Lecture Notes in Computer Science   1811
Edited by G. Goos, J. Hartmanis and J. van Leeuwen

Springer
Berlin
Heidelberg
New York
Barcelona
Hong Kong
London
Milan
Paris
Singapore
Tokyo

Seong-Whan Lee   Heinrich H. Bülthoff
Tomaso Poggio (Eds.)

Biologically Motivated Computer Vision

First IEEE International Workshop, BMCV 2000
Seoul, Korea, May 15-17, 2000
Proceedings

Springer

Series Editors

Gerhard Goos, Karlsruhe University, Germany
Juris Hartmanis, Cornell University, NY, USA
Jan van Leeuwen, Utrecht University, The Netherlands

Volume Editors

Seong-Whan Lee
Korea University
Center for Artificial Vision Research
Anam-dong, Seongbuk-ku, Seoul, 136-701, Korea
E-mail:swlee@image.korea.ac.kr

Heinrich H. Bülthoff
Max-Planck-Institute for Biological Cybernetics
Spemannstr. 38, 72076 Tübingen, Germany
E-mail: heinrich.buelthoff@tuebingen.mpg.de

Tomaso Poggio
Massachusetts Institute of Technology
Department of Brain and Cognitive Sciences
Artificial Intelligence Laboratory, E25-218
45 Carleton Street, Cambridge, MA 02142, USA
E-mail: tp@ai.mit.edu

Cataloging-in-Publication Data

Die Deutsche Bibliothek - CIP-Einheitsaufnahme

Biologically motivated computer vision : proceedings / First IEEE CS
International Workshop BMCV 2000, Seoul, Korea, May 15 - 17, 2000.
Seong-Whan Lee ... (ed.). - Berlin ; Heidelberg ; New York ; Barcelona
; Hong Kong ; London ; Milan ; Paris ; Singapore ; Tokyo : Springer,
2000
  (Lecture notes in computer science ; Vol. 1811)
  ISBN 3-540-67560-4

CR Subject Classification (1998): I.4, F.2, F.1.1, I.3.5, J.2, J.3

ISSN 0302-9743
ISBN 3-540-67560-4 Springer-Verlag Berlin Heidelberg New York

This work is subject to copyright. All rights are reserved, whether the whole or part of the material is
concerned, specifically the rights of translation, reprinting, re-use of illustrations, recitation, broadcasting,
reproduction on microfilms or in any other way, and storage in data banks. Duplication of this publication
or parts thereof is permitted only under the provisions of the German Copyright Law of September 9, 1965,
in its current version, and permission for use must always be obtained from Springer-Verlag. Violations are
liable for prosecution under the German Copyright Law.

Springer-Verlag is a company in the BertelsmannSpringer publishing group.
© Springer-Verlag Berlin Heidelberg 2000

Printed in Germany
Typesetting: Camera-ready by author, data conversion by Steingräber Satztechnik GmbH, Heidelberg
Printed on acid-free paper   SPIN 10720262   06/3142   5 4 3 2 1 0

Preface

It is our great pleasure and honor to organize the First IEEE Computer Society International Workshop on Biologically Motivated Computer Vision (BMCV 2000). The workshop BMCV 2000 aims to facilitate debates on biologically motivated vision system and to provide an opportunity for researchers in the area of vision to see and share the latest developments in state-of-the-art technology. The rapid progress being made in the field of computer vision has had a tremendous impact on the modeling and implementation of biologically motivated computer vision. A multitude of new advances and findings in the domain of computer vision will be presented at this workshop.

By December 1999 a total of 90 full papers had been submitted from 28 countries. To ensure the high quality of workshop and proceedings, the program committee selected and accepted 56 of them after a thorough review process. Of these papers 25 will be presented in 5 oral sessions and 31 in a poster session. The papers span a variety of topics in computer vision from computational theories to their implementation. In addition to these excellent presentations, there will be eight invited lectures by distinguished scientists on "hot" topics. We must add that the program committee and the reviewers did an excellent job within a tight schedule.

BMCV 2000 is being organized by the Center for Artificial Vision Research at Korea University and by the IEEE Computer Society. We would like to take this opportunity to thank our sponsors, the Brain Science Research Center at KAIST, the Center for Biological and Computational Learning at MIT, the Korea Information Science Society, the Korean Society for Cognitive Science, the Korea Research Foundation, the Korea Science and Engineering Foundation, the Max-Planck Institute for Biological Cybernetics, and the Ministry of Science and Technology, Korea. In addition, our special thanks are given to Myung-Hyun Yoo for arranging the workshop, and to Su-Won Shin and Hye-Yeon Kim for their developing and maintaining the wonderful web-based paper submission/review system.

On behalf of the program and organizing committees, we would like to welcome you to BMCV 2000, whether you come as a presenter or an attendee. There will be ample time for discussion inside and outside the workshop hall and plenty of opportunity to make new acquaintances. Last but not least, we would like to express our gratitude to all the contributors, reviewers, program committee and organizing committee members, and sponsors, without whom the workshop would not have been possible.

Finally, we hope that you experience an interesting and exciting workshop and find some time to explore Seoul, the most beautiful and history-steeped city in Korea.

May 2000

Heinrich H. Bülthoff
Seong-Whan Lee
Tomaso Poggio

Workshop Co-Chairs

H. H. Bülthoff	Max-Planck-Institute for Biological Cybernetics, Germany
S.-W. Lee	Korea University, Korea
T. Poggio	MIT, USA

Program Committee

S. Akamatsu	ATR, Japan
J. Aloimonos	University of Maryland, USA
J. Austin	University of York, UK
S.-Y. Bang	POSTECH, Korea
C.-S. Chung	Yonsei University, Korea
L. da F. Costa	University of Sao Paulo, Brazil
R. Eckmiller	University of Bonn, Germany
S. Edelman	University of Sussex, UK
D. Floreano	EPFL, Switzerland
K. Fukushima	University of Electro-Communications, Japan
E. Hildreth	MIT, USA
K. Ikeuchi	University of Tokyo, Japan
A. Iwata	Nagoya Institute of Technology, Japan
C. Koch	California Institute of Technology, USA
M. Langer	Max-Planck-Institute for Biological Cybernetics, Germany
S.-Y. Lee	KAIST, Korea
H. Mallot	Tübingen University, Germany
T. Nagano	Hosei University, Japan
H. Neumann	Ulm University, Germany
E. Oja	Helsinki University of Technology, Finland
L.N. Podladchikova	Rostov State University, Russia
A. Prieto	University of Granada, Spain
W. von Seelen	Ruhr University of Bochum, Germany
T. Tan	Academy of Sciences, China
J. Tsotsos	University of Toronto, Canada
S. Ullman	Weizmann Institute of Science, Israel
H. R. Wilson	University of Chicago, USA
R. Wuertz	Ruhr University of Bochum, Germany
M. Yachida	Osaka University, Japan
D. Young	University of Sussex, UK
J. M. Zurada	University of Louisville, USA

Organizing Committee

H.-I. Choi	Soongsil University, Korea
K.-S. Hong	POSTECH, Korea
W.-Y. Kim	Hanyang University, Korea
I. S. Kweon	KAIST, Korea
Y.-B. Kwon	Chungang University, Korea
Y. B. Lee	Yonsei University, Korea
S.-H. Sull	Korea University, Korea
Y. K. Yang	ETRI, Korea
B. J. You	KIST, Korea
M.-H. Yoo	Korea University, Korea

Organized by

Center for Artificial Vision Research, Korea University
IEEE Computer Society

Sponsored by

Brain Science Research Center, KAIST
Center for Biological and Computational Learning, MIT
Korea Research Foundation
Korea Science and Engineering Foundation
Korean Society for Cognitive Science
Max-Planck-Institute for Biological Cybernetics
Ministry of Science and Technology, Korea
SIG-CVPR, Korea Information Science Society

Table of Contents

Invited Paper (1)

CBF: A New Framework for Object Categorization in Cortex 1
    M. Riesenhuber, T. Poggio

Invited Paper (2)

The Perception of Spatial Layout in a Virtual World 10
    H.H. Bülthoff, C.G. Christou

Segmentation, Detection and Object Recognition

Towards a Computational Model for Object Recognition in IT Cortex 20
    D.G. Lowe

Straight Line Detection as an Optimization Problem:
An Approach Motivated by the Jumping Spider Visual System 32
    F.M.G. da Costa, L. da F. Costa

Factorial Code Representation of Faces for Recognition 42
    S.Choi, O.Lee

Distinctive Features Should Be Learned 52
    J.H. Piater, R.A. Grupen

Moving Object Segmentation Based on Human Visual Sensitivity 62
    K.-J. Yoon, I.-S. Kweon, C.-Y. Kim, Y.-S. Seo

Invited Paper (3)

Object Classification Using a Fragment-Based Representation 73
    S. Ullman, E. Sali

Computational Model

Confrontation of Retinal Adaptation Model
with Key Features of Psychophysical Gain Behavior Dynamics 88
    E. Sherman, H. Spitzer

Polarization-Based Orientation in a Natural Environment 98
  V. Müller

Computation Model of Eye Movement in Reading Using Foveated Vision 108
  Y. Ishihara, S. Morita

New Eyes for Shape and Motion Estimation 118
  P. Baker, R. Pless, C. Fermuller, Y. Aloimonos

Top-Down Attention Control at Feature Space for Robust Pattern Recognition 129
  S.-I. Lee, S.-Y. Lee

A Model for Visual Camouflage Breaking 139
  A. Tankus, Y. Yeshurun

Active and Attentive Vision

Development of a Biologically Inspired Real-Time Visual Attention System 150
  O. Stasse, Y. Kuniyoshi, G. Cheng

Real-Time Visual Tracking Insensitive
to Three-Dimensional Rotation of Objects 160
  Y.-J. Cho, B.-J. You, J. Lim, S.-R. Oh

Heading Perception and Moving Objects 168
  N.-G. Kim

Dynamic Vergence Using Disparity Flux 179
  H.-J. Kim, M.-H. Yoo, S.-W. Lee

Invited Paper (4)

Computing in Cortical Columns:
Curve Inference and Stereo Correspondence 189
  S.W. Zucker

Invited Paper (5)

Active Vision from Multiple Cues 209
  H. Christensen, J.-O. Eklundh

Posters

An Efficient Data Structure for Feature Extraction
in a Foveated Environment 217
  E. Nattel, Y. Yeshurun

Parallel Trellis Based Stereo Matching Using Constraints 227
    H. Jeong, Y. Oh

Unsupervised Learning
of Biologically Plausible Object Recognition Strategies 238
    B.A. Draper, K. Baek

Structured Kalman Filter for Tracking Partially Occluded Moving Objects 248
    D.-S. Jang, S.-W. Jang, H.-I. Choi

Face Recognition under Varying Views 258
    A. Sehad, H. Hocini, A. Hadid, M. Djeddi, S. Ameur

Time Delay Effects on Dynamic Patterns in a Coupled Neural Model 268
    S.H. Park, S. Kim, H.-B. Pyo, S. Lee, S.-K. Lee

Pose-Independent Object Representation by 2-D Views 276
    J. Wieghardt, C. von der Malsburg

An Image Enhancement Technique Based on Wavelets 286
    H.-S. Lee, Y. Cho, H. Byun, J. Yoo

Front-End Vision: A Multiscale Geometry Engine 297
    B.M. ter Haar Romeny, L.M.J. Florack

Face Reconstruction Using a Small Set of Feature Points 308
    B.-W. Hwang, V. Blanz, T. Vetter, S.-W. Lee

Modeling Character Superiority Effect in Korean Characters by Using IAM 316
    C.S. Park, S.Y. Bang

Wavelet-Based Stereo Vision 326
    M. Shim

A Neural Network Model
for Long-Range Contour Diffusion by Visual Cortex 336
    S. Fischer, B. Dresp, C. Kopp

Automatic Generation of Photo-Realistic Mosaic Image 343
    J.-S. Park, D.-H. Chang, S.-G. Park

The Effect of Color Differences
on the Detection of the Target in Visual Search 353
    J.-Y. Hong, K.-J. Cho, K.-H. Han

A Color-Triangle-Based Approach to the Detection of Human Face 359
   C. Lin, K.-C. Fan

Multiple People Tracking
Using an Appearance Model Based on Temporal Color 369
   H.-K. Roh, S.-W. Lee

Face and Facial Landmarks Location Based on Log-Polar Mapping 379
   S.-I. Chien, I. Choi

Biology-Inspired Early Vision System
for a Spike Processing Neurocomputer 387
   J. Thiem, C. Wolff, G. Hartmann

A New Line Segment Grouping Method
for Finding Globally Optimal Line Segments 397
   J.-H. Jang, K.-S. Hong

A Biologically-Motivated Approach to Image Representation
and Its Application to Neuromorphology 407
   L. da F. Costa, A.G. Campos, L.F. Estrozi, L.G. Rios-Filho, A. Bosco

A Fast Circular Edge Detector for the Iris Region Segmentation 417
   Y. Park, H. Yun, M. Song, J. Kim

Face Recognition Using Foveal Vision 424
   S. Minut, S. Mahadevan, J.M. Henderson, F.C. Dyer

Fast Distance Computation with a Stereo Head-Eye System 434
   S.-C. Park, S.-W. Lee

Bio-Inspired Texture Segmentation Architectures 444
   J. Ruiz-del-Solar, D. Kottow

3D Facial Feature Extraction and Global Motion Recovery
Using Multi-modal Information 453
   S.-H. Kim, H.-G. Kim

Evaluation of Adaptive NN-RBF Classifier
Using Gaussian Mixture Density Estimates 463
   S.W. Baik, S. Ahn, P.W. Pachowicz

Scene Segmentation by Chaotic Synchronization and Desynchronization 473
   L. Zhao

Electronic Circuit Model of Color Sensitive Retinal Cell Network 482
    R. Iwaki, M. Shimoda

The Role of Natural Image Statistics in Biological Motion Estimation 492
    R.O. Dror, D.C. O'Carroll, S.B. Laughlin

Enhanced Fisherfaces for Robust Face Recognition 502
    J. Yi, H. Yang, Y. Kim

Invited Paper (6)

A Humanoid Vision System for Versatile Interaction 512
    Y. Kuniyoshi, S. Rougeaux, O. Stasse, G. Cheng, A. Nagakubo

ICA and Space-Variant Imaging

The Spectral Independent Components of Natural Scenes 527
    T.-W. Lee, T. Wachtler, T.J. Sejnowski

Topographic ICA as a Model of Natural Image Statistics 535
    A. Hyvärinen, P.O. Hoyer, M. Inki

Independent Component Analysis of Face Images 545
    P. C. Yuen, J.H. Lai

Orientation Contrast Detection in Space-Variant Images 554
    G. Baratoff, R. Schönfelder, I. Ahrns, H. Neumann

Multiple Object Tracking in Multiresolution Image Sequences 564
    S. Kang, S.-W. Lee

A Geometric Model for Cortical Magnification 574
    L. Florack

Neural Networks and Applications

Tangent Fields from Population Coding 584
    N. Lüdtke, R.C. Wilson, E.R. Hancock

Efficient Search Technique for Hand Gesture Tracking in Three Dimensions 594
    T. Inaguma, K. Oomura, H. Saji, H. Nakatani

Robust, Real-Time Motion Estimation from Long Image Sequences
Using Kalman Filtering 602
    J.A. Yang, X.M. Yang

T-CombNET - A Neural Network Dedicated to Hand Gesture Recognition 613
  M.V. Lamar, M.S. Bhuiyan, A. Iwata

Invited Paper (7)

Active and Adaptive Vision: Neural Network Models 623
  K. Fukushima

Invited Paper (8)

Temporal Structure in the Input to Vision Can Promote Spatial Grouping 635
  R. Blake, S.-H. Lee

Author Index 655

CBF: A New Framework for Object Categorization in Cortex

Maximilian Riesenhuber and Tomaso Poggio

Center for Biological and Computational Learning and Dept. of Brain and Cognitive Sciences, Massachusetts Institute of Technology, Cambridge, MA 02142, USA
{max,tp}@ai.mit.edu,
http://cbcl.mit.edu

Abstract. Building on our recent hierarchical model of object recognition in cortex, we show how this model can be extended in a straightforward fashion to perform basic-level object categorization. We demonstrate the capability of our scheme, called "Categorical Basis Functions" (CBF) with the example domain of cat/dog categorization, using stimuli generated with a novel 3D morphing system. We also contrast CBF to other schemes for object categorization in cortex, and present preliminary results from a physiology experiment that support CBF.

1 Introduction

Much attention in computational neuroscience has focussed on the neural mechanisms underlying object recognition in cortex. Many studies, experimental [2, 7, 10, 18] as well as theoretical [3, 11, 14], support an image-based model of object recognition, where recognition is based on 2D views of objects instead of the recovery of 3D volumes [1, 8]. On the other hand, class-level object recognition, *i.e.*, *categorization*, a central cognitive task that requires to generalize over different instances of one class while at the same time retaining the ability to discriminate between objects from different classes, has only just recently been presented as a serious challenge for image-based models [19].

In the past few years, computer vision algorithms for object detection and classification in complex images have been developed and tested successfully (*e.g.*, [9]). The approach, exploiting new learning algorithms, cannot be directly translated into a biologically plausible model, however.

In this paper, we describe how our view-based model of object recognition in cortex [11, 13, 14] can serve as a natural substrate to perform object categorization. We contrast it to another model of object categorization, the "Chorus of Prototypes" (COP) [3] and a related scheme [6], and show that our scheme, termed "Categorical Basis Functions" (CBF) offers a more natural framework to represent arbitrary object classes. The present paper develops in more detail some of the ideas presented in a recent technical report [15] (which also applies CBF to model some recent results in Categorical Perception).

2 Other Approaches, and Issues in Categorization

Edelman [3] has recently proposed a framework for object representation and classification, called "Chorus of Prototypes". In this scheme, stimuli are projected into a representational space spanned by prototype units, each of which is associated with a class label. These prototypes span a so-called "shape space". Categorization of a novel stimulus proceeds by assigning that stimulus the class label of the most similar prototype (determined using various metrics [3]).

While Chorus presents an interesting scheme to reduce the high dimensionality of pixel space to a veridical low-dimensional representation of the subspace occupied by the stimuli, it has severe limitations as a model of object class representation:

- COP cannot support object class hierarchies. While Edelman [3] shows how similar objects or objects with a common single prototype can be grouped together by Chorus, there is no way to provide a common label for a *group* of prototype units. On the other hand, if several prototypes carry the same class label, the ability to name objects on a subordinate level is lost. To illustrate, if there are several units tuned to dog prototypes such as "German Shepard", "Doberman", *etc.*, they would all have to be labelled "dog" to allow successful basic-level categorization, losing the ability to name these stimuli on a subordinate level.

  This is not the case for a related scheme presented by Intrator and Edelman [6], where in principle labels on different levels of the hierarchy can be provided as additional *input* dimensions. However, this requires that the complete hierarchy is already known at the time of learning of the first class members, imposing a rigid and immutable class structure on the space of objects.

- Even more problematically, COP does not allow the use of different categorization schemes on the same set of stimuli, as there is only one representational space, in which two objects have a certain, fixed distance to each other. Thus, objects cannot be compared according to different criteria: While an apple can surely be very similar to a chili pepper in terms of color (cf. [15]), they are rather different in terms of sweetness, but in COP their similarity would have exactly one value. Note that in the unadulterated version of COP this would at least be a shape-based similarity. However, if, as in the scheme by Intrator & Edelman [6], category labels are added as input dimensions, it is unclear what the similarity actually refers to, as stimuli would be compared using all categorization schemes simultaneously.

These problems with categorization schemes such as COP and related models that define a global representational space spanned by prototypes associated with a fixed set of class labels has led us to investigate an alternative model for object categorization in cortex, in which input space representation and categorization tasks are decoupled, permitting in a natural way the definition of categorization hierarchies and the concurrent use of alternative categorization schemes on the same objects.

3 Categorical Basis Functions (CBF)

Figure 1 shows a sketch of our model of object categorization. The model is an extension of our model of object recognition in cortex (shown in the lower part of Fig. 1), that explained how view-tuned units (VTUs) can arise in a processing hierarchy from simple cell-like inputs. As discussed in [13, 14], it accounts well for the complex visual task of invariant object recognition in clutter and is consistent with several recent physiological experiments in inferotemporal cortex. In the model, feature specificity and invariance are gradually built up through different mechanisms. Key to achieve invariance and robustness to clutter is a MAX-like response function of some model neurons which selects the maximum activity over all the afferents, while feature specificity is increased by a template match operation. By virtue of combining these two operations, an image is represented through an (overcomplete) set of features which themselves carry no absolute position information but code the object through a combination of local feature arrangements. At the top level of the model of object recognition, view-tuned units (VTUs) respond to views of complex objects with invariance to scale and position changes (to perform view-invariant recognition, VTUs tuned to different views of the same object can be combined, as demonstrated in [11]).

VTU receptive fields can be learned in an unsupervised way to adequately cover the stimulus space, *e.g.*, through clustering [15], and can in the simplest version even just consist of all the input exemplars. These view-tuned, or *stimulus space-covering units* (SSCUs) (see caption to Fig. 1) serve as input to *categorical basis functions* (CBF) (either directly, as illustrated in Fig. 1, or indirectly by feeding into view-invariant units [11] which would then feed into the CBF — in the latter case the view-invariant units would be the SSCUs) which are trained in a supervised way to participate in categorization tasks on the stimuli represented in the SSCU layer. Note that there are no class labels associated with SSCUs. CBF units can receive input not only from SSCUs but also (or even exclusively) from other CBFs (as indicated in the figure), which allows to exploit prior category information during training (cf. below).

3.1 An Example: Cat/Dog Categorization

In this section we show how CBF can be used in a simple categorization task, namely discriminating between cats and dogs. To this end, we presented the system with 144 randomly selected morphed animal stimuli, as used in a very recent physiological experiment [4] (see Fig. 2).

The 144 stimuli were used to define the receptive fields (*i.e.*, the preferred stimuli) of 144 model SSCUs by appropriately setting the weights of each VTU to the C2 layer (results were similar if a k-means procedure was used to cluster the input space into 30 SSCUs [15]). The activity pattern over the 144 SSCUs was used as input to train an RBF categorization unit to respond 1 to cat stimuli and -1 to dog stimuli. After training, the generalization performance of the unit was tested by evaluating its performance on the same testing stimuli as used in a

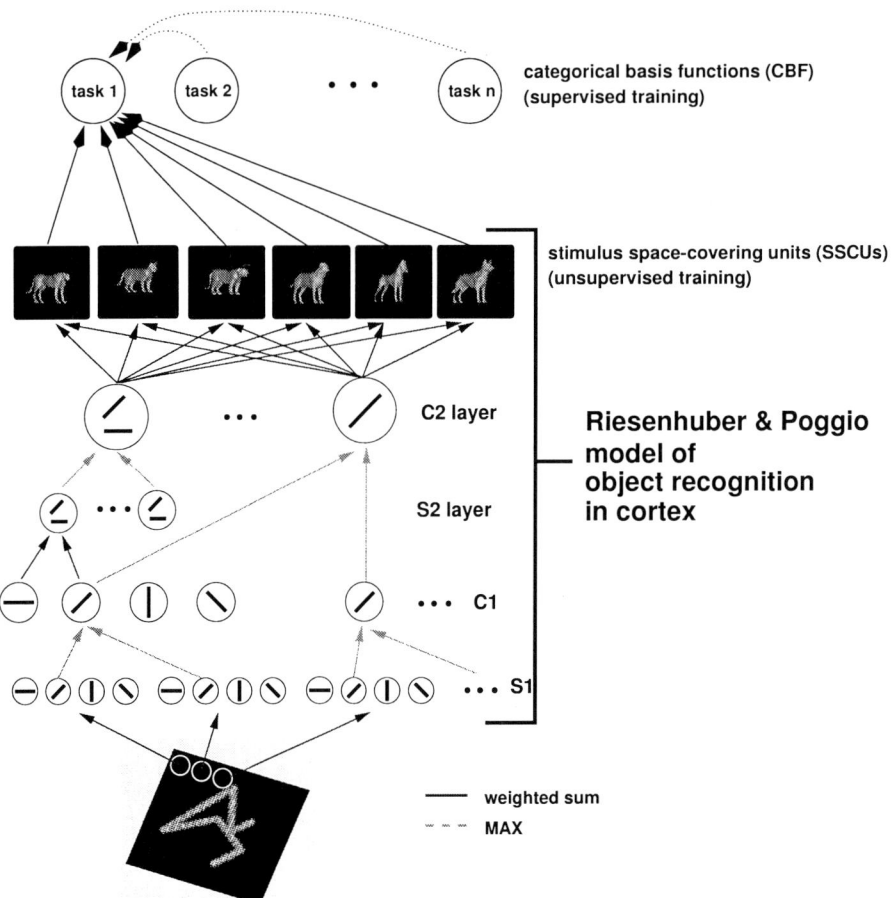

Fig. 1. Our model of object categorization. The model builds on our model of object recognition in cortex [14], which extends up to the layer of view-tuned units (VTUs). These VTUs can then serve as input to categorization units, the so-called categorical basis functions (CBF), which are trained in a supervised fashion on the relevant categorization tasks. In the proof-of-concept version of the model described in this paper the VTUs (as stimulus space-covering units, SSCUs) feed directly into the CBF, but input to CBFs could also come from view-invariant units, that in turn receive input from the VTUs [11], in which case the view-invariant units would be the SSCUs. A CBF can receive inputs input not just from SSCUs but also from other CBFs (illustrated by the dotted connections), permitting the use of prior category knowledge in new categorization tasks, *e.g.*, when learning additional levels of a class hierarchy.

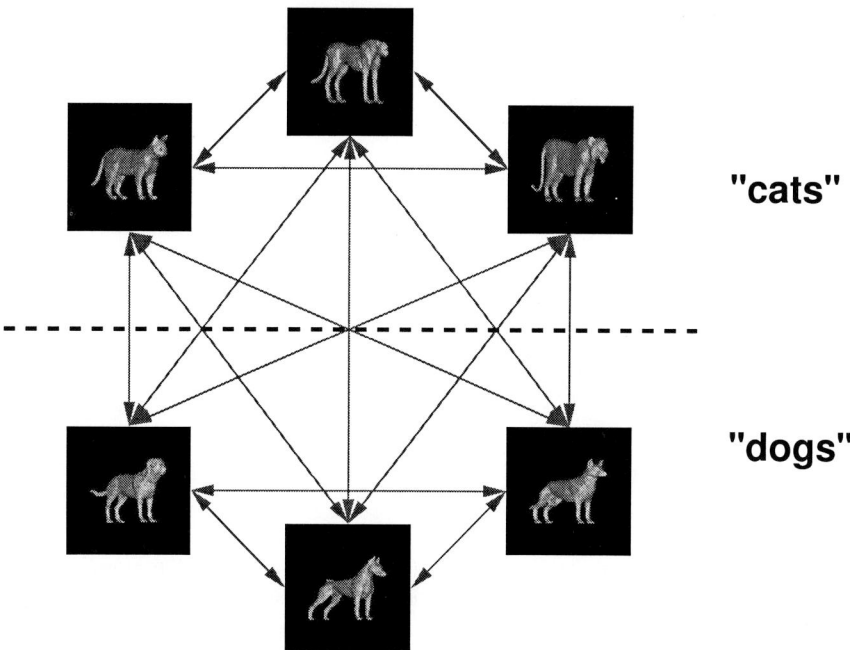

Fig. 2. Illustration of the cat/dog morph space [15]. The stimulus space is spanned by six prototypes, three "cats" and three "dogs". Using a 3D morphing system developed in our lab [17], we can generate objects that are arbitrary combinations of the prototype objects, for instance by moving along prototype-connecting lines in morph space, as shown in the figure and used in the test set. Stimuli were equalized for size and color, as shown (cf. [4]).

recent physiology experiment [4]: Test stimuli were created by subdividing the prototype-connecting lines in morph space into 10 intervals and generating the corresponding morphs (with the exceptions of morphs on the midpoint of each line, which would lie right on the class boundary in the case of lines connecting prototypes from different categories), yielding a total of 126 stimuli.

The response of the categorization unit to stimuli on the boundary-crossing morph lines is shown in Fig. 3. Performance (counting a categorization as correct if the sign of the categorization unit agreed with the stimulus' class label) on the training set was 100% correct, performance on the test set was 97%, comparable to monkey performance, which was over 90% [4]. Note that the categorization errors lie right at the class boundary, *i.e.*, occur for the stimuli whose categorization would be expected to be most difficult.

3.2 Use of Multiple Categorization Schemes in Parallel

As pointed out earlier, the ability to allow more than one class label for a given object, *i.e.*, the ability to categorize stimuli according to different criteria, is a

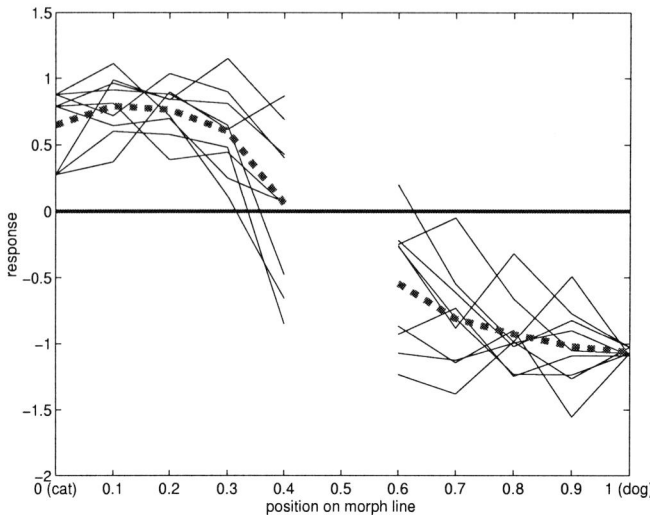

Fig. 3. Response of the categorization unit (based on 144 SSCU, 256 afferents to each SSCU, $\sigma_{SSCU} = 0.7$) along the nine class boundary-crossing morph lines. All stimuli in the left half of the plot are "cat" stimuli, all on the right-hand side are "dogs" (the class boundary is at 0.5). The network was trained to output 1 for a cat and -1 for a dog stimulus. The thick dashed line shows the average over all morph lines. The solid horizontal line shows the class boundary in response space. From [15].

crucial requirement of any model of object categorization. Here we show how an alternative categorization scheme over the same cat/dog stimuli (*i.e.*, using the same SSCUs) used in the previous section, can easily be implemented using CBF.

To this end, we regrouped the stimuli into three classes, each based on one cat and one dog prototype. We then trained three category units, one for each class, on the new categorization task, using 180 stimuli, as used in an ongoing physiology experiment (D. Freedman, M. Riesenhuber, T. Poggio, E. Miller, in progress). Each unit was trained to respond at a level of 1 to stimuli belonging to "its" class and -1 to all other stimuli. Each unit received input from the same 144 SSCU used before.

Performance on the training set was 100% correct (defining a correct labeling as one where the label of the most strongly activated CBF corresponded to the correct class). On the testing set, performance was 74%. The lower performance as compared to the cat/dog task reflects the increased difficulty of the three-way classification task. This was also borne out in an ongoing experiment, where a monkey has been trained on the same task using the same stimuli — psychophysics are currently ongoing that will enable us to compare monkey performance to the simulation results. In the monkey training it turned out to be necessary to emphasize the (complex) class boundaries, which presumably also gives the boundaries greater weight in the SSCU representation, forming a better basis to serve as inputs to the CBFs (cf. [15]).

3.3 Learning Hierarchies

The cat/dog categorization task presented in the previous section demonstrated how CBF can be used to perform basic level categorization, the level at which stimuli are categorized first and fastest [16]. However, stimuli such as cats and dogs can be categorized on several levels. On a superordinate level, cats and dogs can be classified as mammals, while the dog class, for instance, can be divided into dobermans, German shepherds and other breeds on a subordinate level.

Naturally, a model of object categorization should be able not just to represent class hierarchies but also to exploit category information from previously learned levels in the learning of new subdivisions or general categories. CBF provides a suitable framework for these tasks: Moving from, *e.g.*, the basic to a superordinate level, a "cat" and a "dog" unit, resp., can be used as inputs to a "mammal" unit (or a "pet" unit), greatly simplifying the overall learning task. Conversely, moving from the basic to the subordinate level (reflecting the order in which the levels are generically learned [16]), a "cat" unit can provide input to a "tiger" unit, limiting the learning task to a subregion of stimulus space.

4 Predictions and Confirmations of CBF

A simple prediction of CBF would be that when recording from a brain area involved in object categorization with cells tuned to the training stimuli we would expect to find object-tuned cells that respect the class boundary (the CBF) as well as cells that do not (the SSCUs). In COP, or in the Intrator & Edelman scheme, however, the receptive fields of all object-tuned cells would be expected to obey the class boundary.

Preliminary results from Freedman *et al.* [4] support a CBF-like representation: They recorded from 136 cells in the prefrontal cortex of a monkey trained on the cat/dog categorization task. 55 of the 136 cells showed modulation of their firing rate with stimulus identity ($p < 0.01$), but only 39 of these 55 stimulus-selective cells turned out to be category-selective as well.

One objection to this line of thought could be that cells recorded from that are object-tuned but do not respect the class boundary might be cells upstream of the Chorus units. However, this would imply a scheme where stimulus space-tuned units feed into category units, which would be identical to CBF.

Regarding the representation of hierarchies, Gauthier *et al.* [5] have recently presented interesting results from an fMRI study in which human subjects were required to categorize objects on basic and subordinate levels. In that study, it was found that subordinate level classification activated additional brain regions compared to basic level classification. This result is compatible with the prediction of CBF alluded to above that categorization units performing subordinate level categorization can profit from prior category knowledge by receiving input from basic level CBFs. Note that these results are not compatible with the scheme proposed in [6] (nor COP), where a stimulus activates that same set of units, regardless of the categorization task.

5 Discussion

The simulations above have demonstrated that CBF units can generalize within their class and also permit fine distinction among similar objects. It will be interesting to see how this performance extends to even more naturalistic object classes, such as the photographic images of actual cats and dogs as used in studies of object categorization in infants [12]. Moreover, infants seem to be able to build basic level categories even without an explicit teaching signal [12], possibly exploiting a natural clustering of the two classes in feature space. We are currently exploring how category representations can be built in such a paradigm.

Acknowledgments

Fruitful discussions with Michael Tarr and email exchanges with Shimon Edelman are gratefully acknowledged. Thanks to Christian Shelton for MATLAB code for k-means and RBF training and for the development of the correspondence program [17].

References

1. Biederman, I. (1987). Recognition-by-components : A theory of human image understanding. *Psych. Rev.* **94**, 115–147.
2. Bülthoff, H. and Edelman, S. (1992). Psychophysical support for a two-dimensional view interpolation theory of object recognition. *Proc. Nat. Acad. Sci. USA* **89**, 60–64.
3. Edelman, S. (1999). *Representation and Recognition in Vision*. MIT Press, Cambridge, MA.
4. Freedman, D., Riesenhuber, M., Shelton, C., Poggio, T., and Miller, E. (1999). Categorical representation of visual stimuli in the monkey prefrontal (PF) cortex. In *Soc. Neurosci. Abs.*, volume 29, 884.
5. Gauthier, I., Anderson, A., Tarr, M., Skudlarski, P., and Gore, J. (1997). Levels of categorization in visual recognition studied with functional mri. *Curr. Biol.* **7**, 645–651.
6. Intrator, N. and Edelman, S. (1997). Learning low-dimensional representations via the usage of multiple-class labels. *Network* **8**, 259–281.
7. Logothetis, N., Pauls, J., and Poggio, T. (1995). Shape representation in the inferior temporal cortex of monkeys. *Curr. Biol.* **5**, 552–563.
8. Marr, D. (1982). *Vision: a computational investigation into the human representation and processing of visual information*. Freeman, San Francisco, CA.
9. Papageorgiou, C., Oren, M., and Poggio, T. (1998). A general framework for object detection. In *Proceedings of the International Conference on Computer Vision, Bombay, India*, 555–562. IEEE, Los Alamitos, CA.

10. Perrett, D., Oram, M., Harries, M., Bevan, R., Hietanen, J., Benson, P., and Thomas, S. (1991). Viewer-centred and object-centred coding of heads in the macaque temporal cortex. *Exp. Brain Res.* **86**, 159–173.
11. Poggio, T. and Edelman, S. (1990). A network that learns to recognize 3D objects. *Nature* **343**, 263–266.
12. Quinn, P., Eimas, P., and Rosenkrantz, S. (1993). Evidence for representations of perceptually similar natural categories by 3-month-old and 4-month-old infants. *Perception* **22**, 463–475.
13. Riesenhuber, M. and Poggio, T. (1999). Are cortical models really bound by the "Binding Problem"? *Neuron* **24**, 87–93.
14. Riesenhuber, M. and Poggio, T. (1999). Hierarchical models of object recognition in cortex. *Nature Neurosci.* **2**, 1019–1025.
15. Riesenhuber, M. and Poggio, T. (1999). A note on object class representation and categorical perception. Technical Report AI Memo 1679, CBCL Paper 183, MIT AI Lab and CBCL, Cambridge, MA.
16. Rosch, E. (1973). Natural categories. *Cognit. Psych.* **4**, 328–350.
17. Shelton, C. (1996). Three-dimensional correspondence. Master's thesis, MIT, (1996).
18. Tarr, M. (1995). Rotating objects to recognize them: A case study on the role of viewpoint dependency in the recognition of three-dimensional objects. *Psychonom. Bull. & Rev.* **2**, 55–82.
19. Tarr, M. and Gauthier, I. (1998). Do viewpoint-dependent mechanisms generalize across members of a class? *Cognition* **67**, 73–110.

The Perception of Spatial Layout in a Virtual World

Heinrich H. Bülthoff[1] and Chris G. Christou[2]

[1] Max-Planck-Institute for Biological Cybernetics, Tübingen, Germany
heinrich.buelthoff@tuebingen.mpg.de
[2] Unilever Research, Wirral, UK.

Abstract. The perception and recognition of spatial layout of objects within a three-dimensional setting was studied using a virtual reality (VR) simulation. The subjects' task was to detect the movement of one of several objects across the surface of a tabletop after a retention interval during which time all objects were occluded from view. Previous experiments have contrasted performance in this task after rotations of the observers' observation point with rotations of just the objects themselves. They found that subjects who walk or move to new observation points perform better than those whose observation point remains constant. This superior performance by mobile observers has been attributed to the influence of non-visual information derived from the proprioceptive or vestibular systems. Our experimental results show that purely visual information derived from simulated movement can also improve subjects' performance, although the performance differences manifested themselves primarily in improved response times rather than accuracy of the responses themselves.

1 Introduction

As we move around a spatial environment we appear to be able to remember the locations of objects even if during intervening periods we have no conscious awareness of these objects. We are for instance able to remember the spatial layout of objects in a scene after movements and predict where objects should be. This ability requires the use of a spatial representation of the environment and our own position within it. Recent experiments have shown that although people can perform such tasks their performance is limited by their actual experience of the scene. For instance, Shelton & McNamara [11] found that subjects ability to make relative direction judgements from positions aligned with the studied views of a collection of objects were superior to similar judgements made from misaligned positions. In general it has been shown that accounting for misalignments in view requires more effort in as much as response times and error rates are higher than for aligned views. Similar findings were also reported by Diwadkar & McNamara [5] in experiments requiring subjects to judge whether a configuration of several objects was the same as a configuration of the same objects studied previously from a different view. They found that response latencies were a linear function of the angular distance between the test and previously studied views.

These findings and others (e.g., see [9]) have led researchers to conclude that the mental representation of spatial layout is egocentric in nature and encodes the locations of objects with respect to an observer-centred reference frame. However, Simons & Wang [12] found that the manner in which the transformation in view is brought about is important. For instance, people can compensate better for changes in the retinal projection of several objects if these changes are brought about by their own movement. These experiments contrasted two groups of subjects. The first group performed the task when the retinal projection of objects was a result of the (occluded) rotation of the objects themselves while for the second group it was a result of their own movement. In both cases subjects performed equivalent tasks; that is, to name the object that moved when the retinal projection was both the same and different compared to an initial presentation of the objects. Simons & Wang attributed the superior performance by the 'displaced' observers to the involvement of extra-retinal information which, for instance, may be derived from vestibular or proprioceptive inputs. Such information could allow people to continually update their position in space relative to the configuration of test objects. This is known as spatial updating.

This result supports the findings in other spatial layout experiments which contrasted imagined changes in orientation with real yet blind-folded changes in orientation [10,8,6]. The subjects' task was to point to the relative locations of objects from novel positions in space after real or imagined rotations or translations. It was found that translation is less disruptive than rotation of viewpoint and that, when subjects are blindfolded, actual rotation is less disruptive than imagined rotation. Again, this implicates the involvement and use of extra-retinal, proprioceptive or vestibular cues during actual movement of the observers. These cues could be used for instance to specify the relative direction and magnitude of rotation and translation and thus could support spatial updating in the absence of visual cues. Whilst the involvement of non-visual information in the visual perception of spatial layout is interesting in itself, it does not seem plausible that only such indirect information is used for spatial updating. Indeed any information which yields the magnitude and direction of the change in the viewers position could be used for spatial updating including indirect information derived from vision itself. Simons & Wang did test whether background cues were necessary for spatial updating by repeating their experiment in a darkened room with self-luminous objects. The results were only slightly affected by this manipulation. That is, spatial updating still occurred. However, this only means that visually derived movement is not a necessity for spatial updating. It does not imply that visually derived movement cannot facilitate it. To determine whether spatial updating can occur through purely visual sources of information we constructed a simple vision-only based spatial updating experiment using a virtual reality simulation to eliminate unwanted cues and depict the implied movement within a realistic setting. We attempted to replicate the conditions of the original Simons & Wang [12] experiment but with simulated movement rather than real movement and within a simulated environment rather than a real environment.

1.1 The Principles of Virtual Environment Simulation

An alternative to using real-world scenes for studying spatial cognition is to use a virtual environment (or virtual reality) simulation. These are becoming increasingly popular means of simulating three-dimensional (3D) space for studying spatial cognition (e.g., [1,7,4]. Such simulation allows one to actively move through a simulated space with almost immediate visual feedback and also allows objects in the scene to move (either smoothly or abruptly) from one position to another.

Fig. 1. Three views of the simulated environment marking the three stages of the experiment as seen by the viewpoint-change observers (see below). The first image depicts the 5 test objects. The middle image shows the fully lowered curtain. The right hand image shows the new view of the objects after a counter-clockwise rotation in view. In this case the 5 objects and table did not rotate so that the retinal projection is different from that in the initial view. The pig and the torch have exchanged places and the appropriate response would have been to press the 'change' button.

In essence, our simulation consisted of a 3D-modelled polygonal environment in which a virtual camera (the view of the observer) is translated and rotated and whose projection is rendered on a desktop computer monitor in real-time (see Figure 1). That is, the scene was rendered (i.e. projected onto the image plane and drawn on the monitor) approximately 30 times a second. A Silicon Graphics Octane computer performed the necessary calculations. The cameras motion was controlled by the observer using a 6 degrees-of-freedom motion input device (Spacemouse). The initial stages of development involved the construction of the 3D environment, created using graphics modelling software (3DStudio Max from Kinetix, USA.) The illumination in most VR simulations is usually calculated according to a point illumination source located at infinity. The visual effects produced by such a model are unrealistic because a non-extended light source at infinity produces very abrupt changes in illumination which can be confused with changes in geometry or surface reflectance (lightness). Furthermore, in any real scene the light reaching the eye is a function of both direct illumination from the light source and indirect illumination from other surfaces. The latter helps to illuminate surfaces occluded from the light source and produces smooth gradients of illumination across surfaces which can also eliminate the confusion between reflectance and illumination mentioned above (see [3]). Therefore, in our experiments we pre-rendered the surfaces of our virtual environment using software that simulates the interreflective nature of diffuse illumination. This produced realistic smooth shadows and ensured that regions not visible to the source directly could still be illuminated by indirect light.

Once the 3D model was constructed, interactive simulation software could be used to control the simulated observer movement and subsequent rendering of visible field of view onto the screen. On SGI computers this is performed using the IRIS Performer 'C' programming library. This software also provides the functionality for detecting the simulated observers' collision with surfaces in the scene and makes sure the observer stays within the confined bounds of the simulation.

2 Method

2.1 Materials

The experiment utilised the 3D simulated environment described above together with 25 three-dimensional models of common objects. The objects were chosen for their ease of identification and consisted of, for instance, pig, torch, clock, lightbulb and toothbrush (see Figure 1). All objects were scaled to the same or similar size and no surface texturing or colouring was used. The objects could occupy any one of 7 possible evenly spaced positions across the platform and these positions where computed once at the beginning of each experiment. The position of each object for a given trial was chosen at random from the list of all possible locations. The object positioning ensured sufficient distance between each object so that the objects did not overlap and minimized the chances of one object occluding another from any given viewpoint.

2.2 Subjects

Male and female subjects were chosen at random from a subject database. All 28 were naïve as to the purposes of the experiment and had not performed the experiment in the past. They were randomly assigned to one of the two groups of 14 and given written instructions appropriate to the group chosen (see below). They were given an initial demonstration of their task.

Subjects viewed the scene through a viewing chamber that restricted their view to the central portion of the computer monitor and maintained a constant viewing distance of 80 cm. Before each block of trials they were allowed to move themselves freely through the environment for 3 minutes (using the Spacemouse) to get a better impression of its dimensions and to acquaint themselves with the five test objects, which were visible on the platform. Observers' simulated height above the floor was held constant at an appropriate level. The experiment was completely self-initiated and subjects received on-screen instructions at each stage.

2.3 Procedure

In keeping with the original experimental paradigm of Simons & Wang [12] we facilitated a retention interval by the simulation of a cylindrical curtain which could be lowered to completely obscure the objects from view (see Figure 1). Each trial consisted of the following format. The configuration of objects would be viewed for 3 seconds from the start viewpoint. The curtain was then lowered, eventually obscuring the objects completely. The observers' view was then rotated to a new viewpoint (Group B) or rotated half-way to this new viewpoint and then rotated back to the start viewpoint (group A). This retention interval (during which time the objects were not visible) lasted for 7 seconds for both groups. The curtain was then raised revealing the objects for a further 3 seconds. Subjects then had to decide if one of the objects was displaced (to a vacant position) during the intervening period. They responded by pressing one of two pre-designated keys of the computer keyboard. The subject's response together with their response latency (calculated from the second presentation of objects) was stored for later analysis.

The experiment consisted of 5 blocks of 12 trials each. For each block 5 new objects were chosen and a new start position around the platform randomly selected as the start viewpoint. In 50% of the trials there was a displacement of one of the objects. Also in 50% of the trials, the platform was rotated by 57 degrees in the same direction as the observer. Thus, for group B (different observation point), when the platform was rotated and no object was displaced, exactly the same retinal configuration of objects on the table was observed (apart from the background). In the case of group A (same observation point), only when the platform was not rotated and no object displaced was the retinal configuration of objects exactly the same. Thus both groups had to determine displacement of one of the objects when the retinal projection of objects was the same or a fixed rotation away from the observation point. For group B, however, the rotation was the result of the observers simulated movement whereas for group A it was the result of the rotation of the platform and objects. The rotation angle of the simulated observation point and platform was always 57 degrees. For each block of trials the computer chose either a clockwise or anti-clockwise rotation away from the start observation point. This ensured that subjects paid attention to the rotation itself. When the table was rotated both groups were notified by on-screen instructions and received a short audible tone from the computer

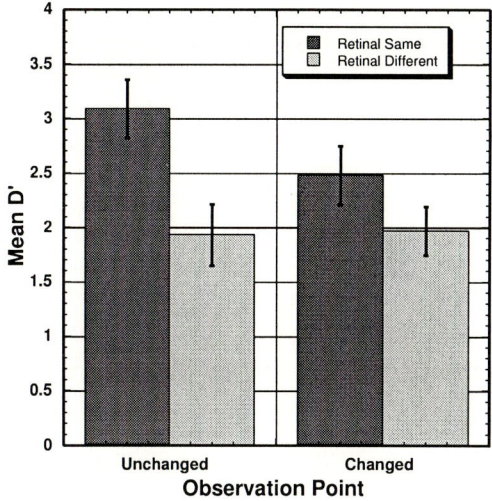

Fig. 2. Shows the mean d' for according to the two subject groups (which differed in terms of changed or unchanged observation) and same or different retinal projection. The error bars are standard errors of the mean.

3 Results

The proportion of correct responses and false alarms were used to calculate a mean d' score for each observer. On the whole mean d' was always above 1.75 which indicates that subjects found the task quite easy. An analysis of variance (ANOVA) with two between-subject factors (Group A or B, i.e. same or different observation point) and two within-subject factors (same or different retinal projection) revealed that the effect of group on d' was not significant [$F(1,26)=.76$, $p=0.39$], that the effect of the within-subject factor retinal projection was significant [$F(1,26)=22.76$, $p<0.00005$] and that the interaction between these two was approaching significance [$F(1,26)=3.56$, $p<0.07$]. These results are plotted in Figure 2 which shows that group B (changed observation point) were least affected by a different retinal projection of the objects. This is more strongly indicated by the response times (RT). A similar ANOVA on RT revealed again that the effect of group was not significant [$F(1,26)=0.65$, $p=0.4$], that the effect of retinal projection was significant [$F(1,26)=21.67$, $p<0.0005$] and also the interaction between these was also significant [$F(1,26)=19.3$, $p<0.0005$]. This interaction is portrayed in Figure 3 which shows that there was no significant difference in response times for group B for identifying configurations viewed from either the original or rotated observation point. This is in sharp contrast to the group A observers who required on average 400 ms more in order to perform this judgement after rotation of the objects.

These results for RT indicates apparently view-independent performance for Group B and view-dependent performance for Group A. However, these response times were averaged for all responses regardless of whether the response was correct or incorrect. It may be that these differences correlate only with the different error rates for each of the individual conditions rather than reflecting differences in mental processing. We therefore extracted the response times corresponding only to correct responses and performed t-tests for related pairs of within-subject data and for each group. The mean RTs for group A were 1447 ms and 2090 ms for same and different retinal projections respectively. For group B, mean RTs were 1544 ms and 1654 ms for same and different retinal projections respectively. We found that these RT measures for group B were not significantly different [$t(13)=-1.08$; $p<0.29$] whereas the means for group A were significantly different [$t(13)=-5.14$; $p<0.0005$]. Therefore, the correct responses for group A reflect a possible difference in the amount of time required to perform the task correctly when the retinal projection was the same or different. For group B, who performed the task from a different observation point, there was no significant cost in response time regardless of whether the same or different retinal projection of objects was viewed.

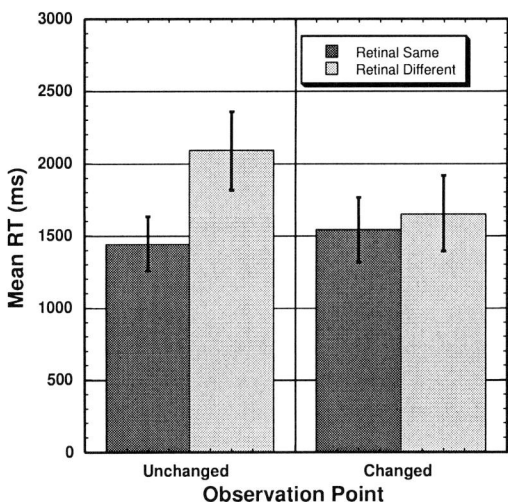

Fig. 3. Shows the mean response times (RT) for all responses (correct and incorrect). Error bars are again standard errors of the mean.

4 Conclusion

The ability to remember the relative spatial locations of several objects in a three-dimensional display has been used to reveal some of the properties of human spatial representation and the perception of spatial layout. One important ability that requires spatial representation (and processing of spatial representations) is the ability to make judgements about relative locations of objects in a display after changes in the observers' viewpoint. Previous experiments suggest that such an ability is view-dependent and becomes view independent when additional information regarding the magnitude of the orientation shift is available from proprioceptive or vestibular sources [12,14]. We wanted to determine if this additional information could be derived solely from visual sources.

Informal responses from our subjects in the current experiment suggest that the principle means of performing the task was to remember a verbalised sequence (such as, pig, torch, clock, lightbulb and toothbrush). However, this method of storage still implies some directional encoding of the next object in each sequence. That is, it must encode the relative direction of the next object in the sequence with respect to the previous object. Furthermore, we may also suggest that the position of the start object in the sequence (i.e. pig) needed to be stored (perhaps in a world-centred reference frame) as well. Otherwise, if the start object was indeed the object that moved then this sequence encoding would be difficult to use. All this suggests that one or more reference frames were in used to encode the positions of objects and that differences either in the mode of encoding or operations on these encodings were the cause of the differences between groups that we have observed.

For conditions where objects rotated and the observers' observation point remained constant we found that performance dropped dramatically compared with trials in which the objects did not rotate and the observation point was also constant (group A). This misalignment of objects resulted in more errors and longer response latencies. This is in keeping with view-dependent results obtained in both object recognition (e.g., [2,13]) and scene recognition studies [5,4]). If either a retinal-centred or body-centred encoding was used these orientation changes required some computational effort or transformation to extract the necessary matching information. One means of performing the task is to imagine looking at the objects from a new position which preserves the original retinal-centred encoding. In this situation, one could determine the movement of an object by comparing two similar representations. If this is the case however, then the results of previous studies on imagined spatial layout would have predicted the loss in performance (e.g., [8]) which we obtained.

For the subjects in the view-change group (group B) the differences between same retinal image and different retinal image were found to be statistically insignificant. Again one strategy for subjects would have been to imagine they were at the original viewpoint before the move and compare relative spatial positions in a retinal-centred reference frame. However, what they may also do here is continually update the possible changes that are occurring to the visual appearance of objects as their viewpoint changes. This may explain why in the original experiments by Simons & Wang viewpoint-change subjects performed better with different retinal projections of objects than with the same retinal projections (we would expect that matching two dissimilar retinal images results in more errors, not less). The operation of this predictive mechanism may be initiated automatically whenever we move around the world. The fact that there was no real movement in our present study may explain why we did not obtain similar elevated performance by Group B subjects. In future experiments, this could be tested by comparing abrupt and smooth changes in view and also by observing the influence of actual walking but with visual feedback derived from a head mounted display.

As mentioned above, our computer simulation differed from the previous experiments in one crucial factor. Namely, that in this simulation observers' movement was implied rather than actual (observers passively viewed their rotation around the table). In our experiment therefore, the key information supporting the task was entirely visual. At first it appears that this is contradictory to the results of Experiment 2 reported by Simons & Wang [12] in which they performed the same task after removing the visual detail in the background. The background detail was reduced by performing the experiment in the dark and with self-luminous objects. Simons & Wang found that this did not affect spatial updating and the superior performance of view-change subjects. However, the fact that the absence of visual background does not affect performance does not imply that visual background cannot provide the necessary information. Furthermore, in our experiment the superior performance of viewpoint-change subjects is reflected principally by differences in their response times, which were not recorded in Simons & Wang's experiments. It appears that the crucial factor is the provision of additional information regarding the motion or change of position of the observer. This additional information may be derived from various sources, both visual and non-visual.

In conclusion, this experiment has revealed a positive benefit of simulated movement even within a simulation of a three-dimensional space and even when only visual input specifies that the observer has moved. This indicates that information from both visual and non-visual modalities can interact in the facilitation of stable perception and view-invariant recognition. The basis of this facilitation appears to be that of providing a stable environment or 3D space and a continual mapping of the position of the observer within this space. Future studies may determine whether this does indeed imply spatial updating of the observers' relative position in space and perhaps suggest some model of how spatial updating functions in the perception and recognition of spatial layout.

References

1. V. Aginsky, C. Harris, R. Rensink, and J Beusmans. Two strategies for learning a route in a driving simulator. *Journal of Environmental Psychology*, 17:317 – 331, 1999.
2. H. H. Bülthoff and S. Edelman. Psychophysical support for a 2-D view interpolation theory of object recognition. *Proceedings of the National Academy of Science*, 89:60 – 64, 1992.
3. C. Christou and A. Parker. *Simulated and Virtual Realities: Elements of Perception*, chapter Visual realism and virtual reality. Taylor & Francis, London, 1995.
4. C. G. Christou and H. H. Bülthoff. View dependency in scene recognition after active learning. *Memory & Cognition*, 27:996 – 1007, 1999.

5. V. A. Diwadkar and T. McNamara. View dependence in scene recognition. *Psychological Science*, 8(4):302 – 307, 1997.
6. M. J. Farrell and I. H. Robertson. Mental rotation and the automatic updating of body-centered spatial relationships. *Journal of Experimental Psychology: Learning, Memory and Cognition*, 24(1):227 – 233, 1998.
7. E. A. Maguire, N. Burgess, J. G. Donnett, R. S. J. Frackowiak, D. D. Frith, and J. O'Keefe. Knowing where and getting there: a human navigation network. *Science*, 280(1):921–924, 1998.
8. C. C. Presson and D. R. Montello. Updating after rotational and translational body movements: Coordinate structure of perspective space. *Perception*, 23(12):1447–1455, 1994.
9. J. J. Rieser. Access to knowledge of spatial structure at novel points of observation. *Journal of Experimental Psychology: Learning, Memory and Cognition*, 15(6):1157–1165, 1989.
10. J. J. Rieser, A. E. Garing, and M. F. Young. Imagery, action, and young childrens spatial orientation - its not being there that counts, its what one has in mind. *Child Development*, 65(5):1262 – 1278, 1994.
11. A. L. Shelton and T. P. McNamara. Multiple views of spatial memory. *Psychonomics Bulletin & Review*, 4(1):102 – 106, 1997.
12. D. J. Simons and R. F. Wang. Perceiving real-world viewpoint changes. *Psychological Science*, 9(4):315 – 320, 1998.
13. M. Tarr and H. H. Bülthoff. *Object recognition in man, monkey and machine*. MIT Press, 1999.
14. R. X. F. Wang and D. J. Simons. Active and passive scene recognition across views. *Cognition*, 70(2):191 – 210, 1999.

Towards a Computational Model for Object Recognition in IT Cortex

David G. Lowe

Computer Science Dept., Univ. of British Columbia,
Vancouver, B.C., V6T 1Z4, Canada,
http://www.cs.ubc.ca/~lowe
lowe@cs.ubc.ca

Abstract. There is considerable evidence that object recognition in primates is based on the detection of local image features of intermediate complexity that are largely invariant to imaging transformations. A computer vision system has been developed that performs object recognition using features with similar properties. Invariance to image translation, scale and rotation is achieved by first selecting stable key points in scale space and performing feature detection only at these locations. The features measure local image gradients in a manner modeled on the response of complex cells in primary visual cortex, and thereby obtain partial invariance to illumination, affine change, and other local distortions. The features are used as input to a nearest-neighbor indexing method and Hough transform that identify candidate object matches. Final verification of each match is achieved by finding a best-fit solution for the unknown model parameters and integrating the features consistent with these parameter values. This verification procedure provides a model for the serial process of attention in human vision that integrates features belonging to a single object. Experimental results show that this approach can achieve rapid and robust object recognition in cluttered partially-occluded images.

1 Introduction

During the past decade, there have been major advances in our understanding of how object recognition is performed in the primate visual system. There is now a broad body of evidence [21] showing that object recognition makes use of neurons in inferior temporal cortex (IT) that respond to features of intermediate complexity. These features are typically invariant to a wide range of changes in location, scale, and illumination, while being very sensitive to particular combinations of local shape, color, and texture properties.

This paper describes a computer vision system for performing object recognition that also makes use of local image features of intermediate complexity that are invariant to many imaging parameters. The approach is called a scale invariant feature transform (SIFT), as it transforms an image into a representation that is unaffected by image scaling and other similarity transforms. Features are located at peaks of a function computed in scale space. The features describe local image regions around these peaks using a representation that emphasizes gradient orientations while allowing small shifts in gradient position, in a manner modeled on the responses of complex cells in primary visual cortex. These features are bound to object interpretations through a process of indexing followed by a best-fit solution for object parameters. This achieves feature integration in a manner similar to the process of serial visual attention that has been shown to play an important role in object recognition in human vision [23]. The result is a system that is able to recognize 3D objects from any viewpoint under varying illumination in cluttered natural images with about 1 second of computation time.

There has been some previous research [14,24] on building computer systems for object recognition that use similar intermediate features to those in IT cortex. One problem has been that these earlier systems use correlation to measure the presence of intermediate features in an image, which has a prohibitive computational cost due to the need to compare each feature at every location, scale, and orientation to the image. This paper describes a staged filtering approach in which stable key points in scale space are first identified, and then feature detection is performed only at these canonical points and with respect to a canonical scale and orientation. This reduces the cost by about a factor of 10,000, making it easily suitable for practical applications. Furthermore, the feature descriptors are modified to make them less sensitive to affine transformation, illumination change and local distortions as compared to image correlation. A final stage of solving for explicit model pose parameters allows for robust verification of object interpretations.

2 Related Research

An understanding of the intermediate-level features in visual cortex was first obtained in a novel approach developed by Tanaka and his associates [7,20]. They recorded from individual neurons in anesthetized monkeys while using a library of objects and a computer graphics editing system to characterize their responses. First, a collection of complex real-world objects were tested to obtain an initial response. An image of the best initial object was then subject to numerous attempts at simplification and modification to obtain an optimal response. Although some neurons in anterior IT cortex responded to very simple line or bar features, in most cases the optimal response was obtained by features of intermediate complexity, such as a dark five-sided star shape, a circle with a thin protruding element at a particular orientation, or a green horizontal textured region within a triangular boundary. Some neurons responded only to more complex shapes, such as moderately detailed face or hand images.

These intermediate-complexity neurons were often highly sensitive to small variations in shape, such as the degree of rounding of corners or relative lengths of elements. On the other hand, the neurons exhibited a wide range of invariance to other parameters, such as retinal location, size, and contrast. Detailed studies of size and position invariance [6,16] have shown that about 55% of the neurons have a size invariance range of greater than 2 octaves and 20% have a range of more than 4 octaves. Most neurons have a receptive field covering a large portion of the image (an average 25 degrees of visual angle and usually including all of the fovea). These properties would seem ideally suited to determining object identity without needing to replicate the neuron for each combination of values of the imaging parameters.

Neurons that were close together in cortex often responded to small variations of the same feature. Based on the average size of these related feature columns as a proportion of the total size of the brain region, Tanaka [20] estimated that there was room for about 1300 such feature columns. However, if the column sizes vary, then the large ones would be preferentially sampled and the total number of feature types could be far greater.

The feature responses have been shown to depend on previous visual learning from exposure to specific objects containing the features. Logothetis, Pauls & Poggio [10] examined the responses of neurons in monkeys that had been trained to classify views of wire-frame and spheroidal shapes. They discovered many neurons that responded only to particular views of these shapes, while exhibiting the usual invariance to large ranges of scale and location. Booth & Rolls [3] found similar results for 10 plastic objects that had been placed in the monkey's cage for a period of weeks or months without any training. In addition to the usual view-dependent neurons, they found a small population of neurons that responded to any view of a particular object (these responded to a conjunction of shape views, rather than to a simple feature such as color that was shared between views). In a dramatic illustration of learning, Tovee, Rolls & Ramachandran [22] showed that a face sensitive neuron could learn to recognize degraded face images (that were previously unrecognizable) by exposure to 5 seconds of training images that showed the transition between normal and degraded images.

It is also known that object recognition in human vision uses a serial process of attention to bind features to object interpretations, determine pose, and segment an object from a cluttered background [23]. A wide range of psychophysical experiments [25] have shown that preattentive object descriptions consist of only a collection of isolated features, and serial attention is necessary to represent shape relationships and integrate features into a common object description. In this paper we will describe the use of a Hough transform to generate object hypotheses, followed by best-fit parameter solving and selection of consistent features to perform binding and model verification.

Within the computer vision field, there has been recent work on using dense collections of local image features for object recognition. One approach has been to use a corner detector (more accurately, a detector of peaks in local image variation) to identify repeatable image locations, around which local image properties can be measured. Schmid & Mohr [18] used the Harris corner detector to identify interest points, and then created a local image descriptor at each interest point from an orientation-invariant vector of derivative-of-Gaussian image measurements. These image descriptors were used for robust object recognition by looking for multiple matching descriptors that satisfied object-based orientation and location constraints. This work was impressive both for the speed of recognition in a large database and the ability to handle cluttered images. However, the corner detector examines an image at only a single scale. As the change in scale becomes significant, the detector responds to different image points. Also, since the detector does not provide an indication of the object scale, it is necessary to create image descriptors and attempt matching at a large number of scales.

Other approaches in computer vision to appearance-based recognition include eigenspace matching [15], color histograms [19], and receptive field histograms [17]. These approaches have all been demonstrated successfully on isolated objects or pre-segmented images, but due to their more global features it has been difficult to extend them to cluttered and partially occluded images.

3 Key Localization

Rather than searching all possible image locations for particular features, we obtain far better efficiency by first selecting key locations and scales of interest and describing the local image region around each location. The key locations are selected in a manner that is invariant with respect to image translation, scaling, and rotation, and is minimally affected by noise and small distortions.

Lindeberg [8] has shown that under some rather general assumptions on scale invariance, the Gaussian kernel and its derivatives are the only possible smoothing kernels for scale space analysis. To achieve rotation invariance and a high level of efficiency, we have chosen to select key locations at maxima and minima of a difference of Gaussian function applied in scale space. This can be computed very efficiently by building an image pyramid with resampling between each level. Furthermore, it locates key points at regions and scales of high variation, making these locations particularly useful for characterizing the image. Crowley & Parker [4] and Lindeberg [9] have previously used the difference-of-Gaussian in scale space for other purposes.

As the 2D Gaussian function is separable, its convolution with the input image can be efficiently computed by applying two passes of the 1D Gaussian function in the horizontal and vertical directions:

$$g(x) = \frac{1}{\sqrt{2\pi}\sigma} e^{-x^2/2\sigma^2}$$

For key localization, all smoothing operations are done using $\sigma = \sqrt{2}$, which can be approximated with sufficient accuracy using a 1D kernel with 7 sample points.

A pyramid of Gaussian images is computed at all scales differing by factors of 1.5, using smoothing followed by resampling. Images at adjacent scales are subtracted prior to resampling to obtain the difference of Gaussian function. See [13] for details on how this can computed very efficiently. Maxima and minima of this scale-space function are determined by comparing each pixel in the pyramid to its immediate neighbours in location and scale.

4 Key Orientation and Stability

To characterize the image at each key location, the smoothed image at each level of the pyramid is processed to extract image gradients and orientations. At each pixel, A_{ij}, the image gradient magnitude, M_{ij}, and orientation, R_{ij}, are computed using pixel differences:

$$M_{ij} = \sqrt{(A_{ij} - A_{i+1,j})^2 + (A_{ij} - A_{i,j+1})^2}$$

$$R_{ij} = \mathrm{atan2}\,(A_{ij} - A_{i+1,j}, A_{i,j+1} - A_{ij})$$

The pixel differences are efficient to compute and provide sufficient accuracy due to the substantial level of previous smoothing.

Robustness to illumination change is enhanced by thresholding the gradient magnitudes at a value of 0.1 times the maximum possible gradient value. This reduces the effect of a change in illumination direction for a surface with 3D relief, as an illumination change may result in large changes to gradient magnitude but will probably have less influence on gradient orientation.

Each key location is assigned a canonical orientation so that the image descriptors are invariant to rotation. In order to make this as stable as possible against lighting or contrast changes, the orientation is determined by the peak in a histogram of local image gradient orientations. The orientation histogram is created using a Gaussian-weighted window with σ of 3 times that of the current smoothing scale. These weights are multiplied by the thresholded gradient values and accumulated in the histogram at locations corresponding to the orientation, R_{ij}.

The stability of the resulting keys can be tested by subjecting natural images to affine projection, contrast and brightness changes, and addition of noise. The location of each key detected in the first image can be predicted in the transformed image from knowledge of the transform parameters. Testing on a collection of natural images [13] shows that about 70% of the key locations will be stable to these changes and will be detected at the correct corresponding location, orientation, and scale.

5 Local Image Description

Given a stable location, scale, and orientation for each key, it is now possible to describe the local image region in a manner invariant to these transformations. In addition, it is desirable to make this representation robust against small shifts in local geometry, such as arise from affine or 3D projection. One approach to this is suggested by the response properties of complex neurons in the visual cortex, in which a feature position is allowed to vary over a small region while orientation and spatial frequency specificity are maintained. Edelman, Intrator & Poggio [5] have performed experiments that simulated the responses of complex neurons to different 3D views of computer graphic models, and found that the complex cell outputs provided much better discrimination than simple correlation-based matching. This can be seen, for example, if an affine projection stretches an image in one direction relative to another, which changes the relative locations of gradient features while having a smaller effect on their orientations and spatial frequencies.

This robustness to local geometric distortion can be obtained by representing the local image region with multiple images representing each of a number of orientations (referred to as orientation planes). Each orientation plane contains only the gradients corresponding to that orientation, with linear interpolation used for intermediate orientations. Each orientation plane is blurred and resampled at fewer locations to allow for larger shifts in positions of the gradients.

This approach can be efficiently implemented by using the same precomputed gradients and orientations for each level of the pyramid that were used for orientation selection. For each keypoint, we use the pixel sampling from the pyramid level at which the key was detected (to maintain scale invariance). The pixels that fall in a circle of radius 8 pixels around the key location are inserted into the orientation planes. The orientations are measured relative to that of the key by subtracting the key's orientation. For our experiments we used 8 orientation planes, each sampled over a 4×4 grid of locations, with a sample spacing 4 times that of the pixel spacing for that pyramid level. The blurring is achieved by allocating the gradient of each pixel among its 8 closest neighbors in the sample grid, using linear interpolation in orientation and the two spatial dimensions. This implementation is much more efficient than performing explicit blurring and resampling, yet gives almost equivalent results.

In order to sample the image at a larger scale, the same process is repeated for a second level of the pyramid one octave higher. However, this time a 2×2 rather than a 4×4 sample region is used. This means that approximately the same image region will be examined at both scales, so that any nearby occlusions will not affect one scale more than the other. Therefore, the total number of samples in the SIFT key vector, from both scales, is $8 \times 4 \times 4 + 8 \times 2 \times 2$ or 160 elements, giving enough measurements for high specificity.

6 Indexing and Matching

For indexing, we need to store the SIFT keys for sample images and then identify matching keys from new images. The problem of identifying the most similar keys for high dimensional vectors is known to have high complexity if an exact solution is required. However, a modification of the k-d tree algorithm called the best-bin-first search method (Beis & Lowe [2]) can identify the nearest neighbors with high probability using only a limited amount of computation.

An efficient way to cluster reliable model hypotheses is to use the Hough transform [1] to search for keys that agree upon a particular model pose. Each model key in the database contains a record of the key's parameters relative to the model coordinate system. Therefore, we can create an entry in a hash table predicting the model location, orientation, and scale from the match hypothesis. We use a bin size of 30 degrees for orientation, a factor of 2 for scale, and 0.25

Fig. 1. Top row shows model images for 3D objects with outlines found by background segmentation. Images below show the recognition results for these objects with superimposed model outlines and image key locations used for matching.

times the maximum model dimension for location. These rather broad bin sizes allow for clustering even in the presence of substantial geometric distortion, such as due to a change in 3D viewpoint. To avoid the problem of boundary effects in hashing, a hypothesis that is close to a bin boundary is hashed into bins on both sides of the boundary for all dimensions (analogous to a feature neuron activating multiple object-selective neurons with overlapping receptive fields).

The hash table is searched to identify all clusters of at least 3 entries in a bin, and the bins are sorted into decreasing order of size. Each such cluster is then subject to a verification procedure in which a least-squares solution is performed for the affine projection parameters relating the model to the image [12,13].

Outliers can now be removed by checking for agreement between each image feature and the model, given the parameter solution. Each match must agree within 15 degrees orientation, $\sqrt{2}$ change in scale, and 0.2 times maximum model size in terms of location. If fewer than 3 points remain after discarding outliers, then the match is rejected. If any outliers are discarded, the least-squares solution is re-solved with the remaining points.

Knowledge of the model pose allows us to perform top-down matching, in which all image keys within the model region are checked for whether they are consistent with the model pose. Features are often added at this stage that were not in the original hash bin due to errors in predicting the model parameters from a single feature or due to previous ambiguity in the object match for a feature.

7 Experiments

The top row of Figure 1 shows three model images of typical objects that were used to test recognition. The models were photographed on a black background, and object outlines were extracted by segmenting out the background region (the outlines are used only for display purposes and play no role in recognition). Examples of recognition in occluded, cluttered images are shown below the model images. The SIFT key locations used for recognition are shown superimposed on the test images. The object outlines are also projected onto the image using the best-fit affine parameter solution. Since only 3 keys are needed for robust recognition, it can be seen that the solutions are often highly redundant and can survive substantial occlusion.

Although the model images and affine parameters do not account for rotation in depth of 3D objects, they are still sufficient to perform robust recognition of 3D objects over about a 20 degree range of rotation in depth away from each model view. The images in these examples are of size 384 × 512 pixels. The computation times for recognition of all objects in each image are about 1.5 seconds on a Sun Sparc 10 processor, with about 0.9 seconds required to build the scale-space pyramid and identify the SIFT keys, and about 0.6 seconds to perform indexing and least-squares verification. This does not include time to pre-process each model image, which would be about 1 second per image, but would only need to be done once for initial entry into a model database.

These first examples used only single training images and tested recognition for views within about 20 degrees of the training view. We have recently developed an approach to integrating images from many different viewpoints. Large changes in pose are handled by noticing when the least-squares solution has a high residual (above 0.1 of model image size) and creating a new model view. Points that match across views are linked so that indexing can consider both possible model views for images near the boundaries. For small changes in view (with a low least-squares residual) the key points from all images are combined into a common model view. This allows a single model to incorporate features from training views that have different illumination or changes in model shape, such as differing expressions on a face.

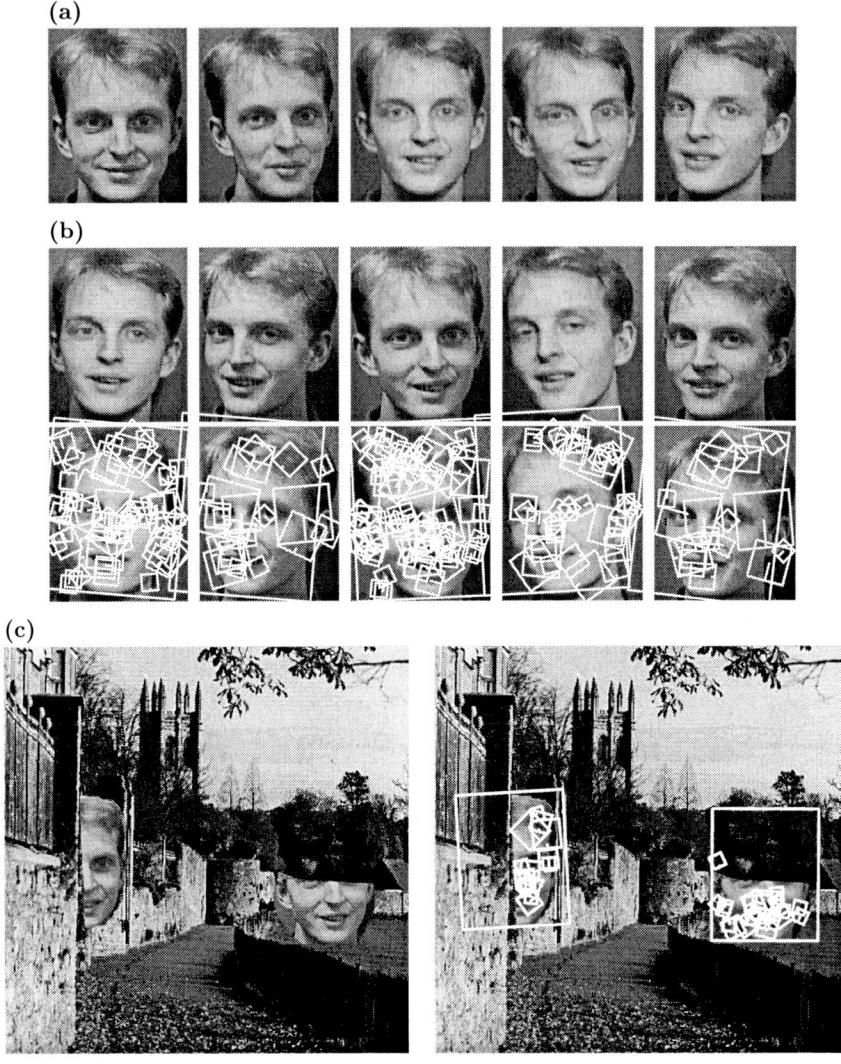

Fig. 2. (a) Five training images of a face were used to build a model. (b) The model was matched to five test images, as shown with the overlay of matched keys below each image. (c) Parts of a test image could be recognized following insertion into a cluttered scene. (Face images are courtesy of AT&T Laboratories Cambridge).

Figure 2 shows this recognition approach being applied to face images. Five training images shown in Figure 2(a) were used to build a model of the face. These were easily matched to one another in spite of variations in pose and illumination, and the features combined into a common model. This model could then be matched to subsequent images, such as the five test images in Figure 2(b). Under each image are shown the keys that were used to match the model to the image, as well as an outer rectangle showing the least-squares solution for the model affine fit. The robustness of the approach is illustrated by taking portions of a test image and inserting them into a natural cluttered image, as shown in Figure 2(c). These face parts are easily identified, as shown by the superimposed keys and solution for the model frame. In fact, the face portion could be considerably smaller and still produce at least 3 key matches. While these experiments are still at a preliminary stage, they show promise for face recognition from arbitrary pose in cluttered scenes.

8 Conclusions and Discussion

We have described an efficient method for the detection of dense local features of intermediate complexity. These SIFT features are sensitive to local image shape properties while being partially invariant to many common imaging transformations. Object matching is performed by identifying clusters of features that agree on a model interpretation, solving for best-fit model parameters, and integrating features that agree with this interpretation. This approach is broadly similar in its use of features and their integration to what is known of object recognition in visual cortex.

One topic that we have not considered is the discrimination and categorization of similar instances within an object class (for example, identifying a particular person or expression from matching a face image). This could probably be performed using the same set of SIFT features by maintaining the correlations between each feature type and the categorizations of interest, but this approach remains to be developed and tested. In IT cortex, neurons are organized in columns that respond to close variations on each feature [20], and these feature columns likely play the role of assisting fine discriminations between similar shapes.

Human vision is sometimes able to identify clear, isolated objects so rapidly that it seems likely that recognition is performed in a purely bottom-up manner [23]. We might model this as consisting of just the Hough transform stage that generates object hypotheses as a conjunction of features without detailed verification of consistent model parameters. However, human object recognition in cluttered scenes appears to require a process of serial attention to one object at a time [25] (a familiar scene can itself be recognized as a single object). This appears to involve the determination of object pose and other parameters, as well as selection and integration of features consistent with these parameters. This process appears to be consistent with the verification component described above, including best-fit parameter solving, outlier detection, and integration of new consistent features that were not sufficiently distinctive to contribute to the initial object hypothesis.

One clear difference to IT cortex is that the SIFT features are invariant to image rotation, while neurons in IT cortex are usually not. However, the rotation invariance brings significant improvements in efficiency and is easily accounted for by the subsequent indexing and least-squares verification. Less is known about how orientation constraints are enforced in primate vision.

Another area for further development is to add new SIFT feature types to incorporate color, texture, and edge groupings, all of which play an important role in primate vision. Scale-invariant edge groupings that make local figure-ground discriminations would be particularly useful at object boundaries where background clutter can interfere with other features. The indexing and verification framework allows for all types of scale and rotation invariant features to be incorporated into a single model representation.

References

1. Ballard, D.H., "Generalizing the Hough transform to detect arbitrary patterns," *Pattern Recognition,* **13,** 2 (1981), pp. 111-122.
2. Beis, Jeff, and David G. Lowe, "Shape indexing using approximate nearest-neighbour search in high-dimensional spaces," *Conference on Computer Vision and Pattern Recognition,* Puerto Rico (1997), pp. 1000–1006.
3. Booth, Michael C.A., and Edmund T. Rolls, "View-invariant representations of familiar objects by neurons in the inferior temporal cortex," *Cerebral Cortex,* **8** (1998), pp. 510–523.
4. Crowley, James L., and Alice C. Parker, "A representation for shape based on peaks and ridges in the difference of low-pass transform," *IEEE Trans. on Pattern Analysis and Machine Intelligence,* **6,** 2 (1984), pp. 156–170.
5. Edelman, Shimon, Nathan Intrator, and Tomaso Poggio, "Complex cells and object recognition," Unpublished Manuscript, preprint at http://kybele.psych.cornell.edu/~edelman/abstracts.html#ccells
6. Ito, Minami, Hiroshi Tamura, Ichiro Fujita, and Keiji Tanaka, "Size and position invariance of neuronal responses in monkey inferotemporal cortex," *Journal of Neurophysiology,* **73,** 1 (1995), pp. 218–226.
7. Kobatake, Eucaly, and Keiji Tanaka, "Neuronal selectivities to complex object features in the ventral visual pathway of the macaque cerebral cortex," *Journal of Neurophysiology,* **71,** 3 (1994), pp. 856–867.
8. Lindeberg, Tony, "Scale-space theory: A basic tool for analysing structures at different scales", *Journal of Applied Statistics,* **21,** 2 (1994), pp. 224–270.
9. Lindeberg, Tony, "Detecting salient blob-like image structures and their scales with a scale-space primal sketch: a method for focus-of-attention," *International Journal of Computer Vision,* **11,** 3 (1993), pp. 283–318.
10. Logothetis, Nikos K., Jon Pauls, and Tomaso Poggio, "Shape representation in the inferior temporal cortex of monkeys," *Current Biology,* **5,** 5 (1995), pp. 552–563.
11. Lowe, David G., "Three-dimensional object recognition from single two dimensional images," *Artificial Intelligence,* **31,** 3 (1987), pp. 355–395.
12. Lowe, David G., "Fitting parameterized three-dimensional models to images," *IEEE Trans. on Pattern Analysis and Machine Intelligence,* **13,** 5 (1991), pp. 441–450.
13. Lowe, David G., "Object recognition from local scale-invariant features," *International Conference on Computer Vision,* Corfu, Greece (September 1999), pp. 1150–1157.
14. Mel, Bartlett W., "SEEMORE: Combining color, shape, and texture histogramming in a neurally-inspired approach to visual object recognition," *Neural Computation,* **9,** 4 (1997), pp. 777–804.

15. Murase, Hiroshi, and Shree K. Nayar, "Visual learning and recognition of 3-D objects from appearance," *International Journal of Computer Vision,* **14,** 1 (1995), pp. 5–24.
16. Perrett, David I., and Mike W. Oram, "Visual recognition based on temporal cortex cells: viewer-centered processing of pattern configuration," *Zeitschrift für Naturforschung C,* **53c** (1998), pp. 518–541.
17. Schiele, Bernt, and James L. Crowley, "Recognition without correspondence using multidimensional receptive field histograms," *International Journal of Computer Vision,* **36,** 1 (2000), pp. 31–50.
18. Schmid, C., and R. Mohr, "Local grayvalue invariants for image retrieval," *IEEE PAMI,* **19,** 5 (1997), pp. 530–534.
19. Swain, M., and D. Ballard, "Color indexing," *International Journal of Computer Vision,* **7,** 1 (1991), pp. 11–32.
20. Tanaka, Keiji, "Neuronal mechanisms of object recognition," *Science,* **262** (1993), pp. 685–688.
21. Tanaka, Keiji, "Mechanisms of visual object recognition: monkey and human studies," *Current Opinion in Neurobiology,* **7** (1997), pp. 523–529.
22. Tovee, Martin J., Edmund T. Rolls, and V.S. Ramachandran, "Rapid visual learning in neurones of the primate temporal visual cortex," *NeuroReport,* **7** (1996), pp. 2757–2760.
23. Treisman, Anne M., and Nancy G. Kanwisher, "Perceiving visually presented objects: recognition, awareness, and modularity," *Current Opinion in Neurobiology,* **8** (1998), pp. 218–226.
24. Viola, Paul, "Complex feature recognition: A Bayesian approach for learning to recognize objects," *MIT AI Memo 1591,* Massachusetts Institute of Technology (1996).
25. Wolfe, Jeremy M., and Sara C. Bennett, "Preattentive object files: shapeless bundles of basic features," *Vision Research,* **37,** 1 (1997), pp. 25–43.

Straight Line Detection as an Optimization Problem: An Approach Motivated by the Jumping Spider Visual System

Felipe Miney G. da Costa and Luciano da F. Costa

Cybernetic Vision Research Group, IFSC-USP, Caixa Postal 369, São Carlos, SP,
13560-970, Brazil
{miney, luciano}@if.sc.usp.br

Abstract. Straight lines are important features in images, and their detection plays a major role in the compression, representation and analysis of visual information. The visual system of spiders from the *Salticidae* family is especially effective for straight line detection, due to their elongated and moveable retinae, which are used to scan the visual field. This paper presents a method for straight line motivated by the visual system of the *Salticidae*, which uses an optimization strategy (namely Nelder and Mead's amoeba with simulated annealing) to find maxima on the continuous ρ-θ parameter space that correspond to straight lines in the image. The method considers the spatially quantized nature of the image spaces and allows unlimited parametric resolution without the need to sample large regions of the parameter space.

1 Introduction

A characteristic that is always present in computational vision, as well as in natural vision, is the large amount of input information to be processed. Since such information always presents a high degree of correlation, natural and artificial vision systems have employed various strategies to reduce such a redundancy, in order to make the analysis and understanding of visual information possible [1]. As most of the relevant data for the recognition of objects resides at the borders of these objects, contour detection is a widely used technique that significantly decreases the amount of information to be processed. Straight line segments represent a further step in the reduction of this information (and increasing abstraction), since each segment can be determined in terms of only its starting and termination points. Straight line detection is also performed by the human visual system [13], as well as by other living beings. A particularly interesting visual system, as far as straight line detection is concerned [3], is that exhibited by the jumping spiders (*Salticidae*), which is one of the most developed among all invertebrates.

Motivated by the visual system of the jumping spiders, a method for straight line detection is proposed in the current article which uses an optimization approach in a continuous parametric (polar) space, where the straight lines are mapped into peaks.

The method allows straight line detection with unlimited accuracy without the need to calculate the entire parameter space, as implied by the Hough Transform. Moreover, the proposed approach allows the issue of straight line segment detection to be treated as an optimization problem, paving the way for the application of many traditional approaches in this area, such as simulated annealing and genetic algorithms. In addition, the straight line parameters are pursued in a continuous space, allowing unlimited precision, without the need to sample large regions in this space.

2 The Jumping Spider Visual System

Jumping spiders have a peculiar and efficient (although highly specialized) visual system, since their hunting behavior and courtship is almost entirely visual. They have four pairs of eyes, but one of them is atrophied (Posterior-Median); two of the other pairs are responsible for wide-angle movement detection (Anterior-Lateral and Posterior-Lateral), and the last pair is responsible for pattern recognition (Anterior-Median). The retinae of the anterior-median pair of eyes are moveable and have elongated shapes, which are almost straight in older species, and boomerang-like in newer ones [10]. The visual system of the jumping spiders is believed to function in the simple, yet effective, way described below.

When the lateral eyes detect movement, the spider positions itself so as to face the source of the movement. The subsequent behavior depends of the following three factors: size, speed and distance to the target. If the target has more than twice the size of the spider, the spider does not approach the target, and tries to escape if it comes towards her. If the target has adequate size, the speed and shape will then be analyzed. When a target is moving fast, i.e. more than 4°/sec, the spider will chase it, guided by her anterior-lateral eyes. Otherwise, the spider will carefully approach the target, and analyze it with her anterior-median eyes. The main purpose of this analysis is to define if the target is a prey or another spider of the same species. If it is considered prey, the spider will attack; if it is a conspecific spider, the behavior will depend on the sex and maturity of the spiders involved, as well as whether it is mating time or not [4].

This analysis is believed to be accomplished through a complex set of rotations and translations of the retinae of the anterior-median eyes, which actively scan the visual field, trying to match their elongated shapes with straight features of the visual scene [4]. This method, which can be understood as a sort of template matching where the template has elongated shape, is effective for straight line detection because when the retina matches a straight feature in the scene, the resulting neural signal is very strong, and little or no further additional processing is necessary to identify the straight line.

The method for straight line segment developed in this paper consists in searching the image with a linear window, trying to match this window with a straight line, such as the jumping spider does.

3 Digital Straight Lines and Continuous Parameter Spaces

Digital straight lines are the straight entities found in spatially quantized or *digital* image spaces. Their discrete (spatially sampled) nature must be carefully taken into account in any attempt to represent and detect them because their characteristics in such discrete spaces are richer and more complex than in continuous spaces, where they are trivial (smallest distance between two points). For instance, many straight lines map into a single digital straight line, implying a degenerated mapping. Even the total number of digital straight lines in an image is not known.

Melter [11] has proposed an interesting criterion for defining a digital straight line, which says that, given a set of points in a digital space, these will correspond to a digital straight line if, when the best line passing through these points is calculated using the least squares method and the resulting line is quantized using the GIQ (Grid Intercept Quantization [5]) method, the resulting points match the original set.

Many applications consider the parameterization of straight lines in terms of the angle of the normal and the distance from the origin (θ and ρ, respectively). A straight line in such a parameter space is the set of points (x, y) obeying:

$$\rho = x.\cos(\theta) + y.\sin(\theta) \qquad (1)$$

In the case of discrete parameter spaces (orthogonal lattices), this formula yields:

$$y.sen(\theta) + x.\cos(\theta) + \frac{\Delta sen(\theta)}{2} < \rho \leq y.sen(\theta) + x.\cos(\theta) - \frac{\Delta.sen(\theta)}{2} \qquad (2)$$

where $\square$ represents the width of a *straight band*, characterized by the fact that all its internal points (x, y) map to the same point (ρ,θ) in the parameter space.

The ρ-θ parameter space has often been adopted in the Hough Transform [8], which maps each of the image points (usually its edge elements) into spatially quantized sinusoidal bands in the discrete parameter space (i.e. the accumulator array) [2].

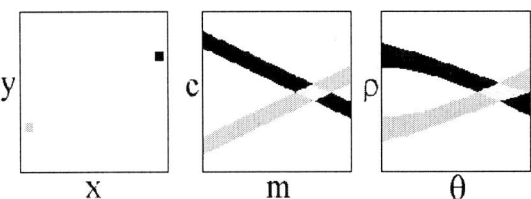

Fig. 1. Pixels in the image (x,y) space, and corresponding digital bands in the (m,c) and (θ,ρ) parameter spaces.

Furthermore, just as sinusoids generated from collinear points intercept over a single point in the continuous ρ-θ space, the sinusoidal bands produced by collinear points intercept over a small polygonal region, inside which all possible (ρ,θ) pairs backmap into the same digital line in the image. In a way analogous to the Hough transform, the regions covered by the bands can be added in order to produce a continuous version of the accumulator array, which is considered henceforth. Figure 2 shows, respective to the polar (a) and slope/intercept (b) parameterizations, these regions on the parameter space generated for all possible straight lines with $\pi/2 < \theta < \pi$, and $\rho > 0$ in a 9x9 image.

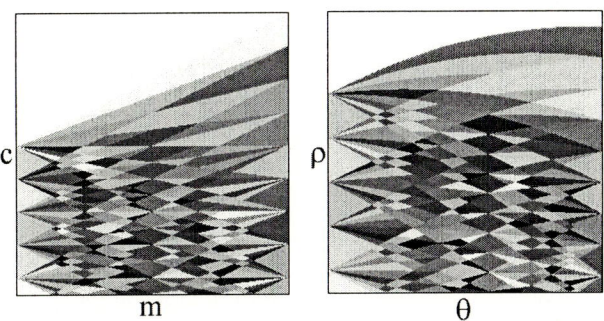

Fig. 2. Polygonal regions corresponding to digital straight lines, for a 9x9 pixels image.

4 Finding Straight Lines with Linear Windows

The Hough Transform has been widely used as a tool for straight line detection. However, at least in the more traditional approaches, a good portion (sometimes the whole) of the parameter space has to be calculated while searching for lines. The possibility of searching the visual space (i.e. the image) by using a linear window, as motivated by the jumping spider visual system, allows adaptive searching schemes to be considered while searching for the straight lines without implying systematic calculation of the parameter space. Indeed, such an approach allows the whole problem of straight line detection to be understood as an optimization approach. More specifically, we will be searching for extrema in the polar parameter space in a sequential and effective fashion, scanning only through a few points in the parameter space. In order to cater for enhanced efficiency, we will henceforth consider the maximization algorithm known as "Downhill Simplex" [14], also to be combined with simulated annealing. This algorithm consists in placing n+1 points in a n-dimensional space, calculating the function values for these points and moving the worst point according to some rules. The detection of straight lines is a two dimensional problem, and the simplex is composed by three points, with the following possible movements: Reflection with Expansion, Simple Reflection, Normal and reflected contractions and *Shrinkage*; which are tried in this order.

More information about the strategies for application of these operations can be found in [14]. While this algorithm can find the nearest maximum in the parameter space, it is not robust to local maxima (i.e. it does not guarantee global extreme) or plateaus on the search surface, which are both commonly found in typical parameter spaces. These effects are discussed in more detail in the following. Figure 3 shows an isolated peak in the parameter space produced by adding the bands defined by a digital straight line in a 9×9 image. Even at this low magnification, plateaus are already visible around the peak. Indeed, the very tip of the peak can be verified to be itself a plateau.

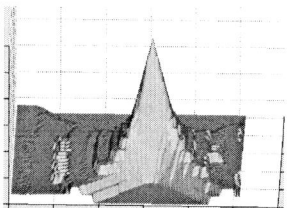

Fig. 3. Plateaus around an isolated peak in the parameter space, produced by the addition of the bands defined by a straight line in a 9x9 image.

In order to prevent the search algorithm from stopping at the plateaus, it is possible to use wider linear windows, which result in smoother surfaces, as shown in Figures 4 and 5 (a).

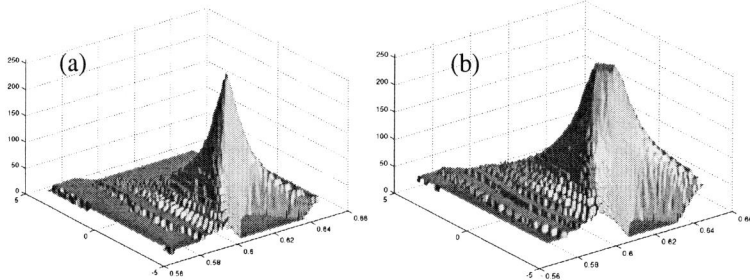

Fig. 4. Peaks corresponding to lines mapped with unitary(a) and double(b) width windows.

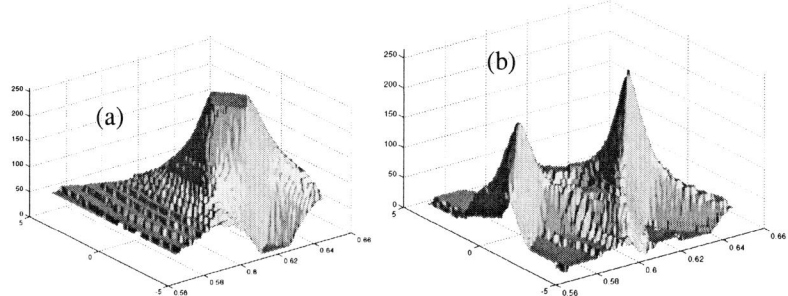

Fig. 5. Peak corresponding to a line mapped with a window with width 3(a) and local maxima produced by interference between the sought curve and other objects in the image (b).

The use of wider windows implies in a loss of precision, as the peak is also broadened. To address this issue, the algorithm starts with a wider window, but proceeds by adaptively narrowing the windows once the number of pixels in the image covered by the current window becomes stable. This process implies little execution time overhead to the algorithm, as the subsequent simplexes are launched close to the peak.

In addition to the plateaus implied by the discrete nature of the image, noise in the image or the presence of close straight lines and interference with other curves and objects can generate local maxima in the parameter space, shown in Figure 5(b). Since the window-scanning algorithm should be capable of avoiding these local maxima, a simulated annealing strategy was incorporated into the downhill simplex.

Simulated annealing strategies [9] were initially proposed for combinatorial minimization, and are widely used where near-optimal solutions are needed and many variables are involved, such as in circuit wiring. Annealing strategies allow non-optimal steps in algorithms with a controlled probability, based on a parameter called "temperature". As this temperature decreases, the algorithm takes less non-optimal steps and reaches a maximum. This allows the algorithm to follow the general topology of a surface at high temperatures, and then to closely analyze the region where the global maximum is more likely to be, as the temperature is decreased. The probability of accepting a given step is based on the function variation caused by the step, f; and on the temperature T, according to the probability: $P(f) = \exp(-f/T)$ [12].

Since there are many parameters controlling the simulated annealing strategy, an effective means for properly setting those parameters is fundamental for a good performance of the algorithm. These parameters are: initial temperature, maximum number of iterations per temperature and temperature reduction factor. The annealing strategy can be incorporated into the Downhill Simplex algorithm, allowing a maxima search in the presence of local maxima. The following strategy adds a fluctuation to the value of the function at the simplex vertexes:

$$f'(x,y) = f(x,y) + K. T. \log(z) \qquad (3)$$

where z is a random number between 0 and 1

With this fluctuation, the classification of the simplex points as best, second best and worst can be modified, with a probability based on the temperature and the difference of the values of the points. Therefore, an effect similar to that of annealing is achieved, and the downhill simplex becomes endowed with an ability to avoid local maxima [15]. This modified algorithm can also benefit from the above discussed adaptive window width, but the searches with narrower windows should not start with a very high temperature, in order to avoid departing from the maximum approached by the previous execution. Thus, when a subsequent search starts, at the point where the previous had ended, it does so at a lower temperature. Although nice, such a strategy implies yet more parameters to the optimization algorithm in addition to the annealing parameters, which are: tolerance of the downhill simplex, initial width of the window and initial temperature reduction with width reduction.

The multidimensional space of parameters for this annealing strategy with adaptive windows is particularly difficult to search, and characterizes a situation suited for the use of Genetic Algorithms. Genetic Algorithms were originally created to study evolution and adaptation in nature, and to apply these mechanisms in computer systems [6,7]. The genetic algorithm used in this paper is a basic version with several parameters as proposed by Holland, using standard values for mutation and crossover rates (0.001 and 0.7, respectively); single point crossover; and fitness proportional selection. The method used also incorporates "elitism", which consists in keeping some of the best individuals from one generation to another [13].

The "chromosome" consists in 48 bits, which describe the values of the 6 involved parameters. The fitness function evaluates the performance of the algorithm using a test image and the set of parameters in an individual. The evaluation is performed taking into account precision and processing time. Thus, the fitness function can be weighed to represent the importance of time/precision in a given application. The adopted population size was 50 individuals, which evolved for 100 generations ("epochs").

5 Experimental Results

Two classes of images have been considered. The first class of images (Figure 6a) consists of a stylized spider profile, with 4 straight line segments; and the second class (Figure 6b) are the borders extracted from a real (as opposed to computer generated) image.

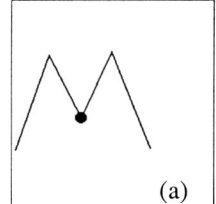 

Fig. 6. (a) spider profile, and (b) borders of a real image.

The first obtained results addressed the selection of an adequate set of parameters for the annealing-based searching method, via Genetic Algorithm. The fitness function incorporated a parameter K in order to weights the achieved precision, in such a way that the higher the parameter K was set, the more relevant became precision (relatively to the processing time). Five runs were performed for each type of image, and results were obtained for two values of K (4000 and 40000), as shown in Tables 1-4.

Table 1. Annealing parameters for the spider profile, K=4000

Evolution	Best fitness	Initial Temp.	Cooling Factor	Max. Iter.	Toler.	Initial width	Cooling w/ width
1	0.9966	154	0.54	50	2.33	1	0.54
2	1.403	334	0.52	18	2.96	1	0.60
3	1.095	156	0.54	35	2.32	1	0.74
4	0.8747	90	0.62	50	3.17	1	0.68
5	1.755	213	0.52	29	0.87	1	0.57

Table 2. Annealing parameters for the spider profile, K=40000

Evolution	Best fitness	Initial Temp.	Cooling Factor	Max. Iter.	Toler.	Initial width	Cooling w/ width
1	1.274	152	0.69	15	0.12	2	0.57
2	1.442	204	0.64	17	0.07	2	0.63
3	1.190	232	0.59	22	0.09	3	0.52
4	1.465	227	0.57	20	0.06	2	0.59
5	0.941	181	0.78	24	0.10	3	0.67

Table 3. Annealing parameters for the real image, K=4000

Evolution	Best fitness	Initial Temp.	Cooling Factor	Max. Iter.	Toler.	Initial width	Cooling w/ width
1	1.439	142	0.56	18	1.24	1	0.74
2	1.520	119	0.58	22	0.93	1	0.66
3	0.983	136	0.53	23	1.72	1	0.53
4	1.266	132	0.56	20	1.33	1	0.85
5	1.412	135	0.59	20	1.27	1	0.64

Table 4. Annealing parameters for the real image, K=40000

Evolution	Best fitness	Initial Temp.	Cooling Factor	Max. Iter.	Toler.	Initial width	Cooling w/ width
1	1.508	240	0.63	23	0.12	2	0.52
2	1.840	276	0.57	21	0.08	2	0.54
3	1.429	311	0.62	31	0.11	1	0.71
4	1.677	259	0.65	28	0.13	2	0.56
5	1.549	217	0.59	26	0.11	1	0.63

For the spider profile, the best parameter configuration used a fast-cooling algorithm, about 20 iterations per temperature and initial temperature around 200. The main differences between the algorithms (K=4000 and K=40000) were the initial width of the window, and the tolerance of the downhill simplex. For the real images, the results were similar, and for all the cases the best fitness corresponded to the lowest tolerance of the simplex.

The annealing-based searching method, considering the best parameter configurations for each type of image, was executed for the spider profile, and the results are shown in table 7, respectively to K=4000 and 40000.

Table 5. Average results for the detection of the lines in a spider profile

Line	(K=4000)	(K=4000)	(K=40000)	(K=40000)
1	1.35	0.0308	0.06	0.0013
2	1.02	0.0113	0.02	0.0010
3	0.94	0.0092	0.04	0.0006
4	0.78	0.0125	0.02	0.0007

The errors in   and   were much smaller for the strategy defined by K=40000, but the average processing time of the algorithm was three times as large (approximately 3000ms for K=4000 and approximately 9000 for K=40000).

For the spider profile and the real image, the parameter space was partitioned in 64 regions, and the algorithm was launched once at each of these regions. Therefore, not all the lines were necessarily detected (e.g. when more than two lines were present in a same window), and alternatives for a more effective identification of the launching points in each image are currently being investigated. For the real image, the lines found are traced over the image in Figure 7, for K=4000 (a) and K=40000 (b).

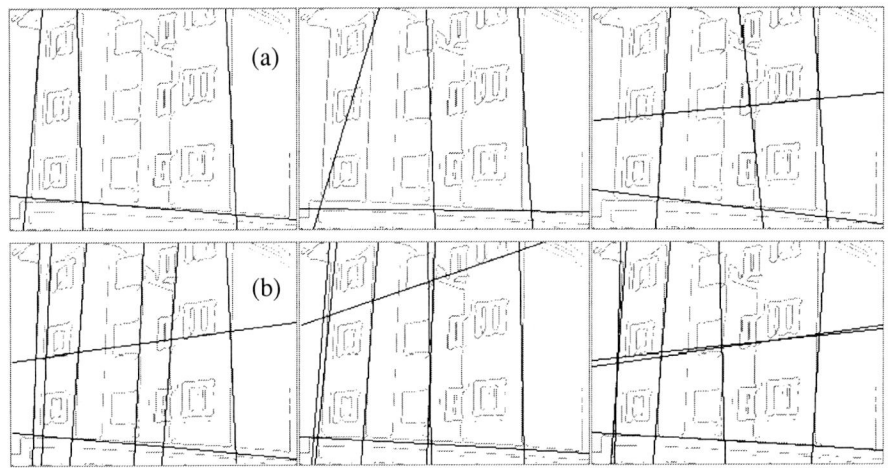

Fig. 7. Detection of straight lines in a real image, K=4000 (a) and K=40000 (b).

6 Concluding Remarks

This paper has presented a method for straight lines detection motivated by the visual system of the *Salticidae*, one of the most sophisticated and powerful amongst the invertebrates. The basic features explored in the current approach are the use of straight scanning windows and the search for straight stimuli in terms of template matching (more specifically the tensorial product between the image and the window). This formulation has allowed the problem of straight line detection to be treated as an optimization problem, in such a way that unlimited resolution can be achieved without calculating a large portion of the parameter space, as it is implied by the Hough transform and related techniques. The specific features appearing when digital straight lines are mapped into continuous parametric spaces (i.e. local extrema and plateaus) have been characterized and explicitly taken into account while devising optimization strategies to control the window scanning. More specifically, the Nelder and Mead's amoeba technique incorporating simulated annealing has been used in order to scan the parameter space while looking for peaks corresponding to the main straight features in the presented image. The determination of a suitable set of values for the parameters controlling the annealing-based scanning has been obtained by using the genetic algorithm while executing the strategy over two distinct types of images, and the respectively obtained parameter configurations have been discussed. The proposed approach to straight line detection has been verified to be effective, especially when the parameters are properly set. It should be observed that the progressive decrease of the window width considered in the presented methodology suggests that the several layer of photoreceptors in the spider retina may perform a similar role.

References

1. Barlow, H. "What does the brain see? How does it understand?" IN: Images and Understanding. Cambridge University press, 1990. (pp. 5-25)
2. Costa, L.F., Effective Detection of Line Segments with Hough Transform. Ph.D. Thesis, King's College, University of London, 1992.
3. Costa L.F., "On jumping Spiders and Hough Transform" IN: Systems and Control 94 - IASTED. Lugano, Switzerland, Jun 1994. (pp. 138 - 141)
4. Forster, L., "Target Discrimination in Jumping Spiders (Araneae: Salticidae)" IN: Neurobiology of Arachnids. Springer - Verlag, Heidelberg, 1985.
5. Freeman, H., "Boundary Encoding and Processing" IN: Picture Processing and Psychopictorics. Academic Press, New York/London, 1970. (pp. 241 - 266)
6. Goldberg, D.E. Genetic Algorithms in Search, Optimization and Machine Learning, Addison-Wesley Publishing Company, 1989.
7. Holland, J.H. Adaptation in Natural and Artificial Systems, University of Michigan Press, 1975.
8. Hough, P V C, Method and Means for Recognizing Complex Patterns. U.S.Patent 3.069.654, Dec 1962.
9. Kirkpatrick, J., Gelatt, C.D. and Vecchi, M.P. "Optimization by Simulated Annealing" IN: Science, Vol. 220, n.4598, May 1983. (pp. 671 - 680)
10. Land, M.F., "Morphology and Optics of Spider Eyes" IN: Neurobiology of Arachnids. Springer - Verlag, Heidelberg, 1985.
11. Melter, R.A., Stojmenovic, I. and Zunic, J. "A New Characterization of Digital Lines by Least Square Fits" IN: Pattern Recognition Letters. Vol. 14, n.2, 1993. (pp. 83 - 88)
12. Metropolis, N., Rosenbluth, A., Rosenbluth, M., Teller, A. and Teller, E. "Equation of State Calculations by Fast Computing Machines" IN: Journal of Chemical Physics, Vol. 21, 1953. (pp. 1087 - 1090)
13. Mitchell, M. An Introduction to Genetic Algorithms, The MIT Press, Massachusetts, 1998.
14. Nelder, J.A. and Mead, R. "A simplex Method for Function Minimization" IN: The Computer Journal, 1965. (pp. 308 - 313)
15. Press, W.H., Teukolski, S.A., Vetterling, W.T. and Flannery, B.P. Numerical Recipes in C, Cambridge University Press, 1992.

Factorial Code Representation of Faces for Recognition

Seungjin Choi and Oyoung Lee

Department of Electrical Engineering
Chungbuk National University
48 Kaeshin-dong, Cheongju
Chungbuk 361-763, Korea
{schoi,loy250}@engine.chungbuk.ac.kr

Abstract. The information-theoretic approach to face recognition is based on the compact coding where face images are decomposed into a small set of basis images. A popular method for the compact coding may be the principal component analysis (PCA) which eigenface methods are based on. PCA based methods exploit only second-order statistical structure of the data, so higher-order statistical dependencies among pixels are not considered. Factorial coding is known as one primary principle for efficient information representation and is closely related to redundancy reduction and independent component analysis (ICA). The factorial code representation exploits high-order statistical structure of the data which contains important information and is expected to give more efficient information representation, compared to eigenface methods. In this paper, we employ the factorial code representation in the reduced feature space found by the PCA and show that the factorial code representation outperforms the eigenface method in the task of face recognition. The high performance of the proposed method is confirmed by simulations.

1 Introduction

The task of face recognition is to identify each person in an unlabeled set of face images (test set), given a set of face images labeled with the person's identity (training set). The information-theoretic approach to face recognition is based on the compact coding (undercomplete data representation) where face images are decomposed into a linear sum of small number of basis images. The feature vectors are obtained by projecting face images onto the subspace spanned by the basis images. A popular method for the compact coding may the factor analysis (FA) or PCA. A number of face recognition algorithms are based on PCA. They include eigenpitures [13], eigenfaces [15], Fisherfaces [5], and local feature analysis (LFA) [11], to name a few.

The compact coding seeks for a linear transformation that allows the input to be represented with a reduced number of basis vectors with minimal reconstruction error. The PCA is a well-known exemplary method of compact coding. The PCA aims at finding a set of mutually orthogonal basis vectors that capture the directions of maximum variance in the data and for which the basis coefficients (features) are uncorrelated [9]. It is known that the PCA is the optimal linear coding in mean square sense. In this sense, the PCA might be an appropriate candidate for dimensionality reduction, but it does not guarantee optimal classification [7]. Factor analysis (FA) is closely related to the PCA and finds a linear data representation that best model the covariance structure of the data. PCA or FA based methods exploit only second-order structure of the data, so higher-order statistical dependencies among pixels are not considered.

Factorial coding is known as one primary principle for efficient information representation and is closely related to redundancy reduction [3] and independent component analysis (ICA). The factorial code representation aims at finding a linear data representation which best model the probability distribution of the data, so higher-order statistical structure is incorporated. In this paper, we investigate the factorial code representation in the framework of latent variable models and apply it to the task of face recognition. Since image space is very high dimensional space, we find a factorial code representation in the reduced feature space that is already found by the PCA method. Both theoretical results and experimental results are presented to confirm the validity of the proposed method.

2 Eigenfaces

In the framework of pattern recognition, each of the pixel values in a sample image is considered as a coordinate in a high dimensional space, i.e., *the image space*. Let us consider a set of $m_r \times m_c$ face images, $\{I_t\}_{t=1}^{N}$. It is often convenient to regard a two-dimensional image I as a vector $\boldsymbol{x}$. The vectorized image $\boldsymbol{x}$ is constructed by the concatenation of rows of I, and a training set of images is $\{\boldsymbol{x}_t\}_{t=1}^{N}$ where each $\boldsymbol{x}_t$ is m-dimensional vector ($m = m_r \times m_c$). In general, face images are very high dimensional space, so it is desirable to represent the images in a low-dimensional space for characterization or classification. The assertion that only a relatively small number of basis vectors should be necessary for face images to be identified, is rooted: (1) in the idea that humans are able to store and recognize enormous number of faces; (2) since recognition is *instantaneous*, it is conceivable that our brain adopts a preprocessing of visual information to encode it in efficient form [13]. From both mathematical and cognitive viewpoints, the PCA might be an efficient method for representing input data in a low-dimensional space. This leads to eigenfaces [13, 15].

The PCA seeks for a linear transformation from m-dimensional image space to n-dimensional feature space ($n < m$). The feature vectors $\{\boldsymbol{z}_t\}$ are defined by

$$\boldsymbol{z}_t = \boldsymbol{U}\left(\boldsymbol{x}_t - \boldsymbol{\mu}\right), \tag{1}$$

where $\boldsymbol{\mu}$ is average face defined by

$$\boldsymbol{\mu} = \frac{1}{N} \sum_{t=1}^{N} \boldsymbol{x}_t, \tag{2}$$

and $U \in \mathbb{R}^{n \times m}$ is a linear transformation matrix with orthonormal rows. In PCA, the matrix U is chosen to minimize the reconstruction error. It is known that the rows of U correspond to the principal eigenvectors of the sample covariance matrix of $\{x_t\}$.

The sample covariance matrix C_x (the scatter matrix) is given by

$$C_x = \frac{1}{N} \sum_{t=1}^{N} (x_t - \mu)(x_t - \mu)^T$$

$$= \frac{1}{N} \Phi \Phi^T, \tag{3}$$

where Φ is a matrix, each column of which is a caricature of image vector x_t, i.e.,

$$\Phi = [x_1 - \mu, \ldots, x_N - \mu]. \tag{4}$$

Since the sample covariance matrix C_x is symmetric, we have the following eigen-decomposition:

$$C_x = V \Lambda V^T, \tag{5}$$

where V is the modal matrix (the column vectors of V correspond to the normalized eigenvectors of C_x) and Λ is the corresponding eigenvalue matrix whose diagonal elements are arranged in descending order of their magnitude. First n column vectors are selected and are assigned to the n row vectors of the matrix U in PCA. The n row vectors of the matrix U are called *eigenfaces* and are served as basis images.

In general, the number of sample images, N is less than the dimension of the image space, m. In this case, the rank of the sample covariance matrix given in (3) is at most N and the $m - N$ remaining eigenvectors belong to the null space of this degenerate matrix. Since only first n eigenvectors ($n \leq N < m$) are important, one can use a simple trick known as *the snap-shot method* [13, 15]. The snap-shot method is to constrain the eigenvectors we search for to be linear combinations of pattern vectors. Instead of finding the eigenvectors of C_x directly, one first find the eigenvectors of the matrix $\Phi^T \Phi$. Let us consider the eigenvectors $\{e_t\}$ of the matrix $\Phi^T \Phi$ such that

$$\Phi^T \Phi e_t = \lambda_t e_t. \tag{6}$$

Premultiplying both sides by Φ, we have

$$\Phi \Phi^T \Phi e_t = \lambda_t \Phi e_t. \tag{7}$$

One can see that Φe_t are the eigenvectors of $\Phi \Phi^T$. Note that $\{\Phi e_t\}$ spans the eigenspace of the ensemble of face images, but they are no longer orthogonal each other. A simple orthogonalization method (for example, Gram-Schmidt orthogonalization) can be employed. As will be shown in experimental results, the orthogonalized eigenvectors improves the classification performance. Alternative computationally efficient method of computing first few eigenvectors of the high dimension covariance matrix is to maximum likelihood PCA using EM optimization [14].

3 Factorial Code Representation

3.1 Learning Linear Generative Models

The approach we take here is to find a factorial code representation for the feature vectors $\{z_t\}$ that are found by the PCA method. In the context of linear generative model, the n-dimensional feature vector z is assumed to be generated by

$$z = a_1 s_1 + \cdots a_n s_n,$$
$$= As, \tag{8}$$

where $A = [a_1, \ldots, a_n] \in \mathbb{R}^{n \times n}$ whose column vectors are basis vectors (basis images) and the elements of $s \in \mathbb{R}^n$ are called basis coefficients (latent variables).

The factorial code representation aims at finding both A and s that best model the probability distribution of observed data (observed density). In other words, we find the linear generative model (8) that best match the model density and the observed density. Let us denote the observed density and model density by $p^o(z)$ and $p(z)$, respectively. We assume that basis coefficients $\{s_i\}$ are statistically independent, i.e., their joint density is factored into the product of marginal densities,

$$p(s) = \prod_{i=1}^{n} p_i(s_i). \tag{9}$$

In order to find a matrix A that best match $p^o(z)$ and $p(z)$, we consider the Kullback-Leibler divergence between them. The risk R that we consider here, is given by

$$R = KL\left[p^o(z) \| p(z)\right]$$
$$= \int p^o(z) \log \frac{p^o(z)}{p(z)} dz. \tag{10}$$

Since the transformation from latent variable vector s to the feature vector z in the model (8) is linear, the Jacobian of the transformation is simply the absolute value of its determinant. Thus, we have the following relation

$$\log p(z) = -\log |\det A| + \sum_{i=1}^{n} \log p_i(s_i). \tag{11}$$

Then the loss function L is given by

$$L = \log |\det A| - \sum_{i=1}^{n} \log p_i(s_i), \tag{12}$$

where $\log p^o(z)$ was neglected since it does not depend on A.

The gradient descent method gives the learning algorithm for A that has the form

$$\Delta A = -\eta_t \frac{\partial L}{\partial A}$$
$$= -\eta_t A^{-T} \left\{ I - \varphi(s) s^T \right\}, \qquad (13)$$

where $\eta_t > 0$ is the learning rate and $\varphi(s)$ is an elementwise function whose ith element $\varphi_i(s_i)$ is given by

$$\varphi_i(s_i) = -\frac{d \log p_i(s_i)}{ds_i}. \qquad (14)$$

Employing the natural gradient [1], we have

$$\Delta A = -\eta_t A A^T \frac{\partial L}{\partial A}$$
$$= -\eta_t A \left\{ I - \varphi(s) s^T \right\}. \qquad (15)$$

At each iteration, the basis coefficient basis vector, s is computed by $s = A^{-1} z$ using the current estimate of A. Then the value of A is updated by (15). This procedure is repeated until A converges.

The function $\varphi_i(s_i)$ depends on the prior $p_i(s_i)$ that has to be specified in advance. Depending to the choice of prior, we will have different data representation. For efficient information representation, we consider the Laplacian prior that is one of super-Gaussian prior which gives heavy-tailed distribution (sparse distribution). For the Laplacian prior, the function $\varphi_i(s_i)$ has the form

$$\varphi_i(s_i) = \text{sgn}(s_i), \qquad (16)$$

where $\text{sgn}(\cdot)$ is the signum function. Sparseness constraint was shown to be useful to describe the receptive field characteristics of simple cells in primary visual cortex [10].

Alternatively it is possible to learn A^{-1} instead of A. The A^{-1} coincides with the ICA filter [6]. If we define $W = A^{-1}$, then the natural gradient learning algorithm for W is given by

$$\Delta W = \eta_t \left\{ I - \varphi(s) s^T \right\} W. \qquad (17)$$

This is a well-known ICA algorithm [2].

3.2 Classification

Once the linear generative model for the reduced feature space is learned, new feature vector in the factorial code representation, s is given by

$$s = A^{-1} U (x - \mu). \qquad (18)$$

Simple similarity measure between two feature vectors might be the Euclidean inner product. For a test image, the corresponding feature vector s_{test} is computed by the relation (18). The feature vectors in the test set is assigned the class label of the feature vector in the training set, s_{train} with the maximal normalized correlation calculated by

$$\text{corr} = \frac{s_{test}^T s_{train}}{||s_{test}||_2 ||s_{train}||_2} \tag{19}$$

Alternatively, various vector norms (for example l_1, l_2, l_∞) of difference between two feature vectors can be used for classification.

One can also consider the Fisher's linear discriminant (FLD) which is a linear transformation that maximizes the ratio of between-scatter matrix and within-scatter matrix. The between-scatter matrix $\boldsymbol{S}_B$ is defined by

$$\boldsymbol{S}_B = \sum_{k=1}^{c} (\boldsymbol{\mu}_k - \boldsymbol{\mu})(\boldsymbol{\mu}_k - \boldsymbol{\mu})^T, \tag{20}$$

where N_k is the number of data in class k, $\boldsymbol{\mu}_k$ is the sample average of the data in class k, and c is the number of class. The within-scatter matrix $\boldsymbol{S}_W$ is defined by

$$\boldsymbol{S}_W = \sum_{k=1}^{c} \sum_{t \in \mathcal{C}_k} (\boldsymbol{x}_t - \boldsymbol{\mu}_k)(\boldsymbol{x}_t - \boldsymbol{\mu}_k)^T. \tag{21}$$

Then the Fisher's linear discriminant $\boldsymbol{W}_{FLD}$ is obtained by

$$\boldsymbol{W}_{FLD} = \arg\max_{\boldsymbol{W}} \text{tr} \left\{ \left(\boldsymbol{W} \boldsymbol{S}_W \boldsymbol{W}^T \right)^{-1} \left(\boldsymbol{W} \boldsymbol{S}_B \boldsymbol{W}^T \right) \right\}, \tag{22}$$

where tr denotes the trace operator. Note that if raw data $\boldsymbol{x}$ is replaced by the feature vector $\boldsymbol{z}$ (from PCA), then the resulting base images (row vectors of $\boldsymbol{W}_{FLD}$) is known as Fisher faces [5].

4 Experimental Results

We evaluated face recognition performance for the eigenface method and the factorial code representation method using the AR face database [8]. The data set contains frontal view faces of 40 people with different facial expressions (neutral expression, smile, anger) and illumination conditions (left-light-on, right-light-on) from two sessions which are separated by two weeks. There were 400 face images (40 people * 5 images * 2 sessions), 200 face images of which were selected as the training set (neutral expression, anger, and right-light-on from first session; smile and left-light-on from second session). The rest of face images (200 face images) were chosen as the test set (smile and left-light-on from first session; neutral expression, anger, and right-light-on from second session). Each face image was cropped to the size of 46 × 50 and the rows of face images were concatenated to produce 2300 × 1 column vectors. Sample images for the training set and the test set are shown in Figure 1.

The PCA method was applied to a set of training face images to find a linear transformation U. Feature vectors $z \in \mathbb{R}^n$ were calculated by

$$z = U(x - \mu). \tag{23}$$

First twenty eigenfaces (which correspond to row vectors of U) are shown in Figure 2 (a). The classification was performed using a nearest neighbor classifier in the reduced feature space ($n = 20$ and $n = 50$). The snap shot method was used to calculate principal eigenvectors. The result with a postprocessing of orthogonalization improved the classification performance (see Figure 3 (a)).

The factorial code representation was applied to the feature vector z. We used the Laplacian prior, so the corresponding nonlinear function $\varphi_i(s_i)$ was $\varphi_i(s_i) = \text{sgn}(s_i)$ where $\text{sgn}(\cdot)$ is the signum function. The basis images and coefficients were learned by the algorithm (15). Twenty basis images learned by the factorial code representation are shown in Figure 2 (b). Principal component feature vector z was projected onto the basis images learned by the factorial code to produce new feature vectors. These new feature vectors were used for classification using a nearest classifier. The factorial code representation method showed much improved recognition performance in contrast to the eigenface method (see Figure 3 (b)).

We have applied the FLD method in the reduced feature space. It also improved the recognition performance, compared to a simple nearest classifier. In the case of 50 features, FLD provided good performance (see Figure 4). The FLD was also applied for classification in factorial code representation. However, there was no improvement in recognition. Although FLD exploits within-class spreads, only second-order statistics is utilized, while the factorial code representation exploits higher-order structure of the data.

In our experiments, factorial code representation always gave better performance, compared to eigenfaces and Fisher faces. The result is summarized in Table 1.

5 Conclusions and Discussions

In the information-theoretic approach to face recognition, face images are decomposed into a set of basis images and feature vectors are obtained by projecting face images into basis images. In this paper, we have applied the factorial coding method to obtain basis images. It was shown that the factorial code representation outperformed the eigenface method where the PCA was used. In contrast to the PCA where only second-order structure of the data was used, the factorial code representation exploited higher-order statistical structure of the data which may contain much of important information. The factorial code representation approach that we consider in this paper is similar to the independent component representation [4] where the nonlinear information maximization based ICA method [6] was applied.

The set of data that we used here consisted of only frontal views. Although the factorial code representation exploits the higher-order structure of the data, its usefulness is limited by global linearity. For face recognition under varying views, it will be desirable to incorporate view-based eigenspaces [12] into the factorial coding. Alternatively local linear model could be adopted.

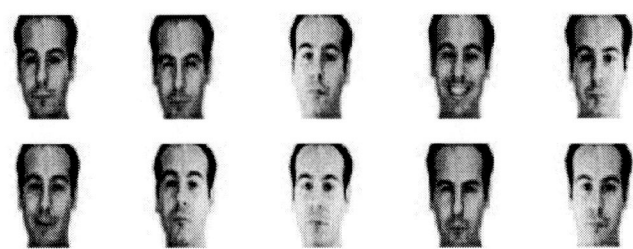

Fig. 1. Sample images in the training set (first row) and in the test set (second row).

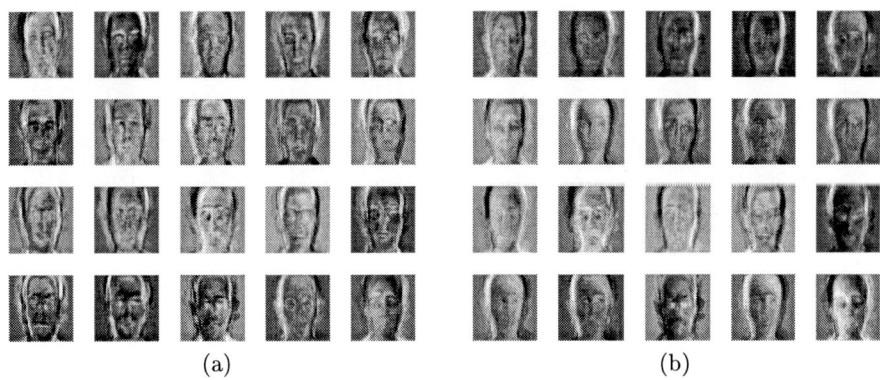

Fig. 2. First 20 basis images: (a) in eigenface method; (b) factorial code. They are ordered by column, then, by row.

Acknowledgments

This work was supported by the Braintech 21, Ministry of Science and Technology in KOREA. The first author is grateful to Prof. A. Cichocki in Brain Science Institute, RIKEN, Japan for helpful discussion.

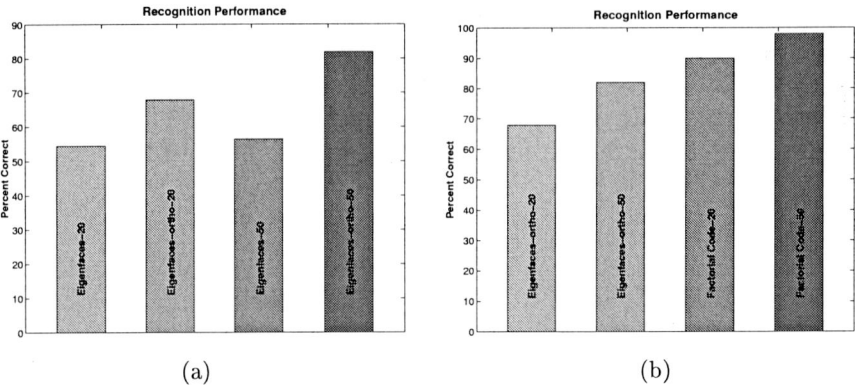

Fig. 3. The comparison of recognition performance: (a) the eigenface method using 20 or 50 principal components without orthogonalization and with orthogonalization in the snap shot method; (b) factorial code representation using 20 or 50 basis images.

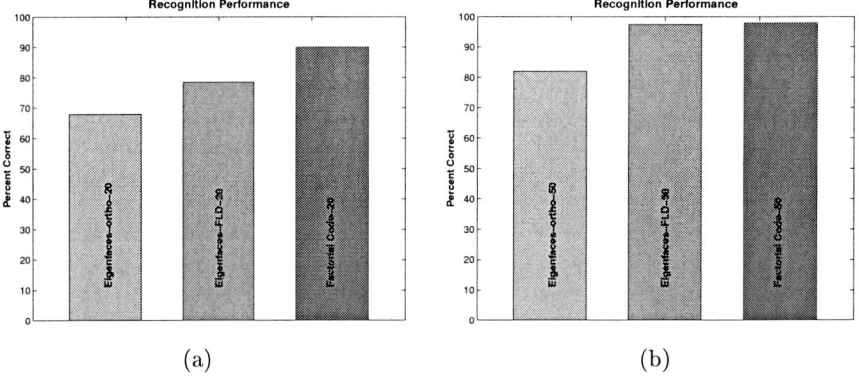

Fig. 4. The comparison of recognition performance: (a) 20 features; (b) 50 features. In both cases, we compare the eigenface method using a nearest classifier, eigenface+FLD, and factorial code representation using a nearest classifier.

Type of method	Percent correct
Eigenfaces-20	54.5 %
Eigenfaces-ortho-20	68 %
Eigenfaces-FLD-20	78.5 %
Factorial code-20	90 %

(a)

Type of method	Percent correct
Eigenfaces-50	56.5 %
Eigenfaces-ortho-50	82 %
Eigenfaces-FLD-50	97.5 %
Factorial code-50	98 %

(b)

Table 1. The recognition performance comparison: (a) for 20 features; (b) for 50 features.

References

1. S. Amari. Natural gradient works efficiently in learning. *Neural Computation*, 10(2):251–276, Feb. 1998.
2. S. Amari, T. P. Chen, and A. Cichocki. Stability analysis of learning algorithms for blind source separation. *Neural Networks*, 10(8):1345–1351, 1997.
3. H. B. Barlow. Unsupervised learning. *Neural Computation*, 1:295–311, 1989.
4. M. S. Bartlett, H. M. Lades, and T. J. Sejnowski. Independent component representations for face recognition. In *Proceedings of the SPIE Symposium on Electronic Imaging: Science and Technology; Conference on Human Vision and Electronic Imaging III*, pages 528–539, San Jose, California, Jan. 1998.
5. P. N. Belhumeur, J. P. Hespanha, and D. J. Kriegman. Eigenfaces vs. Fisherfaces: Recognition using class specific linear projection. *IEEE Trans. Pattern Analysis and Machine Intelligence*, 19(7):711–720, Oct. 1997.
6. A. Bell and T. Sejnowski. An information maximisation approach to blind separation and blind deconvolution. *Neural Computation*, 7:1129–1159, 1995.
7. K. Fukunaga. *An Introduction to Statistical Pattern Recognition*. Academic Press, New York, NY, 1990.
8. A. Martinez and R. Benavente. The AR face database. Technical Report CVC #24, Computer Vision Center, Purdue University, June 1998.
9. E. Oja. Neural networks, principal component analysis, and subspaces. *International Journal of Neural Systems*, 1:61–68, 1989.
10. B. A. Olshausen and D. J. Field. Emergence of simple-cell receptive field properties by learning a sparse code for natural images. *Nature*, 381:607–609, 1996.
11. P. Penev and J. Atick. Local feature analysis: A general statistical theory for object representation. *Network: Computation in Neural Systems*, 7(3):477–500, 1996.
12. A. Pentland, B. Moghaddam, and T. Startner. View-based and modular eigenspaces for face recognition. In *Proc. IEEE Conf. Computer Vision and Pattern Recognition*, 1994.
13. L. Sirovich and M. Kirby. Low-dimensional procedure for the charaterization of human faces. *Journal of the Optical Society of America A*, 4(3):519–524, 1987.
14. M. E. Tipping and C. M. Bishop. Mixtures of probabilistic principal component analyzers. *Neural Computation*, 11(2):443–482, 1999.
15. M. Turk and A. Pentland. Eigenfaces for recognition. *Journal of Cognitive Neuroscience*, 3(1):71–86, 1991.

Distinctive Features Should Be Learned

Justus H. Piater and Roderic A. Grupen

University of Massachusetts, Amherst MA 01003, USA
{Piater, Grupen}@cs.umass.edu
http://www.cs.umass.edu/~piater

Abstract. Most existing machine vision systems perform recognition based on a fixed set of hand-crafted features, geometric models, or eigensubspace decomposition. Drawing from psychology, neuroscience and intuition, we show that certain aspects of human performance in visual discrimination cannot be explained by any of these techniques. We argue that many practical recognition tasks for artificial vision systems operating under uncontrolled conditions critically depend on incremental learning. Loosely motivated by visuocortical processing, we present feature representations and learning methods that perform biologically plausible functions. The paper concludes with experimental results generated by our method.

1 Introduction

How flexible are the representations for visual recognition, encoded by the neurons of the human visual cortex? Are they predetermined by a fixed developmental schedule, or does their development depend on their stimulation? Does their development cease at some point during our maturation, or do they continue to evolve throughout our lifetime? For some of these questions, the answers have been well established. For example, the development of receptive fields in the early visual pathways is influenced by stimulation of the visual system. Some visual functions do not develop at all without adequate perceptual stimulation during a maturational *sensitive period*. Higher-order visual functions such as pattern discrimination capabilities are also subject to a developmental schedule. It is still debated to what extent feature learning for pattern discrimination continues throughout adulthood. Recent psychological studies indicate that humans are able to form new features if required by a discrimination task [11].

In contrast to the human visual system, most work on machine vision has not used learning at the level of feature detectors. In the following section, we briefly discuss visual object recognition by humans and machines, and we argue that low-level learning is an essential ingredient of a robust and general visual system. The remainder of the paper presents our experimental system for learning discriminative features for recognition.

This work was supported in part by the National Science Foundation under grants CISE/CDA-9703217, IRI-9704530 and IRI-9503687, by the Air Force Research Labs, IFTD (via DARPA) under grant F30602-97-2-0032, and by Hugin Expert A/S through a low-cost Ph.D. license of their Bayesian network library.

2 Feature Learning in Humans and Machines

How do humans learn recognition skills? Two principal hypotheses can be identified [9]: According to the Schema Hypothesis, sensory input is matched to internal *representations of objects* that are built and refined through experience. On the other hand, the Differentiation Hypothesis postulates that *contrastive relations* are learned that serve to distinguish among the items. Psychological evidence argues for a strong role of Differentiation learning [9,13]. What exactly the discriminative features are and how they are discovered is unclear. It appears that feature discovery is a hard problem even for humans and takes a long time to learn [5]:

- Neonates can distinguish certain patterns, apparently based on statistical features like spatial intensity variance or contour density.
- Infants begin to note simple coarse-level geometric relationships, but perform poorly in the presence of distracting cues. They do not consistently pay attention to contours and shapes.
- At the age of about two years, children begin to discover fine-grained details and higher-order geometric relationships. However, attention is still limited to "salient" features [15].
- Over much of childhood, humans learn to discover distinctive features even if they are overshadowed by more salient distractors.

There is growing evidence that even adults learn new features when faced with a novel recognition task. In a typical experiment, subjects are presented with computer-generated renderings of unfamiliar objects that fall into categories based on specifically designed but unobvious features. After learning the categorization, the subjects are asked to categorize other objects that exhibit controlled variations of the diagnostic features, which reveals the features learned by the subjects. Schyns and Rodet [12] employed three categories of "Martian cells." The first category was characterized by a feature called X, the second by a feature Y, and the third by a feature XY, which was a composite of X and Y. Subjects were divided into two groups that differed in the order they had to learn the categories. Subjects in one group first learned to discriminate categories X and Y and then learned category XY, whereas the other group learned XY and X first, then Y. Subjects of the first group learned to categorize all objects based on two features (X and Y), whereas the subjects of the second group learned three features, not realizing that XY was a compound consisting of the other two. Evidently, feature generation was driven by the recognition task. For a summary of evidence for feature learning in adults, see a recent article [11].

Feature learning does not necessarily stop after learning a concept. Tanaka and Taylor [14] found that bird experts were as fast to recognize objects at the subordinate level ("robin") as they were at the basic level ("bird"). In contrast, non-experts are consistently faster on basic-level discriminations as compared to subordinate-level discriminations. Gauthier and Tarr [4] trained novices to become experts on unfamiliar objects and obtained similar results. These findings indicate that the way experts perform recognition is qualitatively different than novices. We suggest that experts have developed specialized features, facilitating rapid and reliable recognition in their domain of expertise.

General theories of vision such as those by Marr [6] and Biederman [2] have sparked extensive research efforts in both human and machine vision, and have contributed substantially to our understanding of how visual processes may operate. However, they have not led to artificial vision systems of noteworthy generality. Why is this so? From our point of view, a key reason is that most theories of vision do not address adaptation and learning. The real world is very complex, noisy, nonstationary – too variable for any fixed visual system, too unpredictable for its designer. Today's functional vision systems are highly specialized and operate under well-controlled conditions. They break if the built-in assumptions about task and environment do not hold.

Consider visual recognition. It is easy to see that there is no particular representation that can express all perceivable distinctions between objects or object categories that may later be required of a recognition system. Most existing machine vision systems perform recognition either based on a fixed set of hand-crafted features, eigen-subspace decomposition, or geometric model matching. In the first case, features are chosen in a best effort to express the distinctions required, but not too much more to avoid overfitting. The same is true of geometric models. How much detail should be encoded in the models? On the one hand, the level of detail should be kept low to increase generalization and efficiency; on the other hand, models should contain sufficient detail to express the distinctions required by a given task. Thus, both these methods are restricted to tasks that are well-defined at design time. We call such tasks *closed*. In contrast, almost all human visual learning takes place in *open* settings, where tasks are open-ended and evolve over time. While eigen-subspace representations (or related subspace methods that optimally separate instances by class label) are to some extent consistent with certain aspects of human visual mechanisms (e.g. face recognition), it appears unlikely that such methods can account for all of biological discrimination learning since they can tolerate only a limited degree of occlusion and object variability.

Humans can learn an impressive variety of distinctions ranging from miniscule local features such as a tiny scratch to abstract global features such as symmetry. In light of the evidence cited above, it seems clear that humans are capable of forming new representations of global and local appearance characteristics in a task-driven way. Thus, a key concept for building artificial vision systems of substantially increased generality and robustness is task-driven learning or adaptation. An adaptive system should be able to

- optimize its performance on-line with respect to individual tasks,
- expand its functionality incrementally,
- optimize its performance on-line under the actual working conditions, and
- track a nonstationary environment by adapting its parameters

by building new representations and adapting parameters. In the following sections, we describe our current work on a model of feature learning for recognition that addresses all of these issues, building on our previous work [8].

3 An Infinite Feature Space

We argued above that any fixed object representation is insufficient for learning arbitrary distinctions. Instead, we begin by specifying a small set of *primitive* features that can be combined into *higher-order* features according to a small number of rules that will be discussed below. All features that can be represented in this way form an infinite feature *space*. The structural complexity of a feature, i.e. the number of primitive features that form a compound, naturally provides a partial ordering of this space. Our learning procedure searches the feature space beginning with structurally simple features, and considers more complex features as needed [1]. The underlying assumption is that structurally simple features are easier to discover and have less discriminative potential than complicated features, but are still useful for some aspects of the learning problem.

3.1 Primitive Features

In our current system, primitive features are local appearance descriptors represented as vectors of local filter responses. The filters are oriented derivatives of 2-D Gaussian functions, with orientations chosen such that they form a steerable basis [3]. Here, the steerability property permits the efficient computation of filter responses of Gaussian-derivative kernels at any orientation, given $d+1$ measured filter responses for the dth derivative at specific orientations. Specifically, our system currently uses two specific variants of such descriptors:

- An *edgel* is encoded as a 2-vector containing the filter responses to the two first-derivative basis filters. These values encode the local intensity gradient of horizontal (G_x) and vertical (G_y) orientation. Using the steerability property, the magnitude of gradients in any orientation can be computed.
- A *texel* is represented as an 18-vector comprising the responses to the basis filters of the first three derivatives at two scales. This represents a local texture signature. Like edgels, texels have an associated orientation that is defined by the two first-derivative filter responses. When the orientation of a texel is steered, the entire vector containing all derivatives is rotated rigidly with reference to the first derivative computed at the largest scale [10].

This choice of low-level representations is plausible of biological early vision. While it is unlikely that any biological visual systems exploit steerability, this is an attractive computational alternative in the absence of massively parallel hardware. Steerability leads to rotational invariance which simplifies artificial vision systems at essentially no extra cost. We are not aware of any conclusive evidence for or against the biological faithfulness of our texel representation.

3.2 Higher-Order Features

Primitive features by themselves are not very discriminative. However, spatial combinations of these can express a wide range of shape and texture characteristics at various degrees of specificity or generality. We suggest the following four complementary types of feature composition:

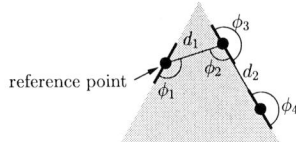

Fig. 1. A geometric feature of order 3, composed of three primitives. The feature is defined by the angles ϕ and the distances d. Each primitive is either an edgel or a texel.

- *Geometric* relations are given by the relative angles and distances between the participating lower-order features (Fig. 1). As long as these are rotation-invariant, so is their geometric composition. Geometric features are useful for representing e.g. corners, angles, and collinearity.
- *Topological* relations here refer to relaxed geometric relationships between component features that allow some degree of variability in angles and distances. Topological compound features are more robust to viewpoint changes than are geometric features, at the expense of specificity.
- *Conjunctive* features assert the presence of all component features without making any statement about their geometric or topological relationship.
- *Disjunctive* features are considered to be present in a scene if at least one component feature is detected. This can express statements such as "If I see a dial *or* a number pad, I may be looking at a telephone."

In contrast to other work, our features can be composed into increasingly complex and specific descriptors of 2-D shape, which is consistent with current models of the inferotemporal cortex. Features are computed at various scales, generated by successively subsampling images by factor two. This achieves some degree of scale invariance. Moreover, many compositions of edgels are inherently tolerant to changes in scale. For example, the arrangement shown in Fig. 1 applies equally to triangles of any size. Another desirable property of these features is that no explicit contour extraction or segmentation is required. This avoids these two difficult open problems in computer vision and should provide robustness to various kinds of image degradation. In contrast, the human visual system detects meaningful contours with remarkable robustness. This capability can probably not be explained entirely as a low-level visual process, but is supported by pre-segmentation recognition and task-dependent top-down processes.

4 Bayesian Networks for Recognition

The presence of a given feature $\mathbf{x}^*$ at a point i in the image is denoted by its *strength* $s \in [0,1]$. For primitive features, $s = \max\{0, r(\mathbf{x}^*, \mathbf{x}(i))\}$, where r is the normalized cross correlation function. The vector quantity $\mathbf{x}^*$ is a model feature, and the function $\mathbf{x}(i)$ returns the corresponding feature at location i. A geometric feature is described by the concatenation of the constituent feature values $\mathbf{x}$. In the case of topological and conjunctive features, the strength of the compound feature is the product of the strengths of its constituents; for disjunctive features, the maximum is used. Recognition is based on the maximum strengths of features found in the scene (or within a region if interest). Mapping feature

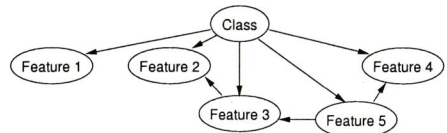

Fig. 2. A Bayesian network for one class. Note some interdependent features. A network such as this is created for each class.

vectors to class (or object) labels is the problem of classification, for which many algorithms exist. We chose Bayesian networks for their attractive properties that are desirable for open-domain recognition problems. In our system, each class is modeled as a separate Bayes net. The presence of an object is modeled as a discrete random variable with two states, *true* and *false*. The presence of an object gives rise to observable features, represented by random variables whose distributions are conditional on the presence of an object of this class. Assuming that the features are conditionally independent given the class, the resulting Bayes net has the topology of a star, with arcs connecting the class node to each of the feature nodes.

If some features are not independent, corresponding arcs must be inserted between the appropriate feature nodes. For example, in Fig. 2, Feature 3 may be a geometric composition with Feature 2, that is also in the feature set. Then, the presence of Feature 3 in an image implies the presence of Feature 2. Thus, in the Bayes net there is an arc from node 3 to node 2. An analogous argument holds for topological and conjunctive features, such as Feature 5 in Fig. 2, that combines Features 3 and 4. In the case of disjunctive features, the direction of the argument (and that of the additional arrows) is reversed.

Our feature strengths are continuous. We split each feature variable into two bins, corresponding to "present" and "not present", using a threshold. This threshold is determined individually for each feature variable such that its discriminative power between its own class and a misrecognized class is maximized. The discriminative power of a feature variable given a threshold is measured in terms of the Kolmogorov-Smirnoff distance (KSD). The KSD between two conditional distributions of a random variable is the difference between the cumulative probabilities at a given value of this variable under the two conditions. This separates the instances of the two conditions optimally, in the Bayesian sense, using a single cutpoint.

To perform recognition, we compute feature values one by one and update the Bayesian network after incorporating each feature. The class with the highest posterior probability gives the recognition result. Features are processed in decreasing order of informativeness. The informativeness of a feature is defined by the mutual information between a feature and the class node, i.e. its potential to reduce the entropy in the class random variable. In practice, only a fraction of all features are computed, because the entropy in the class nodes diminish before all features have been queried. This phenomenon suggests a straightforward, but very effective *forgetting* procedure: We delete any features that cease to be used during recognition.

5 Adaptive Feature Generation

As an agent (e.g. an animal, a human or a robot) interacts with the world, it uses vision (and maybe other sensory modalities) to acquire state information about the world, and performs actions appropriate in this state. This requires that the agent's visual features discriminate relevant aspects of the state of the world. We posit that such features are generated in response to feedback received during interaction with the world. For simplicity, we restrict the following discussion to a conventional supervised-learning scenario: The actions of the agent consist of naming class labels, the sensory input is an image, and the feedback received from the world consists of the correct class label. We further assume that the agent can retrieve random *example images* of known classes. This assumption is realistic in many cases. For example, an infant can pick up a known object and view it from various viewpoints; or a child receives various examples of letters of the alphabet from a teacher.

Initially, the agent does not know about any objects or features. When it is presented with the first object, it simply remembers the correct answer given by the teacher. When it is shown the second object, it will guess the only category it knows about.

When the agent gives a wrong answer, it needs to learn a new feature to discriminate this object category from the mistaken category (or categories). This is done by random sampling, with a bias for structurally simple features. We employ the following heuristic procedure, where each step is iterated up to a constant number of times:

1. Pick a random feature from some other Bayes net (corresponding to another class) that is not yet part of this Bayes net (corresponding to the true class). This promotes the usage of general features that are characteristic of more than one class.
2. Sample a new feature directly from the misrecognized image by either picking two edgels and turning them into a geometric compound, or by picking a single texel.
3. Pick a random feature that is already part of this Bayes net, find its strongest occurrence in the current image, and expand it geometrically by picking an additional edgel or texel close-by.
4. Pick two random features and combine them into a conjunctive feature.
5. Pick two random features and combine them into a disjunctive feature.

After each new feature is generated, it is evaluated on a small set of example images, retrieved from the environment, that contains examples of the true class and the mistaken class(es). If it has any discriminative power, it is then added to the Bayes net of the true class using the conditional probabilities estimated using a small set of example images, randomly chosen from the training set. If the image is now recognized correctly by the expanded Bayes net, the feature learning procedure stops; if not, the feature is removed from the net, and the learning procedure continues. This procedure may terminate without success.

During operation of the learning system, an instance list of all classes encountered and features queried is maintained. Periodically, all feature cutpoints, class priors and conditional probabilities in the Bayes nets are updated according to this list.

Feature learning does not have to stop after learning a training set perfectly. The system can continue to search for better features. The quality of a feature is its discriminative power at a given stage during a recognition procedure, given by the KSD between its own class and the combined set of all other classes. We can train our system to develop better features by imposing a minimum KSD on all features that are used during a recognition procedure. If a feature does not meet this requirement, the system has to learn a new and better feature. The minimum KSD can iteratively be raised, until the system fails to find adequate features. As a consequence, fewer (but superior) features will be queried while recognizing a given image, and many of the inferior features will become obsolete. We suggest this procedure, called *feature upgrade*, as a crude model of expert learning, as outlined in Section 2.

6 Experiments

To illustrate that our algorithm is able to produce discriminative features, we performed pilot experiments on two example tasks (Fig. 3). In the COIL task, the images of the first four objects from the COIL-20 database [7] were split into two disjoint sets such that no two neighboring viewpoints were represented in the same set. As a result, each image set contained 36 images, spaced 10 degrees apart on the viewing sphere, at constant elevation. We performed a 2-fold cross-validation on these two sets: In one run, one set served as a training set and the other as the test set; in a second run, the roles were reversed. In the PLYM task, there were eight geometric objects on 15 artificially rendered images each, covering a small section of the viewing sphere[1]. We performed a 10-fold stratified cross-validation on this data set, with random subdivision of the 15 images of each class into 10 subsets of 1 or 2 images each.

The results of the experiments are summarized in Table 1. While the recognition results fall short of current machine recognition technology, they were achieved by an uncommitted visual system with a strong bias toward few and simple features that had access only to a small number of random training views at any given time during an incremental training procedure. Most of these properties are contrary to current computer vision technology, but are characteristic of biological vision systems.

In accord with our biased search strategy, most learned features were isolated texels and simple geometric compounds of edgels and/or texels. Smaller numbers of the other compound types of features were also found. In most cases, the training set was not learned perfectly. This is because our system currently gives up after 10 iterations through the training set. Clearly, more effective techniques for finding distinctive features are called for.

---

[1] http://www.cis.plym.ac.uk/cis/levi/UoP_CIS_3D_Archive/8obj_set.tar

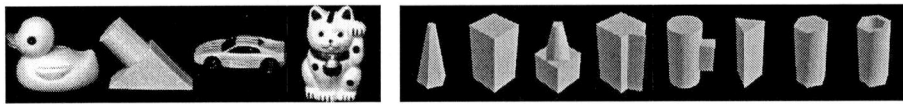

Fig. 3. Objects of the COIL task (left) and the PLYM task (right).

Table 1. Summary of experimental results. The "expert level" column gives the number of feature upgrade iterations. The "other" columns contain cases where the system returned an ambiguous answer, or no answer at all.

Task	expert level	avg. # features queried	Training Set: correct	wrong	other	Test Set: correct	wrong	other
COIL	0	44	0.98	0.02		0.81	0.19	
	1	36	0.85	0.11	0.04	0.73	0.23	0.05
	2	23	0.97	0.03		0.83	0.16	0.01
	3	11	0.83	0.14	0.03	0.67	0.27	0.06
PLYM	0	19	1.00			0.72	0.28	
	1	21	1.00			0.76	0.21	0.03
	5	13	0.95	0.03	0.02	0.71	0.09	0.20

As the minimum KSD required of a feature is increased during feature upgrade, it is increasingly difficult to find appropriate features in order to learn the training set perfectly. However, feature upgrade has the desired effect of decreasing the number of features queried during recognition, and where the training set is learned well, it also tends to reduce the number of false recognitions while marginally increasing the correct recognition rate on the test set.

7 Conclusions

There is overwhelming evidence that humans learn features for recognition in a task-driven manner. Biological learning is on-line and incremental. We have presented an artificial vision system that follows these characteristics, based on an infinite combinatorial feature space and a generate-and-test search procedure for finding discriminative features. Our method successfully learns to discriminate objects. We also proposed that developing visual expertise involves the construction of better features. Our system models this by increasing the minimum KSD required of features during recognition. While our system reflects certain aspects of human vision, it is not a complete model in that it focuses on appearance-based discriminative features. Biological vision systems are probably composed of several complementary algorithms.

As a model of feature learning for discrimination, the main limitation of our system is the undirected search for features in images that is only guided by a few simple heuristics. A more faithful (and more practical) model requires a developmental schedule that initially constrains the search for features to increase the likelihood of finding useful features fast, while temporarily restricting generality. Over time, these restrictions should be relaxed, while the system learns better heuristics from experience. This is an area of further research.

Another critical limitation of our current system is the restricted expressiveness our feature space that encodes only high-contrast edge, corner and texture information. As such, our model roughly corresponds to the human visual system during early infancy [5]. A more complete model should at least encode color and blob-type features. In addition, more sophisticated recognition requires higher-level features such as qualitative ("Gestalt") features (e.g. parallelism, symmetry, continuity, closure) and multiplicity (a triangle has three corners; a bicycle wheel has many spokes). We hope to address these in future work.

References

1. Y. Amit, D. Geman, and K. Wilder. Joint induction of shape features and tree classifiers. *IEEE Trans. Pattern Anal. Mach. Intell.*, 19(11):1300–1305, 1997.
2. I. Biederman. Recognition-by-components: A theory of human image understanding. *Psychological Review*, 94:115–147, 1987.
3. W. T. Freeman and E. H. Adelson. The design and use of steerable filters. *IEEE Trans. Pattern Anal. Mach. Intell.*, 13(9):891–906, 1991.
4. I. Gauthier and M. J. Tarr. Becoming a "Greeble" expert: Exploring mechanisms for face recognition. *Vision Research*, 37(12):1673–1682, 1997.
5. E. J. Gibson and E. S. Spelke. The development of perception. In J. H. Flavell and E. M. Markman, editors, *Handbook of Child Psychology Vol. III: Cognitive Development*, chapter 1, pages 2–76. Wiley, 4th edition, 1983.
6. D. Marr. *Vision: A Computational Investigation into the Human Representation and Processing of Visual Information*. Freeman, San Francisco, 1982.
7. S. A. Nene, S. K. Nayar, and H. Murase. Columbia object image library (COIL-20). Technical Report CUCS-005-96, Columbia University, New York, NY, Feb. 1996.
8. J. H. Piater and R. A. Grupen. Toward learning visual discrimination strategies. In *Proc. Computer Vision and Pattern Recognition (CVPR '99)*, volume 1, pages 410–415, Ft. Collins, CO, June 1999. IEEE Computer Society.
9. A. D. Pick. Improvement of visual and tactual form discrimination. *J. Exp. Psychol.*, 69:331–339, 1965.
10. R. P. N. Rao and D. H. Ballard. An active vision architecture based on iconic representations. *Artificial Intelligence*, 78:461–505, 1995.
11. P. G. Schyns, R. L. Goldstone, and J.-P. Thibaut. The development of features in object concepts. *Behavioral and Brain Sciences*, 21(1):1–54, 1998.
12. P. G. Schyns and L. Rodet. Categorization creates functional features. *J. Exp. Psychol.: Learning, Memory, and Cognition*, 23(3):681–696, 1997.
13. J. R. Silver and H. A. Rollins. The effects of visual and verbal feature-emphasis on form discrimination in preschool children. *J. Exp. Child Psychol.*, 16:205–216, 1973.
14. J. W. Tanaka and M. Taylor. Object categories and expertise: Is the basic level in the eye of the beholder? *Cognitive Psychology*, 23:457–482, 1991.
15. J.-P. Thibaut. The development of features in children and adults: The case of visual stimuli. In *Proc. 17th Annual Meeting of the Cognitive Science Society*, pages 194–199. Lawrence Erlbaum, 1995.

Moving Object Segmentation Based on Human Visual Sensitivity

Kuk-Jin Yoon[1], In-So Kweon[1], Chang-Yeong Kim[2], and Yang-Seok Seo[2]

[1] Department of Electrical Engineering and Computer Science,
KAIST, 373-1 Kusong-dong, Yusong-ku, Taejon, 305-701, Korea.
kjyoon@covra1.kaist.ac.kr, iskweon@kaist.ac.kr
[2] Signal Processing Laboratory, Samsung Advanced Institute of Technology,
P.O. Box 111, Suwon 440-660, Korea.

Abstract. The edge and motion are the main features that human visual system (HVS) perceives intensively. This paper proposes an algorithm for the segmentation of the moving object with accurate boundary using color and motion focusing on the HVS perception in the general image sequence. The proposed algorithm is composed of three parts: color segmentation, motion analysis, and region refinement and merging part. In the color segmentation phase, K-Means algorithm is used in consideration of the sensitivity of the human color perception to get the boundaries that HVS perceives. The global and local motion estimation are performed in parallel with color analysis. After that, Bayesian clustering using color and motion provides more accurate boundary. In the final stage, regions are merged taking into account their motion. The experimental results of the proposed algorithm show the accurate moving object boundary coinciding with the boundary that HVS perceives.

1 Introduction

Object-based video analysis describes the contents of the video focusing on the information about the objects, such as shape, color, intensity and gesture, and etc. Object-based technologies take an important role in video analysis. Moving object segmentation is the heart of the object-based technologies and can be applied to various applications such as object recognition, object tracking, object-based video indexing and object-based video coding. So, it has been studied for the last few decades in computer vision and image coding area. As the result, many algorithms and systems have been proposed and developed. In spite of these efforts, however, it is the one of the most difficult problem yet. The best known mechanism for moving object segmentation is the Human Visual System (HVS). Moving object segmentation is the high level analysis. The performance evaluation is answered by the person. So, in the moving object segmentation process, it is important to extract the object boundary coinciding with HVS because HVS has high sensitivity about motion and edge.

In this paper, we present a motion segmentation algorithm that adaptively combine color and motion based on the sensitivity of the human visual system.

2 Previous Works

It is assumed that the object is composed of a few regions. Also, the region is composed of the pixels that have coherent feature. So, image segmentation is the fist step to extract the objects in image.

Image segmentation algorithm can be classified into three groups according to the used feature: spatial segmentation, temporal segmentation and spatio-temporal segmentation. Spatial segmentation uses the spatial feature only, such as color, intensity, and texture. The result of spatial segmentation shows accurate region boundary without any contextual information. Therefore, it is difficult to extract the contextually meaningful object. Temporal segmentation can segment the contextually meaningful object because it uses temporal information. But, the motion, which is the temporal feature, is difficult to be estimated accurately; the result of temporal segmentation shows inaccurate region boundaries. To define and segment the contextually meaningful moving object in the image sequence, it is necessary to use the spatial feature and temporal feature together. Spatio-temporal segmentation uses the two kinds of features to segment moving objects.

In many previous researches, the spatial features, such as color, intensity and texture, determine object boundary [1, 2, 3]. Input images are segmented into regions by spatial feature. Then motion of each region is estimated to merge the region having similar motion. So, the boundary is determined by the spatial feature only and the location of the moving region is estimated by temporal feature only. In simple images, color boundary is accurate. But, if the contrast of the spatial feature of the object and background is low, some part of object and background may be merged. To overcome this limitation, approaches using spatial and temporal feature simultaneously to determine the boundary and location of moving object have been proposed [4, 5]. These approaches use the joint similarity measure of spatial and temporal feature between the pixel and region to assign the region label to each pixel. The joint similarity measure is assumed the form of the weighted linear combination of spatial similarity and temporal similarity. So, the determination of optimal weight value is difficult and the boundary location may be changed depending upon the weight.

3 Proposed Algorithm

We propose an algorithm that preserves the accurate boundary of moving object in consideration of HVS. Figure (1) shows the overall structure of the proposed algorithm composition. In this paper, we assume that the object is composed with a few color regions having coherent motion. So, we use color as the spatial feature and motion as the temporal feature. The proposed algorithm consists of three components: the color segmentation and motion estimation and region refinement process. Because color and motion are the main cues that HVS has high sensitivity, the preservation of accurate color boundary and accurate motion estimation is important. After the color image segmentation and motion analysis, region boundaries are refined by using the color and motion information of each region together in the post-processing part. So, region boundaries are refined to coincide the boundaries that the HVS perceives. In the final phase, regions that have similar motion vectors are merged.

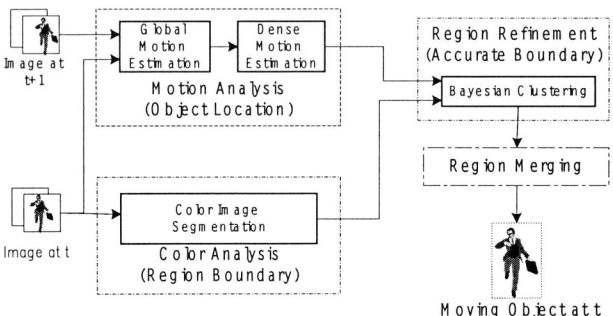

Fig. 1. Proposed Algorithm

3.1 Color Image Segmentation

There are many approaches to color image segmentation. Among the many color segmentation algorithms, we use the K-Means algorithm, which is known as a powerful method to deal with the large color pixel set [7, 8, 9].

In color image segmentation, it is important to preserve the human perceivable color boundary. But, conventional clustering methods don't have reflected the characteristics of the human visual system. First, the Euclidean distance between two color points in a color space cannot accurately reflect the difference of the HVS color perception. According to the position of color points in the color space, the same Euclidean distance does not mean the same color difference. Next, they do not consider the sensitivity of the human color perception. The HVS shows different color perception sensitivity according to the color distribution in the image domain [6]. In conventional approaches, the cluster centers, shapes in the color space and the boundaries of color regions in the image domain are determined by the color values in the image only. As the result, the region boundary doesn't coincide the human perceivable color boundaries and some boundaries are missed. In addition to these problems, the K-Means algorithm has the fatal limitation: how to determine the number of clusters. That is very important for the accuracy and the efficiency of clustering based segmentation.

To solve these problems, we propose a color segmentation algorithm that specifically takes into account the characteristics of the human color perception. We use the CIELab color space, in which Euclidean distance of two colors is proportional to the difference that human visual system perceived. We also introduce the color weight to consider the human color perception. Human visual system is more sensitive to the color in homogeneous region than complex region as mentioned earlier. Figure (2) shows the flower garden image with two test color points: one is in the sky and the other is in the flower garden. Although the colors of test points are similar with colors of neighboring points, the HVS perceives the color of the point in the sky more vividly.

Fig. 2. Flower Garden Image with Test Color Points

To take into account this phenomenon, we assign a weight value to each pixel color. The pixels surrounded by a region with a homogeneous color have larger weights and the pixels in the complex colored region have smaller weights. As the result, the cluster center and shapes are formed to preserve the more sensitive color boundaries that are analogous with HSV. Color weight means the color sensitivity of each pixel color considering the neighboring color distribution.

To assign a color weight to each pixel, we use a local mask centered at each pixel. We first calculate the average color and the color variance in the local mask. The average color is given by

$$[L_{avg} \quad a_{avg} \quad b_{avg}] = \frac{1}{N}[\sum_{j \in M} L_j \quad \sum_{j \in M} a_j \quad \sum_{j \in M} b_j] \tag{1}$$

where N is the number of pixels in the local mask M. In CIELab color space, a small Euclidean distance between two color points is proportional to the difference that HVS perceives. But, a large Euclidean distance has no meaning but only large difference in HVS. Therefore, color difference in CIELab color space can be modeled as

$$D_{i,j} = 1 - e^{-E_{i,j}/\gamma} \tag{2}$$

where γ is the normalized factor and $E_{i,j}$ is the Euclidean distance given by,

$$E_{i,j} = \sqrt{(L_i - L_j)^2 + (a_i - a_j)^2 + (b_i - b_j)^2} \tag{3}$$

The color variance in the local mask is given by

$$v_i = \frac{1}{N}\sum_{j \in M} D_{avg,j}^2 = \frac{1}{N}\sum_{j \in M}(1 - e^{-\sqrt{(L_{avg}-L_j)^2 + (a_{avg}-a_j)^2 + (b_{avg}-b_j)^2}/\gamma})^2 \tag{4}$$

The color variance in the local mask is the measure how the color pattern varies in the local area. A large color variance means that there is frequent color pattern variation in the neighboring area. So, large weight is assigned to the pixel that has a small color variance. We use a Gaussian kernel to generate the weight

$$W_i = e^{\frac{-v_i^2}{2\sigma^2}} \qquad (5)$$

To determine the number of clusters, we use the calculated weight values. The number of clusters must be determined taking into account the distribution of color components in the color space and image domain. If the average value of the color weights is large, it means that there are large parts of homogeneous region in image domain. In this case, a small number of clusters are enough. For the large variance of the weight value, a large number of clusters are needed. We use the following equation to determine the number of clusters.

$$K(m_w, v_w) = M \times [e^{-m_w * \alpha} + (1 - e^{-v_w * \beta})] \qquad (6)$$

where M denotes the maximum number of clusters to be used and α, β are some constants value to adjust the optimal number of clusters. We set α, β as 4, 2 respectively. In the K-Means process, color weight is used to determine the center of each cluster. The objective function and the center of each cluster are given by

$$J = \sum_{i=1}^{K} \sum_{x \in S_i} \|x - m_i\|^2 w(x), \quad m_i^{t+1} = \frac{\sum_{x \in S_i} x \times w(x)}{\sum_{x \in S_i} w(x)} \qquad (7)$$

3.2 Motion Estimation

Motion is also an important cue to HVS. In the image, there are two kinds of motion: global motion and local motion. Global motion is due to the motion of the camera, such as panning, tilting, zoom-in and zoom-out, and local motion is due to moving object in the scene. To extract the moving object, global motion is compensated first. After global motion compensation, the motion in the image is wholly due to object motion.

To estimate and compensate the global motion, we use the 6-parameter affine model as the motion model and the NDIM (Normalized Dynamic Image Model) as the image model [10].

The NDIM is devised to overcome the intensity change due to illumination change or object motion. The NDIM models the intensity change 1st order linear equation. The relationship between corresponding pixels can be simplified using ICA (Intensity Conservation Assumption);

$$\frac{(I_1(\mathbf{x}) - m_1(\mathbf{x}))}{\sigma_1(\mathbf{x})} = \frac{(I_2(\mathbf{x} + \Delta \mathbf{x}) - m_2(\mathbf{x} + \Delta \mathbf{x}))}{\sigma_2(\mathbf{x} + \Delta \mathbf{x})} \qquad (8)$$

where m and σ denote the average and standard deviation of intensity in a small window.

After global motion compensation, there are local motion components only in the image. This is due to the object motion completely. We use the robust method proposed by Black and Anandan [11] to compute dense optical flow.

3.3 Region Boundary Refinement and Region Merging

The initial boundary of each region is determined by color. If the color contrast between object and background is high, color is a good cue to determine the boundary of moving region. In this case, color boundary coincides the boundary that HVS perceives. But, if the contrast between object and background is low, the resulting boundary may be inaccurate and some parts of object may be merged with the background. Region boudnary refinement is done for the accurate object boundary.

We propose the statistical approach, Bayesian clustering, to refine the region boundaies. It is assumed that the pixels in a same region have coherent feature. Therefore, pixels in a same region show similar statistical property when the regions are modeled as the statistical models. Let x_{ij} denote a statistically independent 5×1 feature vector consisting of three color components and two motion components. Also let the pixel set be $S=\{x_{ij}, i=1, 2, ..., W, j=1, 2, ..., H\}$. The number of regions M is already computed by the color image segmentation process. Then, we can partition the set S into the M regions, $R=\{R_1, R_2, ..., R_M\}$. The pixels in the same region k are described by the probability density $p_k(x_{ij}|\theta_k)$, where p_k is known function and θ_k is a parameter vector whose values have to be determined. We model the color and motion as the Gaussian random variables, so p_k is given as Gaussian. If a label of the pixel in the border of the region is incorrect, the probability density value will be low because that pixel has different statistical property from the region. The region label of that pixel must be changed if the pixel has larger probability density value when it is assumed that that pixel belongs to the neighboring region. In other words, the pixel must have a region label in which has the largest probability density value. Therefore, if we assume priors of θ and R be uniform, then the MAP estimates (R*, θ*) are given by

$$(R^*, \theta^*) = \arg \max_{R,\theta} P(S|R,\theta) \quad (9)$$

Since the pixels are independent, the joint density of S has the following form

$$P(S|R,\theta) = \prod_{k=1}^{M} (\prod_{x_{ij} \in R_k} p_k(x_{ij}|\theta_k)) \quad (10)$$

By applying natural logarithm, we can obtain the following relation.

$$\ln(P(S|r,\theta)) = \ln(\prod_{k=1}^{M} (\prod_{x_{ij} \in R_k} p_k(x_{ij}|\theta_k))) = \sum_{k=1}^{M} \sum_{x_{ij} \in R_k} \ln(p_k(x_{ij}|\theta_k)) \quad (11)$$

Therefore, eq. (9) can be rearraged into

$$(R^*, \theta^*) = \arg\max_{R,\theta} \sum_{k=1}^{M} \sum_{x_{ij} \in R_k} \ln(p_k(x_{ij} \mid \theta_k)) \quad (12)$$

For a fixed M, to obtain the MAP estimates of R and θ, we have to minimize the function

$$J(R,\theta) = -\sum_{k=1}^{M} \sum_{x_{ij} \in R_k} \ln(p_k(x_{ij} \mid \theta_k)) \quad (13)$$

For a fixed θ, R which minimizes the eq. (13) can be obtained by

$$R_k = \{x_{ij}, \ln p_k(x_{ij} \mid \theta_k) > \ln p_t(x_{mn} \mid \theta_t), \forall k \neq t, t = 1,2,...,M\} \quad (14)$$

Similarly, for a fixed R, the value of θ is unique and it can be obtained using

$$\theta_k = \max \sum_{x_{ij} \in R_k} \ln(p_k(x_{ij} \mid \theta_k)) \quad (15)$$

Because we choose p_k to be Gaussian, $\theta_k = \{\mu_k, \sigma_k\}$ an explicit expression for θ can be given. The parameter estimates are

$$\mu_k = \frac{1}{N_k} \sum_{x_{mn} \in R_k} x_{mn}, \quad \sigma_k^2 = \frac{1}{N_k} \sum_{x_{mn} \in R_k} (x_{mn} - \mu_k)^2 \quad (16)$$

where N_k denotes the number of pixels in the region.

Because we have initial region boundaries and region labels of the pixels already, we test the statistics of pixels in the border of the regions only. Starting from an initial R_k and θ_k computed in the previous stage, Bayesian clustering steps are like as follows.

Step1. Initialize R_k and θ_k, k=1, 2, ..., M
Step2. Given R_k^i, Compute θ_k^{i+1} using eq.(16)
Step3. Given θ_k^i, Compute R_k^{i+1} using eq.(14)
Step4. If $R_k^i = R_k^{i+1}$, stop the iteration. Else go to step2.

As the result of region refinement, we can get the accurate region boundaries coinciding with the boundaries that HVS. Also, accurate motion information is obtained by refining the boundary of each region.

We assumed that the object is composed with a few color regions having coherent motion. To extract the complete object, regions that have a similar motion are considered to belong to one object. Because we are interested in moving regions only, we must detect the regions that have independent motion first. We determine the moving region by applying a threshold to the magnitude of motion vector. After that, we merge the regions having a similar motion vector into a larger one.

4 Experimental Results

Figure (3) shows the experimental result of color image segmentation. For the conventional algorithm, we assume that all pixel colors have same weights.

    (a) Flower garden   (b) Conventional algorithm   (c) Proposed algorithm

Fig. 3. Color Image Segmentation Results

From the result, we can observe that the proposed methods provides more accurate boundary in uniform color regions such as the sky.

Figure (4) shows the extracted motion for two consequent images, which include sudden intensity changes. As mentioned earlier, there are two kinds of motion. The global motion is due to the panning of the camera and local motion is due to the swing motion of the player. To estimate the global motion, input images are transformed to the NDIM images first. In Figure (4-b), only the moving object parts are detected as outliers after global motion compensation.

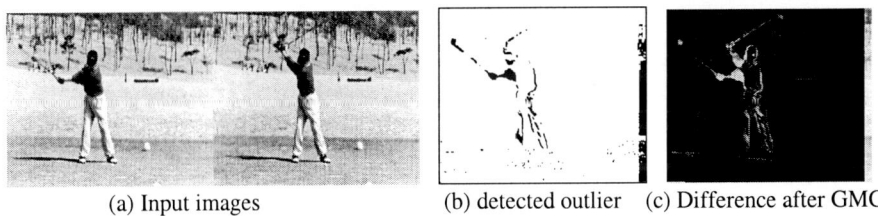

    (a) Input images   (b) detected outlier   (c) Difference after GMC

Fig. 4. Global Motion Estimation Result

Another results for the Foreman sequence are shown in Figure (5). In these input images, the color of helmet is very similar with the background color as shown in Figure (5). As you can see, some parts of the object and the background are merged where the color contrast between the object and the background are vary low. Some parts of the hat of the object in the foreground are merged with the wall in the background because they have the similar color. That is an unavoidable result in color segmentation. In parallel with color segmentation, motion in the image is estimated. Figure (5-c) shows the estimated optical flow and relative disparity. The low intensity denotes the large disparity of the pixel.

Figure (6) shows the region refinement result. We use the color and motion information together to refine the border of the regions. So, we can overcome the limitation of the color segmentation. As the result, we can get the accurate boundaries coinciding the boundaries that HVS perceives because the boundaries are refined by two main visual cues. Also, the accurate motion of each region is obtained.

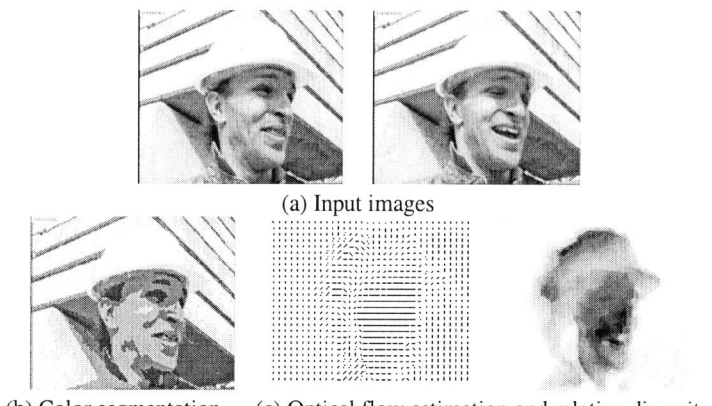

(a) Input images

(b) Color segmentation   (c) Optical flow estimation and relative disparity

Fig. 5. Color Segmentation and Motion Estimation Results

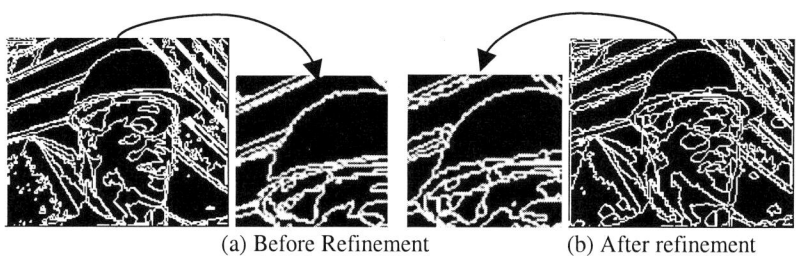

(a) Before Refinement   (b) After refinement

Fig. 6. Region Refinement Result

Figure (7) shows the final result by the proposed algorithm and a conventional algorithm, which uses the spatial feature for boundary information and temporal feature for moving region location respectively. The conventional algorithm shows erroneous boundaries and some parts of foreground region are lost. But, the proposed algorithm shows reasonably accurate boundary that HVS perceives by using two visual cues; motion and color. Also, some experimental results are also shown in Figure (8).

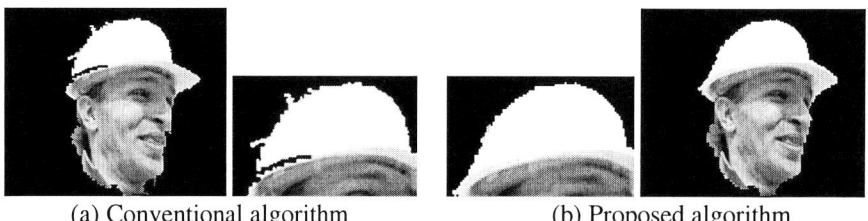

(a) Conventional algorithm   (b) Proposed algorithm

Fig. 7. Moving Object Segmentation Result

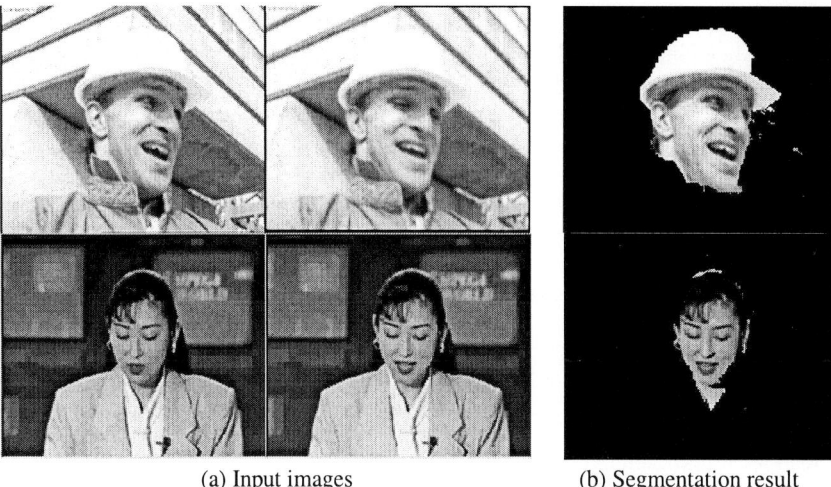

(a) Input images　　　　　　　　(b) Segmentation result

Fig. 8. Moving Object Segmentation Results

5 Conclusion and Further Works

In this work, we proposed a moving object segmentation algorithm with accurate boundary, that human visual system perceivable, focusing on the two main visual cues, color and motion. Human visual system has different sensitivity to color and is also very sensitive the boundary information. We introduced the color weight in clustering processing to assign more weight to the color that HVS shows high sensitivity. The experimental result of the color segmentation shows the more accurate region boundaries in uniform colored regions that HVS perceives more vividly. But, color is insufficient to extract the accurate boundary because the color contrast may be low between foreground and background at the border of the object. Therefore, another visual cue, motion, is used together with color to extract the proper boundary for HVS. The final result of the proposed algorithm gives an accurate segmentation result with HVS perceivable boundaries.

References

1. V. Rehrmann : Object Oriented Motion Estimation in Color Image Sequence, European Conference on Computer Vision, Vol. 1, Page(s) : 704-719, 1998.
2. M. Hoetter and R. Thoma : Image Segmentation Based on Object Oriented Mapping Parameter Segmentation, Signal Processing, Vol. 15, No. 3, pp.315-334, October 1988.
3. J. Guo, J. W. Kim and C-C. J. Kuo : Fast Video Object Segmentation Using Affine Motion And Gradient-Based Color Clustering, IEEE Second Workshop on Multimedia Signal Processing, Page(s): 486 –491, 1998.

4. K. W. Song, E. Y. Chung, J. W. Park, G. S. Kim, E. J. Lee and Y. H. Ha : Video Segmentation Algorithm using a Combined Similarity Measure for Content-based Coding, Proceedings of the Picture Coding Symposium, Page(s) :261-264, 1999.
5. J. G. Choi, S. W. Lee and S. D. Kim : Spatio-temporal video segmentation using a joint similarity measure, IEEE transactions on circuits and systems for video technology, Vol. 7, No. 2. April 1997.
6. B. A. Wandell : Color Appearance: the Effects of Illumination and Spatial Patterns, Proc. Nat. Acad. Sci., USA, v 90, p. 1494-1501, 1993.
7. T. Uchiyang and M.A. Arbib : Color Image Segmentation Using Competitive Learning, IEEE Transactions on Pattern Analysis and Machine Intelligent, vol. 16, No. 12, December 1994.
8. N. Kehtarnavaz, J. Monaco, J. Nimtschek and A. Weeks : Color Image Segmentation Using Multiscale Clustering, Proceedings of the IEEE Southwest Symposium on Image Analysis and Interpretation, 142-147, 1998.
9. L. Lucchese, S. K. Mitra : An Algorithm for Unsupervised Color Image Segmentation, IEEE Second Workshop on Multimedia Signal Processing, Page(s): 33 –38, 1998.
10. Y.S. Moon and I.S. Kweon : Robust Dominant Motion Estimation Algorithms against Local Linear Illumination Variations, Technical Report of Robotics and Computer Vision Lab, Dept. of EE, KAIST, Jun. 1999.
11. M.J. Black and P. Anandan : The Robust Estimation of Multiple Motions: Parametric and Piecewise-smooth Flow Field, Computer Vision and Image Understanding, Vol. 63, No. 1, January, pp. 75-104, 1996

Object Classification
Using a Fragment-Based Representation

Shimon Ullman and Erez Sali

The Weizmann Institute of Science
Rehvot 76100, Israel
shimon@wisdom.weizmann.ac.il

Abstract. The tasks of visual object recognition and classification are natural and effortless for biological visual systems, but exceedingly difficult to replicate in computer vision systems. This difficulty arises from the large variability in images of different objects within a class, and variability in viewing conditions. In this paper we describe a fragment-based method for object classification. In this approach objects within a class are represented in terms of common image fragments, that are used as building blocks for representing a large variety of different objects that belong to a common class, such as a face or a car. Optimal fragments are selected from a training set of images based on a criterion of maximizing the mutual information of the fragments and the class they represent. For the purpose of classification the fragments are also organized into types, where each type is a collection of alternative fragments, such as different hairline or eye regions for face classification. During classification, the algorithm detects fragments of the different types, and then combines the evidence for the detected fragments to reach a final decision. The algorithm verifies the proper arrangement of the fragments and the consistency of the viewing conditions primarily by the conjunction of overlapping fragments. The method is different from previous part based methods in using class specific overlapping object fragments of varying complexity, and in verifying the consistent arrangement of the fragments primarily by the conjunction of overlapping detected fragments. Experimental results on the detection of face and car views show that the fragment-based approach can generalize well to completely novel image views within a class while maintaining low mis-classification error rates. We briefly discuss relationships between the proposed method and properties of parts of the primate visual system involved in object perception.

1 Classification and the Generalization Problem

Object classification is a natural task for our visual system: we effortlessly classify a novel object as a person, dog, car, house, and the like, based on its appearance. Even a three-year old child can easily classify a large variety of images of many natural classes. In contrast, visual classification proved extremely difficult to reproduce in artificial computer vision system. It is therefore natural to study the mechanisms and processes used by biological visual system for object classification, and to examine the applicability of similar methods to computer vision system. Such studies may lead to the development of better artificial systems dealing with natural images, and can also shed light on what appears to be fundamental differences in the processes of visual information by current computer systems on the one hand, and by biological systems on the other.

It is interesting to note in this context the difference between general object classification and the specific identification of individual objects. Classification is concerned with the general description of an object as belonging to a natural class of similar objects, such as a face or a dog, whereas identification involves the recognition of a specific individual within a class, such as the face of a particular person, or the make of a particular car. For human vision, the general classification of an object as a car, for example, is usually easier than the identification of the specific make of the car (Rosch *et al.*1976). In contrast, current computer vision systems can deal more successfully with the task of recognition compared with classification. This may appear surprising, because specific identification requires finer distinctions between objects compared with general classification, and therefore the task appears to be more demanding.

The main difficulty faced by a recognition and classification system is the problem of variability, and the need to generalize across variations in the appearance of objects belonging to the same class. Different dog images, for example, can vary widely, because they can represent different kinds of dogs, and for each particular dog, the appearance will change with the imaging conditions, such as the viewing angle, distance, and illumination conditions, with the animal's posture, and so on. The visual system is therefore constantly faced with views that are different from all other views seen in the past, and it is required to generalize correctly from past experience and classify correctly the novel image. The variability is complex in nature: it is difficult to provide, for instance, a precise definition for all the allowed variations of dog images. The human visual system somehow learns the characteristics of the allowed variability from experience. This makes classification more difficult for artificial system than individual identification. In performing identification of a specific car, say, one can supply the system with a full and exact model of the object, and the expected variations can be described with precision. This is the basis for several approaches to identification, for example, methods that use image combinations (Ullman & Basri 1991) or interpolation (Poggio & Edelman 1990) to predict the appearance of a known object under given viewing conditions. In classification, the range of possible variations is wider, since now, in addition to variations in the viewing condition, one must also contend with variations in shape of different objects within the same class.

In this paper we propose an approach to classification that uses a fragment-based representation. In this approach, images of objects within a class are represented in terms of class-specific fragments. These fragments provide common building blocks that can be used, in different combinations, to represent a large variety of different images of objects within the class. In the next section we discuss the problem of selecting a set of fragments that are best suited for representing a class of related objects, given a set of example images. We then illustrate the use of these fragments to perform classification and deal with the variability in shape between different objects of the same class. We also discuss the problem of coping with variability in the viewing conditions, focusing on the problem of position invariance, with possible application to other aspects of object recognition. Finally, we conclude with some comments about similarities between the proposed approach and aspects of the human visual system.

2 The Selection of Class-Based Fragments

2.1 A Brief Review of Related Past Approaches

Before describing the selection of fragment from a collection of example images, we will review briefly past approaches to recognition, focusing on methods that bear relevance to the approach developed here.

A popular framework to classification is based on representing object views as points in a high-dimensional feature space, and then performing some partitioning of the space into regions corresponding to the different classes. Typically, a set of n different measurements are applied to the image, and the results constitute an n-dimensional vector representing the image A variety of different measures have been proposed, including using the raw image as a vector of grey-level values, using global measures such as the overall area of the object's image, different moments, Fourier coefficients describing the object's boundary, or the results of applying selected templates to the image. Partitioning of the space is then performed using different techniques. Some of the frequently used techniques include nearest-neighbor classification to class representatives using, for example, vector quantization techniques, nearest-neighbor to a manifold representing a collection of object or class views (Murase & Nayar 1995,), separating hyperplanes performed, for example, by Perceptron-type algorithms and their extensions (Minsky & Papert 1969), or, more optimally, by support vector machines (Vapnik 1995). The vector of measurements may also serve as an input to a neural network algorithm that is trained to produce different outputs for inputs belonging to different classes (Poggio & Sung 1995).

More directly related to our approach are methods that attempt to describe all object views belonging to the same class using a collection of fundamental building blocks. The eigenspace approach (Turk & Pentland 1990) belongs to this general approach. In this method, a collection of objects within a class, such as a set of faces, are viewed as a set of vectors constructed from the raw grey level values. A set of principal components is extracted from these images to describe the images economically with minimal residual error. The principal components are used as the building blocks for describing new images within the class, using linear combination of the basic images. For example, a set of `eigenfaces' is extracted and used to represent a large space of possible faces.

In this approach the building blocks are global in nature. Other approaches have used more localized building blocks that represent smaller parts of the objects in question. One well-known scheme is the Recognition By Components (RBC) method (Biederman 1985) and related schemes using generalized cylinders as building blocks (Binford 1971, Marr 1982, Marr & Nishihara 1978). The RBC scheme uses a small number of generic 3-D parts such as cubes, cones, and cylinders. Objects are described in terms of their main 3-D parts, and the qualitative spatial relations between parts. Other part-based schemes have used 2-D local features as the underlying building blocks. These building blocks were typically small simple image features, such as local image patches of 2-5 pixels together with their qualitative spatial relations, (Amit & Geman 1997, Nelson & Selinger 1998), corners the direct output of local receptive fields of the type found in primary visual cortex (Edelman 1993).

2.2 Fragments of Intermediate Complexity in Size and Resolution

Unlike other methods that use local 2-D features, we do not employ universal shape features. Instead, we use object fragments that are specific to a class of objects, taken directly from example views of objects in the same class. That is, the shape fragments used to represent faces, for instance, would be different from shape fragments used to represent cars, or letters in the alphabet. These fragments are used as a set of common building blocks to represent, by different combinations of the fragments, different objects belonging to the class. The fragments we detect are divided into equivalence sets that contain views of the same general region in the objects under different transformations and viewing conditions. As discussed later, the use of fragment views achieves superior generalization capability with a smaller number of example views compared with more global methods.

The use of the combination of image fragments to deal with intra-class variability is based on the notion that images of different objects within a class have a particular structural similarity -- they can be expressed as combinations of common substructures. Roughly speaking, the idea is to approximate a new image of a face, say, by a combination of images of partial regions, such as eyes, hairline etc. of previously seen faces. In this section we describe briefly the process of selecting class-based fragments for representing a collection of images within a class. This procedure will be described in more detail elsewhere. In the following section, we describe the use of the fragment representation for performing classification tasks.

Examples of fragments for the class of human faces (roughly frontal) and the class of cars (sedans, roughly side views) are illustrated in Figure 1. The fragments used as a basis for the representation were selected by the principle of maximizing mutual information $I(C,F)$ between a class C and a fragment F. This is a natural measure to employ, because it measures how much information is added about the class once we know whether the fragment F is present or absent in the image. In the ensemble of natural images in general, prior to the detection of any fragment, there is an a-priori probability $p(C)$ for the appearance of an image of a given class C. The detection of a fragment F adds information and reduces the uncertainty (measured by the entropy) of the image. We select fragments that will increase the information regarding the presence of an image from the class C by as much as possible, or, equivalently, reduce the uncertainty by as much as possible. This depends on $p(F|C)$, the probabilities of detecting the fragment F in images that come from the class C, and on $p(F|NC)$ where NC is the complement of C.

A fragment F is highly representative of the class of faces if it is likely to be found in the class of faces, but not in images of non-faces. This can be measured by the likelihood ratio $p(F|C) / p(F|NC)$. Fragments with a high likelihood ratio are highly distinctive for the presence of a face. However, highly distinctive features are not necessarily useful fragments for face representation. The reason is that a fragment can be highly distinctive, but very rare. For example, a template depicting an individual face is highly distinctive: its presence in the image means that a face is virtually certain to be present in the image. However, the probability of finding this particular fragment in an image and using it for making classification is low. On the other hand, a simple local feature, such as a single eyebrow, will appear in many more face images, but it will appear in non-face images as well. The most informative features are therefore fragments of intermediate size, as can be seen in Figures 2. In selecting and using optimal fragments for classification, we distinguish between what we call the 'merit' of a fragment and its 'distinctiveness'. The merit is defined by the mutual information

$$I(C,F) = H(C) - H(C/F) \qquad (1)$$

where I is the mutual information, and H denotes entropy (Cover & Thomas 1991). The merit measures the usefulness of a fragment F to represent a class C, and the fragments with maximal merit are selected as a basis for the class representation. The distinctiveness is defined by the likelihood ratio above, and it is used in reaching the final classification decision, as explained in more detail below. In summary, fragments are selected on the basis of their merit, and then used on the basis of their distinctiveness.

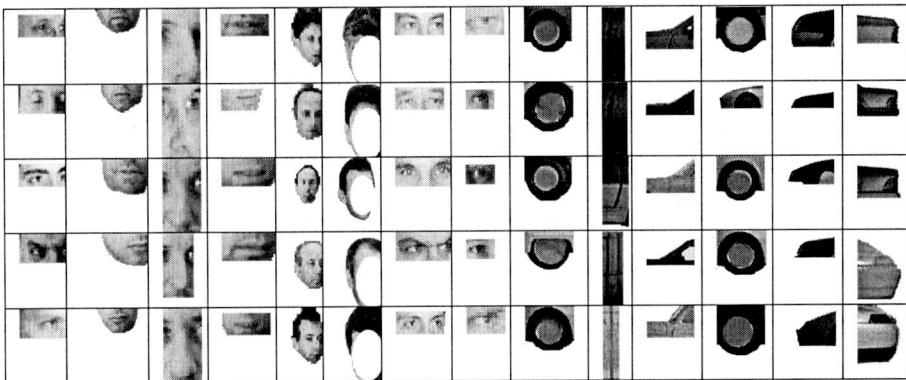

Fig. 1. Examples of face and car fragments

Our procedure for selecting the fragments with high mutual information is the following. Given a set of images, we start by comparing the images in a pairwise manner. The reason is that a useful building block that appears in multiple face images must appear, in particular, in two or more images, and therefore the pairwise comparison can be used as an initial filter for identifying image regions that are likely to serve as useful fragments. We then perform a search of the candidate fragments in the entire database of faces, and also in a second database composed of many natural images that do not contain faces. In this manner we obtain estimations for p(F|C) and p(F|NC) and, assuming a particular p(C), we can compute the fragment's mutual information. For each fragment selected in this manner, we extend the search for optimal fragments by testing additional fragments centered at the same location, but at different sizes, to make sure that we have selected fragments of optimal size. The procedure is also repeated for searching an optimal resolution rather than size.

We will not describe this procedure in further detail, except to note that large fragments of reduced resolution are also highly informative. For example, a full-face fragment at high resolution in non-optimal because the probability of finding this exact high-resolution fragment in the image is low. However, at a reduced resolution, the merit of this fragment is increased up to an optimal value, at which it starts to decrease. In our representation we use fragments of intermediate complexity in either size of resolution, and it includes full resolution fragments of intermediate size, and larger fragments of intermediate resolution.

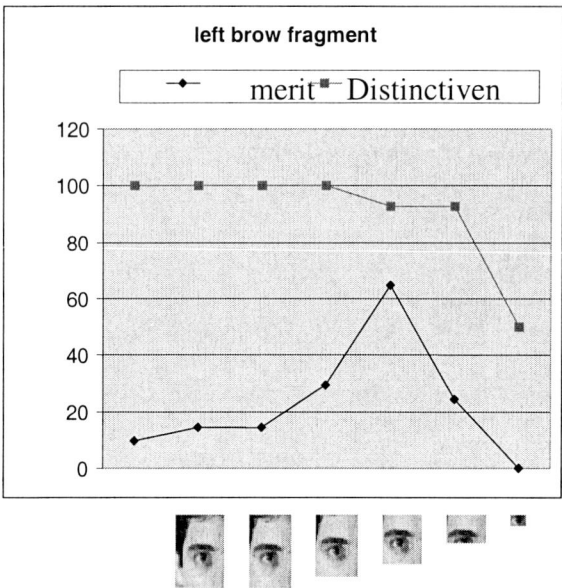

Fig. 2. Selecting optimal fragments by maximizing mutual information. The fragment's merit (based in mutual information) and distinctiveness (based on likelihood ratio) as a function of size. The fragment is optimal at an intermediate size.

3 Performing Classification

In performing classification, the task is to assign the image to one of a known set of classes (or decide that the image does not depict any known class). In the following discussion, we consider a single class, such as a face or a car, and the task is to decide whether or not the input image belongs to this class. This binary decision can also be extended to deal with multiple classes. We do not assume that the image contains a single object at a precisely known position, consequently the task includes a search over a region in the image. We can therefore view the classification task also as a detection task, that is, deciding whether the input image contains a face, and locating the position of the face if it is detected in the image.

The algorithm consists of two main stages. In the first stage basic fragments are detected by comparing the image at each location with several sets of stored fragment views. Each set contains fragments of objects in a class, seen under various viewing conditions. The comparison is performed by combining the results of three comparison criteria: qualitative gray-level based representation, gradient and orientation measures. The second stage combines the results of the individual fragment detectors. It verifies that a sufficient subset of fragment-types have been detected, and enforces the consistency of the fragments viewing parameters. The main tool for verifying the consistency is the use of multiple overlapping fragments. With respect to position in the image, we also incorporate a test for rough position that we found experimentally to be helpful. The consistency of the other viewing parameters such as rotation and illumination is ensured only by the detection of overlapping fragments. The algorithm is performed on the image at several scales so that object views at different scales can be detected. Each level detects objects at scale differences of ±35%. The combination of several scales enables the detection of objects under considerable changes in their size.

In the following sections we describe the details of the algorithm. We begin by describing the similarity measure used for the detection of the basic fragments.

3.1 Similarity between Image Patches

We have evaluated several methods, both known and new, to measure similarity between gray level patches in the stored fragment views and patches in the input image. Many of the comparison methods we tested gave satisfactory results within the subsequent classification algorithm, but we found that a method that combined qualitative image based representation suggested by Bhat and Nayar (1997) with gradient and orientation measures gave the best results. The method measured the qualitative shape similarity using the ordinal order of the pixels in the regions, and measured the orientation difference using gradient amplitude and direction. For the qualitative shape comparison we computed the ordinal order of the pixels in the two regions, and used the normalized sum of displacements of the pixels with the same ordinal order as the measure for the regions' similarity. (See Fig. 3).

The similarity measure $D(F,H)$ between an image patch H and a fragment patch F is a weighted sum of their sum of ordinal displacements d_i, their absolute orientation difference $|\alpha_F - \alpha_H|$ and their absolute gradient difference $|G_F - G_H|$:

$$D(F,H) = k_1 \sum_i d_i + k_2 |\alpha_F - \alpha_H| + k_3 |G_F - G_H| \tag{2}$$

This measure appears to be successful because it is mainly sensitive to the local structure of the patches and less to absolute intensity values.

3.2 The Detection of Fragments

For the detection of fragment views in the images we compared local 5x5 gray level patches in each fragment view to the image using the above similarity measure. Only regions with sufficient variability were compared, since in flat-intensity regions the gradient, orientation and ordinal-order have little meaning. We allowed flexibility in the comparison of the fragment view to the image by matching each pixel in the fragment view to the best pixel in some neighborhood around its corresponding location. Most of the computations of the entire algorithm are performed at this stage. To speed up the application we reduced the search regions for fragments of each type as the search proceeded. We implemented the ordinal measure calculation on ASP's associative processor (ASP, 1998) and achieved a speed factor of approximately 8.5.

An associative processor is especially suitable for such computations since it can process in parallel thousands of pixels.

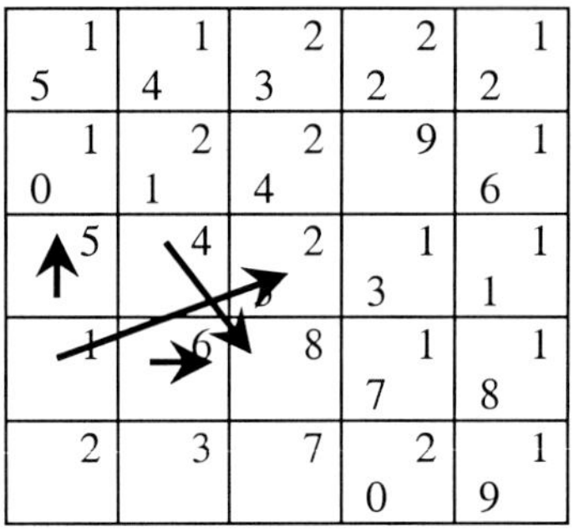

15	14	23	22	11
10	21	24	29	16
↑5	4	2	13	11
↙4	→6	8	17	18
2	3	7	20	19

Fig. 3. Displacement vectors for four pixels with the highest ordinal order. The displacement vectors connect the locations of pixels with similar gray-level ordinal order in the two compared regions. The sum of the four displacements in this case is $1+\sqrt{5}+1+\sqrt{5}$.

3.3 Merging the Detection of the Different Fragment Types

Following the detection of the individual fragments, the final stage of merging the results for the entire object detection is performed. To detect an object only if the fragments are organized properly and are consistent in their viewing conditions, this stage uses the detection of 'binding' fragments as well as the so-called 'pointing' method. It also verifies that fragments from a sufficient subset of fragment-types have been detected, although some occlusion is also allowed for. The 'binding fragments' are fragments with large overlap with other basic fragments such as an eye with a part of a nose, or a lower resolution view of a large part of the object.

In the 'pointing' method each detected fragment (with similarity value above a threshold) 'points' to a common anchor region of a possible object. In face detection, for example, we used the tip of the nose as the anchor. A mouth fragment will therefore point up and a forehead fragment down. The overall contributions of the different fragment types are summed for every location in the image. This procedure is used to integrate the information from all the fragments that are arranged in roughly the expected positions around the anchor location. Each fragment-type contributes to the overall sum associated with a particular location with an associated magnitude of M computed by:

$$M = W_{Type} \cdot \underset{\text{All Part Views} S_i}{Max} (S_i - S_{TH}) \qquad (3)$$

where W_{type} is the weighting factor of the fragment type, S_{TH} a threshold similarity value and S_i the similarity value of all the fragments of that type that point to the location in question. The natural choice for the weighting factor is the distinctiveness defined above based on the likelihood ratio.

Locations that are pointed to by most of the fragment types with high similarity values are candidate object locations. At the final stage we reject some of these locations according to the following rules. First, we reject locations where less then 3/4 of the fragment types were detected. We also compare the collection of image fragments contributing to the candidate location to several low-resolution example views of objects from the class in question. This global filtering proved useful in further enforcing the consistent arrangement of the fragments. After the restrictions are applied to the merged detection results, we mark locations where the merged results exceed a threshold as final detection locations.

4 Experimental Results

We have tested our algorithm on face and car views. For faces we used a set of 1104 part views, taken from a set of 23 male face views under three illuminations and three horizontal rotations. The parts were grouped by 8 types – eye pair, nose, mouth, forehead, low-resolution view, mouth and chin, single eye and face outline. For cars, we used 153 parts of 6 types. Several examples of the fragments used by the algorithm are shown in Figure 1. Figure 4 is the result of the individual fragment detectors that were then merged to yield full-face detection in Figure 5. The images in Figure 6 are additional examples. Note that although the system used only male views in few illuminations and rotations, it detects male and female face views under various viewing conditions. Figure 7 demonstrates the detection of a partly occluded face view.

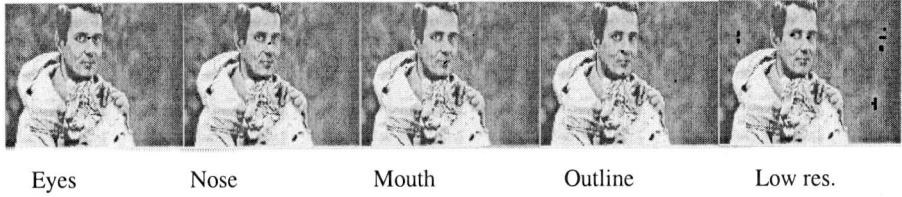

Eyes   Nose   Mouth   Outline   Low res.

Fig. 4. Detection of individual fragments

We have tested the rate of face detection vs. false detection by applying the algorithm to two images – a complex image of a cathedral (Fig. 8A) that does not contain faces, and an image that contains multiple faces (Fig. 8B). We first fixed the system thresholds so that it will not detect any face in the cathedral image while detecting as many faces as possible in the other image. The number of faces that were detected vs. the number of fragments used in each of the eight fragment types is shown in (Fig. 8C). The fragments were taken from the same set of global views and had less overall "image area cover" thenthe global views. We also tried to detect faces by measuring similarity to low resolution (32x22) face views. The number of faces that were detected in Fig. 8B vs. the number of low-resolution example views that were used for the extraction of fragments and as low-resolution patterns is shown in (Fig. 8C). It indicates that the use of multiple fragments performs much better then the use of global views.

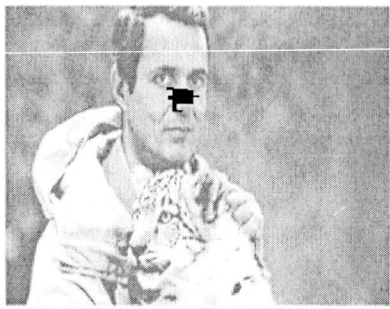

Fig.5. Final face detection

Fig. 6. Examples of face and car detection (Most of the images are taken from the CMU face detector gallery.) Note the detection under different rotations, illuminations and in cluttered scenes.

Fig. 7 Detection of partly occluded face

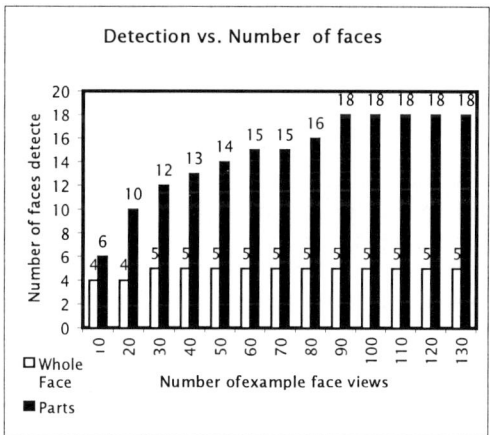

Fig. 8. Sensitivity vs. false detection rate. Image (A) contains 20 faces, (B) contains no faces. System parameters were tuned so that no false detection will occur in image (B), while the maximal number of faces will be detected in (A).

The number of faces detected in (A) vs. the number of training fragments is shown in (C) for the use of fragments (dark bars) and whole views (white bars).

The results of applying the method to these and other images indicate that the fragment-based representation generalizes well to novel objects within the class of interest. Using face fragments obtained from 26 individuals it was possible to classify correctly diverse images of males and females, in both real images and drawings, that are very different from the faces in the original training set. This was achieved while maintaining low false alarm rates on images that did not contain faces. Using a modest number of informative fragments, in different combinations, appears to have an inherent capability to deal with shape variability within the class. The fragment-based scheme was also capable of obtaining significant position invariance, without using explicit representation of the spatial relationships between fragments. The insensitivity to position as well as to other viewing parameters was obtained primarily by the use of a redundant set of overlapping fragments, including fragments of intermediate size and higher resolution, and fragments of larger size and lower resolution.

5 Some Analogies with the Human Visual System

In visual areas of the primate cortex neurons respond optimally to increasingly complex features in the input. A simple cell in the primary visual area (V1) responds best to a line or edge at a particular orientation and location in the visual field (Hubel & Wiesel 1968). In higher-order visual areas of the cortex, units were found to respond to increasingly complex local patterns. For example, V2 units respond to collinear arrangements of features (Von der Heydt 1984), some V4 units respond to spiral, polar and other local shapes (Gallant *et. al.* 1993), TE units respond to moderately complex features that may resemble e.g. a lip or an eyebrow (Tanaka 1993), and anterior IT units often respond to complete or partial object views (Perorate *et. al.* 1982, Rolls 1984, Logothetis *et. al.* 1994). Together with the increase in the complexity of their preferred stimuli, units in higher order visual areas also show increased invariance to viewing parameters, such as position in the visual field, rotation in the image plane, rotation in space, and some changes in illumination (Logothetis *et al.* 1994, Perret, Rolls & Caan 1982, Rolls 1983, Tanaka 1993). The preferred stimuli of IT units are highly dependent upon the visual experience of the animal. In monkeys trained to recognize different wire objects, units are found that respond to specific full or partial views of such objects (Logothetis *et al.* 1994). In animals trained with fractal-like images, units are subsequently found that respond to on or more of the images in the training set (Miyashita & Chang 1988).

These findings are consistent with the view that the visual system uses object representations based on class related fragments of intermediate complexity, constructed hierarchically. The preferred stimuli of simple and intermediate complexity neurons in the visual system are specific 2-D patterns. Some binocular information can also influence the response, but this additional information, which adds 3-D information associated with a fragment under particular viewing conditions, can be incorporated in the fragment based representation. The preferred stimuli are dependent upon the family of training stimuli, and in this sense appear to be class-dependent rather than, for example, a small set of universal building blocks used by other classification schemes. Invariance to viewing parameters such as position in the visual field or spatial orientation appears gradually, possibly by the convergence of more elementary and less invariant fragments onto higher order units. From this theory we can anticipate the existence of two types of intermediate complexity units that have not been reported so far. First, for the purpose of classification, we expect to find units that respond to different types of partial views. As an example, a unit of this kind may respond to different shapes of hairline, but not to a mouth or nose regions. Second, because the invariance of complex shapes to different viewing parameters is inherited from the invariance of the more elementary fragment, we expect to find intermediate complexity units, responding to partial object views at a number of different spatial orientations and perhaps different illumination conditions.

In our implementation we used the notion of a `pointing' mechanism to limit the relative spatial displacements allowed for the different fragments within a more global shape. The particular implementation we used is not intended to be directly related to biological mechanisms. However, similar results can be obtained by processes that are more biological in nature. For example, attentional mechanisms, that limit the processing to restricted regions of visual space, can play a similar role in limiting the combination of fragments and ensure that all the fragments are detected within a common region. Together with the use of overlapping fragments, this can prevent the "illusory conjunction" of fragments that are not properly arranged.

6 Summary

Object classification is a challenging task because individual objects within the same class can have large variations is shape that are difficult to define precisely and to compensate for during the classification process. In our approach, objects within a class are represented by class-specific fragments. These are image fragments that appear in similar shape in multiple individual objects within the class, and that are highly informative about the class in questions. The fragments we obtain are typically of intermediate complexity, either in size or resolution, and they form a redundant, overlapping set of `building blocks' for representing individual objects within the class.

The experimental results support the view that the fragment-based approach can generalize well to novel object images, and it can detect and classify objects of a given class despite large variability in shape. It can also deal with changes in position, pose, and illumination, but these aspects will have to be extended and tested further in future work. The results support the notion that the intra-class variability of views can be expressed, in large part, in terms of different combinations of shared common sub-structures.

The detection algorithm initially detects fragment views of different fragment types, and then combines the results for the detection of the entire object. The consistency in the fragments viewing parameters is ensured by the use of overlapping fragments that "bind" the parts together and by testing directly that the fragments are approximately aligned. An important advantage of this method is that it can compensate well for shape variations by matching a novel shape within a class with a new combination of stored fragments. The formation of a new combination also results in a system that uses less training examples compared with the use of global shapes. The scheme uses high-resolution smaller fragments with coarser larger ones and this increases the efficiency of the scheme, as well as its ability to impose the proper arrangement of the components.

The method is relatively simple because it does not require the estimation of the viewing parameters and does not require the explicit representation and matching of spatial relations. The use of class-specific rather than universal fragments also has limitations, since it implies that dealing with a new class of objects will require extending the set of stored object fragments. This raises interesting learning issues, currently under study, concerning the automatic extraction of useful fragments from a set of novel object views.

A number of natural extensions to the basic classification scheme outlined above will be considered in the future. One is the construction of fragments in a hierarchical manner. In the algorithm described above objects views are represented in terms of intermediate complexity fragments. These fragments in turn can be constructed from simpler fragments. Starting from simple local fragments, at the level of complexity of features found in primary visual cortex, more complex fragments can be constructed hierarchically. It will be of interest to consider the effects of having different number of levels. In the primate visual system, the hierarchy includes about five levels from V1 to anterior IT, and this may be a reasonable guideline to consider. As in our scheme, at each level it will be useful to extract the most informative fragments possible at that level, and to use a redundant, overlapping set of fragments. Fragments that can be considered equivalent, will be grouped together (as in the fragment types used above) to generate increased invariance to shape variations and to changes in viewing conditions. In such a hierarchical scheme, it is likely that the similarity scheme used above for the purpose of detecting fragments in the image will be modified: the similarity between an image region and a stored fragments could be based instead on the more basic sub-fragments they have in common.

A second area of further development has to do with invariance to viewing conditions, including rotations in space and changes in illumination. The current algorithm already exhibits some invariance to these parameters, because each fragment type includes fragments at somewhat different orientations and illuminations. However, additional fragments covering a wider range of changes will be required to reach invariance comparable to human perception. It is not entirely clear whether the mechanism of using multiple overlapping fragments will be sufficient to impose the correct overall configuration when the number of alternative fragments will be substantially larger, and it is conceivable that additional mechanisms, possible including top-down verification, will be required.

Finally, the scheme should be extended to deal effectively with multiple classes. Each class added to the system will be represented by a set of class-specific fragments that are optimal for the class in question. Some of them may be similar to already existing fragments for other classes, but others will be new. The overall scheme will remain similar, however, it remains to be seen how to best organize the system to deal as efficiently as possible with a large number of known classes.

References

ASP associative processor, http://www.asp.co.il

Amit Y.,Geman D., Wilder K., "Joint Induction of Shape Features and Tree Classifiers", IEEE Transactions on Pattern Analysis and Machine Intelligence, Vol. 19, No. 11, November 1997. **

Bhat D., Nayar K. S., "Ordinal measures for image correspondence", *IEEE Trans. on PAMI* Vol. 20 No. 4, (1998) 415-423.

Biederman I., "Human image understanding: recent research and theory", *Computer Vision, Graphics and Image Processing, (1985)* 32:29-73.

Binford T. O. "Visual perception by computer", IEEE conf. on systems and control 1971.

Brooks R., "Symbolic reasoning among 3-D models and 2-D images", Artificial intelligence (17) (1981) 285-348.

Cootes T.F., Taylor C.J., Cooper D.H., Graham J., Active shape models -- their training and applications. *Computer Vision and Image Understanding*, 61 (1995) 38-59.

Cover, T.M. & Thomas, J.A. *Elements of Information Theory.* Wiley Series in Telecommunication, New York, 1991.

Edelman, S. Representing 3D objects by sets of activities of receptive fields 70, 37-45. *Biological cybernitics,* 70, (1993) 37-45.

Grimson W. E. L., Recognition of Object Families Using Parametrized Models, Proc. First International Conference on Computer Vision, (1987) 93-101.

Grimson, E.W.L., & Lozano-Perez, T. Localizing overlapping parts by searching the interpretation tree. *IEEE Trans. On Pattern Analysis and Machine Intelligence*, 9 (1987) 469-482.

Hubel, D. H., Wiesel, T. N. "Receptive fields and functional architecture of monkey striate cortex", *Journal of physiology,* 195 (1968) 215-243.

Logothetis N. K., Pauls J., B•lthoff H. H., Poggio T., "View-dependent object recognition in monkeys", *Current biology*, 4 (1994) 401-414.

Marr D., *Vision*, W.H. Freeman, San Francisco CA, 1982.

Marr D., Nishihara H. K. "Representation and recognition of the spatial organization of three dimensional structure" *Proceedings of the Royal Society of London B,* 200 (1978) 269-294.

Mel W. B., SEEMORE: "Combining color, shape and texture histogramming in a neurally inspired approach to visual object recognition", *Neural computation* 9 (1997) 777-804.

Minsky M. and Papert S., *Perceptrons*, The MIT Press, Cambridge Massachusetts, 1969.

Miyashita, Y. & Chang, H.S. Neuronal correlate of pictorial short-term memory in the primate temporal cortex. Nature, 331, (1988) 68-70.

Murase, H. & Nayar, S.K. Visual learning and recognition of 3-D objects from appearance. *International J. of Com. Vision*, 14 (1995) 5-24.

Nelson C. R., and Selinger A., "A Cubist approach to object recognition", ICCV-98 (1998) 614-621.

Perret D. I., Rolls E. T. Caan W., "Visual neurons responsive to faces in the monkey temporal cortex", *Experimental brain research*, 47 (1982) 329-342.

Poggio T. and Sung K., "Finding human faces with a gaussian mixture distribution-base face model", *Computer analysis of image and patterns* (1995) 432-439.

Poggio, T. \& Edelman, S. A network that learns to recognize three-dimensional objects. Nature, 343 (1990) 263-266.

Rolls E. T., "Neurons in the cortex of the temporal lobe and in the amygdala of the monkey with responses selective for faces", *Human neurobiology,* 3 (1984) 209-222.

Rosch, E. Mervis, C.B., Gray, W.D., Johnson, S.M. & Boyes-Braem, P. Basic objects in natural carogories. *Cognitive Psychology*, 8 (1976) 382-439.

Tanaka, K., "Neural mechanisms of object recognition", *Science*, Vol. 262 (1993) 685-688.

Turk M. and Pentland A., "Eigenfaces for recognition", *Cognitive Neuroscience*, 3 (1990) 71-86.

Ullman, S. \& Basri, R. Recognition by linear combination of models. *IEEE PAMI*, 13(10) (1991) 992-1006.

Vapnik, V. *The Nature of Statistical Learning Theory.* Springer, New York, 1995.

von der Heydt R., Peterhans E., Baumgartner G., "Illusory contours and cortical neuron responses", *Science,* 224 (1984) 1260-1262.

Confrontation of Retinal Adaptation Model with Key Features of Psychophysical Gain Behavior Dynamics

Eilon Sherman and Hedva Spitzer

Department of Biomedical Engineering, Faculty of Engineering,
Tel Aviv University, Tel-Aviv, Israel

Abstract. The study aimed to establish a comprehensive computational model of intensity adaptation mechanisms, which predicts key features of experimental responses (1). We elaborated on a previous adaptation model (2) which presents retinal adaptation mechanisms and predicts responses to aperiodic stimuli. The model suggests that the temporal decline in the response of the retinal ganglion cells is a reflection of the adaptation mechanism ("curve shifting"(3)). This adaptation mechanism is applied to each cell receptive-field (RF) region (center and surround) separately, and only then the subtraction operation between the two regions is performed. The elaborated model was tested by simulating various periodic sinusoidal fields, which varied in DC level, and frequency (1-30 Hz). The model's results are in agreement with various psychophysical and physiological findings and predict most of the psychophysical key features (1). Until now, no existing model has been able to predict the key features of the experimental findings (1).

1 Introduction

The human visual system can distinguish small changes in a large range (over 10^8) of ambient light levels. The visual system counters this challenge mainly by using various adaptation mechanisms. These mechanisms were extensively studied from both physiological and psychophysical aspects (see reviews 3-5). It is common to regard adaptation to light intensity (first order) as a low-level mechanism, which originates at the retinal level (3-5). Light intensity adaptation refers to the changes in the visual system's sensitivity due to changes in light level, and enables it to obtain high sensitivity at a wide range of light levels (3).

A computational model has to predict experimental results, including cardinal features of important physiological and psychophysical adaptation effects. Among these adaptation features are Weber's law; the "curve-shifting" effect (6, 3); and the results of the Crawford paradigm - the change of light sensitivity over a course of time (3). Many of the computational adaptation models emphasized predictions results of either periodic or aperiodic experimental traditions (7), i.e., experiments in which the stimuli were periodic, mainly sinusoidal, or from experiments in which the stimuli were aperiodic, such as step functions.

Since there are a large number of different adaptation models and the physiological adaptation mechanisms are not fully understood, the need to have a common tool of testing the models was raised. Graham and Hood (1,7) confronted a group of models with a set of results from psychophysical experiments of periodic and aperiodic traditions. Hood and his colleagues performed additional psychophysical experiments using a probed-sinewave paradigm (1). They measured the threshold to a probe stimulus (a brief flash) at various temporal phases of a spatially wide background field, which is sinusoidally modulated. They chose this paradigm since this stimulation includes aperiodic (probe) and periodic stimulation and because it enables a method of testing the temporal properties of adaptation. Hood and his colleagues (1) found that the probe threshold amplitude depends on the temporal phase differently at various frequencies. The authors defined several characteristic properties of the analyzed probe-threshold responses as the key features of the data for assessing adaptation models. According to this line of thought, different adaptation models were tested for their ability to predict the key features. These models were derived from the periodic tradition (8), from the aperiodic tradition (7) and included merged models, which stemmed from the two traditions (1,9-11). The study of Wiegand et. al. (11) was the only model which succeeded in the quantitative prediction of phenomena from both traditions, periodic and aperiodic (1). However, all the tested models, including the study of Wiegand et al. (11) failed to predict all the important aspects of the key features.

Our adaptation model assumed that one of the main features of light adaptation is the curve-shifting mechanism (2). A curve-shifting effect is the transition from one response curve to another, due to a change in the light intensity, in order to obtain higher gain in the new light intensity. The change from one response curve to another is reflected in the time decay of a cell response, to a step function. The model also predicts Weber's law as a specific case of equal levels of the adaptation stimulus and the conditional stimulus (2). Since the model takes into account illumination history and the dependence of a temporal filter on it, it was worthwhile to test qualitatively the predictions of the model to the probed-sinewave paradigm.

1.1 The Elaborated Model

For the sake of clarity and self-containment, we have briefly summarized our previous model, whose essentials have been previously published (2). The model assumes that the adaptation process occurs in the retina before the subtraction between the responses of the center and the surround regions of the RF (3). The same equations, that describe the center's RF region response, are applied to the surround region with different parameters. For this reason the subscript "c", which indicates the RF center, has been omitted in all the following equations unless otherwise stated. The block diagram of the model appears in Fig. 1

Let $I(x,y,t)$ be the visual stimulus that falls on the retina, and $R(t)$ the output at the ganglion cell level. The ganglion cell spatial profile is commonly considered to be composed of two opponent regions (for the center and surround RF regions), and it is expressed as a difference of two Gaussians (DOG). Since in our model we assume that adaptation takes place in each of the two RF regions separately, the difference between the two Gaussians should be considered only after the adaptation stage. The spatial output of the RF center region is therefore:

$$G_s(t) = \int_{-\infty}^{\infty}\int_{-\infty}^{\infty} I(x, y, t) f_s(x, y) dx dy \qquad (1)$$

where

$$f_s(x, y) = \exp\left[-(x^2 + y^2)/\rho^2\right]/\pi\rho^2 \qquad (2)$$

, ρ is the estimated radius of the RF center, and I(x,y,t) in this study was taken as varied only in the time domain - as a temporal sinusoidal wave, which creates a homogeneous test field in space.

Next (Fig. 1), a temporal filtering, i.e., a convolution of $Gs(t)$ with the temporal filter $ft(t)$, is carried out.

$$G(t) = Gs(t) * ft(t) \qquad (3)$$

The $ft(t)$, low pass (LP) filter, is given by:

$$f_t(t) = \exp(-t/\tau)/\tau \quad t \Rightarrow 0 \; ; and \; f_t(t) = 0 \quad t < 0 \qquad (4)$$

where τ is the LP filter's time constant.

In accordance with experimental findings we express the cell's sub-region response by the Naka-Rushton (N-R) equation (3). Thus the response of the RF center is:

$$R(t) = R_{max}[G(t)]^n / \{[G(t)]^n + [\sigma(t)]^n\} \qquad (5)$$

where σ is the semi-saturation constant, R_{max} is the maximum response (400 spikes/sec), and n is a constant parameter found in photoreceptors as $n \approx 1$.

The adaptation itself is considered here as a change in the σ parameter [Eq. (5)]. Such a change is equivalent to the curve shift of the "log response vs. log illumination" curve, which has been shown experimentally (3). The model describes the dependence of σ as a function of illumination and the way it develops as a function of time. This is expressed by the output $Gb(t)$, which is termed the "adapting component," and is calculated as a convolution of G with a LP filter- $fb(t)$ (squares in the block diagram, Fig. 1), accordingly:

$$G_b(t) = G(t) * f_b(t) \qquad (6)$$

where $fb(t)$ is:

$$f_b(t) = \exp(-t/\tau_b)/\tau_b \quad t \Rightarrow 0 \; ; and \; f_b(t) = 0 \quad t < 0 \qquad (7)$$

and τ_b is the time constant of the LP filter $fb(t)$. The model takes into account both Weber's law for a specific case (2) and the temporal change in the gain. Consequently, the model suggests that $\sigma(t)$ is described by the following:

$$\sigma(t) = \alpha G_b(t) + \beta \qquad (8)$$

where α and β are constant parameters and their values are based on physiological findings (2,3). According to Eqs. 6-8, the state of adaptation of the RF center depends on the same spatial profile as the response component $G(t)$, this in agreement with physiological findings which showed that the spatial properties of the ganglion cells are common to adaptation and spatial summation (3).

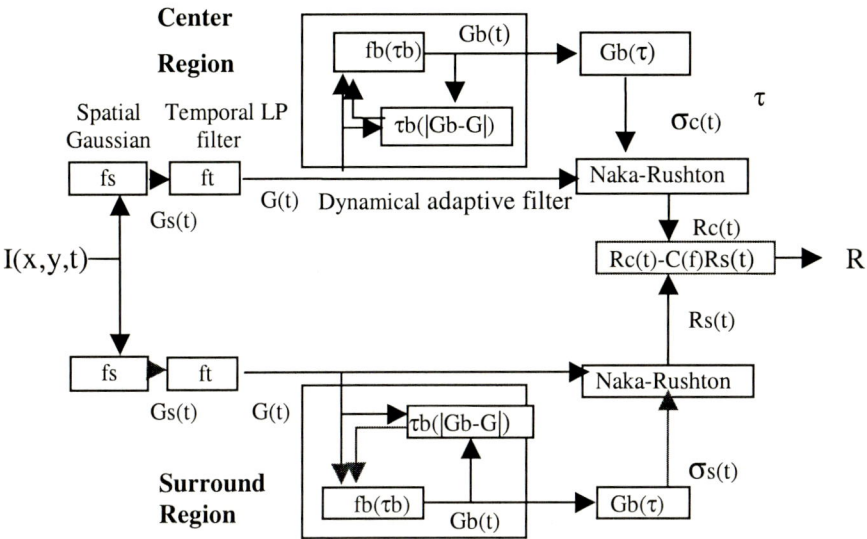

Fig. 1. Schematic block diagram of the adaptation model. Components of the model are detailed in the text.

An important aspect of the model is that the level of response transience is derived from the mechanism that is responsible for the curve shift into a new state of adaptation. This mechanism is expressed mathematically in our model as a change in σ through a change in the temporal parameter $\tau_b(t)$ of the LP filter, fb (Eq. 7). Consequently, the mathematical dynamic temporal parameter is:

$$\tau_b(t) = \tau_m / \left[1 + |G(t) - G_b(t)| / G_n \right] \quad (9)$$

where the constant τ_m is the maximum value of $\tau_b(t)$ and Gn is a normalization constant.

The above development (Eqs. 2-9) describes a procedure to calculate the response of the center region of the RF. The same calculation applies to the surround region $Rs(t)$, changing the parameters to include, for example, its different spatial properties (Tab. 1 in ref. No. 2). The last stage of the model is a subtraction of the center and the surround responses, but with an additional operation of LP filter [11].

Relating the subscripts c and s to the center and the surround regions, respectively, the overall cell response is given by (Fig. 1):

$$R(t) = R_0 + R_c(t) - d_f \cdot R_s(t) \quad (10)$$

where R_0 is the background cell activity at dark illumination, which is taken as constant and d_f is taken as a frequency dependent weight factor:

$$d_f = 0.2 \cdot \tanh(f - 7) + 1.4 \tag{11}$$

where f is the frequency of the input signal.

2 Simulation Methods

Simulation parameters: Our simulations followed the experimental probed-sinewave paradigm of Hood and colleagues (1). The original paradigm included a small (1°) test flash (probe) which was located in the center of the field of view, and a large (18°) symmetric flashing background. However, since our model derived from the physiological mechanism of a single ganglion cell, we calculated its response to a diffused (i.e. an effectively homogeneous) background on the entire RF. Therefore, the stimuli we used (both background and probe) were sinusoidally varied in time, with no spatial changes. The thresholds of noticeable probes were tested with brief (10 msec) probe flashes. The probe flashes were presented at different temporal phases (0, 45, 90, 135, 180, 225, 270 and 315 deg) of the background stimulus, with only one probe present in each trial. The following properties of the background stimulus were varied: the mean luminance and the temporal frequencies.

Procedure: Simulation of the model was performed through a MATLAB program (from Math Works Inc.).

The LP Filter (Eqs. 3, 4) and the dynamical adaptive filter (Eqs. 6, 7) were simulated by solving two-difference equation, following the presentation in the appendix of (7), in order to produce the different outputs G and G_b.

The solving of the difference equations numerically was produced in time steps of 0.5msec. τ, the time constant of both temporal filters, was taken to be 10 msec. τ_b (Eq. 9) was calculated with τ_m=500 msec and G_n=5000 (constants taken roughly from ref. No. 2).

We determine in simulations the probe-threshold by a convergence procedure as described below (see ref. 7, 13). The decision rule requires that:

$$R(I + \Delta I) - R(I) = \Delta R \tag{12}$$

where R is the simulated cell's response (quanta/sec), I is the field intensity (cd/m²), ΔI is the probe's intensity (which is added to the value of I), and ΔR is the increment of the cell's response due to the increment in the field intensity. The response, ΔR, to the just noticeable difference (JND) is the threshold response. The constant decision rule (7) assumes that ΔR has an arbitrary small constant value (usually termed δ, eq.4, ref. 7). Previous simulations showed that this constant threshold value should not affect the probe-threshold results (1). This was tested and verified in our simulations.

The probe intensity is determined once the threshold response is within one-tenth of the threshold's size (i.e. whenever the response is between $R_{ref} + \delta < R < R_{ref} + 1.1 \cdot \delta R_{ref}$).

3 Simulation Results

We mainly focused in the simulations on the model's predictions to the probed-sinewave paradigm, with its resulting key features that were established as a critical test for an adaptation model for psychophysical data from both periodic and aperiodic traditions (1, 5). The key features that were considered (1) are: 1) The probe threshold depends upon the phase at which it is presented for all background frequencies used, 0-16 Hz. These threshold variations are not well described by a sinewave. 2) The peak threshold is over 180° out of phase with the through threshold. 3) The positions of the peaks and troughs shift fairly abruptly at background modulations of 4-8 Hz (Fig. 2IIa). 4) The difference between the peak and trough thresholds vary as a function of temporal frequency in a manner similar to the temporal contrast function. 5) The peak-trough difference dominates at low frequencies of background modulation, while the dc level dominates at higher frequencies (1).

In addition, further simulations were performed in order to test the model's predictions to additional background light intensities (while maintaining the same stimulus contrast), and to additional higher temporal frequencies (13,14). In the following paragraphs we will present the simulation results to probe sine wave paradigm and compare them to the different experimental results.

Our results (Fig. 2II) show an overall good agreement with the experimental findings (Fig. 2I), even though not all the details are similar. It can be seen clearly that the probe-threshold vs. temporal phase curves depend significantly on the stimulus' temporal frequency (the first key feature). The predicted probe-sinewave results show a single peak at the frequency of 8 Hz (Fig. 2IIa), which yields the largest response around the probe phase 230°, as in the experimental results (Fig. 2Ia). The probe-threshold curves follow approximately the stimulus phase at relatively low frequencies (up to 6Hz), with a slight change in shape. At frequencies higher than 6 Hz, there is even a phase reversal of the response curve, with a significant distortion in response-curve shape (second key feature). The Phases of both the simulated (Fig. 2IIb) and the experimental (Fig. 2Ib) probe threshold's peaks and troughs appear to increase as a function of temporal frequency, mainly from 4Hz, with a steep phase shift up to 8Hz, as found experimentally (the third key feature). At higher temporal frequencies, non-monotonic behavior appears in the phase-shift of the simulated peaks and troughs of the probe-thresholds, as in most of the experimental results. The troughs of the probe threshold, which is not considered a key feature, obtain relatively low values in the simulated results, beneath ΔI_0, as opposed to the experimental findings, which were always higher than ΔI_0.

The predictions of our model to the relative differences between the peak and through thresholds and the dc levels show a bell shape with an optimum around 8 Hz, similarly, as experimentally found (Fig. 2Ic). This predicted dependence is similar to the experimental findings, however, the bandwidths of the two predicted curves obtained are somewhat narrower than those found experimentally (the forth key feature). The prediction of the fifth key feature is less agreeable with the empirical data (Fig. 2Ic) than the other key features. It can be seen that the relative values of the dc-levels and the differences between peaks and troughs seem to be lower in the simulated results, than in the experimental results.

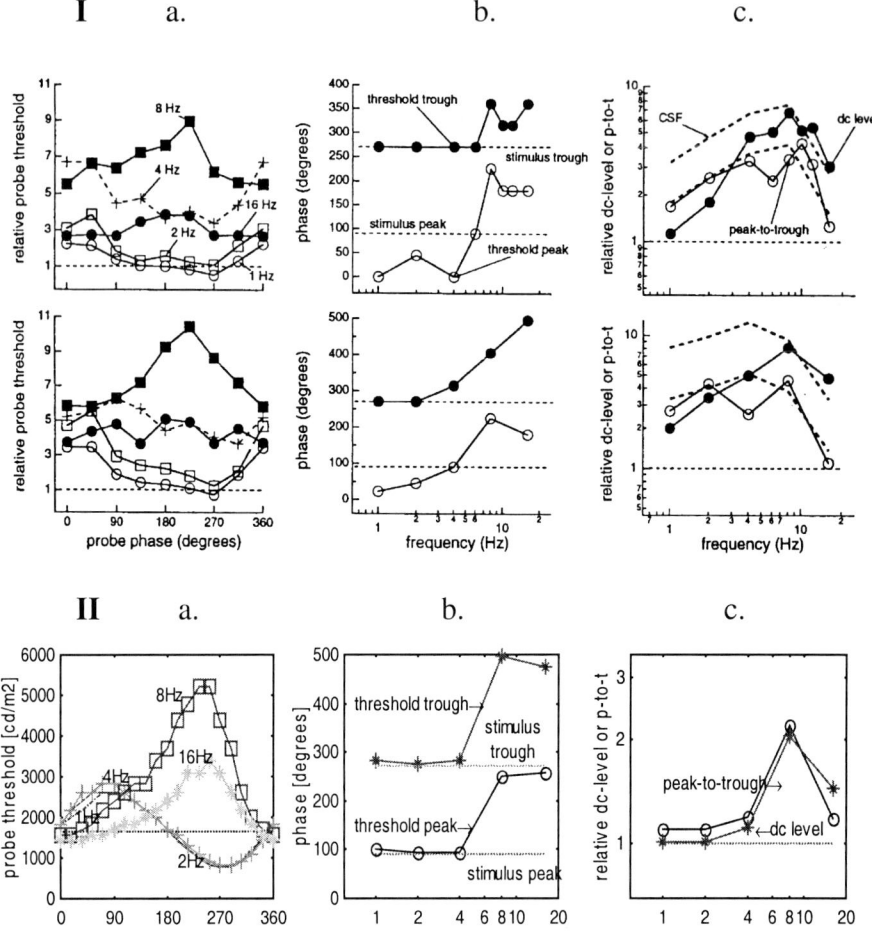

Fig. 2. The probe threshold as a function of the background modulation: To summarize the principal changes in probe threshold with temporal frequency (Fig. 2) we analyzed the results in a similar way to the one done experimentally (Fig. 2I) and theoretically by simulation predictions of other adaptation models (1). Figure 2 presents the simulation results for a mean background intensity of $10^{5.5}$ cd/m^2, 57% in contrast, when the background was modulated at 1- 16 Hz. Note that the relative difference between the peak and the trough values and their relative averaged values (dc-levels) are obtained by normalizing the responses to the response threshold to a steady background of the same luminance (ΔI_0). (Fig. 2c open symbols).

Further predictions: We further tested our model with: a) background which is rapidly modulated (30 Hz), and b) different background intensities.

Boynton and his colleagues (14) found that the probe threshold is varied approximately sinusoidally, and in phase with a rapid temporally square-wave (30 Hz) modulated background. Hood and Graham (13) simulated Wilson's model (12) to predict the probe threshold at 30 Hz and found a sinusoidal dependence and concluded that there is an agreement with the findings. Our results show that the probe threshold varied sinusoidally in phase with the stimulus similarly, as found experimentally, yet without the change of shape (to a square wave, (13)).

We varied intensity levels over 4 log units (10^2, 10^4, $10^{5.5}$ and $10^{6.5}$ cd/m^2) in order to cover a large range of indoor, and outdoor light intensity levels (4). Note that the intensity level of $10^{5.5}$ cd/m^2 (Fig. 2II) roughly equals 250 td, i.e. the light intensity used in the experimental paradigms (1).Note that the contrast was kept constant for all of the different light intensities (57%), following (1).

The results of this simulation show that the shapes of the different curves of the probe thresholds are changed dramatically as a function of the light intensity. At 10^2 cd/m^2 there is almost no dependence of the probe threshold on the probe phase, and at 10^4 cd/m^2 the responses can appear with two peaks (at 8 and 4Hz). At the largest background intensity $10^{6.5}$ cd/m^2 the response curves appear more sinusoidal than in the $10^{5.5}$ cd/m^2, with less distortion in the phase shift. Consequently the above key features are not necessarily obtained in the other background intensities (1).

4 Discussion

The goal of our study, and of other accompanied studies of other researchers, was to explore the temporal dynamics of the light adaptation process (1,13). Our model simulations showed the basic effect of variations in the probe threshold response (Fig. 2II), not because of relatively slow and fast adaptation mechanisms as part of other studies considered (9, 10), but because of adaptive temporal filtering in the model (Eqs. 6-9).

The results predicted 4 out of the 5 key features (1) for background intensity similar to the one used experimentally. The fifth key feature, which is not predicted by our model, is the peak-through difference dominance at low frequencies of background modulation as opposed to dc level dominance at higher frequencies. Beside the agreement of most predicted key features with experimental results there are additional characteristics of the predicted responses, which are in agreement with the experimental ones. The largest relative peak threshold response is obtained at 8Hz and the second largest is obtained at 4 Hz or 16 Hz, while at 1 and 2Hz the responses are much smaller. An additional tested feature, the response to 30Hz (13), was predicted. It should be noted that this feature has been predicted by an additional adaptation model (12), even though several additional assumptions have been added to it (13).

Our predictions to variation in the background intensity show that the key features are mainly maintained for $10^{5.5}$ cd/m^2. The probe threshold responses have major behavior differences at the largest, intermediate and lowest chosen background intensities. Intuitively speaking, the various types of responses for the different background intensities are derived from different gain conditions at the various intensities (Fig. 24 in ref. No. 3, and Eqs. 21-23 in ref. No. 2).

Comparison with other adaptation models: Hood and his colleagues (1) computed the predictions of different adaptation models to the probe threshold paradigm. A summary of the main components and the predictions from these models can be found in Hood and his colleague's study (1).

Our original model was based on physiological mechanisms of adaptation of single cells in the retina (ganglion cells). This model succeeds to predict a large repertoire of findings from physiological experiments (2) and predicted the critical tests, i.e., key features of psychophysical findings (1). Most of the other tested adaptation models (1) mainly considered psychophysical findings, though few of the other models also included modules, which are based on physiological findings (MUNSOL in ref. No. 7). Wilson's model (12), which is also a physiological model, is discussed separately below.

There are structural differences between our model and the other models (1). In our model, adaptation is performed by two separate pathways, for the center and surround receptive field regions, (Fig. 1), whose outputs are subtracted only after the adaptation process (i.e. the curve-shifting effect (3, 6)). This approach is different from the structure of the other models, where subtraction is performed as a mere stage in the adaptation process within a single pathway (1).

In addition, the basic approach to our original study (2) was the modeling of the temporal dynamics of curve shifting (3, 6). Other models use dynamic or static multiplicative and subtractive modules (1) and do not follow the curve-shifting effect. Most models use a static non-linear (SNL) module (mainly Naka-Rushton) and perform the dynamical adaptation in the input to this module, whereas in our model the dynamics of the adaptation lie inside this non-linear Naka Rushton module (Eqs. 5-8), through a dynamic change in the semi-saturation "constant" (σ).

Although there are structural differences, there is some similarity between the components of our model and the other models (1). Among these similar components are low-pass filters (static or dynamic) and adaptive operations (feedback or feedforward). However, our model generally requires a smaller number of stages within each module, and requires no high-pass filtering operation. Our model applied an adaptive operation also to the temporal parameter, τ_b (Eq. 9) in the dynamical adaptive filter (Fig. 1). A similar consideration appears in Sperling and Sondhi (8). Wiegand et al. (11) also considered a dynamical operation in their high order nonlinear filter, but in a different manner through fitting their model's parameters to the experimental results.

Wilson's model (12) was tested for a different feature from the key features (1): predictions to extreme temporal frequencies, 1 and 30 Hz (13). Wilson's physiological adaptation model has a different approach from ours. Wilson modeled the different visual stages in the retina to their details (including the processes at the inter-layer cells, see also remarks in ref. No. 5 and 13). Thus the outcome appears as a multi-layer model which contains a large number of free parameters. The model of Wilson also uses two channels of adaptation, but these channels represent the M_{on} and the M_{off} ganglion cells. These two separated pathways are subtracted to produce a "push-pull" mechanism, which is found to be critical for the successful prediction of the tested features (13). Hood and Graham (13) succeeded to get plausible predictions from Wilson's model (12) only when they assumed that the M_{on} pathway was less sensitive than the M_{off} pathway (5). The significance of two channels was also found in our simulation results. Our model, on the other hand, tried to focus mainly on the relevant adaptation mechanism (that was recorded in the ganglion cells, (2)), and which seemes to us relevant to the relatively fast adaptation in the retina. Therefore, our model appears much simpler and with a much smaller number of parameters (12).

Moreover, our model is the only model so far that has succeeded to predict not only the predictions to extreme temporal frequencies (13), but also many of the psychophysical key features in a wide range of temporal frequencies (1, 13).

References

1. Hood, D.C., Graham, N., Wiegand T.E. and Chase, V.M. (1997). Probed-sinewave paradigm: a test of models of light-adaptation dynamics. Vision Res. 37, 1177-1191.
2. Dahari R. and H. Spitzer (1996). Spatio-temporal adaptation model for retinal ganglion cells. JOSA A 13, 419-439.
3. Shapley, R. and Enroth-Cugell C. (1984). Visual Adaptation and Retinal Gain Controls. Progress in Retinal Res. 3, 263-346.
4. Hood, D.C. and Finkelstein, M.A. (1986). Sensitivity to light. In Boff, K.R. Kaufman, L. & Thomas, J.P. (Eds), handbook of perception and human performance, Vol. I: Sensory processes and perception. New York: John Wiley and Sons.
5. Hood, D.C. (1998). Lower-level visual processing and models of light adaptation. Anu. Rev. Psychol. 503-535.
6. Sakmann, B. and Creutzfeld, O.D. (1969). Scotopic and mesopic light adaptation in the cat's retina. Flugers Arch. 313, 168-185.
7. Graham, N. and Hood D.C. (1992). Modeling the dynamics of light adaptation: the merging of two traditions. Vision Res. 32, 1373-1393.
8. Sperling, G. and Sondhi, M.M. (1968). Model for visual luminance discrimination and flicker detection. JOSA A 58, 1133-1145.
9. Geisler, W.S. (1983). Mechanisms of visual sensitivity: backgrounds and early dark adaptation. Vision Res. 23, 1423-1432.
10. Hayhoe, M.M., Benimoff, N.I. and Hood, D.C. (1987). The time-course of multiplicative and subtractive adaptation process. Vision Res. 27, 1981-1996.
11. Wiegand, T.E., Hood, D.C. and Graham, N. (1995). Testing a computational model of light-adaptation dynamics. Vision Res. 35, 3037 3051.
12. Wilson, H.R. (1997). A neural model of fovial light adaptation and afterimage formation. J. Neurosci. 14, 403-423.
13. Hood , D.C & Graham, N. (1998). Threshold fluctuations on temporally modulated backgrounds: a possible physiological explanation based upon a recent computational model. Vis. Neurosci.15(5), 957-67.
14. Boynton, R., Sturr, J. and Ikeda, M. (1961). Study of flicker by increment threshold technique. JOSA A 51, 196-201.

Polarization-Based Orientation in a Natural Environment

Volker Müller

bbcom GmbH&Co KG, Shanghai Office
Caobao Lu 509, Caohejing Tower 612, 200233 Shanghai, P.R. China
mifu@bbcom.sh.cn

Abstract. Many insects, birds and aquatic animals are sensitive to the polarization state of light and are able to use this information for orientation in their habitat. In this paper we imitate this ability and evaluate the possibility of a computer vision system using a standard CCD-camera to determine directions in outdoor scenes based on the polarization pattern of the sky. Even if the sky is covered with thick clouds, the incident light has a sufficient degree of polarization. A physics-based model is proposed to describe the state of polarization of the skylight, including the depolarizing effect of clouds and dust in the atmosphere. Based on our polarization model we propose features for orientation that are invariant to weather and other environmental conditions. Experiments show that through polarization analysis we can get the same precision in determing a direction as by using a compass.

1 Introduction

So many different animals are sensitive to the state of polarization of light that in natural organs of sight, polarization vision seems to be more the rule than the exception. Use of polarization state of light for orientation in their specific habitat has been reported from insects like honeybees [1], ants [2], crickets [3], dragonflies [4] but also from shrimps and fish [5], sea turtles [6] and migratory birds [7]. In contradiction to most reports, even human are slightly sensitive to polarization, a phenomenon called „Haidinger's brush" [8]. However there is no hint that human use this ability for any purpose, it may just be a relict inherited from early predecessors of man.

Considering the variety of animals using polarization vision we can distinguish two cases: macro orientation, i.e. the ability to navigate over long distances using the state of polarization of the skylight, and micro orientation, i.e. the ability to distinguish certain objects like water pools, enemies or prey by a distinct pattern of polarization caused by reflection of light. In this paper we will focus on the question of macro orientation, how a computer vision system can determine a direction in a natural environment.

A research group at Zurich university made the most comprehensive efforts to study insect vision by designing an „artificial ant", an autonomous agent navigating with polarized light [9], [10]. In their approach three sets of special one-dimensional polarization sensors were used.

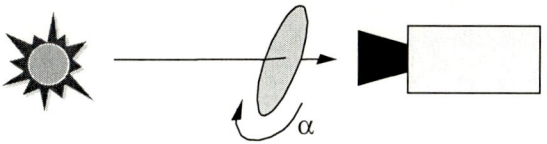

Fig. 1. Analysis of state of polarization of incident light with a CCD camera

Analysis of the state of polarization has been introduced to the computer vision community through the pioneering work of Wolff [11], [12]. As shown in [13], three images taken from an scene with different angles α of a polarization filter in front of a standard CCD-camera provide sufficient information to obtain the degree and orientation of polarization at each single pixel of a two-dimensional image (Fig. 1). Image intensity of each image i is composed of constant term I_c and a variable term I_v depending on the angle α of the polarization filter and the orientation of polarisation β (see fig. 2):

$$I_i = I_c + I_v \cos 2(\alpha_i - \beta) \ . \tag{1}$$

Polarization analysis has been employed in controlled environments to extract physics-based invariants [14], [15]. In outdoor scenes the characteristic polarization pattern of light reflected from man-made objects is used for object recognition [16], a work mainly aiming at military applications.

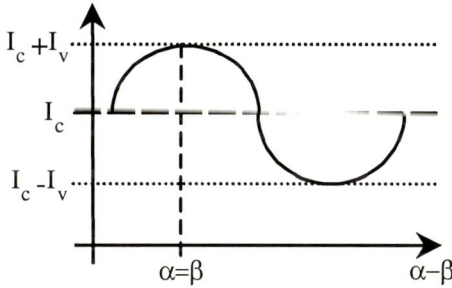

Fig. 2. Image intensity as a function of polarization filter position

The purpose of this paper is to evaluate the possibility of navigation in outdoor scenes relying on the polarization pattern of the sky. The optical sensor is a standard CCD-camera linked with a frame grabber in a PC.

2 A Model for the Polarization State of Skylight

Light is a transverse wave, its oscillations (i.e. the magnitude of its E-vector) occur perpendicular to the direction of propagation of light. While each elementary wave is polarized (i.e. its E-vector has a definite direction), the light of the sun and that of most artifical light sources is a chaotic superposition of a huge number of elementary waves, hence the total influx of light is unpolarized.

Polarization of skylight is caused by scattering. When light from the sun hits a molecule of the atmosphere, it acts like a Hertz dipole [17], each dipol emits light waves with an intensity distribution as shown in Fig. 3. The scattered light is polarized, its orientation is equal to the oscillation of the dipole. The intensity of the scattered light S is proportional to

$$S \sim \frac{1+\cos 2\varphi}{r^2} \omega^4 p_0^2 \,, \tag{2}$$

ω being the angular frequency of the dipole and p_o the dipole moment. Hence we get a maximum polarization for an angle φ of 90^0 between the sun, the scattering molecules in the atmosphere and the observer.

The effective polarization of incident light ρ^i as experienced by an observer on the ground can be approximated by [18]

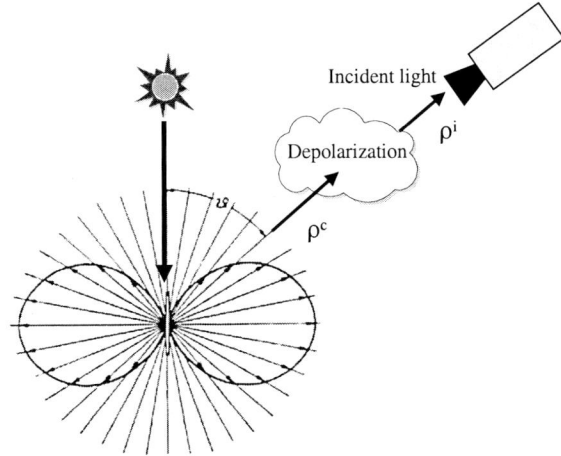

Fig. 3. Scattering and polarization at molecules in the atmosphere

$$\rho^i = \rho^i_{max} \frac{\sin^2 \vartheta}{1+\cos^2 \vartheta} \,. \tag{3}$$

If we define a base-line with the x-axis given by observer C and the projection of a ray from the sun S to C onto the surface of the earth (see fig. 4), then ϑ of Eq. 3 is determined by the polar angle θ_s between C and S, the polar angle θ_i between C and the reflecting molecule M and the azimuth angle ϕ between the x-axis and the direction of observation:

$$\cos \vartheta = \cos \theta_s \cos \theta_i \cos \phi - \sin \theta_s \sin \theta_i \,. \tag{4}$$

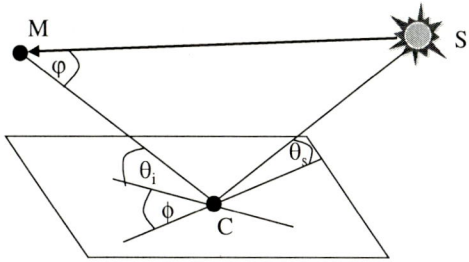

Fig. 4. Geometry of observer C, sun S and scattering molecule M

Fig. 5 (from [9]) shows the relative intensity and direction of polarization for a given position of the sun. The degree of polarisation ρ^c of the celestial hemisphere is defined as

$$\rho^c = \frac{I_{max} - I_{min}}{I_{max} + I_{min}} = \rho^c(\theta_s, \phi, \theta_i) \ . \tag{5}$$

The main characteristics of the polarization distribution are:
- the degree of polarization ρ^c is zero in the direction of the sun and $\vartheta = 180°$ from the sun.
- the polarization has its maximum at an angle of $\vartheta = 90°$ from the sun.
- the pattern of polarization has a mirror symmetry with respect to the plane defined by the solar meridian
- at the solar meridian the orientation of polarization is perpendicular to the meridan.

If we had an ideal dipole, the degree of polarization ρ^c should be 100% at an angle of $\vartheta = 90°$ from the sun (see fig. 3). In [19] a maximum ρ^c of 80% is reported, in a city (Shanghai) with humid climate we measured not more than 30% (see section 3.1), even if the sky is blue. The reason is a depolarization of the light on the way from the dipole to the observer. The polarized light from the celestial hemisphere is scattered by little particles of clouds and dust.

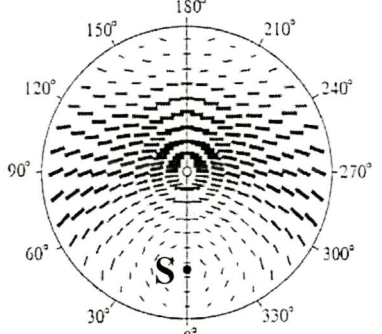

width of black bars proportional to degree of polarization

direction of black bars represents orientation of polarization

Fig. 5. Graphic Representation of the polarization pattern of the celestial hemisphere (from [9]).

Whether scattering causes polarization or depolarization depends on the size of the particles. Molecules in the size of less than 10^{-7}m act as a dipole and cause polarization (Raleight-scattering). Particles of a size of about 10^{-6}m (Mie-scattering) and 10^{-5}m or larger (Fraunhofer scattering) act like rough surfaces and cause depolarization [20].

Depolarization depends on weather conditions (clouds, huminity) and dust. In our model we assume:
- depolarization has a linear characteristics, i.e. it can be expressed by an orientation-dependend factor
- the orientation of the polarization remains unchanged, i.e. there is no optical active medium (like a crystal) between the dipole and the observer.

With this assumptions, the degree of polarization as experienced by the observer becomes

$$\rho^i = \frac{I_{max} - I_{min}}{I_{max} + I_{min}} = \rho^c (\theta_s, \phi, \theta_i) \ d(\phi, \theta_i) \ . \tag{6}$$

Hence we have a reasonably simple model to describe the characteristics of the partially polarized light that hits the surface of the earth, based on the properties of the atmosphere. Much more elaborate models of the process of scattering and polarization in the atmoshere have been proposed [21], however these models fit the requirements of high precision physical measurements, they can not be verified with the limited precision of a standard CCD-camera.

3 Polarization-Based Orientation by a Computer Vision System

In this chapter we will describe a method how a computer vision system can determine a direction based on the model introduced in chapter 2, including the preparatory work to create a data-base, determination of invariants and evaluation of the precision of the estimation of a direction.

3.1 Setup of a Data-Base

Instead of an algebraic solution of eq. (2) to (4), we recorded a comprehensive series of measurements with a polarization filter in front of a lux-meter, i.e. changing the CCD-camera in fig. 1 for a lux-meter, using the larger dynamics and greater precision of the lux-meter.

Fig. 6 shows the degree of polarization ρ^i in relation to the azimuth angle (circular horizontal distance) ϕ from the sun. In this experiment the polar angle θ_i is kept constant. Fig. 6a and Fig. 6b show the results of two consecutive days with different weather conditions. The most surprising result is that the incident light in Fig. 6b is partly polarized despite thick clouds and drizzle. Research on honeybees indicated that they need at least a small patch of blue sky to utilize polarization [22]. The degree of polarization ρ^i is much smaller on cloudy days and the curve of Fig. 6b has a

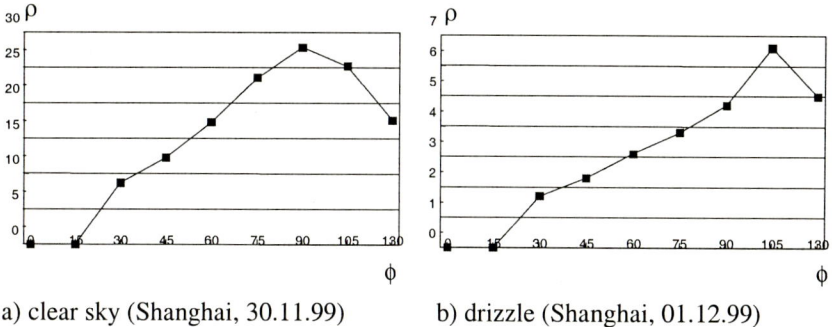

a) clear sky (Shanghai, 30.11.99)   b) drizzle (Shanghai, 01.12.99)

Fig. 6. Degree of polarization at different wheather conditions

slightly different shape than that of Fig. 6a because of the inhomogenity of the clouds. Measurements on other days showed that even in foggy wheater (visibility less than 200m) the incident light has a degree of polarization ρ^i of around 5%.

Fig. 7 shows the orientation of polarization β. As expected, the orientation is independent of the degree of polarization (and the weather), hence the two lines representing different weather conditions are nearly identical. The maxima and minima of polarization are not very distinct, hence there is a certain error in determining the orientation of polarization.

Polarization analysis of insects is sometimes referred to as „polarized light compass" [10], however a more adequate name would be „polarized light sundial", since the polarization pattern of the sky depends very much on the daytime, the time of the year and the geographic position of the observer. Our approach to create a databank of reference polarization values has to take into account that each set of data is only valid for a definite time and location.

3.2 Extraction of Invariants

As shown in Fig. 6, the degree of polarization ρ is a very unreliable feature, since its value depends on weather and other environmental conditions. From eq. 3 we see that the incident light is depolarized by a function $d(\phi,\theta_i)$, d can be assumed as being constant within a small patch of the sky. Hence we can calculate feature f_1 based on two measurements with a slightly different azimuth angle of observation ϕ_i:

$$\rho_1^i(\theta_s,\phi,\theta_i) = d(\phi,\theta_i)\, \rho_1^c(\theta_s,\phi,\theta_i) \tag{7}$$

$$\rho_2^i(\theta_s,\phi+\Delta\phi,\theta_i) = d(\phi,\theta_i)\, \rho_2^c(\theta_s,\phi+\Delta\phi,\theta_i)$$

$$f_1 = \frac{\rho_1^i}{\rho_2^i} = \frac{\rho_1^c}{\rho_2^c}$$

Feature f_1 is independent of the depolarization function d, the stability of this feature under different weather conditions in shown in Fig. 8.

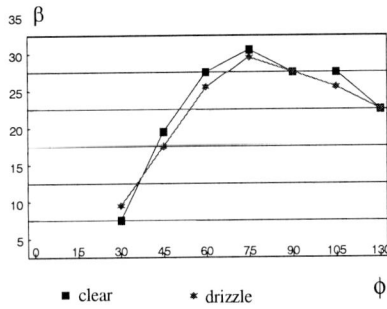 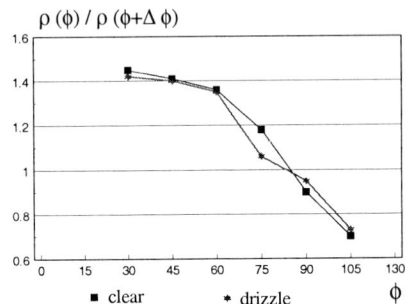

Fig. 7. Orientation of polarization measured under different weather conditions

Fig. 8. Values of feature f_1 at different weather conditions

As shown in Fig. 7, the orientation of polarization is independent of influences from the atmosphere, hence we simply choose the orientation of polarization as a second feature

$$f_2 = \beta\,(\theta_s,\phi,\theta_i) \ . \tag{8}$$

Both features are independent of each other.

The calculation of these features is based on intensity differences between greylevel images, obtained with a polarization filter mechanically rotated in front of the lens. In order to increase the precision of the results, we use the fact that light intensity of a small patch of the sky is homogeneous. Hence we can calculate the average values of any feature f_n in a tiny rectangle of size a * b:

$$\bar{f}_n = \frac{1}{ab} \sum_{x=0}^{a} \sum_{y=0}^{b} f_n \ . \tag{9}$$

Chapter 3.3 shows that in this way the precision can be considerably increased, especially in case of a clowded sky.

3.3 Experimental Results

Analysis of the state of polarization with standard CCD-cameras raises high demands to the experimental setup, mainly for two reasons:
- we have to input three different images with a polarization filter mechanically rotated in front of the lens. This poses the problem of pixel registry. Even small vibrations of the camera may result in a shift between the images, leading to distorted results
- if the degree of polarization is low, the difference of intensity between the three images is low, hence the method is sensitive to noise, especially if illumination is poor, i.e. the signal / noise ratio is small.

Our experiments have shown that analysis of the polarization pattern of the sky is much less sensitive to these distortions than applications in industrial inspection. Small patches of the sky have a homogeneous light intensity, hence an inaccurate pixel registry is less harmful. In addition, the light intensity of the sky is much higher than that of reflected light analysed in industrial inspection, hence the signal / noise ratio noise is larger.

In our experiments we did not simulate a complete 3D-orientation but choose a more practical task: we assumed that the vision system is standing on a plane surface. In vertical direction we fix the orientation of the camera, providing a determined field of view limited by the polar angles $\theta_{i,u}$ and $\theta_{i,d}$. The camera can be rotated in horizontal direction and the vision system shall determine the direction of its line of sight, the azimuth angle ϕ. Fig. 9 shows an original greylevel image of an outdoor scene in Shanghai plus the corresponding degree and orientation of polarization. These images are used to extract features. The results were verified with a compass. Table 1 shows the precision of the estimation at a representative distance from the sun, ϕ_s about 60^0. The results proove that using polarization based orientation we can achieve the same precision as with a compass.

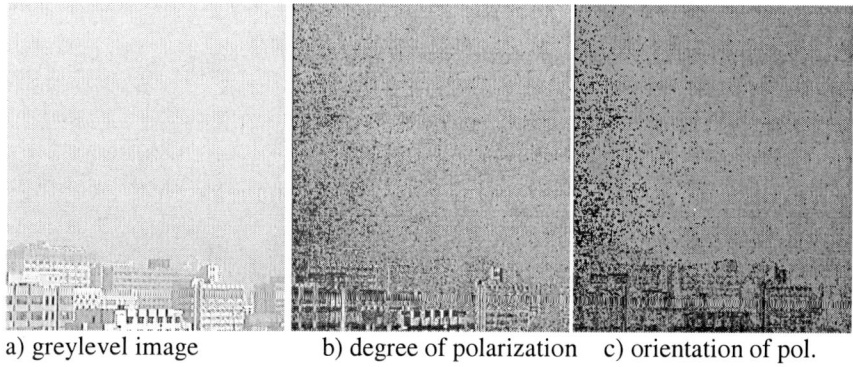

a) greylevel image   b) degree of polarization   c) orientation of pol.

Fig. 9. Exampel of polarization in an urban scene (scaled for printing)

4 Conclusions and Further Work

Our experiments have shown that a computer vision system equiped with a standard CCD-camera and a polarization filter in front of the lens can navigate in outdoor scenes merely based on the state of polarization of skylight. Most important: we were able to determine invariant features that make the system robust against changes of weather and other environmental changes.

Compared with one-dimensional polarimeters, CCD-cameras have certain advantages, they can observe a greater sector of the sky in realtime, hence this method requires no active movement of the vision system (as necessary for the autonomous agent descibed in [9]) to scan the sky and it shows little sensitivity to changes in the atmosphere. In addition, CCD-cameras can determine the direction of changes of the polarization pattern in the sky, hence the mirror-symmetry of the polarization pattern in the celestrial hemisphere does not cause ambigues results.

Table 1. Precision of orientation estimation

Feature	Deviation [°]
Feature 1 (one pixel)	4
Feature 2 (one pixel)	6
Feature 1 (3 x 3 pixel)	< 3
Feature 2 (3 x 3 pixel)	3
Featue 1 & Feature 2 (one pixel)	< 3
Featue 1 & Feature 2 (3 x 3 pixel)	< 2

However, while biology provides clues how to solve vision poblems, artifical systems are likely to follow different approaches. Compass and GPS (global positioning system) being readily available, it is unlikely that a technical system will rely on polarization information to find its way, except (may be) as an auxiliary tool to check the plausiblity of other sources of information.

Hence our further work will concentrate on analysing the depolarization function $d(\phi, \theta_i)$:

- can we match the state of polarization and the colour of the sky in order to obtain colour constancy?
- can we estimate environmental quantities such as humidity and number of particals in the air based on polarization information? Such an analysis may provide online two-dimensional information about pollution of the atmosphere and hence help to detect sources of pollution.

References

1. M. Lehrer: Looking all around: Honeybees use different cues in different eye regions. Journal on Experimental Biology, Vol. 201, pp. 3275-3292.
2. R. Möller, D. Lambrinos, R. Pfeifer, T. Labhart, R. Wehner: Modeling Ant Navigation with an Autonomous Agent. In: From Animals to Animats (SAB98), Proc. 5th International Conference of the Society for Adaptive Behavior, pp. 185-194, MIT Press 1998
3. T. Labhard, B. Hodel, I. Valenzuela: The physiology of the cricket's compound eye with particular reference to the anatomically specialized dorsal rim area. J. Comp. Physiol A, Vol. 155, p. 289-296, 1984
4. G. Horvath, B. Bernath, G. Molnar: Dragonflies find crude oil more attractive than water: multiple choice experiments on dragonfly polarotaxis. In: Naturwissenschaften, Vol. 85, pp. 292-297, 1998
5. G. Horvath, D. Varju: Underwater-Polarization Patterns of Skylight Perceived by Aquatic Animals through Snell's Window on the flat Water Surface. In: Vision Research 35, pp. 1651-1666, 1995
6. H. Fountain, Navigation Satellites have a rival: the sea turtle. In: New York Times, March 12, 1996, B6

7. G. Horvath, I. Pomozi: How celestial polarization changes due to reflection from the deflector panels used in deflector loft and mirror experiments studying avian orientation. Journal of Theoretical Biology, Vol. 184, pp.291-300, 1997
8. C.K. Tri, A.J. Wasik, R.P. Hemenger, R. Garriott: Do wave plates improve the visibility of Haidinger's brushes? In: The American Academy of Optometry, Annual Meeting, San Antonio 1997
9. D. Lambrinos, R. Möller, T. Labhart, R. Peifer, R. Wehner: A mobile robot employing insect strategies for navigation. To appear in: Robotics and Autonomous Systems, special issue on biometric robots, 1999.
10. D. Lambrinos, M. Maris, H. Kobayashi, T. Labhart, R. Peifer, R. Wehner: An autonomous agent navigating with a polarized light compass. Adaptive Behavior, Vol. 6, No. pp 131-161
11. L.B. Wolff: Polarization Methods in Computer Vision, PhD thesis, Columbia University 1990
12. L.B. Wolff, S.A. Shafer, G.E. Healey (Eds.): Physics-Based Vision, Principles and Practice. Jones and Bartlett Publishers, Boston 1992
13. S.K. Nayar, XS.Fang, T.Boult: Removal of Specularities Using Color and Polarization, IEEE Conference on Computer Vision and Pattern Recognition, pp. 583-590, New York, 1993
14. V. Müller: Elimination of Specular Surface-Reflectance Using Polarized and Unpolarized Light. In: B. Buxton, R. Cipolla (Eds.): Computer Vision - ECCV'96, Vol. II, pp. 625-635. Springer, Berlin 1996
15. V. Müller: Photometric Invariants Based on Polarization Analysis, In: M. Frydrych, J. Parkkinen, A. Visa (Ed.): SCIA'97, The 10th Scandinavian Conference on Image Analysis, Lappeenranta, Finland, pp. 855-860, Lappeenranta 1997
16. L.B. Wolff: Reflectance modeling for object recognition and detection in outdoor scenes, ARPA, pp. 799-804, 1996
17. C.Gerthsen, H.O. Kneser, H. Vogel: Physik. Springer. Berlin 1977
18. R. Schwind, G. Horvath: Reflection-Polarization Pattern at Water Surfaces and Correction of a Common Representation of the Polarization Pattern of the Sky. Naturwissenschaften 80 (2/1993), p. 80-81. Springer. 1993
19. J. L. Bahr: Polarization. Internet WWW page, at URL: <www.physics.otago.ac.nz/PHSI110/bahr/wo4/tsld014.htm>, last update: 24.09.98
20. C. Schnörr: Untersuchung zur Farb- und Polarisationsinformation des Streulichtes bei der Partikelkontrolle in Ampullen. In: Procedings of Mustererkennung 1994, 16th symposium of the German society for pattern recognition and 18th workshop of the Austrian society for pattern recognition, pp.650-657, Wien 1994
21. C. F. Bohren (Ed.): Selected papers on scattering in the atmosphere, SPIE Optical Engineering Press, 1989
22. K. Able, J. Swinebroad, J. Waldvogel: Bird Navigation. Internet WWW page, at URL: <http://earthsky.worldofscience.com/1996/esmi961025.html>, last update: 25.10.96

Computation Model of Eye Movement in Reading Using Foveated Vision

Yukio Ishihara and Satoru Morita

Faculty of Engineering, Yamaguchi University,
Tokiwadai Ube 755-8611, Japan

Abstract. Viewpoints tend to be fixed in the middle point of a word in reading, because letter information on the periphery is used for eye movement. Letter information regards letter-shape, and the letter number of a word. In this manuscript, we realize eye movement using the letter number information of a word in reading. In addition, we compare eye movement while apprehending the letter number information of word to eye movement without being given the letter number information of word, and we confirm that viewpoint tends toward the middle point of a word in the former.

1 Introduction

A human captures dense information using the fovea, and captures information thinly at the edge of the fovea[1]. Therefore, we distinguish objects in detail, and try to capture the details of moving objects by moving them onto the fovea via eye movements.

A human can't capture information using the periphery of the fovea densely. But when it senses a moving object, the eyeball immediately moves to the object. These eyeball movements are distinguished into two kinds. One is successive eyeball movement which is performed when the eye tracks an object, and the other is jumpy eyeball movement which is performed when eye captures object which is on the periphery of the fovea. The former is called smooth pursuit movement, and the latter is called saccadic movement. The object which we try to look at moves onto the fovea by using these eyeball movement. The performance that looking at a single position of object is called fixation.

The information of object is captured on the fovea densely and is captured on the periphery of the fovea thinly for a fixation. Therefore, to capture the information of object densely, we require moving eye and looking at the object more than one times. We suppose that this eye movement is different from the eye movements in the other tasks. When we look at the object which doesn't move, only saccadic eye movement is performed. In this paper, task is reading and we realize saccadic eye movement on the computer.

G. Sandini has developed CCD devices which have a high pixel density distribution in the retina center and a lower density away from the center [2]. Like wise, T. Baron has developed a vision system with a high pixel density in the center and a lower away from the center [3][4]. Vision systems with nonuniform sampling have appeared, as well as active vision systems [5] and mechanical systems to support sensory movement [6][7].

In this paper, we use a computer model to explore eye movement in reading, and incorpolate the psychological aspect of eye movement during reading into our computer model.

It is reported that to move viewpoint in reading, letter information on the periphery of the fovea is used. That is letter-shape information obtained for about 10 or 11 positions to the right of the fixation point and the letter number of a word obtained for about 15 positions to the right of the fixation point is used [8]. Relation between viewpoint in reading to voice point is also investigated [9], and it is reported that viewpoints in reading is ahead of voice point and view-voice point span is long at the beginning of sentence and is short at the end. Eye movement in reading and in searching has also been investigated under conditions wherein letter information on the periphery has been both hidden and revealed. The latter isn't affected with variation of letter information on the periphery than the former [10]. In this paper, we realize eye movement using the letter number information of a word.

We discuss fovea and eye movement at next section. At third section, we explain psychological aspect of eye movement and we realize saccadic eye movement using psychological aspect of eye movement. In the fourth section, we compare the above condition with the condition wherein the psychological factor has been incorpolated.

2 Foveated Vision

2.1 Fovea

The log-polar mapping model is widely used as a model of space-variant image sampling in machine vision [11]. Wilson proposes a model for such a space-variant receptive field arrangement and explains how the human contrast sensitivity function arises from it [1].

The center of a receptive field is located on a ring. The eccentricity of the nth ring R_n is the following:

$$R_n = R_0 \left(1 + \frac{2(1 - Ov)Cm}{2 - (1 - Ov)Cm}\right)^n,$$

where R_0 is the radius of the fovea (in degrees). Cm is the ratio of the diameter (in degrees) of the receptive field to the eccentricity (in degrees) of that receptive field from the center of the fovea. Ov is the overlap factor. If the receptive fields touch each other, $Ov=0$. If they extend to the center of the next field, $Ov=0.5$. Also, the radius of the receptive field on the nth ring is $\frac{CmR_n}{2}$, and the number of receptive fields is $\frac{2\pi}{Cm(1-Ov)}$ per ring.

We call the image represented in degree and radius elements with horizon and vertical axis $R\theta$ image. The $R\theta$ image is transformed into a fovea image. The equation which transforms the $R\theta$ image into the fovea image is the following:

$$x = R\cos(\theta) + C_x$$
$$y = R\sin(\theta) + C_y,$$

where (C_x, C_y) is the center of the fovea image.

In this paper, we generate the fovea image using this model.

2.2 Saccade

A human can't capture information using the periphery of the fovea densely. But when it senses object-moving, eyeball immediately moves to move eye to the object. These eyeball movement is classified into two kinds of eyeball movement. One is successive eyeball movement which is performed when eye tracks object, and the other is jumpy eyeball movement which is performed when eye captures object which is on the periphery of the fovea. The former is called smooth pursuit movement, and the latter is called saccadic movement.

The shift to the next viewpoint in the saccade is performed by extracting significant information from the periphery, which is composed of extracted characters. This character extraction is performed by comparing the prepared template of the edge to the $R\theta$ image. The position from which the character is extracted is used to determine the next viewpoint.

3 Eye Movement in Reading

3.1 Reading

Reading involves moving the viewpoint along a sentence, and a human captures visual information at the viewpoint. McConkie, Kerr, Reddiz and Zola reported that viewpoint in reading tends to be in the middle point of a word [8]. Figure 1 shows the proportion of fixations that occur by letter position within words of the different letter number. Fixations tend to be more frequent near the middle. In other words, the distance of eye movement is long on a saccade when a word of the 6-8 letters is to the right of the previous fixation point, and the distance of eye movement is short on a saccard when a word of the 3-5 letters is to the right of the previous fixation point. That is, the information on the periphery is used to move the viewpoint. What information on the periphery is used? How far to the right of the fixation point that information is used? McConkie and Rayner have examined these problems, and have investigated the ease of reading and the length of the saccade under conditions wherein letter information on the periphery has been both revealed and withheld. From this investigations, they found that information is obtained in about 15 positions to the right of the fixation point.

In this paper, we achieve eye movement incorpolating the psychological factor wherein eye movement in reading related to word-length information to the right of the fixation point.

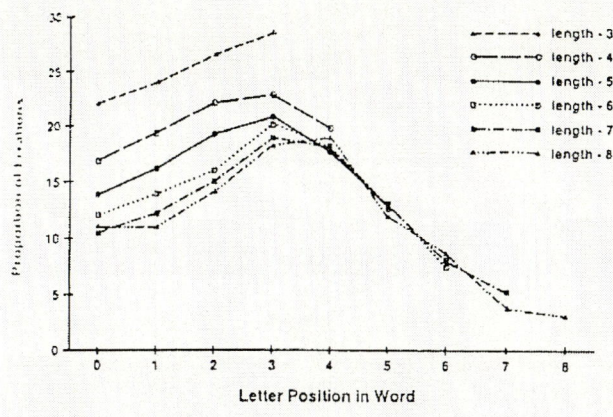

Fig. 1. The proportion of fixations at different letter positions within words of different lengths.

3.2 Eye Movement in Reading

Eye movement in reading varies with understanding and age. Even if a human reads a book several times, eye movement always varies. But our eye movement in reading is similar to other kinds of eye movement. A viewpoint moves from the beginning of a line to the end of a line, and the viewpoint reaches the end of a line and moves from the end of a line to the beginning of a next line. Again, the viewpoint moves along a line. Our eye movement is repeat of these processes. In this paper, following processes are performed to demonstrate this eye movement using computer.

#1. Moving viewpont along a line
#2. Moving viewpoint along a line
#3. Finding the end of a line

Eye movement is performed using these three processes.

Figure 2 shows the flow of these three processes. #1 and #2 are basic loop processes. #3 is executed when the end of a line is found even times in #2.

At first, the viewpoint is at the beginning of a line, and viewpoint moves to the end of a line by repeating #1 and #2. When viewpoint is close to the end of a line, space is found ahead by #2. Then, the viewpoint comes back to the beginning of a line by repeating #2 and #2. When viewpoint is close to the beginning of a line, space is found ahead by #2 . Then, a next line is found by #3 and view point moves to a next line.

Before explaining each process, we define the **attention region**.

When human reads a book, viewpoint moves on a line and doesn't move a previous line and a next line on the way. We think that eye movement is restricted within narrow limits. In this paper, we restrict eye movement using the attention region. The attention region is used to determine a next viewpoint. In the other words, a next viewpoint is determined in the attention region.

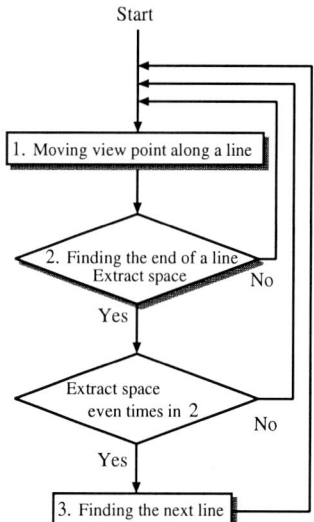

Fig. 2. The flow of three processes.

1. **Moving the viewpoint along a line** This process is performed by moving the attention region.

   At first, we consider moving the attention region in the direction that has maximum edge density around the current viewpoint. To find this direction, we use an edge image (Figure 3(b)) extracted from a fovea image (Figure 3(a)) at the current viewpoint, and rotate the attention region to each direction around the current viewpoint (Figure 4), and examine edge density in the attention region. We then move the attention region in the direction of high edge density.

   But viewpoint does not move evenly along a line and it vibrates on a line, since the attention region lies on a line that contains high density in the direction of both forward eye movement and backward eye movement. To improve this, we put emphasis on the direction to which the attention region moves, and consider both edge density and the direction and thus move the attention region. By this, the viewpoint doesn't vibrate on a line and moves to the direction with high edge density and which is much the same direction as the direction in which the viewpoint has moved. Also because moving the attention region by examining edge density around the current viewpoint, viewpoint can move flexibly even if a line is distorted.

   Then, because eye movement in reading relates to the word-length information at the right of eye movement, consider varying the attention region according to word-length on the direction of eye movement. In this paper, we perform it by finding space between words and covering a word with the attention region(Figure 5).

   We perform eye movement by repeating our movement of attention region and determining the viewpoint of the attention region. Determining the view-

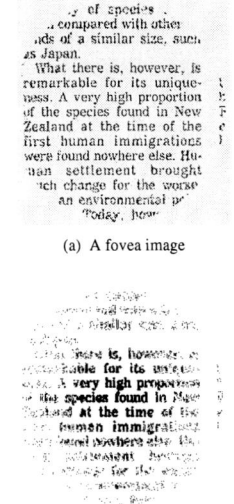

(a) A fovea image

(b) A edge image extracted from a fovea image

Fig. 3. Fovea image and edge.

point of the attention region is performed by giving the low resolution point priority.

2. **Finding the end of a line** This process is performed by finding space ahead.

   With the edge image(Figure 3(b)) extracted from a fovea image(Figure 3(a)) at the current viewpoint, we examine edge density in the direction of eye movement(Figure 6) and determine that the point with low edge density is the end of a line.

3. **Finding the next line** This process is performed by finding space on both sides of the direction of eye movement.

   With the edge image (Figure 3(b)) extracted from a fovea image (Figure 3(a)) at the current viewpoint, examine edge density on both sides of the direction of eye movement(Figure 7), determine that the point with low edge density is the space between line.

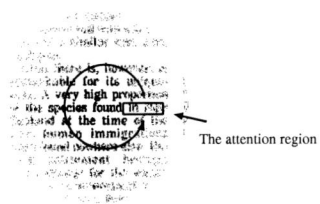

Fig. 4. The calculation of edge density.

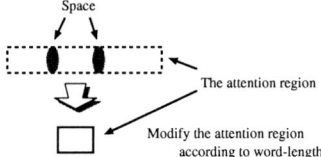

Fig. 5. The modification of the attention region.

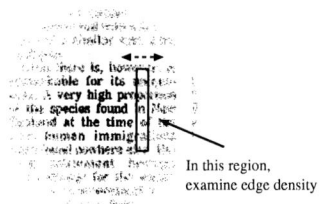

Fig. 6. Space extraction at the end of a line.

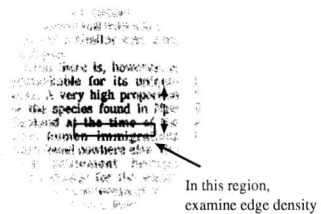

Fig. 7. Space extraction between lines.

4 Experiment

In this paper, we compare eye movement with varying attention regions to eye movement without varying attention regions. The varying of the attention region then proceeds according to word-length in the direction of eye movement in the former. We confirm that the former is more similar to the eye movement which we perform in reading than the latter. To compare both types of eye movement, we examine the viewpoint on a line, making reference to the point that point that viewpoint in reading tends to be in the middle of a word [8].

The fovea image is generated using Wilson's model with parameters: R_0(the radius of the fovea) is 7, Cm(the ratio of the diameter of the receptive field to the eccentricity of that receptive field from the center of the fovea) is 0.3, and Ov(the overlap factor) is 0.96. The field of view is 200 and the maximum length of the attention region within which space can be found in is 54.

We examined eye movement under the following conditions:

- Without varying the attention region.
- While varying the attention region according to the letter numbers of a word in the direction of eye movement.

We experimented using sentences of 1000 words in English. The sentences involve the words of the various letter numbers. Figure 8 is a sample image which we used on experiments. Figure 9 shows a trace of viewpoints. Figures 10 and 11 show the proportion of fixations at different letter positions within words of different lengths under both conditions.

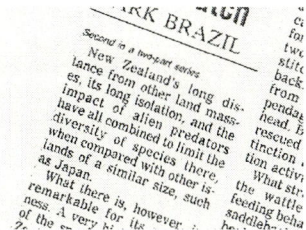

Fig. 8. A sentence image.

Fig. 9. The viewpoint trace.

Figure 9 shows that viewpoint moves to the end of a line and back to the beginning of the line, then moves to the next line. comparing Figure 10 to Figure 11, it is found that the proportion of fixations at a words first letter for eye movement while varying the attention region is smaller than the proportion of eye movement without varying it.

This is because eye movement, which falls on letters at both the beginning and ending of a word, is restricted by variances in the attention region depending on the letter number of a word in the direction of the eye movement. Therefore, in our model, eye movement under the condition of varying attention regions tends to be in the middle point of a word in a manner similar to that of human eye movement.

By comparing Figure 1 to Figure 11, it is found that the proportion of fixations for 6-8 letter words is the biggest at third and fourth letter position in Figure 1 and is the greatest at the second and third letter position in Figure 11. That is, in our experiment, viewpoint tends to be a point in the middle point of a word in the case of long words(6-8 lengths words). Also, in Figure 1 the proportion of fixations at the 6-8 letter position within long words is large.

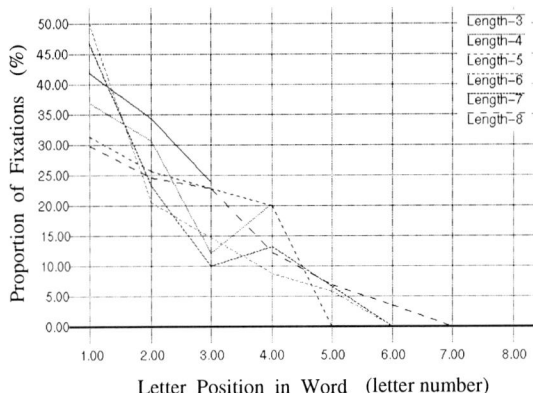

Fig. 10. The proportion of fixations at different letter positions within words of different lengths for eye movement without varying the attention region.

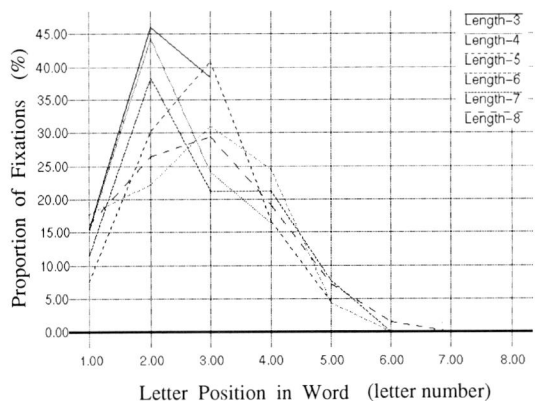

Fig. 11. The proportion of fixations at different letter positions within words of different lengths for eye movement with varying attention regions.

We next consider the relation between the attention region and tendency that viewpoint tends to be the point which is in the middle in the case of long words. If the maximum length of the attention region is short, viewpoint doesn't move to the back of a word and the proportion of fixations on the middle point of a word is large. On the contrary, if the maximum length of the attention region is long, it can cover a long word with the attention region and viewpoint will tend to be on the middle point of a word. Therefore, on our experiment, because the maximum length of the attention region is short and cannot cover long words with the attention region, the viewpoint tended to be the middle point of a word. But this problem can be solved by extending the attention region.

5 Conclusion

Eye movement in reading relates to the word-length in the direction of eye movement. In this paper, we realized eye movements using word-length information in reading. In addition, we compared the eye movements with using word-length information to the eye movement without using word-length information, and we confirmed that the former was similar to human eye movement.

References

1. S. W. Wilson: "On the retina-cortial mapping", Int. J. Man-Machine Stud. 18, pp.361-389, 1983.
2. G. Sandini, P. Dario and F. Fantini, "A Retina like space variant CCD sensor," SPIE 1242, pp. 133-140, 1990.
3. T. Baron, M. D. Levine and Y. Yeshurun, "Exploring with a Foveated robot eye system," 12th ICPR, pp. 377-380, 1994.
4. H. Yamamoto, Y. Yeshurun and M. D. Levine, "An Active Foveated Vision System: Attentional Mechanisms and Scan Path Convergence Measures," Computer Vision and Image Understanding, pp. 50-65, 1996.
5. D. H. Ballard, "Animated Vision," Univ. of Rochester, Dept. of Computer Science, Tech. Rept. 61, TR, 1990.
6. K. Pahlavan, T. Uhlin and J. Eklundh, "Integrating primary ocular processes," proc. ECCV, pp. 526-541, 1992.
7. J. L. Crowery, P. Bobet and M. Mesrabi, "Gaze control for a binocular camera head," proc. ECCV, pp. 588-596, 1992.
8. K. Rayer, "Eye Movements in Reading", pp.7-20, Academic Press, 1983.
9. R. A. Monty and J. W. Senders, "Eye Movements and Psychological Process, "Lawrance elbaum Associates pub., pp.371-395, 1976.
10. R. A. Monty and J. W. Senders, "Eye Movements and Psychological Process, "Lawrance elbaum Associates pub., pp.417-427, 1976.
11. J. J. Koenderink and A. J. VanDoorn, "Visual detection of spatial contrast: Influence of location in the visual field, target extent, and illumination level," Biol. Cybernetics, 30, pp. 157-167, 1978.

New Eyes for Shape and Motion Estimation

Patrick Baker[1], Robert Pless[1], Cornelia Fermüller[1], and Yiannis Aloimonos[1]

Center for Automation Research
University of Maryland
College Park, Maryland 20742, USA
{pbaker,pless,fer,yiannis}@cfar.umd.edu

Abstract. Motivated by the full field of view of insect eyes and their fast and accurate estimation of egomotion, we constructed a system of cameras to take advantage of the full field of view (FOV) constraints that insects use. In this paper, we develop a new ego-motion algorithm for a rigidly mounted set of cameras undergoing arbitrary rigid motion. This egomotion algorithm combines the unambiguous components of the motion computed by each separate camera. We prove that the cyclotorsion is resistant to errors and show this empirically. We show how to calibrate the system with two novel algorithms, one using secondary cameras and one using self calibration. Given this system calibration, the new 3D motion algorithm first computes the rotation and then the 3D translation. We apply this algorithm to a camera system constructed with four rigidly mounted synchronized cameras pointing in various directions and present motion estimation results at www.cfar.umd.edu/~pbaker/argus.html.

1 Introduction

Argus, the guardian of Hera, the goddess of Olympus had one hundred eyes, spread all over his body. He defeated, according to myth, a whole army of Cyclops, one-eyed giants. For Argus, physical space and visual space were the same thing – he could get Euclidean models of the world. Non-mythological creatures also have various methods of sampling the visual world, optimized for their particular environment. Flying insects, for instance, can perform complicated analysis of visual motion and execute navigational tasks more accurately and more quickly than state of the art visual algorithms allow. Lessons from insects can inspire new approaches to creating a visual system to give fast, accurate 3D motion estimates to an agent navigating through the world. This has important consequences: improving the accuracy of motion estimation permits a visual system to create better structural models of the environment and improves "dead reckoning" types of navigation.

However, not all aspects of an insect visual system have a useful computational or machine vision analogy — the biological and artificial capabilities differ with respect to both the representations of the visual images, and the computation or processing of the visual images. Furthermore, a biological visual system may be optimized to solve a diverse set of problems not all of which are required for machine systems. For instance, insects have a nearly full field of view, but rather low resolution, so they are much better at motion estimation than shape recognition. With this in mind, we concentrate on studying one fundamental difference between insect eyes and typical cameras. Specifically, what advantages are gained by using eyes that sample the light intensity from nearly all possible directions?

In the computer vision community, the problem of finding the 3D motion of the camera relative to a scene, the "ego-motion problem", has received a considerable amount of attention (for instance 8, 17, 10). It was discovered that there are inherent ambiguities which render the computation impossible, in particular, for cameras with a small field of view, it is difficult to differentiate between image motion due to translation and rotation 1, 11. Mathematically, these ambiguities manifest themselves as valleys in the function one optimizes to find the motion parameters (the epipolar error, for instance). The position of the optimal solution within this valley is easily obscured by image noise. Larger field of view cameras were subsequently recognized as helping to solve these ambiguities, and algorithms were created to use image motion from the full field of view to constrain the 3D motion 13.

The construction of larger field of view cameras has recently become more common, with the advent of catadioptric cameras, for instance 12 or 7. These cameras have a larger field of view, but their calibration is more difficult than with standard cameras. Often, catadioptric cameras are designed as stereo pairs, they have also been used to simulate a spherical camera 4. The hyperbolic or parabolic mirrors which these cameras use typically introduce sampling distortions which make image motion computations difficult. Other algorithms which use egomotion from many cameras include 14, which uses cameras pointing the *same direction*. Because of this, the advantages of a full field of view are not obtained, although there are advantages in reconstructing a scene using both stereo and motion cues. The fact that we have high resolution cameras oriented in various direction allows the possibility of computing accurate reconstructions in the directions in which we have data. Perhaps the most closely related work to ours is in 16, where they note many of the same advantages of having cameras with non-overlapping fields of view and implement a two camera system for computing egomotion. Ambiguities may remain in the motion estimation problem for two camera, for a particular set of motion directions.

We extend the previous work in the following ways. First, we present a formal proof that solving for the motion parameters using optic flow from a full spherical field of view does not suffer from the ambiguities that cause problems in the small field of view case. Second, we have built a device that has synchronized high-resolution cameras, oriented to cover widely distributed pieces of the viewing sphere. We give an algorithm to quickly solve for the rotation of the camera system, and argue that solving for the motion from multiple cameras pointing in widely distributed directions is just as good at eliminating ambiguities as using a real or virtual spherical imaging surface. Using this rotation, the translation can then be solved for as a linear system, accounting for that fact that the minimum of our error function for each camera lies in a valley. Given cameras which do not share centers of projection, this allows us to solve for the absolute translation. On www.cfar.umd.edu/~pbaker/argus.html we give motion estimation results from the application of our algorithm on an outdoor sequence taken from our constructed camera system.

2 Advantages of a Spherical Camera

The basic geometry of image motion is well understood. As a system moves in its environment, every point of the environment has a velocity vector relative to the system. The projections of these 3D velocity vectors on the retina of the system's eye constitute the motion field. For an eye moving with translation $\mathbf{t}$ and rotation $\boldsymbol{\omega}$ in a stationary environment, each scene point $\mathbf{R} = (X, Y, Z)$ measured with respect to a coordinate system $OXYZ$ fixed to the nodal point of the eye has velocity $\dot{\mathbf{R}} = -\mathbf{t} - \boldsymbol{\omega} \times \mathbf{R}$. Projecting $\dot{\mathbf{R}}$ onto a retina of a given shape gives the image motion field.

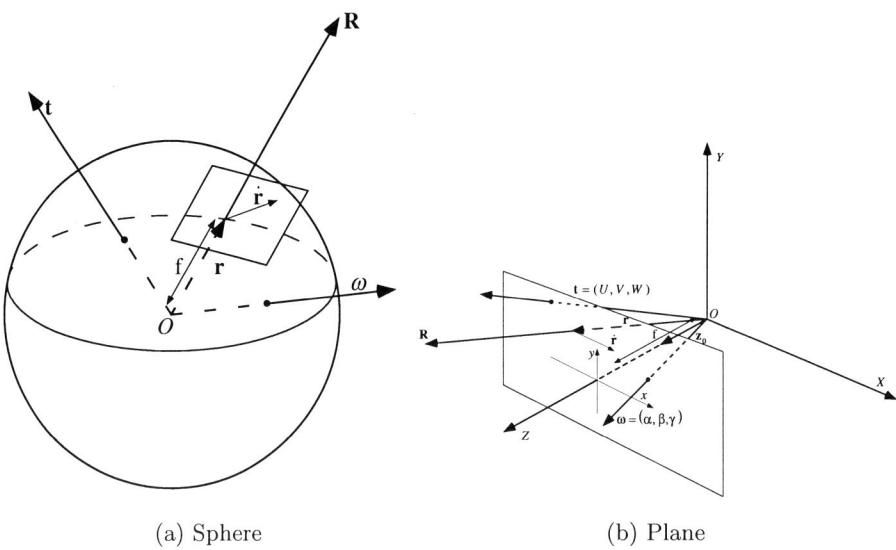

(a) Sphere (b) Plane

Fig. 1. Image formation on the sphere and the plane. The system moves with a rigid motion with translational velocity $\mathbf{t}$ and rotational velocity $\boldsymbol{\omega}$. Scene points $\mathbf{R}$ project onto image points $\mathbf{r}$ and the 3D velocity $\dot{\mathbf{R}}$ of a scene point is observed in the image as image velocity $\dot{\mathbf{r}}$.

If the image is formed on a plane (Fig. 1a) orthogonal to the Z axis at distance f (focal length) from the nodal point, then an image point $\mathbf{r} = (x, y, f)$

and its corresponding scene point $\mathbf{R}$ are related by $\mathbf{r} = \frac{f}{\mathbf{R}\cdot\mathbf{z}_0}\mathbf{R}$, where $\mathbf{z}_0$ is a unit vector in the direction of the Z axis. The motion field becomes

$$\dot{\mathbf{r}} = -\frac{1}{(\mathbf{R}\cdot\mathbf{z}_0)}(\mathbf{z}_0 \times (\mathbf{t} \times \mathbf{r})) + \frac{1}{f}\mathbf{z}_0 \times (\mathbf{r} \times (\boldsymbol{\omega} \times \mathbf{r}))$$
$$= \frac{1}{Z}\mathbf{u}_{\text{tr}}(\mathbf{t}) + \mathbf{u}_{\text{rot}}(\boldsymbol{\omega}), \tag{1}$$

with $Z = \mathbf{R}\cdot\mathbf{z}_0$ representing the depth. If the image is formed on a sphere of radius f (Fig. 1b) having the center of projection as its origin, the image $\mathbf{r}$ of any point $\mathbf{R}$ is $\mathbf{r} = \frac{\mathbf{R}f}{|\mathbf{R}|}$, with R being the norm of $\mathbf{R}$ (the range), and the image motion is

$$\dot{\mathbf{r}} = \frac{1}{|\mathbf{R}|f}\left((\mathbf{t}\cdot\mathbf{r})\mathbf{r} - \mathbf{t}\right) - \boldsymbol{\omega} \times \mathbf{r}$$
$$= \frac{1}{R}\mathbf{u}_{\text{tr}}(\mathbf{t}) + \mathbf{u}_{\text{rot}}(\boldsymbol{\omega}). \tag{2}$$

The motion field is the sum of two components, one, $\mathbf{u}_{\text{tr}}$, due to translation and the other, $\mathbf{u}_{\text{rot}}$, due to rotation. The depth Z or range R of a scene point is inversely proportional to the translational flow, while the rotational flow is independent of the scene in view. As can be seen from (1) and (2), the effects of translation and scene depth cannot be separated, so only the direction of translation, $\mathbf{t}/|\mathbf{t}|$, can be computed. We can thus choose the length of $\mathbf{t}$; throughout the following analysis f is set to 1, and the length of $\mathbf{t}$ is assumed to be 1 on the sphere and the Z-component of $\mathbf{t}$ to be 1 on the plane. The problem of egomotion then amounts to finding the scaled vector $\mathbf{t}$ and the vector $\boldsymbol{\omega}$ from a representation of the motion field.

Equations (1) and (2) show how the motions of image points are related to 3D rigid motion and to scene depth. By eliminating depth from these equations, one obtains the well known epipolar constraint 9; for both planar and spherical eyes it is

$$(\mathbf{t} \times \mathbf{r}) \cdot (\dot{\mathbf{r}} + \boldsymbol{\omega} \times \mathbf{r}) = 0. \tag{3}$$

Equating image motion with optic flow, this constraint allows for the derivation of 3D rigid motion on the basis of optic flow measurements. One is interested in the estimates of translation $\hat{\mathbf{t}}$ and rotation $\hat{\boldsymbol{\omega}}$ which best satisfy the epipolar constraint at every point $\mathbf{r}$ according to some criterion of deviation. The Euclidean norm is usually used, leading to the minimization 6, 11 of the function[1]

$$M_{ep} = \iint_{\text{image}} \left[(\hat{\mathbf{t}} \times \mathbf{r}) \cdot (\dot{\mathbf{r}} + \hat{\boldsymbol{\omega}} \times \mathbf{r})\right]^2 d\mathbf{r}. \tag{4}$$

---

[1] Because $\mathbf{t} \times \mathbf{r}$ introduces the sine of the angle between $\mathbf{t}$ and $\mathbf{r}$, the minimization prefers vectors $\mathbf{t}$ close to the center of gravity of the points $\mathbf{r}$. This bias has been recognized 15 and alternatives have been proposed that reduce this bias, but without eliminating the confusion between rotation and translation.

Expressing $\dot{\mathbf{r}}$ in terms of the real motion from (1) and (2), function (4) can be expressed in terms of the actual and estimated motion parameters $\mathbf{t}, \boldsymbol{\omega}, \hat{\mathbf{t}}$ and $\hat{\boldsymbol{\omega}}$ (or, equivalently, the actual motion parameters $\mathbf{t}, \boldsymbol{\omega}$ and the errors $\mathbf{t}_\epsilon = \mathbf{t} - \hat{\mathbf{t}}$, $\boldsymbol{\omega}_\epsilon = \boldsymbol{\omega} - \hat{\boldsymbol{\omega}}$) and the depth Z (or range R) of the viewed scene. To conduct any analysis, a model for the scene is needed. We are interested in the statistically expected values of the motion estimates resulting from all possible scenes. Thus, as our probabilistic model we assume that the depth values of the scene are uniformly distributed between two arbitrary values $Z_{\min}$ (or $R_{\min}$) and $Z_{\max}$ (or $R_{\max}$) ($0 < Z_{\min} < Z_{\max}$). Parameterizing $\mathbf{n}$ by ψ, the angle between $\mathbf{n}$ and the x axis, we thus obtain the following two functions:

$$E_{ep} = \int_{Z=Z_{\min}}^{Z_{\max}} M_{ep} dZ, \qquad (5)$$

measuring deviation from the epipolar constraint. Function (5) is a five- dimensional surfaces in $\mathbf{t}_\epsilon, \boldsymbol{\omega}_\epsilon$, the errors in the motion parameters.

Epipolar Minimization on the Plane Denote estimated quantities by letters with hat signs, actual quantities by unmarked letters, and the differences between actual and estimated quantities (the errors) by the subscript "ϵ." Furthermore, let $\mathbf{t} = (x_0, y_0, 1)$ and $\boldsymbol{\omega} = (\alpha, \beta, \gamma)$. Since the field of view is small, the quadratic terms in the image coordinates are very small relative to the linear and constant terms, and are therefore ignored.

Considering a circular aperture of radius e, setting the focal length $f = 1$, $W = 1$ and $\hat{W} = 1$, the function in (5) becomes

$$E_{ep} = \int_{Z=Z_{\min}}^{Z_{\max}} \int_{r=0}^{e} \int_{\phi=0}^{2\pi} r \left(\left(\frac{x - x_0}{Z} - \beta_\epsilon + \gamma_\epsilon y + x \right) (y - \hat{y}_0) \right.$$

$$\left. - \left(\frac{y - y_0}{Z} + \alpha_\epsilon - \gamma_\epsilon x + y \right) (x - \hat{x}_0) \right)^2 dr\, d\phi\, dZ$$

where (r, ϕ) are polar coordinates ($x = r\cos\phi, y = r\sin\phi$). Performing the integration, one obtains

$$E_{ep} = \pi e^2 \left((Z_{\max} - Z_{\min}) \left(\frac{1}{3} \gamma_\epsilon^2 e^4 + \frac{1}{4} \left(\gamma_\epsilon^2 (\hat{x}_0^2 + \hat{y}_0^2) + 6\gamma_\epsilon (\hat{x}_0 \alpha_\epsilon + \hat{y}_0 \beta_\epsilon) \right. \right. \right.$$

$$\left. \left. + \alpha_\epsilon^2 + \beta_\epsilon^2 \right) e^2 + (\hat{x}_0 \alpha_\epsilon + \hat{y}_0 \beta_\epsilon)^2 \right) + (\ln(Z_{\max}) - \ln(Z_{\min})) \left(\frac{1}{2} (3\gamma_\epsilon (x_{0_\epsilon} y_0 - y_{0_\epsilon} x_0) + \right.$$

$$\left. x_{0_\epsilon} \beta_\epsilon - y_{0_\epsilon} \alpha_\epsilon) e^2 + 2(x_{0_\epsilon} y_0 - y_{0_\epsilon} x_0)(\hat{x}_0 \alpha_\epsilon + \hat{y}_0 \beta_\epsilon) \right)$$

$$\left. + \left(\frac{1}{Z_{\min}} - \frac{1}{Z_{\max}} \right) \left(\frac{1}{4} (y_{0_\epsilon}^2 + x_{0_\epsilon}^2) e^2 + (x_{0_\epsilon} y_0 - y_{0_\epsilon} x_0)^2 \right) \right) \qquad (6)$$

(a) Assume that the translation has been estimated with a certain error $\mathbf{t}_\epsilon = (x_{0_\epsilon}, y_{0_\epsilon}, 0)$. Then the relationship among the errors in 3D motion at the minima of (6) is obtained from the first-order conditions $\frac{\partial E_{ep}}{\partial \alpha_\epsilon} = \frac{\partial E_{ep}}{\partial \beta_\epsilon} = \frac{\partial E_{ep}}{\partial \gamma_\epsilon} = 0$, which yield

$$\alpha_\epsilon = \frac{y_{0_\epsilon}(\ln(Z_{\max}) - \ln(Z_{\min}))}{Z_{\max} - Z_{\min}}$$
$$\beta_\epsilon = \frac{-x_{0_\epsilon}(\ln(Z_{\max}) - \ln(Z_{\min}))}{Z_{\max} - Z_{\min}}$$
$$\gamma_\epsilon = 0 \tag{7}$$

It follows that $\alpha_\epsilon/\beta_\epsilon = -x_{0_\epsilon}/y_{0_\epsilon}$, $\gamma_\epsilon = 0$, which means that there is no error in γ and the projection of the translational error on the image is perpendicular to the projection of the rotational error. This constraint is called the "orthogonality constraint."

(b) Assuming that rotation has been estimated with an error $(\alpha_\epsilon, \beta_\epsilon, \gamma_\epsilon)$, the relationship among the errors is obtained from $\frac{\partial E_{ep}}{\partial x_{0_\epsilon}} = \frac{\partial E_{ep}}{\partial y_{0_\epsilon}} = 0$. In this case, the relationship is very elaborate and the translational error depends on all the other parameters—that is, the rotational error, the actual translation, the image size and the depth interval.

(c) In the general case, we need to study the subspaces in which E_{ep} changes least at its absolute minimum; that is, we are interested in the direction of the smallest second derivative at 0, the point where the motion errors are zero. To find this direction, we compute the Hessian at 0, that is the matrix of the second derivatives of E_{ep} with respect to the five motion error parameters, and compute the eigenvector corresponding to the smallest eigenvalue. The scaled components of this vector amount to

$$x_{0_\epsilon} = x_0 \quad y_{0_\epsilon} = y_0 \quad \beta_\epsilon = -\alpha_\epsilon \frac{x_0}{y_0} \quad \gamma_\epsilon = 0$$

$$\alpha_\epsilon = 2y_0 Z_{\min} Z_{\max}(\ln(Z_{\max}) - \ln(Z_{\min}))/\left((Z_{\max} - Z_{\min})(Z_{\max}Z_{\min} - 1) + \left((Z_{\max} - Z_{\min})^2(Z_{\max}Z_{\min} - 1)^2 + 4Z_{\max}^2 Z_{\min}^2(\ln(Z_{\max}) - \ln(Z_{\min}))^2\right)^{\frac{1}{2}}\right)$$

As can be seen, for points defined by this direction, the translational and rotational errors are characterized by the orthogonality constraint $\alpha_\epsilon/\beta_\epsilon = -x_{0_\epsilon}/y_{0_\epsilon}$ and by the constraint $x_0/y_0 = \hat{x}_0/\hat{y}_0$; that is, the projection of the actual translation and the projection of the estimated translation lie on a line passing through the image center. We refer to this second constraint as the "line constraint." These results are in accordance with previous studies 2, 11, which found that the translational components along the x and y axes are confused with rotation around the y and x axes, respectively, and the "line constraint" under a set of restrictive assumptions.

Epipolar Minimization on the Sphere The function representing deviation from the epipolar constraint on the sphere takes the simple form

$$E_{ep} = \int_{R_{\min}}^{R_{\max}} \int\int_{\text{sphere}} \left\{ \left(\frac{\mathbf{r} \times (\mathbf{r} \times \mathbf{t})}{R} - (\boldsymbol{\omega}_\epsilon \times \mathbf{r}) \right) \cdot (\hat{\mathbf{t}} \times \mathbf{r}) \right\}^2 dA\, dR$$

where A refers to a surface element. Due to the sphere's symmetry, for each point $\mathbf{r}$ on the sphere, there exists a point with coordinates $-\mathbf{r}$. Since $\mathbf{u}_{tr}(\mathbf{r}) = \mathbf{u}_{tr}(-\mathbf{r})$ and $\mathbf{u}_{rot}(\mathbf{r}) = -\mathbf{u}_{rot}(-\mathbf{r})$, when the integrand is expanded the product terms integrated over the sphere vanish. Thus

$$E_{ep} = \int_{R_{\min}}^{R_{\max}} \int\int_{\text{sphere}} \left\{ \frac{\left((\mathbf{t} \times \hat{\mathbf{t}}) \cdot \mathbf{r}\right)^2}{R^2} + \left((\boldsymbol{\omega}_\epsilon \times \mathbf{r}) \cdot (\hat{\mathbf{t}} \times \mathbf{r})\right)^2 \right\} dA\, dR$$

(a) Assuming that translation $\hat{\mathbf{t}}$ has been estimated, the $\boldsymbol{\omega}_\epsilon$ that minimizes E_{ep} is $\boldsymbol{\omega}_\epsilon = 0$, since the resulting function is non-negative quadratic in $\boldsymbol{\omega}_\epsilon$ (minimum at zero). The difference between sphere and plane is already clear. In the spherical case, as shown here, if an error in the translation is made we do not need to compensate for it by making an error in the rotation ($\boldsymbol{\omega}_\epsilon = 0$), while in the planar case we need to compensate to ensure that the orthogonality constraint is satisfied!

(b) Assuming that rotation has been estimated with an error $\boldsymbol{\omega}_\epsilon$, what is the translation $\hat{\mathbf{t}}$ that minimizes E_{ep}? Since R is uniformly distributed, integrating over R does not alter the form of the error in the optimization. Thus, E_{ep} consists of the sum of the two terms:

$$K = K_1 \int\int_{\text{sphere}} \left((\mathbf{t} \times \hat{\mathbf{t}}) \cdot \mathbf{r}\right)^2 dA$$

$$L = L_1 \int\int_{\text{sphere}} \left((\boldsymbol{\omega}_\epsilon \times \mathbf{r}) \cdot (\hat{\mathbf{t}} \times \mathbf{r})\right)^2 dA,$$

where K_1, L_1 are multiplicative factors depending only on $R_{\min}$ and $R_{\max}$. For angles between $\mathbf{t}, \hat{\mathbf{t}}$ and $\hat{\mathbf{t}}, \boldsymbol{\omega}_\epsilon$ in the range of 0 to $\pi/2$, K and L are monotonic functions. K attains its minimum when $\mathbf{t} = \hat{\mathbf{t}}$ and L when $\hat{\mathbf{t}} \perp \boldsymbol{\omega}_\epsilon$. Consider a certain distance between $\mathbf{t}$ and $\hat{\mathbf{t}}$ leading to a certain value K, and change the position of $\hat{\mathbf{t}}$. L takes its minimum when $(\mathbf{t} \times \hat{\mathbf{t}}) \cdot \boldsymbol{\omega}_\epsilon = 0$, as follows from the cosine theorem. Thus E_{ep} achieves its minimum when $\hat{\mathbf{t}}$ lies on the great circle passing through $\mathbf{t}$ and $\boldsymbol{\omega}_\epsilon$, with the exact position depending on $|\boldsymbol{\omega}_\epsilon|$ and the scene in view.

(c) For the general case where no information about rotation or translation is available, we study the subspaces where E_{ep} changes the least at its absolute minimum, i.e., we are again interested in the direction of the smallest second derivative at 0. For points defined by this direction we calculate $\mathbf{t} = \hat{\mathbf{t}}$ and $\boldsymbol{\omega}_\epsilon \perp \mathbf{t}$.

3 Finding the Rotation

The above shows that if a spherical camera is used, then we can expect to have an accurate rotation, which would then allow the calculation of an accurate translation. Cameras with a common nodal point do exist 12, but their sampling properties are such that flow computation is less well defined. Nevertheless, ego-motion computation has been done, for example in 4. Here a different approach is taken. Given N ordinary cameras, mounted rigidly together, it is possible to simulate the error properties of the sphere in the following way. We assume here that the external calibration between all the cameras is known. More specifically, for every camera i, we know the projection matrix

$$P_i = R_i^T [K_i \mid \mathbf{c}_i] \tag{8}$$

where K_i is the calibration matrix, $\mathbf{c}_i$ is the position of the camera, and R_i is the rotation of camera. The process by which we calibrated is explained at www.cfar.umd.edu/~pbaker/argus.html.

---

1. Calculate $\hat{\boldsymbol{\omega}}_i$ and $\hat{\mathbf{t}}_i$ using a fast algorithm
2. Keep only $\hat{\gamma}_i$ and compute $\hat{\boldsymbol{\omega}}$ using the least squares outlined in (9).
3. Using this accurate rotation, optimize using the epipolar constraint only on translation.

Fig. 2. The γ algorithm for computing $\boldsymbol{\omega}$

First, note that when cameras are mounted rigidly together, the rotation $\boldsymbol{\omega}$ in every camera is the same. Thus information from every camera can be used to compute the $\boldsymbol{\omega}$ for the system. From (7), it can be seen that γ_ϵ is very small when using epipolar minimization. Put in another way, the error in the measurement of the projection of the rotation vector $\boldsymbol{\omega}$ onto the optical axis of each camera is likely to be small. Now for each camera i, let it have rotation γ_i around its optical axis $\mathbf{n}_i$, where $\mathbf{n}_i$ is expressed in the fiducial coordinate system. The calculation of $\boldsymbol{\omega}$ can now be formed as a linear system:

$$\begin{bmatrix} \mathbf{n}_1^T \\ \mathbf{n}_2^T \\ \vdots \\ \mathbf{n}_N^T \end{bmatrix} \boldsymbol{\omega} = \begin{bmatrix} \gamma_1 \\ \gamma_2 \\ \vdots \\ \gamma_N \end{bmatrix} \tag{9}$$

We have assumed here that the $\mathbf{n}_i$ are known, which is a function of the calibration discussed later. Also, note that the cameras must be placed in such a way that the $\mathbf{n}_i$ are linearly independent, that is the optical axes do not all lie in a plane. This implies the need for at least three cameras for this algorithm to remove all ambiguities from the rotation, and four to provide an accuracy measure. Using the rotational calibration, the optic flow, and the accurate γ_i, the algorithm in Fig. 2, called the γ-algorithm, can thus be used to compute $\boldsymbol{\omega}$.

4 Computing the Translation

Once the rotation is obtained, it still remains to calculate the translation. As in the case of the cyclotorsional component of the rotation, part of the translation direction is well defined, even from a small field of view. Geometric results 11, 15, statistical analysis 5, and empirical studies 3 all show that the translational estimate — when projected onto the image plane — lies along a line between the true translational direction and the center of the image. Combined with the scale factor ambiguity, this implies that the translation lies on the plane through this line and the focal point defines a set of possible 3D translations.

If the line from the translational estimate to the image center is in the direction of $\mathbf{q}$, then $\mathbf{m} = \hat{z} \times \mathbf{q}$ is the normal to the plane containing the true translation $\mathbf{t}$. Since there are many cameras, calibrated with one another, these planes can be intersected in the following manner to calculate the true $\mathbf{t}$.

It is assumed that the rotation $\boldsymbol{\omega}$ of the camera system has been calculated. Recall from (8) that the center of projection (in the fiducial coordinate system) is $\mathbf{c}_i$ and the camera is rotated with rotation matrix R_i.

First we consider the case where $\boldsymbol{\omega} = 0$ The translation of the camera system, in the coordinate system of camera i, is $R_i \mathbf{t}$. This translation is constrained by $\mathbf{m}$ in the equation

$$\mathbf{t}^T R_i^T \mathbf{m} = 0 \tag{10}$$

If we then assume that the camera system is rotating with rotation $\boldsymbol{\omega}$, then a compensatory translation of $\boldsymbol{\omega} \times \mathbf{c}_i$ is added to $\mathbf{t}$ in the ith camera, so that the equation becomes

$$(\mathbf{t} - \boldsymbol{\omega} \times \mathbf{c}_i)^T R_i^T \mathbf{m} = 0 \tag{11}$$

This is a linear equation in $\mathbf{t}$ and known variables for each camera. Given n cameras such that the $\mathbf{m}_i^T R_i$ are linearly independent, it is possible to solve for the translation.

$$\begin{bmatrix} \mathbf{m}_1^T R_1 \\ \mathbf{m}_2^T R_2 \\ \vdots \\ \mathbf{m}_n^T R_n \end{bmatrix} \mathbf{t} = \begin{bmatrix} \mathbf{m}_1^T R_1(\boldsymbol{\omega} \times \mathbf{c}_1) \\ \mathbf{m}_2^T R_2(\boldsymbol{\omega} \times \mathbf{c}_2) \\ \vdots \\ \mathbf{m}_n^T R_n(\boldsymbol{\omega} \times \mathbf{c}_n) \end{bmatrix}$$

If the $\mathbf{c}_i$ are small or zero, then this equation becomes homogeneous and only the direction of $\mathbf{t}$ is defined. This is equivalent to the case of a spherical camera with a patchwise field of view. If the $\mathbf{c}_i$ are significant *and* there is a rotation, then the linear system will give an absolute translation, not subject to a scale factor, which concurs with the result found in 16. This allows for metric reconstructions without having to solve a correspondence problem. Thus the system is a multi-camera stereo system with cameras that do not have to be pointing at the same object.

5 Summary and Conclusions

This paper has shown a method for constructing a system of cameras so that egomotion is easier to compute and more stable. We have proved that a spherical camera gives no rotational error; and the gamma algorithm provides the same unbiased motion estimation by combining only the reliable component of rotation from each camera. We use this rotation to solve for the translation; if the cameras do not share a nodal point, the metric 3D translation is computed.

These results have implications beyond the better computation of 3D motion. Using high resolution cameras looking in multiple directions gives advantages for motion estimation, these accurate motion estimates permit higher quality reconstructions and models. Solving directly for all three parameters of the camera translation — instead of just the direction of translation — simplifies the task of linking differential camera motions into trajectories through space.

These considerations also help to explain evolutionary choices for biological eye designs. Primate vision has very high resolution, allowing detailed reconstruction and manipulation of objects. Flying birds and insects, which require fast 3D motion computations, generally have a nearly complete field of view, perhaps because this simplifies the computation problems. The fact that the two hemispheres of an insect eye do *not* share a common nodal point allows, theoretically, a solution to the actual translational velocity, as opposed to only the direction of translation. In general, biological eye designs are constrained by computational and physical limits which make high resolution spherical eyes impractical. Our system uses both high resolution cameras and multiple camera orientations to compute accurate 3D motion and detailed models.

References

G. Adiv. Inherent ambiguities in recovering 3D motion and structure from a noisy flow field. In *Proc. IEEE Conference on Computer Vision and Pattern Recognition*, pages 70–77, 1985.

K. Daniilidis and M. E. Spetsakis. Understanding noise sensitivity in structure from motion. In Y. Aloimonos, editor, *Visual Navigation: From Biological Systems to Unmanned Ground Vehicles*, Advances in Computer Vision, chapter 4. Lawrence Erlbaum Associates, Mahwah, NJ, 1997.

A.M. Earnshaw and S.D. Blostein. The performance of camera translation direction estimators from optical-flow: Analysis, comparison, and theoretical limits. *IEEE Transactions on Pattern Analysis and Machine Intelligence*, 18(9):927–932, September 1996.

Joshua Gluckman and Shree K. Nayar. Egomotion and omnidirectional sensors. In *Proc. International Conference on Computer Vision*, 1998.

D. J. Heeger and A. D. Jepson. Subspace methods for recovering rigid motion I: Algorithm and implementation. *International Journal of Computer Vision*, 7:95–117, 1992.

B. K. P. Horn. *Robot Vision*. McGraw Hill, New York, 1986.

H. Ishiguro, K. Ueda, and S. Tsuji. Omnidirectional visual information for navigating a mobile robot. In *Proceedings of 1993 IEEE International Conference on Robotics and Automation*, volume 1, pages 799–804. IEEE Comput. Soc. Press, 1993.

J. J. Koenderink and A. J. van Doorn. Invariant properties of the motion parallax field due to the movement of rigid bodies relative to an observer. *Optica Acta*, 22:773–791, 1975.

H. C. Longuet-Higgins. A computer algorithm for reconstructing a scene from two projections. *Nature*, 293:133–135, 1981.

D. Marr. *Vision*. W.H. Freeman, San Francisco, CA, 1982.

S. J. Maybank. *Theory of Reconstruction from Image Motion*. Springer, Berlin, 1993.

S. Nayar. Catadioptric omnidirectional camera. In *Proc. IEEE Conference on Computer Vision and Pattern Recognition*, pages 482–488, Puerto Rico, 1997.

R. C. Nelson and J. Aloimonos. Finding motion parameters from spherical flow fields (or the advantage of having eyes in the back of your head). *Biological Cybernetics*, 58:261–273, 1988.

Federico Pedersini, Augusto Sarti, and Stefano Tubaro. Egomotion estimation of a multicamera system through line correspondences. In *Proceedings. International Conference on Image Processing*, volume 2, pages 175–178. IEEE Computer Society, 1997.

M. E. Spetsakis and J. Aloimonos. Optimal motion estimation. In *Proc. IEEE Workshop on Visual Motion*, pages 229–237, 1989.

An-Ting Tsao, Yi-Ping Hung, Chiou-Shann Fuh, and Yong-Sheng Chen. Ego-motion estimation using optical flow fields observed from multiple cameras. In *Proc. IEEE Conference on Computer Vision and Pattern Recognition*, pages 457–462. IEEE Computer Society, 1997.

S. Ullman. *The Interpretation of Visual Motion*. MIT Press, Cambridge, MA, 1979.

Top-Down Attention Control at Feature Space for Robust Pattern Recognition

Su-In Lee and Soo-Young Lee

Brain Science Research Center and Department of Electrical Engineering
Korea Advanced Institute of Science and Technology
373-1 Kusong-dong, Yusong-gu, Taejon 305-701, Korea
leepc@cais.kaist.ac.kr
sylee@ee.kaist.ac.kr

Abstract. In order to improve visual pattern recognition capability, this paper focuses on top-down selective attention at feature space. The baseline recognition system consists of local feature extractors and a multi-layer Perceptron (MLP) classifier. An attention layer is added just in front of the multi-layer Perceptron. Attention gains are adjusted to cope with the top-down attention process and ellucidate expected input features. After attention adaptation, the distance between original input features and expected features becomes an important measure for the confidence of the attended class. The propsoed algorithms improves recognition accuracy for handwritten digit recognition tasks, and is capable of recognizing 2 superimposed patterns one by one.

Introduction

Selective attention is known as a mechanism to overcome limited resources in brain, and improves perception capability in complex visual and auditory scenes. [1-4] There exist two approaches to model the selective attention, one emphasizes bottom-up signal and the other emphasizes top-down signal. The bottom-up approaches look for some saliency from input signal, and usually result in 'spot-light' attended area in the input space. [5] In the top-down attention approaches an attention cue is specified at the output layer and back-propagated to control the attention. [6] Recently it is believed that both "bottom-up" and "top-down" paths exist in human attention process.

Recently a simple but efficient top-down attention algorithm was developed to improve recognition accuracy in noisy and superimposed patterns. [7,8] In Ref. 8 the attention is made at the input space. In this paper we further extend the top-down attention algorithm to incorporate attention at feature space. In early vision systems local orientation features are extracted first, and it may be natural to look for attention at this feature space. [9]

Fukushima developed Neocognitron model which has a structure similar to the hierarchical model of the visual nervous system with selective attention and attention switching capabilities. [6] However, the Neocognitron model has many unknown parameters to be determined heuristically, and requires prohibitively expensive computation for many real-time applications.

In Section 2 we describe the baseline recognition system, i.e., Self-Organizing Feature with Fuzzy Association (SOFFA) neural network model, which incorporates local feature extractors in front of an MLP classifierIn. Section 3 a new simple selective attention algorithm is applied to the baseline model. Also, an attention switching algorithm is introduced to recognize superimposed patterns one by one. A brief concluding remark will follow in Section 4.

2 The Baseline System: SOFFA Neural Network Model

The baseline recognition system, Self-Organizing Feature with Fuzzy Association (SOFFA) neural network model, consists of three stages, i.e., local feature extraction stage, fuzzy abstraction stage, and a multi-layer Perceptron (MLP) neural classifier. The first stage plays a role of extracting local features of the input pattern. This module provides an information about the existence and the location of some feature on the input pattern. In the second stage, the network abstracts the information effectively maintaining the distinctive feature and this function make the network relatively insensitive for spatial error. In fact, these two stages resemble to the S-cells and C-cells of the Neocognitron model and human visual sensory nervous systems. The third stage is a MLP neural network model with more than one hidden layer, which performs classification and recognition. The structure of the SOFFA neural network model is illustrated in Fig. 1.

2.1 Local Feature Extraction

Simple predefined 12 features are used here and shown in Fig 2. Since learning procedure need not be done in this stage, the computationsl requirement is drastically reduced. These are 33 pixels squares and express 12 angles of straight lines. Among 9 pixels of a feature, 3 pixels are black and coded as 1 and white pixels are coded as -0.5, because the background of a input pattern have to be coded as 0 in all feature planes. These are chosen based on the theories which assert that various angles of straight lines are most prevalent features to be extracted in human visual recognition process. There are 12 feature planes corresponding to each local features. Each pixel in a feature plane has connection only from the 3×3 contiguous square with the original input. The location of the 3×3 square shifts by one input pixels neighbors in a certain feature plane. This square is a corresponding feature of the feature plane. For bipolar binary representations, black pixels of input pattern are coded as 1 and white pixels are -1. Each feature plane detects their own feature and displays the results.

Top-Down Attention Control at Feature Space for Robust Pattern Recognition   131

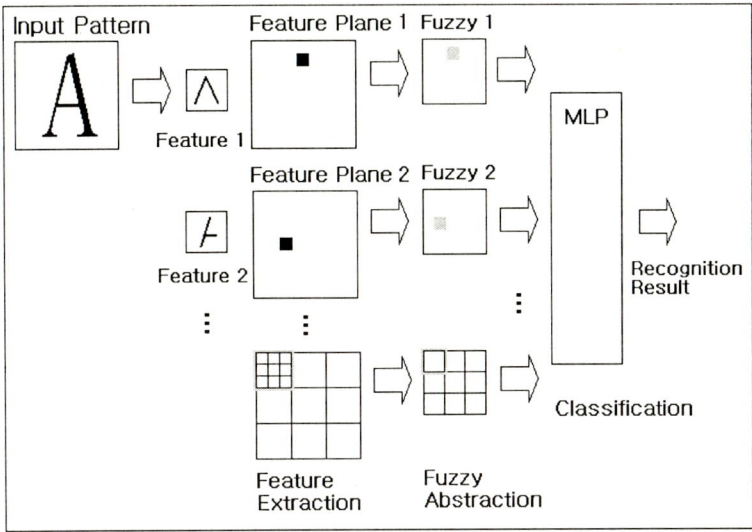

Fig 1. The Structure of SOFFA Neral Nework Model

The size of feature plane becomes 18x18 because the input pattern size is 16x16, which is extended to 20x20 by extending background in order to detect features at the edge of the pattern. All 324 pixels within a feature plane have the same weight, so they all detect the same feature. Thus all feature planes serves as a detector of a specific feature across the entire input field.

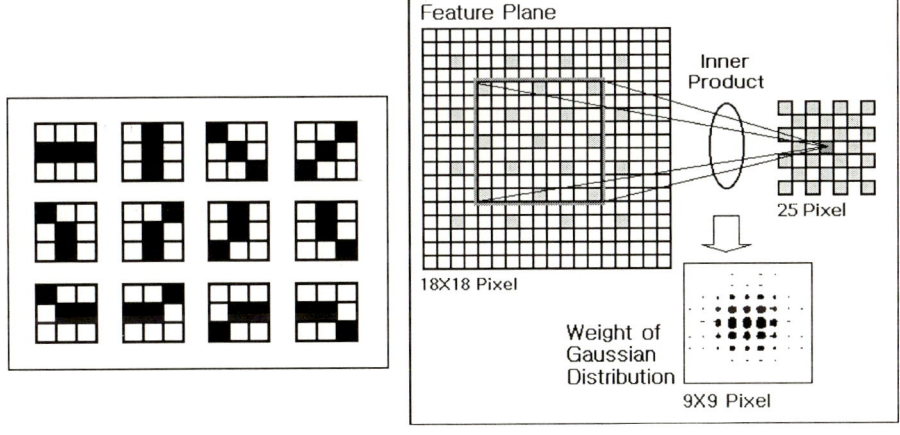

Fig 2. 12 Local Features   **Fig 3.** Fuzzy Abstraction

2.2 Fuzzy Abstraction

In this module, the network abstracts 12 feature planes maintaining the distinctive features of them. Proposed fuzzification stage abstracts the feature plane consisting of 18×18 pixel to 25 pixel. This procedure is illustrated in Figure 3. First of all, 25 pixels in the feature plane have to be selected uniformly. The output of this stage is weighted sum of corresponding 9×9 square on the feature plane of which center is one of the selected pixels. The squared weight consisting of 9×9 square is obtained from Gaussian distribution with a certain variance. This function is applied to all feature planes identically.

2.3 Multi-layer Perceptron (MLP)

An MLP with a single hidden layer is applied in this paper. It is possible for some pixels on the feature planes to have negative values because the input pattern and local features are coded as (1, -1) and (1, -0.5), respectively.

However, these negative values are changed to zeros before fuzzification. Each feature plane displays the existence and location of its own feature on the input pattern. If a pixel on a feature plane has relatively large value, it means that there are local feature which resembles the feature in corresponding location on the input pattern. However, a negative valued pixel means existence of the opposite feature, which is not a big concern here. Therefore, negative values are converted into zeros, which is widely accepted in the visual and auditory models.

2.4 Experimental Results

The proposed SOFFA model was tested on recognition of 480 numeral patterns. The numeral database consists of samples of the handwritten digits collected from 48 people. Each digit is encoded as a 16×16 binary pixel array. Roughly 16% of the pixels are black coded as 1 and rest are white coded as 0. Four experiments were conducted with different training sets of 280 training patterns each, and the rest 200 patterns were test pattern. Standard hyperbolic-tangent bipolar sigmoid function was used as a nonlinear function. A one hidden-layer MLP was trained by error back propagation learning rule. Training progress sank below 10^{-4} of normalized output error or when a maximum number of epochs had been reached and learning rate was 0.01. In order to discuss the effectiveness of local feature extraction and fuzzification stage, experiment was carried out in four cases. First, the number of features was reduced to four as the top row in Fig. 2. Second, the effectiveness of fuzzification was concerned, so we can divide into four cases by matching pairs of feature extraction and fuzzification methods. The definition of these cases and the number of neurons in MLP for each cases are summarized in Table 1.

False recognition rates for test patterns of all cases are summarized in Table 2. The false recognition rate is clearly lower with the fuzzification. It appears that fuzzification stage provides tolerance for spatial error because test patterns are not identical in shape and position to training patterns. Also, it showed better recognition capability when more features were used. Finally, the results shows that recognition capability was achieved by proposed SOFFA model because the false recognition rate of Case 1&3 was much lower than std MLP.

Table 1. Definition of each cases and number of neurons of MLP

Cases	Case1	Case2	Case3	Case4	Std MLP
# of Local Features	12	12	4	4	×
Fuzzy Abstraction	O	×	O	×	×
# of Input Neurons of MLP	300	3888	100	1296	256
# of Hidden Neurons of MLP	15	15	15	15	15
# of Output Neurons of MLP	10	10	10	10	10

Table 2. False recognition Rates of Baseline System

Patterns (Number of Patterns)	epochs	False Recognition Rates [%]				
		Case1	Case2	Case3	Case4	Std MLP
Training Pattens(280)	5000	0.00	0.00	1.50	0.42	0.00
Test Patterns (200)	5000	0.50	28.00	5.50	42.00	9.66

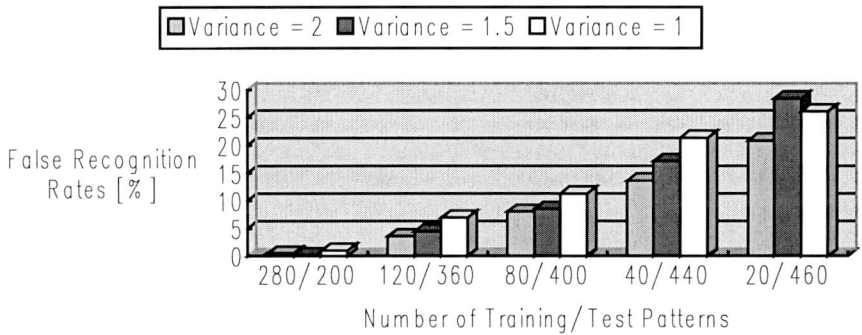

Fig 4. False recognition rates with various number of training patterns and variance values of fuzzification

In Fig. 4, the false recognition rates for Case 1 is plotted as a function of both the number of training/test patterns and variances of fuzzification. In this case, the false recognition rate of training patterns were all 0, so need not to be plotted. As the number of training patterns decreases, the false recognition rate increases. It is because the network trained by more number of training patterns has better generalization capability. Also, the false recognition rate increases as the variance of fuzzification decreases. It shows that the larger the variance of fuzzification is, the more insensitive to the spatial change the network becomes.

3 Selective Attention Model

3.1 Top-Down Attention Model

If the early selection model with top-down control is applied, the network for selective attention is designed as depicted in Fig. 5. [8] The dotted box is a standard one hidden-layer Perceptron and the attention layer with one-to-one connectivity is added in front of the input layer. During training, all attention gains are fixed to 1, causing the network behave as an ordinary MLP. Therefore, proposed selective attention model shares training phase with SOFFA model proposed in the previous work, and selective attention algorithm is applied during the test phase only. In order to apply selective attention, three steps are implemented here. First, top N activation classes are chosen as candidates by trained SOFFA model. And then, the process of attention adaptation follows for each candidate. Attention gains are adjusted to have strong response from a particular class under each candidate–call it the attention class. If a target output vector $t^s = [t^s_1, t^s_2, ... t^s_M]^T$ under a certain candidate is defined and a_k's are set to 1 as an initial value, adaptation starts with the input $x=[x_1, x_2, ... x_N]^T$ and pre-trained synaptic weights W.

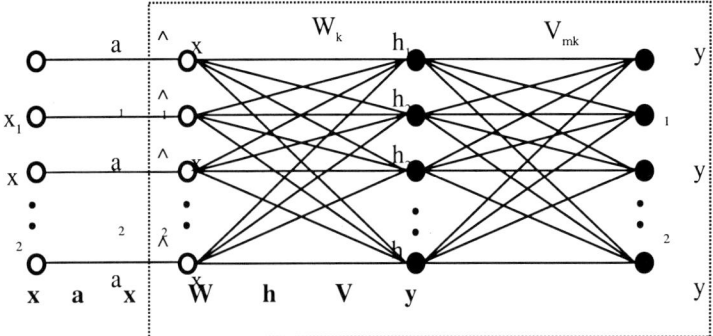

Fig 5. Proposed network architecture for top-down selective attention

For bipolar binary output representations, t_i^s is 1 for the attention class and -1 for the others. Attention adaptation proceeds to minimize an output error based on gradient-descent algorithm with error back-propagation learning rule. At the (n+1)'th iterative epoch, the attention gain a_k is updated as

$$a_k[n+1] = a_k[n] - \eta (\partial E / \partial a_k)[n] = a_k[n] + \eta\, x_k \delta_k^{(0)}[n] \quad (1a)$$

$$\delta_k^{(0)} = \sum_j W_{jk}^{(1)} \delta_j^{(1)} \quad (1b)$$

where

$$E^s \equiv \tfrac{1}{2} \sum_i (t_i^s - y_i)^2$$

and $\delta_j^{(1)}$ and $W_{jk}^{(1)}$ denote the attention output error, the j'th attribute of the back-propagating error at the first hidden-layer, and the synaptic weight between the input $\hat{x}_k$ and the j'th neuron at the first hidden layer, respectively, η is an adaptation rate. After attention adaptation for all candidates, one of N candidates is selected with maximum attention measure. The attention measure is defined as

$$M \equiv \frac{A_O + 1}{E_A^\alpha D_A^\beta} \quad (2a)$$

$$D_A \equiv \frac{1}{2N} \sum_{k=1}^{N} (x_k - \hat{x}_k)^2 = \frac{1}{2N} \sum_{k=1}^{N} x_k^2 (1 - a_k)^2, \quad E_A \equiv \frac{1}{2M} \sum_{i=1}^{M} [t_i - y_i(\hat{x})]^2 \quad (2b)$$

where D_A is the square of Euclidean distance between two input patterns before and after the application of selective attention, E_A is the output error after the application of selective attention, and A_U is the output activation before attention adaptation. Finally, α and β are relative weighting factors between the three parameters. If the attention adaptation proceeded successfully, D_A becomes small. However, if the input pattern is much different from training vectors belongs to the attention class, the attention adaptation converges to a local minimum with a significant output error E_A. In this case, the attention measure become small and E_A becomes an important measure.[8]

3.2 Experimental Results for Test Data

The proposed selective attention algorithm was tested on the same data set in Section 2. Attention adaptation process lasted before the output error falls below a certain value. The values of criterion, i.e., target output error, were defined as 0.005, 0.00125, and 0.0002. These values correspond to normalized output errors when the output vectors are $\{y_{i=k} = 0.9,\ y_{i \neq k} = -0.9\}$, $\{y_{i=k} = 0.95,\ y_{i \neq k} = -0.95\}$, and $\{y_{i=k} = 0.98,\ y_{i \neq k} = -0.98\}$, respectively. In Fig. 6, both results from proposed SOFFA model and selective attention model are compared. Here, SA and SOFFA denote with and without the selective attention process, respectively. In each case, the network with selective attention process reduced false recognition rate drastically. Especially, error reduction rate was higher as the network was trained with more number of training patterns. In other words, as the number of training patterns increases, there was great enhancement in the effectiveness of selective attention process. These results agree with biological concept of selective attention which is based on top-down control. Biological attention may occur being directed based on expectations and object knowledge.

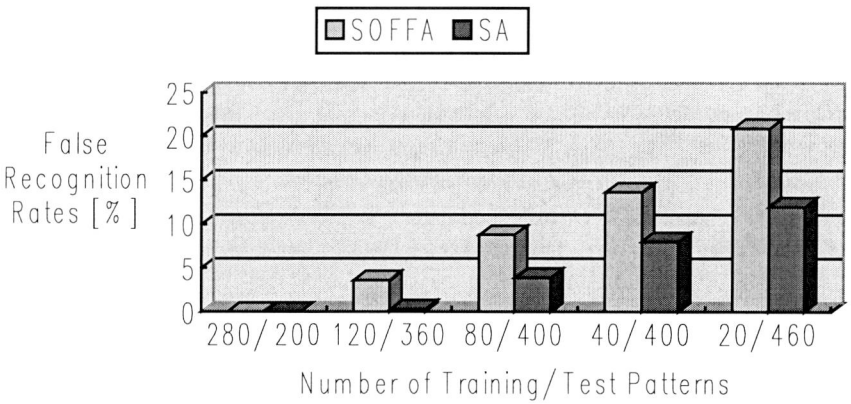

Fig 6. False recognition rates comparison between SOFFA model and selective attention model with various number of training patterns

3.3 Attention Switching for Superimposed Pattern

If we want to recognize superimposed patterns in sequence, an algorithm based on selective attention can be applied. In this case, a visual pattern recognition-where two pattenrs are spatially overlapping is dealt with. This is an extreme case of a situation that is common in visual pattern recognition. Our strategy for recognizing superimposed patterns consists of three steps as follows.

First, one pattern of a superimposed pattern is recognized through selective attention algorithm. Second, attention is switched from the recognized pattern to the remaining pixels by removing attention from the pixels related to the first recognized pattern. During the attention switching, attention gains of which value is larger than 1 after first-stage selective attention are clamped to 0, and all other gains are set to be 1. Finally, the recognition process is performed again with remaining feature image to recognize the second pattern. This step can be done either by proposed SOFFA algorithm or selective attention model. Of course, it will show better performance for selective attention model for the second pattern.

The proposed selective attention and attention switching algorithm were tested on superimposed numeral data. The superimposed patterns were generated using OR operator for two different binary digits. We want to generate two sets of superimposed patterns; one set is made of two training patterns, the other is made of two test patterns. For each set, two groups consisting of 20 patterns were selected and 360 superimposed patterns were generated by superimposing pairs between two groups of superimposed patterns from different output classes. Then, SOFFA(Self-Organizing Feature Extraction with Fuzzy Association) model was tested on two sets of superimposed patterns. We simply use trained networks in last sections. Each network was trained with different number of training patterns and different values of variance in fuzzification process. Two classes which showed highest activity in MLP output were selected as recognized classes. False recognition rates for superimposed patterns generated from training patterns are summarized in Table 3. The more training patterns are used during training procedure and the larger value of variance was used during fuzzification, the more error was occurred.

Table 3. False recognition rates(%) of superimposed numeral patterns constituting of training patterns

Patterns	Variance	Number of Traing/ Test Patterns			
		280/ 200	120/ 360	80/ 400	40/ 440
First Pattern	2	21.4	20.6	18.2	13.9
	1	19.9	19.4	16.9	7.50
Second Pattern	2	57.8	50.0	55.0	39.7
	1	53.6	43.3	41.9	40.6

For superimposed patterns made of training patterns, the network trained by 40 training patterns showed best performance for the first pattern. In fact, if the first pattern was not recognized correctly, attention switching algorithm malfunctions. Therefore, proposed attention switching algorithm was applied for the network trained by 40 training patterns. The false recognition rates are summarized in Table 4. After attention switching, the second pattern was recognized SOFFA model. The recognition performance was improved a little bit for the first pattern, and attention switching algorithm drastically reduced the errors for the second pattern in both cases.

Table 4. False recognition rates(%) of superimposed numeral patterns with variance = 1.0 (1) and variance = 2.0 (2)

Recognition Process	First Pattern	Second Pattern
(1) No Selective Attention or Switching	7.50	40.6
Selective Attention & Switching	7.17	27.0
(2) No Selective Attention or Switching	13.9	39.7
Selective Attention & Switching	11.7	26.11

4 Conclusion

In order to make robust recognition system on positional error or spatial overlapping, this paper proposed Self-Organizing Feature with Fuzzy Association(SOFFA) neural network model with top-down selective attention. The top-down attention illucidates expected input features by error backpropagation algorithm. Proposed model showed a significant improvement in recognition rates both in test patterns and superimposed patterns. It is meaningful that two procedures which are critical during human recognition mechanism, were applied with each other. Also, these are much simpler than other biologically inspired algorithms.

Acknowledgements

S.Y. Lee acknowledges supports from Korean Ministry of Science and Technology as a Brain Science and Engineering Research Program.

References

1. Broadbent, D. E.: Perception and Communication. Pergamon Press. (1958)
2. Cowan, N.: Evolving conceptions of memory storage, selective attention, and their mutual constraints within the human information processing system. Psychological Bulletin **104** (1988) 163-191
3. Parasuraman, R. (Ed.) The Attentive Brain, MIT Press (1998)
4. Anderson, C., Olshausen, B., and Essen, D.: A neurobiological model of visual attention and invariant pattern recognition based on dynamic routing of information. Journal of Neuroscience **13** (1993) 4700-4719
5. Ltti, L., Kock, C., and Niebur, E.: A Model of saliency-Based Visual Attention for rapid Scene Analysis. IEEE Trans. Pattern Analysis and machine Intelligence **20** (1998)
6. Fukushima, K.: Neural network model for selective attention in visual pattern recognition and association recall. Applied Optics **26** (1987) 4985-4992
7. Park, K.Y, Lee, S.Y.: Selective Attention for Noise Robust Speech Recognition, International Joint Conference on Neural Networks, Washington, D.C., July (1999)
8. Lee, S.Y., Michael C.M. Robust Recognition of Noisy and Superimposed Patterns via Selective Attention. Neural Information Processing Systems **12** (2000)
9. Arbib, M.A. (Ed.) The handbook of Brain Theory and Neural Networks. MIT Press (1998)
10. Kim, J.W.: Isolated Word Recognition Using SOFFA Neural Network, Master Thesis, Korea Advanced Institute of Science and Technology (1995)

A Model for Visual Camouflage Breaking

Ariel Tankus* and Yehezkel Yeshurun*

Department of Computer Science
Tel-Aviv University
Tel-Aviv 69978, Israel
{arielt,hezy}@math.tau.ac.il

Abstract. Some animals use counter-shading in order to prevent their detection by predators. Counter-shading means that the albedo of the animal is such that its image has a flat intensity function rather than a convex intensity function. This implies that there might exist predators who can detect 3D objects based on the convexity of the intensity function. In this paper, we suggest a mathematical model which describes a possible explanation of this detection ability. We demonstrate the effectiveness of convexity based camouflage breaking using an operator ("D_{arg}") for detection of 3D convex or concave graylevels. Its high robustness and the biological motivation make D_{arg} particularly suitable for camouflage breaking. As will be demonstrated, the operator is able to break very strong camouflage, which might delude even human viewers. Being non-edge-based, the performance of the operator is juxtaposed with that of a representative edge-based operator in the task of camouflage breaking. Better performance is achieved by D_{arg} for both animal and military camouflage breaking.

1 Introduction

"Camouflage is an attempt to obscure the signature of a target and also to match its background" [1]. Work related to camouflage can be roughly divided into two: camouflage assessment and design (e.g, [1], [2]), and camouflage breaking. Despite the ongoing research, only little has been said in the computer vision literature on visual camouflage breaking: [6], [11], [5], [3], [4]

In this paper, we address the issue of *visual camouflage breaking*. We present biological evidence that detection of the convexity of the graylevel function may be used to break camouflage. This is based on Thayer's principle of counter-shading [12], which observes that some animals use apatetic coloration to prevent their image (under sun light) from appearing as convex graylevels to a viewer. This implies that other animals may break camouflage based on the convexity of the graylevels they see (or else there was no need in such an apatetic coloration).

Our goal is therefore to detect 3D convex or concave objects under strong camouflage. For this task, we employ our proposed operator (D_{arg}), which is

---

* Supported by the Minerva Minkowski center for geometry, and by grant from the Israel Academy of Science for Geometric Computing.

applied directly to the intensity function. D_{arg} responds to smooth 3D convex or concave patches in objects. The operator is not limited by any particular light source or reflectance function. It does not attempt to restore the three dimensional scene. The purpose of the operator is *detection* of convex or concave objects in highly cluttered scenes, and in particular under camouflage conditions.

The robustness and invariance characterizing D_{arg} (see [10]) as well as the biological motivation make it suitable for camouflage breaking, even for camouflages that might delude a human viewer. In contrast with existing attempts to break camouflage, our operator is context-free; its only a priori assumption about the target is its being three dimensional and convex (or concave). In order to evaluate the performance of the operator in breaking camouflage, we juxtaposed D_{arg} with a representative edge-based operator. Due to lack of room only a small portion of the comparison can be given in the paper.

The next section defines the operator D_{arg} for convexity-based detection. Section 2.1 gives intuition for D_{arg} and is of particular importance for understanding its behavior. Section 3 utilizes D_{arg} for camouflage breaking. Section 3.1 brings the biological evidence for camouflage breaking by detection of graylevel convexity. Section 3.2 establishes the connection between the biological evidence and the specific convexity detector D_{arg}. Section 4 delineates a camouflage breaking comparison of an edge-based method with our convexity detector. Concluding remarks are in section 5.

2 Y_{arg}, D_{arg}: Operators for Detection of Convex Domains

We next define an operator for detection of three dimensional objects with smooth convex and concave domains.

Let $I(x, y)$ be an input image, and $\nabla I(x, y) = (\frac{\partial}{\partial x} I(x, y), \frac{\partial}{\partial y} I(x, y))$ the Cartesian representation of the gradient map of $I(x, y)$. Let us convert $\nabla I(x, y)$ into its *polar* representation. The gradient argument is defined by:

$$\theta(x, y) = \arg(\nabla I(x, y)) = \arctan\left(\frac{\partial}{\partial y} I(x, y) , \frac{\partial}{\partial x} I(x, y)\right)$$

where the two dimensional arc tangent is:

$$\arctan(y, x) = \begin{cases} \arctan(\frac{y}{x}), & \text{if } x \geq 0 \\ \arctan(\frac{y}{x}) + \pi, & \text{if } x < 0 , y \geq 0 \\ \arctan(\frac{y}{x}) - \pi, & \text{if } x < 0 , y < 0 \end{cases}$$

and the one dimensional $\arctan(t)$ denotes the inverse function of $\tan(t)$ so that: $\arctan(t) : [-\infty, \infty] \mapsto [-\frac{\pi}{2}, \frac{\pi}{2}]$.

The proposed convexity detection mechanism, which we denote: Y_{arg}, is simply the y-derivative of the argument map:

$$Y_{arg} = \frac{\partial}{\partial y} \theta(x, y) = \frac{\partial}{\partial y} \arctan\left(\frac{\partial}{\partial y} I(x, y) , \frac{\partial}{\partial x} I(x, y)\right)$$

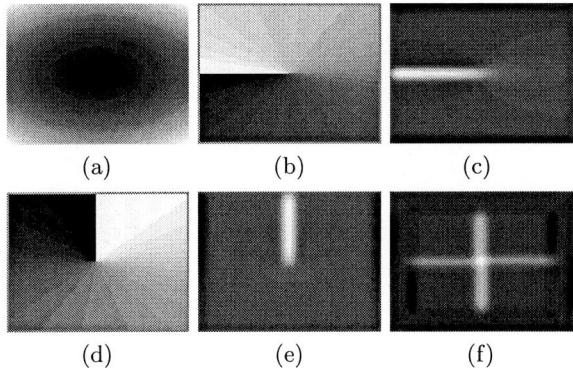

Fig. 1. *(a)* Paraboloidal gray-levels: $I(x,y) = 100x^2 + 300y^2$. *(b)* Gradient argument of (a). Discontinuity ray at the negative x-axis. *(c)* Y_{arg} of (a) ($= \frac{\partial}{\partial y}$ of (b)). *(d)* Rotation of (a) (90° c.c.w.), calculation of gradient argument, and inverse rotation. *(e)* Rotation of (a) (90° c.c.w.), calculation of Y_{arg}, and inverse rotation. *(f)* Response of D_{arg}, the isotropic operator.

To obtain an isotropic operator based on Y_{arg}, we rotate the original image by 0°, 90°, 180° and 270°, operate Y_{arg}, and rotate the results back to their original positions. The sum of the four responses is the response of an operator which we name: D_{arg}.

2.1 Intuitive Description of the Operator

- **What Does Y_{arg} Detect?**
  Y_{arg} detects the zero-crossings of the gradient argument. This stems from the last step of the gradient argument calculation: the two-dimensional arc-tangent function. The arc-tangent function is discontinuous at the negative part of the x-axis; therefore its y-derivative approaches infinity there. In other words, Y_{arg} approaches infinity at the negative part of the x-axis of the arctan, when this axis is being crossed. This limit reveals the zero-crossings of the gradient argument (see [10] for more details).
- **Why Detect Zero-Crossings of the Gradient Argument?**
  Y_{arg} detects zero-crossings of the gradient argument of the intensity function $I(x,y)$. The existence of zero-crossings of the gradient argument enforces a certain range of values on the gradient argument (trivially, values near zero). Considering the intensity function $I(x,y)$ as a surface in $\mathbb{R}^3$, the gradient argument "represents" the direction of the normal to the surface. Therefore, a range of values of the gradient argument means a certain range of directions of the normal to the intensity surface. This enforces a certain structure on the intensity surface itself.
  In [10] we have characterized the structure of the intensity surface as either a paraboloidal structure or any derivable strongly monotonically increasing transformation of a paraboloidal structure (Fig. 1).

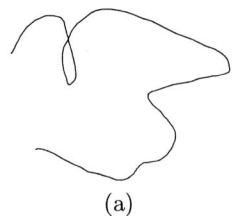 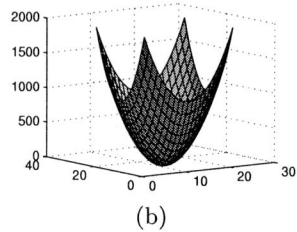

(a)   (b)

Fig. 2. 3D vs. 2D convexity. *(a)* 2D convexity: A contour is a 1D surface in $\mathbb{R}^2$. *(b)* 3D convexity: A paraboloid is a 2D surface in $\mathbb{R}^3$.

Since paraboloids are arbitrarily curved surfaces, they can be used as a local approximation of 3D convex or concave surfaces (Recall, that our input is discrete, and the continuous functions are only an approximation!). The detected intensity surface patches are therefore those exhibiting 3D convex or concave structure. The convexity is three dimensional, because this is the convexity of the intensity surface $I(x, y)$ (= 2D surface in $\mathbb{R}^3$; Fig. 2(b)), and *not* convexity of contours (= 1D surface in $\mathbb{R}^2$; Fig. 2(a)). This 3D convexity of the intensity surface is characteristic of intensity surfaces emanating from smooth 3D convex bodies.

– **Summary**
We detect the zero-crossings of the gradient argument by detecting the infinite response of Y_{arg} at the negative x-axis (of the arctan). These zero-crossings occur where the intensity surface is 3D convex or concave. Convex smooth 3D objects usually produce 3D convex intensity surfaces. Thus, detection of the infinite responses of Y_{arg} results in detection of domains of the intensity surface which characterize 3D smooth convex or concave subjects.

3 Camouflage Breaking

The robustness of the operator under various conditions (illumination, scale, orientation, texture) has been thoroughly studied in [10]. As a result, the smoothness condition of the detected 3D convex objects can be relaxed. In this paper, we further increase the robustness demands from the operator by introducing very strong camouflage.

3.1 Biological Evidence for Camouflage Breaking by Convexity Detection

Next, we exhibit evidence of biological camouflage breaking based on detection of the convexity of the intensity function. This matches our idea of camouflage breaking by direct convexity estimation (using D_{arg}). We bring further evidence, that not only can intensity convexity be used to break camouflage, but also there are animals whose coloring is suited to prevent this specific kind of camouflage breaking.

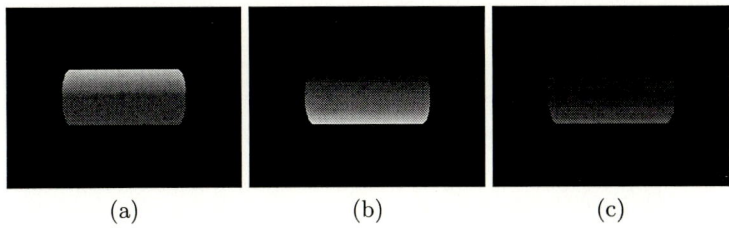

Fig. 3. Thayer's principle of counter shading. **(a)** A cylinder of constant albedo under top lighting. **(b)** A counter-shaded cylinder under ambient lighting (produced by mapping a convex texture). **(c)** Thayer's principle: the combined effect of counter-shading albedo and top lighting breaks up the shadow effect (= convex intensity function).

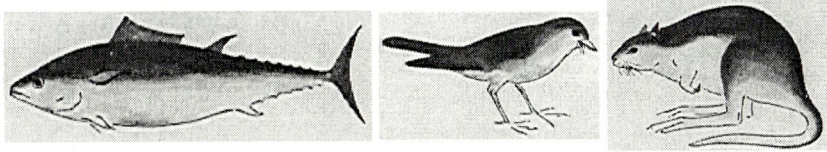

Fig. 4. Thayer's principle. The upper part of the animal is the darker one; transition from the dark part to the bright part is obtained by a gradual change of albedo. When the animal would be in sun light, this coloration would break the convexity of the intensity function.

It is well known that under directional light, a smooth three dimensional convex object produces a convex intensity function. The biological meaning is that when the trunk of an animal (the convex subject) is exposed to top lighting (sun), a viewer sees shades (convex intensity function). As we shall see, these shades may reveal the animal, especially in surroundings which break up shadows (e.g., woods) (see [8]). This biological evidence supports D_{arg} approach of camouflage breaking by detecting the convexity of the intensity function.

The ability to trace an animal based on these shadow effects has led, during thousands of years of evolution, to coloration of animals that dissolves the shadow effects. This counter-shading coloration was first observed at the beginning of the century [12], and is known as Thayer's principle. Portmann [8] describes Thayer's principle: "If we paint a cylinder or sphere in graded tints of gray, the darkest part facing toward the source light, and the lightest away from it, the body's own shade so balances this color scheme that the outlines becomes dissolved. Such graded tints are typical of vertebrates and of many other animals." Figure 3 uses ray tracing to demonstrate Thayer's principle of counter-shading when applied to cylinders. The sketches in Fig. 4, taken from [8], demonstrate how animal coloration changes gradually from dark (the upper part) to bright (the lower part). When the animal is under top lighting (sun light), the gradual change of albedo neutralizes the convexity of the intensity function. Had no counter-shading been used, the intensity function would have been convex (as in Fig. 3(a)), exposing the animal to convexity based detectors (such

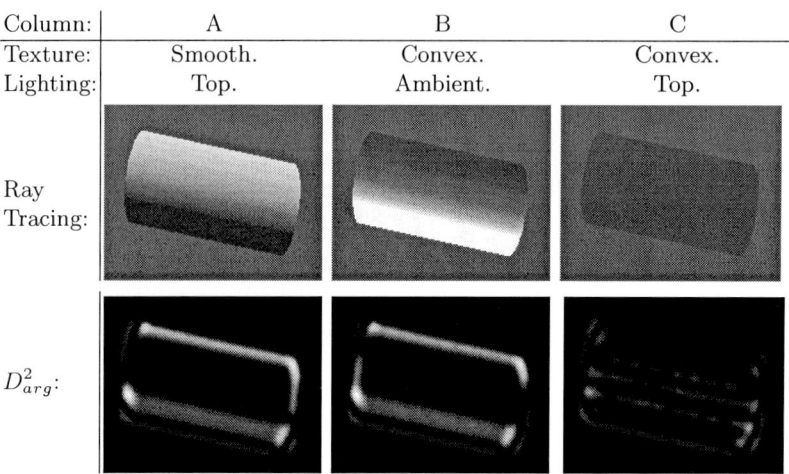

Fig. 5. Operation of D_{arg}^2 on a counter-shaded cylinder. **Column A:** A smooth cylinder under top lighting. **Column B:** The counter-shaded cylinder under ambient lighting. **Column C:** The counter-shaded cylinder under top lighting. The counter-shaded cylinder can barely be noticed under top lighting, due to the camouflage. Under top lighting, the response of D_{arg} is much stronger when the cylinder is smooth than when it is counter-shaded, showing this type of camouflage is effective against D_{arg}.

as D_{arg}). Putting counter-shading into effect neutralizes the convexity of the intensity function thus disabling convexity-based detection.

The existence of counter-measures to convexity based detectors implies that there might exist predators who can use convexity based detectors similar to D_{arg}.

3.2 Thayer's Counter-Shading Against D_{arg}-Based Detection

Let us demonstrate how Thayer's principle of counter-shading can be used to camouflage against D_{arg}-based detectors. In Fig. 5 we once again consider a synthetic cylinder; this time we operate D_{arg} on each of the images of that cylinder. As can be seen, the counter-shaded cylinder under top lighting (Fig. 5, Column C) attains much lower D_{arg} values than the smooth cylinder under the same lighting (Fig. 5, Column A). This is because counter-shading turns the intensity function from convex to (approximately) planar.

To see the transition from a convex intensity function to a planar one due to camouflage, we draw (Fig. 6-left) the vertical cross-sections of the intensity functions of Fig. 5. The smooth cylinder under top lighting (Column A) produces a convex cross-section. The albedo, or the counter-shaded cylinder under ambient lighting (Column B), consists of graded tints of gray (i.e, convex counter-shading). Finally, the counter-shaded cylinder under top lighting (Column C) produces a flat intensity function, which means a lower probability of detection by D_{arg}.

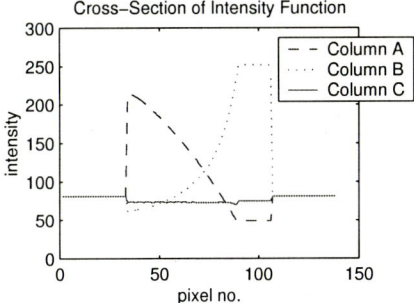 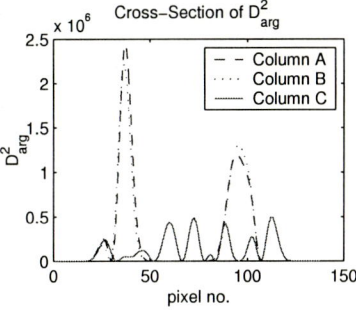

Fig. 6. Cross-sections (parallel to the y-axis, at the center of the image) of: **Left:** The intensity functions of Fig. 5. Thayer's counter-shading yields a flat intensity function for a cylinder. **Right:** D_{arg}^2 of Fig. 5. Under top lighting, the flattened intensity function of the counter-shaded cylinder has a lower D_{arg} response than that of the convex intensity function of the smooth cylinder.

We verify that the flat intensity function is indeed harder to detect using D_{arg} than the convex intensity function: we show that D_{arg} has a lower response to the counter-shaded cylinder under top lighting than it has to the smooth cylinder under the same lighting. This is obvious from Fig. 6-right which shows the vertical cross-sections of the responses of D_{arg} to the various images of the cylinder.

The above demonstrates that Thayer's principle of counter-shading is an effective biological camouflage technique against convexity-based camouflage breakers, and more specifically, against D_{arg}. One can thus speculate that convexity-based camouflage breaking might also exist in nature (or else, the camouflage against it would be unnecessary).

4 Experimental Results

In this section we juxtapose the D_{arg} operator with a typical edge-based operator —the radial symmetry transform [9]— as camouflage breakers. This operator seeks generalized symmetry, and has been shown there to generalize several edge-based operators. We compare D_{arg} with edge-based methods, since camouflage by super-excitation of a predator's edge detectors is evident in the animal kingdom [7] (which implies edges are biologically used for camouflage breaking).

The radial symmetry operator is scale-dependent, while the peaks of D_{arg} are not. Therefore, we have compared D_{arg} with radial symmetry of radii: 10 and 30 pixels (i.e, 2 radial symmetry transformations performed for each original image). In the paper, only one radius is introduced per original, but similar results were obtained for the other radius as well.

Apatetic Coloration in Animal Animals use various types of camouflage to hide themselves, one of which is apatetic coloration. Fig. 7 exhibits a natural

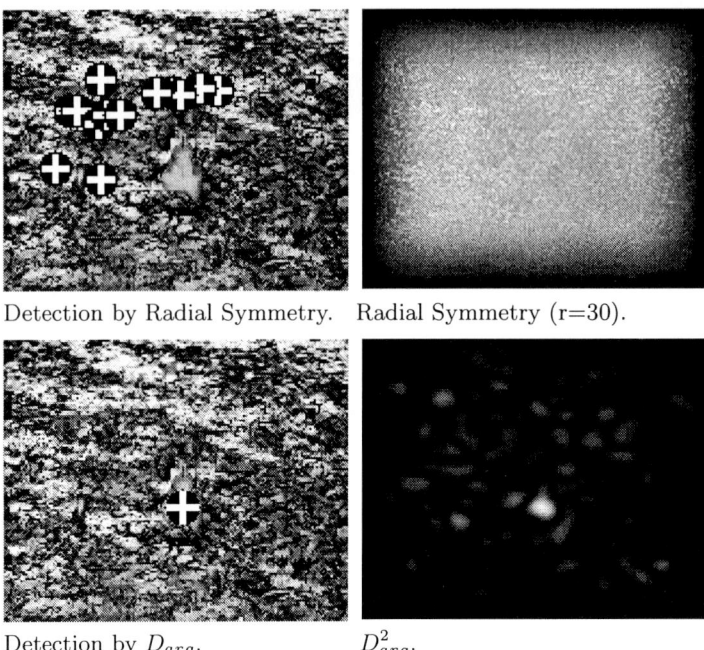

Fig. 7. A hidden squirrel. The squirrel is on a leafy ground shaded by a tree. The shades and leaves form many edges "deluding" edge-based methods. Even human viewers find it difficult to locate the squirrel in the image. D_{arg} detects the squirrel, breaking the camouflage.

camouflage of a squirrel in a leafy environment under the shades of a nearby tree. The camouflaged fur has many edges which mix with the environment, preventing the radial symmetry operator from isolating any specific target. D_{arg}, however, produces a single strong peak, exactly on the squirrel. The convexity of the squirrel (and in particular, its belly) is the reason for its detection by D_{arg}. The only smooth 3D convex region in the image is the belly of the squirrel. Though some of the shades might look similar to the belly of the squirrel (even to a human viewer), they do not possess the property of being a projection of a 3D convex object, so their graylevels introduce no 3D convexity.

Another example of camouflage by apatetic coloration is Fig. 8. The figure shows two Rocky Mountain sheep (*Ovis Canadensis*) in their natural rocky environment. The coloration of the Rocky Mountain sheep fits their habitat (pay attention in particular to the upper sheep in Fig. 8). D_{arg} detects the sheep as the main subject, since they appear smooth (from the photographic distance), and are three dimensional and convex. Due to the apatetic coloring, the rocky background produces much stronger edges than the sheep, thus attracting edge-based methods. Radial symmetry specifies no single target.

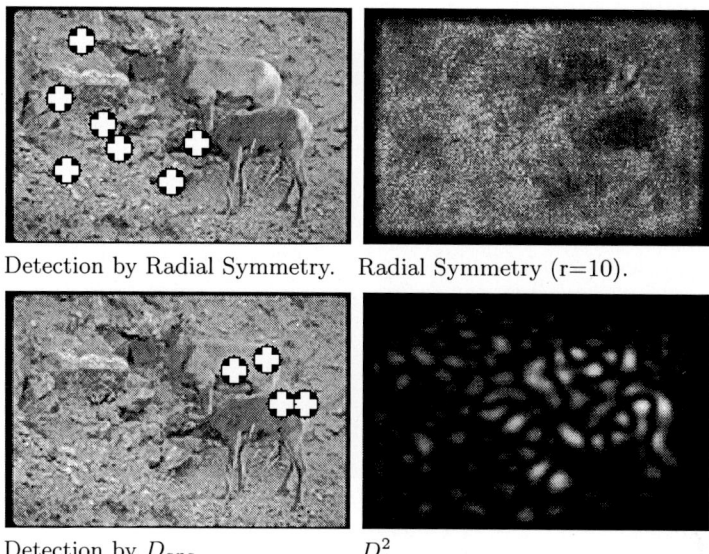

Detection by Radial Symmetry. Radial Symmetry (r=10).

Detection by D_{arg}. D_{arg}^2.

Fig. 8. Rocky mountain sheep in their rocky habitat. Edge based methods fail to detect the sheep due to its apatetic coloration. D_{arg}, however, highly responds to the convexity of the intensity function.

Military Camouflage Breaking camouflage of concealed equipment is of particular interest. Figure 9 presents a tank in camouflage paints in front of a tree. The tree produces edges distracting the radial symmetry operator. The convexity of the intensity function near the wheels of the tank exposes the tank to the D_{arg} detector.

An example of breaking clothes camouflage appears in Fig. 10. Camouflage clothes on the background of dense bushes and a river makes the subject very hard to detect. Edge based detection misses the subject. D_{arg} isolates the camouflaged subject, but with one outlier (out of three detections).

5 Conclusions

Thayer's principle states that various animals use counter-shading as a major basis for camouflage. The observation of such a counter-measure in animals implies that other animals might use convexity detection to break camouflage (or otherwise there was no need for the counter-measure). We illustrate how Thayer's counter-shading prevents detection based on the convexity of the graylevel function (D_{arg}). The effectiveness of camouflage breaking by convexity detection (for subjects which are not counter-shaded according to Thayer's principle) is demonstrated using D_{arg}. The operator D_{arg} is basically intended for detection of image domains emanating from smooth convex or concave 3D objects, but the smoothness assumption can be relaxed. Finally, a comparison between the

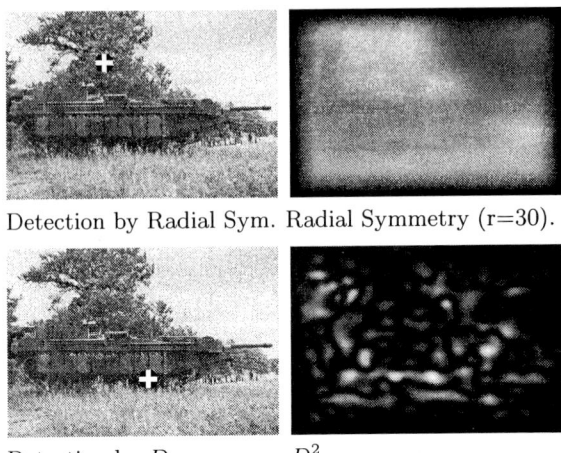

Detection by Radial Sym. Radial Symmetry (r=30).

Detection by D_{arg}. D^2_{arg}.

Fig. 9. A tank in camouflage paints. Convexity-based detection using D_{arg} is *not* distracted by the background (tree, grass). The tree distracts edge-based detection.

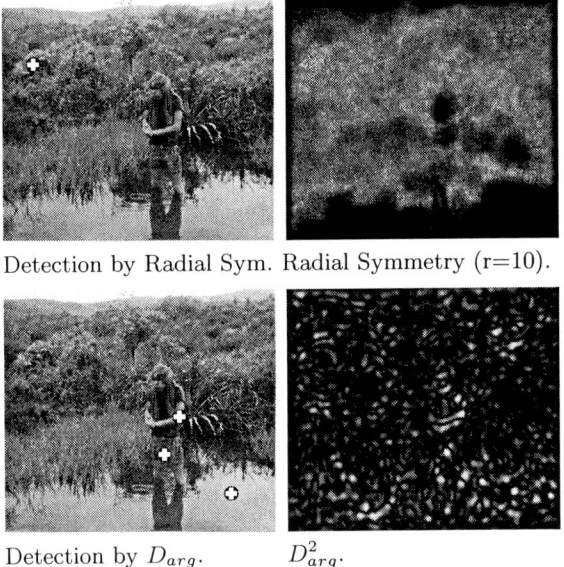

Detection by Radial Sym. Radial Symmetry (r=10).

Detection by D_{arg}. D^2_{arg}.

Fig. 10. A camouflaged soldier on the bank of a river near dense bushes. The edge distribution of the soldier unites with the camouflage (especially the helmet). In spite of the strong camouflage, D_{arg} detects the soldier in 2 out of 3 locations it isolates. The incorrect detection is of clouds reflecting in the water.

convexity-based camouflage breaker (D_{arg}) and an edge-based operator (radial symmetry) has been delineated. Convexity-based camouflage breaking was found to be highly robust and in many cases much more effective than edge-based techniques.

References

1. A. C. Copeland and M. M. Trivedi. Models and metrics for signature strength evaluation of camouflaged targets. *Proceedings of the SPIE*, 3070:194–199, 1997.
2. F. M. Gretzmacher, G. S. Ruppert, and S. Nyberg. Camouflage assessment considering human perception data. *Proceedings of the SPIE*, 3375:58–67, 1998.
3. Song Guilan and Tang Shunqing. Method for spectral pattern recognition of color camouflage. *Optical Engineering*, 36(6):1779–1781, June 1997.
4. Lu Huimin, Wang Xiuchun, Liu Shouzhong, Shi Meide, and Guo Aike. The possible mechanisms underlying visual anti-camouflage: a model and its real-time simulation. *IEEE Trans. on Systems, Man & Cybernetics, Part A (Systems & Humans)*, 29(3):314–318, May 1999.
5. S. Marouani, A. Huertas, and G. Medioni. Model-based aircraft recognition in perspective aerial imagery. In *Proc. of the Intl. Symposium on Comp. Vision*, pages 371–376, USA, 1995.
6. S. P. McKee, S. N. J. Watamaniuk, J. M. Harris, H. S. Smallman, and D. G. Taylor. Is stereopsis effective in breaking camouflage? *Vision Research*, 37:2047–2055, 1997.
7. D. Osorio and M. V. Srinivasan. Camouflage by edge enhancement in animal coloraion patterns and its implications for visual mechanisms. *Proceedings of the Royal Society of London B*, 244:81–85, 1991.
8. Adolf Portmann. *Animal Camouflage*, pages 30–35. The University of Michigan Press, 1959.
9. Daniel Roiofeld, Haim Wolfson, and Yehezkel Yeshurun. Context free attentional operators: the generalized symmetry transform. *Intl. Journal of Computer Vision*, pages 119–130, 1995.
10. Ariel Tankus, Yehezkel Yeshurun, and Nathan Intrator. Face detection by direct convexity estimation. *Pattern Recognition Letters*, 18:913–922, 1997.
11. I. V. Ternovskiy and T. Jannson. Mapping-singularities-based motion estimation. *Proceedings of the SPIE*, 3173:317–321, 1997.
12. Abbott H. Thayer. An arraignment of the theories of mimicry and warning colours. *Popular Science Monthly, N.Y.*, pages 550–570, 1909.

Development of a Biologically Inspired Real-Time Visual Attention System

Olivier Stasse, Yasuo Kuniyoshi, and Gordon Cheng

Humanoid Interaction Laboratory,
Intelligent Systems Division,
Electrotechnical Laboratory,
1-1-4 Umezono Tsukuba Ibaraki 305-8568, Japan
{stasse,kuniyosh,gordon}@etl.go.jp,
http://www.etl.go.jp/~{stasse,kuniyosh,gordon}

Abstract. The aim of this paper is to present our attempt in creating a visual system for a humanoid robot, which can intervene in non-specific tasks in real-time. Due to the generic aspects of our goal, our models are based around human architecture. Such approaches have usually been contradictory, with the efficient implementation of real systems and its demanding computational cost. We show that by using PredN[1], a system for developing distributed real-time robotic applications, we are able to build a real-time scalable visual attention system. It is easy to change the structure of the system, or the hardware in order to investigate new models. In our presentation, we will also present our system with a number of human visual attributes, such as: log-polar retino-cortical mapping, banks of oriented filters providing a generic signature of any object in an image. Additionally, a visual attention mechanism — a psychophysical model — FeatureGate, is used in eliciting a fixation point. The system runs at frame rate, allowing interaction of same time scale as humans.

1 Introduction

This paper describes our attempt to build a biologically inspired visual system. We want to explore concepts created by psychologists, neurophysiologists, and biologists. By creating a humanoid interacting with the real-world, in order to solve new robotic problems, and understand which concepts lead to complex and interesting behaviors. In this context, we have tried to build a visual system, see Figure 1 which is designed for interacting in open situations, i.e. where a task performed by a system is not operationally defined, but is the result of the complex interaction among the parts of the system, and the environment.

One of the main difficulties in trying to build such systems is to deal with the time-scale problem. Our current interaction experiments are with humans, and the system is based on human studies. We believe that the real-time aspect is crucial in the analysis of the emergent behaviors between the robot and the subject. Due to the computational complexity of the processing, trying to deal

---

[1] **P**arallel **R**eal time **E**vent and **D**ata driven **N**etwork

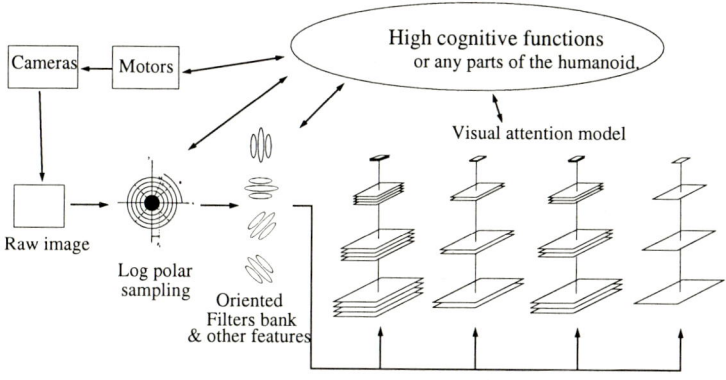

Fig. 1. Structure of our initial visual system

with this aspect in vision is difficult. Finally a main key point is the integration of this system in a complex robotic project. Our humanoid is intended to be a real-world experimental platform for the investigation of biologically inspired models, in any regard, some technical problems must be solved. From a software point of view, problems such as structural modification of a complex system, modification of hardware, as well as updating parts of a system, must be made as easy as possible.

We aimed to build a visual system which integrates an attention system with multiple cues, that is non dedicated to any specific task, and is able to perform active vision tasks. In its current form, the system embraces a number of human visual attributes, such as: log-polar retino-cortical mapping, banks of oriented filters providing a generic signature of any object in an image. Additionally, a visual attention mechanism — a psychophysical model — FeatureGate, is used in eliciting a fixation point. The system runs at frame rate, allowing interaction of the same time scale as humans.

Section 2 describes past works, upon which we build our current visual attention system. Section 3 presents the biological components of our generic visual attention system, which can be integrated into a more complex robotic system. Section 4 describes PredN, especially developed for dealing with the pure software engineering aspect, and the experimental results of our implementation.

2 Previous Work

Numerous visual attention models have been presented in the past, see [1] for a review on psychological, and computer vision models, and [2] for a neurophysiological review. Due to their nature such kinds of models are used typically to predict and mimic off-line biological systems. When visual attention models are proposed for robotic systems [3], and [4], technical problems like time constraints, or computational power reduced the number of features and the complexity of the visual attention problem. Few systems have been proposed to handle both real-time constraints in robotic system and strong biological grounds. The system presented in [5] by Rao and Ballard is one of them. The visual attention model is based on multi scale derivatives of Gaussian filters. Using the steerable properties of such filters [6], the authors have shown that the overall output provides a general signature of a point, robust to size, rotation and view point modifications. In order to tackle the real-time problem, the authors used algorithms adapted to their powerful hardware dedicated to vision processing, $DataCube^{TM}$ $MaxVideo\ MV200$. The increasing power of computers give us now the possibility to consider other technical solution to solve similar problems, and extend the range of algorithms which can be used. Being a test-bed for several experiments, our robot is designed to handle different tasks. Obviously vision is an important cue, and so its integration with the rest of the system must be as open as possible. This excludes hardware which could over-reduce the range of possible algorithms. Another system presented in [7] has real time performance. However this system focus on gaze fixation, and for this reason uses only symmetry as input feature. Also several subtle modifications are introduced, their implementation needs a network of eleven C40 DSPs for one camera. Others robotics system have been construct such [8] which focus on object recognition. Unfortunately this paper does not give results about computational time.

We proposed in this paper the implementation of a general purpose visual attention system. It is strongly biologically inspired, open to structural software modifications, and is able to maintain real-time performances. This is achieved by using off-the-shelf hardware and a software architecture developed at our laboratory.

3 Biological Foundation

3.1 Log-polar Mapping

The log-polar mapping was done according to the scheme given in [9], see Figure 2. In this expression, ρ_0 corresponds to the size of the fovea where we keep a normal image with full resolution. Once a coordinate mapping has been chosen, there are several ways to compute the intensity from the Cartesian image to the cortical plane, see [10] for a review. Biologically, it has been reported that the photoreceptors' spatial integration of the light can be modelized by a Gaussian [11], whose scale parameter is an increasing function of eccentricity. Therefore we choose an overlapping circular receptive field model. Each receptive field is a Gaussian kernel, for which the center follows the scheme depicted in Figure 2, and with a size which increases according to the distance to the center of the image. Figure 2.e shows a cross through log-polar sampling in the Cartesian plane, Figure 2.f shows the same image in the cortical plane. Further to the biological foundation established, we have other reasons in adopting such formulation. Indeed numerous image processing tools have been designed for log-polar representation. However, a large part of them hold only because of the conformal mapping assumption. A complex mapping is said to be conformal if

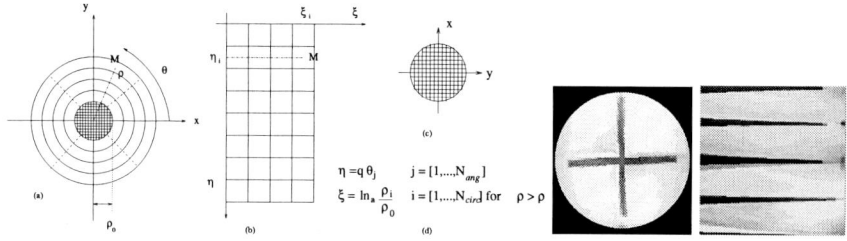

Fig. 2. A point M in the three coordinate systems: Cartesian (x,y), polar (ρ,θ), and log-polar (η,ξ). From [9] (a) for the initial image, (b) for the cortical plane, (c) for the fovea, (d) Discrete formulation (e) **Retinal image**. An image of a cross through log-polar sampling. (f)**Cortical image**. The periphery in the cortical plane.

it has the property of preserving angles. The previous formulation is conformal, but has a singular point at zero. At some point, the radius will become smaller than one, and even if we use sub-pixel precision methods, the singularity will still hold, and a part of the center must be removed. In [12], Schwartz proposed another formulation by adding a constant inside the logarithmic function. The major drawback, is a discontinuity at the center of the image, which needs special treatment during processing.

3.2 Feature Maps

As discussed, the feature maps used in this visual attention system are based on derivatives of Gaussian and includes also optical flows, see Figure 4.b. In [5], Rao and Ballard showed that this is similar to a PCA analysis, or a piecewise decomposition of the local structure of the image. In [13], they showed that a neural network which can reconstruct the input data under a constraint of minimal length coding, converged to similar functions. We would like to use the same scheme both in the fovea and in the periphery. For the fovea, it can be also used to compute the orientation [6]. Our future plan is to use the scaling effect to raise the dimensionality of the signature without the computational cost involved by a pyramid like concept as done in [5]. Also [5] the authors presented some experiences with a log-polar transformation, it is unclear if the scheme has been applied on the retina image and in the Cartesian coordinates, or on the cortical image in the log-polar references. Because the Gaussian derivatives are similar to a PCA, and because the mapping is conformal, we believe that the signature will be still usable in this configuration. However, this assumption needs further investigation.

3.3 FeatureGate

The FeatureGate model has been originally created by Cave [14], and used for the Vanderbilt humanoid project in [15]. This network models a visual attention system based on the spotlight paradigm with top-down and bottom-up integration. Its structure is representative of the psychological approach of visual attention, in the sense that it is possible to define implementation which are equivalent to other models [16], or modify it to integrate subtle refinement such as the ones proposed in [1]. An underlying concept is that selective attention is related to the computational resource allocation of visual systems. In our case, this is a major criterion for the investigation of biological and psychological models for visual attention. At this time, the choice of this model has been mainly driven by computational time constraints, and its structure makes it suitable for distributed implementation. However, we must stress that this is not a definite choice, and we intend to implement other models for example [17].

The structure of FeatureGate. FeatureGate is composed of feature modules and one activation module. A module, indexed by m, contains N levels, each level contains D_m maps. For example, a feature module can be the color, and the maps are its three channels: red, green and blue. The lowest levels contain the data of the features, the others contain the results of a local competitive algorithm. The latest level contains the position of the attention point and the values of the features associated. The local competitive algorithm at each location (x, y) of the image is computed in a 8 point neighborhood.

Two kinds of functions influence the election process. The first function called *salience function*, and noted by $\Phi_B\{U_{m,l}\}(x,y)$, where $U_{m,l}(x,y)$ represents the values contained in the maps of the module m at the level l at the location (x,y), compute the saliency of the information. The second function called *discrimination function*, and noted by $\Phi_T\{U_{m,l}\}(x,y)$, computes the proximity, for the current feature, to a target noted by $f = \{f_m(x,y) | m \in \{1, 2, \ldots, M\}, |f_m(x,y)| = D_m\}$. The election process is done based on the computation of the activation. The activation map is computed by adding the *salience function* and the *discrimination function* for each module together.

Each layer of the activation module is divided in square areas of size a. From this area, the location with the highest activation value is found. Then the corresponding features in all the modules are propagated to the upper layer, and the location of the winner in the initial reference is stored.

Implementation of FeatureGate. In [15] the author proposed the following functions, and emphasized the fact that this is not specific to the model, but was proposed for implementation purposes. First, a metric is defined to compare different values of features, in this case the Euclidean distance, $\rho()$ is used. The bottom-up participation to the activation is computed as follows:

$$\Phi_B\{U_{m,l}\}(x_0, y_0) = \alpha_m \sum_{\overline{\mathcal{N}}_{m,l}(x_0, y_0)} \rho[U_{m,l}(x,y), U_{m,l}(x_0, y_0)], \qquad (1)$$

where $U_{m,l}^k(x,y)$ is the value of the feature m in the map k at the position (x,y) of the layer l, $\overline{\mathcal{N}}_{m,l}(x_0, y_0)$ is the deleted neighborhood of (x_0, y_0), and α_m a positive parameter. The top-down computation is done by comparing the neighborhood of (x_0, y_0) to the feature target f_m. A subset $\mathcal{S}(x_0, y_0) \subseteq \overline{\mathcal{N}}_{m,l}(x_0, y_0)$ is defined as follows:

$$\mathcal{S}(x_0,y_0) = \{U_{m,l}(x,y) \in \overline{\mathcal{N}}_{m,l}(x_0,y_0) | \rho[U_{m,l}(x,y), f_m] < \rho[U_{m,l}(x_0,y_0), f_m]\} \quad (2)$$

We proposed here a version slightly different from [15], for the computation of the top-down activation:

$$\Phi_T\{U_{m,l}\}(x_0, y_0) = -\beta_m \quad \rho[U_{m,l}(x_0, y_0), f_m] - \frac{\beta_m}{\|\mathcal{S}(x_0, y_0)\|} \times \sum_{(x,y) \in \mathcal{S}(x_0, y_0)} \left\{ \rho[U_{m,l}(x_0, y_0), f_m] - \rho[U_{m,l}(x,y), f_m] \right\} \quad (3)$$

where β_m is a positive number. Because they are only negative terms, this function acts only as an inhibition. Indeed the distance of the feature to the target is a penalty. Moreover, if some neighborhood points are closer to the target than the central point, a penalty is given. We added a normalization factor to favor the case where an area of points is close to the target. In that case the set $\mathcal{S}(x_0, y_0)$ is bigger, but the distances are smaller. Since we want to favor such a case, rather than a point that is far away from the target, and with a neighborhood different from the target. It provides a blob-like effect.

It can be noticed that the main way for an external process to influence the result of the computation is to modify the (α_m, β_m) couple. In the initial model those parameters were scalar. To adapt the scheme to our distributed implementation and provide more flexibility, we defined them as vectors.

From an experimental point of view, a major advantage of FeatureGate is the fact it does not over specify in the definition of the features, the distance function, and the bottom-up and top-down functions. This gives the possibility to try different implementations according to the hardware and to the models. This model supports incremental integration of additional features, and distributed computation of the activation function. However, a bottleneck lies in the synchronization of these activations, and the propagation of the result to the layers, in order to update the next stage of FeatureGate. The computation of the distance to a target is also an open issue, the topology of the features space plays a crucial role in the visual attention model. For example, it is well known that the human color perception in the RGB space is non-linear. Finally, the cooperation-competition scheme must be flexible enough to try different configurations (difference of Gaussian, fully connected, one-to-one, module by module, binary association, and so on).

4 Practical Implementation

4.1 PredN – Parallel Real Time Event and Data Driven Network

PredN is a software system developed by our laboratory presented in [18]. One of the main goal has been to make as simple as possible the design of complex

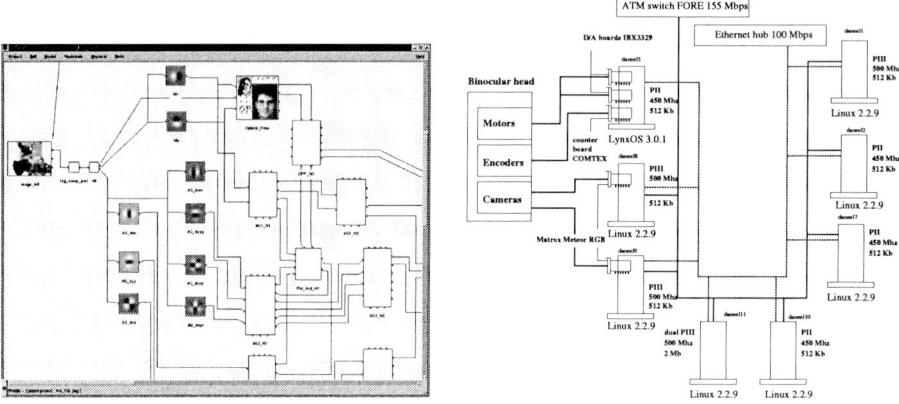

Fig. 3. (a)**XPredN**: Implementation of FeatureGate (b) Our current physical architecture .

robotic application. The main points of PredN are a computation model based on multi-thread architecture, a event driven scheduling, a general modelization of a physical structure, and an integration of code reusability concepts as design patterns. This tool gives us the possibility to separate real-time programming techniques and user code with a minimal loss of performance. It helps to solve problems such platform migration, and structural software modification.

A Graphical User Interface has been designed in order to help the manipulation of structures, codes, and configuration files. The graphical display of the application in Figure3 is similar to the structure of the visual system depicted in Figure1.

The user code is integrated in Nodes like the one on which a camera is represented. This Node is periodic and transmits the image to two others Nodes. One extracts the fovea from the raw image, while the other constructs the log-polar sampling of the periphery. In Figure 3 the log-polar sampling is followed by a temporal filter, providing two information: the filtered image, and its temporal derivative. The former one is used to feed the Gaussian derivative filters, when the temporal derivative is sent to the optical flow node. Optical flow has been synchronized to wait for the information arriving from the first Gaussian derivative filters, and the temporal filter. The outputs of all these features are integrated in the distributed implementation of FeatureGate, which is explained in the next section.

4.2 Physical Platform

For processing we are using a cluster of 7 PCs, connected by ATM (155 Mbps) and Ethernet (100 Maps). For low-level control, it is handled by a computer running the real-time OS, $LynxOS^{TM}$. All the other nodes in our PCs cluster are running the Linux OS, with an extension for ATM support. The topology of this configuration is depicted in Figure 3.b.

Here, we exploit the structure of PredN, by providing only communication and run-time specific libraries for each particular platform. In this way we can easily change the location of a Node, for instance to another machine, without breaking the overall structure of the application. The current implementation of the log-polar transformation takes $23ms$ to perform each projection for a ratio of 10 between the inner receptive fields' size and the outer receptive fields' size.

4.3 Feature Maps

The feature maps for orientation are computed with a bank of 9 filters. They are 5×5 Gaussian derivative convolution masks. The derivative orders are first, second and third see Figure 4.a. We have used the flow capability of the Pentium III^{TM}, by using an implementation in SSE code in order to speed up the computation. A 64×64 image is computed in $4ms$.

For the temporal filter and the optical flow we have used the algorithms developed in [19]. Both of these have also been developed in SSE code. The algorithm to compute the optical flow is initially for Cartesian image, and it is coherent for the fovea. We use it as well in the cortical plane, to gain an approximation of the motion, and in order to simulate the cost of the computation of optical flow specific to a log-polar scheme. Our future work will be to implement an appropriate algorithm such as [20]. The current algorithm applied to the fovea can be seen in Figure 4.b.

4.4 FeatureGate

FeatureGate has been implemented by creating a Node for each feature module layer. One of the advantages is that if a feature module doesn't have the same size as the others, it can also be managed at the activation module level. For example, in Figure 3 the Gaussian filters $dG1\_N1$, $dG2\_N1$ and $dG3\_N1$ respectively are the Gaussian derivatives of first, second and third order in the first layer of FeatureGate, providing vectors with a size of 60×60. The vectors provided by the optical flow feature module layer OPF_N1 have a size of 56×56. By letting the activation module layer, make the coordinates projection from one map to the other, adding or removing a feature will limit the modification of the structure of the application. The activation Node synchronizes all the outputs of the module layers of one level. When all the activations are computed by each module, the activation module elects for each receptive field the winner, and creates a "winner" vector. This vector is sent back to each module to propagate the vector features of the points elected. Therefore each feature module layer sends the vector features for a number of points divided by the square of the undersample. Each Node can have parameters α_m and β_m which can be changed externally. An input port accepts the coordinates of a point in the module feature reference, and updates an output port. The scheduling function of the asking Node must be true when this port is updated, in order to allow the asking Node to wake-up.

The undersample has been fixed to 10, $\beta_m = \alpha_m = 1$, which gives a Feature-Gate implementation of 2 levels. The timing for the first level is the following: first Gaussian derivative, $4ms$, second derivative, $10ms$, third derivative $14ms$, optical flows $3.8ms$. Because the undersample is 10, results for the second level are negligible. In this implementation, the α_m and β_m parameter are specifics

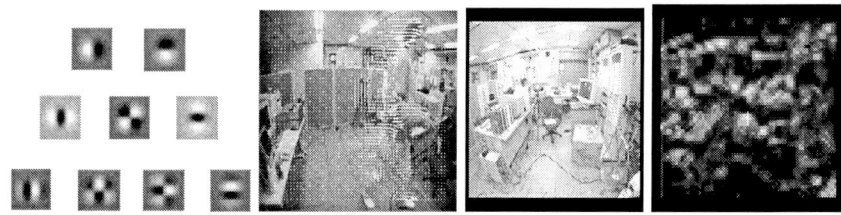

Fig. 4. (a)**Gaussian maps**, (b)**Optical Flow**: 60 × 60, (c)**Gaze fixation point**, (d)**Activation map from the image (c)**

for each module. This makes it possible to weigh the features which participate in the election process. The Figure 4.d gives an example where only the module which integrates the second derivative of gaussian influences the bottom-up process. It is possible to use the selection point as a gaze fixation point, or to use the map provided by one of the modules as a mask. The map's meaning depends on the value of the couple (α_m, β_m) in one module. Therefore this distributed form makes possible a parallel use of those modules by different visual processes. The only explicit constraint of FeatureGate, is the synchronization module which elects a set of data to be processed by the next level.

5 Conclusion

In the context of our humanoid project, we have built a visual attention system which proposes a rich set of visual features, integrates both bottom-up and top-down, is tunable according to the task, practically easy to extend, and is real-time. Also visual attention deals also with spatial memory, object memory, and motor control we believe that it needs further consideration which is beyond this paper, and are under investigation in our laboratory. This is strongly related to our future work which will focus on the interaction between location and image signature through visual attention, this is in order to solve the inhibition problem.

We gratefully acknowledge the COE program of STA and the Brain Science Project of STA for partial funding of this work. We would like to thank Sbastien Rougeaux who developed the SSE code, and David Peeters who tested the FeatureGate model. And also we are grateful to Katrina Cheng for proof reading draft versions of this paper.

References

1. Ruggero Milanese, *Detecting salient regions in an image: from biological evidence to computer implementation*, Ph.D. thesis, Departement of Computer Science, University of Geneva, Switzerland, 1993.
2. Colby, "The neuroanatomy and neurophysiology of attention," *Journal of Child Neurology*, vol. 6, pp. 90–118, 1991.
3. Westelius C.J., *Focus of Attention and Gaze Control for Robot Vision*, Ph.D. thesis, Linkoping University, 1995.
4. Atsuto Maki, *Stereo Vision in Attentive Scene Analysis*, Ph.D. thesis, KTH Computational Vision and Active Perception Laboratory (CVAP), 1996.
5. Rajesh P.N. Rao and Dana H. Ballard, "An active vision architecture based on iconic representations," *Artificial Intelligence Journal*, vol. 78, pp. 461–505, 1995.
6. William T. Freeman, "The design and use of steerable filters," *IEEE Transactions on Pattern Analysis and Machine Intelligence*, vol. 13(9), pp. 891–906, September 1991.
7. Gal Sela and Martin D. Levine, "Real-time attention for robotic vision," *Real-time Imaging*, vol. 3, pp. 173–194, 1997.
8. Barbel Mertsching and Maik Bollmann, "Visual attention and gaze control for an active vision system," *Progress in connectionnist-based information systems. Springer*, vol. 1, pp. 76–79, 1997.
9. C. Capuro, F. Panerai, and G. Sandini, "Dynamic vergence using log-polar images," *International Journal of Computer Vision*, vol. 24, pp. 79–94, 1997.
10. Marc Bolduc and Martin D. Levine, "A review of biologically motivated space-variant data reduction models for robotic vision," *Computer Vision and image understanding*, vol. 69, pp. 170–184, 1998.
11. C. Braccini, G. Gambarella, G. Sandini, and V. Tagliasco, "A model of the early stages of the human visual system: Functional and topological transformations performed in the peripheral visual field," *Biological Cybernetics*, vol. 44, pp. 47–58, 1982.
12. E. L. Schwartz, "Computational anatomy and functional architecture of the striate cortex," *Vision Research*, vol. 20, pp. 645–669, 1980.
13. Rajesh P. N. Rao and Dana H. Ballard, "Efficient encoding of natural time varying images produces oriented space-time receptive fields (technical report)," Tech. Rep., National Resource Laboratory for the Study of Brain and Behavior, Department of Computer Science, University of Rochester, August 1997.
14. K. R. Cave, "The featuregate model of visual selection," *(in review)*, 1998.
15. J. A. Driscoll, R. A. Peters II, and K. R. Cave, "A visual attention network for a humanoid robot," in *IROS 98*, 1998.
16. Jeremy M. Wolfe, "Visual search: A review," *Visual Search, Attention*, 1996.
17. Zhihua Wu and Aike Guo, "Selective visual attention in a neurocomputational model of phase oscillators," *Biological cybernetics*, vol. 80, pp. 205–214, 1999.
18. Olivier Stasse and Yasuo Kuniyoshi, "Predn: Achieving efficiency and code reusability in a programming system for complex robotic applications," *Submitted to the International Conference on Robotics and Automation*, 2000.
19. S. Rougeaux and Y. Kuniyoshi, "Robust real-time tracking on an active vision head," in *Proc. Int. Conf. on Intelligent Robots and Systems (IROS'97)*, September 1997, 7-11.
20. J. Dias Arajo, C. Paredes, and J. Batista, "Optical normal flow estimation on log-polar images. a solution for real-time binocular vision," *Real-Time Imaging*, vol. 3(3), pp. 213–228, June 1997.

Real-Time Visual Tracking Insensitive to Three-Dimensional Rotation of Objects

Young-Joo Cho[1,2], Bum-Jae You[1], Joonhong Lim[2], and Sang-Rok Oh[1]

[1] Intelligent System Control Research Center, Korea Institute of Sci. & Tech.
P.O. Box 131, Cheongryang, Seoul 130-650, Korea
ybj@amadeus.kist.re.kr

[2] School of Electrical Engineering and Computer Science, Hanyang University,
Sa-1 dong 1271, Ansan, Kyungki-do 425-791, Korea
jhlim@aser.hanyang.ac.kr

Abstract. Visual tracking is essential for many applications such as vision-based control of intelligent robots, surveillance, agriculture automation, medical image processing, and so on. Especially, a fast and reliable visual tracking is important since the performance of visual tracking determines the reliability and real-time characteristics of overall system. It is not easy in visual tracking, however, to estimate the configuration of a target object in real-time when the three-dimensional pose of the target object is changing. On the contrary, a human being is able to track an object without the estimation of three-dimensional pose of the object even though three-dimensional rotations and /or occlusion by other objects change the original image of the object.

This paper proposes a fast and reliable SSD-based visual tracker insensitive to three-dimensional rotation of an object as well as translation, two-dimensional rotation, scaling, and shear of an object by proposing a performance measure for distortion of current image with respect to an original reference image. The performance measure is a combination of aspect ratio of a rectangle, variations of four internal angles of the rectangle from ninety degrees, the direction of rotation and the angular velocity of the rectangle. So, the reference image for visual tracking is updated whenever the performance measure is greater than an initialized distortion rate without the estimation of three-dimensional pose of an object. The algorithm is experimented in real-time successfully at a personal computer adopted with a general-purpose frame grabber.

1 Introduction

Visual tracking is essential for many applications such as vision-based control of intelligent robots, surveillance, agriculture automation, medical image processing, and so on. Especially, a fast and reliable visual tracking is important since the performance of visual tracking determines the reliability and real-time characteristics of overall system. It is not easy in visual tracking, however, to estimate the configuration of a target object in real-time when the three-dimensional pose of the target object is changing while illumination condition is varying and other objects occlude the target object. There have been proposed various approaches for visual tracking divided into two categories, two-dimensional versus three-dimensional visual tracking. For two-dimensional visual tracking, there have been proposed several tracking algorithms based on optical flows, regions of uniform brightness, depth information, sum of squared differences, and color [1][2][3][4][5]. Yamane, Shrai, and Miura[1] proposed a high-speed visual tracker based on a specialized hardware including 15 digital signal processors by using optical flows and regions of uniform brightness. It was applied for visual tracking of a walking man successfully. Also, Shirai, Yamane, and Okada[2] introduced a visual tracking algorithm using optical flows, depth information of feature points, and regions of uniform brightness while the tracker was applied for tracking a couple of people in motion including occlusions between two target objects. A number of digital signal processors, however, are used because of expensive calculation cost. In [3], a snake model-based visual tracker was proposed. It resolved occlusion problems in complex backgrounds even though its computational load is heavy. In [4], Hager and Toyama proposed a SSD(sum of squared differences)-based visual tracker by optimizing a performance index measuring the difference between two images. Parameters for affine transform - such as three-dimensional rotation, scale, shear and translation of a plane - between two contiguous images are estimated in real-time by using only a general- purpose computer system such as a workstation or personal computer. Also, Hager and Belhumeur introduced a visual tracking algorithm immune to illumination variations and occlusions with outliers in [5]. Next, several approaches for three-dimensional visual tracking of an object have been proposed [6][7][8][9][10]. Most algorithms estimate the three-dimensional pose of the object after extraction of several features such as line segments, points, optical flows and projective invariance.

A human being, however, is able to track an object without the estimation of three-dimensional pose of the object even though three-dimensional rotations and /or occlusion by other objects change the original image of the object. That is, we can track a person in crowded environments, such as department stores or bus/train stations, even though his or her front face is not shown to our eyes. It is enough just to see one's face at starting time. Also, one can track easily a person rotating oneself at a fixed place like a ballerina.

In this paper, we propose a fast and reliable SSD-based visual tracker insensitive to three-dimensional rotation of an object as well as translation, two-dimensional rotation, scaling, and shear of an object by proposing a performance measure for distortion of current image with respect to an original reference image. The performance measure is a combination of aspect ratio of a rectangle, variations of four internal angles of the rectangle from ninety degrees, the direction of rotation and the angular velocity of the rectangle. So, the reference image for visual tracking is updated whenever the performance measure is greater than an initialized distortion rate without the estimation of three-dimensional pose of an object. The algorithm is experimented in real-time successfully at a personal computer adopted with a general-purpose frame grabber. This paper is composed of as follows. Chapter 2 reviews the concept of SSD-based high-speed visual tracking proposed in [4] while our proposed visual tracker insensitive to three-dimensional rotation is introduced in Chapter 3. Experimental results for a doll and a cylindrical object are proposed in Chapter 4 and this paper is concluded in Chapter 5.

2 Visual Tracking by SSD (Sum-of-Squared Differences)

In [4], the distortion of a reference image under orthographic projection is described by changes of six parameters. Four parameters are components of a positive definite matrix, $\boldsymbol{A}$, representing rotation, scaling, and shear of an object, and another two parameters are components of a translation vector, $\boldsymbol{d}$ in image plane. The six parameters for a target image are estimated in a very fast manner by minimizing the following object function at each time t under the assumption that prospective effects are minimal.

$$O(\boldsymbol{A};\boldsymbol{d}) = \sum_{x \in \mathcal{W}} \left[I(\boldsymbol{A}x + \boldsymbol{d}, t) - I(x, t_0) \right]^2 w(x) \qquad (1)$$

where $I(\boldsymbol{x}, t)$ means the brightness of a position $\boldsymbol{x} = (x, y)^T$ in an image plane, $\mathcal{W}$ represents a target region in the image plane, and $w(\cdot)$ is an arbitrary positive weighting vector. The following state vector is estimated at each time step by minimizing the object function (1) by least square technique.

$$\boldsymbol{S}_t = (\boldsymbol{A}_t, \boldsymbol{d}_t, \gamma_t) \qquad (2)$$

The state vector is composed of six parameters for matrices $\boldsymbol{A}$ and $\boldsymbol{d}$, and residual value, γ, indicating the similarity between a reference image and a current image in an estimated target region in least square sense. Most previous SSD-based visual tracking methods find a tracking area that has the minimum SSD value and the shortest distance from the previous tracking area by searching an estimated tracking window as shown in [11] and [12]. The searching process is a time-consuming one since template-matching techniques are used to measure the similarity between the reference image and a current image. In [4], however, all parameters in the state vector are estimated by solving the optimization problem directly. So, the computational cost becomes cheap and its fast visual tracking is possible without any specialized processing hardware. The optimization is, however, done under the assumption that the reference image is on a plane and the distortion is limited to a slight affine transform where perspective effects are small. That is, the tracker may lose the moving target object when its three-dimensional rotation is greater than a safe region for visual tracking.

3 Proposed Visual Tracking Algorithm Insensitive to Three-Dimensional Rotation

In order to make the visual tracker adaptive to three-dimensional rotation of an object, we consider an approach to update the reference image for visual tracking

Fig. 1. The change of tracking window by three-dimensional rotations

instead of estimating three-dimensional pose of an object. It is more similar with the eye system of human beings in the sense that one can track an object without the three-dimensional model of the object and it is not necessary to measure the pose of the object. A performance measure to update the reference image is defined as a distortion rate of current image with respect to previous reference image. The distortion rate is measured from the shape of a rectangular tracking window in practice under the assumption that the tracking window is always square in the beginning of visual tracking. The shape of the tracking window is changed as the object makes three-dimensional motions with respect to a camera or the camera makes a motion with respect to the object as shown in Fig. 1. So, we define the following state for SSD-based visual tracking by including the performance measure η_t as shown in equation (3).

$$\boldsymbol{S}_t = (\boldsymbol{A}_t, \boldsymbol{d}_t, \gamma_t, \eta_t) \tag{3}$$

The performance measure, η_t, is determined by the aspect ratio and the distance of each internal angle from ninety degrees of the tracking window as shown in the following equation. And, the direction of rotation of the tracking window is used for determining the location of the updated tracking window.

$$\eta_t = w_1(\sigma_t - 1)^2 + w_2 \sum_{i=1}^{4}(\theta_{i,t} - 90°)^2 \tag{4}$$

where σ_t represents the aspect ratio and $\theta_{i,t}$ means each internal angle of the tracking window in degree while w_1 and w_2 are weighting values. The aspect ratio is not enough to represent the distortion of a square rectangle since the aspect ratio is near 1 when the square rectangle is rotated with respect to $y = x$ axis. The reference image for visual tracking and the location of tracking window is updated when the performance measure, η_t , is greater than a given threshold value which is determined experimentally. That is, the matrix A_t, transition vector d_t and residual value γ_t becomes

$$A_t = A_0, \ d = 0, \ \gamma_t = 0 \qquad (5)$$

where A_0 is the initial matrix for A which means no rotation, no scaling and no shear. And, a rectangular tracking window is located at a new position in image plane based on the direction of rotation of the target object by the following rules. When a motion of an object is a combination of motions introduced the following rules, the coordinate of a corner of new tracking window is determined by applying two rules simultaneously.

Rule 1 When the direction of rotation of the current tracking window with respect to its y-axis is clockwise, the x-coordinate of the left corner of new tracking window is the minimum x-coordinate of the current tracking window.

Rule 2 When the direction of rotation of the current tracking window with respect to its y-axis is counter-clockwise, the x-coordinate of the right corner of new tracking window is the maximum x-coordinate of the current tracking window.

Rule 3 When the direction of rotation of the current tracking window with respect to its x-axis is upward, the y-coordinate of the upper corner of new tracking window is the minimum y-coordinate of the current tracking window.

Rule 4 When the direction of rotation of the current tracking window with respect to its x-axis is downward, the y-coordinate of the lower corner of new tracking window is the maximum y-coordinate of the current tracking window.

So, when an object is rotated with respect to y=x axis, the coordinate of the upper-left or lower-right corner is determined by the combination of Rule 1 and Rule 3 or Rule 2 and Rule 4, respectively.

4 Experimental Results

The proposed algorithm is experimented on a IBM-PC with Pentium II 200Mhz CPU adopted with a Matrox Genesis-LC frame grabber. The operating system is Windows-NT and Visual C++ is used for programming. The experimental system is shown in Fig. 2.

Sobel operator is used to find image gradient, ∇I. The size of our reference image is 80(pixels) x 80(pixels) while the resolution is halved in real implementation in order to reduce computational loads by using sub-sampling. Experiments are executed on two objects. One is a cylinder and the other is a doll. The cylinder is rotated in clockwise direction with respect to its main axis while the

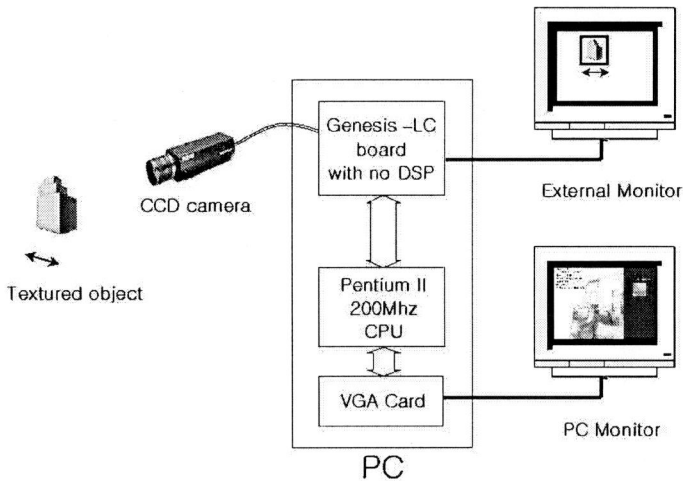

Fig. 2. Experimental System

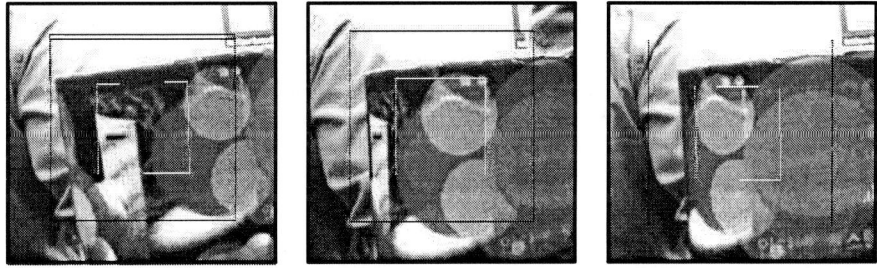

Fig. 3. Experimental Results for a Cylinder

doll has a motion of rotations in clockwise direction and translations. The experimental results are shown in Fig. 3 and Fig. 4. In Fig. 4, the first image is the reference image for visual tracking in the beginning phase and the motion sequence of the doll is upper left, upper right, lower left and lower right. The tracking is executed at 40 Hz in average sense.

Fig. 4 shows the residual value in case of visual tracking of the cylinder. The value is in the range from 0 to 6. By updating the reference image, the residual value exists in a reasonable area maintaining the similarity between the reference image and current image.

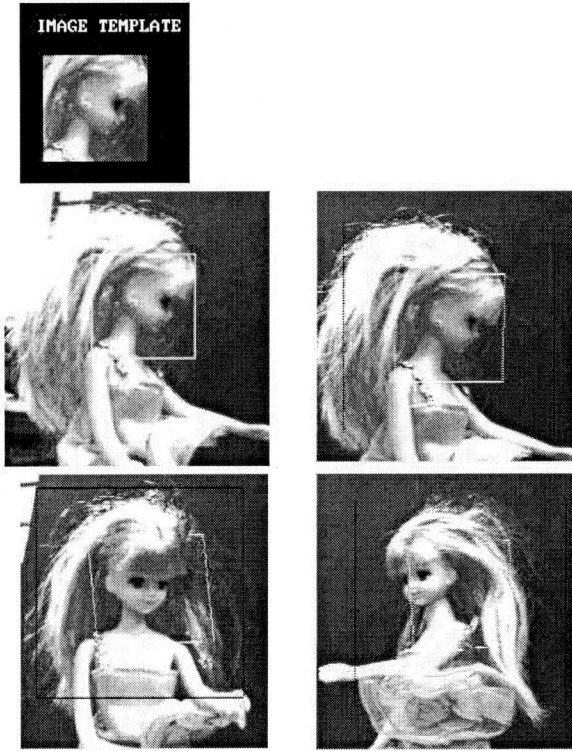

Fig. 4. Experimental Results for a Doll

5 Conclusion

We have proposed a real-time visual tracking algorithm insensitive to three-dimensional rotation of objects. The algorithm is based on SSD but three-dimensional visual tracking is possible by updating the reference image for tracking using the proposed performance measure. It is more similar with the eye system of human beings in the sense that one can track an object without the three-dimensional model of the object and it is not necessary to measure the pose of the object. The proposed algorithm is experimented to two objects, a cylinder and a doll, successfully at 40 Hz speed in average sense. Contrary to other algorithms using optical flows, snake model and depth information, our algorithm maintains the advantage of SSD-based approach in [4] and becomes insensitive to three-dimensional motion of an object. It is expected that the developed visual tracking algorithm can be applied for security systems in department stores or other public places. We are under development several visual tracking algorithms using color information and an active stereo head-eye system with vergence for resolving occlusion problems. In addition, visual tracking in complex changing backgrounds is an interesting one.

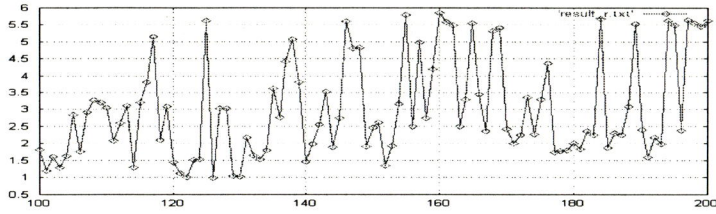

Fig. 5. Residual Value

References

1. T. Yamane, Y. Shirai, and J. Miura, "Person Tracking by Integrating Optical Flows and Uniform Brightness Regions", *Proceedings of IEEE International Conference on Robotics & Automation*, pp. 3267-3272, 1998.
2. Y. Shirai, T. Yamane, R. Okada, "Robust Visual Tracking by Integrating Various Cues", *IEICE Transaction on Information & Systems*, vol. E81-D, no. 9, pp. 951-958, 1998.
3. Natan Peterfreund, "Robust Tracking of Position and Velocity With Kalman Snakes", *IEEE Transaction on Pattern Analysis and Machine Intelligence*, vol. 21, no. 6, pp. 564-569, 1999.
4. Gregory D. Hager and Kentaro Toyama, "X Vision: A Portable Substrate for Real-Time Vision Applications", *Technical Report YALEU/DCS/RR-1078*, 1997.
5. Gregory D. Hager and Peter N. Belhumeur, "Efficient Region Tracking With Parametric Models of Geometry and Illumination", *IEEE Transaction on Robotics and Automation*, vol. 20, no. 10, pp. 1025-1038, 1998.
6. S.-W. Lee, B.-J. You, and G. D. Hager, "Model-based 3-D Object Tracking using Projective Invariance", *Proceedings of IEEE International Conference on Robotics and Automation*, pp. 1589-1594, 1999.
7. A. J. Bray, "Tracking Objects using Image Disparities", *Image and Vision Computing*, vol. 8, no. 1, pp. 4-9, 1990.
8. D. B. Gennery, "Visual tracking of known three-dimensional objects", *International Journal of Computer Vision*, vol. 7, no. 3, pp. 243-270, 1992
9. D. G. Lowe, "Robust model-based motion tracking through the integration of search and estimation", *International Journal of Computer Vision*, vol. 8, no. 2, pp. 113-122, 1992.
10. R. S. Stephens, "Real-time 3D object tracking", *Image and Vision Computing*, vol. 8, no. 1, pp. 4-9, 1990
11. Nikolas P. Papanikolopoulos, Pradeep K. Khosla, and Takeo Kanade, "Visual Tracking of a Moving Target by a Camera Mounted on a Robot: A Combination of Control and Vision", *IEEE Transaction on Robotic and Automation*, vol. 9, no. 1, 1993.
12. Kevin Nickels and Seth Hutchinson, "Measurement Error Estimation for Feature Tracking", *Proceedings of IEEE International Conference on Robotics and Automation*, pp. 3230-3235, 1999.

Heading Perception and Moving Objects

Nam-Gyoon Kim

Center for the Ecological Study of Perception and Action
University of Connecticut, Storrs, CT 06269-1020, USA

Abstract. The present study was directed at understanding the influence of moving objects on curvilinear heading perception. Displays simulated observer movement over a ground plane along a curved (i.e., circular and elliptical) path in the presence of moving objects, depicted as transparent, opaque or black cubes. Objects either moved parallel to or intersected the observer's path, and either retreated from or approached the moving observer. Heading judgments were accurate across all conditions. Discussion focused on the significance of these results for computational models of heading perception and the possible role of dynamic occlusion.

1 Introduction

It has generally been assumed that the visual system recovers the information for self motion from a velocity vector field defined over an image plane. A velocity vector field is described as a function which assigns a vector to each point in the plane. A flow field defined on the retina likewise consists of image vectors uniquely determined by observer translation and concomitant eye rotation. That is, each image vector describes the instantaneous velocity of the projected image of a particular environmental element whose position changes as a result of observer movement.

The preceding holds only if the environment is stationary and the resultant image motion is rigid (Fig. 1a). A moving object violates the rigidity assumption because it introduces image vectors whose properties are not uniquely determined by observer movement (Fig. 1b-d). As a result, a moving object is problematic for models designed to decompose an image flow field into two components corresponding to observer translation and eye rotation.

Recently, two studies concerned with human heading perception addressed the issue of moving objects [1; 2]. The Royden and Hildreth study was an attempt to validate Hildreth's [3] vector subtraction model whereas the Warren and Saunders study was an attempt to validate Hatsopoulos and Warren's [4] template model. Despite their differences, both studies reported accurate perceptions of heading in the presence of a moving object. However, there is a reason to suspect that the models may have circumvented the process involved in recovering the global flow pattern by oversimplifying the experimental setup that incorporated only rectilinear translation and a single moving object. Note that an observer's translation along a straight path yields a global radial pattern that contains information specific to the observer's movement. Specifically, the focus of expansion (FOE) specifies the observer's heading direction. These well-defined flow field properties can simplify the search for information (i.e., the optical concomitants for self motion) in the flow field.

Whereas the flow field corresponding to linear observer translation contains certain well-defined properties, the flow field resulting from an observer's translation along a curved path does not because it lacks a comparable singularity and a formal description for the ensuing global pattern [5]. Accordingly, determining what constitutes the requisite information for self motion during observer movement along a curved path in the presence of moving objects is a major problem for existing models of the visual system.

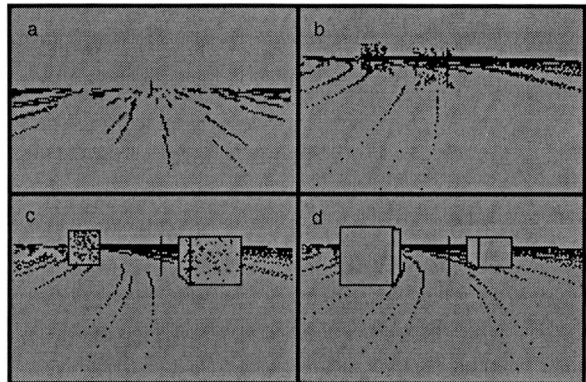

Fig. 1. (a) Optical flow due to observer movement along a linear path. The observer's heading direction is specified by the FOE. (b-d) Displays used in the(a) transparent, (b) opaque, and (c) black objects conditions in the experiment. In (c-d), the contours are shown to enhance the visibility of the objects. In the actual display, the objects were drawn in black, the same color as that of the background. Optical flow lines were generated by observer movement along a circular path with the radius of 160 m. The vertical bar identifies the observer's circular heading.

2 Experiment

The experiment described here examined the effect of moving objects on perceived direction of heading, but in a more generalized context (i.e., locomotion along a curved path in the presence of moving objects). Focus was given specifically to four separate aspects of such a situation.

First, the simulated self-motion varied in terms of the type and curvature of the path. The type of path was either circular or elliptical. The curvature of the path varied from large, corresponding to a sharp turn, to small, corresponding to a very gradual turn. Second, the effect of moving objects on perceived heading was examined by introducing two objects to the flow field. Moreover, the path along which each object moved varied. Either both objects moved in parallel with the observer, or one crossed the observer's future path from inside to outside or crossed from outside to inside while the other remained in a parallel path with the observer's.

Third, the direction in which the objects moved varied. Either both objects receded away from the observer into the distance or they approached the observer.

Last, the transparencies of objects varied: transparent, opaque, or blank. As noted earlier, two computational models, Warren and Saunders' [2] template model and Hildreth's [3] vector subtraction model, differ with respect to how moving objects are treated. Whereas image vectors corresponding to moving objects are pooled together with image vectors corresponding to the stationary environment in Warren and Saunders' model, moving objects are segmented based on relative motion in Hildreth's model. A third possibility is that observers may segment moving objects based on dynamic occlusion and then determine heading. To test these hypotheses, Warren and Saunders specifically manipulated the discrepancy between image vectors corresponding to moving objects and those corresponding to the stationary environment in terms of object transparency. If the object is depicted as transparent, there will be strong relative motion between object elements and the background within the object's contour (Fig. 1b). If the object is depicted as opaque, object motion will yield dynamic occlusion of the background surface with additional relative motion along the boundary of the object contour (Fig. 1c). If the object is depicted as black, its motion will yield only dynamic occlusion of the background surface with no relative motion (Fig. 1d).

Hence, by manipulating the way in which objects were depicted, *viz.*, transparent, opaque, or black, Warren and Saunders [2] were able to test various hypotheses about how the visual system copes with moving objects. If the visual system simply pools over the entire velocity vector field without segmenting the object first, observers' judgments should be biased in the presence of transparent and opaque objects, but not in the presence of black objects. Further, the hypothesis predicts poorer performance with opaque objects than transparent objects. With occlusion of the background dots by the image of the moving object, the overall signal-to-noise ratio (SNR) would be lower in the opaque condition than in the transparent condition. Alternatively, as Hildreth [3] contends, the visual system could rely on the relative motion of neighboring elements at different depths to segment moving objects and determine heading. Thus, the relative motion hypothesis predicts that perceived heading would be most accurate with transparent objects and less accurate for opaque and black objects due to the weaker or non-existent relative motion, respectively. Finally, if the visual system utilizes dynamic occlusion rather than relative motion to segment moving objects, more accurate heading judgments would be predicted with black or opaque objects than with transparent objects.

It is expected that the performance of human observers under such conditions will provide a means for evaluating existing models of heading perception in the presence of moving objects, but more importantly, it will enhance our understanding of the problem of visual navigation.

2.1 Method

Forty-eight undergraduate students participated in partial fulfillment of a course requirement. Eight participants were randomly assigned to each Transparency x Object Motion Direction condition. All had normal or corrected-to-normal vision.

Displays were simulations of observer movement parallel to a random-dot ground plane along a curved path. Simulated observer velocity was held constant at 13.2 m/s in terms of tangential velocity. The ground plane consisted of 150 dots placed randomly within each cell of an appropriately scaled grid (see [5] for further details). The dots were single white pixels drawn on a black background. Approximately the same number of dots were maintained during the duration of each trial.

The moving object appeared as 80 white dots randomly positioned in a hexahedron. For transparent objects, the projected images of these dots were simply displayed on the screen. For opaque and black objects, the overall contour of the object was drawn to yield the effect of dynamic occlusion of the background surface.

Each display was generated in real time on a Silicon Graphics Indigo 2 High Impact R10000 for the opaque and black object conditions and on a Silicon Graphics Indigo 2 R4400 for the transparent object condition. Displays were presented on a 19-inch screen refreshed at 60 frames/s. The display screen had a pixel resolution of 1280 H x 1080 V pixels and subtended 42û H x 32û V when viewed from 50 cm from a chin rest. Each display lasted about 2.5s (150 frames).

Added to the flow field were two minivan size objects whose paths were varied in three ways: In object path A, both objects moved parallel to the observer's path. In object path B, the outside object crossed the observer's path while the inside object continued to move along its own path. In object path C, the inside object crossed the observer's path while the outside object continued to move along its own path.

For the approaching objects, objects lay between 55 and 70 m in front of the observer and moved toward the observer at a tangential velocity of 6.2 m/s regardless of the curvature of the observer's path. Because objects approached the observer, the relative speed was 19.8 m/s. For the retreating objects, objects lay between 15 and 30 m in front of the observer and receded away from the observer at two different angular velocities ranging from 0.35 û/s to 2.8 û/s.

Heading accuracy was assessed in terms of heading angle, the visual angle defined by a target on the ground, the point of observation, and the point at which the observer's path passed the target (see [6] for further details). Heading angle varied randomly among values of ±0.5û, 1û, 2û and 4û. The curvature of the observer's path varied randomly among ±160/320 m, 320/160 m, 160/160 m and 320/320 m (see [5] for further details). This yielded a 2 (object motion direction) x 3 (transparency) x 3 (object path) x 4 (curvature) x 2 (turn direction) x 4 (heading angle) x 2 (target location) mixed design with 192 trials. All variables were within-subjects except object motion direction and transparency, which were between-subjects.

Participants initiated the displays by pressing the space bar, then watched the display until it stopped and a blue vertical bar appeared on the ground surface. At that time, participants pressed one key if the path of travel seemed to be to the left of the vertical bar and another key if the path of travel seemed to be to the right of the bar. A short practice session was provided prior to the experiment.

Heading accuracy was evaluated in terms of heading thresholds [6]. However, four participants for the retreating objects and three participants for the approaching objects failed to reach threshold in one of the curvature conditions. Consequently, only thresholds for object paths were analyzed.

2.2 Results and Discussion

For analysis, the results were collapsed over the signs of turn direction and target location and entered into a mixed design analysis of variance (ANOVA) with object motion direction, transparency, object path, curvature, and heading angle as independent variables. The ANOVA showed the main effects of curvature, $F(3, 132) = 24.96, p < .0001$, and heading angle, $F(3, 132) = 289.63, p < .0001$. The effect of curvature was consistent with prior findings [5;6] in which curvilinear heading was shown to be more difficult to determine for larger curvature of the movement trajectory. The ANOVA also revealed an object path x curvature interaction, $F(6, 264) = 2.44, p < .05$, a third order interaction involving object path, curvature, and motion direction, $F(6, 264) = 3.60, p < .01$, and a fourth order interaction involving object path, curvature, heading angle, and transparency, $F(36, 792) = 1.70, p < .01$.

However, the main effects of motion direction, $F(1, 44) = 1.89, p > .05$, object path, $F < 1$, and transparency, $F < 1$, were all insignificant. Unreliable effects of these factors suggest that the different patterns of disturbances induced by the variations indirection of object motion, object path, and object transparency had a minimal effect on perceived curvilinear heading. Overall performance in each condition of transparency was 71% for transparent objects, 72% for opaque objects, and 70% for black objects, respectively. Except for the fourth order interaction reported above, there were no significant interactions involving transparency.

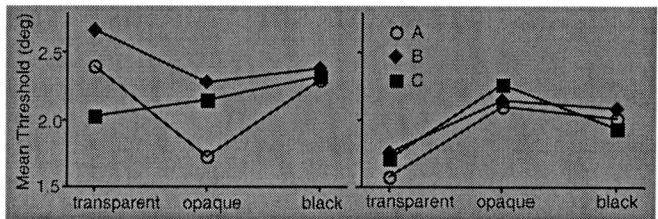

Fig. 2. Mean heading threshold for the retreating (left) and approaching (right) objects, for each condition of object's path (bottom), as a function of transparency.

Thresholds for the three object paths are presented as a function of object transparency for each condition of object motion direction in Fig. 2. The mean heading thresholds were 2.3û for the retreating objects and 2.0û for the approaching objects. Two separate ANOVAs on threshold were performed for each condition of object motion direction with transparency and object path as independent variables. Both ANOVAs confirmed the results of the preceding analysis. Neither of the two main effects nor their interaction was reliable.

The overall mean threshold of 2.2û was slightly higher than the overall thresholds reported in [5] and [6], but accurate enough to support successful navigation through the environment. Note that a heading accuracy of 4.2û is needed for drivers to maintain a speed of 13.2 m/s to brake properly for a pedestrian or animal [7]. Taken together, different object paths, different types of transparency, and different directions of object motion did not appear to affect heading judgments.

Because curvilinear paths introduce asymmetries in the resulting flow field, Warren et al. [6] examined a possible constant error in terms of heading bias measured as the total percentage of "outside" responses. A 50% outside response rate indicates no bias. For the retreating objects, the percentage of outside responses was 39% for transparent objects, 51% for opaque objects, and 56% for blank objects, respectively. Of these, an inside tendency was significant for transparent objects, $t(7) = -4.83, p < .01$. For the approaching objects, it was 49% for transparent objects, 43% for opaque objects, and 56% for blank objects. None reached statistical significance.

To further examine the contrasting pattern of heading bias, two mixed design ANOVAs were conducted for each condition of object motion direction on percentage of outside responses with transparency, object path, and curvature as independent variables. The results for the retreating objects revealed significant main effects of object path, $F(2, 42) = 22.97, p < .001$, and curvature, $F(3, 63) = 72.99, p < .001$. The main effect of transparency was not significant, $F(2, 21) = 2.67, p > .05$, but was qualified by a significant interaction with object path, $F(4, 42) = 4.22, p < .01$.

With respect to the main effect of curvature, participants showed inside preferences for two trajectories with small curvature, 160/320, 39%, $t(23) = -3.53, p < 0.01$, and 320/320, 42%, $t(23) = -2.44, p < .05$, respectively. With respect to the main effect of object path, participants showed an outside tendency for object path C, 60%, $t(23) = 3.61, p < .01$. With respect to the transparency x object path interaction, a simple effects analysis showed that the effect of object path was significant for all conditions of transparency, whereas the effect of transparency was significant only for object path A, $F(2, 21) = 7.95, p < .01$, but not for object path B or C, $F < 1$. It appears that a strong inside tendency, especially, for object path A, 24%, $t(7) = -10.40, p < .0001$, was the source of this interaction.

The same analysis for the approaching objects, however, confirmed only the main effects of object path, $F(2, 42) = 5.12, p < .01$, and curvature, $F(3, 63) = 106.75, p < .001$, but not the object path x transparency interaction, $F(4, 42) = 1.26, p > .05$. Further analyses revealed no particular pattern of bias at each condition of object path. More importantly, the insignificance of transparency in conjunction with the absence of the object path x transparency interaction further corroborates the suspicion that the seeming inside tendency with transparent objects at object path A observed in the retreating object condition may have been an anomaly.

In sum, performance was quite accurate despite the various perturbations introduced to the flow field. Of particular interest was that the effect of transparency was virtually negligible, not only on the accuracy of perceived heading, but also on bias, except for some anomalous inside tendency with transparent objects, especially at object path A. Also notable were virtually negligible effects of object path and direction of object motion. The absence of a particular pattern of bias further corroborates these results. Participants' judgments were influenced by curvature of the observer's path and were consistent with the results of the previous studies without moving objects [5; 6]. That is, inside bias for observer paths of small curvature and outside bias for observer paths of larger curvature. Taken together, there is virtually no evidence that indicates that the presence of moving objects affected human observers' capacity to perceive direction of curvilinear heading.

3 General Discussion

Two computational models of the visual system [2; 3] have been noted for their ability to handle noise induced by moving objects. Except for the fact that Hildreth's [3] model employs an extra computational procedure to segment objects from the flow field, both models achieve robust performance by exploiting the redundancy of the flow field (i.e., pooling together velocity vectors from all regions of the image plane).

The results of the present study, however, question the underlying strategies of these models. As noted, a unique pattern (i.e., the radial pattern of expansion or the FOE) characterizes the flow field resulting from rectilinear translation. With the additional assumptions that the stationary environment with respect to moving objects occupies the largest area of the visual field and/or consists of the largest number of coherent optical elements, the idiosyncrasies of the flow field can be exploited to simplify the process involved in recovering the global flow pattern. In other words, the apparent effectiveness of these models may lie within a very circumscribed condition involving the observer's linear translation with the additional constraints limiting the range of object motion and the number of object features.

The manipulations employed in the present study were specifically geared to compensate for these limitations. First of all, the flow field corresponding to observer translation along a curved path lacks identifiable features without a single encompassing description (see [5] for further details). The manipulations of path and surface transparency all contributed to produce flow fields that violated the preceding assumptions. In particular, the variation in object path induced changes in the projected shape, size, and location of the moving object, yielding varying patterns of perturbation in the flow field. Because these conditions eliminate the qualitative properties characteristic of the flow field, any scheme operating at the level of image vectors will be ineffective as a means to segment the moving object from the stationary environment. Indeed, reliable performance across all these variations confirms the suspicion that the effectiveness of Hildreth's [3] model is based on an a priori distinction of the flow field utilizing its distinctive properties that is available only in rectilinear translation. On the other hand, variations in the transparency of the object surface yielded different levels of SNR, a condition that would force a model such as Warren and Saunders' [2], which relies on a spatial pooling scheme, to perform differently. The insignificance of transparency clearly undermines its validity. It maybe the case that the model is effective only when image vectors corresponding to the stationary environment make up the majority and those of the moving object the minority of the flow field, as in the condition employed by Warren and Saunders.

Instead, the results corroborate the findings of Kim et al. [8]. In particular, in their Experiments 1 and 2, Kim et al. examined specifically the perceptual consequences of optical flow wherein the rigidity of the environment was violated either with respect to the ground or with respect to the medium through the manipulation of noise. Their results demonstrated that the basis of (in)accurate perception is not noise density, but the way in which noise affects the global pattern of the flow field. One notable difference between these two studies, however, is the scale at which noise perturbed the flow field. In Kim et al.'s study, noise was scattered, perturbing the flow field globally. In the present study, noise was concentrated on the surfaces of the moving object, perturbing the flow field locally. Moreover, the two types of noise used in Kim et al.'s study produced different effects on perceived heading. Whereas the flow field rendered by the nonrigid ground degraded the performance, annihilating perception eventually when SNR reached 0.5; the flow field rendered by the nonrigid medium had a minimal effect even at an SNR level as low as 0.08. The performance observed in the present study is comparable to the performance in Kim et al.'s medium condition. Based on their results, Kim et al. concluded that the performance demonstrated by their human participants could only be accounted for by the basic informational structure, which they called the ordinal relations following Gibson [9]. When the ordinal relations among spatial elements were preserved optically, accurate perception ensued; when absent, perception failed. Hence, to the extent that the ordinal properties are available, human observers can register the requisite information such as that specifying direction of heading.

It can be argued, then, that the specific perturbation induced by the moving objects left the ordinal properties of the flow field intact. However, how does the visual system extract the requisite information from a flow field perturbed by moving objects? Based on the results of the present study, a brute-force based strategy such as a spatial pooling scheme employed in [2] can be easily ruled out as a possible candidate. The strategy is susceptible to the effect of noise density. A consistency criterion which played a key role in [3] can be also utilized under rectilinear translation, but may not be applicable to a more generalized context involving curvilinear translation. A possible resolution will besought from J. J. Gibson whose concern on visual control of locomotion pioneered the research on optical flow. In particular, Gibson [9; 10; 11] postulated two optical bases for the perception of object motion. These two candidates are discussed in the following.

3.1 Informational Candidates for Object Motion

Any rigid motion can be analyzed in terms of six motion parameters, three for translation and three for rotation. The corresponding optical motion can be described by perspective transformations under projective geometry. In contrast, objects in nature are voluminous, hence, made of many faces (i.e., polyhedrons). When an object turns, its projected faces become unprojected while unprojected faces become projected. Optically, however, continuous deletion of the optical structure corresponding to the background texture along the leading edge of the moving object's image and continuous accretion along the trailing edge destroys a one-to-one correspondence between environmental elements and their optical counterparts. Hence, the resultant optical disturbance is a breaking of its adjacent order, not a transformation [11]. Indeed, this problem poses immense difficulties for any computational model that extracts certain properties of the flow field by performing computations defined over velocity vectors. Unprojection of the occluded surfaces means no corresponding vectors, hence, no computation. To make matters worse, not only is there a void (i.e., an absence of velocity vectors) in the flow field, but the location of that void is continuously shifting while, at the same time, annihilating and creating vectors along its borders.

Static occlusion, or superposition, has been well recognized as a pictorial cue for depth. For Gibson [9], however, *dynamic occlusion* provides more; that is, it specifies not only "the existence of an edge in the world, and the depth at the edge, but É *also specifies the existence of one surface behind another, that is, the continued existence of a hidden surface*" (p. 204). To that extent, dynamic occlusion becomes an important source of information for the perception of observer movement: "The point to be emphasized about motion perspective É is that it does *not* elicit a perception of objective motion but a detection of subjective movement" [10, p. 340].

Indeed, it is this information, the dynamic occlusion that specifies the existence of the background surface despite the unprojection of its parts due to the moving object, which supported reliable performance in the opaque and blank object conditions as observed in the present study. In other words, an accretion and deletion at the borders of the moving object specified the continuity of the background surface, thereby preserving the ordinal relations among its surface elements. To that extent, perceiving the consequences of observer movement was equally reliable with or without the presence of moving objects.

Reliable performance in the transparent object condition demands a different explanation. Perhaps an explanation can be found from a typical description of optical flow, one that involves a moving point of observation in an open environment. Unlike in a cluttered environment, in an open environment all surfaces are projected at the point of observation; and a one-to-one correspondence between environmental elements and their optical counterparts is preserved in the optical transformation. Accordingly, locomotion in an open environment is specified by flow of the optic array, "a change in perspective structure, a change in the perspectives of the ground if outdoors and of the floor, walls, and ceiling if indoors" [11, p. 122]. Moreover, for Gibson, "The optical flow of the ambient array is almost never perceived as motion; it is simply *experienced as kinesthesis*, that is *ego locomotion*" [11, p. 123].

Whereas optical flow, or a global transformation of the optic array, specifies observer locomotion, a *local transformation* of the optic array specifies object motion. Or, as Gibson [10] put it, "motion perspective [optical flow in Gibson's later terminology] caused by locomotion entails change in the *whole* of the textured ambient array whereas the alteration of perspective caused by an objective motion entails only change in *part* of the ambient array" (p. 341). It is this whole-part or global-local distinction that provides the information for the segregation of observer movement from object motion. And it is this distinctive pattern of disturbances that accounts for the reliable performance in the transparent object condition.

In the present study, each object traced a complicated path of its own; collectively, they yielded a non-distinct pattern of perturbation. Nonetheless, the depiction of optical flow (i.e., the patterning of the transformation of the optic array corresponding to observer locomotion) was unequivocal because the transformation was manifested as a change in the perspectives of the ground. In fact, the significance of the rigid ground plane has been well documented in the self-motion estimation literature, especially in a flow field with concomitant eye rotation [12; 13]. Thus, the degraded performance reported in various studies may well be accounted for by the non-distinct nature of optical flow because the simulated flow field was either with respect to a fronto-parallel plane [14] or with respect to a random dot cloud [12; 15].

Similar reasoning can be also applied to the apparent discrepancy between the results of the present study and those reported in [1] and [2]. Note that the latter two studies observed a small, but nonetheless systematic, bias in their participants' judgments; whereas the participants of the present study demonstrated virtually little bias. The major difference between those two studies and the present study, ignoring the difference in the type of the simulated path of locomotion (i.e., linear vs. curvilinear), is the type of background. In [1] and [2], the flow field was defined with respect to fronto-parallel planes. In the present study, however, it was with respect to a ground plane. As a land-based animal, most of our locomotion is restricted to the ground surface. It appears that this physical constraint plays a more important role by reducing the degrees of freedom embedded in the flow field, thereby simplifying the description of the problem. In other words, the flow field simulating observer locomotion along a ground plane renders the flow pattern distinct, further simplifying the perceiving of the consequences of observer movement.

3.2 Concluding Remarks

For Gibson [9; 11], the optic array specifies the permanent properties of the environment. However, what is specified in the ambient optical structure is neither surfaces nor objects, but the underlying *adjacent order* of the structure. It is important to note that the order characterizing the optical structure reflects the corresponding order of the layout of the environment. As Gibson [11] put it, "the reality underlying the dimensions of space is the adjacent order of objects or surface part" (p.101). Hence, for Gibson, perception is specific to ordinal stimulation, which is, in turn, specific to the adjacent and successive order of the environmental layout.

To the extent that ordinal properties of the environmental layout are specified in the flow field, it appears that the consequences of observer movement can be perceived. The results of the present study corroborate this observation. Performance was reliable and consistent, even when a significant portion of the flow field was either unprojected at the point of observation (in the black condition; Fig. 1d)or was occupied by discrepant vectors (in the opaque condition; Fig. 1c).

In contrast, locomotion engendered in an open, uncluttered environment is an abstract case, suitable only for a geometrical analysis of a flow field. The environment we live in is typically cluttered with opaque objects and the light reflected from the surfaces of the objects travels in a straight line. Because not all surface elements are projected to the point of observation, there is no projective correspondence between the environmental features and the corresponding optical elements for unprojected hidden surfaces. The "true problem," then, as Gibson [16] noted, "is how surfaces are perceived when they are temporally occluded, or hidden, or covered, that is, when they are not projected in the array at a fixed point of observation" (p. 162). Nonetheless, the results of the present study clearly demonstrate that the effect of dynamic occlusion on perception is quite robust and human observers are quite reliable in exploiting that effect to detect the requisite information from optical flow.

References

1. Royden C.S., Hildreth E.C.: Human heading judgments in the presence of moving objects. Percept Psychophys **58** (1996) 836-856
2. Warren W.H., Saunders J.A.: Perceiving heading in the presence of moving objects. Perception **24** (1995) 315-331
3. Hildreth E.C.: Recovering heading for visually-guided navigation. Vision Res **32** (1992) 1177-1192
4. Hatsopoulos N.G., Warren W.H.: Visual navigation with a neural network. Neural Networks **4** (1990) 303-317
5. Kim N.-G., Turvey M.T.: Visually perceiving direction of heading on circular and elliptical paths. J Exp Psychol Hum Percept Perf **24** (1998) 1690-1704
6. Warren W.H., Mestre D.R., Blackwell A.W., Morris M.W.: Perception of circular heading from optical flow. J Exp Psychol Hum Percept Perf **17** (1991) 28-43
7. Cutting J.E.: Perception with an eye for motion. MIT Press, Cambridge MA (1986)
8. Kim N.-G., Fajen B.R., Turvey M.T.: Perceiving circular heading in noncanonical flow fields. J Exp Psychol Hum Percept Perf (in press)
9. Gibson J.J.: The senses considered as perceptual systems. Houghton Mifflin, Boston (1966)
10. Gibson J.J.: What gives rise to the perception of motion. Psychol Rev **75** (1968) 335-346
11. Gibson J.J.: The ecological approach to visual perception. Houghton Mifflin, Boston (1979)
12. van den Berg A.V.: Robustness of perception of heading from optic flow. Vision Res **32** (1992) 1285-1296.
13. van den Berg A.V., Brenner, E.: Humans combine the optic flow with static depth cues for robust perception of heading. Vision Res **34** (1994) 2153-2167
14. Warren W.H., Hannon D.J.: Eye movements and optical flow. J Opt Soc Am **A 7** (1990) 160-169
15. Royden C.S., Crowell J.A., Banks M.S.: Estimating heading during eye movements. Vision Res **34** (1994) 3197-3214
16. Gibson J.J.: On the analysis of change in the optic array. Scan J Psychol **18** (1977) 161-163

Dynamic Vergence Using Disparity Flux

Hee-Jeong Kim, Myung-Hyun Yoo, and Seong-Whan Lee*

Center for Artificial Vision Research, Korea University,
Anam-dong, Seongbuk-ku, Seoul 136-701, Korea
{hjkim, mhyoo, swlee}@image.korea.ac.kr

Abstract. Vergence movement enables human and vertebrates, having stereo vision, to perceive the depth of an interesting visual target fixated by both left and right eyes. To simulate this on a binocular robotic camera head, we propose a new control model for vergence movement using disparity flux. Experimental results showed that this model is efficient in controlling vergence movement in various environments. When the perception-action cycle is short enough to approach to the real-time frame rate, the precision of disparity flux increases, and then a more accurate control of vergence movements on the stereo robotic head is possible.

1 Introduction

Human and many vertebrates, having stereo vison, can perform eye movements consisting of saccade, pursuit, tremor and vergence movement. The critical ecological reason for the existence of eye movements is the necessity to shift the high-resolution foveal area onto the most interesting, important and informative parts of a visual scene, that is, stable fixation on the visual target. Among four basic eye movements, the vergence movement plays a significant role in fulfilling this necessity, directly related to the perception of depth using stereoscopic information.

In the view of machine vision research, vergence movement is defined as the motion of turning both cameras of a stereo head-eye system symmetrically so that their image centers display the same point or feature in the real world within a limited number of perception-action cycles. The binocular geometry of vergence is illustrated in Figure 1. The left and right optical center points and fixation point are the vertices of a triangle inscribed in the geometrical horopter circle. Here the angle between the left and right optical axes on its intersection with the horopter, that is, the fixation point, is called a vergence angle θ_{verge}. The angle parameters including the vergence angle θ_{verge} of the camera axes for controlling the vergence movement can be obtained using disparities estimated from correspondence matching in stereo image pairs. To estimate disparities for the control of vergence movement, vision researchers have developed ideas and

---

* To whom all correspondence should be addressed. This research was supported by Creative Research Initiatives of the Ministry of Science and Technology, Korea.

techniques such as the correlation method, phase analysis method, and so on [1,2,4,5,6]. Here, we choose the phase analysis method to estimate disparities in stereo image pairs [1,2]. With these disparity maps we propose a new control model for vergence movement on a binocular robotic head using disparity flow and flux.

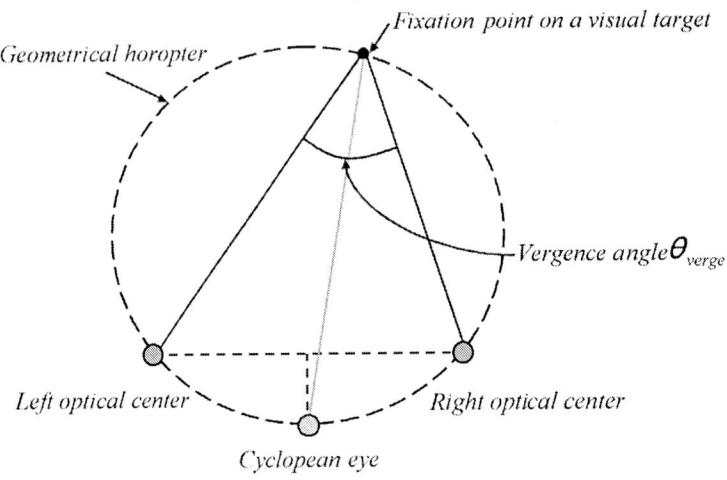

Fig. 1. Binocular geometry of vergence

2 The Proposed Architecture for Vergence Control

In Figure 2 we represent the outline of our proposed control model for vergence movement on a head-eye system. It shows the role of the disparity flow and flux in the entire control loop.

This section is organized as follows. Section 2.1 introduces the basic idea about the phase-based method to estimate disparities in stereo image pairs. In Section 2.2 and 2.3, the concepts of disparity flow and flux are described in detail, respectively. Then, the vergence-adjustment mapping process for real vergence movement of a binocular robotic head is presented in Section 2.4. Finally, the overall flow chart of the proposed control architecture for vergence movement is presented in Section 2.5.

2.1 Disparity Estimation

The basic idea of the phase-based algorithm for estimating disparities is conceptualized to compute a local disparity as a spatial shift from a local phase difference observed in the frequency domain, obtained by convolving both left

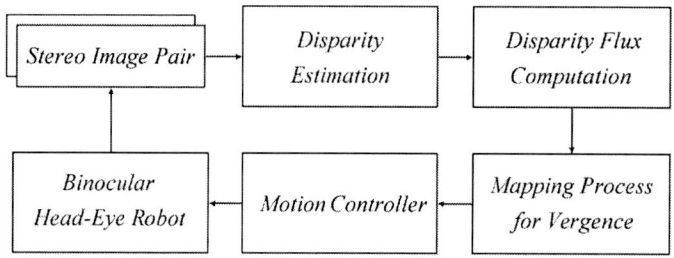

Fig. 2. The outline of the proposed control model

and right spatial images with a complex filter such as the Gabor filter. This procedure is illustrated in Figure 3. Many researchers in stereo vision laboratories, for example, the Computer Science Institute at the University of Kiel and the CVAP Laboratory at the Royal Institute of Technology, have used the phase-based method to estimate disparities and to construct disparity map [1,2]. These ideas are formulated as:

$$V_l(x,y) \cong e^{j\omega D(x,y)} \cdot V_r(x,y) \qquad (1)$$

$$D(x,y) \cong (argV_l - argV_r)/\omega, \qquad (2)$$

where $V_l(x,y)$ and $V_r(x,y)$ are the left and right images convolved with the Gabor filter respectively. $D(x,y)$ is a disparity value at (x,y) in a spatial image. And ω is the central filter frequency(radian/pixel). Also, in order to improve the performance of this method we adopt the coarse-to-fine strategy as in [1,2].

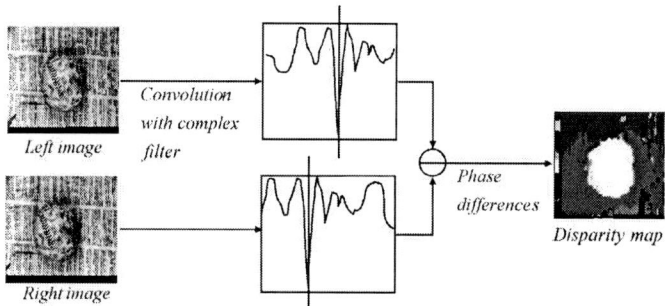

Fig. 3. The basic idea of the phase-based algorithm

2.2 Disparity Flow

To utilize disparity maps obtained from the phase-based disparity estimation process, we propose a simple concept of disparity flow. The *disparity flow* is defined as the signed quantity of change of the disparity value D at a given spatial location (x, y) for one perception-action cycle Δt. This simple concept is illustrated in Figure 4. And the disparity flow F_D is formulated as follows:

$$F_D = \frac{dD(x,y)}{dt} = D_{t-\Delta t}(x,y) - D_t(x,y). \tag{3}$$

The shorter the perception-action cycle is, the higher the precision of disparity flow is. That is, a small change in a disparity value is reflected to the control mechanism of vergence when the perception-action cycle approaches to the real-time frame rate.

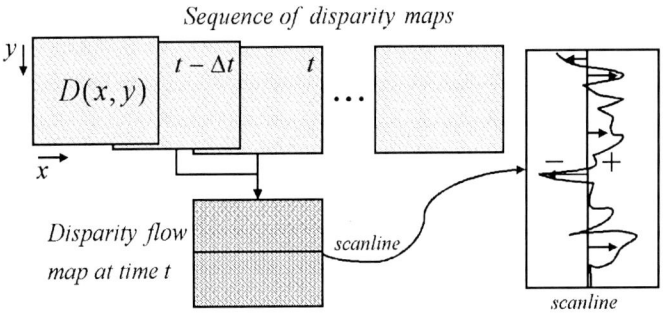

Fig. 4. The concept of disparity flow

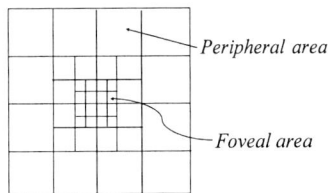

Fig. 5. An example of a multi-level attention window; here, 3-levels of attention window are displayed.

2.3 Flux of Disparity Flows

After the process of computing disparity flow maps we use the *flux of disparity flows* to reflect them to our control mechanism of vergence. The flux of disparity flows is defined as the total volume of disparity flows F_D, passing through an *attention window* W_A, per perception-action cycle Δt. This can be expressed as the mathematical formula given below:

$$\Phi_{F_D} = \int_{W_A} F_D dW_A \cong \sum_{j,k=0}^{n} F_D(j,k), \tag{4}$$

where n is the total number of squares, as j times k, in the attention window W_A. An attention window[1] [3] is a pick-up box which is made by a logarithmic subdivision toward the center point of the window as shown in Figure 5. To compute the flux of disparity flows at a given time we simply pick up one disparity flow value in every square of the attention window whose center point coincides with the center of the disparity flow map and all the disparity flow values picked up in this manner are then summed. This is illustrated in Figure 6 (a) and (b).

It can be easily noticed that the computed flux using the attention window at a specific time will flow out of the window or flow into the window or remain unchanged. Namely, the total sum of the disparity flow values in the attention window at a specific time may indicate the direction of the flow, the sign (+) or (-) or zero. This signed property of the sum can be utilized to control the vergence axes on a binocular robotic head at any specific moment.

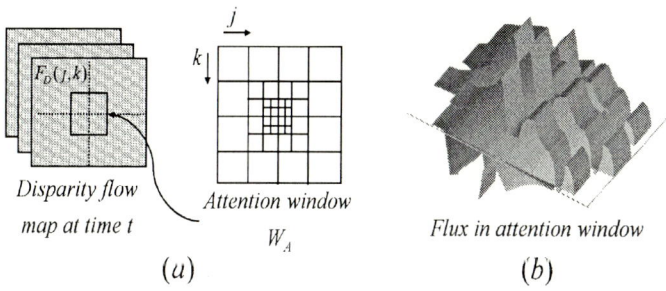

Fig. 6. Flux of disparity flows

2.4 Vergence-Adjustment Mapping Process

When motion occurs with a change of depth on the fixation point in a visual scene, disparity flows in the attention window are generated. If the flux at any given time is a small value near to zero or exactly equal to zero, that is, the current flux value is *in the range of stable vergence*, then it is said that the current state of vergence is very *stable*. Instantly, when disparity flows are detected in the attention window, or the current flux value is not in the stable range close to zero, the left and right vergence axes are controlled to converge or diverge symmetrically as the vergence-adjustment angle through the mapping process

---

[1] As we see in Figure 5, the attention window consists of logarithmic subdivided squares, similar to the fovea-peripheral log-polar map in space-variant sensing. In Figure 5 the total number of the levels of the window is 3.

until the flux value is included in the range of stable vergence. The mapping process consists of two steps: one step to quantize the flux value into 7-levels and the other step to link that level with an appropriate adjustment angle in the table[2]. For selecting the action of vergence, convergence or divergence, the signed property of the summed disparity flows is used.

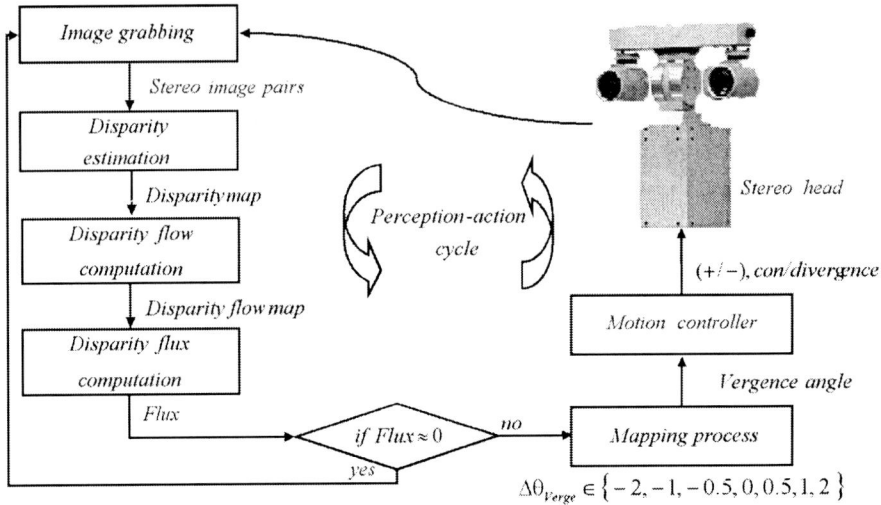

Fig. 7. Overview of the proposed control architecture

2.5 Overview of the Proposed Control Architecture

The overall flow chart of the proposed control architecture for vergence movement is illustrated in Figure 7. In the figure the control flow forms a closed-loop, a compact perception-action cycle.

3 Experimental Results

3.1 Experimental Environment

Experiments were carried out to test the performance of the proposed control model for vergence movement on our binocular head-eye robot.[3] The head-eye

---

[2] The vergence-adjustment table can be seen like {-2, -1, -0.5, 0, 0.5, 1, 2}. An element of this table represents how many degrees of the relative angle between the left and right vergence axes the vergence is to be adjusted.

[3] This head-eye robot is a commercial component made by Helpmate Robotics Inc. It is a BiSight system, mounted on a UniSight system, equipped with FUJINON H10x11E-X41 lenses connected to PULNiX TMC-7 series color CCD cameras on each vergence axis of the robot.

robot is equipped with 4 degrees of freedom: independent vergence for each left and right camera and a common tilt and a pan. In our experiments for vergence movement we use only the left and right vergence axes of the head-eye robot. To capture stereo image pairs for estimating disparities, two image grabbers[4] connected with CCD cameras were used. And a motion controller[5] was used to control the left and right axes of our head-eye robot.

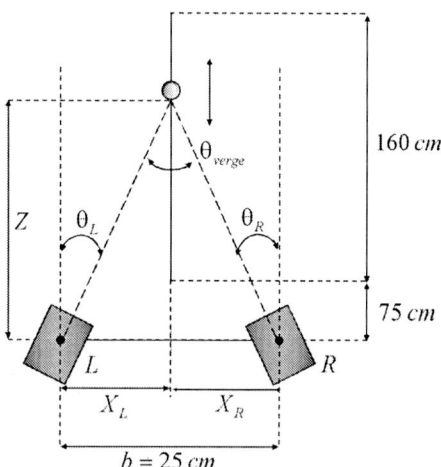

Fig. 8. Set up used for experiments: The cameras are verging on a pole object on the extended line from the center of the baseline. The line extends over 160 cm in length, starting at 75 cm from baseline.

In order to be able to accurately quantify the performance of the proposed vergence system, the experimental set-up[6] shown in Figure 8 was used to create controlled vergence stimuli in a realistic environment, similar to the one used in [4]. This set-up is composed of a pole object on the extended line from the center of the baseline[7] b of our binocular camera head, as shown in Figure 8. The length of the baseline b of our head-eye robot is a constant value(25 cm). θ_L and θ_R are the left and right vergence motor angles, respectively.

From this simple figure, we can develop a relationship between a vergence angle θ_{verge} and a relative depth Z of a verged object. First, the vergence angle is defined as the angle between the optical axes of the two cameras, that is equivalent to the sum of θ_L and θ_R. This is expressed as follows:

---

[4] Matrox Meteor-II image grabber made by Matrox Electronic Systems Ltd.
[5] A PMAC(Programmable Multi-Axes Controller, IBM PC-ISA bus version) made by DELTA TAU data systems Inc. is used for our head-eye system. This PMAC board is connected to a PC-based host computer(Pentium III MMX 500Mhz) via ISA bus.
[6] In this setup the tilt angle is exactly zero, that is, the head-eye robot looks straight ahead.
[7] i.e., the distance between the rotation centers of two cameras

$$\theta_{verge} = |\theta_L| + |\theta_R|. \tag{5}$$

In this set-up, the left and right vergence motor angles are equal, *i.e.*, the horizontal distances X_L and X_R from the center of the baseline are the same as half of the baseline length. Thus, the depth Z of the fixation point on the verged object from the center of the baseline can be obtained by:

$$\theta = |\theta_L| = |\theta_R| = \frac{\theta_{verge}}{2}, X = X_L = X_R = b/2 \tag{6}$$

$$Z = \frac{X}{tan\theta} = \frac{b}{2tan\theta} = \frac{b}{2tan(\theta_{verge}/2)}. \tag{7}$$

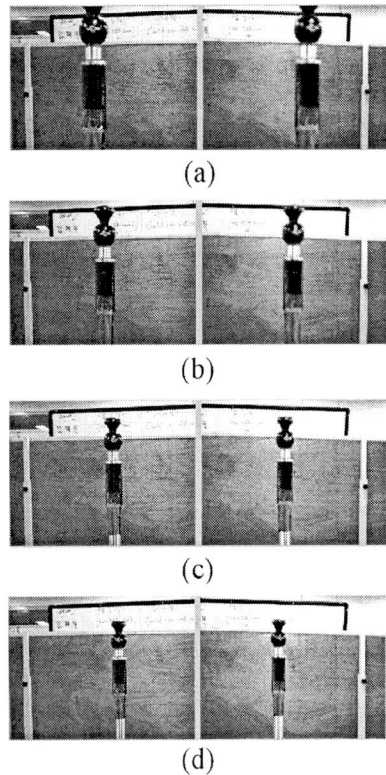

(a)

(b)

(c)

(d)

Fig. 9. Constant vergence angle(open-loop) experiment: (a) object: 75 cm, fixation: 95 cm, flux = -780 (b) object: 95 cm, fixation: 95 cm, flux ≈ 0(verged) (c) object: 115 cm, fixation: 95 cm, flux = 832 (d) object: 135 cm, fixation: 95 cm, flux = 1524

3.2 Results and Analysis

In order to measure the performance of the proposed control system in different situations, some experiments were conducted as described below.

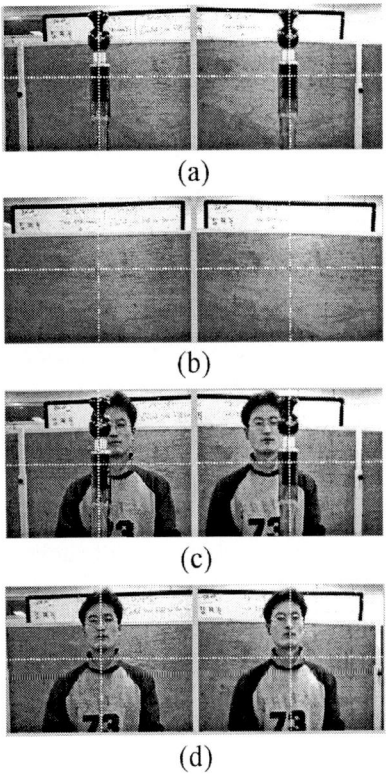

Fig. 10. Saccadic vergence experiment: (a) object: 95 cm, fixation: 95 cm (b) object: 235 cm, fixation: 235 cm (c) object: 95 cm, fixation: 95 cm (d) object: 135 cm, fixation: 135 cm

- **constant vergence angle experiment**

The objective of this experiment was to test the open-loop response of the vergence system to change of the flux and direction of motion. The vergence angle was fixed so that the left and right optical axes were verging midway along the extended line from the center of the baseline(95 cm). Figure 9 was obtained by moving the pole object back and forth(by constant step off 20 cm) in a range 75~135 cm from the center of the baseline. In Figure 9 (b) the disparity flux is almost zero at the 95 cm from the baseline. It shows that the flux reaches its minimum value, which indicates stable verging, when the visual target crosses the fixation point(at 95 cm from the baseline in this experiment).

• **saccadic vergence experiment**

To measure the response of the proposed control system on unexpected depth changes with large vergence errors in the visual scene, the saccadic vergence experiment was conducted at indoor environment ranging approximately from 95 cm to 235 cm. In Figure 10 (a) and (c), the fixation point was initially located at the pole object at 95 cm with stable vergence. When the pole object disappears in the visual field, the flux abruptly changes and vergence movement of the head-eye robot is issued. When the flux value approached to zero, vergence movements are turned off and the flux is in the stable vergence range in Figure 10 (b) and (d).

4 Conclusions and Further Research

We proposed a control model for dynamic vergence eye movement of stereo head-eye system using disparity flux. Experiments under various conditions show that this model makes it possible a precise control of vergence movement. Also, this model can be applied to the control of vergence tracking.

As the perception-action cycle approaches to the real-time frame rate, the precision of disparity flux as well as the accuracy of the vergence movement control on the stereo robotic head increases. And a more efficient disparity estimation method robust to the variation of illumination and the noise would contribute to the more accurate control of dynamic vergence. This control model will be further studied to include smooth pursuit movement.

References

1. A. Maki, T. Uhlin, J. O. Eklundh: Disparity Selection in Binocular Pursuit. In Proc. of IAPR Workshop on Machine Vision Applications, Kawasaki, Japan. (1994) 182–185
2. M. Hansen, G. Sommer: Real-Time Vergence Control using Local Phase Differences. Machine Graphics and Vision. **5** (1996) 51–63
3. J. H. Piater, R. A. Grupen, K. Ramamritham: Learning Real-Time Stereo Vergence Control. In Proc. of the 1999 IEEE International Symposium on Intelligent Control, Cambridge, MA. (1999)
4. C. Capurro, F. Panerai, G. Sandini: Dynamic Vergence Using Log-Polar Images. International Journal of Computer Vision. **24** (1997) 79–94
5. J. Batista, P. Pexixoto, H. Araujo: Real-Time Vergence and Binocular Gaze Control. IEEE/RSJ International Conference on Intelligent robotic system. (1997)
6. D. Coombs, C. Brown: Real-Time Binocular Smooth Pursuit. International Journal of Computer Vision. **11** (1993) 147–164

Computing in Cortical Columns:
Curve Inference and Stereo Correspondence

Steven W. Zucker*

Center for Computational Vision and Control
Dept. of Computer Science and Electrical Engineering
Yale University
New Haven, CT U.S.A.
steven.zucker@yale.edu

Abstract. In stereoscopic images, the behavior of a curve in space is related to the appearance of the curve in the left and right image planes. Formally, this relationship is governed by the projective geometry induced by the stereo camera configuration and by the differential structure of the curve in the scene. We propose that the correspondence problem– matching corresponding points in the image planes– can be solved by relating the differential structure in the left and right image planes to the geometry of curves in space. Specifically, the compatibility between two pairs of corresponding points and tangents at those points is related to the local approximation of a space curve using an osculating helix. This builds upon earlier work on co-circularity, or the transport of tangents along the osculating circle. To guarantee robustness against small changes in the camera parameters, we select a specific osculating helix. A relaxation labeling network demonstrates that the compatibilities can be used to infer the appropriate correspondences in a scene. Examples on which standard approaches fail are demonstrated.

1 Introduction

Visual cortex is organized largely around orientation; that is, around selective responses to local oriented bars. In a classical observation of Hubel and Wiesel, recordings along a tangential penetration encounter a sequence of cells with regular shifts in orientation preference, while normal penetrations reveal cells with similar orientation and position preferences but different receptive field sizes. Together they define an array of orientation columns, and these columns provide a representation for visual information processing.

The question is thus raised: how can visual information processing be structured on orientation hypercolumns? Specifically, does the columnar architecture suggest which processing should take place within columns, and which between? It is commonly held that, if receptive fields were considered as oriented filters,

---

* Research supported by AFOSR; L. Iverson computed the co-circular compatibilities and S. Alibhai implemented the stereo computations.

then intra-columnar processing amounts to scale-spaces of filters. But what information processing tasks are these filters solving, and how does the scale-space notion arise from within the task? Physiologically, in what sense does this require elaboration across different layers, rather than just additional connections? Do certain non-linearities arise naturally?

A clue derives from considering the inter-columnar interactions, for which it is commonly held that, for "edge detection" and curve inference, co-aligned receptive fields should be mutually supportive. This suggests a role for long-range horizontal processes, and the anatomical and physiological evidence is accumulating ([12, 18, 14, 21]). However, how can the results of such "edge detectors" be evaluated (Fig. 1)? Furthermore, major exceptions already exist in the literature (e.g., Kapadia et al, 1995; Fig. 10), which show facilitation between cells with about 50 deg orientation differences. How can such exceptions be explained, and should the computational structure of boundary detection be elaborated to include them?

Finally, we note that, although undoubtedly important, the inference of boundary segments is only one of the tasks that might be located partly or completely in V1; given the ocular-dominance columns, stereo also seems a likely candidate. Is this possible?

We attempt to answer both of these larger questions from an information processing viewpoint. We first review aspects of our model for curve inference, and then sketch a model for stereo correspondence that explicitly takes advantage of the columnar architecture. To our knowledge this is the first formal stereo model that uses monocular orientation to support 3-D differential and projective geometry. The results open up a new class of curve-based stereo algorithms, and suggest why stereo computations are distributed across V1 and V2 in primates.

2 Curve Inference and Tangent Maps

We have developed a computational model based on the formal observation that visual orientation is a substrate for representing those tangents that approximate the curve bounding an object, the set of curves participating in a texture flow (e.g., a hair pattern), and other visual patterns. Formally horizontal interactions between orientations are necessary to reduce the errors inherent in locally estimating orientation, in localizing corners and discontinuities (as occur at the point where one object occludes another in depth), and for grouping and completing contour fragments obscured by highlights and specularities [23]. The relevant mathematics comes from differential geometry, which dictates that any such interactions must involve curvature [19]. The analysis is not unlike driving a car, in the following sense. At each instant of time the axis of the car defines its orientation, and the relationship between the orientation of the car at one instant with that at the next depends on how much the road curves; in operational terms, it depends on how much the steering wheel has to be turned during transport. This requires that curvature must be represented systematically with respect to orientation in cortex, and we (and others) have established that an-

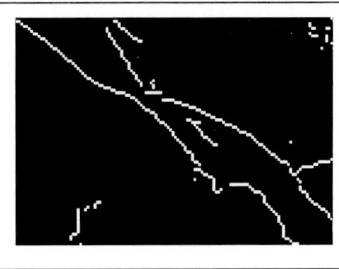

Fig. 1. An illustration of the geometric problems in interpreting standard approaches to edge detection. The edge map (middle) is obtained from the Canny operator, Matlab implementation, scale=3. This operator is designed with normalization and a hysteresis stage to enforce co-aligned facilitation thus, in some sense, realizing the two basic postulates for filters. While the result seems solid at first glance, a detail from the shoulder/arm region (right) shows the incorrect topology; neither the detail of the shoulder musculature, nor the separation from the arm, are localized correctly. A more careful examination of the full edge map in (middle) confirms the incorrect topology. We contend that a suffficent approach to edge detection should be topologically correct, which casts doubt on the above result.

other property of cortical neurons – "endstopping" – is sufficent for achieving this [4]. The majority of superficial, interblob orientation selective cells in V1 are also endstopped to some extent, and these bi-selective dimensions of orientation and endstopping are precisely what is required to represent tangent and curvature. We have used such notions of transport to derive the strength of horizontal interactions [23], which agree with available data for straight situations (curvature = 0) ([18, 14, 21], but generalize as well to explain data such as Kapadia et al[12]; see Fig. 2. Other more recent models ([13, 22]) adopt only our curvature = 0 case, and cannot explain the non-co-linear data.

We now suggest why the results in Fig. 3 are desirable. Rather than comparisons with "intuitive" notions of edge, we appeal to the basic mathematics of the situation. Whitney has classified maps from smooth surfaces into smooth surfaces, and has shown that only two situations can occur generically (i.e., without changing under small changes in viewpoint): the fold and the cusp (the position where the fold disappears into the surface); see Fig. 4.

Folds clearly indicate boundaries when viewed from a given position; in fact, the word implies that the tangent plane to the surface "folds" away from the viewer's line of sight. They thus become singular, the 2-D tangent space collapses to a 1-D tangent, and this is the object we seek. With the tangent direction established, the intensity profiles in the tangent and the normal directions can be examined separately. Note that linear receptive fields would average these together. We observe immediately that,

- *Normal direction:* The fold condition can take on a different intensity profile for a bounding edge (which involves a dark-to-light transition) from an interior fold (which involves a light-to-dark-to-light transition) or vice versa.

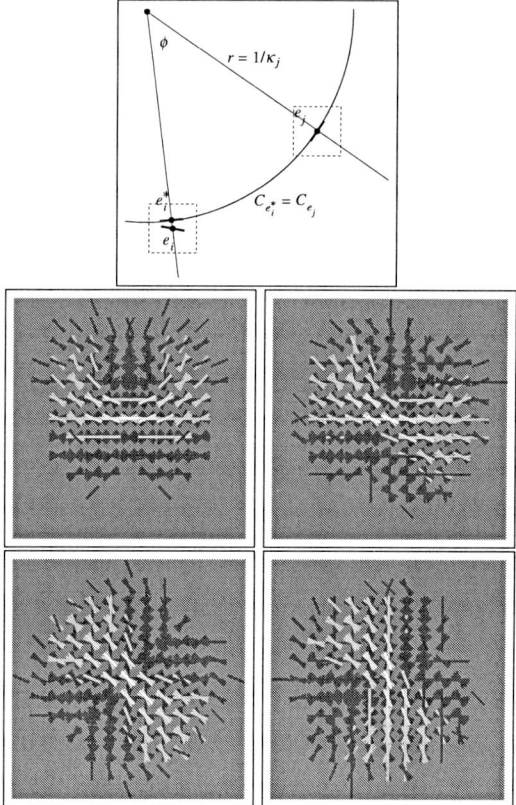

Fig. 2. The geometry of inter-columnar interactions. (top) Co-circularity indicates how consistent a neighboring tangent e_j is with a given tangent e_i, based on local transport along the osculating circle. The relaxation labeling network [8, 17] selects those tangents that minimize the mismatch between e_i and e_i^*, weighted by the initial strength of match. (bottom) Four examples of the compatibilities derived from co-circularity. Conceptually you can think of each bar as indicating the orientation preference of a single layer II-III pyramidal cell; multiple bars at the same position indicate several cells in the same orientation hypercolumn. Connections are relative to the center bright bar, with bright contrast indicating excitatory and dark inhibitory. The connections are intended to model long-range horizontal interactions, so they extend about 3 columns in each direction. Four cases are shown, clockwise from lower-left: co-aligned facilitation ($\theta_i = 45\,\text{deg}$ and $\kappa_i = 0.0$), curved a large amount in the negative sense ($\theta_i = 0\,\text{deg}$ and $\kappa_i = -0.2$), curved a small amount in the negative sense ($\theta_i = 22.5\,\text{deg}$ and $\kappa_i = -0.1$), and curved a small amount in the positive sense ($\theta_i = 67.5\,\text{deg}$ and $\kappa_i = 0.1$). Notice in particular that many of the excitatory connections would appear to be between co-aligned cells, given the loose definition of alignment commonly used in the physiological literature (e.g., $\pm 15°$); however, in the high curvature example there are cases of excitatory connections with approx. 50 deg relative orientation. This is precisely the outlier data from Kapedia et al discussed earlier.

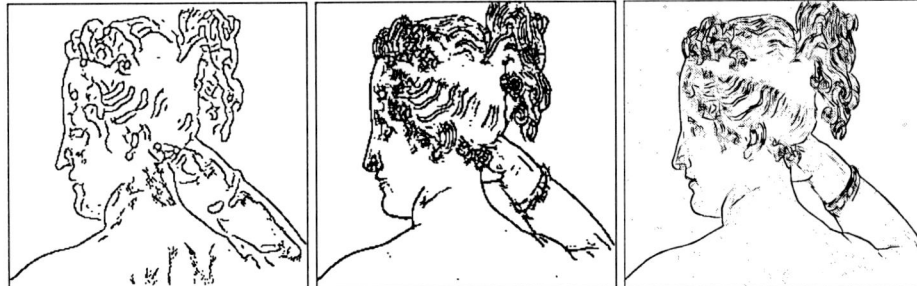

Fig. 3. Performance of our model for boundary and edge detection. (left) The Canny output at a scale larger than Fig. 1(center). Note how the topological problems remain despite the scale variation. (middle) The tangent map obtained from our logical/linear operators [10]. Note differences in the edge topology, with the shoulder musculature clearly indicated and proper "T" junctions around the neck and chin. (right) The result of our relaxation process using the co-circularity compatibilities (5 iterations of [8]). Note how the isolated responses through the hair have been removed, and how the details in high curvature regions (such as the ear) have been improved.

This latter profile is often called a line. Standard linear operators blur both together.
– *Tangential direction:* The definition of a tangent demands that continuity conditions exist (that is, that the limit of one point approaching another must exist). This corresponds to continuity contraints on the intensity pattern.

A necessary condition for a tangent to exist is that one or the other of the above intensity and continuity conditions must be satisfied. We have developed a class of non-linear local operators, called logical/linear operators [10], that use Boolean conditions to test whether the above structural criteria are met; if so, they return the average; if not, they veto to zero. "Edge operators" are separated from "line operators", and lines can arise either in light-dark-light conditions (typical of a crack or a crease) or dark-light-dark conditions (typical of a highlight). Note that both of these latter conditions refer to surface markings, rather than to surface boundaries.

3 The Geometry of Stereo Correspondence

The objects in our visual environment weave through space in an endless variety of depths, orientations and positions (Figs. 5, 6), and this suggests the need to extend the techniques from the first part of this paper to general space curves. Nevertheless, in computer vision the dominant theme has been to develop region-based methods to solve the correspondence problem [6, 7, 11, 15, 16, 20], or else to focus on long, straight lines. Edge features have been introduced to reduce the complexity of matching, but ambiguities along the epipolar lines (Fig. 5) are dealth with by a distance measure, e.g. similarity of orientation, over the possible

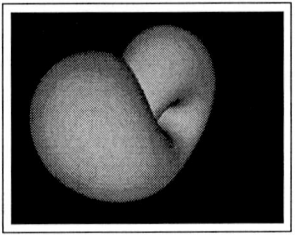 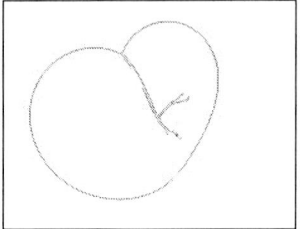 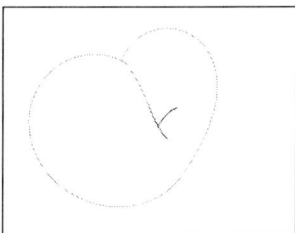

Fig. 4. The motivation from differential topology for early vision. The image of the Klein bottle in (left) shows how "T" junctions can arise from occlusion relationships (e.g., at the top of the figure), and how certain interior edges can end (e.g., where the fold smoothly joins the body). The Canny edge structure shown in (middle) is inconsistent with both of these topological observations. Notice how the boundary "T"-junction is not connected, how it smooths the outline, and how the interior folds blur into the shading. In (right) is shown the output of the logical/linear operator. Notice how the "T"-junctions are maintained, and how the contours end at cusps. Such configurations resemble the shoulder musculature in the Paolina image. Dark tangents signify lines; gray ones edges; see text.

edge matches. While this filters the number of potential matches, it does not solve the problem of which edge elements in each figure should correspond.

Fig. 5. A stereo pair of twigs. There are many places in the stereo pair where the ordering constraint breaks down. The images were taken using a verged stereo rig and a baseline of 9cm.

To resolve these remaining ambiguities, global (heuristic) constraints have been introduced, including the uniqueness constraint [6, 15, 20]; the ordering constraint (the projections of two objects in an image must have the same left to right ordering as the objects in the scene) [2, 6, 20]; the smoothness constraint [3, 6, 7, 15, 16], and so on. However, each of these is heuristic, and breaks down for the above images; see Figs. 7, 8. It is our goal to develop a stereo correspondence system that can work for scenes such as these. It is a second goal to determine whether such computations could be carried out in primate visual cortex.

There is a natural sense in which the occluding contour of an object weaves through space as does the object with which it is associated. The projection of the occluding contour is an edge that winds across the image plane. We focus on the relationship between the differential-geometric properties of curves in a scene and the projection of those curves into the left and right image planes. For details, see [1].

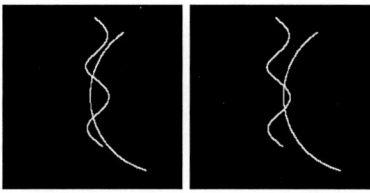

Fig. 6. A synthetic stereo pair consisting of a y-helix and a circular arc.

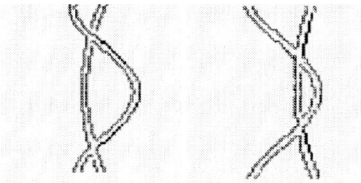

Fig. 7. The false matches obtained using the PMF algorithm of Pollard, Mayhew, and Frisby in the TINA system are indicated by the darkened pixels in the left and right images. This is a zoom on the previous example. The correct match for a pixel in the left image occurs on the same raster line in the right but is not darkened.

Specifically, we will relate the local approximation of a curve in three space to its projection in the left and right image planes. As previously utilized, the behavior of a planar curve can be described in terms of its tangent and curvature. Similarly, a curve in three space can be described by the relationships between its tangent, normal and binormal. As the curve moves across depth planes, there exists a positional disparity between the projection of the curve in the left image and the projection in the right image. However, there also exist higher order disparities, for example disparities in orientation, that occur. It is these types of relationships that can be capitalized upon when solving the correspondence problem. Rather than defining a distance measure that compares the positions or orientations of a left/right edge-element pair, we will require that neighboring pairs be locally consistent. That is to say, (i) we shall interpret edge elements as signaling tangents; and (ii) we shall require that there exists a curve in three space whose projection in the left and right image planes is commensurate with the locus of tangent pairs in a neighborhood of the proposed match.

Clearly, the role of projective geometry must not be ignored. One would not want correspondence relations that depended in minute detail on camera

Fig. 8. The reconstruction using PMF in the TINA system of the scene from the stereo pair in Fig. 2. It is shown from three separate orthogonal projections. The rightmost "birds-eye view" illustrates the complete merging of two curves in depth. The incorrect reconstruction is due to the failure of the ordering constraint in the image.

parameters and precise camera calibration, otherwise the solution would be too delicate to use in practice. We seek those relationships between the differential structure of a space curve and the differential structure of image curves on which stereo correspondence is based to be invariant to small changes in the camera parameters.

3.1 Differential Formulation of the Compatibility Field for Correspondence

Let $\alpha(s)$ be a curve in $\Re^3$. Assuming measurements of the curvature, torsion and Frenet frame of the curve at $\alpha(0)$, we obtain a local approximation of the curve by taking the third order Taylor expansion of α at $s = 0$. Using the Frenet equations and keeping only the dominant term in each component, we obtain the Frenet approximation of α at $s = 0$ (see Fig. 9):

$$\hat{\alpha}(s) = \alpha(0) + sT_0 + \kappa_0 \frac{s^2}{2} N_0 + \kappa_0 \tau_0 \frac{s^3}{6} B_0 \tag{1}$$

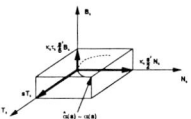

Fig. 9. Shown is the Frenet approximation to a curve in $\Re^3$ at a point $\alpha(0)$. Without loss of generality, the point $\alpha(0)$ is assumed to be coincident with the origin.

Another approximation can be derived for the case of a planar curve. Let α be unit speed planar curve with $\kappa > 0$. There is one and only one OSCULATING CIRCLE ψ which approximates α near $\alpha(s)$ up to second-order in the Frenet sense.

TRANSPORT PROBLEM Given a unit speed curve α, determine a unit vector field, $\boldsymbol{u}$, on α such that $\alpha'(s) \cdot \boldsymbol{u}(s) = \mathrm{const}$.

For any unit speed space curve, α, we can define the vector field $\boldsymbol{u} = T$ such that the pointwise relation $\alpha' \cdot \boldsymbol{u} = 1$ holds. If a curve is reconstructed from an image, then the reconstructed curve must satisfy this property. This assumes that α was known around the point $\alpha(0)$. Unfortunately, this assumption is unrealistic. However, we can use the observation that α can be locally approximated at 0 via the osculating circle, ψ: if the tangent at the position $\psi(\delta s)$ is transported along the osculating circle from $\psi(0)$ to $\psi(\delta s)$, it should approximate the tangent at $\alpha(\delta s)$ to first order. This is the co-circularity constraint that was used earlier in this paper; see also [19].

For the stereo correspondence problem, we are given two edge maps (one for the left camera and one for the right); each of these will be consistent (in the sense that they satisfy the transport constraint); our goal now is to make them consistent with a local approximation to the space curve from which they project. The osculating notion for space curves requires an osculating helix. To begin, recall that a unit speed circular helix is a space curve $h(s)$ where $h(s) = \left(a \cos\left(\frac{s}{c}\right), a \sin\left(\frac{s}{c}\right), \frac{bs}{c}\right)$ where $c = \sqrt{a^2 + b^2}$.

OSCULATING HELIX Let α be a unit speed space curve with $\kappa > 0$. There is one and only one unit speed circular helix, $\tilde{h}$, that locally approximates α at s in the sense of Frenet: $\tilde{h}(0) = \alpha(s); \boldsymbol{T}_{\tilde{h}}(0) = \boldsymbol{T}_\alpha(s); \boldsymbol{N}_{\tilde{h}}(0) = \boldsymbol{N}_\alpha(s); \kappa_{\tilde{h}} = \kappa_\alpha(s); \tau_{\tilde{h}} = \tau_\alpha(s)$.

Now, consider a point $M \in \Re^3$. There exists a family of unit speed smooth space curves that pass through M. For each such curve, it is possible to construct an osculating helix that locally approximates the curve at M. Each such approximation is only valid in a small neighborhood around M (denoted $\mathcal{N}(M)$). The projection of M onto the left image plane is $m_l = P_l(M)$. Similarly, the projection of the neighborhood around M is a neighborhood around m_l, $\mathcal{N}(m_l) = P_l(\mathcal{N}(M))$, where $\mathcal{N}(m_l)$ is a connected subset of the image plane. The neighborhood of the stereo pair $(m_l, m_r) = (P_l(M), P_r(M))$ is the set of corresponding points in the space $\mathcal{N}(m_l) \times \mathcal{N}(m_r)$.

Let M be a point in $\Re^3$ and $\boldsymbol{T}_M$ the tangent vector at M. The mapping S is the projection of M and $\boldsymbol{T}_M$ onto the left and right image planes:

$$S : E^3 \times T(E^3) \mapsto E^2 \times E^2 \times T(E^2) \times T(E^2)$$

We refer to the space $E^2 \times E^2 \times T(E^2) \times T(E^2)$ as the stereo tangent space and denote it with the symbol ϑ. A point $i \in \vartheta$ can be represented as $(x_l, y_l, x_r, y_r, \theta_l, \theta_r)$ where x and y are the projection of M in the image plane and θ is the orientation of the projected tangent in $T(E^2)$. The mapping S is invertible everywhere except for the line between the optical centers and the focal planes.

Two image points, one from the left image plane and one from the right, are referred to as *corresponding points* if the inverse projection of the pair is a point on some object in the scene.

We also define the projection map:
$$\pi : E^2 \times E^2 \times T(E^2) \times T(E^2) \mapsto E^2 \times E^2$$
$$\pi(x_l, y_l, x_r, y_r, \theta_l, \theta_r) = (x_l, y_l, x_r, y_r)$$

The map $P^{-1}(\pi(i))$ maps the stereo tangent pair, i, to the corresponding point in three space. The map $S^{-1}(i)$ maps the stereo tangent pair, i, to a point in three space along with its associated tangent.

Since α at M can be approximated by an osculating helix at M, the image of the curve α in a neighborhood around $(m_l, m_r) = (P_l(M), P_r(M))$ is commensurate with the projection of the osculating helix into the left and right image planes. Furthermore, the direction of corresponding tangents in the left and right image planes is approximately equal to the direction of the projected tangents along the osculating helix. *Each pair of corresponding points and tangents is referred to as a stereo tangent pair.* If i is a stereo tangent pair, $i \in \vartheta$, associated with a point on an arbitrary curve in three space, then in a neighborhood around i, the set of stereo tangent pairs will be commensurate with the positions and tangents of the projected osculating helix. We denote the neighborhood around i as $\mathcal{N}(i)$, where $\mathcal{N}(i) = \{j \in \vartheta | \pi(j) \in \mathcal{N}(\pi(i))\}$. Formally:

COMPATIBLE: Let $i \in \vartheta$ and $j \in \mathcal{N}(i)$. $\{i,j\}$ are *compatible* under S if and only if $\exists h(s)$, a circular helix, such that $S(h(s)) \supset \{i,j\}$. We denote the compatibility of i and j under the helix h as $(i \stackrel{h}{\sim} j)$.

Figure 10 depicts the projection of several helices onto the left and right image planes, each of which is locally consistent (according to the co-circularity transport constraint). Equivalently, for a pair of nearby points that lie on the curve in the left image, for example m_l and n_l, the position and tangent at n_l satisfy the transport constraints at m_l. The same property holds for m_r and n_r, two nearby points on the curve in the right image plane. In addition, for each pair of neighboring stereo pairs, $m = (m_l, m_r)$ and $n = (n_l, n_r)$, the inverse projection of the corresponding points lie on the original helix and the inverse projection of the tangents at m and n have the same orientation as the tangent in three space. This is true for every pair of corresponding points that lie on the perspective projection of the helix. Since the inverse projections $P^{-1}(m)$ and $P^{-1}(n)$ lie on a common helix, the transport constraints at m are necessarily satisfied, when we transport along the image of the helix from n to m. Therefore, we can express the compatibility between a pair of corresponding points as a compatibility relationship between stereo tangent pairs. However, *one cannot arbitrarily pick a curve in the left image and a curve in the right image and expect there to be a helix in $\Re^3$ that projects to the given curves.* Only those helices that satisfy the compatibility relation for a given stereo tangent pair are admissible pairings of curves in left and right images. It is the difference in the projection of the helices that allows us to solve the correspondence problem. In fact, we have the UNIQUENESS LEMMA:

Let $i_1, i_2 \in \vartheta$ such that $P^{-1}(\pi(i_1)) = P^{-1}(\pi(i_2))$ and $S^{-1}(i_1) \neq S^{-1}(i_2)$. There does not exist a circular helix h and $j \in \mathcal{N}(i_1) = \mathcal{N}(i_2)$, such that $(i_1 \stackrel{h}{\sim} j)$ and $(i_2 \stackrel{h}{\sim} j)$.

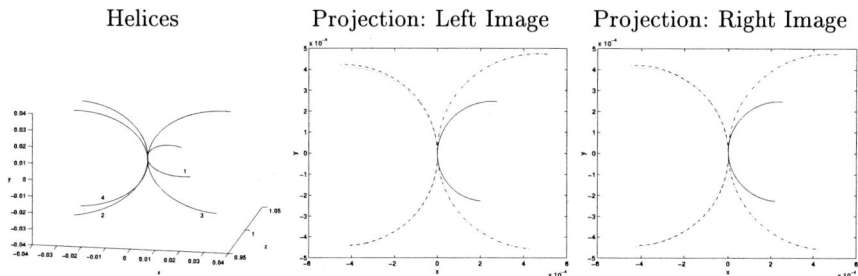

Fig. 10. (top) Four helices in three space whose projection pass through the point $i = (0, 0, 0, 0, \pi/2, \pi/2)$. (Bottom) Projections in the left and right image planes.

Thus for every tangent element (tangel) in the left tangent field, there is only one matching tangel in the right tangent field.

The COMPATIBILITY FIELD for $i \in \vartheta$ under S is the set of all pairs (j, h_j), where $j \in \mathcal{N}(i)$ and h_j is a unit speed circular helix, such that the relation $(i \stackrel{h_j}{\sim} j)$ holds for the camera model S. We denote the compatibility fields for a point i as $\text{CF}_S(i)$.

The solution to the correspondence problem can now be framed in terms of the local approximation of curves. Consider a point M on a smooth curve in three space and a pair of points, not necessarily corresponding points, in the image plane $m = \{m_l, m_r\}$ (the subscripts, respectively, indicate the left and right components of the pair). Earlier, it was shown that there exists an osculating helix to the curve at M. Therefore, m_l and m_r are a match if and only if the image of the osculating helix in the left and right image planes is coincident with the locus of edge points around m_l and m_r.

Formally, solving the correspondence problem is equivalent to solving the transport problem. Consider two arbitrary curves, one in the left image plane and one in the right. Pick two tangents, (m_l, θ_l) and (m_r, θ_r) such that they lie on matching epipolar lines in the left and right image planes. The two curves can be expressed as single curve β in ϑ. Let $\beta(0) = (m_l, m_r, \theta_l, \theta_r)$ and $\beta(\delta s) = (\hat{m}_l, \hat{m}_r, \hat{\theta}_l, \hat{\theta}_r)$ be two points along β. (m_l, θ_l) and (m_r, θ_r) are corresponding tangents if there exists a helix in E^3 that satisfies the transport criteria at $\beta(0)$ and $\beta(\delta s)$. That is to say, there exists a unit speed osculating helix, $h \in \mathcal{H}$, such that $S(h(0), \boldsymbol{T}_h(0)) = \beta(0)$ and $S(h(\delta t), \boldsymbol{T}_h(\delta t)) \approx \beta(\delta s)$. These transport criteria are expressed as the position and tangent direction of the projected helix.

The compatibility field around a point $i = (m_l, m_r, \theta_l, \theta_r) \in \vartheta$ is the set of all stereo tangent pairs in a neighborhood of i for which the transport constraints can be satisfied. To draw this in a figure we separate the left components of i, (m_l, θ_l), from the right components (m_r, θ_r). The result is what appears to be a tangent field surrounding the point for which the compatibility field was constructed. Each tangent in the left compatibility field has a matching tangent in the right field. However, there may be a disparity in both the position and orientation between the matching tangents. It is the differences in position and orientation between corresponding tangents in the left and right compatibility fields that are the key to solving the correspondence problem. This relationship is captured directly in the compatibility fields.

However, we are now at the place where the projective geometry (the camera model) interacts with the differential geometry:

OBSERVATION: Given a stereo tangent pair i, the equations used to determine the osculating helix to the curve α at $\alpha(S^{-1}(i))$, are underdetermined.

The osculating helix is characterized by nine variables: three for position, two for the direction of the tangent, two for the direction of the normal, one for curvature and one for torsion. Given a stereo tangent pair, it is possible to invert the perspective projection matrices to obtain the position, M, and tangent of the curve in $\Re^3$. However, given only a stereo tangent pair, it is not possible to determine the direction of the normal, curvature or torsion. Therefore, there does not exist enough information to uniquely determine the helix passing through M. The family of possible helices passing through M is homeomorphic to the space $S^1 \times \mathbb{R}^+ \times \mathbb{R}$ (where S^1 denotes the unit circle). In the next section, we will provide an argument based on the principle of differential invariance to consider only a subset of the helices in $S^1 \times \mathbb{R}^+ \times \mathbb{R}$.

3.2 The Invariant Compatibility Field

The construction of a compatibility field requires a precise characterization of the camera (intrinsic/extrinsic) parameters. It would be exhausting if the compatibility fields had to be recomputed for every set of camera parameters; it would be unrealistic to assume that the camera parameters could be obtained with arbitrary precision. We seek an INVARIANT COMPATIBILITY FIELD: Consider the mappings S_1 and S_2 where each mapping is constructed using a different set of camera parameters. Let $i \in \vartheta$. For every $j \in \mathrm{CF}_{S_1}(i)$ $\exists h_j \in \mathcal{H}$ such that $S_1(h_j) \supset \{i,j\}$. The $\mathrm{CF}_{S_1}(i)$ is invariant to changes in the camera parameters if for all h that satisfy $S_1(h) \supset \{i,j\}$, $S_2^{-1} S_1(h) \in \mathcal{H}$.

To simplify the analysis, we assume that only the camera parameters θ, dx and dy are variable. Further we assume that the fixation point of the camera, (x_o, y_o, z_o) is such that $z_o > x_o$, $z_o > y_o$ and $z_o \gg$ dx. The last restriction constrains the value of θ to be close to zero. Lastly, we assume that the camera are symmetrically verged at the fixation point.

If the compatibility field is computed for a small neighborhood around a point $i \in \vartheta$, then the affine projection equation can be used to study the invariance of the compatibility fields under changes in the camera parameters. Specifically, we determine if a helix that satisfies the compatibility criteria between two points in ϑ is preserved under changes in the camera parameters by studying the deformation of the spherical image of a circular helix under $F \triangleq A_{2*}^{-1} \circ A_{1*}$, where A_{1*} is the affine derivative map of S_1. We refer to F as the affine stereo transformation.

We can use the spherical image of a helix, σ, to define a circular cone in R^3, $\mathbf{x}_\sigma(u,v) = u\sigma(v)$. Let α be a unit speed cylindrical helix. For every parameterization β of α, the image of β' lies on a cone defined by the spherical image of α. Thus an equivalence relation can be defined on the family of cylindrical helices.

Each equivalence class consists of all helices that are the reparameterizations of a given unit speed cylindrical helix. The spherical image of every helix in the equivalence class lies on the cone defined by the spherical image of the unit speed helix. With additional lemmas [1] we have:

Theorem The compatibility field is invariant if $\forall h \in \mathcal{H}$, $F(\mathbf{x}_\sigma) \in \mathcal{C}$, where $\mathcal{C}$ is the family of all circular cones, $\mathbf{x}_\sigma \in \mathcal{C}$ and σ is the spherical image of the circular helix.

A global property of a circular cone is that there exists a vector $\boldsymbol{a}$ such that the angle between $\mathbf{x}(u,v)$ and $\boldsymbol{a}$ is a constant, $\angle(\mathbf{x}, \boldsymbol{a}) = k$, $\forall u > 0, v \in \mathbb{R}$. If the cone $F(\mathbf{x})$ is a circular cone then:

$$\angle(F(\mathbf{x}(u,v)), F(\boldsymbol{a})) = k$$

$$\lambda(u,v) \triangleq \frac{\partial \angle(F(\mathbf{x}(u,v)), F(\boldsymbol{a}))}{\partial v} = 0 \qquad (2)$$

The function λ is a smooth continuous function because the surface of the cone is smooth. Since λ is a function of the derivative map A_*, it is also a function of the camera parameters, e.g. θ_l and θ_r. If the camera parameters defined by A_1 are fixed, and if initially $A_2 = A_1$, the derivative of λ with respect to some camera parameter of A_2 measures the invariance of the compatibility field to changes in that camera parameter. If for example $\frac{d\lambda}{d\theta_l}$ is non-zero, then the helix is deformed by the mapping $A_2^{-1} A_1$ as θ_l varies. Therefore, λ can be used to study which unit speed circular helices are invariant to changes in the camera parameters.

Equation 2 can be used to study the invariance of the helix h_1 by considering its spherical image σ_1. Similarly, it is possible to study the invariance of h_2 using the spherical image $\sigma_2 = R(\sigma_1)$, where R is a rotation matrix. Therefore, it is possible to study the invariance of an entire family of helices, each with a different direction by considering different rotation matrices. There is a natural symmetry on S^2 (unit sphere) based on the fact that the deformation associated with a point on S^2 and its conjugate are equal. Figure 11 shows an example of the surface that is generated by considering the deformation for a family of helices, a subset of $\mathcal{H}_0$, that have the same value of curvature and torsion but whose directions vary. The locus of points along the bottom of the valley represent the helix whose axis is defined by the vector $(0,0,1)$. We will denote this helix as the z-helix. The figure demonstrates that the z-helix is most stable to small changes in the camera parameter θ_l. A similar observation is made when other camera parameters are varied.

Figure 12:Top shows three helices which are oriented along three different and mutually perpendicular axes. The degree of invariance to changes in θ_l for each of the helix directions are shown at the bottom of the figure. Of the three helices, the z-helix is the most stable to changes in the camera parameters for most ratios of $\frac{\tau}{\kappa}$. A similar observation is made for small changes in θ_r, dx and dy.

It was stated earlier that any solution to the stereo correspondence problem must exhibit some invariance to camera parameter uncertainty. In figure 12 the

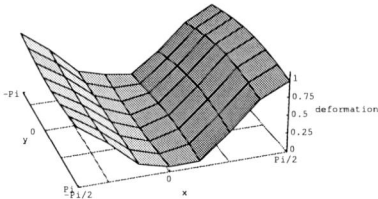

Fig. 11. The degree of deformation, normalized between 0 and 1, for helices pointing in different directions. The direction of each helix is represented by a rotation around the x and y axis of the standard Euclidean frame. The variables x and y represent a rotation about the x and y axis respectively. The helix with 0 deg of rotation around the x-axis is referred to as the z-helix. The value of curvature and torsion for each helix in the family is $\frac{1}{\sqrt{2}}$ and $\frac{1}{\sqrt{2}}$. See text for details.

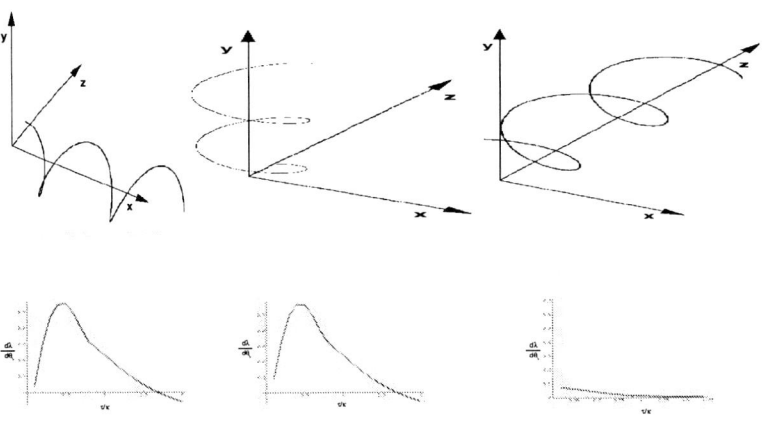

Fig. 12. Top: Shown are three helices in E^3, denoted x-helix, y-helix, z-helix from left to right respectively. Bottom: The degree of deformation of each helix type as the left camera angle is slightly perturbed. The maximum amount of deformation of each helix type is shown as a function of the ratio τ/κ. The line of sight is along the z-axis. The z-helix, the rightmost helix, is significantly more stable under camera perturbations than the other two.

magnitude of the maximum deformation of the x- and y-helix are significantly larger compared to the z-helix. Therefore, of the three helices, the z-helix comes closest to satisfying the principle of differential invariance.

The orientation disparity encoded by a stereo tangent pair can now be interpreted in terms of the physical behavior of a z-helix in three space. If the amount of torsion is equal to zero, the helix is confined to the plane of fixation and the orientation disparity is necessarily zero [9]. As the torsion increases, the z-helix begins to twist out of the plane of fixation. The tangent at the point where the helix crosses the plane of fixation is no longer parallel to the plane. This difference induces an orientation disparity between the projection of the tangent in the left eye and the tangent in the right eye (Fig. 13).

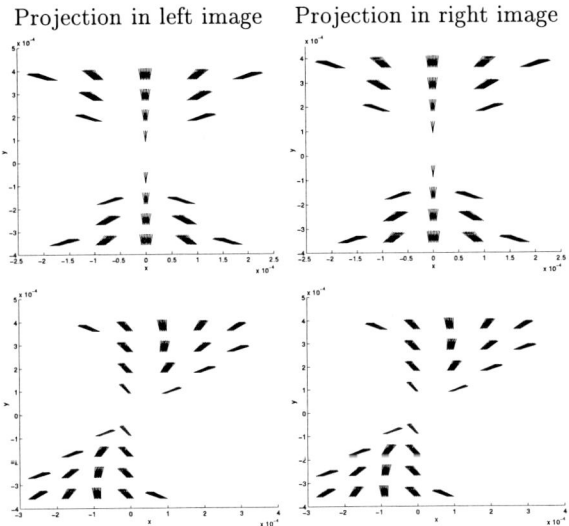

Fig. 13. Top: The positive discrete compatibility field for the point $\hat{\imath} = (0,0,0,0,\pi/2, pi/2)$ projected in the left and right image planes. Bottom: The positive discrete compatibility field for the point $\hat{\imath} = (0,0,0,0,\pi/3, pi/3)$ projected in the left and right image planes.

4 Results

The first step in our stereo process is to obtain a discrete tangent map representation of the stereo pair. Figure 14 is the reconstruction of the scene from the synthetic stereo pair using our approach to correspondence matching and without curvature information. In those few places where the differential geometric constraints are inadequate, we select those matches that lie closest to the fixation plane. The most significant observation with respect to the reconstruction is that the change in the ordering of primitives, in this case discrete

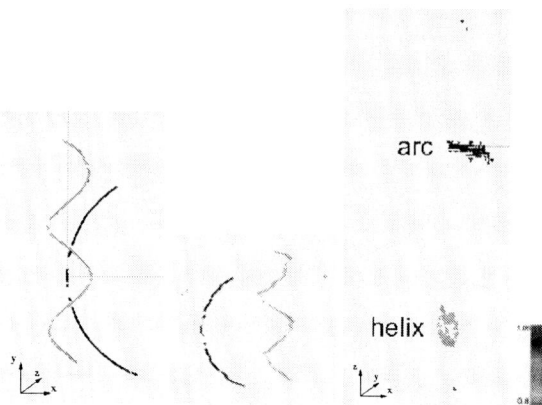

Fig. 14. The reconstruction of the stereo scene from the stereo pair in figure 6 is shown from three separate projections. The color of each space tangent represents its depth with red being the closest to the camera and blue the farthest away. Note that the approximate geometry of the curves in three space is reconstructed and that the curves are localized at distinct depths. The result is obtained after 5 iterations of relaxation labelling and $\sigma_c = 1.75$.

tangent elements, does not effect the matching algorithm. Two curves, a y-helix and an arc, are both successfully segmented and rendered at the correct depth. Another significant difference between the TINA reconstruction and ours is that the our curves are both isolated in space. This difference is due in part to the failure of the PMF matching algorithm near junctions in the image and our use of the discrete tangent map to represent multiple distinct tangents at a given point. At the image junctions, the tangent map encodes two separate tangents directions, one for the arc and the other for the helix. The multiple tangents at the junction in the left image participate in the matching process with the appropriate tangents at the junction in the right image. This ensures that the parts of the helix in the scene that correspond with the junction areas in the image have valid matches in the left and right image planes. The same fact also applies for the arc.

The errors in matching induced by the lack of ordering are further exaggerated when the helix is farther out in front of the circular arc. This is because the extent of the image where the ordering constraint is invalid increases. Figure 6 contains the exact same curves as the previous stereo pair but exaggerates the parts of the image where the ordering constraint breaks down. Predictably, the number of pixels that are falsely matched using the PMF algorithm increases as is shown in figure 7. The reconstruction of the scene as computed by TINA is shown in figure 8.

In contrast to the stereo reconstruction created by TINA, our stereo algorithm returns a significantly better reconstruction of the original scene (see figure 14). As in the previous stereo pair, the only constraint used in the computation was a bias for matches close to the fixation plane. The stereo reconstruction results in two curves, an arc and a y-helix that are separated in depth.

Fig. 15. (top) A stereo pair of Asiatic lilies. The cameras had a 9cm baseline. The stereo pair can be viewed using uncrossed fusing. (bottom) Discrete tangent map computed as in the first part of this paper.

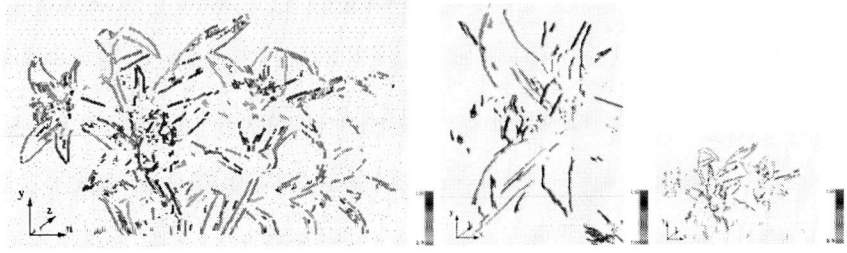

Fig. 16. The depth map associated with the stereo lily pair. The color bars indicate depth scales associated with the image to their left. Each point in the lily image is a tangent in three space that is geometrically consistent with its neighbors. (Left) Full lily image. (Middle) Flower detail magnified. (Right) A cropping plane 1.01m away from the cyclopean center to show the details of objects in the foreground.

The gaps in the stereo reconstructions are due mainly to the inability match horizontal tangents. image tangents whose orientations are less than 20deg from the horizontal are not used in the stereo reconstruction. There are some false matches as shown in the left panel of the above figures.

5 Summary and Conclusions

We have shown how differential-geometric and -topological ideas can be used to specify local measurements of orientation (via logical/linear operators), refinement of local measurements by co-circular transport, and how these can be

Fig. 17. Left: The stereo reconstruction of the twigs scene. There are some mismatches near the base of the twigs and along the branch receding in depth. The reconstruction is not affect by those areas where there is failure of the ordering constraint. Middle: A cropping plane 1.03m from the cyclopean point was placed to highlight the twigs in the foreground. Right: A birdseye view of the reconstructed scene. The twigs are arranged in roughly three layers as can be seen by fusing the stereo pair.

extended for contour-based stereo correspondence. In other work we have shown how segmentation processes can operate within columns, and how endpoints and textures can be defined [5]. Furthermore, we have shown how the relaxation computation can be implemented by layer II-III pyramidal cells [17], at least to a continuous approximation, on physiological time scales of about 5 spikes in about 25 msec. This implementation can also provide additional orientation resolution. We close with the observation that if orientation is sampled too coarsely, the correspondence algorithm performs poorly (Fig. 18); perhaps this is why the physiological realization involves V1 to V2 projections.

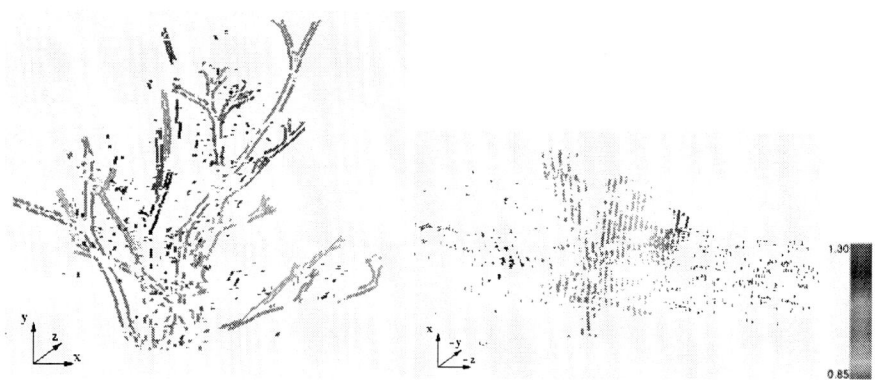

Fig. 18. (Left) A stereo reconstruction of the twigs using a coarse tangent map. (Right) A birdseye view of the scene. When compared with Fig. 17 (right), the lack of separation into depth layers can be seen. The additional orientation resolution is required for the torsion (or, equivalently, the orientation disparity) information. We speculate that this additional requirement for orientation resolution explains why the stereo computation is distributed across V1 and V2.

References

1. S. Alibhai and S. W. Zucker. Contour-based correspondence for stereo. *CVC Technical Report, Yale University*, 1999.
2. H. H. Baker and T. O. Binford. Depth from edge and intensity based stereo. In *Proceeding of the 7th Joint Conference of Artificial Intelligence*, pages 631–636, 1981.
3. S.T. Barnard and W.B. Thompson. Disparity analysis of images. *PAMI*, PAMI-2(4):333–340, July 1980.
4. A. Dobbins, S. W. Zucker, and M. Cynader. Endstopped neurons in the visual cortex as a substrate for calculating curvature. *Nature*, 329:438–441, 1987.
5. B. Dubuc and S. W. Zucker. Indexing visual representations through the complexity map. *Proc. Fifth International Conf. on Computer Vision*, 1995.
6. J.P. Frisby and S.B. Pollard. Computational issues in solving the stereo correspondance problem. In M.S. Landy and J.A. Movshon, editors, *Computational Models of Visual Processing*, chapter 21, pages 331–357. MIT Press, Cambridge, MA, 1991.
7. W.E. Grimson. A computer implementation of a theory of human stereo vision. In *Philisophical Transactions of the Royal Society of London*, volume B292, pages 217–253. 1981.
8. Robert Hummel and Steven Zucker. On the foundations of relaxation labelling processes. *IEEE Trans. Pattern Analysis and Machine Intelligence*, 5(3):267–287, 1983.
9. B.J. Rogers I.P. Howard. *Binocular Vision and Stereopsis*. Oxford Psychology Series No. 29. Oxford, 1995.
10. Lee Iverson and Steven Zucker. Logical/linear operators for image curves. *IEEE Trans. Pattern Analysis and Machine Intelligence*, 17(10):982–996, October 1995.
11. Jesse Jin, W.K. Yeap, and B.G. Cox. A stereo model using log and gabor filtering. *Spatial Vision*, 10(1):3–13, 1996.
12. M. Kapadia, M. Ito, C. Gilbert, and G. Westheimer. Improvement in visual sensitivity by changes in local context: Parallel studies in human observers and in v1 of alert monkeys. *Neuron*, 15:843–856, 1995.
13. Z. Li. A neural model of contour integration in the primary visual cortex. *Neural Computation*, 10:903–940, 1998.
14. R. Malach, Y. Amir, M. Harel, and A. Grinvald. Relationship between intrinsic connections and functional architecture revealed by optical imaging and in vivo targeted biocytin injections in primate striate cortex. *Proc. Natl. Adad. Sci. (USA)*, 90:10469–10473, 1993.
15. David Marr and T. Poggio. A theory of human stereo vision. *Proceedings of the Royal Society of London*, B 204:301–328, 1979.
16. G. Medioni and R. Nevatia. Segment-based stereo matching. *Computer Vision, Graphics and Image Processing*, 31:2–18, July 1985.
17. D. Miller and S. W. Zucker. Computing with self-excitatory cliques: A model and an application to hyperacuity-scale computation in visual cortex. *Neural Computation*, 11(1):21–66, 1999.
18. J. Nelson and B. Frost. Intracortical facilitation among co-oriented, co-axially aligned simple cells in cat striate cortex. *Exp. Brain Research*, 61:54–61, 1985.
19. Pierre Parent and Steven Zucker. Trace inference, curve consistency and curve detection. *IEEE Trans. Pattern Analysis and Machine Intelligence*, 11(8):823–839, August 1989.

20. Doron Sherman and Shmuel Peleg. Stereo by incremental matching of contours. *IEEE Trans. Pattern Analysis and Machine Intelligence*, 12(11):1102–1106, November 1990.
21. D. T'So, C. Gilbert, and T. Wiesel. Relationships between horizontal interactions and functional architecture in cat striate cortex as revealed by cross-correlation analysis. *J. Neuroscience*, 6:1160–1170, 1986.
22. S.-C. Yen and L. Finkel. Salient contour extraction by temporal binding in a cortically-based network. *Adv. in NIPS*, 9, 1997.
23. S. W. Zucker, A. Dobbins, and L. Iverson. Two stages of curve detection suggest two styles of visual computation. *Neural Computation*, 1:68–81, 1989.

Active Vision from Multiple Cues

Henrik Christensen and Jan-Olof Eklundh

Computational Vision and Active Perception
Numerical Analysis and Computational Science
Royal Institute of Technology
SE-100 44 Stockholm, Sweden
{hic,joe}@nada.kth.se

Abstract. Active vision involves processes for stabilisation and fixation on objects of interest. To provide robust performance for such processes it is necessary to consider integration and processing as closely coupled processes. In this paper we discuss methods for integration of cues and present a unified architecture for active vision. The performance of the approach is illustrated by a few examples.

1 Introduction

A fundamental problem in computational vision is figure ground segmentation, where objects of interest are separated from the remainder of the visual input. Traditionally computer vision methods have considered recovery of basic visual characteristics in a bottom-up driven manner. The characteristics are then combined as a basis for recognition and tracking. Such approaches actually presuppose that what is (an object) of interest is known a priori and independent of the task at hand. Furthermore, since the selected cues, often just a single one, are given, the possibility to use or capitalize on *available* information is limited. In an active perception approach to vision the full process of computation is considered task oriented. In a such an approach models of the physical world (in terms of dynamics, geometry, and optics) provide constraints that can be used to control the overall process and turn it into a tractable problem. Moreover the system can sample the world and acquire information in relation to the task(s) at hand.

Fundamental to use of active vision is the ability to control the image acquisition process and the associated set of processes. An active vision head such as the KTH head [1] enables control of intrinsic and extrinsic parameters. Also the body motion is important, although it often is less dependent on the task. Still it provides additional constraints and information. This control in turn provides the basis for extraction of visual features such as depth from accommodation, and control of key parameters such as depth of field of view. The control of extrinsic parameters enable also parameterisation of disparity estimation, stabilisation, and reduction or introduction of image parallax. Utilizing these characteristics is by no means new, but the controlled use of such features provides a basis for robust extraction of features that facilitate figure-ground segmentation. In this paper extraction and fusion of features in the context of active vision is considered as a basis for robust figure-ground segmentation.

The rest of the paper is organised with an outline of prior work on cue integration and active vision with pointers to the biological and machine vision literature. Various approaches to integration are discussed and an overall architecture for an active vision system is presented. The performance of the presented approach is illustrated through a number of application examples and finally a number of conclusions are presented and issues for future research are outlined.

2 Cue Integration

It is well-known from the physiological litterature that biological systems provide a rich set of features through the visual areas of V1, V2, V3, V4 and MT. The visual cortex provides access to a rich set of features in terms of band-pass filtered version of the scene (V1), optical flow channels (V3 and V5) and a range of different disparity estimators (V2) in terms of correlation type matching of visual patterns. There are also indication for colour constancy response in V4 [2]. An excellent description of the wide range of different features can be found in [3,4]. In the biological literature various models for cue integration have been proposed. Many of these approaches [5,6,2] exploit a weighted averaging approach to fusion, i.e.

$$f(\boldsymbol{p}) = \sum_i w_i f_i(\boldsymbol{p})$$

Where $f(\boldsymbol{p})$ is the feature value at the location $\boldsymbol{p}$ and the parameters w_i are weights that may change in response to visual input (i.e. texture, contrast, etc). Such approaches rely on the fact that an abundance of visual features are available from a 'visual front-end' in combination with methods that allow estimation of the 'utility'/confidence for these features (a task dependent characteristic).

In a computer vision context in particular Bayesian optimization has been applied as a basis for fusion of cues, as elegantly illustrated by Blake and Zisserman [7]. Variational approaches as regularisation have been presented by e.g. Aloimonos and Shulman [8]. These approaches rely on the formulation of an optimization approach that includes two components: feature - phenomena models that explicitly model the relationship between scene objects and features as for example modelled by conditional probabilities, and models of scenes represented by a priori probabilities. A problem with these approaches is that it might be difficult to derive or learn such models in a robust manner.

More recently coincidence of features has also been used as a basis for integration, as for example reported by Uhlin et al. [9,10]. In these approaches the co-occurance of features that corroborate a particular hypothesis are interpreted as evidence. They allow use of qualitative methods such as voting as a basis for fusion of cues. These methods rely on much weaker models of the underlying scene. The methods are thus more robust to variations in the scene and the feature set. Voting has been widely used in reliable computing as a basis for integration of information, as for example reported by Parhami [11]. In voting based fusion three different concepts are needed: a) a voting domain (Θ), b) a fusion operator (Λ) and c) a selection mechanism. For each different cue a voting function (v_i) is defined that maps visual input to the voting space.

One can then think of each cue estimator v_i as a mapping:

$$v_i : \Theta \to [0;1]. \tag{1}$$

The voting domain may for example be 3D space or the image plane. I.e., we can estimate the presence of a particular feature at a given location in the image, as used in many image segmentation methods. Alternatively the voting domain could be a control space like the control space for the camera head. In term of selection there are several possible methods which can be used. If each, of the n, cue estimators (v_i) produce a binary vote for a single class (i.e. present or not present) a set of thresholding schemes can be used ($v_i : \Theta \to \{0,1\}$):

Unanimity: $\sum v_i(\theta) = n$
Byzantine: $\sum v_i(\theta) > \frac{2}{3}n$
Majority: $\sum v_i(\theta) > \frac{n}{2}$

The fusion operator defines the relative weighting of votes, unless a uniform weight is used. In the application section a few examples of use of voting will be provided.

3 An Architecture for Active Vision

An important mechanism of integrated systems is the fusion of cues, but to provide robust results it is not enough to simply perform data-driven integration. In general the integration must be considered in a systems context. With inspiration from the physiology, component based computer vision methods and successful integration of systems, it is possible to propose an architecture. Important components of an active vision system includes:

Visual Front-end. that provides a rich set of visual features in terms of region characteristics (homogeneity, colour, texture), discontinuities (edges and motion boundaries), motion (features (discontinuities) and optical flow (regions)) and disparity (binocular filtering). The estimation of features is parameterised based on "the state of the imaging system", the task and non-visual cues. Particular care is here needed to use theoretically sound methods for feature estimation, as for example reported by [12].

Non-visual cues. The active vision system is typically embodied and the motion of the active observer influences the visual input. This information can be fed-forward into the visual process to compensate / utilize the motion [13].

Tracking. facilitates stabilisation and fixation on objects of interest. The motion of objects can be fed-back into the control system to enable stabilisation of objects, which in turn simplifies segmentation as for example zero-disparity filtering can be utilized for object detection [14,15]. In addition motion descriptions can be delivered to higher level processes for interpretation [16], etc.

Recognition. A fundamental part of any vision system is the ability to recognize object. Recognition defines the objective for the figure ground segmentation and provides at the same time the basis for interpretation etc of visual scenes.

Attention. All visual systems utilize a computational substrate that is resource bounded and in addition low-level processes must be tuned to the task at hand. To accommodate this there is a need for attention mechanism that can schedule and tune the involved processes to the current task. Attention involves both selection [17] and tuning [18]. In addition attention is also responsible for initiation of saccades for the head-eye system.

Cue Integration. Finally cue integration is key to achieve robustness as it provides the mechanism for fusion of information. Integration of cues can be divided into two different components a) fusion for stabilisation and b) fusion for recognition. In fusion for stabilisation/tracking the basic purpose it combine visual motion, disparity and basic features to enable frame-by-frame estimation of motion. For recognition a key aspect is figure-ground segmentation to filter out features of relevance for the recognition.

Head-Eye Control Generation. Based on visual motion, attention input and (partial) recognition results the control of extrinsic and intrinsic parameters must be generated (both in terms of continuous control and gaze-shifts). Through consideration of basic imaging characteristics the control can be generated from the above information, see for example [19,18,1,14]

One approach to organisation of these "modules" is shown in figure 1. This architecture is one example of a system organisation, it focusses primarily on the major processing modules and their interconnection. The model thus fails to capture the various representations (short term memory, 2D vs 3D motion models, etc.), and the use of feedback is only implicitly represented by the double arrows. The models is first and foremost used as a basis for formulation of questions such as: "should attention be a separate module or part of several other modules?", "How is the channels to other bodily functions accomplished?", "Is memory a distributed entity organised around modules or is it more efficient to use a representation specific memory organisation?". We can of course not answer all these questions (in this paper).

In the above architecture the 'visual front-end' is divided into two components (2D features and disparity estimation). These processes are tightly coupled through the cue integration modules. The non-visual cues are integrated directly into the stabilisation / tracking part of the system. As can be seen from the architecture attention plays a central role in the control of feature estimators, integration of cues and control generation (saccadic motion). The architecture is intentionally divided into a part that considers temporal aspects (stabilisation and ego-motion compensation) and a recognition part that utilizes figure ground information for segmentation. In practise these two parts are of course closely linked to satisfy the need for stabilisation and recognition, which is accomplished through attention and task control.

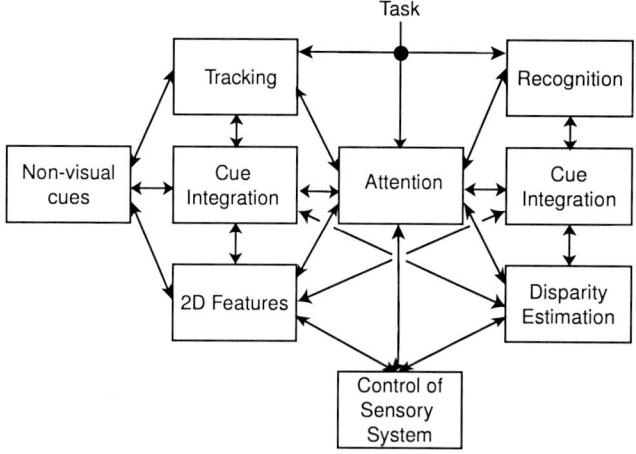

Fig. 1. Architecture for active vision system

4 Application Examples

Over the last few years a central part of the research within the CVAP group has focussed on the design and implementation of the active visual observer, which is our testbed for research on biologically plausible methods for (active) computational vision. The implementation of the system outlined in figure 1 is thus work in progress. In the following a few examples of the use of the above mentioned approach are provided to illustrate the adopted methodology.

A fundamental part of figure ground segmentation is fusion of disparity and motion information to enable detection and tracking of objects. Several have reported detection and tracking of objects in a stationary setting. In the context of an active vision system the assumption of a stationary background can not be utilized. There is here a need for stabilisation of the background (compensation for ego-motion) and subsequent computation of disparity and motion to allow detection and tracking of objects. In many scenes it can be assumed that the background is far away compared to the object of interest. For many realistic scenes it is further possible to assume that the background can be approximated by a plane (i.e. the depth variation of the background is small compared to the depth of the foreground). One may then approximate the motion of the background by a homography (planar motion). Using motion analysis over the full image frame the homography can be estimated and the image sequence can be warped to provide stabilisation on an image by image basis (to compensate for ego-motion of the camera). To aid this process it is possible to feed-forward the motion of the cameras to the warping process. As the depth (in general) is unknown there is still a need for estimation of the homography. Once the background has been stabilised it is possible to use standard motion estimation techniques for detection of independently moving objects. Using a fixation mechanism for the camera head further enables simple estimation of disparity as the search region can be reduced and thus made tractable. Using motion and disparity estimation it is possible to setup a voting method. The voting space is here the motion / disparity space. Through clustering of pixels that exhibit similar motion at a given depth it is possible to segment independently moving objects and track them over time as reported by Pahlavan et al. [20] and Nordlund [21]. In figure 2 an example image, the voting space and the corresponding segmentation is shown.

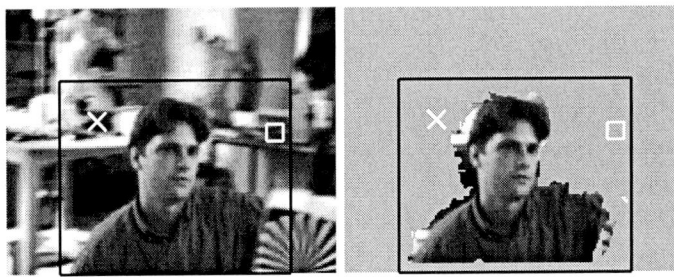

Fig. 2. Example of figure ground segmentation using disparity and motion. The left figure shows an image from the sequence, while the right figure shows segmentation of the moving person

For tracking of general objects individual cues do not provide the needed robustness. We have thus investigated use of combination of regions (colour/homogeneity), edges, correlation and motion as a basis for maintenance of fixation. In this case the different cues were integrated using majority voting directly in the image space (after stabilisation). The method allow simple recognition of the moving object(s). The method was evaluated for tracking of a robot manipulator (that enables controlled experiments). The performance of individual feature detectors and the combined system was compared. Example images from the experiments are shown in figure 3. A summary of the results obtained are shown in table 1.

Fig. 3. Example images used for evaluation of voting based fusion for stabilisation

Table 1. Performance of different cues when used in stabilisation experiment

Module	$\bar{X}$	STD X	$\bar{Y}$	STD Y	(X,Y)	STD
Color	-0.44	3.84	-7.86	1.64	**8.74**	**1.59**
Motion	-0.44	2.29	-4.82	2.72	**5.52**	**2.31**
Disparity	7.03	4.46	13.79	2.32	**15.92**	**3.30**
Edges	-1.65	2.39	-4.10	2.54	**5.00**	**2.55**
NCC	-3.75	3.31	1.82	0.65	**4.70**	**2.56**
Voting	-2.17	4.09	-0.20	1.42	**4.01**	**2.65**

The above mentioned results are reported in more detail in [22].

5 Summary and Issues for Future Research

Task orientation and integration in a systems context is fundamental to detection and tracking of objects. Two processes that are fundamental to the design of active vision systems. Methods for integration of cues have been discussed as a basis for segmentation in an active vision system. A unified architecture for active vision has been outlined and it has been demonstrated how such an approach can be utilized for construction of systems that allow detection and tracking of objects in cluttered scenes.

One aspect that has not been addressed is recognition of objects, a fundamental problem in any computer vision systems. In terms of recognition current research is focussed on methods for fusion of cues to enable efficient indexing into a scene database and the use of such methods for maintenance of a scene model, which in turn requires consideration of issues related to attention and memory organisation.

Acknowledgements

The experiemntal results reported in this paper were generated by T. Uhlin and D. Kragic. In addition the paper has benefitted from discussion with M. Björkman, P. Nordlund, and K. Pahlavan. The work has been sponsored the the Foundation for Strategic Research and the Swedish Technical Research Council.

References

1. K. Pahlavan, "Active robot vision and primary ocular processes," Tech. Rep. PhD thesis, Dept. of Num. Anal. and Comp. Sci., Royal Inst. Technology, Stockholm, 1993.
2. S. Zeki, *A vision of the brain.* Oxford, UK: Oxford: Blakcwell Scientific, 1993.

3. M. J. Tovée, *An Introduction to the visual system.* Cambridge, UK: Cambridge University Press, 1996.
4. S. E. Palmer, *Vision Science: Photons to Phenomology.* MIT Press, 1999.
5. H. Bülthoff and H. Mallot, "Interaction of different modules in depth perception," in *First International Conf. on Computer Vision* (A. Rosenfelt and M. Brady, eds.), (London, UK), pp. 295–305, IEEE, Computer Society, June 1987.
6. J. Clark and A. Yuille, *Data fusion for sensory information processign systems.* Boston, Ma. – USA: Kluwer Academic Publishers, 1990.
7. A. Blake and A. Zisserman, *Visual Reconstruction.* Cambridge, MA: MIT Press, 1987.
8. J. Aloimonos and D. Shulman, *Integration of Visual Modules.* Academic Press, Inc, 1989.
9. J.-O. Eklundh, P. Nordlund, and T. Uhlin, "Issues in active vision: attention and cue integration/selection," in *Proc. British Machine Vision Conference 1996*, pp. 1–12, September 1996.
10. C. Bräutigam, J.-O. Eklundh, and H. Christensen, "Voting based cue integration," in *5th Euopean Conference on Computer Vision* (B. Neumann and H. Burkhardt, eds.), LNCS, (Heidelberg), Springer Verlag, May 1998.
11. B. Parhami, "Voting algorithms," *IEEE Transactions on Reliability*, vol. 43, no. 3, pp. 617–629, 1994.
12. J. Gårding and T. Lindeberg, "Direct estimation of local surface shape in a fixating binocular vision system," in *European Conference on Computer Vision* (J.-O. Eklundh, ed.), vol. 1 of *Lecture Notes in Computer Science*, (Stockholm), pp. 365–376, Springer Verlag, May 1994.
13. F. Panarai, G.Metta, and G. Sandini, "An artificia vestibular system for reflex control of robot eye movements," in *3rd International Conference on Cognitive and Neural Systems*, (CNS Boston University, USA), May 1999.
14. D. J. Coombs and C. M. Brown, "Cooperative gaze holding in binocular vision," *IEEE Control Systems Magazine*, vol. 11, pp. 24–33, June 1991.
15. D. H. Ballard, "Animate vision," *Artificial Intelligence*, vol. 48, pp. 57–86, Feb. 1991.
16. M. Black, "Explaining optical flow events with parameterized spatio-temporal models," in *Proc. IEEE Computer Vision and Pattern Recognition*, (Fort Collins, CO), pp. 326–332, 1999.
17. J. Tsotsos, S. Culhane, W. Wai, Y. Lai, N. Davis, and F. Nuflo, "Modelling visual attention via selective tuning," *Artificial Intelligence*, vol. 78, no. 1-2, pp. 507–547, 1995.
18. K. Pahlavan, T. Uhlin, and J.-O. Eklundh, "Dynamic fixation and active perception," *Intl. Journal of Computer Vision*, vol. 17, pp. 113–136, February 1996.
19. C. S. Andersen and H. I. Christensen, "Using multiple cues for controlling and agile camera head," in *Proceedings from The IAPR Workshop on Visual Behaviours, Seattle 1994.*, pp. 97–101, IEEE Computer Society, June 1994.
20. K. Pahlavan, T. Uhlin, and J. Eklund, "Integrating primary ocular processes," in *Proceeding of Second European Conference in Computer Vision*, pp. 526 – 541, 1992.
21. P. Nordlund and J.-O. Eklundh, "Figure-ground segmentation as a step towards deriving object properties," in *Proc. 3rd Workshop on Visual Form*, (Capri), p. (To appear), May 1997.
22. D. Kragic and H. Christensen, "Integration of visual cues for active tracking of an end effector," in *Proc. IEEE/RSJ international Conference on Intelligent Robots and Systems*, (Kyongju, Korea), pp. 362–368, October 1999.

An Efficient Data Structure for Feature Extraction in a Foveated Environment

Efri Nattel and Yehezkel Yeshurun [*]

The Department of Computer Science
The Raymond and Beverly Sackler Faculty of Exact Sciences
Tel-Aviv University
Tel-Aviv, 69978, Israel
{nattel, hezy}@math.tau.ac.il

Abstract. Foveated sampling and representation of images is a powerful tool for various vision applications. However, there are many inherent difficulties in implementing it. We present a simple and efficient mechanism to manipulate image analysis operators directly on the foveated image; A single typed table-based structure is used to represent various known operators. Using the Complex Log as our foveation method, we show how several operators such as edge detection and Hough transform could be efficiently computed almost at frame rate, and discuss the complexity of our approach.

1 Introduction

Foveated vision, which was originally biologically motivated, can be efficiently used for various image processing and image understanding tasks due to its inherent compressive and invariance properties ([14,13,16]). It is not trivial, however, to efficiently implement it, since we conceptualize and design algorithms for use in a Cartesian environment. In this work, we propose a method that enables implementation of image operators on foveated images that is related to ([14]), and show how it is efficiently used for direct implementation of feature detection on foveated images. Following the classification of Jain (in [3]), we show how local, global, and relational (edge detection, Hough Transform and Symmetry detection, respectively) are implemented by our method.

Both our source images and the feature maps are foveated, based on Wilson's model ([16,15]). To achieve reasonable dimensions and compression rates, the model's parameters are set in a way that follows biological findings – by imitating the mapping of ganglions between the retina and V1. In order to use camera-made images, we reduced the field of view and the foveal resolution. Our simulations are done from initial uniform images with 1024×682 pixels, which are mapped to a *logimage* $L = u_{\text{Max}} \times v_{\text{Max}} = 38 \times 90$ *logpixels*.

---

[*] This work was supported by the Minerva Minkowski center for Geometry, and by a grant from the Israel Academy of Science for geometric Computing.

2 The Complex Log Mapping

The complex log mapping was suggested as an approximation to the mapping of visual information in the brain ([10,11] and others). The basic complex-log function maps a polar coordinate to a Cartesian point by $(u(r,\theta), v(r,\theta)) = (\log r, \theta)$. We follow [12,13,15,16] and remove a circular area from the center of the source image, assuming that it is treated separately.

In addition to [15,16], we select the relevant constants in the model in a way that follows biological findings. In order to achieve meaningful dimensions and compression rates, we follow the path and number of ganglions.

2.1 Modeling the Human Retina

. **The sampling model:** According to [15,16] the retinal surface is spanned by partially overlapping, round-shaped receptive fields; Their centers form a complex log grid. The foveal area is excluded from the model. Using these assumptions – the eccentricity of the nth ring ($0 \leq n \leq u$) R_n is $R_0 e^{\frac{w \cdot n}{2p(1-o_v)}}$, where $w = \log\left(1 + \frac{2(1-o_v)c_m}{2-(1-o_v)c_m}\right)$, R_0 is the radius of the fovea (in degrees), c_m is the ratio between the diameter (in degrees) of the receptive field and its eccentricity (in degrees), o_v is an overlap factor and p is the number of photoreceptors in a radius of one receptive field. The radius of the receptive field on the nth ring is $\frac{c_m \cdot R_n}{2}$ and the number of receptive fields per ring is $v = \frac{2\pi}{c_m(1-o_v)}$. Ganglion cells appear to obey these assumptions ([9,16]). Following [11] and others, and extrapolating the model towards the $> 30°$ periphery, we can use ganglions as the modeled elements and let $o_v = 0.5$.

. **Field of view:** The retinal field of a single eye is $208° \times 140°$ ([5,8]). The foveal area is a circle with a radius of $2.6°$ in the center of this field. We therefore let $R_0 = 2.6°$, and note that $R_u = 104°$ (The number of ganglions in the extreme periphery is very small, thus we neglect the fact that the field of view is not really circular).

. **Number of modules:** There are $\approx 10^6$ neurons in the main optic nerve, 75% of them are peripheral. The number of modules (halves of hypercolumns) in one hemifield of V1 is 2500, 1875 of them represent the periphery ([1]).

. **Computing u, c_m and p:** Following [16] we first model the spatial mapping of modules. For u and c_m, we solve $R_u = 104$ and $u \cdot v = 1875$, which gives $u = 33$, and $c_m = 0.221$, or a grid of 33×56.8 modules. If 750000 ganglions are equally divided to the receptive fields, we get 400 cells per receptive field. Roughly assuming that these cells are equally distributed inside every field in a 20×20 matrix, we get a grid of 660×1136 ganglions for the peripheral area.

2.2 Modeling Foveated Images

. **Foveal resolution:** Polyak's cones density in the foveaola matches Drasdo's foveal resolution of up to 30000 $ganglions/deg^2$, assuming a $ganglions/cones$

ratio of 2 ([6,2]). We therefore define $F = 28648\ ganglions/deg^2$ to be the maximal ganglions density (we ignore the density of $22918\ cones/deg^2$ in the very central $20'$ of the foveaola).

. **Selecting the portion of the view:** In practice images with a field of view of $208°$ are rarely used. Our suggested model uses the central $74° \times 53°$ of the human visual field. This portion has the same field as a $24mm$ camera lens, projected on a $35mm$ film ([4]). It is wider than the human eye's portion (around $39° \times 26°$, achieved using a $50mm$ lens), but not too wide to create distortions.

Viewing this portion with the resolution F we get 12525×8970 pixels; Excluding the foveal area leaves $\approx 1.11 \cdot 10^8$ pixels. The logarithmic mapping of the portion results $u \times v = 24 \times 56.8$ modules, or $480 \times 1136 \approx 545280$ ganglions (The v parameter remains the same and u is the minimal n such that $R_n > \frac{74}{2}$).

. **Lowering the foveal resolution:** We now adjust the above portion to the dimensions of a typical screen, which has a lower resolution. Dividing the uniform portion by 12.66 in each dimension we fit it into 1024×682 pixels – a reasonable size for input images. On a 14" monitor such an image takes $\approx 74° \times 53°$, when viewed from about $20cm$. We also reduce the sampling rate of the modules by the same factor – from 20^2 ganglions to 1.6^2. The 24×56.8 modules will now be composed of about $38 \times 89.8 \approx 3408$ ganglions.

2.3 Summary and Example

Table 1 summarizes the dimensions of the human retina and our model.

Table 1. Numerical data about the human eye and our model. Additional constants are: $w = 0.11061$, $o_v = 0.5$, $c_m = 0.221$, $R_0 = 2.6$.

Parameter	Human eye	Portion of the eye	Model
Fov. resolution ($units/deg^2$)	28648	28648	180
Horizontal field	208	74	74
Vertical field	140	53	53
Horiz. field (w_u)	35205	12525	1024
Vert. field (h_u)	23696	8970	682
Uniform units	$8.34 \cdot 10^8$	$1.12 \cdot 10^8$	706910
Uniform units (ex. fovea)	$8.33 \cdot 10^8$	$1.11 \cdot 10^8$	703080
Log units (ex. fovea)	750000	545280	3408
Logmap's width (ex. fovea)	660	480	38
Logmap's height	1136	1136	89.8
Peripheral compression (x:1)	1110	203	203
p	10		0.8
R_N	104		37

The following pictures (Figure 1) illustrate the sampling models we presented above. The left image shows our 780×674 input image and the middle image

Fig. 1. Left: An input image; Middle: Its Logimage; Right: Reverse logmap

shows its 38×90 logimage. The center of the image was chosen to be the origin for our mapping. The right image shows the reverse mapping of that logimage (the central black area is the foveal area that is not mapped).

3 The Logmap and Operator Tables

A *logmap* is a table with $u_{\text{Max}} \times v_{\text{Max}}$ entries, each containing a list of the uniform pixels that constitute the receptive field of this logpixel entry. Given that table, and if we assume (for simplicity) that the receptive field's kernel is just averaging, we can transform a uniform image with pixel values $\bar{z}$ for each z to a *logimage*. For every l with an attached list $\{z_1, \ldots z_n\}$, we assign the logpixel value $\bar{l}$ to be $\frac{1}{n} \sum \bar{z_i}$.

An *operator table* $Op_k(l)$ is defined $\forall l \in L$ as

$$\left\{ \left(w_i, L_i^{(k)} \right) \right\}_{i=1..N} \quad (1)$$

where $L_i^{(k)} = \left(l_i^{(1)}, \ldots, l_i^{(k)} \right)$ is a k-tuple of $l_i^{(j)} \in L$, and k is constant.

Suppose that F is a function of k parameters. For any $l \in L$ we define Apply$(Op_k(l))$ as

$$\sum_{i=1}^{N} w_i F(\overline{l_i^{(1)}}, \ldots, \overline{l_i^{(k)}}) \quad (2)$$

For $k = 1$ we use $F(\overline{l_i^{(1)}}) = \overline{l_i^{(1)}}$. Applying is the process of "instantiating" the operator on a source logimage, resulting in a target logimage. Preparing the tables can be done in a preprocessing stage, leaving minimal work to be done at applying time. Usually the preprocessing and applying stages can be carried out in parallel for each logpixel.

An operator table of l can be normalized by multiplying every weight in the list of $Op_k(l)$ by a given constant. Since each l has a different index list, we can multiply each list by a different constant, thus normalizing a global operator.

Two operator tables can be added. The resulting table for each l will contain all the k-tuples from the source tables; k-tuples that appeared on both tables will appear once in the resulting table with a summed weight.

4 Edge and Phase Maps

Let (u_0, v_0) be an arbitrary logpixel, with a corresponding field center of (x_0, y_0) in the uniform space. Let $(s, t) = (\varphi(x, y), \psi(x, y))$ be the transformation defined by a translation of a Cartesian coordinate by $(-x_0, -y_0)$ and a rotation by $(-\arctan(y_0/x_0))$. Projecting a logimage to the uniform space, we can express the result as functions of both the x-y and the s-t coordinate systems, by $F(x, y) = f(\varphi(x, y), \psi(x, y))$. It can be easily shown that $\left(\frac{\partial F}{\partial x}\right)^2 + \left(\frac{\partial F}{\partial y}\right)^2 = \left(\frac{\partial f}{\partial s}\right)^2 + \left(\frac{\partial f}{\partial t}\right)^2$. Thus the *magnitude* of the gradient vector ∇F can be approximated using the s-t system as follows: We define G_s, G_t and $\nabla L(u, v)$ to be

$$\begin{aligned} G_s(u, v) &= \overline{(u+1, v)} - \overline{(u-1, v)} \\ G_t(u, v) &= \overline{(u, v+1)} - \overline{(u, v-1)} \\ \nabla L(u, v) &= (G_s^2 + G_t^2)^{0.5}. \end{aligned} \quad (3)$$

(G_t/G_s) approximates the tangent ∇F, relative to the s-t coordinates. Since this system is rotated, the phase of L will be

$$\psi L(u, v) = \arctan\left(\frac{G_t}{G_s}\right) + \frac{2\pi v}{v_{Max}} \quad (4)$$

5 Spatial Foveated Mask Operators

We suggest operators that use different masking scales: Logpixels that reside on peripheral areas will have a larger effective neighborhood than central logpixels (using fixed-scaled neighborhood yields poor peripheral detection). Suppose that we are given a spatial mask g with $M \times N = (2\hat{M}+1) \times (2\hat{N}+1)$ elements, and let $\lambda \in \Re^+$ be an arbitrary constant. We mark the rounded integer value of $x \in \Re$ by $\lceil x \rceil$, the set of pixels in l's receptive field by R_l, and its size by $|R_l|$. For any uniform pixel (x, y) we can find the closest matching logpixel l, and define a neighborhood P around (x, y) as $\left\{\left(\lceil x + m\lambda\sqrt{|R_l|}\rceil, \lceil y + n\lambda\sqrt{|R_l|}\rceil\right)\right\}$, where $|m| \leq \hat{M}, |n| \leq \hat{N}$ are reals.

Every $p \in P$ corresponds to a logpixel $l_p \in L$. We add the logpixels l_p to l's operator list; Each addition of l_p that corresponds to a pixel $(x + m\lambda\sqrt{|R_l|}, y + n\lambda\sqrt{|R_l|})$ will have a weight of $\left(g(\lceil m \rceil, \lceil n \rceil)/(\lambda^2|R_l|^2)\right)$. λ is used to control our masking area (usually $\lambda = 1$). Normalizing the weight compensates for our sampling method: $|R_l|$ compensates for the several additions to the same l_p that may occur using different (x, y) pairs. An additional $\lambda^2 |R_l|$ compensates for the different sizes of P. Note: In [14] this normalization is done in the "applying" stage.

Mask building and applying can be viewed in terms of translation tables (see [14]). A translation by (x, y) can be viewed as a spatial mask $T_{(x,y)}$, with $T(-x, -y) = 1$ and $T(i, j) = 0$ for every other (i, j). For each of the possible offsets we build $\bar{T}_{(x,y)} = T_{(x\lambda\sqrt{|R_l|}, y\lambda\sqrt{|R_l|})}$. Any mask operator G can now be defined as $G(u, v) = \sum_{m=-M}^{M} \sum_{n=-N}^{N} g(m, n) \cdot \bar{T}_{(m,n)}(u, v)$, giving $\text{Apply}(G(u, v)) = \sum_{m=-M}^{M} \sum_{n=-N}^{N} g(m, n) \cdot \text{Apply}(\bar{T}_{(m,n)}(u, v))$.

6 The Foveated Hough Transform

We construct a Hough map that detects lines in a given edge-logimage. Let Γ be a set of k angles, $\Gamma = \{\gamma_i | 1 \leq i \leq k, \gamma_i = \left(\frac{2\pi i}{k}\right)\}$, let z be an arbitrary pixel in a uniform image Z, and q the respective logpixel in L.

For each $z \in Z$ we find the parameterization of the k lines λ_i passing through z and having angles of γ_i, respectively. We define (ρ_i, θ_i) (the coordinates of the normal vector of λ_i that passes through 0) as these parameterizations. For an arbitrary i we observe $(u(\rho_i, \theta_i), v(\rho_i, \theta_i))$. For $u_0 \leq \rho_i \leq u_{\text{Max}}$ and θ_i there is a logpixel $p \in L$ with these coordinates. Thus we can add the coordinates of q to p's operator table, which will function as the voting plane of the Hough transform. Note that the actual results of the voting depend on the logpixels' *values*. They can be calculated once a logimage with the values $\bar{q}$ is given.

The operator table of a single p is made of a series of logpixels which lie on a "band" in Z. That band passes through its parameterizing source logpixel, and is orthogonal to the line (ρ_i, θ_i). We define the *thickness* of that band as the number of pixels along (ρ_i, θ_i) that intersect p. As the u-value of a logpixel p gets larger, more parameterizations fall into the same p: The number of these contributers increases linearly with p's thickness, which can be shown to be proportional to the diameter of its receptive field. We therefore normalize the table during its construction, by dividing every contribution to a logpixel p by p's diameter.

7 The Foveated Symmetry Operator

We show a foveated version of the Generalized Symmetry Transform ([7]), that detects corners or centers of shapes. As in the case of mask operators, our operator is *scale dependent*; It detects both corners and centers, smaller near the fovea and larger in the periphery. Our input is a set of logpixels $l_k = (u_k, v_k)$, from which an edge logmap $(r_k, \theta_k) = (\log(1 + \|\nabla(e_k)\|), \arg(\nabla(e_k)))$ can be obtained. We define $L^{-1}(l)$ as the uniform pixel that corresponds to the center of l's receptive field. Our operator table for a logpixel l will be of the form $\{w_i, (l_{ai}, l_{bi})\}_{i=1..N}$ (assigning $k = 2$, $F(\overline{l_{ai}}, \overline{l_{bi}}) = \overline{l_{ai}} \cdot \overline{l_{bi}}$ in (1), (2)).

For any l we find all the pairs (l_a, l_b) of its circular neighborhood $\Gamma(l)$: We traverse all the possible logpixels-pairs (l_a, l_b), find $mid(l_a, l_b)$ and add this pair to l's list if $mid(l_a, l_b) = l$. $mid(l_a, l_b)$ is defined to be the closest matching logpixel to the pixel $((L^{-1}(l_a) + L^{-1}(l_b))/2)$. When adding a pair, we compute σ and add a weight of $\mathcal{D}_\sigma^\star$ to that pair in l's list. σ and $\mathcal{D}_\sigma^\star$ are defined as –

$$\sigma(l_a, l_b) = \lambda\sqrt{|R_{mid(l_a,l_b)}|}$$

$$\mathcal{D}_\sigma^\star(l_a, l_b) = \begin{cases} e^{-\frac{\|L^{-1}(l_a) - L^{-1}(l_b)\|^2}{2\sigma^2(l_a,l_b)}} & \text{if it is } > e^{-1} \\ 0 & \text{otherwise} \end{cases}$$

λ is a constant that acts as the symmetry radius (σ in [7]). After the construction is done, we divide each weight by the number of elements in its respective list. The number of elements in each list is approximately the same for all the lists in a single table. Thus the normalization is done only to equalize weights of tables with different λ values.

Applying the symmetry operator for l results in $\sum_{i=1}^{N} w_i P(a_i, b_i) r_{a_i} r_{b_i}$, where α_{ij} is the angle between the x-axis and the line (a_i, b_i); and $P(a_i, b_i) = (1 - \cos(\theta_{a_i} + \theta_{b_i} - 2\alpha_{ij}))(1 - \cos(\theta_{a_i} - \theta_{b_i}))$ (for a detailed definition, see [7]).

8 Complexity

Suppose that we are given an operator Op_k. We define the difference operator $Op\Delta_{nk}(u, v)$ as $Op_k(u, v\hat{+}n) - \text{Shift}(Op_k(u, v), n)$ where $\hat{+}$ means $+_{mod\, v_{\text{Max}}}$. We also define the $\text{Shift}(Op_k(u, v), n)$ of Op_k by n as the operator which is defined by $\left\{\left(w_i, (u_i, v_i\hat{+}n)^{(1)}, \ldots, (u_i, v_i\hat{+}n)^{(k)}\right)\right\}_{i=1..N}$

We say that Op_k has *radial invariance* if for every (u, v), the squared sum of weights in $Op\Delta_{nk}(u, v)$ is $\ll$ from the one of Op_k. If an operator is radial invariant, only one representative for each u in the table needs to be stored. This cuts the storage complexity and preprocessing time by a factor of v_{Max}.

For the Hough operator, it can be shown that a list with length of $O(u_{\text{Max}} + v_{\text{Max}})$ is attached for each l. Since the Hough operator is radial-invariant, the space complexity of the operator is $O(u_{\text{Max}} \cdot (u_{\text{Max}} + v_{\text{Max}}))$.

Similarly, it can be shown for the symmetry operator that the total space complexity is $O(u_{Max} \cdot v_{Max})$. The symmetry operator is radial invariant, thus it can be represented by $O(u_{Max})$ elements and the bound on the preprocessing time can be tightened to $O(u_{Max})$.

The preprocessing time for the mask operators is large (computations for each mask element is proportional to the size of the uniform image). However the resulting operator requires only $O(u_{Max} \cdot v_{Max})$ space.

In our model, a relatively small number of elements in each logpixel's list (< 25 for each translation mask, ≈ 440 for the Hough operator, $\approx 70/295$ for the symmetry with $\lambda = 2/4$) enables almost a frame rate processing speed. Note that since we build feature *maps*, the number and quality of interesting points we may extract from a map does not affect the extraction complexity.

Fig. 2. Foveated edge detection

9 Results

Figure 2 demonstrates our edge detector. We transferred the left image to a logimage and applied the edge operator. The back-projection of the result is shown in the right image, and can be compared with the uniform equivalent (middle).

To demonstrate a spatial mask detector, we set a 5×5 mask that detects "Γ"-shaped corners. The image in Fig. 3(a) is used as our input; It is a logmap of uniform squares with sizes that are proportional to the eccentricity of their upper-left corners. We constructed a set of translation operators and used them to construct a corner-detector. The result of applying the operator (along with the original image) is shown in Fig. 3(b).

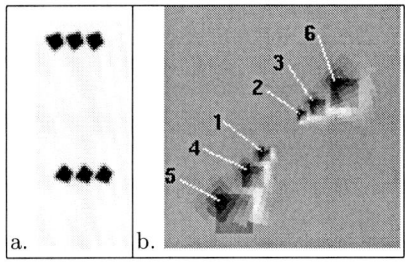

Fig. 3. Corner detection using a foveated spatial mask. Highest peaks designated by their order

The Hough operator is demonstrated on an image with 5 segments that was transferred into logimages using two fixation points (Fig. 4, left). The edge and hough operators were applied (middle) and the 5 highest local maxima were marked. The derived lines are drawn over the edge-logimage (right), they all had similar votes.

For symmetry extraction we created a uniform image with 6 squares. The applied operator when $\lambda = 2$ and 4 are shown in Fig. 5. The two leftmost figures show the resulting logimages, the right figure shows the projection along with

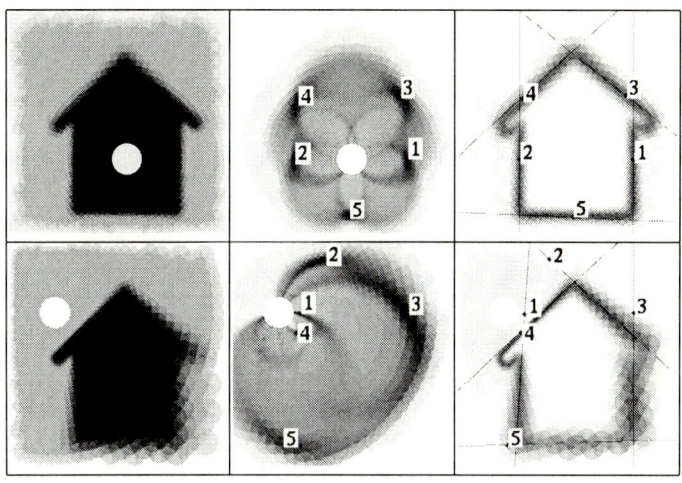

Fig. 4. The foveated Hough operator. See text for details

the source image, for $\lambda = 2$. The symmetry operators indeed detect the corners of the squares (for $\lambda = 2$) or their centers (for $\lambda = 4$).

10 Conclusions

In this paper we have presented an efficient mechanism that can be used to implement various image analysis operators directly on a foveated image. The

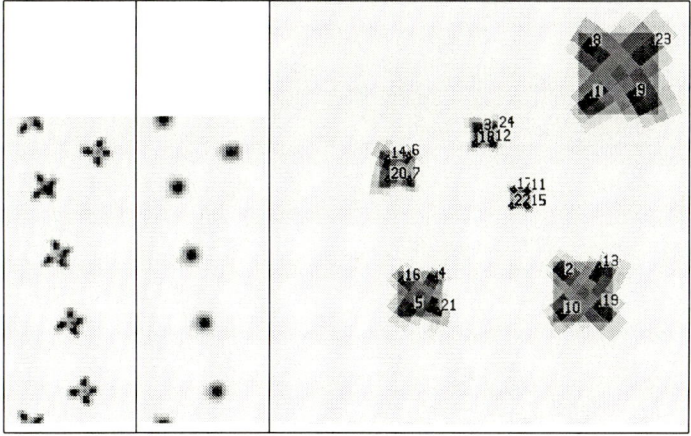

Fig. 5. Foveated corner and center detection using the symmetry operator. Highest peaks designated by their order

method is based on a space variant weighted data structure. Using this approach, we show how some common global and local operators can be implemented directly, at almost frame rate, on the foveated image.

References

1. N.R. Carlson. *Physiology of behavior, forth edition.* Allyn and Bacon, Maryland, 1991.
2. N. Drasdo. Receptive field densities of the ganglion cells of the human retina. *Vision Research*, 29:985–988, 1989.
3. R. Jain, R. Kasturi, and B.G. Schunk. *Machine Vision.* McGraw Hill, New York, 1995.
4. M. Langford. *The Step By Step Guide to Photography.* Dorling Kindersley, London, 1978.
5. M.D. Levine. *Vision in Man and Machine.* McGraw-Hill, New York, 1985.
6. S. Polyak. *The Vertebrate Visual System.* The university of Chicago press, 1957.
7. D. Reisfeld, H. Wolfson, and Y. Yeshurun. Context free attentional operators: the generalized symmetry transform. *International Journal of Computer Vision*, 14:119–130, 1995.
8. A. Rojer and E. Schwartz. Design considerations for a space-variant visual sensor with complex logarithmic geometry. In *Proceedings of the 10th IAPR International Conference on Pattern Recognition*, pages 278–285, 1990.
9. B. Sakitt and H. B. Barlow. A model for the economical encoding of the visual image in cerebral cortex. *Biological Cybernetics*, 43:97–108, 1982.
10. E.L. Schwartz. Spatial mapping in the primate sensory projection: Analytic structure and relevance to perception. *Biological Cybernetics*, 25:181–194, 1977.
11. E.L. Schwartz. Computational anatomy and functional architecture of striate cortex: A spatial mapping approach to perceptual coding. *Vision Research*, 20:645–669, 1980.
12. M. Tistarelli and G. Sandini. Dynamic aspects in active vision. *Journal of Computer Vision, Graphics, and Image Processing: Image Understanding*, 56(1):108–129, July 1992.
13. M. Tistarelli and G. Sandini. On the advantages of polar and log-polar mapping for direct estimation of time-to-impact from optical flow. *IEEE Transactions Pattern Analysis and Machine Intelligence*, 15(4):401–410, April 1993.
14. R. Wallace, P. Wen Ong, B. Bederson, and E. Schwartz. Space variant image processing. *International Journal of Computer Vision*, 13(1):71–90, 1994.
15. S.W. Wilson. On the retino-cortial mapping. *International journal of Man-Machine Studies*, 18:361–389, 1983.
16. H. Yamamoto, Y. Yeshurun, and M.D. Levine. An active foveated vision system: Attentional mechanisms and scan path covergence measures. *Journal of Computer Vision and Image Understanding*, 63(1):50–65, January 1996.

Parallel Trellis Based Stereo Matching Using Constraints

Hong Jeong and Yuns Oh

Dept. of E.E., POSTECH, Pohang 790-784, Republic of Korea

Abstract. We present a new center-referenced basis for representation of stereo correspondence that permits a more natural, complete and concise representation of matching constraints. In this basis, which contains new occlusion nodes, natural constrainsts are applied in the form of a trellis. A MAP disparity estimate is found using DP methodsin the trellis. Like other DP methods, the computational load is low, but it has the benefit of a structure is very suitable for parallel computation. Experiments are performed under varying degrees of noise quantity and maximum disparity, confirming the performance.
Keywords: Stereo vision, constraints, center-reference, trellis.

1 Introduction

An image is a projection that is characterized by a reduction of dimension and noise. The goal of stereo vision is to invert this process and restore the original scene from a pair of images. Due to the ill-posed nature [11] of the problem, it is a very difficult task. The usual approach is to reduce the solution space using *a prior* models and/or natural constraints of stereo vision such as the geometrical characteristics of the projection.

Markov random fields (MRFs) [12] can be used to describe images and their properties, including disparity. Furthermore, the Gibbsian equivalence [3] makes it possible to model the prior distribution and thus use maximum a posteriori (MAP) estimation. Geman and Geman [8] introduced this approach to image processing in combination with annealing [9], using *line processes* to control the tradeoff between overall performance and disparity sharpness. Geiger and Girosi [7] extended this concept by incorporating the mean field approximation. This approach has become popular due to the good results. However, the computational requirements are very high and indeterministic, and results can be degraded if the emphasis placed on the line process is chosen poorly.

An alternative class of methods is based upon dynamic programming (DP) techniques, such as the early works by Baker and Binford [1] and Ohta and Kanade [10]. In both of these cases, scan lines are partitioned by feature matching, and area based methods are used within each partition. DP is used in both parts in some way. Performance can be good but falls significantly when an edge is missing or falsely detected. Cox et al. [5] and Binford and Tomasi [4] have also used DP in methods that consider the discrete nature of pixel matching. Both employ a second smoothing pass to improve the final disparity map.

While the DP based methods are fast, their primary drawback is the requirement of strong constraints due to the elimination or simplification of the prior probability. However, the bases used to represent the disparity in the literature have been inadequate in sufficiently incorporating constraints in a computationally efficient manner. Post processing is often used to improve the solution, but this can greatly increase the computation time and in some cases only provides a modest improvement.

In this paper we examine the basic nature of image transformation and disparity reconstruction in discrete space. We find a basis for representing disparity that is concise and complete in terms of constraint representation. A MAP estimate of the disparity is formulated and an energy function is derived. Natural constraints reduce the search space so that DP can be efficiently applied in the resulting disparity trellis. The resulting algorithm for stereo matching is well suited for computation by an array of simple processor nodes.

This paper is organized as follows. In Sec. 2, the discrete projection model and pixel correspondence is presented. Section 3 deals with disparity with respect to the alternative coordinate system and the natural constraints that arise. The problem of finding the optimal disparity is defined formally in Sec. 4 and reduced to unconstrained minimization of an energy function. In Sec. 5, this problem is converted to a shortest path problem in a trellis, and DP is applied. Experimental results are given in Sec. 6 and conclusions are given in Sec. 7.

2 Projection Model and Correspondence Representation

We begin by defining the relationships of the coordinate systems between the object surfaces, and image planes in several discrete coordinate systems. We also introduce representation schemes for denoting correspondence between sites in the two images. While some of the material presented here has been discussed before, it is included here for completeness.

For the 3-D to 2-D projection model, it is assumed that the two image planes are coplanar, the optical axes are parallel, the focal lengths are equal, and the epipolar lines are the same. Figure 1(a) illustrates the projection process for epipolar scan lines in the left and right images through the focal points p^l and p^r respectively and with focal length l. The finite set of points that are reconstructable by matching image pixels are located at the intersections of the dotted projection lines represented by solid dots. We call this the *inverse match* space.

The left image scan line is denoted by $\boldsymbol{f}^l = \begin{bmatrix} f_1^l & \ldots & f_N^l \end{bmatrix}$ where each element is any suitable and dense observed or derived feature. We simply use intensity. The right image scan line $\boldsymbol{f}^r$ is similarly represented.

Scene reconstruction is dependent upon finding the correspondence of pixels in the left and right images. At this point, we define a compact notation to indicate correspondence. If the true matching of a pair of image scan lines is known, each element of each scan line can belong to one of two categories:

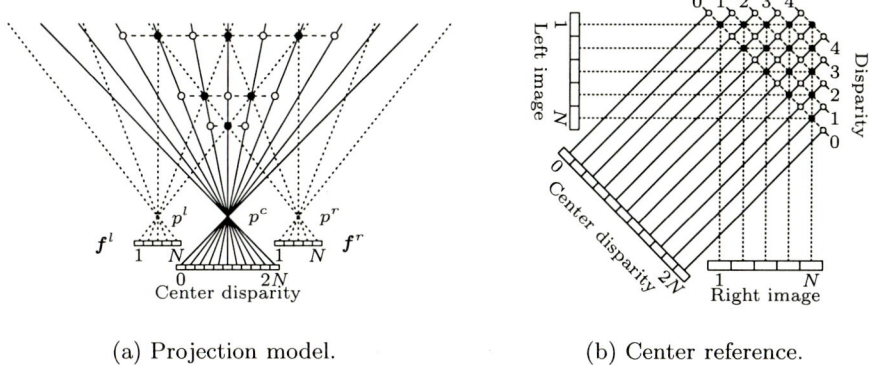

Fig. 1. Discrete matching spaces.

1. The element has a corresponding element in the other image scan line and the two corresponding points, f_i^l and f_j^r, form a *conjugate pair* denoted (f_i^l, f_j^r). The pair of points are said to match with disparity $d = i - j$.
2. The element is not visible in the other image scan line and is therefore not matched to any other element. The element is said to be *occluded* in the other image scan line and is indicated by $(f_i^l, \emptyset)$ if the element came from the left image (*right occlusion*) and $(\emptyset, f_i^r)$ if the element came from the right image (*left occlusion*). This case was also labeled *half occlusion* in [2].

Given a set of such associations between elements of the left and right image vectors, the *disparity map* with respect to the left image is defined as

$$\boldsymbol{d}^l = \begin{bmatrix} d_1^l & \ldots & d_N^l \end{bmatrix} , \qquad (1)$$

where a disparity value d_i^l denotes the correspondence $(f_i^l, f_{i+d_i^l}^r)$. The disparity map with respect to the right image $\boldsymbol{d}^r$ is similarly defined, and a disparity value d_j^r denotes the correspondence $(f_{i-d_j^r}^l, f_j^r)$.

While the disparity map is popular in the literature, it falls short in representing constraints that can arise in discrete pixel space in a manner that is complete and analytically concise.

3 Natural Constraints in Center-Referenced System

Using only left- or right-referenced disparity, it is difficult to represent common constraints such as pixel ordering or uniqueness of matching with respect to *both* images. As a result, the constraints are insufficient by themselves to reduce the solution space and post processing is required to produce acceptable results [10,4]. Some have used the discrete inverse match space directly [6,5] to more fully incorporate the constraints. However, they result in a heavy computational load or still suffer from incomplete and unwieldy constraint representation.

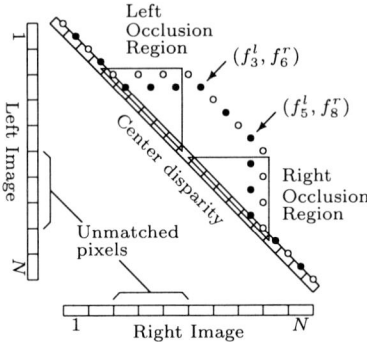

Fig. 2. Occlusion representation.

We propose a new *center-referenced* projection that is based on the focal point p^c located at the midpoint between the focal points for the left and right image planes. This has been described in [2] as the *cyclopean eye* view. However, we use a projection based on a plane with $2N+1$ pixels of the same size as the image pixels and with focal length of $2l$. The projection lines are represented in Fig. 1(a) by the solid lines fanning out from p^c. The inverse match space is contained in the discrete *inverse* space $\mathcal{D}$, consisting of the intersections of the center-referenced projection lines and the horizontal dashed *iso-disparity* lines. The inverse space also contains an additional set of points denoted by the open dots in Fig. 1(a) which we call *occlusion* points.

The 3-D space can be transformed as shown in Fig. 1(b) where the projection lines for p^l, p^r, and p are now parallel. The iso-disparity lines are given by the dashed lines perpendicular to the center-referenced projection lines. We can now clearly see the basis for a new center referenced disparity vector

$$\boldsymbol{d} = \begin{bmatrix} d_0 & \ldots & d_{2N} \end{bmatrix}, \tag{2}$$

defined on the center-referenced coordinate system on the projection of $\mathcal{D}$ onto the center image plane through p. A disparity value d_i indicates the depth of a real world point along the projection line from site i on the center image plane through p. If d_i is a match point $(o(i+d_i) = (i+d_i) \mod 2 = 1)$ it denotes the correspondence $(f^l_{\frac{1}{2}(i-d_i+1)}, f^r_{\frac{1}{2}(i+d_i+1)})$ and conversely (f^l_i, f^r_j) is denoted by the disparity $d_{i+j-1} = j - i$.

There are various ways of representing occlusions in the literature and here we choose to assign the highest possible disparity. Fig. 3 shows an example of both left and right occlusions. The conjugate pair (f^l_5, f^r_8) creates a right occlusion resulting in some unmatched left image pixels. If visible, the real matching could lie anywhere in the area denoted as the Right Occlusion Region (ROR), and assigning the highest possible disparity corresponds to the locations that are furthest to the right. Using only the inverse match space, these are the solid dots in the ROR in Fig. 3. However, the center-referenced disparity contains additional occlusion points (open dots) that are further to the right and we use

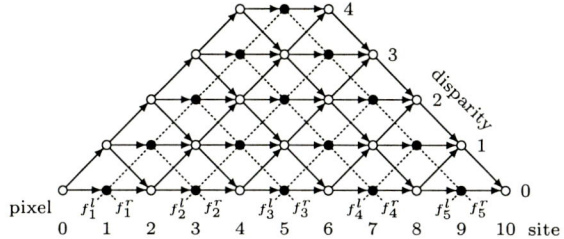

Fig. 3. Disparity trellis for $N = 5$.

these to denote the disparity. This new representation of occlusion simplifies the decisions that must be made at a node.

Now we can evaluate how some natural constraints can be represented in the center-referenced discrete disparity space.

For parallel optical axes, a negative disparity would imply that the focal point is behind rather than in front of the image plane. This violates our projection model, thus $d_i \geq 0$.

Since disparity cannot be negative, the first pixel of the right image can only belong to the correspondence (f_1^l, f_1^r) or the left occlusion $(\emptyset, f_1^r)$. Likewise, pixel N of the left image can only belong to the correspondence (f_N^l, f_N^r) or the right occlusion $(f_N^l, \emptyset)$. This gives the endpoint constraints $d_0 = d_{2N} = 0$.

The assumption that the image does not contain repetitive narrow vertical features [6], i.e. the objects are cohesive, is realized by bounding the disparity difference between adjacent sites: $-1 \leq d_i - d_{i-1} \leq 1$

The uniqueness assumption [6], that is, any pixel is matched to at most one pixel in the other image, is only applicable to match points. At such points, this assumption eliminates any unity disparity difference with adjacent sites. The discrete nature of $\mathcal{D}$ means that match points are connected only to the two adjacent points with identical disparity values.

In summary the constraints are:

$$\begin{aligned}
&\text{Parallel axes: } d_i \geq 0 , \\
&\text{Endpoints: } d_0 = d_{2N} = 0 , \\
&\text{Cohesiveness: } d_i - d_{i-1} \in \{-1, 0, 1\} , \\
&\text{Uniqueness: } o(i + d_i) = 1 \Rightarrow d_{i-1} = d_i = d_{i+1} .
\end{aligned} \quad (3)$$

If the disparity is treated as a path through the points in $\mathcal{D}$ then the constraints in (3) limit the solution space to any directed path though the trellis shown in Fig. 3. In practice the maximum disparity may be limited to some value $d_{\max}$ which would result in the trunctation of the top of the trellis.

4 Estimating Optimal Disparity

In this section we define stereo matching as a MAP estimation problem and reduce it to an unconstrained energy minimization problem

For a scan line, the observation is defined as g^l and g^r which are noise-corrupted versions of f^l and f^r respectively. A MAP estimate $\hat{d}$ of the true disparity is given by

$$\hat{d} = \arg\max_d P(g^l, g^r | d) P(d) , \qquad (4)$$

where Bayes rule has been applied and the constant $P(g^l, g^r)$ term has been removed. Equivalently, we can minimize the energy function

$$U_t(d) = -\log P(g^l, g^r | d) - \log P(d) , \qquad (5)$$
$$= U_c(d) + U_p(d) . \qquad (6)$$

To solve (6), we need the conditional $P(g^l, g^r | d)$, and the prior $P(d)$. First we introduce the notations $a(d_i) = \frac{1}{2}(i - d_i + 1)$, $b(d_i) = \frac{1}{2}(i + d_i + 1)$, and $\Delta g(d_i) = (g^l_{a(d_i)} - g^r_{b(d_i)})^2$ in order to simplify the expressions hereafter.

The conditional expresses the relationships between the two images when the disparity is known. Since $o(i + d_i) = 1 \Rightarrow (g^l_{a(d_i)}, g^r_{b(d_i)})$, if the corrupting noise is Gaussian, then

$$P(g^l, g^r | d) = \frac{1}{(\sqrt{2\pi}\sigma)^\eta} \prod_{i=1}^{2N} \exp\{-\frac{1}{2\sigma^2}\Delta g(d_i) o(i + d_i)\} , \qquad (7)$$

where σ^2 is the variance of the noise and $\eta = \sum_{i=0}^{2N} o(i + d_i)$ is the number of matched pixels in the scan line. The energy function is given by

$$U_c(d) = -\log\{\frac{1}{(\sqrt{2\pi}\sigma)^\eta}\} - \log \prod_{i=0}^{2N} \exp\{-\frac{1}{2\sigma^2}\Delta g(d_i) o(i + d_i)\} ,$$
$$= \sum_{i=1}^{2N} \left[\frac{1}{2\sigma^2}\Delta g(d_i) - \log \frac{1}{\sqrt{2\pi}\sigma}\right] o(i + d_i) . \qquad (8)$$

An occlusion occurs whenever $d_i \neq d_{i-1}$ and every two occlusions means one less matching. Since there are a maximum of N matchings that can occur (8) can be rewritten as

$$U_c(d) = -Nk + \sum_{i=1}^{2N} \left[\frac{1}{2\sigma^2}\Delta g(d_i) o(i + d_i) + k\Delta d_i\right] , \qquad (9)$$

where $k = \frac{1}{2}\log\frac{1}{\sqrt{2\pi}\sigma}$ and $\Delta d_i = (d_i - d_{i-1})^2$.

The use of complex prior probability models, such as the MRF model [8], can be used to reduce the ill-posedness of disparity estimation. We use constraints to reduce the solution space so a very simple binomial prior based on the number of occlusions or matches is used:

$$P(d) = \prod_{i=1}^{2N} \exp\{\alpha\frac{1}{2}(1 - \Delta d_i)\} \exp\{\beta\frac{1}{2}\Delta d_i\} , \qquad (10)$$

where $\alpha = \log(1 - P_o)$, $\beta = \log P_o$, and P_o is the probability of an occlusion in any site.. The energy equation for the prior is

$$U_p(d) = -\log \prod_{i=1}^{2N} \exp\{\alpha \frac{1}{2}(1 - \Delta d_i)\} \exp\{\beta \frac{1}{2}\Delta d_i\} ,$$

$$= N\alpha + \frac{1}{2}\sum_{i=1}^{2N}(\beta - \alpha)\Delta d_i .$$

(11)

Substituting (8) and (11) into (6) we get the total energy function

$$U_t(d) = N(k + \alpha) + \sum_{i=1}^{2N} \left[\frac{1}{2\sigma^2}\Delta g(d_i)o(i + d_i) + (2k + \beta - \alpha)\Delta d_i\right] .$$

Removing the constant additive terms and factors, the final form of the energy function becomes

$$U(d) = \sum_{i=1}^{2N} [\Delta g(d_i)o(i + d_i) + \gamma \Delta d_i] ,$$

(12)

where all the parameters are combined into $\gamma = -\log[\sqrt{2\pi}\sigma(1 - P_o)/P_o]$.

Thus the final optimization problem is to find the disparity vector that minimizes the energy represented by (12). The constraints are implemented by restricting the disparity to valid paths through the trellis in Fig. 3. The use of the simple prior in (10) results in an energy function that has the same form as that in (8); only the parameter is changed. The final energy function is similar to that presented in [5].

5 Implementation

The optimal disparity is the directed path through the trellis in Fig 3 that minimizes (12). Here we use DP techniques to efficiently perform this search. The resulting algorithm is suitable for parallel processing.

The trellis contains two types of nodes, occlusion nodes and match nodes. Occlusion nodes are connected to two neighboring occlusion nodes, with cost γ, and one neighboring match node. Match nodes have only one incoming path from an occlusion node, and associated with the match node or the incoming path is the matching cost for a pair of pixels.

We apply DP techniques progressing through the trellis from left to right. An occlusion node (i, j) chooses the best of the three incoming paths after adding γ to the diagonal paths, and updates its own energy. Each match node merely updates the energy of the incident path by adding the matching cost of the left and right image pixels. At initialization, the cost of the only valid node $(0, 0)$ is set to zero and the other costs are set to infinity. The algorithm terminates at node $(2N, 0)$ and the optimal disparity path $\hat{d}$ is found by tracing back the best path from $P(2N, 0)$.

The shortest path algorithm for disparity is formally given by:

1. Initialization: Set all costs to infinity except for $j = 0$.

$$U(0,j) = \begin{cases} 0 & j = 0 , \\ \infty & \text{otherwise.} \end{cases}$$

2. Recursion: For $i = 1$ to $2N$ find the best path and cost into each node j.

    (a) $i = j$ even: $U(i,j) = \min_{\alpha \in [-1,1]} U(i-1, j+\alpha) + \gamma\alpha^2$,

    $P(i,j) = \arg\min_{\alpha \in [-1,1]} U(i-1, j+\alpha) + \gamma\alpha^2$,

    (b) $i = j$ odd: $U(i,j) = U(i-1,j) + (g^l_{\frac{1}{2}(i-j+1)} - g^r_{\frac{1}{2}(i+j+1)})^2$,

    $P(i,j) = j$.

3. Termination: $i = 2N$ and $j = 0$.

$$\hat{d}_{2N} = P(2N, 0) .$$

4. Backtracking: Find the optimal disparity by tracing back the path.

$$\hat{d}_{i-1} = \hat{d}_i + P(i, \hat{d}_i), \quad i = 2N, \ldots, 1 .$$

At each step i, each node uses accumulated cost or decision information only from neighboring nodes in the previous step $i-1$ (or $i+1$ in Phase 4), and matching and occlusion costs for a given scan line are fixed. Thus the recursion equations can be calculated in parallel at each step, making the algorithm suitable for solution with parallel processor architectures. The computational complexity is $\mathcal{O}(N^2)$, or $\mathcal{O}(N)$ if the maximum disparity is fixed at $d_{\max}$.

6 Experimental Results

The performance of this algorithm was tested on a variety of synthetic and real images. To assess the quantitative performance, we used synthetic images to control the noise and the maximum disparity. Qualitative assessments were performed on both synthetic and real images.

Fig. 4 shows four test samples. The top row is the left image and the bottom row is the calculated disparity. The first column is a 256×256 binary random dot stereogram (RDS). The estimated disparity shows great accuracy with relatively few errors located in or near the occlusion regions. The second column is a large disparity synthetic image of a sphere above a textured background. The result is good with both sharp and gradual disparity transitions are reproduced well. However, vertical disparity edges are somewhat jagged. The third column is the Pentagon image and again the results are good with small features in the building and in the background (such as the road, bridge and some trees) being detected. As with the sphere image, there is some breakup of vertical disparity edges.

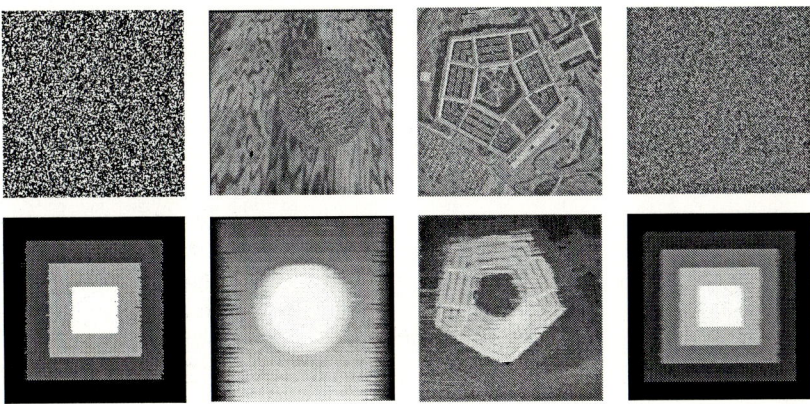

Fig. 4. Sample test sets: From left to right - binary RDS, sphere, Pentagon and gray RDS, and from top to bottom - left and disparity image.

The Pentagon disparity is similar to that of other DP methods with strong constraints, including [5,10]. However this method has the benefit of a highly concurrent simple single-pass algorithm with low computational complexity. The MRF based methods [8,7] tend to blur any sharp disparity discontinuities, but vertical disparity boundaries are more coherent.

To quantitatively assess noise performance, 256 gray level RDS image pairs were used. Each pixel was generated from a Gaussian distribution $N(128, \sigma_s^2)$ to which Gaussian noise $N(0, \sigma_n^2)$ was added. Defining SNR $= 10\log(\sigma_s^2/\sigma_n^2)$, matching was performed on a variety of these RDS images with various SNRs. The performance was quantified in terms of pixel error rate, that is, the fraction of sites where the calculated disparity did not match the real disparity. A test sample containing 5 disparity levels with a step of 12 pixels between each level and SNR of 9 dB is shown in the fourth column of Fig. 4.

The PER with respect to SNR is shown in Fig. 5(a) for three image pairs; RDS1 with two disparity levels and a 16 pixel step between the levels, and RDS2 and RDS3 with 5 disparity levels and 8 and 12 pixel steps respectively. The graph shows that a very high fraction of the pixels are correctly matched at high SNR, and that the performance is robust with respect to noise.

The PER performance verses maximum disparity is shown in Fig. 5(b) for two image pairs; RDS4 with 2 disparity levels and RDS5 with 5 disparity levels. The postfix 'a' indicates that no noise was added and 'b' indicates that the SNR was 9 dB. Again, we see that the PER degrades gracefully with respect to maximum disparity. The reason for this degradation is that the disparity path is based on fewer matchings as maximum disparity increases. Both RDS4 and RDS5 have similar performance, indicating that the number of occlusions, and not how those occlusions are distributed, is the dominant factor. However, distributing the occlusions into many small groups rather than a few large groups improves performance slightly.

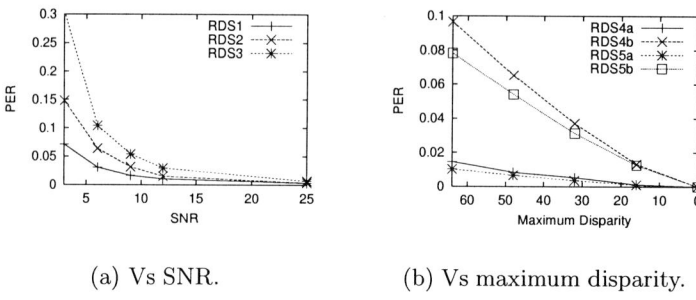

(a) Vs SNR. (b) Vs maximum disparity.

Fig. 5. Error performance for synthetic images.

The overall performance of this method is quite good with disparity estimates equal to that of the best DP methods. Large disparity and finely textured disparity patterns are detected well. The quantitative tests on synthetic images shows good error performance with graceful degradation with respect to noise and disparity. The significant benefit of the center-referenced disparity is the very high degree of concurrency and simplicity of the computational structure. The computation time for these images was significantly better than for MRF-based and most other DP-based techniques, ranging from about 4 s for 256x256 images to about 18 s for 512x512 images on a 350 MHz Pentium-II based PC.

7 Conclusion

We have created a center-referenced projection to represent the discrete match space for stereo correspondence. This space contains additional occlusion points which we exploit to create a concise representation of correspondence and occlusion. Applying matching and projection constraints, a solution space is obtained in the form of a sparsely connected trellis.

The efficient representation of the constraints and the energy equation in the center-referenced disparity space result in a simpler DP algorithm with low computational complexity that is suitable for parallel processing. Systolic array architectures using simple processing elements with only nearest neighbor communication can be used and one is currently being implements using ASICs.

The algorithm was tested on real and synthetic images with good results. The disparity estimate is comparable to the best DP methods. Matching errors were found to degrade gracefully with respect to SNR and maximum disparity.

The occlusion cost was estimated heuristically but an automated mechanism would permit adaption to a variety of images. Also inter-line dependence could smooth vertical disparity edges, using existing techniques or developing a new one to exploit the center-referenced disparity space.

The current model does not fit transparent objects as it is a violation of the uniqueness constraint in (3). However, efficient techniques exist for finding the k best paths through a trellis and these could possibly be applied to find *two* paths through an image region, one for the transparent surface (self or reflection image) and one for the surface behind it. The current algorithm could be applied to each path with the matching cost being a composite function of the two paths.

References

1. H. H. Baker and T. O. Binford. Depth from edge and intensity based stereo. In *Proceedings of the International Joint Conference on Artificial Intelligence*, pages 631–636, Vancouver, Canada, 1981.
2. Peter N. Belhumeur. A Bayesian approach to binocular stereopsis. *International Journal of Computer Vision*, 19(3):237–260, 1996.
3. J. Besag. Spatial interaction and the statistical analysis of lattice systems (with discussion). *Journal of the Royal Statistical Society*, 36(2):192–326, 1974.
4. Stan Birchfield and Carlo Tomasi. Depth discontinuities by pixel-to-pixel stereo. In *Proceedings of the IEEE International Conference on Computer Vision*, pages 1073–1080, Bombay, India, 1998.
5. Ingemar J. Cox, Sunita L. Hingorani, Satish B. Rao, and Bruce M. Maggs. A maximum likelihood stereo algorithm. *Computer Vision and Image Understanding*, 63(3):542–567, May 1996.
6. M. Drumheller and T. Poggio. On parallel stereo. In *Proceedings of the IEEE International Conference on Robotics and Automation*, pages 1439–1448, April 1986.
7. Davi Geiger and Frederico Girosi. Parallel and deterministic algorithms from MRF's: Surface reconstruction. *IEEE Transactions on Pattern Analysis and Machine Intelligence*, PAMI-13(5):401–412, May 1991.
8. Stuart Geman and Donald Geman. Stochastic relaxation, Gibbs distributions, and the Bayesian restoration of images. *IEEE Transactions on Pattern Analysis and Machine Intelligence*, PAMI-6(6):721–741, November 1984.
9. S. Kirkpatrick, C. D. Gelatt Jr, and M. P. Vecchi. Optimization by simulated annealing. *Science*, 220(4598):671–680, May 1983.
10. Y. Ohta and T. Kanade. Stereo by intra- and inter-scanline search. *IEEE Transactions on Pattern Analysis and Machine Intelligence*, PAMI-7(2):139–154, March 1985.
11. Tomaso Poggio and Vincent Torre. Ill-posed problems and regularization analysis in early vision. Artifial Intelligence Lab. Memo 773, MIT Press, Cambridge, MA, USA, April 1984.
12. John H. Woods. Two-dimensional discrete Markovian fields. *IEEE Transactions on Information Theory*, IT-18(2):232–240, March 1972.

Unsupervised Learning of Biologically Plausible Object Recognition Strategies

Bruce A. Draper and Kyungim Baek

Colorado State University, Fort Collins CO 80523, USA
{draper,baek}@cs.colostate.edu

Abstract. Recent psychological and neurological evidence suggests that biological object recognition is a process of matching sensed images to stored iconic memories. This paper presents a partial implementation of (our interpretation of) Kosslyn's biological vision model, with a control system added to it. We then show how reinforcement learning can be used to control and optimize recognition in an unsupervised learning mode, where the result of image matching is used as the reward signal to optimize earlier stages of processing.

1 Introduction

Traditionally, object recognition has been thought of as a multi-stage process, in which every stage produces successively more abstract representations of the image. For example, Marr proposed a sequence of representations with images, edges, $2\frac{1}{2}D$ sketch, and a 3D surfaces [7]. Recently, however, biological theories of human perception have suggested that objects are stored as iconic memories [6], implying that object recognition is a process of matching new images to previously stored images. If so, object recognition is a process of transformation and image matching rather than abstraction and model matching.

Even according to the iconic recognition theory, object recognition remains a multi-stage process. In the iconic approach, it is not reasonable to assume that the target object is alone in the field of view, or that the viewer has previously observed the object from every possible perspective. As a result, there is still a need for focus of attention (e.g. segmentation) and transformation/registration steps prior to matching.

At the same time, there is another psychologically-inspired tradition in computational models of biological vision that suggests that object recognition should not be viewed as a single, hard-wired circuit. Instead, there are many techniques for recognizing objects, and biological systems select among them based on context and properties of the target object. Arbib first described this using a slide-box metaphor in 1970's [1]. Since then, Ulllman's visual routines [17] and the "purposive vision" approach [2] could be viewed as updated versions of the same basic idea.

This paper tries to synthesize Kosslyn's iconic recognition theory with the purposive approach. In particular, it builds on the author's previous work on using reinforcement learning to acquire purposive, multi-stage object recognition strategies [4]. Unlike in previous work, however, this time we assume that memory is iconic and that object recognition is therefore an image matching task. We then use the match score between the stored image (memory) and the sensed image (input) as a reward signal for optimizing the recognition process.

In this way, we build a prototype of an iconic recognition system that automatically develops specialized processes for recognizing common objects. In this way, we not only combine two biologically motivated theories of biological perception, we also avoid the need for hand-labeled training images that limiting our earlier work. Instead, we have an unsupervised rather than supervised system for learning object recognition strategies.

At the moment, our prototype system is extremely simple. This paper presents a demonstration in which the image match score is used as a reward signal and fed back to earlier stages of processing. The goal is to show that this reward signal can be used to make object recognition more efficient. More sophisticated versions, with hopefully higher over-all recognition rates, are under development.

2 Previous Work

Kosslyn has argued since at least 1977 that visual memories are stored essentially as images [5]. This idea received critical neurological support in 1982, when researchers were able to show a retinotopic map of a previously viewed stimulus stored in the striate cortex of a monkey [14]. Since then, the psychological and neurological evidence for iconic memories has grown (see chapter 1 of [6] for an opinionated overview). At the same time, SPECT and PET studies now show that these iconic memories are active during recognition as well as memory tasks [6]. The biological evidence for image matching as a component of biological recognition systems is therefore very strong.

More specifically, Kosslyn posits a two-stage recognition process for human perception, where the second stage performs image transformation and image matching. Although he does not call the first stage "focus of attention", this is essentially what he describes. He proposed pre-attentive mechanisms that extract nonaccidental properties and further suggests that these nonaccidental properties serve as cues to trigger image matching. Beyond hardwired, pre-attentive features, Kosslyn also suggests that biological systems learn object-specific features called signals (see [6] pp.114-115) to predict the appearance of object instances and that these cues are used as focus of attention mechanisms.

Kosslyn's description of image transformation and image matching is imprecise. Much of his discussion is concerned with image transformations, since the image of an object may appear at any position, scale or rotation angle on the retina. He argues that our stored memories of images can be adjusted "to cover different sizes, locations, and orientations" [6], although he never gives a mathematical description of the class of allowable transforms. Tootell's image [14] suggests, however, that at least 2D perspective transformations should be allowed, if not non-linear warping functions. With regard to the matching process itself, Kosslyn doubts that a template-like image is fully generated and then compared to the input. Without further explanation, one is left with a vague commitment to an image matching process that is somehow more flexible than simple correlation.

It should be noted that Kosslyn's model is not the only model of biological object recognition. For example, Rao and Ballard propose a biological model in which images are transformed into vectors of filter responses, and then matched to vectors representing specific object classes [9]. Biederman presents a model based on pre-attentive figure completion [3]. Nonetheless, neither of these theories explain the strikingly iconic nature of Tootell's striate cortex image [14].

In addition, there has been a great deal of interest recently in PCA-based appearance matching [8,16]. While powerful, appearance matching should be understood as one possible technique for the image matching stage of object recognition. In fact, appearance matching is a computationally efficient technique for approximating the effect of correlating one test image to a large set of stored model images (see [15] Chapter 10 for a succinct mathematical review). Thus appearance matching is a potentially useful component of an object recognition system, but it is not by itself a model of biological object recognition.

Previously, we developed a system that learns control strategies for object recognition from training samples. The system, called ADORE, formalized the object recognition control problem as a Markov decision problem, and used reinforcement learning to develop nearly optimal control policies for recognizing houses in aerial images [4]. Unfortunately, the use of this system in practice has been hindered by the need to provide large numbers of hand-labeled training images.

Kosslyn's theory suggests that hand-labeled training images may not be necessary. By using the result of image matching as the training signal, we can move ADORE into an unsupervised learning mode. This removes the need to hand-label training instances, thereby creating an unsupervised system that learns and refines its recognition policies as it experiences the world.

3 The Proposed System

Fig. 1 shows our computational model of biological vision. It can be interpreted as adding a control and learning component to our instantiation of Kosslyn's model [6]. At an abstract level, it has three recognition modules: focus of attention, transformation/registration, and matching[1]. At a more detailed level, each module has multiple implementations and parameters that allow it to be tuned to particular object classes or contexts. For example, the final image matching stage can be implemented many ways. If the goal is to match an image of a specific object instance against a single template image in memory, then image correlation remains the simplest and most reliable comparison method. Alternatively, if the goal is to match an image against a set of closely related templates

---

[1] The focus of attention module has both a pre-attentive and attentive component, although we will not concentrate on this distinction in this paper.

in memory (e.g. instances of the same object under different lighting conditions), then principle components analysis (PCA) may be more efficient. In more extreme cases, if the goal is to find an object that may appear in many different colors (e.g. automobiles), then a mutual information measure may be more effective [18], while a chi-squared histogram comparison may be most appropriate for certain highly textured objects such as trees [12].

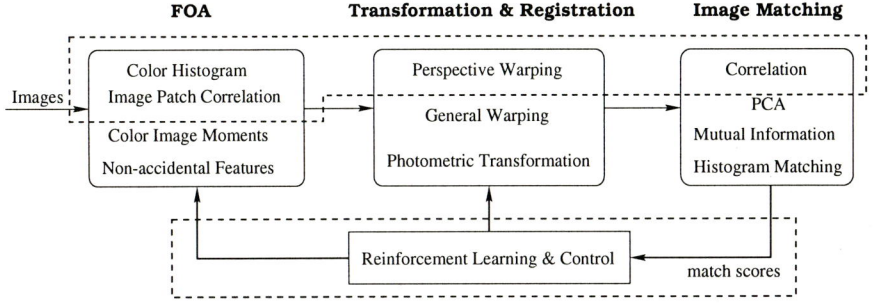

Fig. 1. The proposed system architecture.

In other words, the top-level modules in Fig. 1 represent roles in the recognition process that must be executed in sequence. Each role, however, can be filled by a variety of techniques, depending on the scene and object type. In this way, we address Kosslyn's need for a flexible image matching system, and at the same time incorporate all of Kosslyn's suggestion for focus of attention mechanisms [6].

However, the presence of options within the modules creates a control problem: which technique(s) should be applied to any given input? This is a dynamic control decision, since the choice of technique may be a function of properties of the input image. More importantly, it involves delayed rewards, since the consequences of using a particular FOA or transformation technique are not known until after image matching. We therefore use a reinforcement learning module to generate control policies for optimizing recognition strategies based on feedback from the image matching module (see Fig. 1).

4 The Implemented System

The dashed lines in Fig. 1 outline the parts of the system that have been implemented so far. It includes two technique for FOA module and one technique for each of the others: color histogram matching and image patch correlation for focus of attention; matching four image points to four image points for perspective image transformation and registration; and correlation for image matching. While this is clearly a very limited subset of the full model in terms of its object recognition ability, our goal at this point is to test the utility of the image match score as a reinforcement signal, not to recognize objects robustly.

Currently, the system is "primed" to look for a specific object by a user who provides a sample image of the target. This sample image is then used as the template for image matching. The user also provides the locations of four distinctive intensity surface patches within the template image, for use by our (primitive) focus of attention mechanism. The FOA mechanism then extracts templates at nine different scales from each location, producing nine sets of four surface patches[2].

When an image is presented to the system at run-time, a local color histogram matching algorithm is applied for pre-attentive FOA. If the control module decides that the resulting hypothesis is good enough to proceed further, the FOA system uses a rotation-free correlation algorithm to match the surface patches to the new image. (As described in [10], the rotation-free correlation algorithm allows us to find an object at any orientation without doing multiple correlations.) The result is nine sets of four points each, one set for each scale. Thus the focus of attention mechanism produces nine point set hypotheses for the image transformation step to consider.

Under the control of the reinforcement learning module, the image transformation module selects one of these nine hypotheses to pursue. It then uses the four point correspondences between the input image and model template to compute the perspective transformation that registers the input image to the template. In principle, this compensates not only for changes in translation, rotation and scale, but also for small perspective distortions if the target object is approximately planar.

After computing the image transformation, the system has a third control decision to make. If the transformation is sensible, it will apply the transformation to the input image and proceed to image matching. On the other hand, if the selected set of point matches was in error the resulting image transformation may not make sense. In this case, the control system has the option to reject the current transformation hypothesis, rather than to proceed onto image matching.

The final image matching step is trivial. Since the transformed input image and the object template are aligned, simple image correlation generates the reward signal. In general, if we have found and transformed the object correctly, we expect a greater than 0.5 correlation.

4.1 Optimization through Unsupervised Learning

The proposed system casts object recognition as a unsupervised learning task. Users prime the system by providing a sample image of the target object and the location of unique appearance patches. The system then processes images, rewarding itself for high correlation scores and penalizing itself for low scores. In this way, the system optimizes performance for any given template.

To learn control strategies, the system models object recognition as a reinforcement learning problem. As shown in Fig. 2, the state space of the system has four major "states": two for pre-attentive/attentive FOA hypotheses, one

---

[2] Each set of four points includes patches at a single scale.

for image transformation hypotheses, and one for image matching hypotheses. The algorithms in Fig. 1 are the actions that move the system from one state to another and/or generate rewards, and the system learns control policies that select which action to apply in each state for the given task.

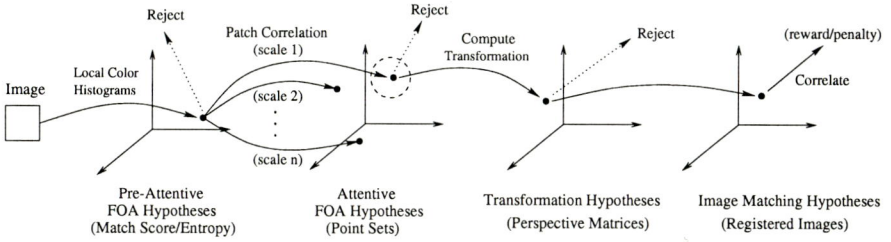

Fig. 2. The iconic view of state spaces and actions. At each state space, the control system decides whether to proceed further or reject current hypothesis.

Two necessary refinements complicate this simple system description. The first is that the four states are actually state *spaces*. For each type of hypothesis, the system has a set of features that measure properties of that type of hypothesis. For example, for the FOA point set hypotheses the system defines nine features: the correlation scores for each point (4 features); the difference between the best and second best correlation scores for each point (to test for uniqueness; 4 more features); and a binary geometric feature based on the positions of the points that tests for reflections[3].

The control policy therefore maps points in these four feature spaces onto actions. In particular, every action takes one type of hypothesis as input, and the control policy learns a Q function for every action that maps points in the feature space of the input data onto expected future rewards. At run-time, these Q functions are evaluated on the features associated with hypotheses, and the action with the highest expected reward is selected and run.

The second refinement is that some actions may return multiple hypotheses. For example, our current attentive FOA algorithm returns nine different point sets corresponding to nine different scales. When this happens, the control system must select the best state/action pair to execute next, where each hypothesis is one state. Again, this is done by selecting the maximum expected reward.

With these refinements, we have a reinforcement learning system defined over continuous state spaces. This implies that we need function approximation techniques to learn the Q functions. Currently we are using backpropagation neural networks for this, although there is some evidence that memory-based function approximation techniques may provide faster convergence [11] and we will experiment with other techniques in the future. Given a function approxi-

---

[3] An image reflection would imply looking at the back of the surface, which we assume to be impossible.

Fig. 3. A *cgw_tree* magazine (left), a *cgw_watch* magazine (middle), and a *wired* magazine image (right).

mation technique, the system can be trained as usual using either $TD(\lambda)$ [13] or Q-Learning [19].

5 Experimental Results

As mentioned earlier, we do not claim at this point to have a robust object recognition system, since much of Fig. 1 remains unimplemented. Instead, the goal of the experiments is to test whether the image match score can be used as a reinforcement signal to make object recognition more efficient. To this end, we consider a "control-free" baseline system that exhaustively applies every routines in Fig. 1, and then selects the maximum correlation score. We then compare this to the strategy learned by reinforcement learning. While the reinforcement learning system obviously cannot create a higher correlation score than exhaustive search, the ideal is that the controlled system would produce nearly as high correlation scores while executing far fewer procedures.

The experiment was performed with a set of color images of magazines against a cluttered background. The dataset has 50 original images with different viewing angles and variations in illumination and scale. There are three types of magazine images in the dataset: 30 *cgw_tree* images, 10 *cgw_watch* images, and 10 images containing the *wired* magazine. Fig. 3 shows these three types of magazines. Each original image was scaled to 14 different resolutions at scales ranging from 0.6 to 1.5 times the original. As a result, the dataset contains 700 images. We define a good sample to be any image that produces a maximum correlation score of 0.5 or higher under exhaustive search.

The first row of Table 1 shows the number of good and bad samples in each dataset. Table 1 also contains several measures of the performance of our control system. The number of rejected samples indicates the number of false negatives for good samples and the number of true negatives for bad samples. Therefore, it is better to have small values for the former and large values for the latter. According to the values in the table, the control system works well on rejecting bad samples for all three cases. The true negative samples classified by the system include all the images without target object and all the false positive samples have final matching scores less that 0.5, which means that, eventually,

Table 1. Results obtained by the policy learned without backtracking.

	CGW_TREE		CGW_WATCH		WIRED	
	good	bad	good	bad	good	bad
# of samples	129	571	31	669	59	641
# of rejected samples	25(19.4%)	550(96.3%)	11(35.5%)	668(99.9%)	20(33.9%)	640(99.8%)
operation_count/sample	2.92	1.97	2.94	1.33	2.85	1.54
#of optimal prediction	53(41.1%)		15(48.4%)		22(37.3%)	
average prediction/optimal	0.936343		0.954375		0.956907	

the system will not consider those samples as positive samples. Therefore, we can say that the control system generates no false positives.

When it comes to the false negatives, however, the control system had a somewhat harder time. One of the reasons is inaccurate predictions by the neural network trained on pre-attentive FOA hypothesis. Once the mistake is made, there is no way to recover from it. Also, even one weak image patch out of four can easily confuse the selection of point matches and the resulting transformed image gets low match score. This is happened most of the times in *cgw_watch* and *wired* cases. These concerns lead us to the need for defining more distinctive feature set.

As a point of reference for efficiency of object recognition, there are few enough control options in the current system to exhaustively pursue all options on all hypotheses for any given image, and then select the maximum match score. This approach executes 28 procedures per image, as opposed to less than three procedures per image on average for *cgw_tree* and even less that two on average for *cgw_watch* and *wired* under the control of the reinforcement learning system. On the other hand, the average reward (i.e. match score) generated by the reinforcement learning system is 94% of the optimal reward generated by exhaustive search for *cgw_tree*, 95% of optimal for *cgw_watch*, and 96% of optimal for *wired*. Thus there is a trade-off: slightly lower rewards in return for more than 93% savings in cost.

We also noticed that the neural networks trained to select point matches are less accurate than the networks trained to decide whether to reject or match transformation hypothesis. This creates an opportunity to introduce another refinement to the reinforcement learning system: since we are controlling a computational (rather than physical) process, it is possible to backtrack and "undo" previous decisions. In particular, if the system chooses the wrong point set, it may notice its mistake as soon as it measures the features of the resulting transformation hypothesis. The system can then abandon the transformation hypothesis and backtrack to select another point set.

Table 2 shows the results with backtracking. The additional operation per sample is less than one on average, and it greatly reduced the false negative rates with no significant increase in the false positive rates. With backtracking, the false negative rate drops from 19.4% to 7.0% on *cgw_tree*, from 35.5% to 19.4% on *cgw_watch*, and from 33.9% to 10.2% on *wired* case, while the largest increase in the false positive rates is only 0.9%. It is open to interpretation, however, whether backtracking could be part of a model of biological vision.

Table 2. Results obtained by the policy learned with backtracking.

	CGW_TREE		CGW_WATCH		WIRED	
	good	bad	good	bad	good	bad
# of samples	129	571	31	669	59	641
# of rejected samples	9(7.0%)	545(95.4%)	6(19.4%)	668(99.9%)	6(10.2%)	640(99.8%)
operation_count/sample	3.72	2.15	3.35	2.11	3.53	1.55
#of optimal prediction	69(53.5%)		17(54.8%)		33(58.9%)	
average prediction/optimal	0.963196		0.939729		0.975222	

6 Conclusions and Future Work

We have described a system that learns control strategies for object recognition tasks through unsupervised learning. The basic recognition process of the system is biologically motivated, and uses reinforcement learning to refine the system's overall performance. We tested the system for three different target objects on a set of color images with differences in viewing angle, object location and scale, background, and the amount of perspective distortion. The learned control strategies were within 94% of optimal for the worst case and 98% of optimal for the best case.

In the future, we would like to complete our implementation of the broader system outlined in Fig. 1. For example, histogram correlation and mutual information can also be used to match images, thereby allowing the system to recognize a broader range of objects. Even more improvements can be made in the focus of attention module. As the set of procedures grows, there will be more actions that can be applied to every type of intermediate data. How accurately can we predict the expected rewards of these actions is, therefore, one of the most important issues in the system implementation. To increase the prediction power, we need to define more distinctive features than those used at the moment.

References

1. M. Arbib. *The Metaphorical Brain: An Introduction to Cybernetics as Artificial Intelligence and Brain Theory*. Wiley Interscience, New York, 1972.
2. J. Aloimonos. Purposive and Qualitative Active Vision. *IUW*, pp.816-828, Sept. 1990.
3. I. Biederman and E. Cooper. Priming contour-deleted images: Evidence for intermediate representations in visual object recognition. *Cognitive Psychology*, **23**:393-419

4. B. A. Draper, J. Bins, and K. Baek. ADORE: Adaptive Object Recognition. *International Conference on Vision Systems*, Las Palmas de Gran Canaria, Spain, 1999.
5. S. K. Kosslyn and S. P. Schwartz. A Simulation of Visual Imagery. *Cognitive Science*, **1**:265–295, 1977.
6. S. K. Kosslyn. *Image and Brain: The Resolution of the Imagery Debate*. MIT Press, Cambridge, MA, 1994.
7. D. Marr. *Vision: A Computational Investigation into the Human Representation and Processing of Visual Information*, W.H. Freeman & Co., San Francisco. 1982.
8. H. Murase and S. K. Nayar. Visual Learning and Recognition of 3-D Objects from Appearance. *International Journal of Computer Vision*, **14**:5–24, 1995.
9. R. P. N. Rao and D. Ballard. An Active Vision Architecture based on Iconic Representations. *Artificial Intelligence*, 78:461–505, 1995.
10. S. Ravela, *et al.*. Tracking Object Motion Across Aspect Changes for Augmented Reality. *Image Understanding Workshop*, Palm Springs, CA, 1996.
11. J. Santamaria, R. Sutton, A. Ram. "Experiments with Reinforcement Learning in Problems with Continuous State and Action Spaces", *Adaptive Behavior* **6**(2):163-217, 1998.
12. B. Schiele and J. L. Crowley. *Recognition without Correspondence using Multidimensional Receptive Field Histograms.* MIT Media Laboratory, Cambridge, MA, 1997.
13. R. Sutton. Learning to Predict by the Models of Temporal Differences. *Machine Learning*, **3**(9):9–44, 1988.
14. R. B. H. Tootell, *et al.*. Deoxyglucose Analysis of Retinotopic Organization in Primate Striate Cortex. *Science*, **218**:902–904, 1982.
15. E. Trucco and A. Verri.*Introductory Techniques for 3-D Computer Vision,* Prentice Hall, Upper Saddle River, NJ., 1998.
16. M. Turk and A. Pentland. Eigenfaces for Recognition. *Journal of Cognitive Neuroscience*, **3**(1):71–86, 1991.
17. S. Ullman. Visual Routines. *Cognition*, **18**:97–156, 1984.
18. P. Viola and W. M. Wells. Alignment by Maximization of Mutual Information. *ICCV*, Cambridge, MA, 1995.
19. C. Watkins. *Learning from Delayed Rewards*. Ph.D. thesis, Cambridge University, 1989.

Structured Kalman Filter for Tracking Partially Occluded Moving Objects

Dae-Sik Jang, Seok-Woo Jang, and Hyung-Il Choi

Soongsil University, 1-1, Sangdo-5 Dong, Dong-Jak Ku, Seoul, Korea
hic@computing.soongsil.ac.kr

Abstract. Moving object tracking is one of the most important techniques in motion analysis and understanding, and it has many difficult problems to solve. Especially estimating and tracking moving objects, when the background and moving objects vary dynamically, are very difficult. The Kalman filter has been used to estimate motion information and use the information in predicting the appearance of targets in succeeding frames. It is possible under such a complex environment that targets may disappear totally or partially due to occlusion by other objects. In this paper, we propose another version of the Kalman filter, to be called Structured Kalman filter, which can successfully work its role of estimating motion information under such a deteriorating condition as occlusion. Experimental results show that the suggested approach is very effective in estimating and tracking non-rigid moving objects reliably.

1 Introduction

There has been much interest in motion understanding with the progress of computer vision and multimedia technologies. Especially estimating and tracking moving objects have received great attention because they are essential for motion understanding [1][2][3]. However, such tasks become very difficult to solve when the background and moving objects vary dynamically and there have been reported no significant solutions for them.

We can find various approaches for tracking moving objects in the literature [4][5]. Most of them make some efforts to estimate motion information, and use the information in predicting the appearance of targets in succeeding frames. The Kalman filter has been used for this purpose successfully [4]. It presumes that the behavior of a moving object could be characterized by a predefined model, and the model is usually represented in terms of its state vector [6]. However, in a real world environment, we often face a situation where the predefined behavior model falls apart. It is possible that a target may disappear totally or partially due to occlusion by other objects. In addition, the sudden deformation of a target itself, like articulated movements of human body, can cause the failure of the predefined behavior model of the Kalman filter.

We propose in this paper another version of the Kalman filter, to be called Structured Kalman filter, which can overcome the above mentioned problems and successfully work its role of estimating motion information under such a deteriorating condition as occlusion. The Structural Kalman filter utilizes the relational information among sub-regions of a moving object. The relational information is to supplement the unreliable measurements on a partially occluded sub-region, so that *a priori* estimate of the next state of the sub-region might be obtained based on the relational information as well as the actual measurements.

2   Organization of Structural Kalman Filter

The Kalman filter is one of the most popular estimation techniques in motion prediction because it provides an optimal estimation method for linear dynamic systems with white Gaussian noise. It also provides a generalized recursive algorithm that can be implemented easily with computers. In general, the Kalman filter describes a system with a system state model and a measurement model as in (1).

$$\vec{s} = \Phi(k-1)\vec{s}(k-1) + w(k)$$
$$\vec{m} = H(k)\vec{k} + v(k) \tag{1}$$

The system state $\vec{s}(k)$ at the k-th time frame is linearly associated with the state at k-1 th time frame, and there is also a linear relationship between the measurement $\vec{m}(k)$ and the system state $\vec{s}(k)$. The random variables $w(k)$ and $v(k)$ represent the state and measurement noise, respectively. They are assumed to be independent of each other, and have white Gaussian distribution. In (1), $\Phi(k)$ is called the state transition matrix that relates the state at time frame k to the state at frame k+1, and $H(k)$ is called the observation matrix that relates the state to the measurement.

Our Structural Kalman filter is a composite of two types of the Kalman filters: *Cell Kalman filters* and *Relation Kalman filters*. The Cell Kalman filter is allocated to each sub-region and the Relation Kalman filter is allocated to the connection between two adjacent sub-regions. Figure 1 shows the framework of the Structural Kalman filter with four Cell Kalman filters (KF_1,KF_2,KF_3,KF_4) and four Relation Kalman filters (KF_{12},KF_{23},KF_{13},KF_{34}). It is the adjacency of sub-regions that requires the assignment of the Relation Kalman filter. Figure 1 thus implies that the sub-region 3 is adjacent to all the other sub-regions. The Cell Kalman filter is to estimate motion information of each sub-region of a target, and the Relation Kalman filter is to estimate the relative relationship between two adjacent sub-regions. The final estimate of a sub-region is obtained by combining the estimates of involved Kalman filters. For example, it is (KF_1,KF_2,KF_3,KF_{12},KF_{13}) that may affect the final estimate of sub-region 1 of Figure 1. When the sub-region 1 is judged not to be occluded, KF_1 is enough to estimate its motion information. However, when the sub-region 1 is judged to be occluded, we rely on the estimates of (KF_2,KF_3,KF_{12},KF_{13}) to supplement the corrupted estimate of KF_1.

Though our Structural Kalman filter is not limited to some specific form of a measurement vector and state vector, we define in this paper a measurement

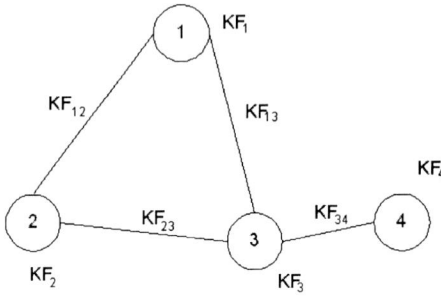

Fig. 1. A sample framework of Structural Kalman filter

vector and system state of the j-th Cell Kalman filter as in (2). That is, the measurement vector contains the position of a sub-region, and the state vector includes the position and its first derivative.

$$\vec{s}^{\,j}_c = \begin{bmatrix} \vec{m}^{\,j}_c(k) \\ \vec{m}^{\,j'}_c(k) \end{bmatrix} \qquad \vec{m}^{\,j}_c(k) = \begin{bmatrix} x^j \\ y^j \end{bmatrix} \qquad (2)$$

The Relation Kalman filter is to predict the relative relationship of adjacent sub-regions. We define its measurement vector and state vector to represent the relative position between the i-th sub-region and the j-th sub-region as in (3).

$$\vec{s}^{\,ij}_r(k) = \begin{bmatrix} \vec{m}^{\,ij}_r(k) \\ \vec{m}^{\,ij'}_r(k) \end{bmatrix} \qquad \vec{m}^{\,ij}_r(k) = \begin{bmatrix} \Delta x^{ij} \\ \Delta y^{ij} \end{bmatrix} \qquad (3)$$

where Δx^{ij} and Δy^{ij} depict the positional difference of the j-th sub-region from the i-th sub-region with respect to x-axis and y-axis, respectively. We need to define another measurement vector, called the inferred measurement vector, which is to indirectly express the measurement of a sub-region through the relative relationship with its pair associated by the connection. Equation (4) tells about it.

$$\begin{aligned} \tilde{m}^j(k) &= \vec{m}^{\,i}_c(k) + \vec{m}^{\,ij}_r(k) \\ \tilde{m}^i(k) &= \vec{m}^{\,j}_c(k) - \vec{m}^{\,ij}_r(k) \end{aligned} \qquad (4)$$

where $\tilde{m}^j(k)$ is computed by adding the measurement vector of $\vec{m}^{\,ij}_r(k)$ to the measurement vector of $\vec{m}^{\,i}_c(k)$. In a similar manner, we can compute $\tilde{m}^i(k)$ by subtracting the measurement vector of $\vec{m}^{\,ij}(k)$ from the measurement vector of $\vec{m}^{\,j}(k)$. When no occlusion happens and the Kalman filters work perfectly, the inferred measurement $\tilde{m}^j(k)$ must be exactly equal to $\vec{m}^{\,j}(k)$.

3 Estimation of Filter Parameters

Our Structural Kalman filter estimates the system state in a way of feedback control that involves the *prediction step* and *correction step*. This strategy is very similar to that of the typical Kalman filter, though the detailed operations in each step are quite different. Especially, the correction step combines the predicted information of Cell filters with that of Relation filters to provide an accurate estimate for the current state.

The prediction step determines *a priori* estimates for the next state based on the current state estimates. This step is applied to each Cell and Relation filter, treating each one independently. Under the linear model as in (1), we ignore the noise terms and get the recursive form of estimates like (5).

$$s_c^{i-}(k+1) = \Phi_c(k) \cdot s_c^{i^*}(k)$$
$$m_c^{i-}(k+1) = H_c(k) \cdot s_c^{i-}(k+1)$$
$$s_r^{ij-}(k+1) = \Phi_r(k) \cdot s_r^{ij^*}(k) \tag{5}$$
$$m_c^{ij-}(k+1) = H_c(k) \cdot s_c^{ij-}(k+1)$$

where the superscript - denotes *a priori* estimate for the next state and the superscript * denotes *a posteriori* estimate. Equation (5) shows how to compute the best estimate of the next measurement after the current measurement has been made. Once we get *a posteriori* estimate of a system state, we can obtain *a priori* estimate of a system state and then in turn *a priori* estimate of a measurement. It provides the answer to our problem of predicting the position of a sub-region. Now the remaining problem is how to get *a posterior* estimate of the system state. This is the task of the correction step.

The correction step of the Structural Kalman filter computes *a posteriori* estimate of the system state as a linear combination of *a priori* estimate and a weighted sum of differences between actual measurements and predicted measurements. When taking the weighted sum of differences, this step requires the participation of all the related Cell filters and Relation filters. For example, to obtain the estimate $s_c^{1^*}(k)$, a posteriori estimate of KF_1 in Figure 1, we need to consider actual and predicted measurement of KF_2, KF_3, KF_{12} and KF_{13}. It is because our Structural Kalman filter is conceived to supplement the unreliable measurement of some sub-region with the measurements of its adjacent sub-regions. Equation (6) tells how to compute a posteriori estimate of KF_1.

$$s_c^{j^*}(k) = s_c^{j-}(k) + (1 - \alpha^j(k)) \cdot \{K_c^j(k)(m_c^j(k) - m_c^{j-}(k))\}$$
$$+ \alpha^j(k) \cdot \{\frac{1}{N_j} \sum_i (1 - \alpha^i(k)) K_r^{ij}(k)(\tilde{m}(k) - m_c^{j-}(k))\} \tag{6}$$

Equation (6) computes *a posteriori* estimate of the j-th Cell filter with three terms. The first term denotes *a priori* estimate that is estimated at the time of k-1. It is adjusted by the second and third term. The second term contains two types of weights, $\alpha^j(k)$ and $K_c^j(k)$, together with the difference between the actual and predicted measurements on the j-th sub-region. The Kalman gain of $K_c^j(k)$ is the typical weight that emphasizes or de-emphasizes the difference. It is computed by manipulating the error covariance [6][7]. Another weight of

$\alpha^j(k)$ is a newly introduced term that reflects the degree of occlusion of the j-th sub-region. As the region is judged occluded more and more, the weight becomes higher and higher so that the j-th sub-region contributes less and less to the determination of the *a posteriori* estimate. The remaining problem is how to determine the measure of occlusion. For the time being, we leave it as it is. We will discuss it in detail later.

The third term of Equation (6) is to reflect the influence of sub-regions adjacent to the j-th sub-region. The amount of influence is controlled by the weight of $\alpha^j(k)$. In other words, as the j-th sub-region is judged occluded more and more, the computation of the *a posteriori* estimate relies on its adjacent sub-regions more and more. However, if some adjacent sub-region is heavily occluded also, the sub-region may not be of help much in the estimation. This consideration is handled by the use of $1 - \alpha^i(k)$ divided by N_j that denotes the number of sub-regions adjacent to the j-th sub-region. The third term also contains the Kalman gain and the difference between the inferred and predicted inferred measurements on the j-th sub-region. The inference is made by (4). That is, it involves the adjacent Cell filter and the associated Relation filter.

The correction step also computes *a posteriori* estimate of the system state of the Relation filter. The estimation is achieved by adding to *a priori* estimate the weighted difference between the actual measurement and predicted measurement as in (7).

$$s_r^{ij^*}(k) = s_r^{ij^-}(k) + K^{ij}(k)(m_r^{ij}(k) - m_r^{ij^-}(k)) \quad (7)$$

In the Relation filter, the measurement depicts the relative difference between measurements of two associated sub-regions. In this paper, we consider only the positional difference for the measurement. The predicted measurement is computed by (5). However, when we compute the actual measurement, we concern the possibility of occlusion and reflect the possibility in the form of (8).

$$m_r^{ij}(k) = \{(1-\alpha^j)(k)m_c^j(k) - (1-\alpha^i)(k)m_c^i(k)\} \\ + \{\alpha^j(k)m_c^{j^-}(k) - \alpha^i(k)m_c^{i^-}(k)\} \quad (8)$$

Equation (8) is the weighted sum of two different types of differences. One is the difference between actual measurements, and the other is the difference between the predicted measurement of associated Cell filters. Each measurement is properly multiplied by the weights representing the amount of occlusion or the amount of not-occlusion of the corresponding sub-region.

So far we have postponed the discussion about the measure of $\alpha^j(k)$ representing the degree of occlusion. In order to judge whether a sub-region is occluded or not, we need a model about the sub-region. If a sub-region matches well with its model, there is a high possibility that the sub-region is preserved without occlusion. However, the problem of modeling a sub-region is not our main concern. One can find various types of approaches for modeling a target in the literature. Some examples are active models and a deformable templates [5].

We assume that we are given a good model of a sub-region. In fact, to define a model of a sub-region in the current time frame, we use its corresponding sub-region of a previous time frame. We compare some predefined features of the pair of corresponding sub-regions. The amount of dissimilarity between the features is computed, and it is then used to determine the measure.

$$\alpha^j(k) = 1 - e^{-|Features^j(k-1) - Features^j(k)|} \tag{9}$$

4 Structural Kalman Filter in Action: Predicting the Motion Information

In the previous two sections, we presented the basic form of the Structural Kalman filter and its operational characteristics, especially in regard to estimating parameters. This section applies the Structural Kalman filter to the specific problem of prediction motion information of moving objects. Figure 2 shows the sequence of images to be used for the experiment. A person is moving around the room. During the movement, its body is partially occluded from time to time.

We obtained an initial model of a target by taking a difference operation between two successive frames [5]. The difference operation yields very naïve areas of moving objects. After some preprocessing like noise removing and region filling, we get a moving area as in Figure 3 (a) which represent a target. The extracted moving area is then partitioned into several sub-regions as in Figure 3 (b) by the criterion of regional homogeneity [5]. We do not go into detail of

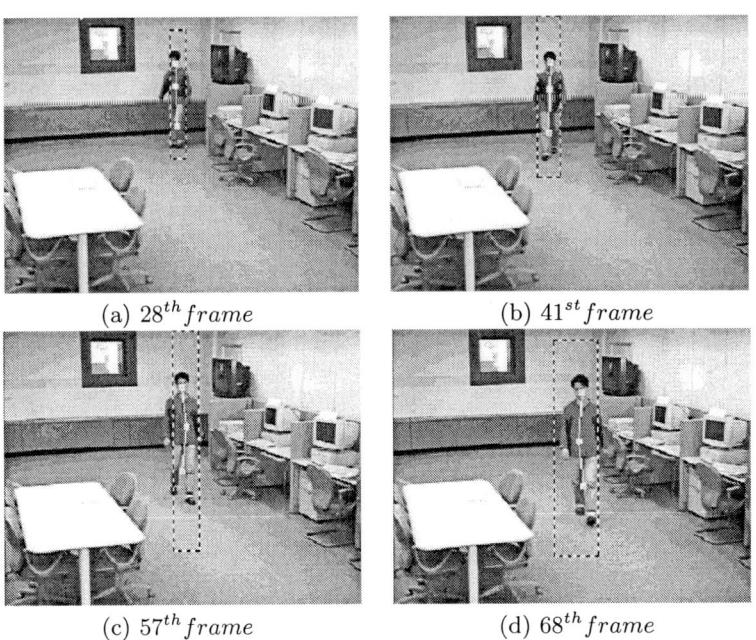

(a) $28^{th} frame$ (b) $41^{st} frame$

(c) $57^{th} frame$ (d) $68^{th} frame$

Fig. 2. Sequence of test images

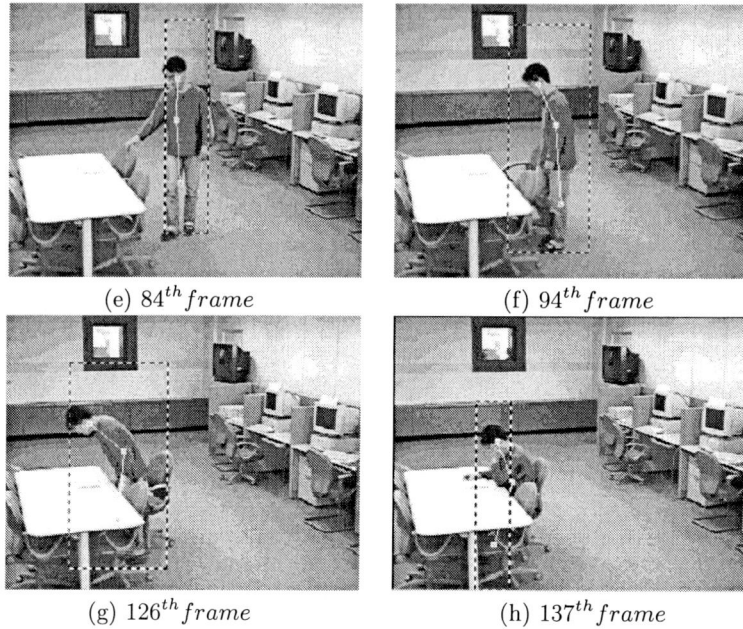

(e) $84^{th} frame$ (f) $94^{th} frame$

(g) $126^{th} frame$ (h) $137^{th} frame$

Fig. 2. Continued

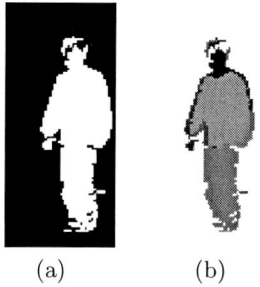

(a)   (b)

Fig. 3. Forming an initial model

the process of forming an initial model, since it is not the main concern of this paper.

In this example, our Structural Kalman filter has 3 cell filters, KF_1, KF_2, and KF_3, and 2 relation filters, KF_{12} and KF_{23}. The center coordinates of sub-regions are used for measurements of the corresponding cell filters, while we use for the measurements of relation filters the disparities between measurements of associated cell filters. The system state involves the measurement and its first order derivative. Under this setting, the system transition matrix and measurement matrix have the simple form like (10).

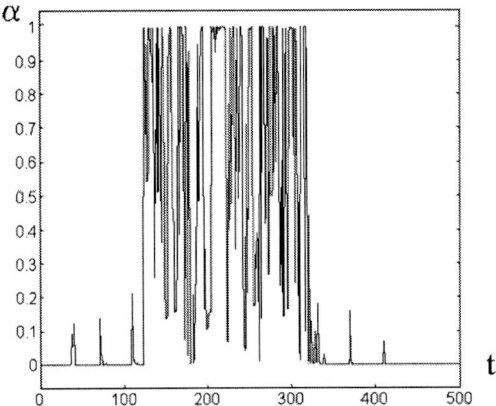

Fig. 4. Amount of occlusion

$$\Phi^j = \begin{bmatrix} 1 & 0 & \Delta t & 0 & \Delta t^2 & 0 \\ 0 & 1 & 0 & \Delta t & 0 & \Delta t^2 \\ 0 & 0 & 1 & 0 & \Delta t & 0 \\ 0 & 0 & 0 & 1 & 0 & \Delta t \\ 0 & 0 & 0 & 0 & 1 & 0 \\ 0 & 0 & 0 & 0 & 0 & 1 \end{bmatrix}$$

$$H^j = \begin{bmatrix} 1 & 0 & 0 & 0 & 0 & 0 \\ 0 & 1 & 0 & 0 & 0 & 0 \end{bmatrix}$$ (10)

We compare sub-regions of one frame, to each of them a cell filter is associated, against sub-regions of successive frame, so that the best matching pair could be found. The comparison is performed with regional features like color, texture and shape. Once the best matching pairs are found, we can obtain $\alpha^j(k)$, the measure representing the degree of occlusion, by using (9). Figure 4 shows values of $\alpha^j(k)$ of 3rd sub-region in a sequence of frames of Figure 2. We can observe that the measure gets very high value at frame 126 due to the severe occlusion.

Figure 5 quantitatively illustrates how the *a posteriori* estimate of $s_c^{j^*}$ is computed by adjusting the *a priori* estimate of $s_c^{j^-}$ with the second and third term of (6). In Figure 5, the solid line depict the second term and the dotted line depict the third term of (6) respectively. We can notice that the contribution of the third term increases as that of the second term decreases. In other words, when a sub-region is judged occluded much, the role of relation filters becomes much important in determining the *a posteriori* estimate.

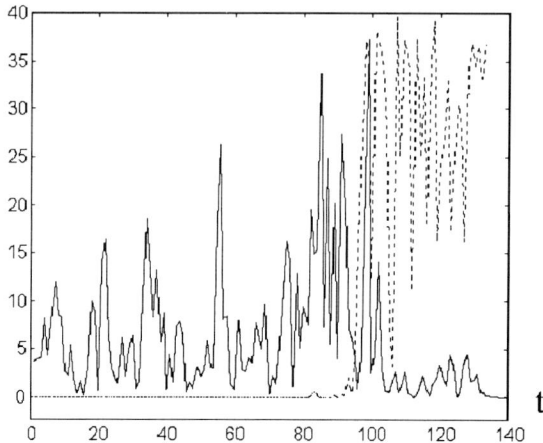

Fig. 5. Variation of influence of relation filters

5 Conclusion and Discussions

In this paper, we proposed a new framework named Structural Kalman filter that can estimate information on occluded region with the help of relation information among adjacent regions. When no occlusion happens, our Structural Kalman filter operates exactly like the typical Kalman filter. It is because relation filters become inactive under such a situation. However as an object gets occluded more and more, its dependency on relation information becomes increased and the role of relation filters becomes crucial in determining the *a priori* estimate of an object.

The Structural Kalman filter was designed to estimate motion information of a moving object. It requires a sort of model of an object, so that the amount of occlusion could be computed through matching the model against corrupted input data. The computed information is to determine how much the outcomes of relation filters are to be involved in the overall outcome. We did not evaluate at present the effects of the matching metric on the performance of the Structural Kalman filter. The model of an object presented in this paper is very simple and coarse. Depending on applications, one may employ more elaborate models like snake models or deformable temples suggested in [8][9]. The Structural Kalman filter requires the partition of an object into several sub-cells. It is because the relation filters are defined among adjacent sub-cells. It is not clear at present how large the number of sub-cells should be. The compulsory partition of a homogeneous object may cause a severe problem when we perform the process of model matching, unless the model can distinguish the partitioned sub-cells. However the judicious design and implementation of the Structural Kalman filter may help solving this problem. For example, we may include some hypothetical objects as parts of a target object or several objects may be combined as one entity. The experiments show very promising results in tracking the moving object that can be occluded partially. The relation information among adjacent sub-regions turns out to be well used in supplementing the unreliable measurement of a partially occluded sub-region. We may draw as conclusions that our Structural Kalman filter provides another framework that can overcome the problems of the Kalman filter caused by corrupted measurements.

Acknowledgement

This work was partially supported by the KOSEF through the AITrc and BK21 program (E-0075)

References

1. Lee, O., Wang, Y.: Motion-Compensated Prediction Using Nodal-Based Deformable Block Matching. J. Visual Communication and Image Representation, Vol. 6, No. 1, March (1995) 26-34
2. Gilbert, L., Giles, K., Flachs, M., Rogers, B., U, Y.: A Real-Time Video Tracking System. IEEE Transaction on Pattern Analysis and Machine Intelligence, Vol. 2 (1980) 47-56
3. Uno, T., Ejiri, M., Tokunaga, T.: A Method of Real-Time Recognition of Moving Objects and Its Application. Pattern Recognition 8 (1976) 201-208
4. Huttenlocher, P., Noh, J., Ruchlidge, J.: Tracking Non-Rigid Objects in Complex Scenes Fourth International Conf. on Computer Vision (1993) 93-101
5. Jang, D., Kim, G., Choi, H.: Model-based Tracking of Moving Object. Pattern Recognition, Vol. 30, No. 6 (1997) 999-1008
6. Haykin, S.: Adaptive Filter Theory, Prentice-Hall. Prentice-Hall, Englewood Cliffs, New Jersey (1986)
7. Minkler, G., Minkler, J.: Theory and Application of Kalman Filtering. Magellan (1994)
8. Kass, M., Withkin A., Terzopoulos, D.: Snakes : Active Contour Models. International J. of Computer Vision (1988) 321-331
9. Williams, J., Shah, M.: A Fast Algorithm for Active Contours. Third International Conf. on Computer Vision (1990) 592-595

Face Recognition under Varying Views

A.Sehad[1], H.Hocini[1], A.Hadid[1], M. Djeddi[2], and S. Ameur[3]

[1] Centre de Développement des Technologies Avancées
128, Chemin Mohamed Gacem, BP. 245, El-Madania. Alger. Algérie
{sehad,hocini,ahadid}@cdta.dz
[2] IHC Université de Boumerdes.
[3] Université de Tizi-ouzou Institut d'Electronique.

Abstract. In this paper "EIGENFACES" are used to recognize human faces. We have developed a method that uses three eigenspaces. The system can identify faces under different angles, even if considerable changes were made in the orientation. First of all we represent the face using the Karhunen-Loeve transform. The face entered is automatically classified according to its orientation. Then we applied the rule of decision of the minimal distance for the identification. The system is simple, powerful and robust.

1 Introduction

Face recognition is not a simple problem since a new image of an object (face) seen in the recognition phase is usually different from the images previously seen by the system in the learning phase. There are several sources for the variations between images of the same face. The image depends on viewing conditions, device characteristic and environment. This including viewing position which determines the orientation, location and size of the face in the image; imaging quality which influences the resolution, blurring and noise in the picture; the light source which influences the reflection. Another source for differences between images take in changes of the faces over time. The face is a dynamic object: it changes according to expressions, mood and age. In addition, the face image may also contain features that can change altogether as hairstyle, beard or classes. An automatic face recognition system should be able to solve these problems. Thus, developing a computational model of face recognition is quite difficult. But such systems can contribute not only to theoretical insights but also to practical applications. Computers that recognize faces could be applied to a wide variety of problems, including criminal identification, security systems and human-computer interaction. Many researches in this field have been done and several systems were developed. In the beginning, much of the work in computer recognition of faces has focused on detecting individual features. In these models, faces are represented in terms of distances, angles and area between the features such as the eyes, the nose, the chin...etc. [1][2][3]. Recent research on computational modeling of faces has found it useful to employ a simpler representation of faces that consists of a normalized pixel-based representation of faces. One of the thoroughly approach in this field is extraction of a global features using PCA (Principal Component Analysis). In this approach, a set of faces is represented using a small number of global eingenvectors known as "eigenfaces". Much of the effort going into the face recognition problem has been concentrated on the processing of frontal (or nearly frontal) 2D pixel-based face representation. Consequently, their performance is sensitive to substantial variations in lighting conditions, size and position in the image. To avoid these problems a preprocessing of the faces is necessary. However, this can be done in a relatively straightforward manner by using automatic algorithms to locating the faces in the images and normalizing them for size, lighting and position. Most face recognition systems operate under relatively rigid imaging conditions: lighting is controlled, people are not allowed to make facial expressions and facial pose is fixed at a fall frontal view. In this paper we explore the eigenface technique of TURK & PENTLAND [4] [5] [6] to generalize face recognition under varying orientations.

2 Eigenfaces

The idea of using the eigenfaces was motivated by a technique developed by SIROVICH & KIRBY (1987) for efficiently representing pictures of faces using Principal Component Analysis (PCA) [7][8]. TURK & PENTLAND used this method for face detection and recognition. A simple approach to extracting the information contained in an image of face is somehow to capture the variation in collection of face images, independent of any feature judgement, and use this information to encode and compare individual face images. In mathematical terms, we consider the principal components of the face distribution or the eigenvectors of the covariance matrix of the set of face images and we treat an image as a point (or vector) in very high dimensional space. These eigenvectors can be thought of as a set of features that together characterize the variation between face images. Each image location contributes more or less to each eigenvector, so that we can display the eigenvector as a sort of ghostly face called "Eigenface". An example of an eigenface is shown in figure. 1.

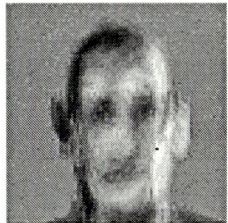

Fig. 1. Example of an eigenface

Each individual face can be represented exactly in terms of a linear combination of the eigenfaces. See figure 2. Given the eigenfaces, every face in the

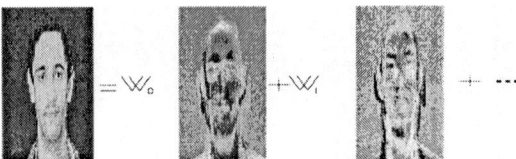

Fig. 2. A face as combination of eigenfaces

database can be represented as a vector of weights (wi). The weights are obtained by projecting the image into eigenface components by a simple product operation. When a new test image whose identification is required is given, the new image is also represented by its vector of weights. The identification of the test image is done by locating the image in the database whose weights are the closest to the weights of the test image. By using the observation that the projection of a face image and a no face image are quite different, a method for detecting the presence of a face in a given image is obtained. It is reported that this approach is fairly robust to changes in lighting condition but degrades as the scale or the orientation change. In this work we generalize this method to deal with the problem of the head orientation in the image.

3 Eigenfaces and Multi-views

As discussed in the previous section, not much work has taken face recognizers beyond the narrow imaging conditions of expressionless and frontal views of faces with controlled lighting. More research is needed to enable automatic face recognizer to run under less stringent imaging conditions. Our goal is to build a face recognition system that work under varying orientations of the face. Our system is a particular application focussed on the use of the eigenfaces in face recognition. It is built with an architecture that allow to recognize a face under most viewing conditions. First, we have to take the face under a compact representation. So we apply a compact representation of a face image using Karhunen-Loeve Transform. The result of this transformation is a weight vector called " feature vector". This is performed by building a projection space called "Face space". The extraction of the feature vector consists then on the projection of the face onto the face space. Therefore, the recognition is done by using feature vector and decision rule of the minimum distance. To solve the problem of the orientation of the face in the image, we have thought of building three spaces of projection. Each space characterizes one possible orientation of a face. Three fundamental orientations had been chosen: frontal face (90^0), diagonal face (45^0) and profile face (0^0). The others orientations will be brought to a nearest mentioned classes of orientation. The achieved spaces of projection are Space of frontal faces, space of diagonal faces and space of profile faces. Figure 3 illustrate examples of eigenfaces of each space.

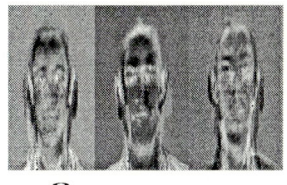  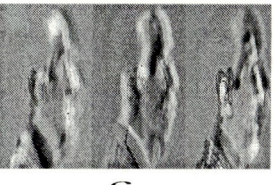

Fig. 3. a): Example of frontal eigenfaces. b):Example of diagonal eigenfaces. c):Example of profile eigenfaces

4 The General Configuration of the System

The system encloses the two following processes: the initialization process and the classification and identification process.

4.1 The Initialization Process

The initialization process includes the following steps:

1. Acquire an initial set of face images (the training set)
2. Calculate the eigenfaces of this set, keeping only the M' images that correspond to the highest eingenvalues.
3. Extract the feature vectors of known faces.
4. For each class (person) determine the class vector by averaging the eigenface vectors.
5. Choose the threshold for each class and faces space.

Results of steps 2, 4 and 5 are stored in the system. This process is executed for the three face spaces.

4.2 The Classification and Identification Process

Once the system is initialized, it is ready for the identification. The classification and identification process includes:

1. Projection of the new face onto the three face spaces.
2. Classification of the face following its orientation (choose the closest orientation).
3. Identification of the face using the feature vectors and determine if the face is known or not.

5 Initialization Process

Let a face image I(x,y) be a two-dimensional N by N array of intensity values. An image may also be considered as a vector of dimension N^2. This vector represents a point in a N^2 dimensional space. A set of images maps a collection of points in this huge space ($N^2 = 128*128 = 16384$). Images of faces, being similar in overall configuration, will not be randomly disturbed in this huge image space and thus can be described by a relatively low dimensional subspace. The main idea of the principal analysis component is to find vectors that best account for the distribution of face images within the entire image space. These vectors define the subspace of face images which is called "face space". Each vector is of length N^2, describes an N by N image, is a linear combination of the original face images. Because these vectors are the eigenvectors of the covariance matrix corresponding to the original face images and because they are face-like in appearance, we refer to them as "eigenfaces". Let the training set of the face images be : $\Gamma_1, \Gamma_2, \ldots, \Gamma_m$.

5.1 Average Face

The average face of the set is defined by:

$$\Psi = \frac{1}{M} \sum_{i=1}^{M} \Gamma_i \qquad (1)$$

with M : number of images used for each orientation.

Each face differs from the average by the vector:

$$\Phi_i = \Gamma_i - \Psi \qquad i = 1, \ldots, M \qquad (2)$$

This set of very large vectors is then subject to principal component analysis. Thus, we obtain a set of M vectors U_n which best describes the distribution of the data.

$$U_l^T . U_k = \begin{cases} 1 \text{ if } l = k \\ 0 \text{ else} \end{cases} \qquad (3)$$

The vectors U_k are the eigenvectors of the covariance matrix:

$$C = \sum_{n=1}^{M} \Phi_n \Phi_n^T = A.A^T \qquad (4)$$

where: $A = [\Phi_1, \Phi_2, \ldots, \Phi_M]$

The matrix C, however is N^2 by N^2 and determining the N^2 eigenvectors is an intractable task for typical image sizes. We need a computationally method to find these eigenvectors. We know that if the number of data points in the image space is less than the dimension of the space (M $<$ N^2), so there will be only M rather N^2 meaningful eigenvectors. Fortunately we can solve for the N^2 dimensional eigenvectors in this case by first solving for the eigenvectors of an M by M matrix, and then taking appropriate linear combination of the face images Φ_i.

Consider the eigenvectors V_i of $L = A^T.A$ such that :

$$A^T.A.V_i = \lambda_i.Vi \qquad (5)$$

Pre multiplying both sides by A we have :
$$A.A^T.A.V_i = \lambda_i.A.V_i$$
From which we see that $A.V_i$ are the eigenvectors of $C = A.A^T$. Following this analysis, we construct the M by M matrix $L = A^T.A$ where

$$L_{m,n} = \Phi_m^T.\Phi_n \tag{6}$$

and find the M eigenvectors V_i of L. These vectors determine linear combinations of the M training set face images to form the eigenfaces U_l

$$U_l = \sum_{k=1}^{m} V_{lk}\Phi_k \quad l = 1,\ldots,m \tag{7}$$

With this analysis, the calculations are greatly reduced from the order of the number of pixels in the image (N^2=128*128=16384) to the order of the number of images in the training set (M=40). In practice, a smaller set of eigenvectors (M') is sufficient for identification. These M' eigenvectors are chosen at those with the largest associated eigenvalues.

5.2 Feature Vector Extraction

A new face imageΓi is transformed into eigenface components (projected onto face space) by a simple operation:

$$W_k^{(i)} = u_k^T(\Gamma_i - \Psi) = u_k^T\Phi i \quad i = 1,\ldots,M \quad k = 1,\ldots,M'. \tag{8}$$

Where $W_k^{(i)}$ represents the projection of the ith face image into k^{th} eigenface. Thus, the feature vector of the ith face image will be :

$$\Omega^{(i)} = \{w_1^{(i)}, w_2^{(i)}, \ldots, w_{M'}^{(i)}\} \tag{9}$$

Each face is represented by a feature vector as shown in 2. We can reconstitute the initial face image using the feature vector. The ratio of reconstitution is given by

$$\gamma = \frac{\sum_{i=1}^{M'} \lambda_i}{tr(L)} = \frac{\sum_{i=1}^{M'} \lambda_i}{\sum_{i=1}^{M} \lambda_i} \tag{10}$$

where λ_i : i^{th} eigenvalue. $tr(L)$: represents the sum of the L diagonal components. Figure. 4 shows a projection of a face and its reconstitution.

5.3 Learning

In this phase, we compute the feature vectors of all the known faces and determine the thresholds of each class and those of the face space.

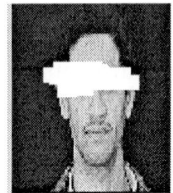 

Fig. 4. An example of reconstitution of a face image

A class feature vector. To each person corresponds one class. In our case, each class is composed by 4 images of the same person taken under different lighting and expressions conditions. Hence the class feature vector is defined by :

$$\Omega^k_{classe} = \frac{1}{4}\sum_{i=1}^{4}\Omega^{(k)}_i \quad k=1,\ldots,NI \tag{11}$$

with NI : number of persons.

The class threshold $DCI^{(k)}$. The threshold of each class (person) is defined by:

$$DCI^{(k)} = max_{i=1,\ldots,4}(\|\Omega^{(k)}_i - \Omega^k_{classe}\|^2) \quad k=1,\ldots,NI \tag{12}$$

The face space class threshold "DEV". The DEV represents the maximum allowed distance from the face space:

$$DEV = max_{i=1,\ldots,m}(d_i) \tag{13}$$

$$d_i = \|\Phi_i - \Phi_{fi}\| \tag{14}$$

$$\Phi_{fi} = \sum_{i=1}^{M'} w_k^{(i)} u_k \quad i=1,\ldots,m \tag{15}$$

6 Classification and Identification Process

The projection of a new face. Before classifying a new face into one of the three orientation classes ($0^0, 45^0 \& 90^0$), we compute the following feature vectors: $\Omega F(frontal), \Omega D(diagonal) and \Omega P(profile)$.

6.1 Classification

The aim of classifying the face is to increase the performances of the system under the changes in the orientation of the face in the image. We use three basic orientations: $0^0, 45^0 and 90^0$. The others will be related to the nearest one. Thus, a profile face will not be compared to the frontal and diagonal images. Only the information related to the orientation of the face will used for identification. The classification is performed as following:

1. Projection of the new face onto the three face spaces (frontal, diagonal and profile), thus we obtain Φ_{fF}, Φ_{fD} and Φ_{fP} (Using equation15).
2. Determine the nearest orientation (or class) to attribute to the face. We compute:
$$DistFrontal = \|\Phi_F - \Phi_{fF}\|$$
$$DistDiagonal = \|\Phi_D - \Phi_{fD}\|$$
$$DistProfil = \|\Phi_P - \Phi_{fP}\|$$

DistFrontal: means the distance between the new face and the frontal face space.

The classification is then done by the following algorithm:

```
Near= min(DistFrontal,DistDiagonal,DistProfil)
Swittch(Near){
Case Distfrontal    Write " frontal face"
                    Lancer-identication(frontal)
                    Break;
Case DistDiagonal   Write " Diagonal face"
                    Lancer-identication(Diagonal)
                    Break;
Case Distprofil     Write " Face of profile"
                    Lancer-identication(Profil )
                    Break;
default             Break
                    }
```

6.2 Identification

After classifying the face into one orientation, we compare the feature vector of this face to these of the faces in the corresponding orientation class. This is done by :

1. Compute the distance which separates the feature vector of the new face $\Omega^{(p)}$ with feature vectors of each class $\Omega^{(p)}_{classe(K)}$:

$$d_k = \|\Omega^{(p)} - \Omega^{p}_{classe(k)}\| \qquad (16)$$

with $k : k^{th} person. P \in \{Frontal, Diagonal, profil\}$

2. Choose the closest class (person) K' that minimizes the distance d_c . thus

$$d_c = min_{i=1,...,NI}(d_i) \qquad (17)$$

3. Identification : if $d_{k'} > DCI^p_{k'}$ then "unknowface" else "knownface" (k^{th}person).

7 Results

We have tested our system in a face database which contains variations in scale, in expressions and in orientation. The first step is to classify the face into the appropriate class orientation (profile, frontal and diagonal). The system performs well this phase, among the entire test set, only three incorrect classifications have been detected. The table(1)illustrate the results of the classification step. After classification, a new face will be identified in its corresponding class of orientation. The tables (2,3,4) show the results of the identification process. Thus we have obtained a rate of 91,1% of a correct identification and a rate of 88,88 % in the discrimination case.

Table 1. Results of classification

	Familiar Faces		unfamiliar Faces
	learned faces	unlearned faces	
Classification	120/120	37/39	8/9
Rate	100%	94.87%	88.88%

Table 2. Results of the identification of frontal faces.

	Frontal Faces			
		Familiar		Unfamiliar
	Learned	Unlearned		Same Scale
		Same Scale	Different Scale	
Recognition	40/40	8/10	2/3	-
Discrimination	-	-	-	3/3

Table 3. Results of the identification of the diagonal faces

	Diagonal Faces			
		Familiar		Unfamiliar
	Learned	Unlearned		Same Scale
		Same Scale	Different Scale	
Recognition	40/40	6/10	1/3	-
Discrimination	-	-	-	2/3

Table 4. results of the identification of the faces of profile

	Faces of profile			
	Familiar			Unfamiliar
	Learned	Unlearned		Same Scale
		Same Scale	Different Scale	
Recognition	40/40	7/10	1/3	-
Discrimination	-	-	-	3/3

8 Conclusion

We have developed a system based on eigenfaces. We have generalized this technique to deal with the problem of the orientation of the face in the image. It seems to be that more the number of classes of orientation increases more the system is efficient, but it is not the case. Because by increasing the number of classes, we will have difficulties to discriminate between classes and easiness to identify faces in one class. While decreasing this number, we will have more easiness for the discrimination between classes but also more difficulties to identify faces in one class. Therefor, it is necessary to find a compromise. In our case, we have used three classes what gives variations of about 20^0 in each class. However, it is necessary to signal that the performance of the system decrease quickly as soon as changes in scale and inclination are signaled. Our approach treated the problem of the face orientation in the image. In the first stage, the face is assigned to a certain class of orientation according to its orientation in the image. Identification is then launched in the appropriate class. We can improve the system by introducing others treatments in the second stage which consider the scale and inclination changes.

References

1. T.Kanade. : Picture Processing by Computer Complex and Recognition of Human Faces. PHD. Univ. Kyoto Japan.(1973)
2. R.Brunnelli, T.Poggio.: Faces Recognition : Features versus Templates. Trans. IEEE. Pattern Anal. Machine Intell., **15** (OCT 1993) 1042-1052
3. I.J.Cox .: Features-Based Face Recognition using Mixture-Distance. In Computer Vision and Pattern Recognition Piscataway, NJ, IEEE (1996)
4. M. Turk, A. Pentland.: Eigenfaces for Recognition. Journal of Cognitive Neuroscience. **Vol 3. N** 1(1991) 71-86
5. M. Turk, A. Pentland.: Face Recognition Using Eigenfaces. IEEE. Proc. Int. Conf. On Pattern Recognition. (1991) 586-591
6. A. Pentland, B. Moghaddam, T. Starner, M. Turk.: View-Based and Modular Eigenspaces for Face Recognition. Proceeding IEEE, Conf. Computer Vision and Pattern Recognition **245** (1994) p84
7. L. Sirovich, M. Kirby.: Low-Dimensional Procedure for the Characterization of Human Faces. Optical Society of America. **Vol 4, N 3** (Mar.1987) p519
8. M. Kirby, L. Sirovich.: Application of the Karhunen-Love Procedure for the Characterization of Human Faces". IEEE. Trans. Pattern Analysis and Machine Intelligence. **Vol.12 , N** 1 (Jan.1990) p103

Time Delay Effects on Dynamic Patterns in a Coupled Neural Model

Seon Hee Park, Seunghwan Kim, Hyeon-Bong Pyo, Sooyeul Lee, and Sang-Kyung Lee

Telecommunications Basic Research Laboratories
Electronics and Telecommunications Research Institute
P.O. Box 106, Yusong-gu, Taejon, 305-600, Korea

Abstract. We investigate the effects of time delayed interactions in the network of neural oscillators. We perform the stability analysis in the vicinity of a synchronized state at vanishing time delay and present a related phase diagram. In the simulations it is shown that time delay induces various phenomena such as clustering where the system is spontaneously split into two phase locked groups, synchronization, and multistability. Time delay effects should be considered both in the natural and artificial neural systems whose information processing is based on the spatio-temporal dynamic patterns.

1 Introduction

When an information is processed in natural neural systems, there is a time delay corresponding to the conduction time of action potentials along the axons. With this physiological background the time delay effects have been theoretically investigated in several neural oscillator models. In [1] a delay has been introduced to investigate the phase dynamics of oscillators in two dimensional layer in the context of temporal coding. The two neuron model with a time delay has been analytically studied by Schuster et. al. [2] focusing on the entrainment of the oscillators due to delay. A delay has been shown to influence the existence and the stability of metastable states in two dimensional oscillators with nearest neighbor coupling [3]. Revently, it has been shown that the time delay induces the multistability in coupled oscillator systems which may provide a possible mechanism for the perception of ambiguous or reversible figures [4].

In this paper we analytically and numerically investigate the time delay effects on dynamic patterns in globally coupled oscillators. Dynamic patterns such as phase locking and clustering represent the collective properties of neurons participating in the information processing. To investigate the time delay effects on these patterns occurring in the neural systems we choose a phase oscillator model. In particular, the phase interactions with more than first Fourier mode will be considered to describe the rich dynamic patterns. The significance of higher mode phase interactions may be found in the nontrivial dynamics in coupled neurons [5]. The phase model with first and second Fourier interaction modes which will be considered in this paper has been introduced to understand the pattern-forming dynamics in the brain behavior [6]. The existence and the stability of clustering state of coupled neural oscillators have been studied in the same model [7].

2 Stability Analysis of Synchronized States

We consider the overdamped oscillator model with first and second Fourier interaction which is given by the following equation of motion [7]

$$\frac{d\phi_i}{dt} = \omega + \frac{g}{N}\sum_{j=1}^{N}[-\sin(\phi_i(t) - \phi_j(t-\tau)) + a) + r\sin2(\phi_i(t) - \phi_j(t-\tau))] \quad (1)$$

$\phi_i, i = 1, 2, ..., N, 0 \leq \phi_i < 2\pi$, is the phase of the i-th oscillator. ω is the uniform intrinsic frequency of the oscillators. $\frac{g}{N}$ is the global coupling of the oscillators scaled down by the number of the oscillators. τ is the time delay. The interaction of the system is characterized by parameters a and r as well. With the symmetry of the system we safely take a in the range $[0, \pi]$. We also consider $g > 0$ (the excitatory coupling) in this paper.

The first Fourier mode in Eq. (1) without time delay is an attractive interaction which yields the synchronization of the system, while the second tends to desynchronize the system. The competition between these two interactions generates nontrivial dynamic patterns. Without time delay the synchronized system bifurcates into two cluser state at critical parameter values. For any coupling constant $2r = \cos a$ defines the critical line where the instability of the phase locked state occurs.

We assume a synchronized state $\phi_i(t) = \Phi(t) = \Omega t$. Then eq. (1) gives

$$\Omega = \omega + g\left[-\sin(\Omega\tau + a) + r\sin(2\Omega\tau)\right]. \quad (2)$$

To analyze the stability of this synchronized state, we deviate ϕ_i and linearize the system around the synchronized state. I.e., for $\phi_i(t) = \Omega t + \delta\phi_i(t)$,

$$\frac{d(\delta\phi_i)}{dt} = \frac{g}{N}[2r\cos(2\Omega\tau) - \cos(\Omega\tau + a)]\sum_{j=1}^{N}(\delta\phi_i(t) - \delta\phi_j(t-\tau)). \quad (3)$$

Converting the above equation to the eigenvalue problem, one obtains

$$\lambda a_i = \frac{g}{N}[2r\cos(2\Omega\tau) - \cos(\Omega\tau + a)]\sum_{j=1}^{N}(a_i - a_j\exp(-\lambda\tau)), \quad (4)$$

where $\delta\phi_i(t) = a_i\exp(\lambda t)$. If $\sum_{i=1}^{N}a_i \neq 0$, one obtains for $\lambda < 0$

$$2r > \cos a \quad (5)$$

in the $\tau \to 0$ limit. This contradicts to the stability condition when $\tau = 0$. Therefore, $\sum_{i=1}^{N} a_i = 0$, and

$$\lambda = g\left[2r\cos(2\Omega\tau) - \cos(\Omega\tau + a)\right]. \tag{6}$$

To visualize the time delay effects on the stability of the synchronized state, we take a small τ. In the $\tau \to 0$ limit, Eq.'s (2) and (6) give

$$\lambda = g\left[2r - \cos a + \sin a(\omega - g\sin a)\tau\right] + O(\tau^2). \tag{7}$$

Therefore, for $\omega > g\sqrt{1 - 4r^2}$, $\lambda = 0$ in Eq. (6), the critical line separating the stable region of the synchronized state from the unstable one, is in the parameter range of

$$2r < \cos a. \tag{8}$$

However, for strong coupling, $\omega < g\sqrt{(1 - 4r^2)}$, the critical line lies in the realm of

$$2r > \cos a. \tag{9}$$

Therefore, for strong coupling, the stability condition is drastically changed even when τ is very small. In Fig. 1, we plot the phase diagram of Eq. (7) in the parameter space of $\cos a$, and τ for several values of ω with fixed coupling constant. $\cos a = 0.5$ in Fig. 1 is the critical line when $\tau = 0$.

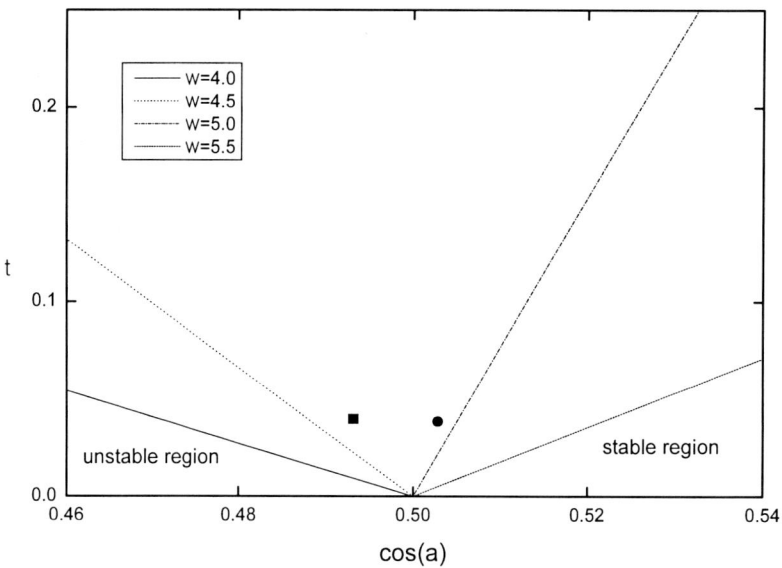

Fig. 1. Plot of critical lines defining the stability of synchronized state when $g = 5.6$ and $r = 0.25$. It can be seen that at some nonzero time delay values the synchronized state for $\tau = 0$ can be unstable when $2r < \cos a$ and the dephased state for $\tau = 0$ can be synchronized when $2r > \cos a$. The corresponding examples are denoted as the solid circle and square, respectively.

3 Numerical Results

So far, we performed the stability analysis around the synchronized ansatz, $\phi_i(t) = \phi(t) = \Omega t$, whose results are independent of N, the number of the oscillators. In this section we investigate numerically the time delay effects on the dynamical patterns of the system. To this end, we choose parameter values where the system is realized at a synchronized state for vanishing time delay. For fixed parameter values the system exhibits various dynamical patterns as time delay values change. We study also the time delay effects at parameter values where the system is desynchronized at zero time delay. In the simulations, we have used the fourth-order Runge-Kutta method with discrete time step of $\Delta t = 0.005$ with random initial conditions.

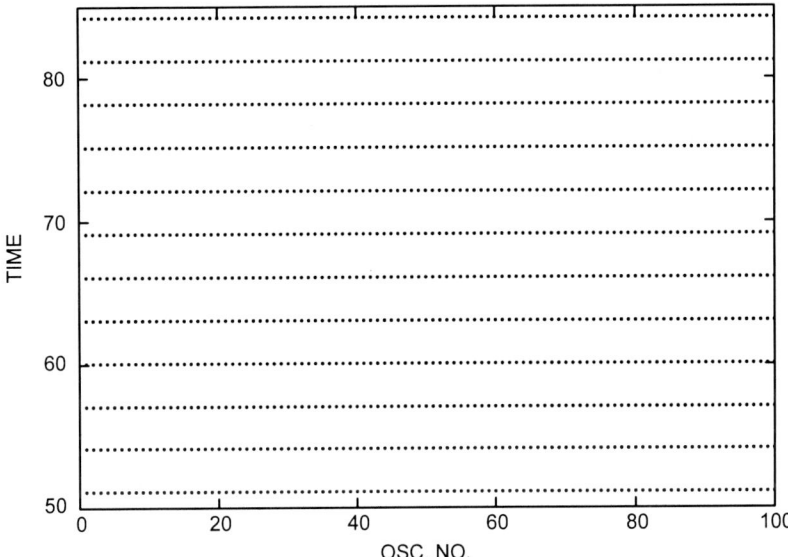

Fig. 2. Times evolution of oscillator phases when $\omega = 5.0$, $g = 5.6$, $a = 1.04$, $r = 0.25$ and $\tau = 0$. The dots represent 0 phase timing of each oscillator. The system is at the synchronized state.

We consider the system with the parameter values given by $\omega = 5.0, g = 5.6, a = 1.04$ and $r = 0.25$, where the synchronized ansatz is stable for vanishing time delay value. In Fig. 2 we plot the time evolution of the system at $\tau = 0$ for $N = 100$. The system is always realized as the synchronized state when there is no time delay.

At $\tau = 0.15$ the system is at clustered state, whose time evolution is shown in Fig. 3. The average ratio of the oscillator population in the two groups is 1:1. Clustering in this paper is induced by time delay which results in the inhibitory

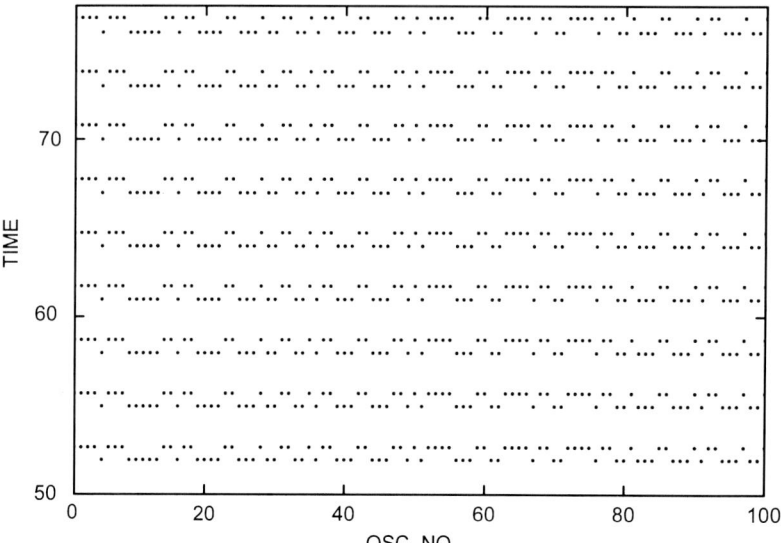

Fig. 3. Time evolution of oscillator phases when $\tau = 0.15$ with the same parameter values as in Fig. 2. The average ratio of the oscillator populations in the two clusters is 1:1.

coupling effects. The phase difference between the two separated groups of oscillators depends on the time delay values. In Fig. 4 we plot the order parameter defined by

$$O(t) = \frac{2}{N(N-1)} \sum_{i,j}^{N} \sin\left(\frac{|\phi_i - \phi_j|}{2}\right) \qquad (10)$$

for three different values of τ where the system is at clustered state.

At $\tau = 0.5$ the system may dwell on a wholly synchronized state or clustered state according to the initial conditions. This is a manifestation of multistability. We plot $O(t)$ in Fig. 5 showing the multistability. The dotted line in Fig. 5 represent the clustering state where the clusters are not uniformly moving. Therefore, the phase difference between the two clusters is not fixed but oscillatorically changed. In Fig. 6 we plot the evolution of ϕ_i's corresponding to the dotted line in Fig. 5. The motion of the clusters is not uniform but the velocity of the clusters depends on the phase value of the clusters. While in [4] the multistability exists either between the synchronized state and a desynchronized state where the oscillators are distributed almost uniformly between the synchronized states with different moving frequencies, the multistability in this paper is realized between the synchronized state and the clustered state.

For $0.6 < \tau < 2.5$, the system exists always in the perfectly synchronized state.

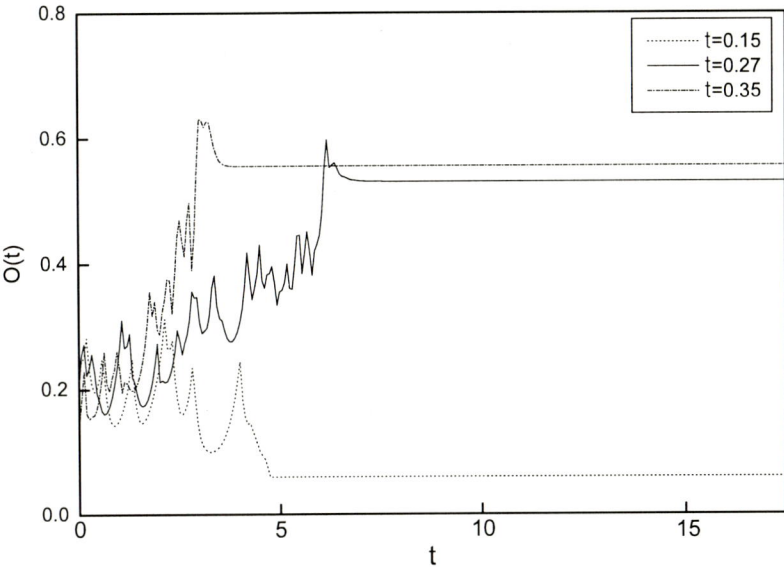

Fig. 4. Plot of the order parameter $O(t)$ for three values of time delay. $O(t)$ represents the distance of the two clusters. The graphs show that the clusters are uniformly moving at the steady states so that $O(t)$ is constant.

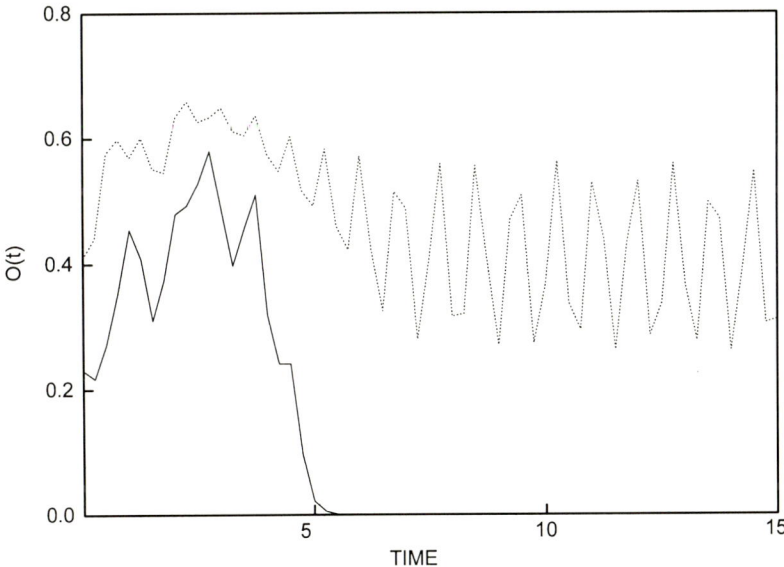

Fig. 5. Plot of the order parameter $O(t)$ for $\tau = 0.5$ showing the multistability of the system. The dotted line represents the clustered state where the distance between the two clusters is changing because of the nonuniform motion of the clusters.

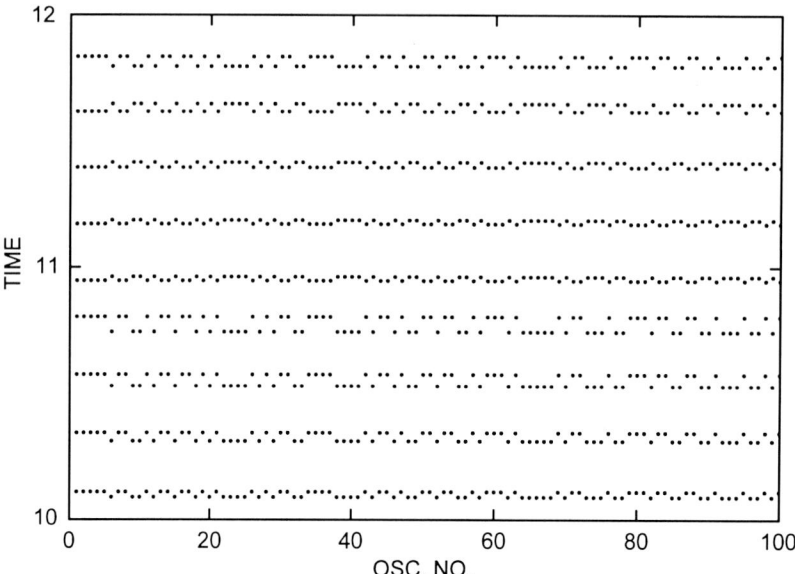

Fig. 6. Time evolution of oscillator phases corresponding to the dotted line in Fig. 5. The distance between the two clusters changes periodically.

We study the time delay effects on the desynchronized state when there is no time delay. We choose $a = 1.25$ with the other paramenter values same as above. When $\tau = 0.49$, the system shows multistability between the synchronized state and the clustered state. For $0.5 < \tau < 2.5$ oscillators are perfectly synchronized. This is a synchronization induced by time delay.

4 Conclusions

In this paper we investigated the time delay effects on the dynamic patterns in the coupled phase oscillators with first and second Fourier mode interactions. The analytical study shows that the time delay drastically changes the stability of the synchronized state. To investigate the time delay effects on dynamical patterns numerically, we fixed all the other parameters than the time delay. The introduction of time delay into a fully synchronized state induces the clustering, and the multistability. Time delay also induces synchronization of desynchronized state when there is no time delay. This shows that the time delay may play an important role in the information processing based on the spatio-temporal structure of neuronal activities [8]. The results in this paper suggest that the time delay is a new route leading to such dynamic patterns. It is expected that the time delay may provide a rich structure of dynamics which may be used to facilitate memory retrieval when the informations are stored on the basis of dynamics of the system [9].

References

1. Konig, P., & Schillen, T. B., (1991) *Neural Computation* **3**, 155-166.
2. Schuster, H. G., & Wagner, P., (1989) *Prog. Theor. Phys.* **81**, 939-945.
3. Niebur, E., Schuster H. G., & Kammen, D. M. (1991) *Phys. Rev. Lett.* **67**, 2753-2756.
4. Kim, S., Park, S. H., & Ryu, C. S., (1997) *Phys. Rev. Lett* **79**, 2911-2914.
5. Okuda, K. (1993) *Physica* **D 63**, 424-436.
6. Jirsa, V., Friedrich, R., Haken, R., & Kelso, J. A. S., (1994) *Biol. Cybern*, **71**, 27-35.
7. Hansel, D., Mato, G., & Meunier, C. (1993) *Phys. Rev.* **E 48**, 3470-3477.
8. Fujii, H., Ito, H., & Aihara, K. (1996). Dynamical cell assembly hypothesis - theoretical possibility of spatio-temporal coding in the cortex. *Neural Networks*, **9**, 1303-1350 and references therein.
9. Wang, L. (1996) *IEEE Trans. Neural networks*, **7**, 1382-1387; Wang, D., Buhmann, J., & von der Malsburg, C. (1990) *Neural Comp.*, **2**, 94-106.

Pose-Independent Object Representation by 2-D Views

Jan Wieghardt[1] and Christoph von der Malsburg[1,2]

[1] Institut für Neuroinformatik, Ruhr-Universität Bochum
D-44780 Bochum, Germany
[2] Lab. for Computational and Biological Vision
University of Southern California, Los Angeles, USA

Abstract. We here describe a view-based system for the pose-independent representation of objects without making reference to 3-D models. Input to the system is a collection of pictures covering the viewing sphere with no pose information being provided. We merge pictures into a continuous pose-parameterized coverage of the viewing sphere. This can serve as a basis for pose-independent recognition and for the reconstruction of object aspects from arbitrary pose. Our data format for individual pictures has the form of graphs labeled with Gabor jets. The object representation is constructed in two steps. Local aspect representations are formed from clusters of similar views related by point correspondences. Principal component analysis (PCA) furnishes parameters that can be mapped onto pose angles. A global representation is constructed by merging these local aspects.

1 Introduction

A major unsolved issue in the neurosciences and in computer vision is the question how three-dimensional objects are represented such as to support recognition invariant to pose. In computer vision, 3-D models have often been proposed, with the argument that with their help object recognition from arbitrary pose can be based on rotation of the model and projection onto the image plane. As biological argument for this view the mental rotation phenomenon [14] is often cited. Also, 3-D models would make it easy to model our brain's ability to derive information on depth rotation from image deformations. But it turned out that it is a difficult and very unreliable process to obtain 3-D models from the available data, which have the form of 2-D views, and there are conflicts with psychophysical results [3].

On the other hand, models based on 2-D views are attractive and successful because of is their close relation to the actual sensory input. Most experimentally successful computational models for object recognition and classification are view-based. On the other hand such models face problems when view-point invariance is required. Image variation due to rotation of the object in depth cannot be easily interpreted or reproduced and is largely neglected in recognition systems. To deal with such difficulties, systems have been proposed that

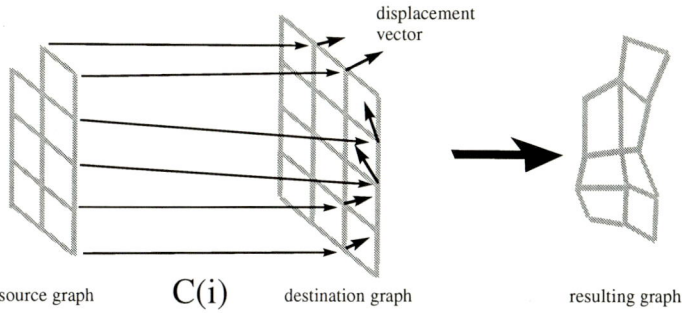

Fig. 1. Matching graphs by estimating displacements for pairs of jets.

are based on the combination of several views, e.g., [18]. However, such systems usually require additional information from non-visual interaction with the object [13,12], or image sequences are required to be continuous to permit connection of different views [2], and there is indeed some psychophysical evidence that temporal continuity is used to understand the transformations of the 2-D projection of 3-D dimensional objects and thus implicitly the underlying 3-D structure [6,7,20].

We try here to shed some light on the question where and to what extent object manipulation and temporal continuity are needed to build a functional object representation from 2-D views by exploring how far we can get without it. It is our goal be to construct an object representation from nothing but single static images and yet have the representation do justice to the topological ordering of views in terms of 3-D parameters. Such a topological ordering of single views is in good agreement with the biological findings by *Tanaka et al.*[21]. Our work is based on raw data in the form of a collection of gray-scale images (collected by G. Peters [13]) of a plastic object. The images cover the upper half of the viewing sphere with a spacing of approximately 3.6^o, resulting in a total of 2500 images. The pose for each image in terms of elevation and declination was recorded but this information is not made use of in this work. Individual images have the format of 128×128 pixels. Our work proceeds in two stages. We first extract clusters of images with high mutual similarity. Clusters contain pictures corresponding to poses within small solid angles. Pictures within a cluster are merged into a *local representation*. By exploiting the overlap between local representations these are in a second step connected to form a *global representation*.

2 Local Representation

First of all, a local understanding of a 3-D object needs to be created, i.e. a representation that allows recognition of distinct views from image data and the interpretation of slight variations in viewing angle.

We assume that during the learning phase the system has access to a number of views of the object which differ only slightly (by less than 20°) in viewing angle. Each single view is represented by a labeled graph. To construct this initial description segmentation is needed to distinguish object from background. We used the method developed in [5] [19], which worked well on our images. To obtain the image description a graph is placed onto the object's segment. The graph has a grid structure of nodes with an equal spacing of 10 pixels. Adjacent nodes are connected by an edge and each node i is labeled with its position $\boldsymbol{x}_i$ in the image and a feature vector $\boldsymbol{J}$. The latter is composed of the complex-valued results of a convolution of the image with a set of Gabor kernels. The vector's n-th entry is given by

$$e_n^T \boldsymbol{J}(\boldsymbol{x}_i) = a_{\boldsymbol{k}_n}(\boldsymbol{x}_i) \exp(i\phi_{\boldsymbol{k}_n}(\boldsymbol{x}_i)) \tag{1}$$

$$e_n^T \boldsymbol{J}(\boldsymbol{x}_i) = \int I(\boldsymbol{x}_i - \boldsymbol{x}') \frac{k_n^2}{\sigma^2} \exp\left(-\frac{k_n^2 x'^2}{2\sigma^2}\right) \left[\exp(i\boldsymbol{k}_n \boldsymbol{x}') - \exp\left(\frac{\sigma^2}{2}\right)\right] d^2 x' \tag{2}$$

with $n = \mu + 8\nu$, $\nu \in \{0,\ldots,5\}$, $\mu \in \{0,\ldots,7\}$, $k_\nu = 2^{-\frac{\nu+2}{2}}$, $\alpha_\mu = \mu\frac{\pi}{8}$ and $\boldsymbol{k}_n = (k_\nu \cos\alpha_\mu, k_\nu \sin\alpha_\mu)$. Throughout this paper these graphs will serve as representations of single views. Within the framework of elastic graph matching this representation has proven to be efficient in recognition tasks for humans and other objects [9,16,17,22,1]. Because its performance in recognition tasks is well established, we concentrate on our main focus, how to organize this collection of single views to yield an understanding of 3-D structure.

In order to derive the transformation properties that constitute the 3-D structure in terms of vision, we need to relate views to each other. The correspondence problem between two graphs $\mathcal{G}^{src}$ and $\mathcal{G}^{dest}$ needs to be solved, i.e., for each node in $\mathcal{G}^{src}$ its corresponding position in the image encoded by $\mathcal{G}^{dest}$ must be established. To achieve this a modified version of elastic graph matching is employed [9]. For the two graphs $\mathcal{G}^{src}, \mathcal{G}^{dest}$ a displacement vector $\boldsymbol{d}_{ij}$ is calculated for all pairs of nodes i^{src}, j^{dest} by maximizing the following jet-similarity function [4]:

$$s_{ij} = \frac{\sum_m a_{\boldsymbol{k}_m}^{src}(\boldsymbol{x}_i) a_{\boldsymbol{k}_m}^{dest}(\boldsymbol{x}_j)(1 - 0.5(\phi_{\boldsymbol{k}_m}^{dest}(\boldsymbol{x}_j) - \phi_{\boldsymbol{k}_m}^{src}(\boldsymbol{x}_i) - \boldsymbol{k}_m \boldsymbol{d}_{ij})^2)}{|J^{src}(\boldsymbol{x}_i)||J^{dest}(\boldsymbol{x}_j)|}. \tag{3}$$

The displacement vector $\boldsymbol{d}_{ij}$ estimates the relative positions of two jets, assuming that they were derived from the same image (or very similar images) and in close neighborhood; the resulting similarity s_{ij} reflects how well these assumptions are met.

In order to match $\mathcal{G}^{src}$ into $\mathcal{G}^{dest}$ we determine the configuration with minimal distortion, i.e. the transformation $j = C(i)$ which yields an injective mapping of the node indices i^{src} in $\mathcal{G}^{src}$ to the node indices in $\mathcal{G}^{dest}$ and minimizes

$$D = \frac{1}{N_{edges}} \sum_i^{N_{nodes}^{src}} \sum_j^{N_{nodes}^{src}} e_{ij} |\eta_{ij}^{src} - \eta_{ij}^{dest'}| \tag{4}$$

with

$$\eta_{ij}^{dest'} = x_{node\ C(i)}^{dest} + \boldsymbol{d}_{i,C(i)} - \left(x_{node\ C(j)}^{dest} + \boldsymbol{d}_{j,C(j)}\right) \tag{5}$$

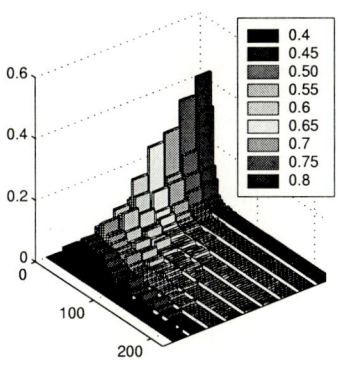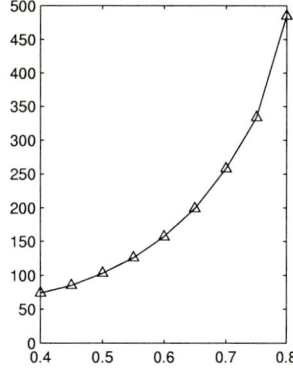

Fig. 2. On the left the normalized histograms on the distribution of cluster sizes are shown for different thresholds. On the right the absolute number of clusters are displayed over the threshold values.

$$\eta_{ij}^{src} = x_{node\ i}^{src} - x_{node\ j}^{src} \tag{6}$$

$$e_{ij} = \begin{cases} 1 & : \quad \text{if there is an edge between node } i \text{ and } j \\ 0 & : \quad \text{else} \end{cases} \tag{7}$$

With this procedure we can match graphs $\mathcal{G}^{src}$ and $\mathcal{G}^{dest}$ onto each other without having to extract Gabor jets with pixel density as is necessary in the conventional method [9] (see fig. 1). The quality of the match is measured by

$$S = \frac{1}{N_{nodes}^{src}} \sum_{i}^{N_{nodes}^{src}} s_{i,C(i)} \tag{8}$$

We now form clusters of similar views. We first pick at random an individual picture as "canonical view", compare it in terms of the similarity (equ. 8) to all other views, and include in the cluster all views above a certain threshold similarity. The procedure is repeated with randomly chosen "canonical views" not yet incorporated in a cluster until all pictures are exhausted (all views were subject to thresholding in each turn, so that each view can belong to two or more clusters). It turned out that clusters always comprised all views from a more or less solid angle on the viewing sphere. The absolute number and size distribution of clusters varied with the similarity as shown in fig. 2. In all experiments the similarity threshold was set to 0.6 (similarities can range from -1 to 1). In this region of values the model is fairly insensitive to the threshold's actual value .

As our next step we will merge the views in a cluster into a continuous local representation that can serve as a basis for object recognition for arbitrary intermediate poses, from which the pose angles of a new image of the object can be estimated accurately, and which represents view transformations under the control of pose angles. To this end we match the graph of the canonical view of a cluster to all other graphs in the cluster, resulting in deformed versions.

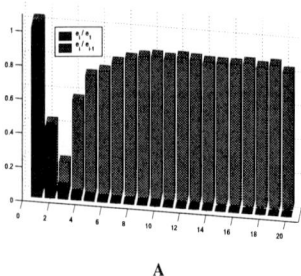
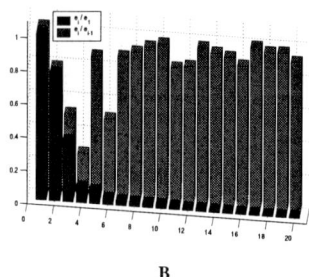

A   B

Fig. 3. *A:* Mean ratios of the eigenvalues to the first eigenvalue (front) and the mean ratios to the preceding eigenvalue (back) for the first 20 eigenvalues. Most of the data variation can be attributed to the first two eigenvalues, i.e., the problem is essentially two-dimensional.
B: Ratios of the eigenvalues to the first eigenvalue (front) and the ratios to the preceding eigenvalue (back) for the first 20 eigenvalues as used by the metric multi-dimensional scaling approach. Most of the data variation can be attributed to the first three eigenvalues, i.e. the problem is essentially three-dimensional.

Following an idea originally proposed by *Lanitis et al.* [10] for hand-labeled selected data for face recognition and interpretation, for each of the deformed versions of the reference graph the node position vectors are concatenated to form a $2 \times N$-tuples ($t_i = (x_{node\ 1}, y_{node\ 1}, \ldots, x_{node\ N}, y_{node\ N})$), and on the set of these we perform principal component analysis (PCA).

The resulting principal components (PCs) are prototypes of graph deformations. The eigenvalue distribution averaged over several clusters is dominated by the two leading ones, see fig. 3A. These correspond to the two viewing angles, azimuth and elevation, which are the main sources of variation in our pictures. When positioning the views within single clusters in a two-dimensional plot by taking the amplitudes of the first two PCs as coordinates, the topology of the pose angles is reproduced faithfully and with little distortion, see fig. 4.

Starting from raw image data we thus have determined the correct topological arrangement of individual views on the viewing sphere, and we have constructed a pose angle-parameterized representation for image transformations in terms of prototypical graph deformations. However, so far only small pose variations have been dealt with. This restriction is imposed by two reasons. First, the system is able only to link pairs of views that are similar enough to permit establishment of meaningful point correspondences and jet similarities. And second, our linear representation of graph transformations breaks down for large pose variations. To represent an object in its entirety by covering the whole viewing sphere, large variations must also be accommodated. This we achieve by patching together the local clusters, making use of their mutual overlaps.

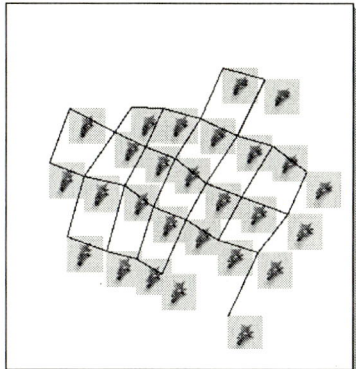 

Fig. 4. *Examples of local representations:* The views within a cluster are plotted with the amplitudes of the first two principal components as coordinates. Views neighboring in terms of pose angle are connected by a line. The projections onto the principal components reflect the local topology given by the viewing angles very well.

3 Global Representation

Although the whole viewing sphere is covered by overlapping clusters we are still lacking a global coordinate system. To construct this we make use of the overlap relations between many pairs of clusters, on the basis of which inter-cluster distances can be approximated as

$$\Delta_{ij} = \frac{1}{N_{overlap}} \sum_{n}^{N_{overlap}} (\delta_{in} + \delta_{jn}), \qquad (9)$$

where the sum runs over all graphs n in the overlap of clusters i and j, and δ_{in} is the distance of graph n to the center of cluster i (distances within a cluster are computed by taking the amplitudes of the first two PCs as Euclidean coordinates). The distance between two non-overlapping clusters is approximated by the smallest sum of distances of a chain of consecutively overlapping clusters that connects the two. If there is no such chain between two clusters they are not assumed to be part of the same object. On the basis of the distance matrix Δ metric multi-dimensional scaling [11] is performed, which yields a global arrangement of the cluster centers. Multi-dimensional scaling is well suited to to retrieve such large scale ordering. The local ordering, which is sometimes neglected by multi-dimensional scaling [8], is still preserved in the known neighborhood relations of the clusters and their associated local coordinate systems.

The way local neighborhood relations are used to retrieve the global structure of the manifold given by a number of is very similar to work done by J.B. Tenenbaum [15].

The eigenvalues derived from the metric multi-dimensional scaling show that the distance matrix can be fairly well interpreted as an embedding of the cluster centers in a three-dimensional Euclidean representation space (see figure 3B).

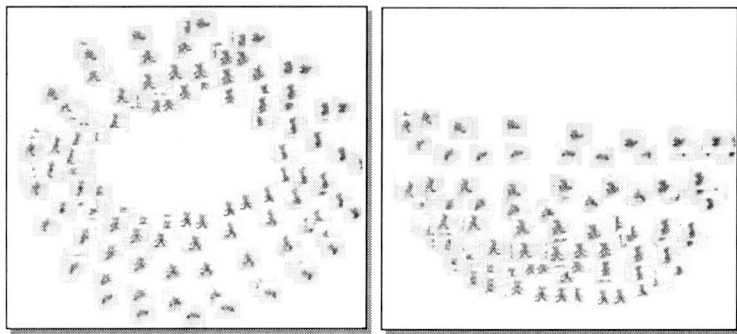

Fig. 5. A subset of 100 image of the 2500 images used in the learning procedure is displayed according to their global coordinates. On the left the projection onto the x-y plane is shown, on the right the projection onto the x-z plane.

All dimensions beyond the third contribute little to the distance matrix and we will therefore ignore them when referring to the global coordinate system.

In order to align the local representations with the global coordinate system, we linearly transform them to fit the overlap data in the following way. The center of the overlap region between clusters i and j in coordinates of cluster i is estimated as

$$o_{ij}^i = \frac{1}{N_{overlap}} \sum_n^{N_{overlap}} p_{ijn}^i, \qquad (10)$$

where $N_{overlap}$ is the number of views in the overlap between cluster j and i, and p_{ijn}^i the position of the n-th of these, in the coordinates of cluster i. In global coordinates this center is estimated as

$$o_{ij}^g = c_j^g + \frac{c_i^g - c_j^g}{1 + \sqrt{\frac{size_i}{size_j}}}, \qquad (11)$$

where c_i^g is the center of cluster i in global coordinates as given by the multi-dimensional scaling and $size_i$ is the size of cluster i in terms of the area it covers in its own two-dimensional PC space. For each neighbor j of cluster i we get a linear equation in 9 variables:

$$o_{ij}^g = \mathbf{A}_i o_{ij}^i + \mathbf{t}_i, \qquad (12)$$

where $\mathbf{A}_i$ and $\mathbf{t}_i$ specify the linear mappings between global and local coordinates. If the number of neighbors is sufficient the linear equation system can be solved in a least square sense. If only insufficient data is available for the cluster under consideration, the cluster is discarded.

Figure 5 displays the global order of views for our test object. Out of the 2500 pictures used for training, 100 sample images are projected into the global coordinate system. It is evident that the views are nicely ordered on a two-dimensional manifold according to the two pose angles. The manifold has a

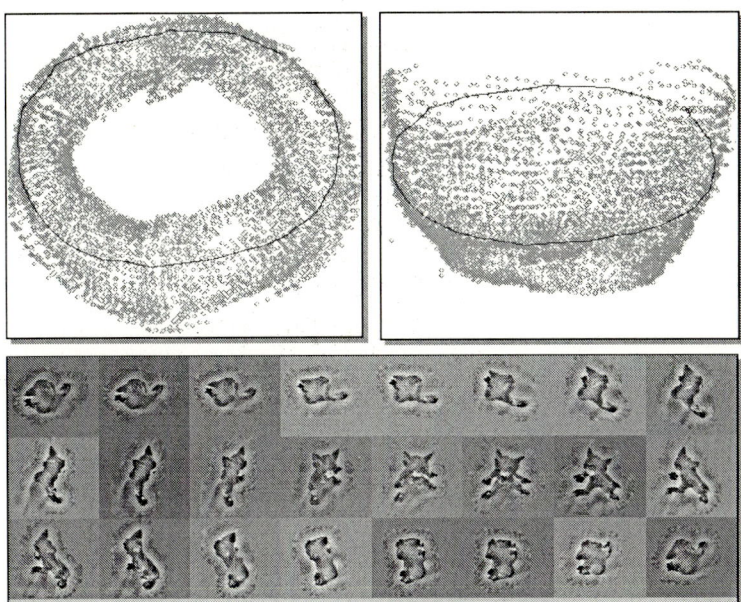

Fig. 6. The top two figures show a path on the manifold. The manifold is visualized by drawing a circle for each view in the training set. Below reconstructions of sample points on the path are shown. The reconstruction was done by picking the cluster closest to the current point and deforming the center-graph according to the local PC coordinates. The reconstruction was then done by a wavelet reconstruction from the unmanipulated jets.

cylindric topology and not, as one might have expected, that of a hemisphere. This is due to the fact that our matching compensates for translation and slight deformation, but not for rotation in the image plane. Thus, the views from the zenith, which differ only by in-plane rotation, are taken as very distinct by the system.

The composite model we have constructed purely from 2-D views and without any reference to a 3-D model of the object is a basis a) for associating an arbitrary new picture of the object with a unique coordinate in representational space and b) for associating an arbitrary point in that space with a concrete picture:

a) Given a picture, a cluster is first determined by finding the best-matching canonical view. The correspondence points resulting from that match constitute a deformed grid. This deformation pattern is projected onto the PCs of the cluster, giving the coordinates within that cluster and correspondingly within the global three-dimensional coordinate system. This procedure is most efficient if a sequence of images is given that corresponds to a continuous trajectory of the object, which can thus be estimated.

b) Given a point in the three-dimensional coordinate system (derived, for instance, from extrapolation of an estimated trajectory) an image can be derived from the model for that point. For that, the closest cluster center is determined and the point is projected onto the linear 2-D subspace of that cluster to determine its position in that coordinate system. For a smooth path on the manifold we have generated those views (see figure 6), showing that a smooth path in 3-D space translates into smooth deformation of the image.

Thus, movements of the object can be associated with paths in representational space and vice versa. This may prove sufficient for interpretation, comparison, and planning of object trajectories.

4 Discussion

Our work shows that it is possible to autonomously generate parameterized object descriptions. Although we set it up as off-line learning, the statistical instruments used all have a biologically plausible on-line equivalent. So why do psychophysical experiments suggest that temporal context is essential in learning object representations?

Possible answers might include:

1. Although not necessary it greatly facilitates the task.
2. The sheer number of views required can only be gained from image sequences.
3. It resolves ambiguities in symmetric objects, that cannot be dealt with in a static and solely vision-based framework.

We believe points 2 and 3 to be the crucial ones. By its very nature, temporal context only provides information on the ordering along a one-dimensional path. But the manifolds involved in acquiring a global understanding of 3-D objects are at least two-dimensional. So a mechanism independent of such context is required anyway to yield correspondences and order between such paths.

Acknowledgments

We would like to thank Gabriele Peters for making here object database available for this study. We acknowledge the support of ONR, contract No.: N00014-98-1-0242.

References

1. Mark Becker, Efthimia Kefalea, Eric Maël, Christoph von der Malsburg, Mike Pagel, Jochen Triesch, Jan C. Vorbrüggen, Rolf P. Würtz, and Stefan Zadel. GripSee: A Gesture-controlled Robot for Object Perception and Manipulation. *Autonomous Robots*, 6(2):203–221, 1999.
2. Suzanna Becker. Implicit learning in 3d object recognition: The importance of temporal context. *Neural Computation*, 11(2):347–374, 1999.
3. H.H. Bülthoff and S. Edelman. Psychophysical support for a 2-d view interpolation theory of object recognition. *Proceedings of the National Academy of Science*, 89:60–64, 1992.

4. D.J.Fleet, A.D.Jepson, and M.R.M.Jenkin. Phase-based disparity measurement. *Image Understanding*, 53(2):198–210, 1991.
5. Christian Eckes and Jan C. Vorbrüggen. Combining Data-Driven and Model-Based Cues for Segmentation of Video Sequences. In *Proceedings WCNN96*, pages 868–875, San Diego, CA, USA, 16–18 September, 1996. INNS Press & Lawrence Erlbaum Ass.
6. Philip J. Kellman. Perception of three-dimensional form by human infants. *Perception & Psychophysics*, 36(4):353–358, 1984.
7. Philip J. Kellman and Kenneth R. Short. Development of three-dimensional form perception. *Journal of Experimental Psychology: Human Perception and Performace*, 13(4):545–557, 1987.
8. J. B. Kruskal. The relationship between multidimensional scaling and clustering. In J. Van Ryzin, editor, *Classification and clustering*, pages 17–44. Academic Press, New York, 1977.
9. M. Lades, J. C. Vorbrüggen, J. Buhmann, J. Lange, C. von der Malsburg, R. P. Würtz, and W. Konen. Distortion invariant object recognition in the dynamic link architecture. *IEEE Transactions on Computers*, 42:300–311, 1993.
10. Andreas Lanitis, Chris J. Taylor, and Timothy F. Cootes. Automatic interpretation and coding of face images using flexible models. *IEEE Trans. PAMI*, 19 7:743 – 756, 1997.
11. K.V. Mardia, J.T. Kent, and J.M. Bibby. *Multivariate Analysis*. Academic Press, 1989.
12. K Okada, S Akamatsu, and C von der Malsburg. Analysis and synthesis of pose variations of human faces by a linear pcmap model and its application for pose-invariant face recognition system. In *Fourth International Conference on Automatic Face and Gesture Recognition, March 26-30, Grenoble, 2000*, 2000.
13. Gabriele Peters. The Interpolation Between Unsimilar Views of a 3-D Object Increases the Similarity and Decreases the Significance of Local Phase. In *Proceedings of the International School of Biophysics*, Naples, Italy, October 11-16, 1999.
14. RN Shepard and J Metzler. Mental rotation of three dimensional objects. *Science*, 171:701–703, 1971.
15. J. B. Tenenbaum. Mapping a manifold of perceptual observations. In M. I. Jordan, M. J. Kearns, and S. A. Solla, editors, *Advances in Neural Information Processing*, volume 10, pages 682–688. MIT Press, 1998.
16. J. Triesch and C. Eckes. Object recognition with multiple feature types. In *ICANN'98, Proceedings of the 8th International Conference on Artificial Neural Networks*, pages 233–238. Springer, 1998.
17. J. Triesch and C. v.d. Malsburg. Robust classification of hand postures against complex backgrounds. In *Proceedings of the Second International Conference on Automatic Face and Gesture Recognition 1996, Killington, Vermont, USA*, 1996.
18. S Ullman and R Basri. Recognition by linear combinations of models. AI Memo 1152, Artificial Intelligence Laboratory, MIT, 1989.
19. Jan C. Vorbrüggen. *Zwei Modelle zur datengetriebenen Segmentierung visueller Daten*, volume 47 of *Reihe Physik*. Verlag Harri Deutsch, Thun, Frankfurt am Main, 1995.
20. Guy Wallis and Heinrich Bülthoff. Learning to recognize objects. *Trends in Cognitive Sciences*, 3(1):22–31, 1999.
21. Gang Wang, Keiji Tanaka, and Manabu Tanifuji. Optical imaging of functional organization in the monkey inferotemporal cortex. *Science*, 272:1665–1668, 1996.
22. L. Wiskott, J.-M. Fellous, N. Krüger, and C. v.d. Malsburg. Face recognition by elastic graph matching. *IEEE Trans. PAMI*, 19 7, 1997.

An Image Enhancement Technique Based on Wavelets

Hae-Sung Lee[1], Yongbum Cho, Hyeran Byun, and Jisang Yoo[2]

[1] Yonsei University, Shinchon-Dong 134, Seodaemun-Koo, Seoul, Korea,
geneel@aipiri.yonsei.ac.kr,
http://wavelets.yonsei.ac.kr
[2] Kwangwoon University, Wolke-Dong 447-1, Nowon-Koo, Seoul, Korea

Abstract. We propose a technique for image enhancement, especially for denoising and deblocking based on wavelets. In this proposed algorithm, a frame wavelet system designed as an optimal edge detector is used. And our theory depends on Lipschitz regularity, spatial correlation, and some important assumptions. The performance of the proposed algorithm is tested on three popular test images in image processing area. Experimental results show that the performance of the proposed algorithm is better than those of other previous denoising techniques like spatial averaging filter, Gaussian filter, median filter, Wiener filter, and some other wavelet based filters in the aspect of both PSNR and human visual system. The experimental results also show that our algorithm has approximately same capability of deblocking as those of previous developed techniques.

1 Introduction

1.1 Previous Denoising Techniques

There exists various noise in most images. So the development of denoising technique was one of the most popular research topics in image processing and computer vision areas during the past thirty years[1]. Although there are a lot of types in noises, the most frequently occurred in images and researched noise is AWG(Additive White Gaussian) noise[2]. In general, AWG noise can be removed by linear filters such as spatial averaging filter, Gaussian smoothing filter, and Wiener filter.

David Donoho has proposed good image enhancement techniques based on wavelets [3][4][5]. And Mallat also proposed a new excellent technique for edge detection and denoising based on wavelets. But his technique has high computational complexity[6][7]. The Mallat's technique was developed by using the concept named Lipschitz regularity. And this Lipschitz regularity plays an important role in this paper also.

After these techniques, a new denoising technique based on wavelets and spatial correlation was developed by Xu et al.[8]. Although this technique showed good performance, it also had high computational complexity.

1.2 Previous Deblocking Techniques

The essential technique, block based DCT(Discrete Cosine Transform) has its theoretical origin to STFT(Short Time Fourier Transform). So there are some defect such as Gibbs phenomenon or blocking effect in JPEG and MPEG[10]. As the compression ratio becomes higher, the effects of blocking is greatly enhanced. Near the compression ratio of 0.25bpp, most people can obtain the blocking effect very easily[9].

There are a lot of postprocessing techniques of deblocking for JPEG and MPEG, such as image adaptive filtering, projection on convex sets, Markov random fields, wavelet transform, and nonlinear filtering[11]. Especially projection on convex sets shows very good deblocking ability, but it's defect is also high computational complexity[12][13]. Another deblocking technique which is based on wavelet transform is developed by Xiong et al.[13]. It shows nearly same deblocking ability as and lower computational complexity than the projection on convex sets. And real time deblocking technique developed by Chou et al. shows nearly the same ability as the projection on convex sets[11]. But this technique can not be adapted to denoising areas which are similar to deblocking areas, because this technique only manipulates the specific pixels which cause the blocking effect.

1.3 Features of Proposed Image Enhancement Technique

We use frame wavelet system, especially the biorthogonal frame wavelet system developed by Hsieh et al.[14]. This is because it satisfies the essential point in denoising and deblocking technique which is to distinguish the noise and block elements which should be removed from the edge elements which should be preserved. Hsieh et al. developed their wavelet filter based on Canny's three criteria for edge detection. The performance of this filter to suppress AWG noise and to detect edges showed superior results to that of Canny's[14][15].

We use spatial correlation in wavelet transformed image to promote the ability of distinguishing noise and block elements from edge elements. This concept was previously used by Xu et al.[8]. But we could analyze the meaning of spatial correlation precisely by using Lipschitz regularity which has not been used in the technique by Xu et al. The proposed technique showed superior or similar denoising and deblocking ability to previous techniques in the aspect of computational complexity and PSNR. Especially our technique could be well adapted to both areas of denoising and deblocking.

In chapter 2, we'll introduce wavelet transform and Lipschitz regularity. We'll explain the our technique for image enhancement in chapter 3. The performance comparison will be described in chapter 4, and the conclusion could be found in chapter 5.

2 Theoretical Bases of Wavelet Transform

2.1 Wavelet Transform

We define wavelet as follows[14][16][17][18].

Definition 1. *If the Fourier transform $\hat{\psi}(\omega)$ of a function $\psi(x)$ satisfies (1), we define $\psi(x)$ as a wavelet,*

$$\int_0^\infty \frac{|\hat{\psi}(\omega)|^2}{\omega} d\omega = \int_{-\infty}^0 \frac{|\hat{\psi}(\omega)|^2}{|\omega|} d\omega = C_\psi < +\infty \qquad (1)$$

Formula (1) means the following.

$$\int_{-\infty}^\infty \psi(u)\, du = 0 \qquad (2)$$

We call (2) as admissibility condition, and define $\psi_s(x) \equiv \frac{1}{s}\psi\left(\frac{x}{s}\right)$ as the dilation of $\psi(x)$ about scale s.

Definition 2. *We define $W_s f(x) = f \star \psi_s(x)$ as the (continuous) wavelet transform of a function $f(x) \in L^2(R)$.*

In Definition 2, $\star$ is the convolution operator. The Fourier transform of $W_s f(x)$ to variable x is simply represented by $\hat{W}_s f(\omega) = \hat{f}(\omega)\hat{\psi}(s\omega)$. By using formula (3), we can get the original function $f(x)$ again from the wavelet transformed result of $f(x)$.

$$f(x) = \frac{1}{C_\psi} \int_0^{+\infty} \int_{-\infty}^{+\infty} W_s f(u)\, \bar{\psi}_s(u-x)\, du\, \frac{ds}{s} \qquad (3)$$

In formula (3), $\bar{\psi}_s(x)$ is the complex conjugate of $\psi_s(x)$.

2.2 Lipschitz Regularity

Lipschitz exponent and Lipschitz regularity are defined as follows[6][7].

Definition 3. *We assume n is positive integer, and $n \leq \alpha \leq n+1$ is satisfied. If there exists two constants, $A, h_0 > 0$ and a polynomial of degree n, $P_n(x)$ for a function $f(x)$, and $|f(x_0 + h) - P_n(h)| \leq A|h|^\alpha$ is satisfied for $h < h_0$, then we say that this function is Lipschitz α at x_0. If a function $f(x)$ satisfies $|f(x_0 + h) - P_n(h)| \leq A|h|^\alpha$ in the interval $x_0 + h \in (a,b)$, we say that this function is uniformly Lipschitz α in the interval (a,b), and the upper limit of α which satisfies Lipschitz α at x_0 for $f(x)$ is the Lipschitz regularity of this function.*

2.3 Wavelet Transform and Image Enhancement Technique

Mallat assumed the derivatives of any smoothing function $\theta(x)$ as wavelets. That is, he assumed $\psi(x) = \frac{d\theta(x)}{dx}$ or $\psi(x) = \frac{d^2\theta(x)}{dx^2}$. And he showed that the three approaches for edge detection, transform based approach, first derivative approach, and second derivative approach, could be unified in the mathematical scheme of wavelet analysis[6][7].

3 Proposed Image Enhancement Technique

3.1 Algorithm Overview

Image enhancement technique consists of wavelet transform part, processing part, and inverse wavelet transform part as shown in figure 1. In figure 1, X is the noised and block effected image, and $\tilde{X}$ is the denoised and deblocked image.

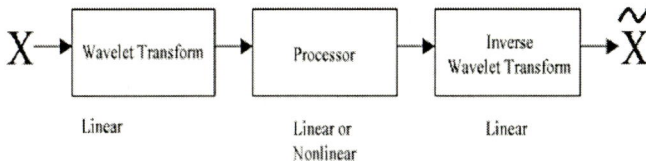

Fig. 1. Image enhancement technique based on wavelet transform

We can find the wavelet transform of a one dimensional signal in figure 2. The first signal in figure 2 is noiseless original signal. The second signal is the result of inserting AWG noise to the original signal. The third signal is the wavelet transform of the second signal at level 1(scale $j = 0 \Leftrightarrow s = 2^j = 1$). The fourth signal is the wavelet transform of the second signal at level 2(scale $j = 1 \Leftrightarrow s = 2^j = 2$). From the fifth signal to the eighth signal means the wavelet transform of the second signal at each level 3, 4, 5, 6 (scale $j = 2, 3, 4, 5 \Leftrightarrow s = 2^j = 4, 8, 16, 32$), respectively[6][7][8].

There are many LMM(Local Modulus Maxima) in the third signal of figure 2. In this paper, $LMM(f(x))$ is the local maximal value of $|f(x)|$. We can find two kinds of LMM in figure 2. In general, the LMMs due to edge elements are much larger than the LMMs due to noise and block elements. And this fact could be the hint for distinguishing edge elements from noise and block elements, though they are all high frequencies. Donoho used the amplitude value of the wavelet transformed coefficient to distinguish edge elements from noise elements[3][4].

In figure 2, as the wavelet transform level grows, the LMMs by noise elements perishes whereas the LMMs by edge element is maintained. So we could elaborate the distinguishing ability by using the spatial correlation of the wavelet transformed coefficients at level 1 and the wavelet transformed coefficients at level 2. In this paper we define the spatial correlation at specific position as the

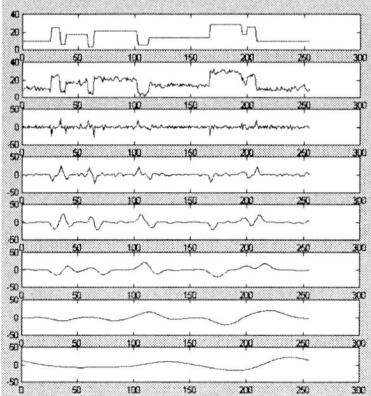

Fig. 2. Original and AWG noised signal, wavelet coefficients according to levels

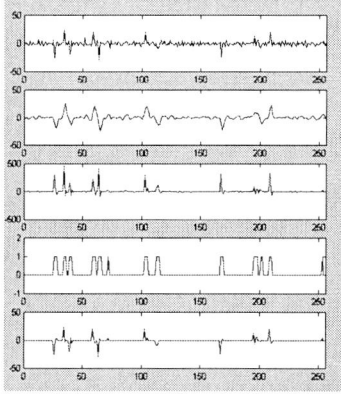

Fig. 3. Our image enhancement algorithm

multiplication of level 1 wavelet transformed coefficient at that position and level 2 wavelet transformed coefficients at that position.

The first signal in figure 3 is the level 1 wavelet transformed coefficients of a AWG noised signal in one dimension. The second signal in figure 3 is the wavelet transformed coefficients at level of the AWG noised signal in one dimension. The third signal is the spatial correlation of the first and second signal. The fourth signal is the mask for setting the coefficients which are less than threshold to zero. The last signal is the result of masking the first signal. So if we inverse wavelet transform of the last signal, then we can get the denoised signal. The most important part of our algorithm is finding the threshold value.

3.2 Analysis of Spatial Correlation

Xu et al. already used spatial correlation and wavelet transform in their research [8]. We derive a different result from that of [8] by using the concept of Lipschitz regularity. As a matter of convenience, we derive our theory for one dimensional signal and assume noise and blocking effect which should be removed are all AWG noises.

We assume $I(x) = f(x) + n(n)$ as input signal, and $O(x)$ as the wavelet transformed coefficient of $I(x)$. In this notation, $f(x)$ is the original signal and $n(x)$ is AWG noise which will be inserted to the original signal. We can find the relation of wavelet coefficient and input signal in the following formulas.

$O_1(x) \equiv W_2^0 I(x) = W_2^0 (f(x) + n(x)) = W_2^0 f(x) + W_2^0 n(x)$
$O_2(x) \equiv W_2^1 I(x) = W_2^1 (f(x) + n(x)) = W_2^1 f(x) + W_2^1 n(x)$

$W_2^0 I(x), W_2^1 I(x)$ are the wavelet transformed coefficients of input signal $I(x)$ at level 1 and 2, respectively. We also define $a(x), b(x), a'(x), b'(x)$ as follows.

$$a(x) \equiv W_2^0 f(x), b(x) \equiv W_2^0 n(x) \quad (4)$$

$$a'(x) \equiv W_2^1 f(x), b'(x) \equiv W_2^1 n(x) \quad (5)$$

And we define $SO_1(x), SO_2(x)$, normalized wavelet transformed coefficient of minimum and maximum as follows.

$SO_1(x) \equiv \frac{255}{max(O_1(x))} O_1(x) = \frac{255 a(x)}{max(a(x)+b(x))} + \frac{255 b(x)}{max(a(x)+b(x))}$

$SO_2(x) \equiv \frac{255}{max(O_2(x))} O_2(x) = \frac{255 a'(x)}{max(a'(x)+b'(x))} + \frac{255 b'(x)}{max(a'(x)+b'(x))}$

We also define k, k'. $k \equiv \frac{255}{max(a(x)+b(x))}, k' \equiv \frac{255}{max(a'(x)+b'(x))}$

Definition 4. *If $LMM\left(W_2^0 I(x)\right)$ has a 0 or a positive Lipschitz regularity at $\tau = \tau_0$ and also has a 0 or a positive Lipschitz regularity for larger scales $(j = 1, 2, 3, \ldots)$, then we define the following set of x as the edge component of $I(x)$ at scale j.*

$\left\{ x \mid \exists_{(\alpha,\beta)}, \forall_{x \in \{x \mid x_0 - \alpha < x < x_0 + \beta\}}, \left(\left| W_2^j I(x) \right| > 0 \right) \wedge \left(W_2^j I(x_0 - \alpha) = W_2^j I(x_0 + \beta) = 0 \right) \right\}$

Theorem 1. *If $x \notin EdgeComponent$ then $a(x) = a'(x) = 0$.*

We define spatial correlation $D(x)$ by the following formula.
$D(x) \equiv SO_1(x) \times SO_2(x)$
$= kk' [a(x) a'(x) + b(x) b'(x) + a'(x) b(x) + b'(x) a(x)]$

Theorem 2. *If $x \in EdgeComponent, D(x) = kk' a(x) a'(x)$*
If $x \notin EdgeComponent, D(x) = kk' b(x) b'(x)$

3.3 Image Enhancement Technique by Using Spatial Correlation

Theorem 3. *If $x \in EdgeComponent$, then $O_1(x) = a(x), O_2(x) = a'(x)$. And if $x \notin EdgeComponent$, then $O_1(x) = b(x), O_2(x) = b'(x)$.*

Theorem 4. *$\forall^{LMM}, LMM(kk' a(x) a'(x)) > LMM(kk' b(x) b'(x))$*

3.4 Proposed Image Enhancement Technique

We propose a new image enhancement technique based on the previous theoretical discussions.

1. Perform wavelet transform to noised and block-effected image at level 1 and 2, in the directions of x and y, respectively. The results are denoted as $O_{1,x}(x,y), O_{1,y}(x,y), O_{2,x}(x,y)$, and $O_{2,y}(x,y)$, respectively. And calculate $O_1(x,y)$ and $O_2(x,y)$ by using the following formula.
$$O_j(x,y) = \sqrt{(O_{j,x}(x,y))^2 + (O_{j,y}(x,y))^2}, j = 1, 2$$
2. Normalize $O_1(x,y)$ and $O_2(x,y)$ to 0 as minimum and 255 as maximum. The results are denoted as $SO_1(x,y)$ and $SO_2(x,y)$.
3. Calculate $D(x,y) = SO_1(x,y) \times SO_2(x,y)$. And find $Max(kk'b(x,y)b'(x,y))$ from $D(x,y)$.
4. Set the values of $O_{1,x}(x,y), O_{1,y}(x,y), O_{2,x}(x,y)$, and $O_{2,y}(x,y)$ to 0 in the regions where $D(x,y) \leq Max(kk'b(x,y)b'(x,y))$ is satisfied. And store these modified regions in $Set(x,y)$
5. Recover the modified values of border edge elements by the method which is described in section 3.3. The results are denoted as $NewO_{1,x}(x,y)$, $NewO_{1,y}(x,y), NewO_{2,x}(x,y)$, and $NewO_{2,y}(x,y)$ respectively.
6. Perform inverse wavelet transform to $NewO_{1,x}(x,y), NewO_{1,x}(x,y)$, $NewO_{2,x}(x,y)$, and $NewO_{2,y}(x,y)$. The result is denoised and deblocked image of the noised and block-effected image.
7. Perform median filtering to the region of $Set(x,y)$ to get final denoised and deblocked image. If an image is heavily noised and block-effected, then do median filtering to all the region of this image.

4 Experiments and Results

4.1 Experimental Results for Denoising

We did experiments for various images and degrees of noises, and have got uniformly good results. We present the results of images of fruit, lena, and peppers for 4.5%(standard deviation 11.5), and 9.5%(standard deviation 23.5) AWG noises. In figure 4, we compare the PSNR results of our proposed algorithm from those of Xu et al., Wiener filter, spatial averaging filter, Gaussian smoothing filter, and median filter. Our denoising algorithm shows higher PSNR than any other algorithms.

And we also confirm that our proposed algorithm needs lower computational complexity than that of Xu et al[8]. The method by Xu et al. has recorded minimum , and maximum CPU clock cycle in experiments, whereas our algorithm has recorded minimum , and maximum CPU clock cycle in the same experimental environment.

Algm. / Images	Noised	Our	Xu	Wiener	Averaging	Gaussian	Median
Lena4.5	22.27	27.73	26.91	27.15	27.05	23.06	26.77
Lena9.5	16.47	24.5	20	22.28	23.71	22.59	22.4
Peppers4.5	22.29	27.9	27.2	27.25	26.93	21.92	27.07
Peppers9.5	16.48	24.81	20.28	22.36	23.81	21.82	22.58
Fruit4.5	22.23	30.16	27.46	28	29.82	25.55	28.35
Fruit9.5	16.51	26.08	20.14	22.87	24.84	24.31	22.95

Fig. 4. Performance comparisons of each denoising algorithms(PSNR:dB)

(a) Original peppers (b) 4.5% noised image (c) 9.5% noised image

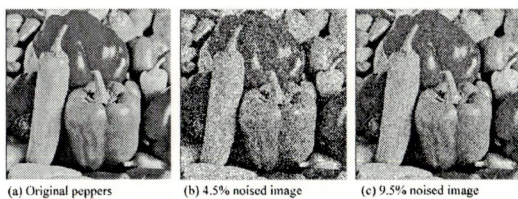

Fig. 5. Original peppers image and noised images for experiments

4.2 Experimental Results for Deblocking

We have used the 512 × 512 pixel sized images of Lena which are compressed at bpp, bpp, and bpp by JPEG to compare the deblocking ability with previous deblocking algorithms.

5 Conclusion

We have proposed a new image enhancement algorithm based on wavelet analysis. Here the image enhancement means the denoising technique for AWG (additive white gaussian) noise and the deblocking technique for blocking effect caused by JPEG and MPEG.

We have developed our image enhancement theory based on Lipschitz regularity, spatial correlation, and wavelets. The essence of our theory was the analysis to the spatial correlation of wavelet transformed coefficients. And we could develop some different results from previous researches by using some reasonable assumptions. We have confirmed that our new algorithm is better than the previous ones by experiments. The denoising ability of our algorithm was superior to those of any other previous denoising algorithms in the aspect of PSNR, and the deblocking ability of our algorithm was similar to those of the best previous deblocking algorithms. Our algorithm could be well adapted to both denoising and deblocking. This is one of the most significant features of our algorithm.

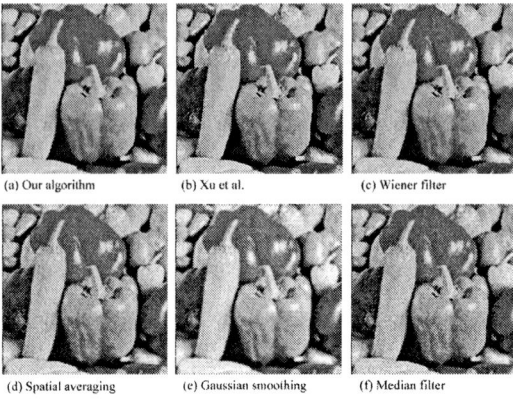

Fig. 6. Denoising algorithms for 4.5 percent AWG noise

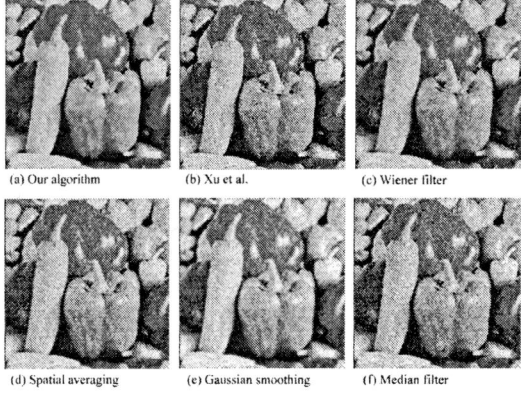

Fig. 7. Denoising algorithms for 9.5 percent AWG noise

This feature could make it possible to implement optimal hardware system like VLSI.

References

1. Gonzalez, R., Woods, R.: Digital Image Processing. Addison Wesley. (1993) 183–450
2. Weeks, A.: Fundamentals of Electronic Image Processing. SPIE/IEEE. (1996) 121–227
3. Donoho, D., Johnstone, I.: Ideal spatial adaptation via wavelet shrinkage. Biometrika **81** (1994) 425–455
4. Donoho, D.: De-noising by soft-thresholding. IEEE Transactions on Information Theory **41(3)** (1995) 613–627
5. Burrus, C., Gopinath, R., Guo, H.: Introduction to Wavelets and Wavelet Transforms:A Primer. Prentice-Hall. (1998) 196–218

Algorithms	0.13bpp	0.15bpp	0.24bpp
JPEG	24.52	26.44	29.58
Our algorithm	25.57	27.55	30.38
Yang et al.[12]	–	27.58	30.43
Xiong et al.[13]	–	27.58	30.37
Chou et al.[11]	–	27.50	30.37

Fig. 8. Deblocking performance comparison(PSNR:dB)

 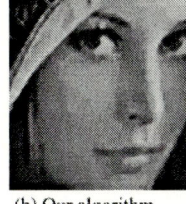 

(a) 0.24bpp compressed   (b) Our algorithm   (c) Xiong et al.

Fig. 9. Comparison of deblocking performance to 0.24bpp JPEG compressed Lena

6. Mallat, S., Hwang, W.: Singularity detection and processing with wavelets. IEEE Transactions on Information Theory **38(2)** (1992) 617–643
7. Mallat, S., Zhong, S.: Characterization of signals from multiscale edges. IEEE Transactions on Pattern Analysis and Machine Intelligence **14(7)** (1992) 710–732
8. Xu, Y., Weaver, J., Healy, D., Lu, J : Wavelet Transform Domain Filters : A Spatially Selective Noise Filtration Technique. IEEE Transactions on Image Processing **3(6)** (1994)
9. Rao, K., Hwang, J.: Techniques & Standards for Image, Video & Audio Coding. Prentice-Hall. (1996) 127–386
10. Vetterli, M., Kovacevic, J.: Wavelets and Subband Coding. Prentice-Hall. (1995) 1–86
11. Chou, J., Crouse, M., Ramchandran, K.: A simple algorithm for removing blocking artifacts in block-transform coded images. Proc. ICIP'98 **1** (1998) 377–380
12. Yang, Y., Galatsanos, P., Katsaggelos, A.: Projection-based spatially adaptive reconstruction of block-transform compressed images. IEEE Trans. Image Processing **4** (1995) 896–908
13. Xiong, Z., Orchard, M., Zhang, Y.: A simple deblocking algorithm for JPEG compressed images using overcomplete wavelet representations. IEEE Trans. on Circuits and Systems for Video Technology (1996)
14. Hsieh, J., Liao, H., Ko, M., Fan, K.: A new wavelet-based edge detector via constrained optimization. Image and Vision Computing **15(7)** (1997) 511–527
15. Canny, J.: A computational approach to edge detection. IEEE Trans. Pattern Anal. Machine Intell. **8(6)** (1986) 679–697
16. Grossmann, A., Morlet, J.: Decomposition of Hardy functions into square integrable wavelets of constant shape. SIAM J. Math. **15** (1984) 723–736

17. Mallat, S.: A theory for multiresolution signal decomposition : The wavelet representation. IEEE Trans. on Pattern Recognition and Machine Intelligence **11(7)** (1989) 674–693
18. Grossman, A.: Wavelet transform and edge detection. Stochastic Processes in Physics and Engineering (1986)
19. Davis, E.: Machine Vision. Academic Press. (1997) 248–249
20. Daubechies, I.: Ten Lectures on Wavelets. SIAM. (1992) 24–63

Front-End Vision: A Multiscale Geometry Engine

Bart M. ter Haar Romeny[1], and Luc M.J. Florack[2]

Utrecht University, the Netherlands
[1] Image Sciences Institute, PO Box 85500, 3508 TA Utrecht
[2] Department of Mathematics, PO Box 80010, 3508 TA Utrecht
`B.terHaarRomeny@isi.uu.nl`, `Luc.Florack@math.uu.nl`

Abstract. The paper is a short tutorial on the multiscale differential geometric possibilities of the front-end visual receptive fields, modeled by Gaussian derivative kernels. The paper is written in, and interactive through the use of Mathematica 4, so each statement can be run and modified by the reader on images of choice. The notion of multiscale invariant feature detection is presented in detail, with examples of second, third and fourth order of differentiation.

1 Introduction

The front end visual system belongs to the best studied brain areas. Scale-space theory, as pioneered by Iijima in Japan [10,17] and Koenderink [11] has been heavily inspired by the important derivation of the Gaussian kernel and its derivatives as regularized differential operators, and the linear diffusion equation as its generating partial differential equation. To view the front-end visual system as a 'geometry-engine' is the inspiration of the current work. Simultaneously, the presented examples of applications of (differential) geometric operations may inspire operational models of the visual system.

Scale-space theory has developed into a serious field [8, 13]. Several comprehensive overview texts have been published in the field [5, 7, 12, 15, 18]. So far, however, this robust mathematical framework has seen impact on the computer vision community, but there is still a gap between the more physiologically, psychologically and psychophysically oriented researchers in the vision community. One reason may be the nontrivial mathematics involved, such as group invariance, differential geometry and tensor analysis.

The last couple of years symbolic computer algebra packages, such as Mathematica, Maple and Matlab, have developed into a very user friendly and high level prototyping environment. Especially Mathematica combines the advantages of symbolic manipulation and processing with an advanced front-end text processor. This paper highlights the use of Mathematica 4.0 as an interactive tutorial toolkit for easy experimenting and exploring front-end vision simulations. The exact code can be used rather then pseudocode. With these high level programming tools most programs can be expressed in very few lines, so it keeps the reader at a highly intuitive but practical level. Mathematica notebooks are portable, and run on *any* system equivalently. Previous speed limitations are now well overcome. The full (1400 pages) documentation is available online, (see www.wri.com).

1.1 Biological Inspiration: Receptive Field Profiles from First Principles

It is well known that the Gaussian kernel, $G(\vec{x}, \sigma) = \frac{1}{\sqrt{2\pi\sigma^2}} \exp(-\frac{\vec{x}.\vec{x}}{2\sigma^2})$ as a front-end visual measurement aperture can be uniquely derived in quite a number of ways (for a comprehensive overview see [17]). These include the starting point that lower resolution levels have a higher resolution level as cause ('causality' [11]), or that there is linearity and no preference for location, orientation and size of the aperture ('first principles' [2]). This Gaussian kernel is the Green's function of the linear, isotropic diffusion equation $\frac{\partial^2 L}{\partial x^2} + \frac{\partial^2 L}{\partial y^2} = L_{xx} + L_{yy} = \frac{\partial L}{\partial s}$, where $s = 2\sigma^2$ is the variance. Note that the derivative to scale is here the derivative to σ^2 which also immediately follows from a consideration of the dimensionality of the equation. All *partial derivatives* of the Gaussian kernel are solutions too of the diffusion equation.

The Gaussian kernel and all of its partial derivatives form a one-parameter *family* of kernels where the scale σ is the free parameter. This is a general feature of the biological visual system: the exploitation of *ensembles* of aperture functions, which are mathematically modeled by families of kernels for a free parameter, e.g. for all scales, derivative order, orientation, stereo disparity, motion velocity etc.

The Gaussian kernel is the unique kernel that generates no *spurious resolution* (e.g. the squares so familiar with zooming in on pixels). It is the physical *point operator*, the Gaussian derivatives are the physical *derivative operators*, at the scale given by the Gaussian standard deviation.

The receptive fields in the primary visual cortex closely resemble Gaussian derivatives, as was first noticed by Young [19] and Koenderink [11]. These RF's come at a wide range of sizes, and at all orientations.

Below two examples are given of measured receptive field sensitivity profiles of a cortical *simple cell* (left) and a Lateral Geniculate Nucleus (LGN) *center-surround* cell, measured by DeAngelis, Ohzawa and Freeman [1], http://totoro.berkeley.edu/.

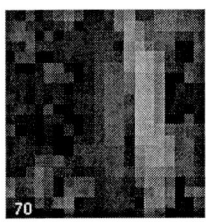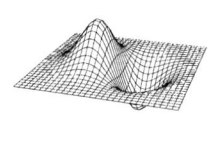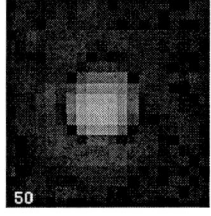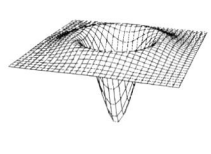

Fig. 1. a. Cortical simple cell, modeled by a first order Gaussian derivative. From[1]. **b.** Center-surround LGN cell, modeled by Laplacean of Gaussian. From [1].

Through the center-surround structure at the very first level of measurement on the retina the *Laplacean* of the input image can be seen to be taken. The linear diffusion equation states that this Laplacean is equal to the first derivative to scale: $\frac{\partial^2 L}{\partial x^2} + \frac{\partial^2 L}{\partial y^2} = \frac{\partial L}{\partial s}$. One conjecture for its presence at this first level of observation might be that the visual system actually measures L_s, i.e. the change in signal ∂L when the aperture is changed with ∂s: at homogeneous areas there is no output, at highly textured areas there is much output. Integrating both sides of $\partial L = (L_{xx}+L_{yy})\, \partial s$ over all scales gives the measured intensity in a robust fashion.

The derivative of the observed (convolved) data $\frac{\partial}{\partial x}(L \otimes G) = L \otimes \frac{\partial G}{\partial x}$ shows that differentiation and observation is accomplished in a single step: convolution with a Gaussian derivative kernel. Differentiation is now done by *integration*, i.e. by the convolution integral.

The Gaussian kernel is the physical analogon of a mathematical point, the Gaussian derivative kernels are the physical analogons of the mathematical differential operators. Equivalence is reached for the limit when the scale of the Gaussian goes to zero: $\lim_{\sigma \to 0} G(x,\sigma) = \delta(x)$, where is the Dirac delta function, and $\lim_{\sigma \to 0} \otimes G(x,\sigma) = \frac{\partial}{\partial x}$. Any differention blurs the data somewhat, with the amount of the scale of the differential operator. There is no way out this increase of the inner scale, we can only try to minimize the effect. The Gaussian kernel has by definition a strong *regularizing* effect [16,12].

2 Multiscale Derivatives

It is essential to work with descriptions that are *independent of the choice of coordinates*. Entities that do not change under a group of coordinate transformations are called *invariants* under that particular group. The only geometrical entities that make physically sense are invariants. In the words of Hermann Weyl: any invariant has a specific geometric *meaning*. In this paper we only study orthogonal and affine invariants. We first build the operators in Mathematica:

The function **gDf[im,nx,ny, σ]** implements the convolution of the image with the Gaussian derivative for 2D data in the Fourier domain. This is an exact function, no approximations other then the finite periodic window in both domains. Variables: **im** = 2D image (as a list structure), **nx, ny** = order of differentiation to *x* resp. *y*, σ = scale of the kernel, in pixels.

```
gDf[im_,nx_,ny_,σ_]:= Module[{xres, yres, gdkernel},
    {yres, xres} = Dimensions[im];
    gdkernel = N[Table[Evaluate[
        D[1/(2 Pi σ²) Exp[-((x²+y²)/(2σ²))],{x,nx},{y, ny}]],
        {y,-(yres-1)/2,(yres-1)/2}, {x,-(xres-1)/2,(xres-1)/2}]];
    Chop[N[Sqrt[xres yres] InverseFourier[Fourier[im]
    Fourier[RotateLeft[gdkernel, {yres/2, xres/2}]]]]]];
```

This function is rather slow, but is exact. A much faster implementation exploits the separability of the Gaussian kernel, and this implementation is used in the sequel:

```
gD[im_,nx_,ny_,σ_] := Module[{x,y,kx,ky,tmp},
    kx ={N[Table[Evaluate[D[(Exp[-(x²/(2σ²))])]/
(σ Sqrt[2π]),{x,nx}]],{x,-4σ, 4σ}]]};
    ky = {N[Table[Evaluate[D[(Exp[-(y²/(2σ²))])]/
(σ Sqrt[2π]), {y, ny}]], {y, -4σ, 4σ}]]};
tmp = ListConvolve[kx, im, Ceiling[Dimensions[kx]/2]];
Transpose[ListConvolve[ky,
Transpose[tmp],Reverse[Ceiling[Dimensions[ky]/2]]]];
```

An example is the gradient $\sqrt{L_x^2 + L_y^2}$ at a range of scales:

```
im = Import["mr128.gif"][[1,1]];
p1 = Table[grad = Sqrt[gD[im,1,0,σ]² + gD[im,0,1,σ]²];
{ListDensityPlot[grad,PlotRange→{0, 40}, DisplayFunction→
Identity],
 ListDensityPlot[σ grad, PlotRange→{0, 40},
DisplayFunction→Identity]}, {σ, 1, 5}];
Show[GraphicsArray[Transpose[p1]]];
```

3 Gauge Coordinates

In order to establish differential geometric properties it is easiest to exploit intrinsic geometry. I.e. that we define a new coordinate frame for our geometric explorations which is related to the local isophote structure, so it is different in every different point. A straightforward definition of a new local coordinate frame in 2D is where we cancel the degree of freedom of rotation by defining gauge coordinates: we locally 'fix the gauge'.

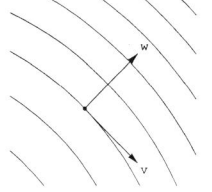

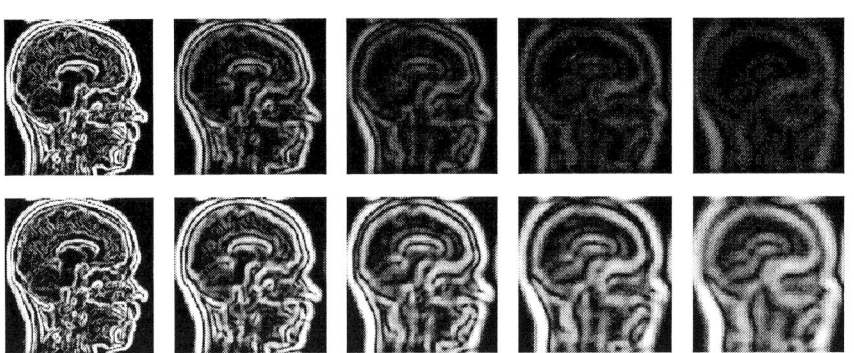

Fig. 2. Top row: Gradient of a sagittal MR image at scales 1, 2, 3, 4 and 5 pixels. Lower row: same scales, gradient in natural dimensionless coordinates, i.e. x'→x/σ, so $\partial/\partial x' \mapsto \sigma \partial/\partial x$. This leaves the intensity range of differential features more in a similar output range due to scale invariance. Resolution 128^2.

The 2D unit vector frame of gauge coordinates $\{v,w\}$ is defined as follows: v is the unit vector in the direction tangential to the isophote, so $L_v \equiv 0$, w is defined perpendicular to v, i.e. in the direction of the intensity gradient.

The derivatives to v and w are by definition features that are invariant under orthogonal transformations, i.e. rotation and translation. To apply these gauge derivative operators on images, we have to convert to the Cartesian $\{x,y\}$ domain. The derivatives to v and w are defined as:

$$\partial_v = \frac{-L_y \partial_x + L_x \partial_y}{\sqrt{L_x^2 + L_y^2}} = L_i \varepsilon_{ij} \partial_j \quad ; \quad \partial_w = \frac{L_x \partial_x + L_y \partial_y}{\sqrt{L_x^2 + L_y^2}} = L_i \delta_{ij} \partial_j$$

The second formulation uses *tensor notation*, where the indices i,j stand for the range of dimensions. δ_{ij} is the nabla operator, δ_{ij} and ε_{ij} are the symmetric Kronecker tensor and the antisymmetric Levi-Civita tensor respectively (in 2D). The definitions above are easily accomplished in Mathematica:

```
δ = IdentityMatrix[2]
ε = Table[Signature[{i,j}],{j,2},{i,2}]
jacobean = Sqrt[Lx^2 + Ly^2];
dv = 1/jacobean*{Lx,Ly}.ε.{D[#1,x],D[#1,y]}&
dw = 1/jacobean*{Lx,Ly}.δ.{D[#1,x],D[#1,y]}&
```

The notation (...#)& is a 'pure function' on the argument #, e.g. D[#,x]& takes the derivative. Now we can calculate any derivative to v or w by applying the operator dw or dv repeatedly. Note that the Lx and Ly are constant terms.

```
rule1 = {Lx → ∂_x L[x,y], Ly → ∂_y L[x,y]};
rule2 = Derivative[n_,m_][L][x,y] → L <>
    Table[x,{n}] <> Table[y, {m}];
Lw = dw[L[x, y]] /. rule1 /. rule2 // Simplify
```

$$\sqrt{Lx^2 + Ly^2}$$

```
Lww = dw[dw[L[x, y]]] /. rule1 /. rule2 // Simplify
```

$$\frac{Lx^2 Lxx + 2 Lx Lxy Ly + Ly^2 Lyy}{Lx^2 + Ly^2}$$

Due to the fixing of the gauge by removing the degree of freedom for rotation, we have an important result: *every derivative to v and w is an orthogonal invariant*, i.e. an invariant property where translation or rotation of the coordinate frame is irrelevant. It means that polynomial combinations of these gauge derivative terms are invariant. We now have the toolkit to make gauge derivatives to any order.

3.1 Examples to 2nd, 3rd and 4th Order

The definitions for the gauge differential operators ∂_v and ∂_w need to have their regular differential operators be replaced by Gaussian derivative operators. To just show the textual formula, we do not yet evaluate the derivative by using temporarily HoldForm (/. means 'apply rule'):

```
gauge2D[im_, nv_, nw_, σ_] := (Nest[dw,Nest[dv,L[x,y],nv],nw]
/. {Lx -> ∂_x L[x,y], Ly -> ∂_y L[x,y]})
/. ((Derivative[n_, m_])[L])[x,y] -> HoldForm[gD[im, n, m,
σ]] // Simplify)
```

3.2 Ridge Detection

L_{vv} is a good ridge detector. Here is the Cartesian (in $\{x,y\}$) expression for L_{vv}:

```
Clear[im,σ]; gauge2D[im, 2, 0, 2]
```

```
(gd[im,0,2,2] gd[im,1,0,2]² - 2 gd[im,0,1,2] gd[im,1,0,2]
gd[im,1,1,2] + gd[im,0,1,2]² gd[im,2,0,2]) /
(gd[im,0,2,2]²+gd[im,0,2,2]²)

im = Import["hands.gif"][[1, 1]];
Lvv = gauge2D[im, 2, 0, 3] // ReleaseHold;
Block[{$DisplayFunction = Identity},p1 = ListDensityPlot[im];
p2 = ListDensityPlot[Lvv];]; Show[GraphicsArray[{p1, p2}]];
```

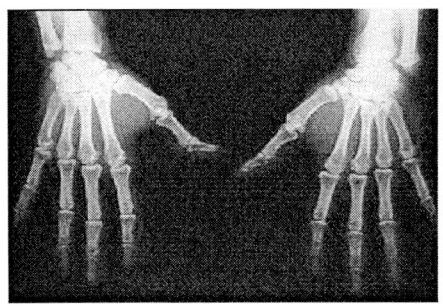 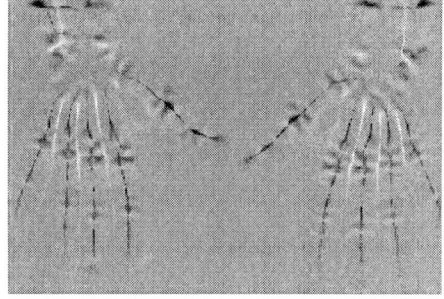

Fig. 3. a. Input image: X-ray of hands, resolution 439x138 pixels. **b.** Ridge detection with L_{vv}, scale 3 pixels. Note the concave and convex ridges.

3.3 Affine Invariant Corner Detection

Corners can be defined as locations with high isophote curvature κ and high intensity gradient L_w. Isophote curvature κ is defined as the change w'' of the tangent vector w' in the gauge coordinate system. When we differentiate the definition of the isophote (L = Constant) to v, we find $\kappa = -Lvv/Lw$:

```
D[L[v, w[v]] == Constant, v]; κ=
w''[v]/.Solve[D[L[v,w[v]]==Constant,{v,2}]/.w'[v]→0,w''[v]]
```

$-L^{(0,2)}[v,w[v]]/L^{(1,0)}[v,w[v]]$

Blom proposed as corner detector [6]: $\Theta^{[n]} = -\frac{L_{vv}}{L_w} L_w^n = \kappa L_w^n$. An obvious advantage is invariance under as large a group as possible. Blom calculated n for invariance under the *affine* transformation $\begin{pmatrix} x' \\ y' \end{pmatrix} \rightarrow \begin{pmatrix} a & b \\ c & d \end{pmatrix} (x\ y)$. The derivatives transform as $\begin{pmatrix} \partial/\partial x \\ \partial/\partial y \end{pmatrix} \rightarrow \begin{pmatrix} a & c \\ b & d \end{pmatrix} \begin{pmatrix} \partial \\ \partial x' \end{pmatrix} \begin{pmatrix} \partial \\ \partial y' \end{pmatrix}$. The corner detectors $\Theta^{[n]}$ transform as $\Theta^{[n]}$ = $(ad-bc)^2 \{(a\ L_{x'} + c\ L_{y'})^2 + (b\ L_{x'} + d\ L_{y'})^2\}^{(n-3)/2} \{2\ L_{x'}\ L_{y'}\ L_{x'y'} - L_{y'}^2 L_{x'x'} - L_{x'}^2 L_{y'y'}\}$. This is a relative *affine invariant* of order 2 if $n=3$ with the determinant $D=(ad-bc)$ of the affine transformation as order parameter. We consider here *special* affine transformations ($D=1$).

So a good corner-detector is $\Theta^{[3]} = -\frac{L_{vv}}{L_w} L_w^3 = L_{vv} L_w^2$. This feature has the nice property that is is not singular at locations where the gradient vanishes, and through its affine invariance it detects corners at all 'opening angles'.

```
im = N[Import["utrecht256.gif"][[1, 1]]];
corner1 = (gauge2D[im, 2, 0, 1] \ gauge2D[im, 0, 1, 1]²) //
ReleaseHold;
corner2 = (gauge2D[im, 2, 0, 3] \ gauge2D[im, 0, 1, 3]²) //
ReleaseHold; Block[{$DisplayFunction = Identity},
p1 = ListDensityPlot[im]; p2 = ListDensityPlot[corner1];
p3 = ListDensityPlot[corner2];]; Show[GraphicsArray[{p1, p2,
p3}]];
```

 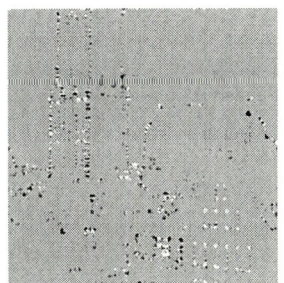 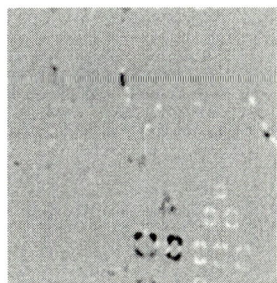

Fig. 4. a. Input image, resolution 256x256. **b.** Corner detection with $L_{vv} L_w^2$, $\sigma=1$ pixel. **c.** idem, $\sigma=3$ pixels.

3.4 T-Junction Detection

An example of third order geometric reasoning in images is the detection of T-junctions. T-junctions in the intensity landscape of natural images occur typically at occlusion points. When we zoom in on the T-junction of an observed (i.e. blurred) image and inspect locally the isophote structure at a T-junction, we see that at a T-junction the change of the isophote curvature κ in the direction perpendicular to the isophotes (the *w*-direction) is high. So a candidate for T-junction detection is $\frac{\partial \kappa}{\partial w}$. We saw before that the isophote curvature is defined as $\kappa = -L_{vv}/L_w$. Thus the Cartesian expression for the T-junction detector becomes

```
κ= Simplify[-(dv[dv[L[x, y]]]/dw[L[x, y]]) /.
  {Lx -> D[L[x, y], x], Ly -> D[L[x, y], y]}];
τ = Simplify[dw[κ] /. {Lx -> D[L[x, y], x],
  Ly -> D[L[x, y], y]}]; % /. Derivative[n_, m_][L][x, y] ->
StringJoin[L, Table[x, {n}], Table[y, {m}]]
```

$1/(Lx^2 + Ly^2)^3$ $(-Lxxy\ Ly^5 + Lx^4\ (2Lxy^2 - Lx\ Lxyy + Lxx\ Lyy) +$
$Ly^4(2Lxy^2 + Lx(-Lxxx + 2Lxyy) + Lxx\ Lyy) + Lx^2\ Ly^2\ (3Lxx^2 -$
$Lx\ Lxxx - 8Lxy^2 + Lx\ Lxyy - 4\ Lxx\ Lyy + 3Lyy^2) +$
$Lx\ Ly^3\ (Lx\ Lxxy + 6\ Lxx\ Lxy - 6Lxy\ Lyy - Lx\ Lyyy) +$
$Lx^3\ Ly\ (2Lx\ Lxxy - 6\ Lxx\ Lxy + 6\ Lxy\ Lyy - Lx\ Lyyy))$

To avoid singularities at vanishing gradients through the division by $(L_x^2+L_y^2)^3 = L_w^6$ we use as our T-junction detector $\tau = \frac{\partial \kappa}{\partial w} L_w^6$, the derivative of the curvature in the direction perpendicular to the isophotes:

```
τ= Simplify[dw[κ]\dw[L[x,y]]⁶ /. {Lx→∂ₓL[x,y], Ly→∂ᵧL[x, y]}];
```

Finally, we apply the T-junction detector on our blocks at a scale of $\sigma=2$:

```
τ= τ /. Derivative[n_,m_][L][x,y]→HoldForm[gD[blocks,n, m,
σ]]; σ = 2; ListDensityPlot[τ // ReleaseHold];
```

Fig. 5. a. Input image: some T-junctions encircled. Resolution 317x204 pixels. **b.** T-juction detection with $\tau = \frac{\partial \kappa}{\partial w} L_w^6$ at a scale of 2 pixels.

3.5 Fourth Order Junction Detection

As a final fourth order example, we give an example for a detection problem in images at high order of differentiation from algebraic theory. Even at orders of differentiation as high as 4, invariant features can be constructed and calculated for discrete images through the biologically inspired scaled derivative operators. Our example is to find in a checkerboard the crossings where 4 edges meet. We take an algebraic approach, which is taken from Salden et al. [14].

When we study the fourth order local image structure, we consider the fourth order polynomial terms from the local Taylor expansion:

```
pol4=(Lxxxx x⁴+4Lxxxy x³ y+6Lxxyy x² y²+4Lxyyy x y³+Lyyyy y⁴)/4!
```

The main theorem of algebra states that a polynomial is fully described by its roots: e.g. $ax^2 + bx + c = (x-x_1)(x-x_2)$. Hilbert showed that the 'coincidenceness' of the roots, i.e. how well all roots coincide, is a particular invariant condition. From algebraic theory it is known that this 'coincidenceness' is given by the *discriminant*:

```
Discriminant[p_, x_] := With[{m = Exponent[p, x]},
  Cancel[((-1)^(1/2*m*(m - 1)) Resultant[p, D[p, x], x])/
  Coefficient[p, x, m]]];
```

The resultant of two polynomials a and b, both with leading coefficient one, is the product of all the differences a_i-b_j between roots of the polynomials. The resultant is always a number or a polynomial. The discriminant of a polynomial is the product of the squares of all the differences of the roots taken in pairs. We can express our function in two variables (x,y) as a function in a single variable x/y by the substitution $y \rightarrow 1$. Some examples:

```
Discriminant[(Lxx x²+2 Lxy x y+Lyy y²)/2!, x] /. {y -> 1}
```

Lxy² + Lxx Lyy

The discriminant of second order image structure is just the determinant of the Hessian matrix, i.e. the Gaussian curvature. Here is our fourth order discriminant:

```
Discriminant[pol4, x] /. {y -> 1}
```

```
(497664 Lxxxy² Lxxyy² Lxyyy² - 31104 Lxxxx Lxxyy³ Lxyyy² -
884736 Lxxxy³ Lxyyy³ + 62208 Lxxxx Lxxxy Lxxyy Lxyyy³ - 648
Lxxxx² Lxyyy⁴ - 746496 Lxxxy² Lxxyy³ Lyyyy + 46656 Lxxxx Lxxyy⁴
Lyyyy + 1492992 Lxxxy³ Lxxyy Lxyyy Lyyyy - 103680 Lxxxx Lxxxy
Lxxyy² Lxyyy Lyyyy - 3456 Lxxxx Lxxxy² Lxyyy² Lyyyy + 1296
Lxxxx² Lxxyy Lxyyy² Lyyyy - 373248 Lxxxy⁴ Lyyyy² + 31104 Lxxxx
Lxxxy² Lxxyy Lyyyy² - 432 Lxxxx² Lxxyy² Lyyyy² - 288 Lxxxx²
Lxxxy Lxyyy Lyyyy² + Lxxxx³ Lyyyy³)/54
```

A complicated polynomial in fourth order derivative images. Through the use of Gaussian derivative kernels each separate term can easily be calculated as an intermediate image. We change all coefficients into scaled Gaussian derivatives:

```
discr4[im_, σ_] :=
  Discriminant[pol4, x] /. {y → 1, Lxxxx → gD[im, 4, 0, σ],
  Lxxxy → gD[im, 3, 1, σ], Lxxyy → gD[im, 2, 2, σ],
  Lxyyy → gD[im, 1, 3, σ], Lyyyy → gD[im, 0, 4, σ]};
ListDensityPlot[noisycheck, ImageSize -> {200, 100}];
ListDensityPlot[discr4[noisycheck,5],ImageSize->{200, 100}];
```

The detection is rotation invariant, robust to noise, and no detection at corners:

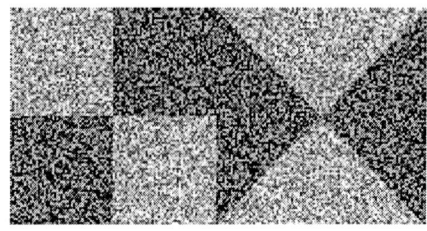 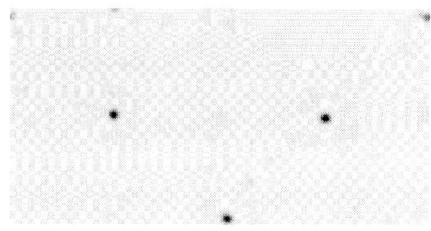

Fig. 6. a. Noisy input image. Resolution 200 x 100 pixels.   **b.** Four-junction detection with the algebraic discriminant D4 at σ=4 pixels.

References

1. Gregory C. DeAngelis, Izumi Ohzawa, and Ralph D. Freeman, "Receptive-field dynamics in the central visual pathways", Trends Neurosci. 18: 451-458, 1995.
2. L. M. J. Florack, B. M. ter Haar Romeny, J. J. Koenderink, and M. A. Viergever, Scale and the differential structure of images, Im. and Vision Comp., vol. 10, pp. 376-388, 1992.
3. L. M. J. Florack, B. M. ter Haar Romeny, J. J. Koenderink, and M. A Viergever, Cartesian differential invariants in scale-space, J. of Math. Im. and Vision, 3, 327-348, 1993.
4. L. M. J. Florack, B. M. ter Haar Romeny, J. J. Koenderink, and M. A. Viergever, The Gaussian scale-space paradigm and the multiscale local jet, International Journal of Computer Vision, vol. 18, pp. 61-75, 1996.
5. L.M.J. Florack, Image Structure, Kluwer Ac. Publ., Dordrecht, the Nether-lands, 1997.
6. B. M. ter Haar Romeny, L. M. J. Florack, A. H. Salden, and M. A. Viergever, Higher order differential structure of images, Image & Vision Comp., vol. 12, pp. 317-325, 1994.
7. B. M. ter Haar Romeny (Ed.), Geometry Driven Diffusion in Computer Vision. Kluwer Academic Publishers, Dordrecht, the Netherlands, 1994.
8. B. M. ter Haar Romeny, L. M. J. Florack, J. J. Koenderink, and M. A. Viergever, eds., Scale-Space '97: Proc. First Internat. Conf. on Scale-Space Theory in Computer Vision, vol. 1252 of Lecture Notes in Computer Science. Berlin: Springer Verlag, 1997.
9. B. M. ter Haar Romeny, L. M. J. Florack, J. J. Koenderink, and M. A. Viergever, Invariant third order properties of isophotes: T-junction detection, in: Theory and Applications of Image Analysis, P. Johansen and S. Olsen, eds., vol. 2 of Series in Machine Perception and Artificial Intelligence, pp. 30-37, Singapore: World Scientific, 1992.
10. T. Iijima: Basic Theory on Normalization of a Pattern - in Case of Typical 1D Pattern. Bulletin of Electrical Laboratory, vol. 26, pp. 368-388, 1962 (in Japanese).
11. J.J. Koenderink: The Structure of Images, Biol. Cybernetics, vol. 50, pp. 363-370, 1984.
12. T. Lindeberg: Scale-Space Theory in Computer Vision, Kluwer Academic Publishers, Dordrecht, Netherlands, 1994.
13. M. Nielsen, P. Johansen, O.F. Olsen, J. Weickert (Eds.), Proc. Second Intern. Conf on Scale-space Theories in Computer Vision, Lecture Notes in Computer Science, Vol. 1682, Springer, Berlin, 1999.
14. A. H. Salden, B. M. ter Haar Romeny, L. M. J. Florack, J. J. Koenderink, and M. A. Viergever, A complete and irreducible set of local orthogonally invariant features of 2-dimensional images, in: Proc. 11th IAPR Internat. Conf. on Pattern Recognition, I. T. Young, ed., The Hague, pp. 180-184, IEEE Computer Society Press, Los Alamitos, 1992.

15. J. Sporring, M. Nielsen, L. Florack: Gaussian Scale-Space Theory, Kluwer Academic Publishers, Dordrecht, the Netherlands, 1997.
16. O. Scherzer, J. Weickert, Relations between regularization and diffusion filtering, J. Math. Imag. in Vision, in press, 2000.
17. J. Weickert, S. Ishikawa, A. Imiya, On the history of Gaussian scale-space axiomatics, in J. Sporring, M. Nielsen, L. Florack, P. Johansen (Eds.), Gaussian scale-space theory, Kluwer, Dordrecht, 45-59, 1997.
18. J. Weickert, Anisotropic diffusion in image processing, ECMI, Teubner Stuttgart, 1998.
19. R.A. Young, Simulation of Human Retinal Function with the Gaussian Derivative Model, Proc. IEEE CVPR CH2290-5, 564-569, Miami, Fla., 1986.

Face Reconstruction Using a Small Set of Feature Points

Bon-Woo Hwang[1], Volker Blanz[2], Thomas Vetter[2] and Seong-Whan Lee[1]*

[1] Center for Artificial Vision Research, Korea University
Anam-dong, Seongbuk-ku, Seoul 136-701, Korea
{bwhwang, swlee}@image.korea.ac.kr
[2] Max-Planck-Institute for Biological Cybernetics
Spemannstr. 38, 72076 Tuebingen, Germany
{volker.blanz, thomas.vetter}@tuebingen.mpg.de

Abstract. This paper proposes a method for face reconstruction that makes use of only a small set of feature points. Faces can be modeled by forming linear combinations of prototypes of shape and texture information. With the shape and texture information at the feature points alone, we can achieve only an approximation to the deformation required. In such an under-determined condition, we find an optimal solution using a simple least square minimization method. As experimental results, we show well-reconstructed 2D faces even from a small number of feature points.

1 Introduction

It is difficult for traditional bottom-up, generic approaches to reconstruct the whole image of an object from parts of its image or to restore the missing space information due to noise, occlusion by other objects, or shadow caused by illumination effects. In contrast to such approaches, top-down, object-class-specific and model-based approaches are highly tolerant to sensor noise and incompleteness of input image information[2]. Hence, the top-down approaches to the interpretation of images of variable objects are now attracting considerable interest[1-4, 6]. Kruger et al. proposed a system for the automatic determination of the position, size and pose of a human head, which is modeled by a labeled graph[3]. The nodes of the graph refer to feature points of a head. However, the location and texture information are not utilized for face reconstruction. Lanitis et al. implemented a face reconstruction using a 'Flexible Model'[4]. About 150 feature points inside a face and its contour were used for reconstruction. The system performs well, but it requires more than 100 feature points for shape reconstruction, and it also requires full texture information.

In this paper, we propose an efficient face reconstruction method from a small set of feature points. Our approach is based on the 2D shapes and textures of a

---

* To whom all correspondence should be addressed. This research was supported by Creative Research Initiatives of the Ministry of Science and Technology, Korea.

data set of faces. Shape is coded by a dense deformation field between a reference face image and each individual face image. Given only a small number of feature points on a novel face, we use the database to reconstruct its full shape and texture. It is a combination of stored shapes and textures that best matches the positions and gray values of the feature points. We anticipate that the proposed method will play an important role in reconstructing the whole information of an object out of information reduced for compressing or partially damaged due to occlusion or noise. In Section 2 and 3, we describe a 2D face model where shape and texture are treated separately, and a method for finding coefficients for face reconstruction, respectively. Experimental results for face reconstruction are given in Section 4. Finally, in Section 5, we present conclusive remarks and discuss some future work.

2 2D Face Model

On the assumption that the correspondence on the face images has already been established[1], the 2D shape of a face is coded as the deformation field from a reference image that serves as the origin of our space. The texture is coded as the intensity map of the image which results from mapping the face onto the reference face[6].

Let $S(x)$ be the displacement of point x, or the position of the point in the face that corresponds to point x in the reference face. Let $T(x)$ be the gray value of the point in the face that corresponds to point x in the reference face. With shape and texture information separated from the face image, we fit a multivariate normal distribution to our data set of faces according to the average of shape $\bar{S}$ and that of texture $\bar{T}$, covariance matrices C_S and C_T computed over shape and texture differences $\tilde{S} - S$ $\bar{S}$ and $\tilde{T} - T$ $\bar{T}$. By Principal Component Analysis(PCA), a basis transformation is performed to an orthogonal coordinate system formed by eigenvectors s_i and t_i of the covariance matrices on our data set of m faces.

$$S = \bar{S} + \sum_{i=1}^{m-1} \alpha_i s_i, \quad T = \bar{T} + \sum_{i=1}^{m-1} \beta_i t_i, \tag{1}$$

where $\alpha, \beta \in \Re^{m-1}$. The probability for coefficients α is defined as

$$p(\alpha) \sim exp\left[-\frac{1}{2}\sum_{i=1}^{m-1}(\frac{\alpha_i}{\sigma_i})^2\right], \tag{2}$$

with σ_i^2 being the eigenvalues of the shape covariance matrix C_S. Likewise, the probability $p(\beta)$ can be computed.

3 Face Reconstruction

In this section, we describe a method for finding coefficients for face reconstruction. First, we define an energy function as the sum of normalized coefficients and

3.1 Problem Definition

Since there are shape and texture elements only for feature points, we achieve an approximation to the deformation required. Our goal is to find an optimal solution in such an underdetermined condition. We define an energy function as the sum of the normalized coefficients. We also set a condition that the given shape or texture information at the feature points must be reconstructed perfectly. The energy function, $E(\alpha)$, describes the degree of deformation from the the average face. The problem(Equation 3) is to find α which minimizes the energy function, $E(\alpha)$, which is given as:

$$\alpha^* = \arg\min_{\alpha} E(\alpha), \tag{3}$$

with the energy function,

$$E(\alpha) = \sum_{i=1}^{m-1} (\frac{\alpha_i}{\sigma_i})^2. \tag{4}$$

under the condition,

$$\tilde{S}(x_j) = \sum_{i=1}^{m-1} \alpha_i s_i(x_j), \ (j = 1, \cdots, n), \tag{5}$$

where $x_1, \cdots, x_n$ are the selected feature points. Since we select only a small number of feature points, n is much smaller than $m - 1$.

3.2 Solution by Least Square Minimization

According to Equation 3~5, we can solve this problem using general quadratic programming. In order to make this problem simpler, we reduce it to a least square problem. Equation 5 is equivalent to the following:

$$\begin{pmatrix} s_1(x_1) & \cdots & s_{m-1}(x_1) \\ \vdots & \ddots & \vdots \\ s_1(x_n) & \cdots & s_{m-1}(x_n) \end{pmatrix} \begin{pmatrix} \alpha_1 \\ \vdots \\ \alpha_{m-1} \end{pmatrix} = \begin{pmatrix} \tilde{S}(x_1) \\ \vdots \\ \tilde{S}(x_n) \end{pmatrix}. \tag{6}$$

To exploit the inherent orthogonal nature of the problem, we rewrite Equation 6 as:

$$\mathbf{S}\,\alpha' = \tilde{\mathbf{S}}, \tag{7}$$

where

$$\mathbf{S} = \begin{pmatrix} \sigma_1 s_1(x_1) & \cdots & \sigma_{m-1} s_{m-1}(x_1) \\ \vdots & \ddots & \vdots \\ \sigma_1 s_1(x_n) & \cdots & \sigma_{m-1} s_{m-1}(x_n) \end{pmatrix},$$

$$\alpha' = (\frac{\alpha_1}{\sigma_1}, \cdots, \frac{\alpha_{m-1}}{\sigma_{m-1}})^T,$$

$$\tilde{\mathbf{S}} = (\tilde{S}(x_1), \cdots, \tilde{S}(x_n))^T, \tag{8}$$

and the row vectors of $\mathbf{S}$ are assumed to be linearly independent. α' can be computed by

$$\alpha' = \mathbf{S}^+ \tilde{\mathbf{S}}, \tag{9}$$

where $\mathbf{S}^+$ is the pseudoinverse of the matrix $\mathbf{S}$, and can be obtained easily using a singular value decomposition as follows[5].

Supposing the singular value decomposition of $\mathbf{S}$ is

$$\mathbf{S} = U\, W\, V^T, \tag{10}$$

the pseudoinverse of $\mathbf{S}$ is

$$\mathbf{S}^+ = V\, W^+\, U^T. \tag{11}$$

The columns of U are eigenvectors of $\mathbf{S}\mathbf{S}^T$, and the columns of V are eigenvectors of $\mathbf{S}^T\mathbf{S}$. The main diagonals of W are filled with the square roots of the nonzero eigenvalues of both. In W^+, all nonzero elements of W are replaced by their reciprocals.

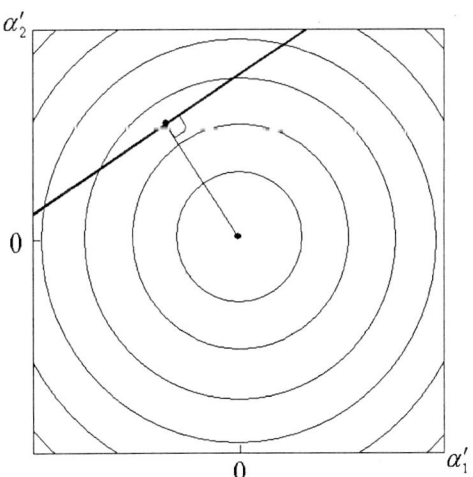

Fig. 1. The Example of least square minimization in 2D-α' space

Figure 1 represents the process described above in 2D-α' space. This is the case that, as we previously assumed, the row vectors of $\mathbf{S}$ are linearly independent, and the number of the bases m-1 and the feature point n are 2 and 1, respectively. Circular contour lines designate two-dimensional probability plots of Gaussian, $P(\alpha)$. The solid line represents the condition given in Equation 5. The point at which the energy function is minimized can be obtained by finding the point which has the minimum distance from the origin.

Using Equations 1, 8 and we obtain

$$S = \bar{S} + \sum_{i=1}^{m-1} \alpha'_i \sigma_i s_i. \qquad (12)$$

By using Equation 12, we can get correspondence of all points. Similarly, we can construct full texture information T. We previously made the assumption that the row vectors of **S** in Equation 8 are linearly independent. Otherwise, Equation 5 may not be satisfied. In other words, the correspondence obtained from Equation 12 may be inaccurate not only between the feature points but also at the feature points. Therefore, for our purpose of effectively reconstructing a face image from a few feature points, selecting the feature points that are linearly dependent would not be appropriate. However, this is unlikely to happen.

4 Experimental Results

For testing the proposed method, we used 200 two-dimensional images of human faces that were rendered from a database of three-dimensional head models recorded with a laser scanner($Cyberware^{TM}$)[1] [6]. The face images had been collected for psychophysical experiments from males and females between twenty and forty years old. They wear no glasses and earrings. Males must have no beard. The resolution of the images was 256 by 256 pixels and the color images were converted to 8-bit gray level images. The images were generated under controlled illumination conditions and the hair of the heads was removed completely from the images. PCA is performed on a random subset of 100 face images. The other 100 images are used to test the algorithm.

4.1 Selection of Feature Points

For face reconstruction from feature points, we first set the number and location of the feature points to be used in the reference face. The feature points from all faces can be automatically extracted by using the given correspondence once the feature points have been selected in the reference face. In Figure 2, the white cross points represent the feature points selected for shape reconstruction. For reconstruction of texture information, additional points are selected, and they are represented by black cross points in the figure. In order to reduce errors caused by noise(e.g. salt and pepper noise), the mid value of the p by p mask that runs on each point is obtained and collected as texture information. In our experiments, 22 feature points are chosen for shape reconstruction and 3 more for texture reconstruction, and A 3 by 3 mask is used for error reduction.

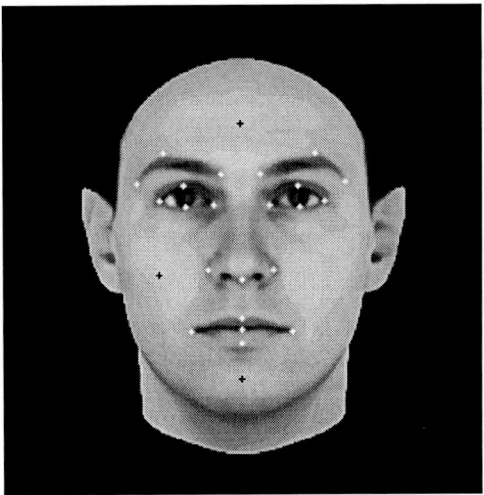

Fig. 2. The selected feature points

4.2 Reconstruction of Shape and Texture

For testing the algorithm, we use the fact that we already have the correct correspondence for the test images, and assume that we knew only the location of the feature points. Using the correspondence we can automatically extract the feature points from all faces, once the feature points have been selected in the reference faces. As mentioned before, 2D-shape and texture of face images can be treated separately. Therefore, A face image can be constructed by combining shape and texture information after reconstruction of both information. In Figure 3, the shape information of the face images is reconstructed from correspondence at the feature points. Instead of texture extraction, we used the standard texture of the reference face image. The images on the left are the original face images and those on the right are the face images reconstructed by the proposed method. Only the images of the inside of the face region are presented in the figure for the convenience of comparison of the reconstructed face images with the original ones. Figure 4 shows face images reconstructed both from shape and texture information at feature points. In contrast to the reconstructed face images in Figure 3, where texture is the same as that of the reference face image, it can be easily noticed that the brightness of the skin is successfully reconstructed.

5 Conclusions and Future Work

In this paper, we have proposed an efficient face reconstruction method from a small set of feature points. The proposed method uses a strategy that minimizes face deformation, provided that the shape and texture of feature points are perfectly reconstructed. As experimental results, well-reconstructed 2D face images

314   B.-W. Hwang et al.

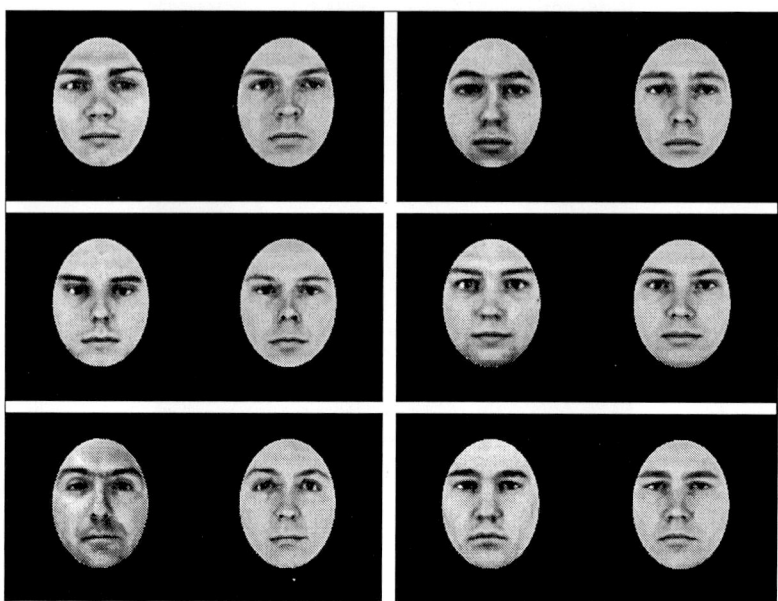

Fig. 3. Examples of shape reconstruction(left: originals, right: reconstructions)

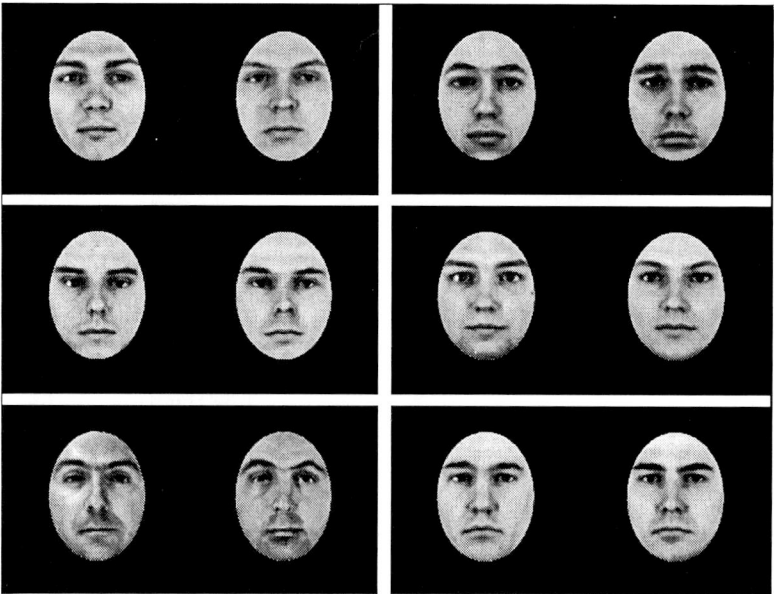

Fig. 4. Examples of shape and texture reconstruction(left: originals, right: reconstructions)

similar to original ones are obtained. However, the feature points are selected heuristically. They need to be automatically chosen linearly independent for the proposed method. Therefore, an elegant algorithm of selecting feature points is required for efficient reconstruction. The proposed method is expected to be useful for reconstruction of information reduced or partially damaged.

Acknowledgement

The authors would like to thank Max-Planck-Institute for providing the MPI Face Database.

References

1. Blanz, V. and Vetter, T.: A morphable model for the synthesis of 3D faces. Proc. of SIGGRAPH'99, Los Angeles (1999) 187–194
2. Jones, M., Sinha, P., Vetter, T. and Poggio, T.: Top-down learning of low-level vision tasks[Brief communication]. Current Biology **7** (1997) 991–994
3. Kruger, N., Potzsch, M., von der Malsburg, C.: Determination of face position and pose with a learned representation based on labelled graphs. Image and Vision Computing **15** (1997) 665–673
4. Lanitis, A., Taylor, C. and Cootes, T.: Automatic interpretation and coding of face images using flexible models. IEEE Trans. on Pattern Anaysis and Machine Intelligence **19:7** (1997) 743–756
5. Press, W., Teukolsky, S., Vetterling, W. and Flannery, B.: Numerical reciples in C, Cambridge University Press, Port Chester, NY (1992)
6. Vetter, T. and Troje, N.: Separation of texture and shape in images of faces for image coding and synthesis. Journal of the Optical Society of America A **14:9** (1997) 2152-2161

Modeling Character Superiority Effect in Korean Characters by Using IAM

Chang Su Park and Sung Yang Bang

Dept. of Computer Engineeing, POSTECH, Pohang, Kyoungbuk, Korea
bylkuse@postech.ac.kr

Abstract. Originally the Interactive Activation Model(IAM) was developed to expalin Word Superiority Effect(WSE) in the English words. It is known that there is a similar phenomena in Korean characters. In other words people perceive a grapheme better when it is presented as a component of a character than when it is presented alone. We modified the orginal IAM to explain the WSE for Korean characters. However it is also reported that the degree of WSE for Korean characters varies depending on the type of the character. Especially a components was reported to be hard to perceive even though it is in a context. It was supposed that this special phenomenon exists for WSE of Korean characters because Korean character is a two-dimensional composition of components(graphemes). And we could explain this phenomenon by introducing weights for the input stimulus which are calculated by taking into account the two-dimensional shape of the character.

1 Introduction

Word Superiority Effect(WSE) means a phenomenon that human can perceive an alphabet better when it is given in the context of a word that when it is given alone or in the context of non-word [1]. In 1969 Reicher showed by using an enforced selection method that WSE really exists and it is not the result of guess after the cognitive process [2]. Since then there have been many researches done on WSE. Especially in 1981 McClelland and Rumelhart proposed a cognitive model called Interactive Activation Model(IAM) in order to explain various aspects of WSE [3]. The key idea of IAM is that the human perception depends on the context. This model is important because not only it explains well many aspects of WSE but also it gives a cognitive model of character or word recognition.

Recently Kim proposed a cognitive model of Korean character recognition [4]. Among his extensive experiments he performed psychological experiments about WSE on Korean characters. Before we summarize his research results, a brief summary on Korean characters is due.

A Korean character consists of three components: the first grapheme, the middle grapheme and the last grapheme which is optional. A Korean character is of a two dimensional composition of these three graphemes. The six structural types of Korean characters with their examples are given in Table 1. The first grapheme is a consonant and located on the left, the top or the left upper corner of a character. The middle grapheme is a vowel. There are three types of the graphemes as seen in the Table 1. It is either a horizontal one, a vertical one or the combination of them. The last grapheme is a consonant and always located in the bottom of a character if it exists.

Table 1. Six types of Korean characters

Type	1	2	3	4	5	6
Structure	C V	C / V	C / V	C V / C	C / V / C	C / V / C
Example	가	고	과	각	곡	곽

Naturally we have words, in Korean language, which are sequences of characters. But forget words for the time being and correspond Korean graphemes to English letters and Korean characters to English words. As seen in the Table 1, the structure of a Korean character is different from that of an English word. An English Word is a one-dimensional arrangement of alphabets but the structure of a Korean character is a two-dimensional composition of graphemes.

It is known that WSE exists also for Korean characters. But in case of Korean characters the phenomenon is called Character Superiority Effect(CSE) rather than WSE because of the obvious reason mentioned above. But in this paper we will use these two terms interchangeably whenever no confusion is worried. The Kim's experiment result is given in Table 2. Now we will describe how to use IAM in order to explain CSE in Korean characters.

Table 2. CSE on Korean characters(%)

Charcter Type	Type of Grapheme Presented	First grapheme	Middle grapheme	last grapheme
Type 4	alone	77	72	77
	in character	91	86	81
Type 5	alone	65	79	66
	in character	77	79	78

2 Modified IAM for Korean Characters

Here we proceed assuming the reader is familiar with the original IAM because of the lack of space. The overall structure of the original IAM is given Fig. 1 a). In order to adapt this model to Korean characters we have to redefine a node of each level. The task is straightforward. By referring to the explanation about Korean characters, the word level in English should corresponded to the character level in Korean, the letter level to the grapheme level. But the meaning of the feature level is essentially the same. The redefined model is given in Fig. 1 b).

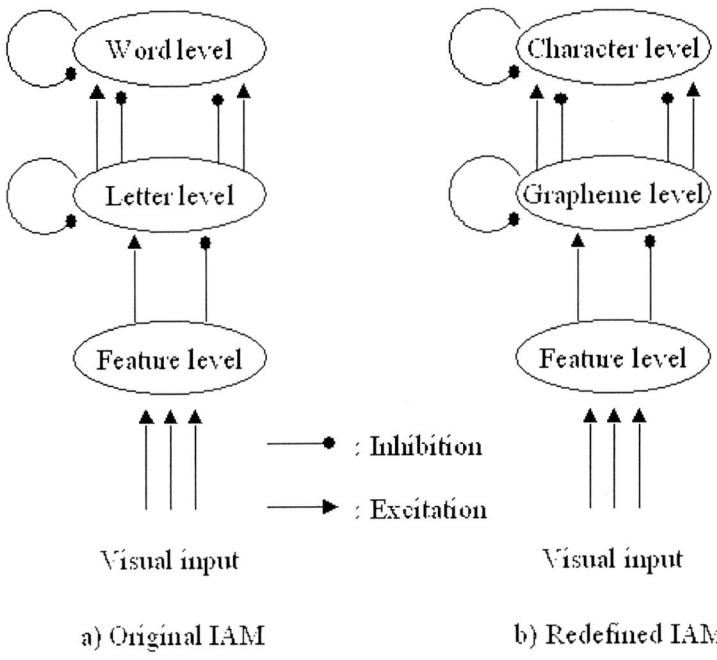

Fig. 1. Basic structure of IAM

In the original IAM each input alphabet is formalized and those given in Fig. 2 are used. In other words any alphabet is represented by a combination of 14 lines segments shown in the rectangular of Fig. 2. Their assumption as that each alphabet is correctly perceived and therefore each line segment has either 1 or 0 as the stimulus to the letter level. We developed a similar formalization of Korean graphemes which are given in Fig. 3. Since the purpose of the current study is to see whether or not the original IAM can be used in order to explain WSE on the Korean characters, we decided to use just 5 consonant graphemes, 5 vertical vowel graphemes and 5 horizontal vowel graphemes as shown in Fig. 3 a), b) and c), respectively. These are obviously a subset of Korean consonant and

Fig. 2. Formalized alphabets in English

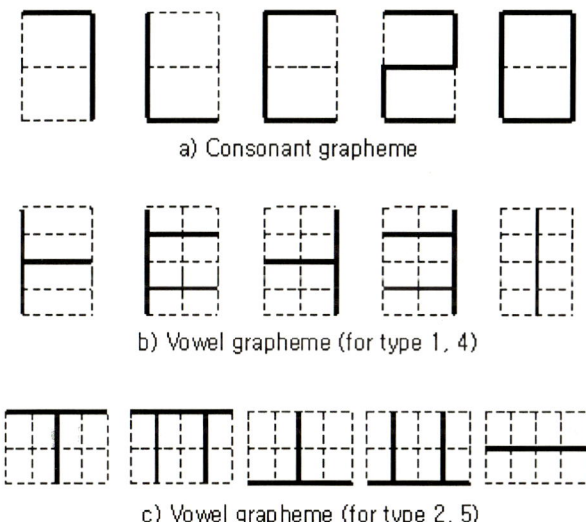

Fig. 3. Formalized Korean graphemes

vowel graphemes. However we don't think this limitation of the set of graphemes makes useless our effort to use IAM to explain WSE in Korean characters.

Further in this study we used only the character types 1, 2, 4 and 5. In other words we excluded the types 3 and 6. We think the types 1, 2, 4 and 5 are enough to represent the compositional property of Korean characters since the type 3 is the combination of the types 1 and 2 and the type 6 of the types 4 and 5. Further it should be noted that the number of the characters of types 3 and 6 is only 6% of the most frequently used 600 characters. In the original IAM the information about the frequency of the each word is faithfully reflected at the word level node. But in our IAM we simply divided the set of all characters included in the model into two groups: high frequent and low frequent characters. Also note that in our simulation actually there are four different model: one for each character type.

3 Simulation I

Table 3. Result of simulation I (%)

Character Type	Type of Grapheme Presented	First grapheme	Middle grapheme	Last grapheme
Type 1	Alone	82.3	80.0	
	In character	83.9	82.6	
Type 2	Alone	82.3	80.2	
	In character	83.9	82.5	
Type 4	Alone	83.1	80.6	82.9
	In character	88.9	86.5	88.9
Type 5	Alone	82.9	80.3	82.9
	In character	88.7	86.6	88.7

Our IAM revised for Korean characters is essentially the same as the original IAM with minor adjustments of parameters. We performed the simulation of our IAM for various inputs. The activity of each node is approximately represented as follows.

$$a_i(t+1) = a_i(t) - \theta_i(a_i(t) - r_i) + f\left(\sum c_j a_j(t)\right) \quad (1)$$

where
- $a_i(t)$: the activation value of the node i at time t
- θ_i : decay constant
- r_i : resting value
- c_j : connection weight with node j which takes into account either the excitation or the inhibition
- f : function to control the influence from the other nodes.

We measured the value of each node after 30 iterations which correspond to the time duration during which a stimulus is exposed to a subject.

The simulation results are summarized in Table 3. As seen in the Table 3, we have the same WSE which was reported in Table 2. In case of the character types 1 and 2, however, the performance differences are less obvious than those in case of the character types 4 and 5.

As in the original IAM paper, we tried the simulation of a case where one of the line segment in the input is unseen. For example, suppose that the position of a line segment of the last grapheme be covered by ink(see Fig. 4). Therefore it is not sure whether the line exists or not. The activities of the possible graphemes in this case are shown in Fig. 5. Since only the one line segment is unknown, only two grapheme are actually possible. Then, as time goes, one of them wins because the character with the last grapheme is marked as more frequently use.

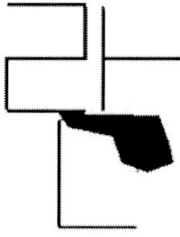

Fig. 4. Incomplete input

4 Modification of the Input Stimulus

So far we showed that we can explain CSE of Korean characters by using IAM with some adaptation. But there is a phenomena of CSE which is unique Korean characters and cannot be explained by using the original IAM implementation.

Kim reported that there is some meaningful difference between CSE on the types 4 and that on the type 5 [5]. As seen in Table 2 CSE appears less definitely in case of the vowel graphemes of the type 5 than in case of those of the type 4. In order to explain this phenomena he hypothesized that the stimulus strength differs depending on the distance from the character shape boundary. In other words line segments closer to the character shape boundary are seen better while line segments closer to the center of the character shape are not seen well. This difference cannot be found in the results of the simulation I. It seems that this is unique to Korean characters which have a two-dimensional composition.

Based on his hypothesis we changed the stimulus strength of each line segment in the input. First we divided an input space into small rectangles and we assigned to each a value v which depends on the distance from the input space boundary as follow.

$$v = \sum_{b \in B} e^{-(|x-x'|+|y-y'|)^c} \qquad (2)$$

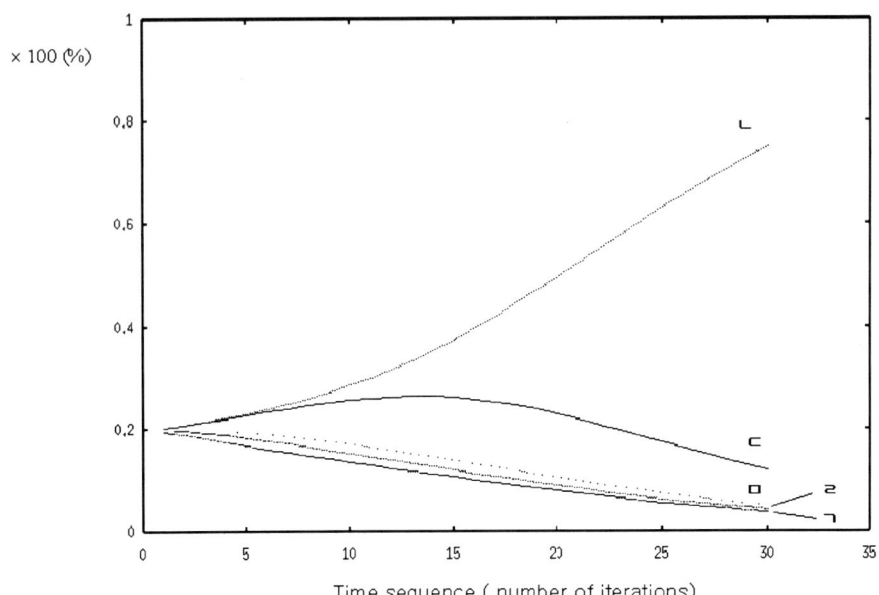

Fig. 5. Result of incomplete input

where

    c : constant
    b : rectangles comprising the character boundary B
    x : x position of the rectangle
    y : y position of the ractangle
    x' : x position of the b
    y' : y position of the b

The visualization graph of v by this formula is shown Fig. 6.

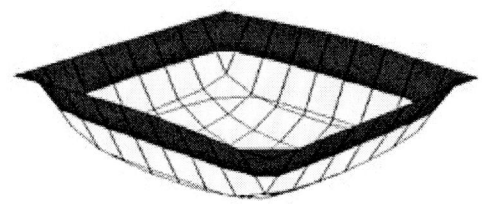

Fig. 6. The value v and the character boundary(shaded)

Then the stimulus strength F of each line segment is given by adding the value v's of the rectangles which are covered by the line segment as follows.

$$F = \beta + K \times \sum_{s \in L} v(s) \qquad (3)$$

where
  s : rectangles comprising the line segment L
  K : constant
  β : base value

The visualization of the v value of the rectangle covered by a line segment is shown Fig. 7.

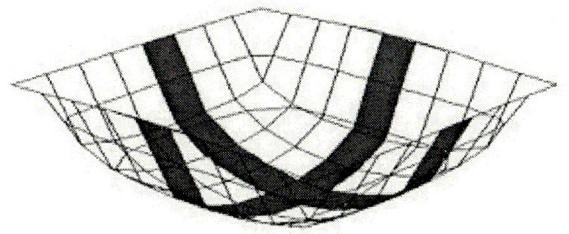

Fig. 7. Example of the rectangles covered by line segment

5 Simulation II

We performed the simulation again using the same IAM but in case using the different input stimulus strengths. The result of the simulation is given in Table 4. As seen in Table 3 the simulation I shows no significant difference between the CSE of the character types 4 and 5. But in case of simulation II we can observe the same difference as in Table 2.

The results for various values of K and /beta are given in Table 5. As seen in the Table 5 the results are not much sensitive to the values of K and /beta.

6 Conclusion

A Korean character consists of two or three graphemes. It is known that we have Character Superiority Effect for Korean characters. In order to use IAM to explain CSE we modified the original IAM by replacing English letters by Korean graphemes and English words by Korean characters. And we were successful in doing so with minor parameter adaptation.

Table 4. Result of simulation II (%)

Character Type	Type of Grapheme Presented	First grapheme	Middle grapheme	Last grapheme
Type 1	Alone	82.8	79.4	
	In character	84.5	81.4	
Type 2	Alone	82.8	79.5	
	In character	84.8	81.3	
Type 4	Alone	83.3	79.8	83.2
	In character	88.7	85.2	89.9
Type 5	Alone	83.2	79.5	83.2
	In character	90.4	81.2	90.3

Table 5. Result with various values of β and K (%)

Character Type	Value of β and K Presented	$\beta = 0.25$ $K = 0.05$ F.G. M.G. L.G.	$\beta = 0.20$ $K = 0.10$ F.G. M.G. L.G.	$\beta = 0.10$ $K = 0.20$ F.G. M.G. L.G.
Type 1	Alone	82.7 79.5	82.8 79.4	82.8 79.3
	In character	84.4 81.7	84.5 81.4	84.6 81.2
Type 2	Alone	82.7 79.7	82.8 79.5	82.8 79.4
	In character	84.6 81.6	84.8 81.3	84.9 81.1
Type 4	Alone	83.3 79.9 83.2	83.3 79.8 83.2	83.4 79.6 83.3
	In character	88.7 85.5 89.8	88.7 85.2 89.9	88.6 85.0 90.0
Type 5	Alone	83.2 79.6 83.1	83.2 79.5 83.2	86.2 79.3 83.2
	In character	90.2 82.4 90.0	90.4 81.2 90.3	90.6 80.3 90.4

- F.G. : first grapheme, M.G. : middle grapheme, L.G. : last grapheme

But it is also known that there is a minor deviation in CSE which is unique to Korean characters. It was conjectured that this special phenomena come from the fact that a Korean character has a two-dimensional composition. In other words, the perception accuracy of each line segment in a grapheme depends on the distance from the character shape boundary. We took this view and assigned a different stimulus strength to each line segment of the input based on the input based on a formula. Then we could obtain the same simulation result as that of the human experiment.

As the original IAM provided a cognition model of word recognition, it is hoped that the IAM presented here can provide a base for a cognition model of Korean character recognition.

References

[1] Robert J. Sternberg, Cognitive Psychology, Harcourt Brace & company, 1996.
[2] Reicher G. M., Perceptual recognition as a function of meaningfulness of stimulus material, Journal of Experimental Psychology, 1969, 274-280.

[3] McClelland J. L., Rumelhart D. E., An Interactive Activation Model of Context Effects in Letter Perception: Part 1. An Account of Basic Findings, Psychological Review, volume 88, number 5, september, 1981, 375-407.
[4] J. K. Kim, J. O. Kim, Grapheme cognition in Korean character context , Ph. D. Thesis (in Korean), Seoul National University, 1994.
[5] Peter T. Daniels, William Bright, The Worlds Writing Systems, Oxford University Press, 1996, 218-227.
[6] Baron J., Thurston I., An analysis of the word-superiority effect, Cognitive Psychology, 1973, 4, 207-228.
[7] Kathryn T. Spoehr, Stephen W. Lehmkuhle, Visual Information Processing, W. H. Freeman and Company, 1982, 133-161.
[8] Florin Coulman, TheBlackwell Encyclopedia of Writing Systems, Blackwell, Oxford, 273-277.
[9] Geoffrey Sampson, Writing Systems, Hutchinson, London, 1985, 120-144.
[10] M. S. Kim, C. S. Jung, Cognitive of grapheme and character by grapheme composition form in Korean, Cognitive Science, Korean Cognitive Science Association, 1989, 1, 27-75.

Wavelet-Based Stereo Vision

Minbo Shim

General Dynamics Robotic Systems, Inc.
1234 Tech Court, Westminster MD 21157, USA
mshim@gdrsi.com

Abstract. Multiresolution frameworks have been embraced by the stereo imaging community because of their human-like approach in solving the correspondence problem and reconstructing density maps from binocular images. We describe a method to recover depth information of stereo images based on a multi-channel wavelet transform, where trends in the coefficients provide overall context throughout the framework, while transients are used to give refined local details into the image. A locally adapted lifting scheme is used to maximize the subband decorrelation energy by the transients. The coefficients in each channel computed from the lifting framework are combined to measure the local correlation of matching windows in the stereogram. The combined correlation yields higher cumulative confidence in the disparity measure than using a single primitive, such as LOG, which has been applied to the traditional area-based stereo techniques.

1 Introduction

Stereo vision is a key component for unmanned ground vehicles (UGV). In order to have autonomous mobility, the vehicle must be able to sense its environment. Generating a dense range estimate to objects in world is critical for UGV's to maneuver successfully to the target. Fig. 1 shows a prototype of the DEMO III Experimental Unmanned Vehicle (XUV) with the autonomous mobility sensors labeled. The DEMO III XUV is a small, survivable unmanned ground vehicle capable of autonomous operation over rugged terrain as part of a mixed military force containing both manned and unmanned vehicles.

In this paper, we propose a stereo algorithm that has an inter-scale backtracking method as well as area-based intra-scale cross correlation algorithm. The proposed stereo algorithm may facilitate the obstacle detection and avoidance mechanisms embedded in the autonomous UGV's under developing. Many traditional multiresolution stereo algorithms use inter-scale area-based correlation measure to generate disparity maps. Certainty maps at each scale are computed and passed to next finer levels in a constrained and interpolated form. However, they do not have a coarse-to-fine backtracking mechanism between images at two contiguous scales. It has been known that the inter-scale backtracking of coefficients in the transform domain improves the performance and the reliability of the intra-scale matching since the backtracking in a wavelet scale space avoids the ambiguities by precisely localizing large-scale elements, and exploiting a mathematical bijection between

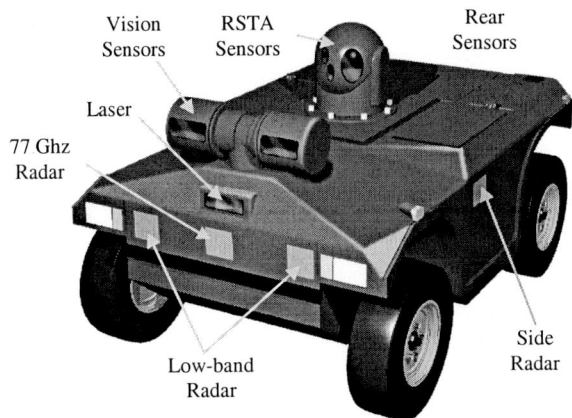

Fig. 1. A prototype of DEMO III XUV with autonomous mobility sensors.

levels to identify corresponding inter-scale elements. This enables us to reduce the complexity of matching and consequently decrease the chances of errors.

We present in Section 2 the wavelet representations and the lifting scheme, and in Section 3 we propose a new multiresolution stereo algorithm.

2 Wavelet Transform and Representations

Both multiorientation and multiresolution are known features of biological mechanisms of the human visual system. There exist cortical neurons that respond specifically to stimuli within certain orientations and frequencies. The retinal image is decomposed into several spatially oriented frequency channels [5]. Wavelet analysis has emerged in the image processing community to give a precise understanding to the concept of the multiresolution.

Wavelet transform is a decomposition of a signal into a wavelet family $\psi_{a,b}(x)$ that is comprised by translations and dilations of a base wavelet $\psi(x)$. In addition to the time-frequency localization, the wavelet transform is able to characterize a signal by its local regularity [7]. The regularity of a function is decided by the decay rate of the magnitude of the transform coefficients across scales. Tracking the regularity helps to identify local tunes (signatures) of a signal. The variation of resolution enables the transform to focus on irregularities of a signal and characterize them locally. Indeed, wavelet representations provide a natural hierarchy to accomplish scale space analysis.

The lifting scheme was introduced as a flexible tool for constructing compactly supported second generation wavelets which are not necessarily translates and dilates of one wavelet function [14]. The lifting allows a very simple basis function to change their shapes near the boundaries without degrading regularities and build a better performing one by adding more desirable properties. Traditionally multiresolution analysis is implemented through a refinement relation consisting of a scaling function and wavelets with finite filters. Dual functions also generate a multiresolution

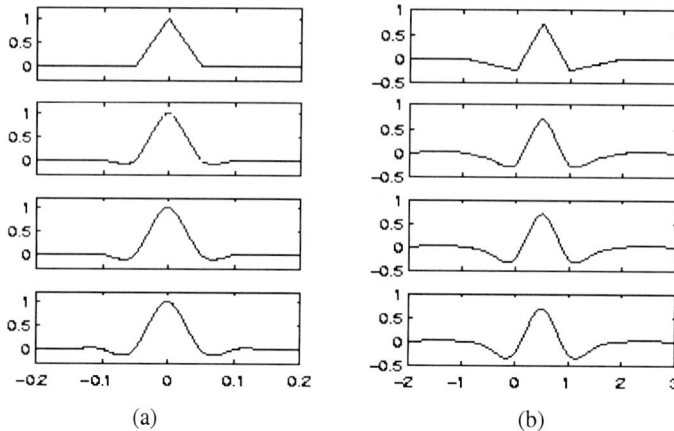

Fig. 2. (a) Interpolating scaling functions resulting from interpolating subdivisions of order 2, 4, 6 and 8 (top to bottom), and (b) the associated wavelet functions with two vanishing moments.

analysis with dual filters. A relation of biorthogonal filters established from the Vetterly-Herley lemma leads to defining the lifting scheme:

$$\varphi(x) = 2\sum_k h_k \varphi(x-k)$$

$$\psi(x) = \psi^o(x) - \sum_k s_k \varphi(x-k)$$

$$\tilde{\varphi}(x) = \tilde{\varphi}^o(x) + \sum_k s_{-k} \tilde{\psi}(x-k)$$

$$\tilde{\psi}(x) = 2\sum_k \tilde{g}_k \tilde{\varphi}(x-k)$$

1)

where $\{h, \tilde{h}^o, g^o, \tilde{g}\}$ is an initial set of finite biorthogonal filters, $\{\varphi, \tilde{\varphi}^o, \psi^o, \tilde{\psi}\}$ the associated biorthogonal wavelets, and s denotes a trigonometric polynomial. The lifting formula 1) implicates that the polynomial s controls the behavior of the wavelet and the dual scaling function. Fig. 2 (a) and (b) show interpolating scaling functions with subdivision of order 2, 4, 6 and 8, and associated wavelet functions, respectively. The standard wavelet transforms usually suffer from boundary artifacts if the length of input is not a power of two, a common problem associated with digital filters operating on a finite set of data. There have been many attempts at minimizing this boundary effect such as zero padding, symmetric extension, and periodization. These workaround approaches are effective in theory but are not entirely satisfactory from a practical viewpoint. However the polynomials reproduced by the lifting scheme are able to adapt them to the finite interval without degrading their regularities. The boundary artifacts appeared thus disappear in the lifting framework. Fig. 3 (a) and (b) respectively show examples of cubic polynomial scaling functions and the associated wavelets accommodating the left boundary with the same regularity throughout the interval. The biorthogonal limitation that the original lifting scheme had was lifted and generalized by factorizing the wavelet transform into a finite number of lifting

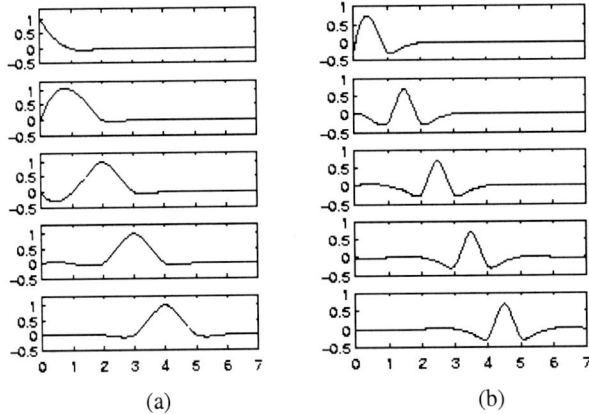

Fig. 3. Examples of the adaptability near the left boundary for **(a)** the cubic interpolating scaling function and **(b)** the associated wavelet.

steps based on Euclidean algorithm [2]. The generalized lifting algorithm can thus be utilized to construct any wavelet and wavelet transform. The studies [2], [14], [13] also showed that the lifting scheme enhances the performance of the wavelet transforms by a factor of two over the standard transforms.

Fig. 4. shows the wavelet coefficients of one of a stereo snapshot taken from our XUV.

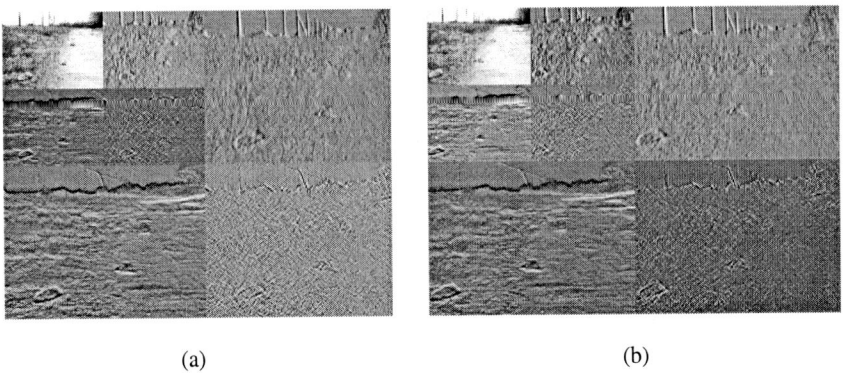

Fig. 4. Two-level wavelet coefficients of **(a)** left and **(b)** right stereo image.

3 Multiresolution Stereo Algorithm

One of the fundamental problems in the stereopsis is finding corresponding points between two stereo images taken from slightly differing viewpoints. Most matching approaches can be classified into two categories: area-based and feature-based matching. Area-based techniques rely on the assumption of surface continuity, and often involve some correlation-measure to construct a disparity map with an estimate of disparity for each point visible in the stereo pair. In contrast, feature-based techniques focus on local intensity variations and generate depth information only at points where features are detected. In general, feature-based techniques provide more accurate information in terms of locating depth discontinuities and thus achieve fast and robust matching. However they yield very sparse range maps and may have to go through perhaps expensive feature extraction process. Area-based methods on the other hand produce much denser results, which is critical in the obstacle detection and avoidance. This is why most real-time stereo algorithms utilized in the modern unmanned ground vehicles (UGV's) are the area-based techniques. In addition, the area-based approaches can be much easily optimized because of its structural regularity.

The depth information of the traditional stereo system to an object in the world is calculated by $r = Bf/D$, where B denotes the baseline between two cameras, f the focal length, D the disparity and r the depth to the object. In our current stereo camera setup, $B = 1219.2$ mm and $f = 8$ mm. Fig. 5 plots the relationship between the range from 3 to 50 meters in front of the vehicle and the disparity that varies from 230 to 8 pixels. Traditional area-based stereo algorithms would traverse in the reference image both 222 pixels left and right from the reference point in the correlation support window. However zero-disparity methods: horopter-based stereo [1], [3], allow us to limit the search range to certain interval, which reduces the computational complexity by the factor of reduction ratio of the disparity searching window size. the tilted horopter that provides more sensitive observation window in this type of stereo geometry can be used in order to extract more distinctive depth information of an object seen in a stereogram [1]. One other element that could be used in the stereo analysis is the range resolution. It provides differential information of the current range map. The range resolution is defined by

$$\nabla r = \left| \frac{d}{dD}\left(\frac{Bf}{D}\right) \right| = \frac{r^2}{Bf}\nabla D \qquad 2)$$

Fig. 6 displays the range resolution. From the result, the range resolution at 3 *m* is about 9 *mm*, and at 50 *m* it is about 5225 *mm*.

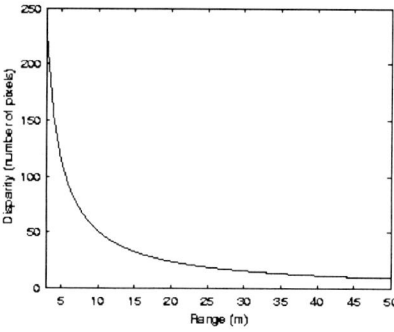

Fig. 5. Disparity in number of pixels as a function of range.

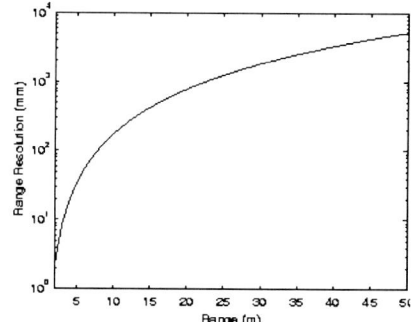

Fig. 6. Range resolution in logarithmic scale.

Our stereo algorithm is motivated by the biological mechanisms of the human visual processing: multiresolution and multiorientation [9], [10], described in the previous section. A common thread among most approaches is Marr-Poggio's model where a set of filters, Laplacian of Gaussian (LOG) of different bands, is applied to each stereo image and then some matching primitives are extracted. The LOG filtering is known to enhance image features as well as alleviating external distortions. Matching is carried out, and the disparities obtained from coarse filtering are used to align the images via vergence of eye movements, thereby causing the finer images to come into the range of correspondence. Instead of using LOG, the proposed algorithm uses wavelet coefficients to accomplish the coarse-to-fine incremental matching in the scale space. Wavelet analysis is known to give a mathematically precise definition to the concept of multiresolution.

The multiresolution stereo algorithm has two main features to improve the reliability of matching by applying both 1) intra-scale correlation measure based on the wavelet multimodality and 2) inter-scale backtracking. They make the best of the properties inherited within multiresolution representations. There have been various correlation-based methods to solve the intra-scale matching problem such as SAD (sum of absolute difference) [11], [3], SSD (sum of squared difference), WSSD (weighted SSD) [1], and SSSD (sum of SSD) [12], [15]. The SSSD was especially designed for multi-baseline camera models. For the intra-scale matching, we adopt a multi-modality sum of SSD (MSSSD). The 2-D dyadic wavelet transform has specific

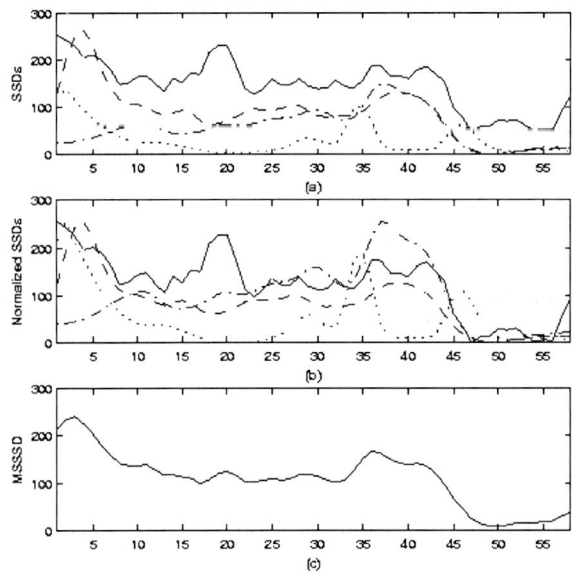

Fig. 7. (a) Each graph plots SSD's of characteristic wavelet coefficients. (b) Normalized version of (a). (c) MSSSD (Multimodality Sum of SSD). In (a) and (b), the dashed dot line plots the wavelet coefficients in the horizontal direction, the dotted line the coefficients in the vertical direction, the solid line the modulus, and the dashed line the angle. The horizontal axis in all these plots is disparity.

orientation tunings, horizontal and vertical, by a set of wavelets $\Psi^k(x, y) \in \mathbf{L}^2(\mathbf{R}^2)$, where $1 \le k \le 2$. The two directional wavelet coefficients form a gradient vector from which we define the angle and the modulus [8]. We call each of the characteristic wavelet representations a modality. Thus MSSSD is to compute SSSD from the distinctive wavelet coefficients of a pair of stereo images. Fig. 7 demonstrates that the MSSSD locates the global minimum that corresponds to the true disparity (48 pixel) much better than each modality alone.

The inter-scale backtracking takes advantages of the fact that the wavelet coefficients are related between scales by regularity properties [7]. Fig. 8 is an example of how the wavelet coefficients from a 1-D scan profile of the left stereo image (shown in Fig. 6(a)) are organized and tracked over a scale space. Fig. 8(b) shows the wavelet coefficients evolving as the scale increases from bottom to top. We observe here that there are spatial distortions such as dislocation, broadening and flattening of features that survive within a level. These features are tracked from their coarse representations and remain localized between levels. For this inter-scale tracking, we adopt the area-based correlation method used in the intra-scale matching. However the way to implement the correlation between two contiguous scales has to take into account the property of the multiresolution framework that is used to generate the transform coefficients.

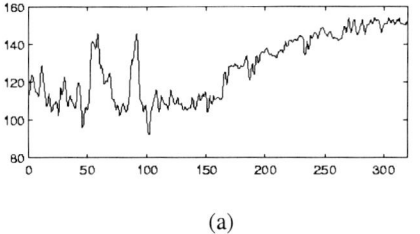
(a)

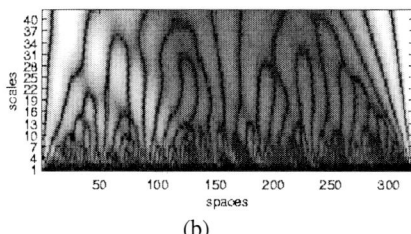
(b)

Fig. 8. Example of scale-space contours of wavelet coefficients. (a) A scan line profile of signal. (b) Evolution of wavelet coefficients within a scale space.

The refinement relations defining the multiresolution are based on the interpolating subdivision [14], a recursive refining procedure to find the value of an interpolating function at the dyadic points. With this property in mind, the inter-scale correlation between the scale j and $j+1$ at a position (x, y) and a tracking offset (u, v) is thus be given by

$$C(x, y; u, v) = \sum_{k=-m}^{m}\sum_{l=-n}^{n} \gamma_{k,l}^{j}(x, y) \gamma_{k,l}^{j+1}(x+u, y+v) \qquad 3)$$

where γ^j is the coefficients at the scale j and $\gamma_{k,l}^{j}(x, y) = r(x + 2^j k, y + 2^j l)$. The upper bounds m and n define the resolution of the inter-scale correlation support window. Although Equation 3) is in the form of cross correlation, the MSSSD used in the intra-scale matching can also be applied here to provide accumulated confidence of the inter-scale backtracking. Fig. 9 illustrates an example that given 10 by 3 (in pixels) inter-scale tracking window and 16 by 5 searching area, the inter-scale

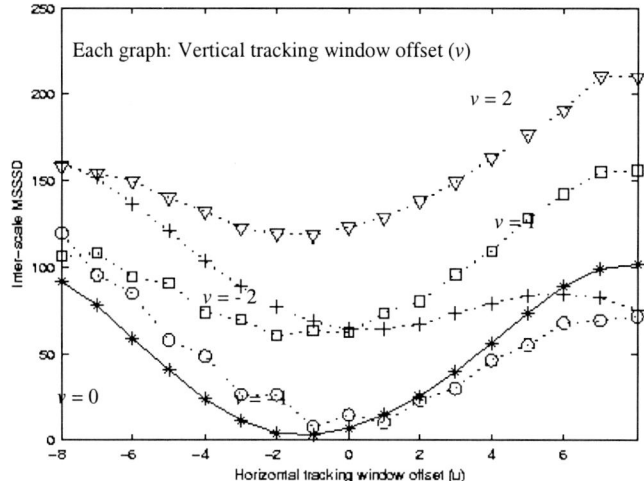

Fig. 9. Example of inter-scale backtracking method with normalized MSSSD.

MSSSD decides with the highest confidence that the offset $(u,v) = (-1,0)$ at the finer scale has the best match to the current reference point at the coarser level. Thus the wavelet coefficients extracted in the lower resolutions are precisely tracked back to the next finer level. Tracking in a wavelet frame scale space obviously avoids the ambiguities introduced by traditional methods and consequently reduces the chance of mismatching.

We show in Fig. 10 the schematic diagram of our proposed multiresolution stereo algorithm. First we rectify one image to adjust some changes introduced by sufficiently high angular and radial distortion of mechanical camera brackets. Then we find the horopter from the rectified images that defines the plane of zero disparity. We warp one image based on the horopter. We apply to the warped image wavelet

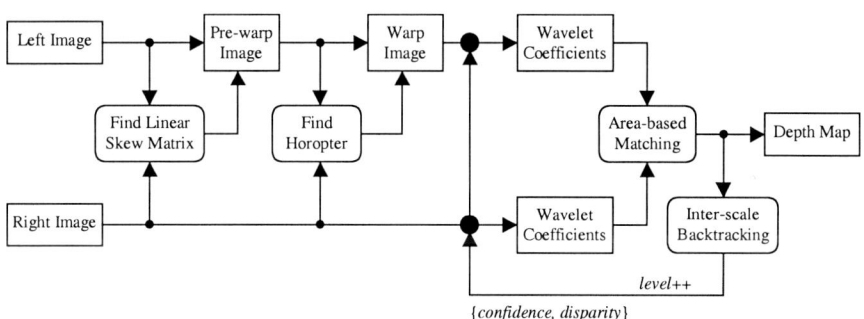

Fig. 10 Proposed Stereo Algorithm

certainty map is passed to the next finer resolution of processing along with the inter-scale correlation map, and the same steps are carried out until it reaches the finest level of resolution. Fig. 11 shows some preliminary results of the proposed stereo algorithm.

(a)   **(b)**   **(c)**

Fig. 11. Preliminary results of the proposed stereo algorithm using a locally adapted lifting scheme. **(a)** Rectified left image to adjust changes introduced by sufficiently high angular and radial distortion of mechanical camera brackets, **(b)** warped left image using two-plane horopter, and **(c)** disparity map.

4 Conclusion

This paper presents a new multiresolution stereo algorithm with wavelet representations. We introduced an intra-scale correlation method and an inter-scale backtracking technique using the Multimodality Sum of Sum of Squared Difference (MSSSD) on the transform coefficients. The results showed that the proposed stereo framework provides cumulative confidence in selecting corresponding points at two contiguous analyzing levels as well as within a scale. The inter-scale backtracking helped the intra-scale matcher to avoid unwanted chances of mismatching by removing the successfully tracked trusty points from the intra-scale matching candidacy. The proposed stereo algorithm as a result would enhance the matching performance and improves reliability over the traditional algorithms.

References

1. P. Burt, L. Wixson and G. Salgian. *Electronically directed focal stereo*, Proceedings of the Fifth International Conference on Computer Vision, pp. 94-101, 1995.
1. C. K. Chui, *Wavelets: A Tutorial in Theory and Applications*, Academic Press, San Diego, CA, 1992.
2. I. Daubechies and W. Sweldens, *Factoring Wavelet Transforms into Lifting Steps*, Preprint, Bell Laboratories, Lucent Technologies, 1996.
3. K. Konolige, Small Vision Systems: *Hardware and Implementation*, Proc. ISRR, Hayama, 1997.

4. K. Konolige, *SRI Stereo Engine and Small Vision Module*, Unpublished notes, http://www.ai.sri.com/~konolige/svm/.
5. S. Mallat, Time-frequency channel decompositions of image and wavelet models, IEEE Trans. ASSP, Vol 37, No. 12, pp. 891-896, 1989.
6. S. Mallat, *A theory for multiresolution signal decomposition: the wavelet representation*, IEEE Trans Pattern Analysis and Machine Intelligence, Vol. 11, pp. 674-693, 1989.
7. S. Mallat and W. Hwang, *Singularity detection and processing with wavelets*, IEEE Trans on Info Theory, Vol 38, No. 2, pp. 617-643, 1992.
8. S. Mallat and S. Zhong, *Characterization of signals from multiscale edges*, IEEE Trans. on Pattern Analysis and Machine Intelligence, Vol. 14, pp. 710-732, 1992.
9. D. Marr, *Vision*, W. H. Freeman and Company, 1982.
10. D. Marr and T. Poggio, A Theory of Human Stereo Vision, in Proceedings of Royal Society London, Vol B-204, pp. 301-328, 1979.
11. L. Matthies and P. Grandjean, *Stereo Vision for Planetary Rovers: Stochastic Modeling to Near Real Time Implementation*, Int. Journal of Computer Vision, Vol. 8, No. 1, pp. 71-91, 1992.
12. M. Okutomi and T. Kanade, *A multiple-baseline stereo*, IEEE Trans. on Pattern Analysis and Machine Intelligence, Vol. 15, No.4, pp. 353-363, 1993.
13. M. Shim and A. Laine, *Overcomplete Lifted Wavelet Representations for Multiscale Feature Analysis*, Proceedings of the IEEE International Conference on Image Processing, Chicago, IL, October, 1998.
14. W. Sweldens, *The Lifting Scheme: A New Philosophy in Biorthogonal Wavelet Constructions*, Wavelet Applications III, Proc. SPIE, San Diego, Vol 2569, pp. 68-79, 1995.
15. T. Williamson and C. Thorpe, *A Specialized Multibaseline Stereo Technique for Obstacle Detection*, Proceedings of the International Conference on Computer Vision and Pattern Recognition (CVPR '98), June, 1998.

A Neural Network Model for Long-Range Contour Diffusion by Visual Cortex

Stéphane Fischer, Birgitta Dresp, and Charles Kopp

Laboratoire de Systèmes Biomécaniques & Cognitifs
I.M.F., U.M.R. 7507 U.L.P.-C.N.R.S.
Ecole Nationale Supérieure de Physique
Boulevard Sebastien Brant
67400 Illkirch - Strasbourg
France
{stephane.fischer, birgitta.dresp, charles.kopp}@ensps.u-strasbg.fr
http://www.neurochem.u-strasbg.fr/fr/lsbmc/unite.html

Abstract. We present a biologically plausible neural network model for long-range contour integration based on current knowledge about neural mechanisms with orientation selectivity in the primate visual cortex. The network simulates diffusive cooperation between cortical neurons in area V1. Recent neurophysiological evidence suggests that the main functional role of visual cortical neurons, which is the processing of orientation and contour in images and scenes, seems to be fulfilled by long-range interactions between orientation selective neurons [5]. These long-range interactions would explain how the visual system is able to link spatially separated contour segments, and to build up a coherent representation of contour across spatial separations via cooperation between neurons selective to the same orientation across collinear space. The network simulates long-range interactions between orientation selective cortical neurons via 9 partially connected layers: one input layer, four layers selecting image input in the orientation domain by simulating orientation selectivity in primate visual cortex V1 for horizontal, vertical, and oblique orientations, and four connected layers generating diffusion-cooperation between like-oriented outputs from layers 2, 3, 4, and 5. The learning algorithm uses standard backpropagation, all processing stages after learning are strictly feed-forward. The network parameters provide an excellent fit for psychophysical data collected from human observers demonstrating effects of long-range facilitation for the detection of a target orientation when the target is collinear with another orientation. Long-range detection facilitation is predicted by the network's diffusive behavior for spatial separations up to 2.5 degrees of visual angle between collinear orientations.

1 Introduction

Mechanisms with orientation selectivity in the primate visual cortex (V1) are assumed to use diffusive cooperation between neurons to generate the perception of continuous visual contours [5] in images where only isolated segments, or fragments, of a contour can be seen (see Figure 1). Hubel & Wiesel (1968)[8] were the first to give a detailed account for orientation selective receptive field structures and related functional properties such as sensitivity to contrast and contrast polarity of visual neurons in monkey striate cortex. Recent neurophysiological evidence suggests that the main functional role of visual cortical neurons, which is the processing of orientation and contours in images, seems to be fulfilled by long-range interactions between orientation selective neurons [5]. These long-range interactions would explain how the visual system is able to link spatially separated contour segments, and to build up a coherent representation of contour across gaps by cooperation between neurons selective to the same orientation across collinear space, as well as to discard irrelevant segments by competition between neurons selective to different orientations. Such a hypothesis is confirmed by psychophysical data showing that collinear orientations in the image facilitate the detection of target-contours, whereas orthogonal orientations in the image suppress the detectability of these targets [2,11]. Kapadia, Ito, Gilbert, & Westheimer [9] have shown that these psychophysical effects correlate with the firing behaviour of orientation selective neurons in V1 of the awake behaving monkey.

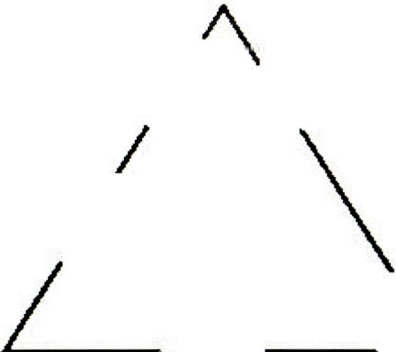

Fig. 1. The perceptual completion of visual contour in images where only isolated segments are presented is presumably based on long-range spatial interactions between visual cortical detectors [6].

2 A Micro-circuit Neural Network for Long-Range Contour Diffusion

Here, we present a micro-circuit-based neural network that simulates long-range interactions between orientation selective cortical neurons. The network has four partially connected layers: one input layer, a second layer selecting image input in the orientation domain by simulating orientation selectivity in primate visual cortex V1, a third layer generating diffusion-cooperation between like-oriented outputs from layer 2, and competition between orthogonally oriented outputs. The fourth layer generates the final output image. The simulation of collinear diffusion-cooperation (see Figure 2) and competition of orthogonally oriented inputs as modeled in layer 3 of our network is based on the neurophysiological evidence for orientation cooperation/competition between cortical cells in V1 [5] which can be seen as the neural substrate of contour filling-in across spatial gaps in the image [9,4]. The learning algorithm uses standard backpropagation, all processing stages after learning are strictly feed-forward. In comparison to more sophisticated neural networks for contour vision across gaps which use competitive learning algorithms via selection of inputs through winner-take-all rules [7], our model uses the most simple computational rule possible, based on the functional properties of the cortical micro-circuit that is needed to simulate local orientation diffusion. This simplicity in architecture and mechanisms

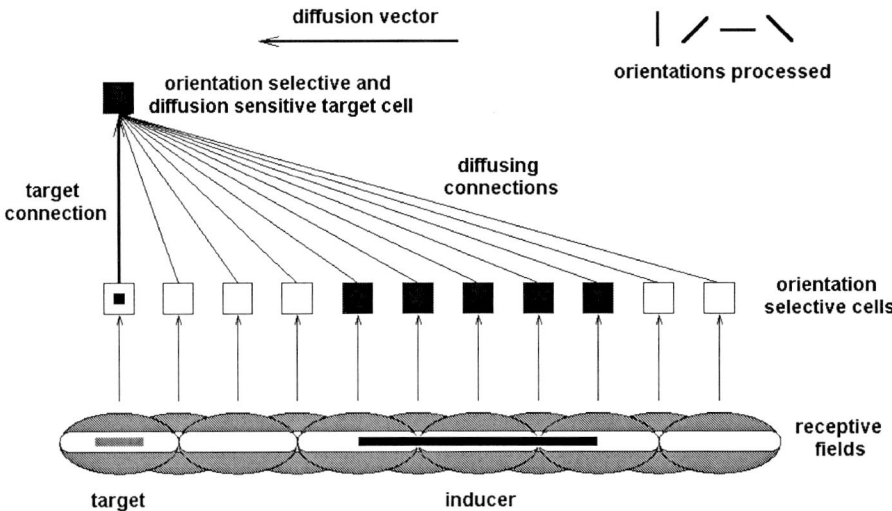

Fig. 2. Neural cooperation achieved by our network model via lateral connectivity is based on the neurophysiological evidence for cooperation between orientation selective cortical cells in V1, which can be seen as the neural substrate of contour completion across spatial gaps between oriented stimuli. A psychophysical correlate of the diffusive neural mechanism has been identified in detection facilitation of collinear targets and inducers [2,9].

allows the network to predict visual behavior with high precision at a strictly local level, as shown by the following psychophysical experiments reproducing contour detection procedures from previously published work on long-range contour integration [3].

3 Psychophysical Test of the Model

Psychophysical thresholds for the detection of line targets have been found to decrease significantly when these targets are presented in a specific context of spatially separated, collinear lines [12]. The findings thus far reported, based on either luminance detection or orientation discrimination procedures [1], imply the existence of two distinct spatial domains for the integration of visual orientation: a local zone producing classic within-receptive field effects [10], and a long-range domain producing effects " beyond the classic receptive field " [1,12]. In the psychophysical experiments designed to probe our model predicitons for long-range contour diffusion, small line targets had to be detected by human observers. In the test conditions, the target was presented at positions collinear with a context line, and the spatial spearation between target and context was varied. In the control conditions, thresholds for the detection of the target line without context line were measured.

3.1 Procedure

The bright target lines were flashed briefly (30 milliseconds), with five different luminance intensities in the threshold region according to the method of constant stimuli, upon a uniform, grey background generated on a high-resolution computer screen. In a two-alternative temporal forced-choice (2afc) procedure, observers had to press one of two possible keys on the computerkeyboard, deciding whether the target was presented in the first or the second of two successive temporal intervals. Each experimental condition corresponded to a total of 200 successive trials. In the test condition, the target was flashed simultaneously with a white, collinear context line, in the control condition the target was presented alone (see Figure 3). Three psychophysically trained observers were used in the experiments.

3.2 Results

Detection thresholds of one of the three observers as a function of the spatial separation between target and context line are shown in Figure 4 for one of the three observers. Results of the other two subjects were very similar. The horizontal line in the graph indicates the level of the detection threshold in the control condition without context. Up to about 25 arcminutes of spatial separation from the context line, detection of the target is dramatically facilitated, the effect decreasing steeply, which describes the short-range spatial domain of

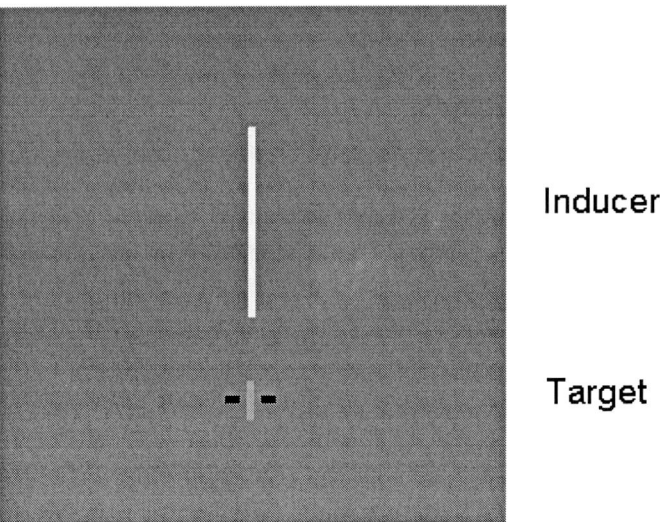

Fig. 3. In the psychophysical task, small line targets had to be detected by human observers. In the test conditions, the target was presented at positions collinear with a context line, as shown here. The spatial spearation between target and context was varied. In the control conditions, thresholds for the detection of the target line without context line were measured.

contour diffusion. The long-range domain is characterized by a constant detection facilitation up to a target inducer distance of about 150 arcminutes (2.5 degrees of visual angle). Predictions of orientation diffusion simulated by our neural network are shown to provide an almost perfect fit to the psychophysical data.

4 Conclusions

The neural network model presented here simulates neural cooperation via intracortical connectivity based on the neurophysiological evidence for cooperation between orientation selective cortical cells in V1, which can be seen as the neural substrate of contour completion across spatial gaps in visual stimuli [6]. The diffusive behavior of the orientation selective cells in the network predicts psychophysical data describing line contour detection performances of human observers with collinear targets and inducers for spatial separations describing short-range and long-range spatial domains [1]. It is concluded that micro-circuit processing characteristics can be used to efficiently simulate intracortical diffusion in the orientation domain, predicted by macro-circuit approaches such as the one by Grossberg and colleagues [7].

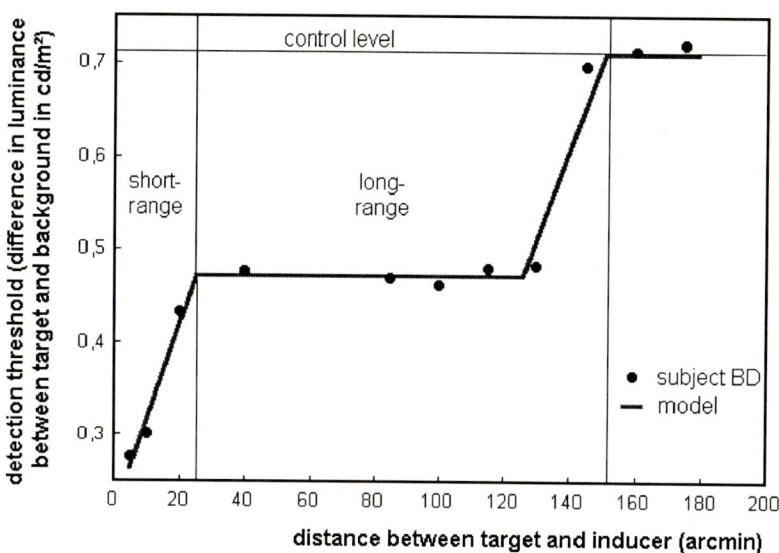

Fig. 4. Detection thresholds of one of the three observers as a function of the spatial separation between target and context line. The horizontal line in the graph indicates the level of the detection threshold in the control condition without context. Up to about 25 arcminutes of spatial separation from the context line, detection of the target is dramatically facilitated, the effect decreasing steeply, which describes the short-range spatial domain of contour diffusion. The long-range domain is characterized by a constant detection facilitation up to a target inducer distance of about 150 arcminutes (2.5 degrees of visual angle). Performance in the detection task, which is indicated by the dots here in the graph, is predicted with high precision by our model parameters (continuous line).

References

1. Brincat, S. L., Westheimer, G. (in press) Integration of foveal orientation signals: Distinct local and long-range spatial domains. *Journal of Neurophysiology*.
2. Dresp, B. (1993) Bright lines and edges facilitate the detection of small light targets. *Spatial Vision*, **7**, 213-225.
3. Dresp, B. (1999) Dynamic characteristics of spatial mechanisms coding contour structures. *Spatial Vision*, **12**, 129-142.
4. Dresp, B., Grossberg, S. (1997) Contour integration across polarities and spatial gaps: From local contrast filtering to global grouping. *Vision Research*, **37**, 913-924.
5. Gilbert, C. D., Wiesel, T. N. (1990) The influence of contextual stimuli on the orientation selectivity of cells in the primary visual cortex of the cat. *Vision Research*, **30**, 1689-1701.
6. Gilbert, C. D. (1998) Adult cortical dynamics. *Physiological Reviews*, **78**, 467-485.
7. Grossberg, S., Mingolla, E. (1985) Neural dynamics of perceptual grouping: textures, boundaries and emergent segmentations. *Perception & Psychophysics*, **38**, 141-171.

8. Hubel, D. H., Wiesel, T. N. (1968) Receptive field and functional architecture of monkey striate cortex. *Journal of Physiology*, **195**, 215-243.
9. Kapadia, M. K., Ito, M., Gilbert, C. D., Westheimer, G. (1995) Improvement in visual sensitivity by changes in local context: parallel studies in human observers and in V1 of alert monkeys. *Neuron*, **15**, 843-856.
10. Morgan, M. J., Dresp, B. (1995) Contrast detection facilitation by spatially separated targets and inducers. *Vision Research*, **35**, 1019-1024.
11. Polat, U., Sagi, D. (1993) Lateral interactions between spatial channels : suppression and facilitation revealed by lateral masking experiments. *Vision Research*, **33**, 993-999.
12. Wehrhahn, C., Dresp, B. (1998) Detection facilitation by collinear stimuli in humans: Dependence on strength and sign of contrast. *Vision Research*, **38**, 423-428.

Automatic Generation of Photo-Realistic Mosaic Image

Jong-Seung Park, Duk-Ho Chang, and Sang-Gyu Park

Virtual Reality R&D Center, ETRI
161 Kajong-Dong, Yusong-Gu, Taejon, 305-350, Korea
{park,dhchang}@etri.re.kr
sgpark@video.etri.re.kr
http://vrcenter.etri.re.kr/~parkjs/

Abstract. This paper presents a method that generates a mosaic image from multiple images or a sequence of images without any assumptions on camera intrinsic parameters or camera motion. The most appropriate transform model is automatically selected according to the global alignment error. Given a set of images, the optimal parameters for a 2D projective transform model are computed using a linear least squares estimation method and then iterative nonlinear optimization method. When the error exceeds a predefined limit, a more simple transform model is applied to the alignment. Our experiments showed that false alignments are reduced even in the case of 360° panoramic images.

1 Introduction

The construction of mosaic images on computer vision applications has been an active research area in recent years. The primary aim of mosaicing is to enhance image resolution and enlarge the field of view. In computer graphics applications, images of real world have been used as environment maps at which images are used as static background of graphic objects. In image based rendering applications, mosaic images on smooth surfaces allow an unlimited resolution also avoiding discontinuities. Such immersive environments provide the users an improved sense of presence in a virtual scene.

In this paper, we present a mosaic generation algorithm using a multi-model approach. The problem is, given a sequence of spatially overlapped images, finding the transformation between two adjacent images, aligning all the images, and constructing a large mosaic view. The major issues of image mosaicing are finding the relative transformation between any two images and generating the global view.

A 2D projective transform algorithm proposed by Szeliski [1] is appropriate for a planar view. For a long sequence of images or a panoramic view images, other transform model is required, e.g., cylindrical mapping [2,7], spherical mapping [9]. To project to a cylinder or to a sphere, the camera parameters are required.

This paper presents a multi-model mosaic method that tries several models of different complexity and chooses a transform model that generates the best mosaic view.

2 Inter-Frame Transform Models

A common model for perspective cameras is the projective mapping from a 3D projective space, $\mathcal{P}^3$, to a 2D projective space $\mathcal{P}^2$:

$$\mathbf{u} \sim P\mathbf{x}$$

where $\mathbf{x} = (x, y, z, 1)^T$ represents a point in $\mathcal{P}^3$, $\mathbf{u} = (u, v, 1)^T$ represents the projection of $\mathbf{x}$ onto the retinal plane defined in $\mathcal{P}^2$, and P is a 3×4 matrix of rank 3 and $\sim$ means equality up to a non-zero scale factor. The matrix P is called a camera projection matrix and it is expressed by

$$P = KT, \quad \text{where} \quad K = \hat{K}\left[I|\mathbf{0}_3\right], \quad \hat{K} = \begin{bmatrix} \alpha_u & -\alpha_u \cot\theta & u_0 \\ 0 & \alpha_v \csc\theta & v_0 \\ 0 & 0 & 1 \end{bmatrix}, \quad T = \begin{bmatrix} R & -R\mathbf{t} \\ \mathbf{0}_3^T & 1 \end{bmatrix}. \tag{1}$$

The camera calibration matrix K describes the transformation of the image coordinates to the pixel coordinates and contains all information about characteristics of the specific camera. The unknowns in the projection matrix are the scale factors, α_u and α_v, the coordinates of the principal point, (u_0, v_0), the angle between the two image axes, θ. The 4×4 matrix T specifies a transform of the world coordinates into the camera coordinates. There are six unknowns in T, three for the rotation matrix R and three for the translation vector $\mathbf{t}$.

Our interest is to recover the geometric relationship of two images containing the same part of a scene. Let $\mathbf{x}$ be a homogeneous 3D world coordinate, and P and P' are 3×4 camera matrices for two images. Then the homogeneous 2D screen locations in the two images are represented by

$$\mathbf{u} \sim P\mathbf{x}, \quad \mathbf{u}' \sim P'\mathbf{x}. \tag{2}$$

We assume that the world coordinate system coincides with the first camera position, i.e., $T = I_{4\times 4}$. Then $\mathbf{u} = K\mathbf{x}$ and, since K is of rank 3, we have

$$\mathbf{x} = \begin{bmatrix} \hat{K}^{-1}\mathbf{u} \\ w \end{bmatrix}$$

where w is the unknown projective depth which controls how far the point is from the origin.

Let $P' = K'T'$ be the camera projection matrix of the second image, then

$$\mathbf{u}' = K'T'\mathbf{x} = \begin{bmatrix} \hat{K}'|\mathbf{0}_3 \end{bmatrix} \begin{bmatrix} R' & \mathbf{t}' \\ \mathbf{0}_3^T & 1 \end{bmatrix} \begin{bmatrix} \hat{K}^{-1}\mathbf{u} \\ w \end{bmatrix} = \begin{bmatrix} \hat{K}'|\mathbf{0}_3 \end{bmatrix} \begin{bmatrix} R\hat{K}^{-1}\mathbf{u} + w\mathbf{t} \\ w \end{bmatrix}$$

and we obtain:
$$\mathbf{u}' = \hat{K}'R\hat{K}^{-1}\mathbf{u} + w\hat{K}'\mathbf{t}. \quad (3)$$

Equation (3) expresses the relationship of two cameras. There are too many unknowns to solve the equation. Although there are some approaches to estimate the camera matrices P and P' such as Amstrong et al [4], or Pollefeys and Gool [5], direct estimation of the camera matrices is impossible without strong assumptions.

If a point $\mathbf{x} = [\hat{\mathbf{x}}^T, w]^T$ lies on a plane with a normal vector $\mathbf{n} = [\hat{\mathbf{n}}^T, -1]^T$, then $\mathbf{n} \cdot \mathbf{p} = 0$ and we obtain $w = \hat{\mathbf{n}}\hat{\mathbf{x}} = \hat{\mathbf{n}}\hat{K}^{-1}\mathbf{u}$. Hence two images are related by
$$\mathbf{u}' = \hat{K}R\hat{K}^{-1}\mathbf{u} + \hat{\mathbf{n}}K^{-1}\mathbf{u}\hat{K}\mathbf{t} = \hat{K}(R + \mathbf{t}\hat{\mathbf{n}}^T)\hat{K}^{-1}\mathbf{u}. \quad (4)$$

Equation (4) shows that the mapping of two planar scene views can be described by a 2D projective transform matrix. Using the idea, Szeliski [1] proposed a method of computing a 3×3 matrix, M_{2D}, that describes the relationship of two planar scene views:
$$\mathbf{u}' = M_{2D}\mathbf{u} \quad (5)$$

or, equivalently,
$$\begin{bmatrix} u' \\ v' \\ 1 \end{bmatrix} \sim \begin{bmatrix} m_1 & m_2 & m_3 \\ m_4 & m_5 & m_6 \\ m_7 & m_8 & 1 \end{bmatrix} \begin{bmatrix} u \\ v \\ 1 \end{bmatrix}. \quad (6)$$

Equation (6) represents a *full projective transform (8-parameter model)* and has eight degree of freedom in the 3×3 matrix M_{2D}.

Several assumptions about 2D planar transformation are possible to reduce the number of free variables:

Pure translation (2-parameter model): The transform allows only translation. The matrix is represented as $M_{2D} = I_3 + [\mathbf{0}_3|\mathbf{0}_3|[t_u, t_v, 1]^T]$ and it has two degree of freedom for t_u and t_v. It is the simplest form of image mosaicing and valid in the case with satellite images.

– *Rigid transform (3-parameter model)*: The transform allows translation plus rotation. A rotation parameter θ is added to the pure translation parameters, t_u and t_v, and the matrix is represented by $m_1 = m_5 = \cos\theta$, $m_4 = -m_2 = \sin\theta$, $m_3 = t_u$, $m_6 = t_v$. It has three degree of freedom, t_u, t_v and θ. This 2D rigid transform model is useful for panoramic image, e.g., Zhu [7].

– *Similarity transform (4-parameter model)*: The transform allows rigid transform plus scaling. A scaling parameter s that is included in m_1, m_2, m_4, m_5 is estimated as well as θ. It has four degree of freedom, t_u, t_v, θ, and s.

– *Affine transform (6-parameter model)*: The transform assumes $m_7 = m_8 = 0$, i.e., the transform allows scaling, rotation and shear but does not allow perspective correction. It has six degree of freedom.

The unknown parameters of the above models can be solved using only correspondences of image points. The perspective transformation is valid for all image points only when there is no motion parallax between frames, e.g., the case of a planar scene or a camera rotating around its center of projection.

3 Estimation of Transform Parameters

To estimate transform parameters, we first find corner points and then match the corners between two images using the correlation method of Faugeras [8]. False match pairs are removed using the epipolar constraint. Among the final match pairs, only the best twenty or less pairs are selected to compute a linear solution of a transform model.

3.1 Linear Least Squares Formulation

For the case of M unknown parameters and given N point correspondences, the problem is to find a parameter vector $\mathbf{m} = [m_1, ..., m_M, 1]^T$ that minimizes $||A\mathbf{m}||$ where $A = [A_1^T, ..., A_N^T]^T$. The submatrix A_i depends on the type of transform. If there is no additional constraint, the solution $\mathbf{m}$ is the eigenvector corresponding to the smallest eigenvalue of A.

For the case of rigid transform (M=4), the matrix A_i is given by

$$A_i = \begin{bmatrix} u_i & -v_i & 1 & 0 & -u_i' \\ v_i & u_i & 0 & 1 & -v_i' \end{bmatrix} \tag{7}$$

for a single match pixel pair (u_i, v_i, u_i', v_i'). There is a constraint to be satisfied: $m_1^2 + m_2^2 = 1$. To solve the problem, we decompose $\mathbf{m}$ into two parts, $\mathbf{m}' = [m_3, m_4, 1]^T$ and $\mathbf{m}'' = [m_1, m_2]^T$, then (7) is expressed as

$$A\mathbf{m} = C\mathbf{m}' + D\mathbf{m}''$$

where

$$C_i = \begin{bmatrix} 1 & 0 & -u_i' \\ 0 & 1 & -v_i' \end{bmatrix}, \quad D_i = \begin{bmatrix} u_i & -v_i \\ v_i & u_i \end{bmatrix}.$$

Using the constrained optimization technique, $\mathbf{m}''$ is given by the eigenvector corresponding to the smallest eigenvalue of matrix E:

$$E = D^T(I - C(C^TC)^{-1}C^T)D .$$

Then $\mathbf{m}'$ is given by $\mathbf{m}' = -(C^TC)^{-1}C^TD\mathbf{m}''$.

For the case of projective transform (M=8), the matrix A_i is given by

$$A_i = \begin{bmatrix} u_i & v_i & 1 & 0 & 0 & 0 & -u_iu_i' & -v_iu_i' & -u_i' \\ 0 & 0 & 0 & u_i & v_i & 1 & -u_iv_i' & -v_iv_i' & -v_i' \end{bmatrix}.$$

For N match pairs of image pixels, the matrix A of dimension $2N \times M$ is constructed. The solution $\mathbf{m}$ is the eigenvector corresponding to the smallest eigenvalue of A.

3.2 Iterative Refinement

Further reduction of transform error is possible using iterative non-linear optimization for the linear solution. Since each corresponding point gives two equations of unknowns, four or more corresponding points between the two views solve for the eight unknowns in the 2-D projective transformation. A parameter computation method from the corresponding points is as follows. For a pixel correspondence (u_i, v_i, u'_i, v'_i) where (u_i, v_i) is a pixel position in image I and (u'_i, v'_i) is the corresponding position in image I', a measure of transform error is defined as the differences of intensity between the corresponding pixels in the two images:

$$e_i = I'(u'_i, v'_i) - I(u_i, v_i). \tag{8}$$

To find a solution that minimizes the error, we compute the partial derivative of e_i with respect to m_k using

$$\frac{\partial e_i}{\partial m_k} = dI'_u \frac{\partial u'}{\partial m_k} + dI'_v \frac{\partial v'}{\partial m_k} \quad \text{where} \quad dI'_u = \left.\frac{\partial I'}{\partial u'_i}\right|_{u'_i, v'_i} \quad \text{and} \quad dI'_v = \left.\frac{\partial I'}{\partial v'_i}\right|_{u'_i, v'_i}.$$

The derivatives are used to constitute an approximate Hessian matrix that is required to optimization. For the case of M unknown parameters, the derivative vector is represented as

$$\frac{\partial e_i}{\partial \mathbf{m}} = \left[\frac{\partial e_i}{\partial m_1}, \cdots, \frac{\partial e_i}{\partial m_M}\right] = dI'_u \mathbf{v}_u + dI'_v \mathbf{v}_v \tag{9}$$

where $\mathbf{v}_u$ and $\mathbf{v}_v$ depend on the type of transform.

For the case of full projective transformation (M=4), the derivatives are given by (9) with

$$\mathbf{v}_u = [u_i, -v_i, 1, 0]^T \quad \text{and} \quad \mathbf{v}_v = [v_i, u_i, 0, 1]^T.$$

For the case of full projective transformation (M=8), the derivatives are given by (9) with

$$\mathbf{v}_u = (1/D_i) \cdot [u_i, v_i, 1, 0, 0, 0, -u_i u'_i, -v_i u'_i]^T,$$
$$\mathbf{v}_v = (1/D_i) \cdot [0, 0, 0, u_i, v_i, 1, -u_i v'_i, -v_i v'_i]^T$$

where $D_i = m_6 u_i + m_7 v_i + 1$.

The total error is defined as just the sum of pixel error:

$$e = \sum_i e_i^2 \tag{10}$$

where e_i is given by (8). The solution of (10) is computed using the Levenberg-Marquardt algorithm with partial derivatives described in (9). The detail steps of the optimization process are described in the work of Szeliski [6].

The 2D image alignment would fail to produce mosaic images in cases of motion parallax between foreground and background, and when the 2D alignment does not consistently align either the foreground or the background.

4 Global Alignment

Let $I_1, \ldots, I_n$ be a sequence of images to be used to generate a mosaic image and $T_{1,2}, \ldots, T_{n-1,n}$ be the corresponding transforms. Once the transform matrices $T_{1,2}, \ldots, T_{n-1,n}$ are obtained for n images, we construct a global map which can be quickly viewed. Let $T_{i,j}$ be the transformation matrix from image I_j to image I_i. To generate a mosaic image with the coordinate of a chosen reference image I_i, we must obtain transform $T_{i,j}$ from image I_j to image I_i for each $j \neq i$. The transform $T_{i,j}$ is computed from a simple composition of the obtained transforms:

$$T_{i,j} = \begin{cases} T_{i-1,i}^{-1} T_{i-2,i-1}^{-1} \cdots T_{j,j+1}^{-1} & \text{if } j < i \\ T_{i,i+1} T_{i+1,i+2} \cdots T_{j-1,j} & \text{if } j > i \end{cases} \quad (11)$$

The reference image is selected as the mid-frame of the given sequence. Then, for each image that is not a reference image, the transform to the reference image is computed. Considering the mapping of image boundary, the 8-parameter transform model is discarded when a mapped image area exceeds double of the original area or below the half of the original area. We try the same process using the 4-parameter transform model, and then 3-parameter transform model when the 4-parameter transform model is discarded.

In the 3-parameter model, if the sweeping view is narrow and long we consider the possibility of panoramic image sequence and try to align the last several images to the first image. If there is an overlapped image to the first image, we change transform model to the 2-parameter model and interpolate the translation vectors so that the first image and the overlapped image reside at the same horizontal altitude.

When the image sequence is decided as a panoramic sequence and the alignment error exceeds a predefined bound, we try image warping process to reduce the error. If there are sufficiently many images to cover the panoramic view in a dense way with a small amount of camera panning between each adjacent pair of images, the 2-parameter model is good enough to generate a fine panoramic view.

When the panning angle is big, an image warping process is required to make a panoramic image. If there is only camera panning motion, where the camera motion is a pure rotation around the y-axis, a rectification algorithm such as the algorithm proposed by Hartley [3] is enough to generate a panorama.

In general, there are camera tilting or translation for the non-exact motion of hand-held camera. A simple solution is to mapping image plane to cylindrical surface. The axis of the cylinder passes through the camera's optical center and has a direction perpendicular to the optical axis. A rotation of the camera around the cylinder's axis is equivalent to a translation on the cylinder. Figure 1 shows the projection warping that corrects the distortion caused by perspective projection. The horizontal pixel position, u', in the cylindrical surface is the same as the horizontal pixel position, u, in the original image. The vertical pixel position, v', in the cylindrical surface is related to the pixel position, (u, v), in

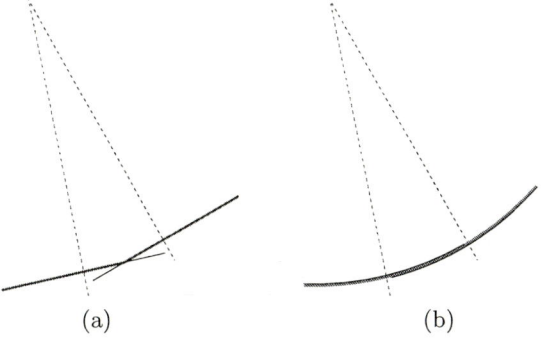

Fig. 1. Projection warping to a cylindrical surface according to the camera parameters.

Fig. 2. Examples of the 8-parameter model. (a) There is only camera rotation. (b) The ground scene is planar.

the original image by:

$$v' = \frac{v}{\cos(\tan^{-1}(u \cdot r_u/f))}$$

where f is the camera focal length and r_u is the horizontal pixel spacing. After the warping, the same mosaicing procedure is applied to generate a global panoramic image.

5 Experimental Results

Image blending process is required to overcome the intensity difference between images. The camera gain is changed frame by frame depending on the light condition and viewing objects.

A common mosaic method is using the 8-parameter projective transform model without any warping. The projective transform is the best model when there is only rotation or the viewing scene is on a plane. Figure 2 shows such examples of the two cases. The projective transform can be directly used to a

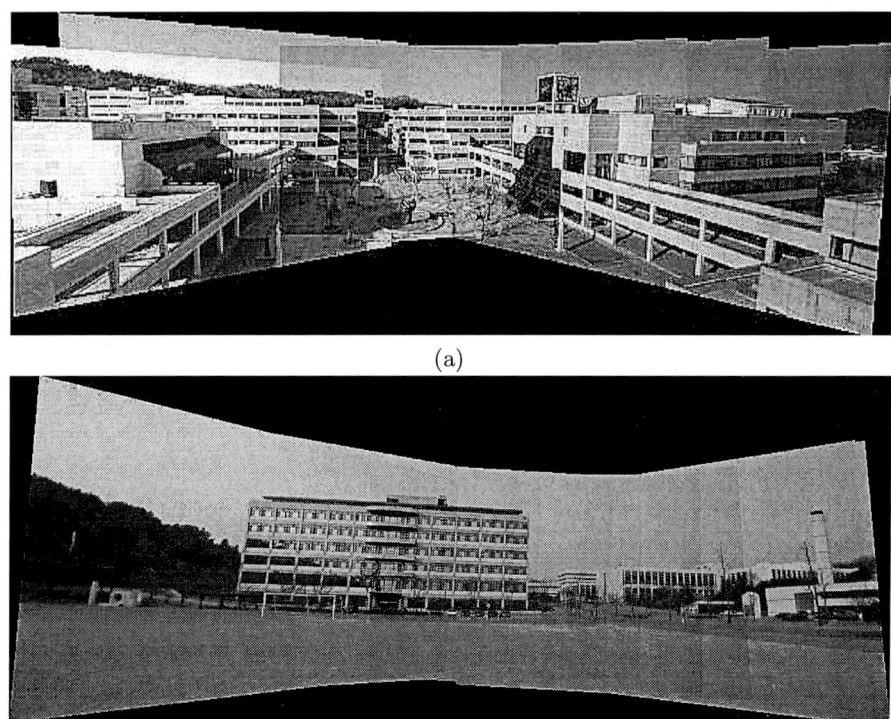

Fig. 3. (a) Mosaic image generated from 7 campus images of size 384×256 using the 8-parameter 2D projective transform model. (b) Mosaic of 10 playground images of size 320×240 using the 8-parameter model.

general set of images only when the images do not cover wide areas of a scene. Figure 3 shows an example using the 8-parameter model to some general cases. However the maximum field of view is restricted to less than 180°. If further images are aligned to the mosaic map, the images are skewed too much and an awful view is generated. To avoid the extreme perspective effect, a change of transform model to lower degree of freedom is indispensable.

Figure 4 shows an example of panoramic mosaic. The 33 images represent a 360° view of a playground. The panoramic view cannot be represented as a single map using the 8-parameter model. By reducing the degree of freedom, we obtain more realistic view to human perception.

Further reduction of alignment error is possible when we know basic characteristics of the camera. If we know the input images contain a panoramic view, we can warp each image according to the camera parameters. Using the warped images, the same mosaic process is applied. Figure 5 shows a comparison of mosaic without warping and mosaic with warping. The photos are acquired using the Fujifilm digital camera MX-700 having 1/1.7-inch CCD of maximum resolu-

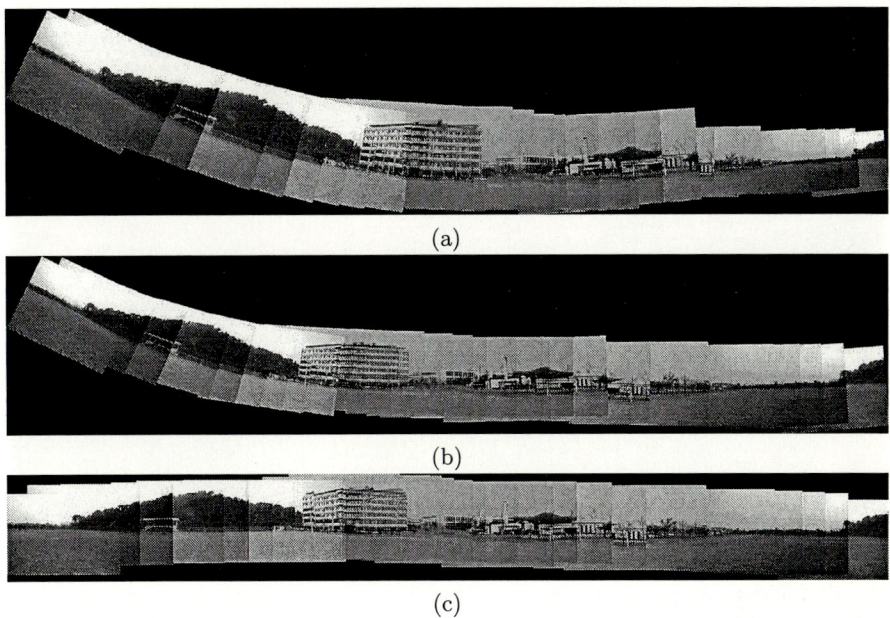

Fig. 4. Mosaic images from 33 playground images of size 320×240 using different transform models. (a) The 4-parameter translation model, (b) the 3-parameter rigid transform model, (c) the 2-parameter similarity transform model.

tion 1280×1024, 35mm focal length. We reduce the images to size 320×240 for the fast computation.

6 Conclusion

We presented a method that generates a mosaic image from multiple images or a sequence of images without assumptions on camera intrinsic parameters or camera motion. The transform model is automatically selected according to the global alignment error. When the error exceeds a predefined limit, more simple transform model is applied to the alignment, in the order of an affine transform model, a similarity transform model, a rigid transform model and a pure translation model. As well as the mosaicing of a planar view, panoramic mosaicing is also possible by decreasing the model complexity. Our experiments show that false alignment from extreme perspective effects is avoidable by selecting appropriate transform model.

The weakness of our method is in the mosaicing of images having moving objects. The intensity change in viewing image area of moving object is regarded as alignment error.

Our further research will be focused on the automatic extraction of camera parameters to construct an environment map that is useful in image-based ren-

(a)

(b)

Fig. 5. Comparison of mosaic images with/without pre-warping from 18 playground images. (a) The 4-parameter model without warping, (b) the 4-parameter model with warping.

dering applications. We are also going to develop an image blending method that is required to overcome the intensity difference between images.

References

1. Szeliski, R.: Image mosaicing for tele-reality applications, Technical Report CRL 94/2, DEC Cambridge Research Lab (1994)
2. Sawhney, H. S. et al.: VideoBrush^{TM}: Experiences with consumer video mosaicing, WACV'98 (1998) 56-62
3. Hartley, R. I.: Theory and Practice of Projective Rectification, International Journal of Computer Vision, **35(2)**, nov (1999) 1-16
4. Amstrong, M., Zisserman, A., Beardsley, P. A.: Euclidean structure form uncalibrated images, BMCV'94 (1994) 509-518
5. Pollefeys, M., Koch, K., Gool, L. V.: Self-calibration and metric reconstruction in spite of varying and unknown internal camera parameters, Proc. IEEE Int. Conf. on Comp. Vis. (1998) 90-95
6. Szeliski, R.: Video mosaics for virtual environments, IEEE Computer Graphics and Applications, **16(2)** (1996) 22-30
7. Zhu, Z.: Xu, G., Riseman, E. M., Hanson, A. R.: Fast generation of dynamic and multiresolution 360° panorama from video sequences, Proceedings of IEEE Int'l Conf. On Multimedia Computing and Systems (1999) 7-11
8. Faugeras, O. et al.: Real time correlation-based stereo: algorithm, implementation and applications, INRIA Technical Report (1993)
9. Gümüştekin, S., Hall, R. W.: Mosaic image generation on a flattened gaussian sphere, WACV'96 (1996) 50-55

The Effect of Color Differences on the Detection of the Target in Visual Search *

Ji-Young Hong, Kyung-Ja Cho, and Kwang-Hee Han

Yonsei Univ., Graduate Program in Cognitive Science,
134 Shinchon-dong, Seodaemun-gu 120-749, Seoul, Korea
{hongcom, chokj, khan}@psylab.yonsei.ac.kr
http://psylab.yonsei.ac.kr/~khan

Abstract. In order for users to find necessary information more easily, the targets should be more salient than the non-targets. In this study, we intended to find the critical color differences that could be helpful in locating the target information. Previous studies have focused on the critical color differences in the C.I.E. color system though the H.S.B color model is often the choice for a color design. Given the popularity of the H.S.B. color, the shift of a focus from the C.I.E. color system to the H.S.B. color model, which is widely used for computer graphic software, seemed quite inevitable. As a result, our experiments revealed that over 60-degree differences in hue dimension and over about 40% difference in brightness dimension resulted in faster search. However, no significant differences in the search time were found along with the saturation dimension.

1 Introduction

It is almost inconceivable that we live without our computers today. In that computers have made all sorts of information more accessible than ever, they exerts positive effects over our every day lives. This, however, is not without a cost. Considering the tremendous amount of the information computers can offer us, it is very likely that we are overwhelmed by the flood of the information even before we can do anything with it. As such it is of vital importance to find the very information we need as quickly and as accurately as possible, and this is even more so when it comes to a visual display.

One way to display the information users want to find effectively is to assign such attributes as proximity, contrast, orientation selectively to the target information. Of many possible attributes, the brightness contrast and the color contrast are instrumental in singling out target information quickly. In particular, the color contrast is deemed crucial for it can draw clear distinction among various information.

Human can distinguish roughly among 2 million different colors[2], and modern computers are capable of displaying even more numbers of colors than human

---

* This work was supported by Yonsei university research fund of 1999.

can tell the differences. However, a close look at the color usage on the widely-used software programs reveals that using all of the 16 million true colors in a given display is hardly the case. This supports the notion that we can display information more effective way using a set of distinctive colors rather than using too many colors simultaneously.

2 Background

In a related study concerning about display various colors simultaneously in a given display, Healey(1996) suggested three different factors in color visualization : color distance, linear separation, and color category[3]. His method can be applied to the selection of a set of colors that will provide clear distinction among the data elements in the data visualization. Healey's study showed that to facilitate a fast and accurate search for information, no more than 7 different colors should be used. Carter and Carter(1989) found out that the critical color difference between targets and non-targets must be at least more than 40 CIELUV units[1]. Nagy and Sanchez(1992) reported that search time was a function of the interaction between the luminosity and the chromaticity difference[4].

Brilliant as these previous studies might be, because they were based on the psychophysical CIELUV color model and obtained through specific apparatus that was scarcely used outside the Lab, there is a problem of generalization in the natural computing situation.

Therefore, with the natural computing situation in mind, our selection of colors was based upon the three dimensions of hue, saturation and brightness which consist of HSB color model, and we intended to find the critical color differences for a faster search. With these three dimensions we found the conditions under which people can search target information effectively and the critical color differences for a faster search.

3 Experiments

In this study, we displayed many non-targets and a single target with various color differences in hue, saturation and brightness by measuring the minimum color differences, we tried to find out the optimal conditions for a faster search.

3.1 Subject

Eighteen college students with normal or corrected acuity served as subjects for this study. They participated in all three conditions (hue, saturation, brightness) the display order was randomized.

3.2 Material

We selected 6 representative scales (0%, 20%, 40%, 60%, 80% and 100%) for brightness and saturation condition, and also 6 representative scales (0°, 60°,

120°, 180°, 240°, 300°) for hue condition within monitor's color gamut. And we arranged these colors as three relative angles like Fig. 1.

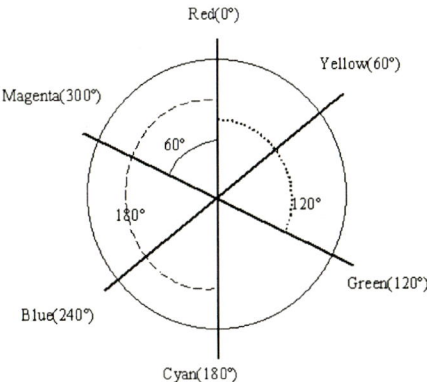

Fig. 1. The selected degrees in accordance with hue differences

For brightness condition, the saturation of the stimuli fixed at 50%, the hue at Red(60°), Green(120°), and Blue(0°). For saturation condition, the brightness of the stimuli is fixed at 50%, and the hue in the same way. For hue condition, the brightness and saturation of the stimuli were fixed at 50%, respectively.

In brightness and saturation condition, the target colors were only two levels : 0% and 100%, and the non-target colors were given all sets of colors except for the target color. In hue condition, we assigned one of the colors randomly to the target's color, and the one of the rests was assigned to the non-target's color. It is necessary to modify the saturation of 0% to the saturation of 10%, because background color was 0% of saturation. The number of stimuli was selected at 6, 12, and 24, and the size of the stimulus was a filled circle of 30 pixel's diameter, and the stimuli were presented at the 128 pixel's distance from the fixation point. The distance from the viewpoint of the subject was 390mm. We set the Monitor's brightness and contrast control knob to 50% and 100% respectively.(This setting was regarded as a natural screen display situation.)

3.3 Equipment

All color search experiments reported were performed on IBM compatible computer with a AGP graphic board(8 bits of resolution for each color gun, and an 17" color monitor with 0.28 dot pitch, and standard 106 key Korean keyboard. We used Microsoft Windows 98 as operating system, and set display to 1024×768 true color resolution. We made a program for the experiment using Microsoft Visual Basic 6.0.

3.4 Procedure

The task was to detect whether there was a target among several non-targets. Subjects were asked to respond to the stimuli after staring at the fixation point on the monitor for 1,000ms. Thirty times of practices preceded the actual task. Subject's task was to answer "Y" if there was a target, "N" if not. The total trials were 540 times. They were given break after every 100 trials. It was recorded the response time in detecting the target along with the dimension of brightness, saturation, hue, respectively.

4 Result and Discussion

In the dimension of brightness, the search time for 20% stimuli was longer than others($F(4,68)=18.485$, $p<0.01$). This implies that there is critical color difference between 20% and the rests. In the dimension of saturation, the slight linear like decrease of search time was observed as difference levels. But analysis of reaction time did not reveal significant differences between levels. This shows that it is difficult to detect the target fast between 40% and 100%, 0% and 60% of saturation difference. The results were superimposed in Fig. 5.

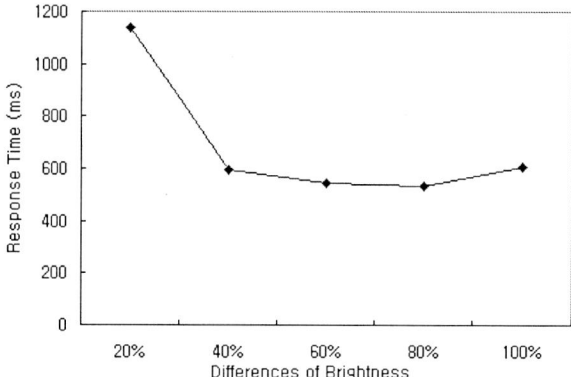

Fig. 2. Response time for the differences of brightness

In dimension of hue, there was no significant difference in search time. However, the mean search time in hue dimension showed tendency of shorter than other dimensions.

Accordingly, we can conclude that a faster search arises from more than 20% of the difference at brightness dimension. In hue dimension, it is safe to say that 60-degree of the hue differences are sufficient to facilitate the search.

The result shows, first, that the differences in hue and brightness dimension are more effective factors than saturation dimension generally. Second, for faster search, brightness difference has to be more than 40% at least, otherwise, the effect is less effective than the effect from the difference of saturation.

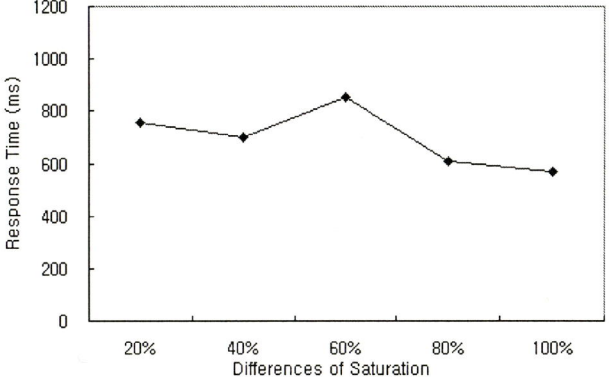

Fig. 3. Response time for the differences of saturation

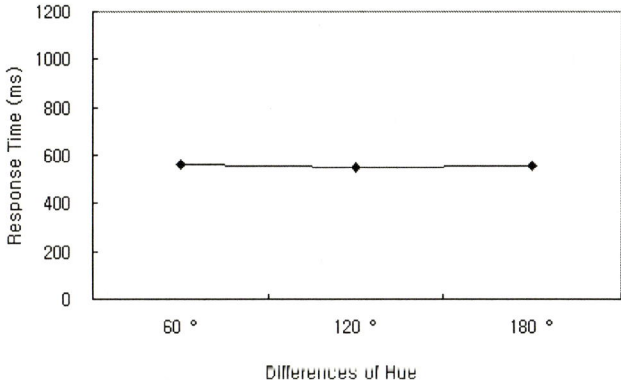

Fig. 4. Response time for the differences of hue

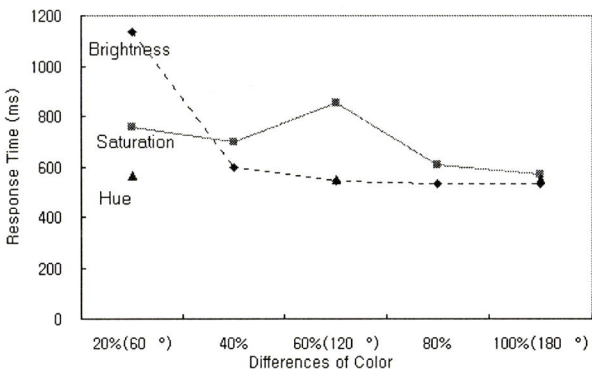

Fig. 5. Comparison of response time for the differences of hue, brightness and saturation

It must be noted that this study needs to investigate the effect of targets within 60 degrees in order to get more information for the hue condition. In order to find further information, as for saturation and brightness, it is required to investigate about more various background colors since background color might be have influence on detecting the targets. Furthermore, it is necessary for more two dimensions, because there might be another possibility to interact between them.

References

1. Carter, R. C.: Calculate the effects of symbol size on usefulness of color. Proceedings of the Human Factors Society 33rd Annual Meeting, (1989) 1368–1372
2. Goldstein, E. B.: Sensation and Perception, California, Brooks/Cole Publishing Co., 4th ed. (1996)
3. Healey, C. G.: Choosing Effective Colours for Data Visualization, Proceedings of IEEE Visualization '96 (1996)
4. Lagy, A. L.,Sanchez, R.R,: Chromaticity and Luminance as Coding Dimensions in Visual Search, Human Factors, **34** (1992) 601–604

A Color-Triangle-Based Approach to the Detection of Human Face

Chiunhsiun Lin and Kuo-Chin Fan

Institute of Computer Science and Information Engineering National Central University,
Chung-Li, Taiwan, 32054, R.O.C.
Kcfan@ncuee.ncu.edu.tw

Abstract. A robust and efficient human face detection system that can detect multiple faces in various surroundings is presented in this paper. The proposed system consists of two principal parts. The first part is to search for the potential face regions. The second part is to perform face verification task. Our system can conquer different size, different lighting condition, varying pose and expression, and noise and defocus problems. In addition to overcome the problem of partial occlusion of mouth and sunglasses, the color segmentation process can successfully extract skin color from complex backgrounds. Experimental results demonstrate that an approximately 98 % success rate is achieved. The experimental results reveal that the proposed method is better than traditional methods in terms of efficiency and accuracy.

1 Introduction

Automatic recognition of human faces is one of the most intricate and important problems in the areas of pattern recognition and computer vision. It can be used in criminology to find out the possible criminals. It can also be used as the security mechanism to replace metal key, plastic card, and password or PIN number. However, we should not deal with face detection problem merely as a preprocessing of a face recognition system because a successful face detection process is the prerequisite to facilitate later face recognition task. If we do not have successful face detection method, the goal of successful face recognition can not be achieved. Therefore, we should deal with face detection problem as important as the face recognition problem.

Abundant of researches have been conducted on human face detection. Some successful systems have been developed and reported in the literature, such as Leung et al. (1995) [1] coupled a set of local feature detectors via a statistical model to find the facial components for face positioning. Rowley et al. (1995) [2] presented a neural network-based face detection system. Lee et al. [3] extracted the face part from the homogeneous background by tracking the face boundary. In their work, they assume that the location of face part is approximately at the center of a captured image. Juell et al. (1996) [4] proposed a hierarchical neural network to detect human faces in gray scale images. An edge enhancing preprocessor and four backpropagation neural networks arranged in a hierarchical structure were devised to find multiple faces in the scene. Instead of gray scale images, Sobottka et al. (1996) [5] located the poses of human faces and facial features from color images. They presented a system that used an ellipse in Hue-Saturation-Value (HSV) color space to express the shape of a face. Chen et al. (1996) [6] proposed a skin color distribution function on an unchanged color space to detect the face-like region. The skin color regions in color images were modeled as several 2-D patterns and verified with the built face model by a fuzzy pattern matching approach. Han et al. (1997)[7] proposed a system using a morphology-based technique to perform eye-analogue segmentation. Then, a trained backpropagation neural network performs the face verification task.

In these aforesaid approach only [5] and [6] use the color information. Color information has not been used widely due to the huge storage and computational requirements. With the advancement in computer technology, color images are now widely used in many applications. However, most of the aforementioned approaches limit themselves to deal with human faces in frontal view, and require the input face to be free of environment and the size to be approximately consistent. Moreover, they cannot tolerate varying pose, expression, and noise and defocus problems. These constraints greatly hinder the usefulness of the system. In order to remove these aforementioned restrictions, a robust system that is suitable for real time face detection is developed in this paper. Speedy and accurate detection of faces in images is a crucial goal to be pursued in any face detection system. Our system is proposed to use this valuable component — color, and use triangle-based approach to solve the problems in aforementioned approaches. The designed system is composed of two principal parts as shown in Figure 1. The first part of the designed system is to search the potential face regions. The second part is to perform the task of face verification for each potential face region. The proposed face detection system can detect multiple faces oriented in complicated background automatically. In addition to cope with the problem of partial occlusion of mouth and sunglasses, it can deal with different size, dissimilar lighting condition, varying pose and expression, and noise and defocus problems.

The rest of the paper is organized as follows. In section 2, segmentation of potential face regions based on color-triangle-based approach is described. In section 3, each of the normalized potential face regions is fed to the weighting mask function to verify whether the potential face region really contains a face. Experimental results are demonstrated in section 4 to verify the validity of the proposed face detection system. Finally, conclusions are given in section 5.

2 Segmentation of Potential Face Regions

The main purpose of this process is to find the regions in an input image that might potentially contain faces. The segmentation process consists of four steps. Firstly, read in a color image. Then, convert this input color image to a binary image. Secondly, label all 4-connected components in the image to form blocks and find out the center of each block. Thirdly, detect any 3 centers of 3 different blocks that form an isosceles triangle. Fourthly, clip the blocks that satisfy the triangle criteria as the potential face region.

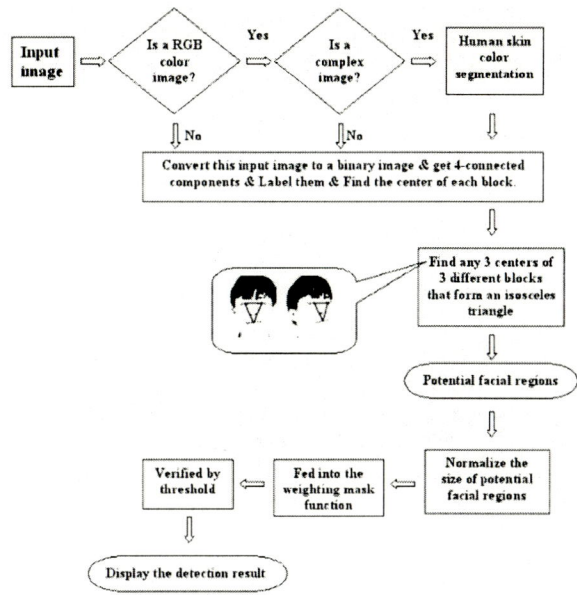

Fig. 1. Overview of our system 2.1 Preprocess the Inputted Image to a Binary Image

When we read in a RGB color image, we will confirm whether the input RGB color image contains complicated backgrounds by checking the number of blocks in the image with threshold $T = 100$. If the number of blocks in the image with threshold $T = 100$ is larger than 30, we conclude that the input RGB color image are with complicated backgrounds. Then we will proceed *human skin color segmentation task*. Our principle is that the output of human skin color segmentation must include all the similar skin color. The output may have other light colors as well, which is not very important. The important thing is that the similar ones should not be missed in the human skin color segmentation process because the other light colors will be eliminated during the binarization process.

In order to use the color options effectively, it is necessary to understand the meanings of the color parameters. Initially, we will explain the so-called Red-Green-Blue (or RGB) color model. For the purpose of color distribution, an image is formed by a set of (n * m) pixels with each pixel being a point in the RGB color space. RGB color space is a 3-dimensional vector space, and each pixel, $p(i)$, is defined by an ordered triple of red, green, and blue coordinates, $(r(i), g(i), b(i))$, which represent the intensities of red, green, and blue light in he color respectively. Each primary color component (red, green, and blue) represents an intensity that varies linearly from 0 to 255 corresponding to the full saturation of that color. The detail of RGB color space can be found in Rafael C. et al. [8].

By carefully observing the results as shown in Figure 2, where Figure 2(a) depicts the r, g, b values of white color, black color, and the colors of skin. From the observation, we realize that the absolute values of r, g, b is totally different with the altered illumination conditions. On the other hand, the relative value between r, g, b is almost similar with the various illumination circumstances. Shown in Figure 2(b) is the original 32-bit-color map. Figure 2(c) is the result of human skin color segmentation of Figure 2(b). If the first rule: the value of *r (i)* *(the intensity of red light) is smaller than* α *(α = 80)*, then the pixel will be assigned to <u>pure red color</u> that we can see about 45 % non-skin colors are replaced by <u>pure red color</u> instead of their original colors. If the value of *(r (i) – g (i)) ((the intensity of red light) - (the intensity of green light))* is *not* between 0 and β (β = 56), then the pixel will be assigned to <u>pure green color</u> that we can comprehend about 50 % more non-skin colors are changed to <u>pure green color</u> instead of their original colors. If the value of *(r (i) – b (i)) ((the intensity of red light) - (the intensity of blue light))* is *not* between 0 and γ (γ = 98), then the pixel will be assigned to <u>pure blue color</u> that we can realize the last 5 % non-skin colors are substituted by <u>pure blue color</u> instead of their original colors. Finally, only the pixels with skin color are not modified. Figure 2(d) is the result of human skin color segmentation of Figure 2 (b). If any one of the above 3 rules is not satisfied, then the pixel is assigned to <u>pure black color</u>. In other words, only the pixels with skin color are remained, and the others are assigned to <u>pure black color</u>. From Figures 2(b), 2(c) and 2(d), we discover that the relative value between r, g, b is roughly similar with the various illumination circumstances. Next, we conduct numerous experiments (testing by real images), and discover that the relative value between r, g, b is approximately similar. This discovery is really useful to succeed the task of human skin color segmentation (It's also valuable for finding the other colors by using different values of (α, β, γ)). Shown in Figure 2(e) to 2(i) illustrate one example of our experiment. Figure 2(e) is the original input RGB color image, 2(f) is the result of human skin color segmentation of 2(e) — only the skin color will be remained, 2(g) is the binary image of the original input RGB color image by $T = 100$ and the number of blocks is 167, (h) is the binary image of the result of human skin color segmentation by $T = 100$ and the number of blocks is 32, 2(i) is the binary image of the result of human skin color segmentation by $T = 100$, then perform the opening operation (erosion first, then dilation) to remove noise, and then the closing operation (dilation first, then erosion) to eliminate holes, and the number of blocks is 21. Due to the decrease of the number of blocks (from 167 to 21), we can expect that the speed will be improved drastly in the complicated background case.

From the experiments of real images, we obtain 3 rules to aid the task of human skin color segmentation:
1. $r(i) > \alpha$: means that the primary color component (red) should be larger than α.
2. $0 < (r(i) - g(i)) < \beta$: means that the primary color component (red – green) should be between 0 and β.
3. $0 < (r(i) - b(i)) < \gamma$: means that the primary color component (red – blue) should be between 0 and γ.

The first rule means that the value of *r (i)* — *the intensity of red light* should be larger than α. The second rule means that the value of *(r (i) – g (i))* — *(the intensity of red light) - (the intensity of green light)* should be between 0 and β. The third rule means that the value of *(r (i) – b (i))* — *(the intensity of red light) - (the intensity of blue light)* should be between 0 and γ. In other words, if the pixels of the input image satisfy the above 3 rules, then the pixels are regarded as skin color and will keep the original color. Otherwise, we will treat the pixels as non-skin color and assign them to pure black color.

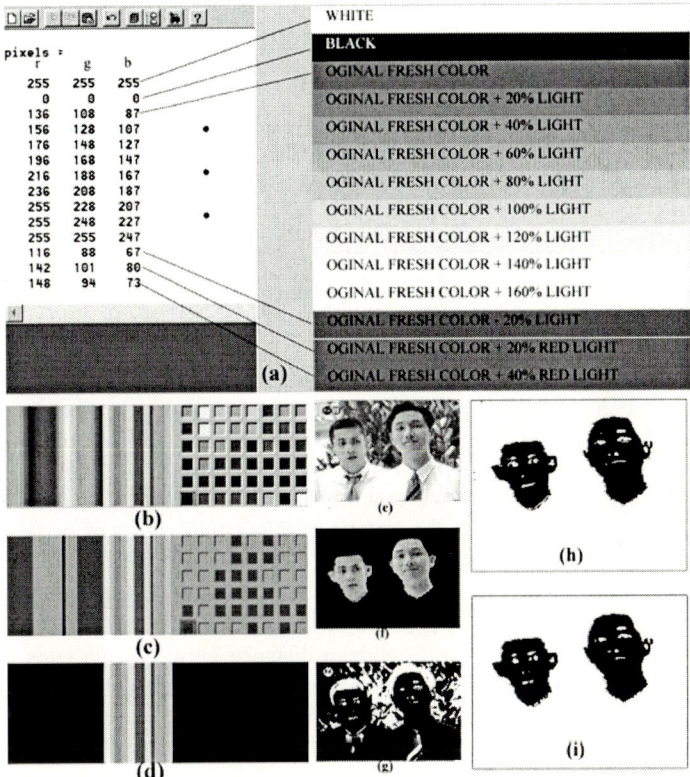

Fig. 2. The human skin color segmentation.

Then, we will use the result of human skin color segmentation as the inputted image, and preprocess it to a gray-level image by eliminating the hue and saturation information while retaining the luminance. If the input image is a gray-level image or a RGB color image with uncomplicated backgrounds, we will skip the step of finding the human skin color segmentation. Then, binarize the gray level image to a "binary image" by simple global thresholding with threshold T because the objects of interest in our case are darker than the background. Pixels with gray level $\leq T$ are labeled black, and pixels with gray level $> T$ are labeled white. Hence, the output binary image has values 1 (black) for all pixels in the input image with luminance less than threshold T and 0 (white) for all other pixels. Before proceeding to the next step, we perform the opening operation (erosion first, then dilation) to remove noise, and then the closing operation (dilation first, then erosion) to eliminate holes. From Figure 2(h), the number of blocks is gotten as 21. Since it can decrease the number of blocks, opening and closing operation can save a lot of executing time in complex background cases. The detail of opening and closing operations can be found in [8].

2.2 Label All 4-Connected Components and Find the 3 Centers of 3 Different Blocks

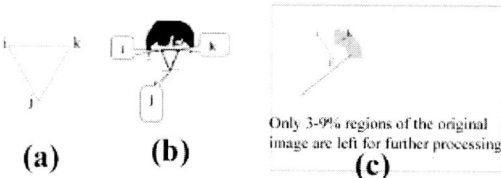

Fig. 3. (a) The isosceles triangle $i\,j\,k$.;(b) Three points (i, j, and k) satisfy the matching rules, so we think them form an isosceles triangle. The search area of the third block center $\underline{k}$ is only limited in the dark area instead of all area of the image; (c) The search area is only limited in the dark area instead of all area of the image.

Here, we use raster scanning (left-to-right and top-to-bottom) to get 4-connected components, label them, and then find the center of each block. The detail of raster scanning can be found in [8]. From careful observation, we discover that two eyes and one mouth in the frontal view will form an isosceles triangle. This is the rationale on which the finding of potential face regions is based. We could search the potential face regions that are gotten from the criteria of *"the combination of two eyes and one mouth (isosceles triangle)"*.

If the triangle $i\,j\,k$ is an isosceles triangle as shown in Figure 3(a), then it should possess the characteristic of "the distance of line $i\,j$ = the distance of line $j\,k$". From observation, we discover that the Euclidean distance between two eyes (line $i\,k$) is about 90% to 110% of the Euclidean distance between the center of the right/left eye and the mouth. Due to the imaging effect and imperfect binarization result, a 25% deviation is given to absore the tolerance. The first matching rule can thereby be stated as (abs(D(i, j)-D(j, k)) < 0.25*max(D(i, j), D(j, k))), and the second matching rule is (abs(D(i, j)-D(i, k)) < 0.25*max(D(i, j), D(j, k))). Since the labeling process is operated from left to right then from top to bottom, we can get the third matching rule as "$i < j < k$". Here, "abs" means the absolute value, "D (i, j)" denotes the Euclidean distance between the centers of block i (right eye) and block j (mouth), "D (j, k)" denotes the Euclidean distance between the centers of block j (mouth) and block k (left eye), "D (i, k)" represents the Euclidean distance between the centers of block i (right eye) and block k (left eye). For example, as shown in Figure 3(b), if three points (i, j, and k) satisfy the matching rules, then i, j, and k form an isosceles triangle. If the Euclidean distance between the centers of block i (right eye) and block j (mouth) is already known, then block center k (left eye) should be located in the area of 75% to 125% of Euclidean distance between the centers of block i (right eye) and block j (mouth), which will form a circle. Furthermore, since the third matching rule is "$i < j < k$", the third block center k is only limited in the up-right part of the circle which is formed by rules 1 and 2. In other words, the search area is only limited in the dark area instead of all area of the image as show in Figure 3(c). As a result, it is not really a selection of C_3^n (select any 3 combination from n blocks). In this way, the triangle-based segmentation process can reduce the background part of a cluttered image up to 97%. This process significantly speeds up the subsequent face detection procedure because only 3-9% regions of the original image are left for further processing.

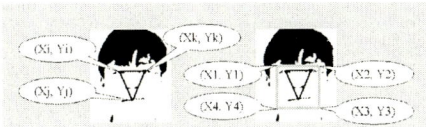

Fig. 4. Assume that (Xi, Yi), (Xj, Yj) and (Xk, Yk) are the three center points of blocks i, j, and k, respectively. The four corner points of the face region will be $(X1, Y1)$, $(X2, Y2)$, $(X3, Y3)$, and $(X4, Y4)$.

After we have found the isosceles triangle, it is easy to obtain the coordinates of the four corner points that form the potential facial region. Since we think the real facial region should cover the eyebrows, two eyes, mouth and some area below mouth [7], the coordinates can be calculated as follows: Assume that (Xi, Yi), (Xj, Yj) and (Xk, Yk) are the three center points of blocks i, j, and k, that form an isosceles triangle. $(X1, Y1)$, $(X2, Y2)$, $(X3, Y3)$, and $(X4, Y4)$ are the four corner points of the face region as shown in Figure 4. X1 and X4 locate at the same coordinate of $(Xi - 1/3*D (i, k))$; X2 and X3 locate at the same coordinate of $(Xk + 1/3*D (i, k))$; Y1 and Y2 locate at the same coordinate of $(Yi + 1/3*D (i, k))$; Y3 and Y4 locate at the same coordinate of $(Yj - 1/3*D (i, k))$; where $D (i, k)$ is the Euclidean distance between the centers of block i (right eye) and block k (left eye).

$$X1 = X4 = Xi - 1/3*D (i, k). \quad (1)$$

$$X2 = X3 = Xk + 1/3*D (i, k). \quad (2)$$

$$Y1 = Y2 = Yi + 1/3*D (i, k). \quad (3)$$

$$Y3 = Y4 = Yj - 1/3*D (i, k). \quad (4)$$

3 Face Verification

The second part of the designed system is to perform the task of face verification. In the previous section, we have selected a set of potential face regions in an image. In this section, we propose an efficient weighting mask function that is applied to decide whether a potential face region contains a face. There are three steps in this part. The first step is to normalize the size of all potential facial regions. The second step is to feed every normalized potential facial region into the weighting mask function and calculate the weight. The third step is to perform the verification task by thresholding the weight obtained in the precious step.

3.1 Normalization of Potential Facial Regions

Normalization of a potential face region can reduce the effects of variation in the distance and location. Since all potential faces will be normalized to a standard size (e.g. 60 * 60 pixels) in this step, the potential face regions that we have selected in the previous section are allowed to have different sizes. Here, we resize the potential facial region using "bicubic" interpolation technique.

3.2 Weighting Mask Function and Weight Calculation

If the normalized potential facial region really contains a face, it should have high similarity to the mask that is formed by 10 binary training faces. Every normalized potential facial region is fed into a weighting mask function that is used to compute the similarity between the normalized potential facial region and the mask. The computed value can be utilized in deciding whether a potential region contains a face or not. The method for generating a mask is to read in 10 binary training masks that are cut manually from the facial regions of images, then add the corresponding entries in the 10 training masks to form an added mask. Next, binarize the added mask by thresholding each entry. Take Figure 5 as an example, we have 10 masks and the size of each mask is 3*3. The first mask is formed by 7 "zero" ("zero" represents white pixel), so the first mask is almost a white 3*3 block. The 9th mask is produced by 7 "one" ("one" represents black pixel), so the 9th mask is almost a black 3*3 block. We added these 10 masks together and get an added mask with the value of each entry as [7 5 7 5 4 7 5 4 1]. If we select the threshold value 5 (if the value of each entry is larger than 4, then we assign its value as 1; otherwise we assign its value as 0.), we can get the final mask that has the values of [1 1 1 1 0 1 1 0 0]. 10 masks and the size of each mask is 3*3. We added them together and get an added mask as [7 5 7 5 4 7 5 4 1]. If we select the threshold value 5 (if the value of each entry is larger than 4, then we assign its value as 1; otherwise we assign its value as 0.), then we can get the final mask of [1 1 1 1 0 1 1 0 0].

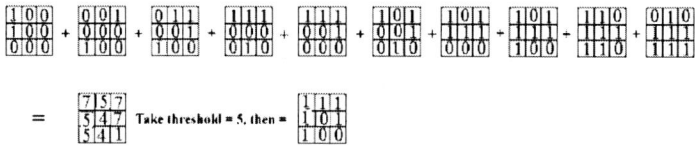

Fig. 5. There are 10 masks and the size of each mask is 3*3. We added them together and get an added mask as [7 5 7 5 4 7 5 4 1]. If we select the threshold value 5 (if the value of each entry is larger than 4, then we assign its value as 1; otherwise we assign its value as 0.), then we can get the final mask of [1 1 1 1 0 1 1 0 0].

In our system, the mask is created by 60*60 pixels. Since the ratio of "the area of two eyes, the nose, and the mouth" and "the facial area excepting two eyes, the nose, and the mouth" is about 1/3, we design the algorithm for getting the weight of the potential facial region as follows.

For all pixels of the potential facial region and the mask: If both of the potential facial region and the mask contain those pixels at the same location of the parts of two eyes, the nose, and the mouth (both of them are black pixels.), then weight = weight + 6. If both of the potential facial region and the mask contain those pixels at the same location of the parts of the skin of the face (both of them are white pixels.), then weight = weight + 2. If the pixel of potential facial region is black and the pixel of the mask is white, then weight = weight - 4. If the pixel of potential facial region is white and the pixel of the mask is black, then weight = weight − 2. The experimental results show that the values of + 6, + 2, - 4, and − 2 as given above can obtain the best match of the facial region. In other words, the range of threshold values is the narrowest.

3.3 Verification

After we have calculated the weight of each potential facial region, then a threshold value is given for decision making. Once a face region has been confirmed, the last step is to eliminate those regions that overlap with the chosen face region, then exhibit the result. Verification of the frontal view - A set of experimental results demonstrates that the threshold values of the frontal view should be set between 4,000 and 5,500.

4 Experimental Results and Discussion

There are 600 test images (include 480 different persons) containing totally 700 faces which are used to verify the effectiveness and efficiency of our system. Some test images are taken from digital camera, some from scanner, and some from videotape. The sizes of the test images range from 10*10 to 640*480 pixels (actually our system could handle any size of image.) In these test images, human faces were presented in various environments.

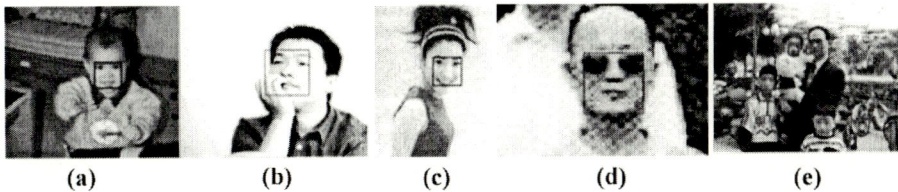

Fig. 6. Experimental results of color images with complex backgrounds: (a) need less than 9.6 seconds to locate the correct face position using PII 233 PC, (b) with partial occlusion of mouth, (c) with noise; (d) with noise and wearing sunglasses, (e) multiple faces.

The execution time required to locate the precise locations of the faces in the test image set is dependent upon the size, resolution, and complexity of images. For example, the color image with complex backgrounds which is 120*105 pixels as shown in Figure 6(a) need less than 9.6 seconds to locate the correct face position using PII 233 PC. The experimental results are showed as Figure 6: Experimental results of color images with complex backgrounds: (a) need less than 9.6 seconds to locate the correct face position using PII 233 PC, (b) with partial occlusion of mouth, (c) with noise, (d) with noise and wearing sunglasses, (e) multiple faces.

5 Conclusions

In this paper, a robust and effective face detection system is presented to extract face in various kinds of face images. The color and triangle-based segmentation process can reduce the background part of a cluttered image up to 97%. Since human skin color segmentation can replace the complex background with the simple/black background, human skin color segmentation can reduce a lot of executing time in the complicated background case. This process significantly speeds up the subsequent face detection procedure because only 3-9% regions of the original image are left for further processing. The experimental results reveal that the proposed method is better than traditional methods in terms of altered circumstance: (1) We can detect the face which is smaller than 50 * 50 pixels. (2) We detect multiple faces (more than 3 faces) in complex backgrounds. (3) We can handle the defocus and noise problems. (4) We can conquer the problem of partial occlusion of mouth or wearing sunglasses. In the future, we plan to use this face detection system as a preprocessing for solving face recognition problem.

References

1. T. K. Leung, M. C. Burl, and P. Perona, "Finding faces in clustered scenes using random labeled graph matching", in Proc. Computer Vision and Pattern Recognition,Cambridge, Mass., Jun. 1995, pp. 637-644.
2. H. A. Rowley, S. Baluja, and T. Kanade, "Human face detection in visual scenes", Tech. Rep. CMU-CS-95-158R, Carnegie Mellon University, 1995. (http://www.cmu.edu/~har/faces.html)
3. S. Y. Lee, Y. K. Ham, and R. H Park, "Recognition of human front faces using knowledge-based feature extraction and neuro-fuzzy algorithm", Pattern Recognition, vol. 29, no. 11, pp. 1863-1876, 1996.
4. P. Juell and R. Marsh, "A hierarchical neural network for human face detection", Pattern Recognition, vol. 29, no. 5, pp. 781-787, 1996.
5. K. Sobottka and I. Pitas, "Extraction of facial regions and features using color and shape information", in Proc. 13th International Conference on Pattern Recognition, Vienna, Austria, Aug. 1996, pp. 421-425.
6. H. Wu, Q. Chen, and M. Yachida, "A fuzzy-theory-based face detector", in Proc. 13th International Conference on Pattern Recognition, Vienna, Austria, Aug. 1996.
7. C. C. Han, H. Y. Mark Liao, G. J. Yu, and L. H. Chen, "Fast face detection via morphology-based pre-processing", in Proc. 9th International Conference on Image Analysis and Processing, Florence, Italy, September 17-19, 1997
8. Rafael C. Gonzalez and Richard E. Woods, "Digital Image Processing", copyright © 1992 by Addison-Wesley Publishing Company, Inc.

Multiple People Tracking Using an Appearance Model Based on Temporal Color

Hyung-Ki Roh and Seong-Whan Lee*

Center for Artificial Vision Research, Korea University,
Anam-dong, Seongbuk-ku, Seoul 136-701, Korea
{hkroh, swlee}@image.korea.ac.kr

Abstract. We present a method for the detection and tracking of multiple people totally occluded or out of sight in a scene for some period of time in image sequences. Our approach is to use time weighted color information, i.e., the temporal color, for robust medium-term people tracking. It assures our system to continuously track people moving in a group with occlusion. Experimental results show that the temporal color is more stable than shape or intensity when used in various cases.

1 Introduction

A visual surveillance system is a common application of video processing research. Its goal is to detect and track people in a specific environment. People tracking systems have many various properties from input camera type to detailed algorithm of body parts detection. Darrell et al. [4] used disparity and color information for individual person segmentation and tracking. Haritaoglu et al. introduced the W4 system [1] and the Hydra system [2] for detecting and tracking multiple people or the parts of their bodies. The KidsRoom system [5] is an application of *closed-world* tracking and utilizes of contextual information to simultaneously track multiple, complex, and non-rigid objects. The Pfinder system [3] used a multi-class statistical model of a person and the background for person tracking and gesture recognition.

The human tracking process is not so simple, in that it does more than merely predict the next position of the target person. The information for each person must be maintained although sometimes he/she is occluded by others or leaves the scene temporarily. To address this problem, each person has to be represented by an appearance model [2]. An appearance model is a set of features with which one person can be discerned from the others. The color, shape, texture or even face pattern can be one of the features of the appearance model.

In this paper, we propose a new people tracking method using an appearance model using the temporal color feature in the image sequences. The temporal color feature is a set of pairs of a color value and its associated weight. The

---

* To whom all correspondence should be addressed. This research was supported by Creative Research Initiatives of the Ministry of Science and Technology, Korea.

weight is determined by the size, duration, frequency, and adjacency of a color object. The duration and frequency of the color object are temporal and play very important roles in the model, but the size and adjacency are neither temporal properties nor directly used in the model. However, the size and adjacency have influence on the calculation of the duration and frequency weights. For this reason, we will include them for the temporal color.

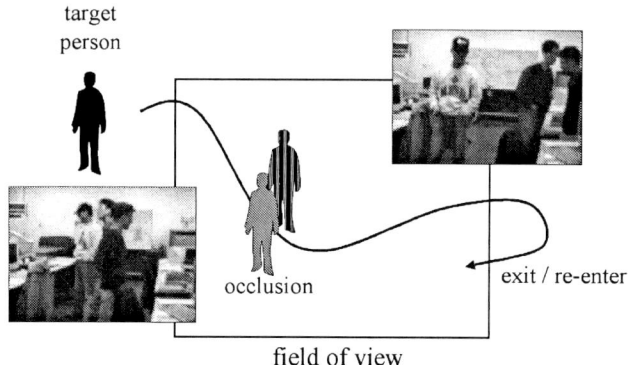

Fig. 1. Difficulties in continous tracking of people

The basic motivation of the temporal color is that the color of the target person's clothes is relatively consistent. In traditional approaches, a temporal template is used for continuous tracking, but it is applicable only to the consecutive frame or a short time occlusion. Since the temporal color is a statistical color information with respect to time, it is not susceptible to shape change or noise. This advantage of the temporal color enables a continuous tracking of each person even when a lengthy occlusion occurs.

2 Appearance Model and Temporal Color

2.1 Tracking in Various Temporal Scales

The person tracking process is a complex one, if there are multiple people in a scene, and even more so if they are interacting with one another. The tracking process can be classified into three types by the temporal scale, namely short-term, medium-term and long-term tracking [4].

The *short-term tracking* process is applied as long as the target stays in the scene. In this type of tracking, we use the second order motion estimation for the continuous tracking of the target person.

The *medium-term tracking* process is applied after the target person is occluded, either partially or totally, or if they re-enter the scene after a few minutes. Short-term tracking will fail in this case, since the position and size correspondences in the individual modules are unavailable. This type of tracking is very important because the meaningful activity of the tracked person may last for several minutes. The statistical appearance model is essential to accomplish medium-term tracking. The appearance model has many features for representing a person, such as shape, texture, intensity, color, and face pattern.

The *long-term tracking* process is the temporal extension of medium-term tracking. The temporal scale is extended to hours or days. Most of the features are unsuitable for this type of tracking, because they become unstable in such a large temporal scale. The only stable, and therefore useful, features are facial pattern and skin color. Since the face recognition module is not installed in our system, the long-term tracking cannot yet be supported.

2.2 Tracking with an Appearance Model

The appearance model plays an important role in the multiple people tracking system. Figure 2 shows the position of the appearance model within the tracking module.

The appearance model is used only for the multiple people tracking module because its main function is the proper segmentation and identification of the multiple people group. However, the appearance model can be used in short-term tracking to make it more robust with the positional information available from the model.

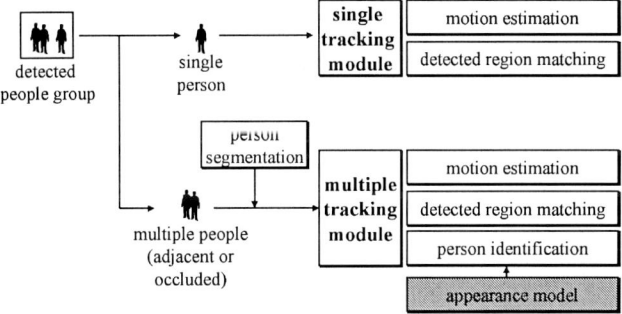

Fig. 2. Appearance model in tracking

In multiple tracking module, the person segmentation process is required. For this purpose, the person segmentation method, which is introduced in [2], is used in our system.

Table 1 shows candidates for features which can be used in the appearance model. Each feature has its own characteristics. The discriminating power (DP) of a feature is its ability of identifying each person. The long-term, the mid-term, and the short-term stability (LS, MS, SS) of a feature indicate their reliability in the corresponding tracking. The time cost (TC) is the complexity of time that it takes in extracting feature information.

Table 1. Candidates for features

features	DP	LS	MS	SS	TC
body position	−	−	−	+	+	+
body shape	+	−	−	+	−
depth	−	−	−	+	−	−
shirt color	+	−	+	+	+
texture	+	−	+	+	−
skin color	−	+	+	+	+
face pattern	+	+	+	+	+	−	−

'+' means that the feature has benefit in the criterion and '−' means that the feature has difficulty in the criterion

Since the color feature are most feasible for fast and robust medium-term tracking, we propose color-based features, namely, the temporal color feature.

2.3 Tracking with Temporal Color Feature

The main idea of temporal color feature is to add shirt color information to the appearance model. The temporal color (F) is defined with the set of color values (C) and its temporal weights (w) as Equation (1).

$$F = \{(C_1, w_1), (C_2, w_2), ..., (C_n, w_n)\} \tag{1}$$

The variable n is determined by the number of color clusters. The first step for calculating the temporal color is the clustering of color space. For each person, a color histogram is constructed and the mean color value of each histogram bin is calculated. Avoiding the quantization error, two bins are merged if their color values are similar. In this way, two or three color clusters for each person can be obtained.

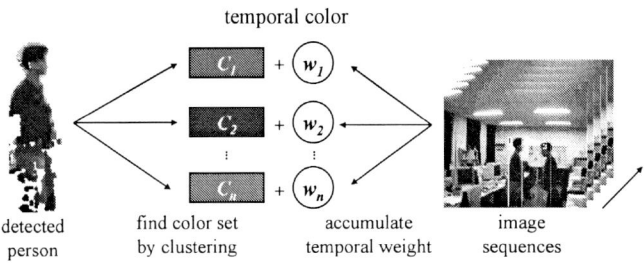

Fig. 3. Temporal color

The color clustering algorithm and the color distance measure are directly related to the accuracy and efficiency of the proposed method. There are many color clustering algorithms and color representations. The considerations for choosing the appropriate algorithms in the proposed method are:

- whether the number of color clusters is known or not
- the color clustering algorithm must have as little time cost as possible
- the color similarity measure must be robust to illumination and noise

In the proposed method, a color cluster is selected by histogram analysis. Each pixel determined as part of the foreground region is represented in the YUV color space. The YUV color representation has an advantage that it separates the luminance component (Y) and the chrominance components (U, V). This advantage makes it possible to minimize the effect of white illumination, because white illumination affects only to the luminance component. To satisfy this condition, we construct a YUV color distance measure between pixel i and pixel j, shown as:

$$d = \alpha(Y_i - Y_j)^2 + \beta((U_i - U_j)^2 + (V_i - V_j)^2) \quad (2)$$

where α and β are weights for luminance components and chrominance components. For minimizing the effect of white illumination, the β value must be larger than the α value.

To construct histogram bins, the Y, U, and V components are divided equally and the histogram bin whose value is over the threshold is selected as the representative color cluster. The average color value of all pixels in a selected histogram bin is the color feature of the proposed method. By the division of histogram bins of the same size, an error can occur where one cluster is divided into two. To overcome this problem, all color values of a selected cluster are compared to each other the two closest color clusters are merged.

Temporal weight is determined by the size, duration, frequency of its associated color and the existence of adjacent objects. The color information of a bigger size, longer duration, and with no adjacent object is more reliable. The relation between weight (w), size (S), duration and frequency (TA), and adjacency function (Γ) can be shown as:

$$w_n^t \propto \frac{S_n^t \cdot TA_n^t}{\Gamma(t)} \quad (3)$$

The higher the temporal weight, the more reliable its associated color value. The S is a constant value for each person and the TA (time accumulation variable) is increased when the same color is detected continuously.

To model the temporal weight in a computational system, it is simplified into a concrete form. The size and adjacency affect only the incremental degree of TA. TA represents the duration and frequency of the appearance of the pixel, the color of which is the same as TA's associated color. The basic notion is to increase its weight when the color appears continuously in the frames, and to decrease its weight when the color disappears frequently or does not appear for long frames.

To simulate the effect of the shape size, the increasing and decreasing degrees of TA are adjusted to the size. The increasing degree (D_i) and decreasing degree (D_d) can be calculated by

$$D_i = \lfloor \frac{|S^t - S_m^{t-1}|}{S_m^{t-1}} \rfloor + 1 \quad and \quad D_d = \lfloor \frac{S^{t-1}}{S_m^{t-1}} \rfloor + 1$$

$$where \quad S_m^t = \frac{S_m^{t-1}(t-1) + S^t}{t} \quad (4)$$

In Equation (4), S^t is the shape size at time t and S_m^t is the average shape size until time t. When TA is increased, only D_i is employed, and when TA is decreased, only D_d is employed.

In addition, the adjacency function is simplified to the adjacency test described below:

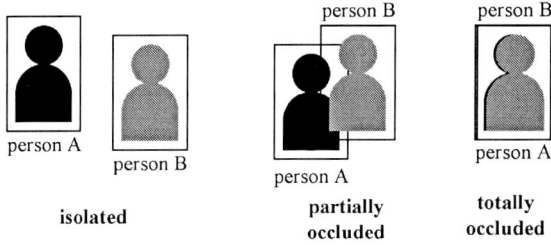

Fig. 4. Cases which are considered in adjacency test

- **isolated** : accumulate its color information
- **partially occluded** : accumulate its color information except for the emerging or vanishing color
- **totally occluded** : do not accumulate its color information for an occluded person, but accumulate its color information for an occluding person

The temporal weight calculation of the proposed method is summarized as follows:

- at $t = 0$, represent each color cluster by the temporal color value and its associated weight and the initial value is set to the base weight.
- at $t = n$, determine whether to increase the temporal weight or to decrease it, according to the result of the adjacency test.
- at $t = n$, increase each weight of the color feature by D_i, when there is a similar color feature at $t = n - 1$; otherwise, decrease it by D_d.

To identify each person after occlusion, the elements whose weight is over the base weight are only used for similarity computation.

3 Experimental Results and Analysis

Our tracking system was implemented and tested on a Pentium II – 333MHz PC, under the Windows 98 operating system. The test video data was acquired using SGI O_2 and a digital camera system. Two video data were collected that contained two or three moving persons in an indoor environment. The frame rate of the video data is varied from 10 Hz to 30 Hz and the size of the input image was 320 by 240 pixels.

Table 2. Experimental results

	NOP	FPS	SUM	DE	TE
scene one (30fps)	2	30	183	2	0
scene one (10fps)	2	10	61	4	0
scene two	3	10	158	14	2

NOP : number of persons, FPS : frame per second, SUM : sum of frames, DE : detection error, TE : tracking error

Table 2 shows parameters and the number of errors for each experiment and Figure 5 shows an the tracking result for scene2 which contains three persons with very complex patterns of movements.

The detection error and tracking error were the number of false detection and false tracking. Because the number of false detection was determined for each frame, one detection error determined that there was one frame which our system could not detect. However, the tracking error was determined for each identification, rather than for individual frames. Therefore, if there was one tracking error, it could affect the next frame, and then all affected errors in the frame are regarded as one tracking error.

To confirm the robustness of the temporal color feature, the variance of the temporal color feature for each person is analyzed. If it is small, the temporal color could be considered as having a robust feature. We compare the color feature used in our system and the shape feature from the temporal template used in [2,5]. Because shape information includes local information, it is more sensitive to small changes. Therefore we use shape size for fair comparison.

In our experiments, the effect of illumination and the complexity of background objects make it difficult to detect the region of a person precisely. Therefore, the variation of shape size is very large, as expected.

Figure 6 shows the variation of each feature for one tracked person. The target person for this graph is the person marked with 'A' in Figure 5. If a set of foreground pixels of the target person (P), the image pixel value $(I(x, y);$ it can be a color value, or 1 for shape), and the number of pixels of the target person (N) is given, the graphs in Figure 6 can be obtained by:

$$\frac{\sum_{x,y \in P}^{N}(M_P - I(x,y))}{N}$$

Fig. 5. Experimental result (scene two)

$$\text{where } M_P = \frac{\sum_{x,y \in P}^{N} I(x,y)}{N} \tag{5}$$

The number of the foreground pixels of the target person is considered for Figure 6 (a), and the Y (luminance) component of the first color cluster of the target person is considered for Figure 6 (b). For each graph in Figure 6, the horizontal axis is the frame number and the vertical axis is the normalized variance.

For the graphs shown in Figure 7, the temporal information is considered. In these graphs, the horizontal axis is frame number and the vertical axis is the normalized variance. The color value is not temporal because the temporal information is not in the color but in its associated weight. The shape is also not temporal. The shape is in the information from the temporal template, excluding intensity information. In order to simulate the variance of the temporal color, the temporal weight is encoded into the color value. This can simulate the variance of the temporal color. The graphs shown in Figure 7, can be drawn from Equation (6).

$$I_t = \frac{I_{t-1} * (t-1) + C(t)}{t} \tag{6}$$

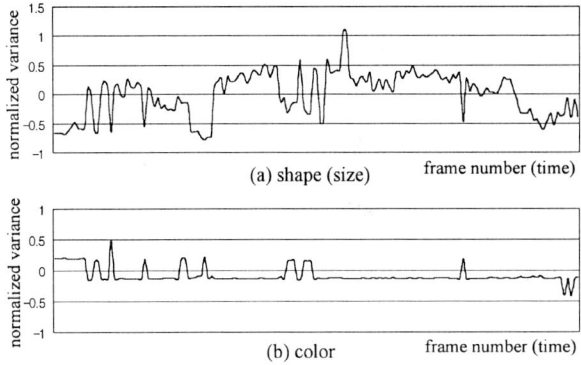

Fig. 6. Comparison of the stability of each feature

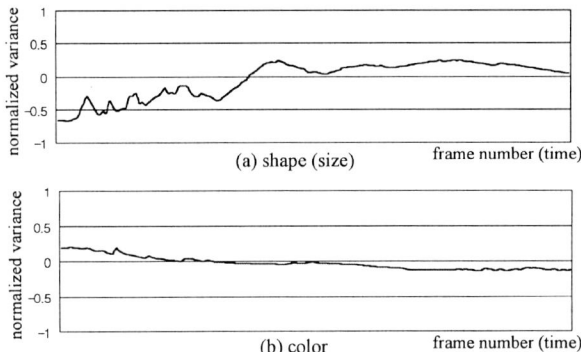

Fig. 7. Comparison of stability of each feature with temporal information

In Equation (6), I_t is the color value with temporal information at time t, and $C(t)$ is the color value at time t, which is equal to the value used in Figure 6.

Figure 7 shows that the color value of temporal weight is more stable than the shape information in an image sequence.

4 Conclusions and Further Research

In this paper, we proposed an appearance model based on temporal color for robust multiple people tracking in short- and mid-term tracking. Although the position, shape, and velocity are suitable features in tracking of consecutive frames, they cannot track the target continuously when the target disappears temporarily. Because the proposed temporal color is accumulated with its associated weight, the target can be continuously tracked when the target is occluded or leaves the scene for a few seconds or minutes.

Our experimental results reveal the stability of the proposed method in the medium-term tracking. The proposed temporal feature can be generalized with any kind of features in the appearance model for person identification in the people tracking process.

Problems with temporal color only occur when people are in uniforms or clothes with similar color. To solve this problem, other features need to be used simultaneously for the tracked target, or face recognition of the target person is required.

Further research will be concentrated on the integration of the tracking system with face pattern recognition for long-term tracking.

References

1. I. Haritaoglu, D. Harwood and L. S. Davis. : W4: Who? When? Where? What? A Real Time System for Detecting and Tracking People. Proc. of International Conference on Face and Gesture Recognition. Nara, Japan, April (1998).
2. I. Haritaoglu, D. Harwood and L. S. Davis. : Hydra: Multiple People Detection and Tracking Using Silhouettes. Proc. of 2nd IEEE Workshop on Visual Surveillance. June (1999) 6-13.
3. C. Wren, A. Azarbayejani, T. Darrell and A. Pentland. : Pfinder: Real-Time Tracking of the Human Body. Trans. on Pattern Analysis and Machine Intelligence. 19 (1997) 780-785.
4. T. Darrell, G. Gordon, M. Harville and J. Woodfill : Integrated Person Tracking Using Stereo, Color, and Pattern Detection. Proc. of Computer Vison and Pattern Recognition. (1998) 601-608.
5. S. S. Intille, J. W. Davis and A. F. Bobick : Real-Time Closed-World Tracking. Proc. of Computer Vison and Pattern Recognition. (1997) 697-703.

Face and Facial Landmarks Location Based on Log-Polar Mapping

Sung-Il Chien and Il Choi

School of Electronic and Electrical Engineering
Kyungpook National University
1370, Sankyuk-dong, Pook-gu, Taegu 702-701, Korea
sichien@ee.knu.ac.kr

Abstract. In this paper, a new approach using a single log-polar face template has been adopted to locate a face and its landmarks in a face image with varying scale and rotation parameters. The log-polar mapping which simulates space-variant human visual system converts scale change and rotation of input image into constant horizontal and cyclic vertical shift in the log-polar plane. The intelligent use of this property allows the proposed method to eliminate the need of using multiple templates to cover scale and rotation variations of faces and thus achieves efficient and reliable location of faces and facial landmarks.

1 Introduction

Human face recognition has attracted many researchers because of its potential application in areas such as security, criminal identification, man-machine interface, etc. Locating a face and its landmarks automatically in a given image is a difficult yet important first step to a fully automatic face recognition system [1]. Since the position, the size, and the rotation of a face in an image are unknown, its recognition task often becomes very difficult.

To overcome these problems, multiple eye templates from Gaussian pyramid images had been used in [2], whereas multiple eigentemplates for face and facial features had been chosen in [3]. Recently, retinal filter banks implemented via self-organizing feature maps had been tested for locating such landmarks [4]. A common feature found in these methods is the use of multitemplate and multiresolution schemes, which inevitably give rise to intensive computation involved.

In this paper, we develop a scale and rotation invariant approach to locate face and its landmarks using a single log-polar face template. The input frontal face images with varying scales and rotations are mapped into the log-polar plane, in which the input candidate faces arising from various fixation points of mapping will be matched to a template. The log-polar mapping [5][6] which simulates space-variant human visual system converts scale change and rotation of input image into constant horizontal and cyclic vertical shift in the output plane. Intelligent use of this property allows us to adopt a single template over a log-polar plane instead of adapting a variety of multiple templates to scale changes and rotations.

2 System Implementation and Experiments

The first step is to construct a face log-polar template in the off-line mode. A face image is histogram-equalized and then smoothed using a Gaussian function to enhance contrast and suppress noise. Then the face region is modeled as an ellipse as shown in Fig. 1a and its most interesting part is segmented out to be a template as shown in Fig. 1b.

Log-polar mapping is the next step. A point (x, y) in the input image is mapped to a point (u, v) in the log-polar image by the following equation detailed elsewhere [5, 6].

$$(u, v) = (\log_b r, \theta), \tag{1}$$

where $r(= \sqrt{x^2 + y^2})$ is the distance from a fixation point and θ is given as $\tan^{-1}\left(\frac{y}{x}\right)$. The logarithmic base b acts as a scale factor and can be determined by matching the given maximum values of r, r_{max} to that of u, u_{max}. Fig. 2a shows an illustration diagram of a log-polar mapping. To obtain a pixel value at the point (u, v) on the log-polar plane, the point is reversely transformed to the corresponding point (x, y) on the Cartesian coordinate by

$$\begin{bmatrix} x \\ y \end{bmatrix} = \begin{bmatrix} x_c \\ y_c \end{bmatrix} + \lfloor b^u \begin{bmatrix} \cos(\theta v) \\ \sin(\theta v) \end{bmatrix} \rfloor, \tag{2}$$

where (x_c, y_c) is the fixation point in the Cartesian coordinate. Here, a crucial decision is choice of fixation points. It has been found from various simulations that a midpoint linking the centers of two pupils performs best when compared to other points such as tip of a nose and the center of mouth. To estimate fixation points more robustly under general environmental conditions such as

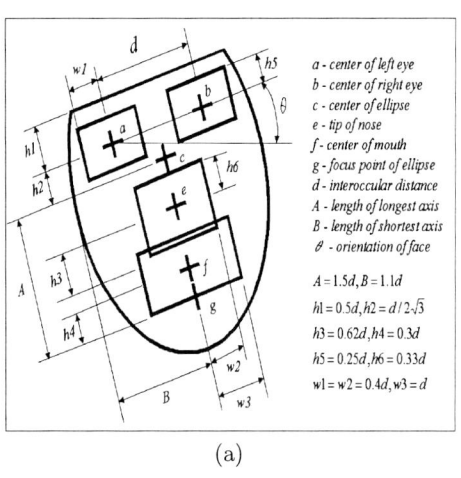
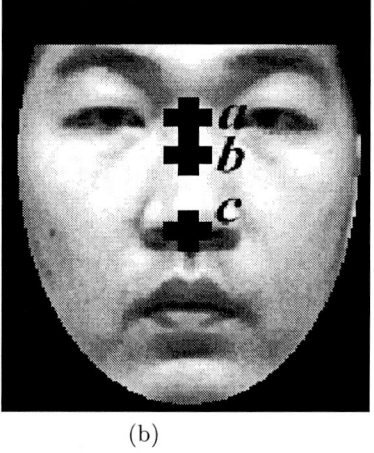

(a) (b)

Fig. 1. (a) Geometry of a face model. (b) Ellipse face model with fixation points **a**, **b**, and **c**.

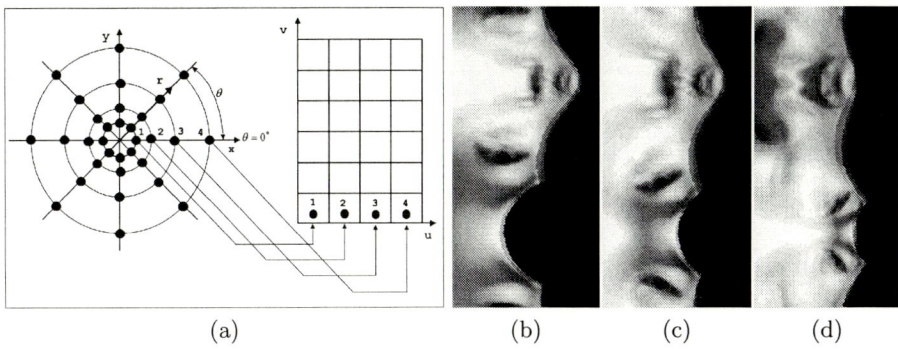

Fig. 2. (a) Schematic diagram of Log-Polar mapping. (b)-(d) Log-polar face templates of at fixation points **a**, **b**, and **c** in Fig. 1b respectively.

uncontrolled illumination, complex background, glasses, several faces, etc, the raster-scan based method using the generalized symmetry transform [10] is being under studied.

Figs. 2b, 2c, and 2d show the resultant log-polar template images of at fixation point **a**, **b**, and **c** in Fig. 1b respectively. These log-polar face templates are somewhat blurred because a template is obtained by realigning and averaging three images of different scales and rotations, diminishing its sensitivity to a particular scale or orientation. Note that the log-polar mapping accentuates a central or fovea region near the fixation point while coarsely sampling and drastically deemphasizing a peripheral region. It thus can provide distinctive facial features and yet offer relative invariance against facial expression, hair styles, and especially mouth movement [11] which is often harmful for the template matching in the Cartesian coordinates.

As a testing database, we collected 795 images from 53 persons of both genders. For each person, 15 images were obtained covering 5 variations of scale, i.e., their interoccular distances from 35 to 130 pixels and 3 angles of in-plane rotation. The maximum face area is about four times larger than the minimum face area and head-tilting is allowed within ± 25 degree. Sample images are shown in Fig. 3.

Now, we detail the on-line location procedure. A given image is also histogram-equalized and smoothed and then segmented using well-known Otsu method [7]. Regions of interest, in which the dark regions are surrounded by the white background can be detected using a typical region labeling algorithm as shown in Fig. 4. As our strategy for selecting candidate fixation points, we focus on detecting two pupils from the segmented regions, because the eye regions are quite prominent and thus can be relatively easy to identify. Moreover, when automatic face recognition system is intended for the purpose of security check, illumination is usually well controlled and it is highly probable that the two shapes of segmented eye regions are quite similar to each other. By the use of shape descriptors such as the boundary length and area for each region, several candidate eye pairs can be located and some of them are marked as in Fig. 5.

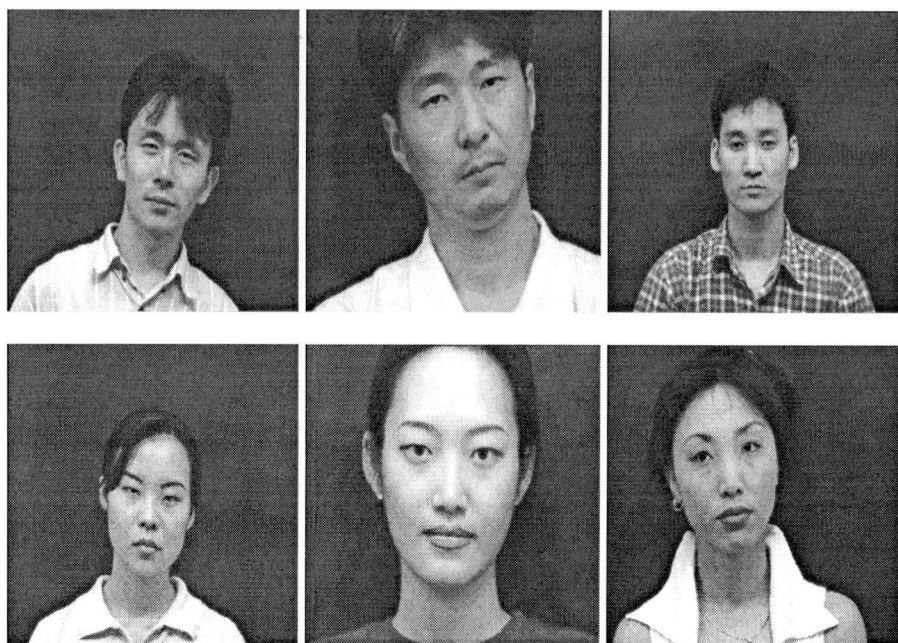

Fig. 3. Sample images of our face database.

Fig. 4. Labeled regions of interest.

Fig. 5 also represents their candidate faces from the elliptic model of a face and their corresponding log-polar images. Then fixation points of mapping are midpoints between the two centers of eye regions. The final step is matching the log-polar images to the templates. As explained previously, scale and rotation variations of an object are reflected as shifts along the corresponding axes of the log-polar plane, which can be corrected by sliding the template to respective axes

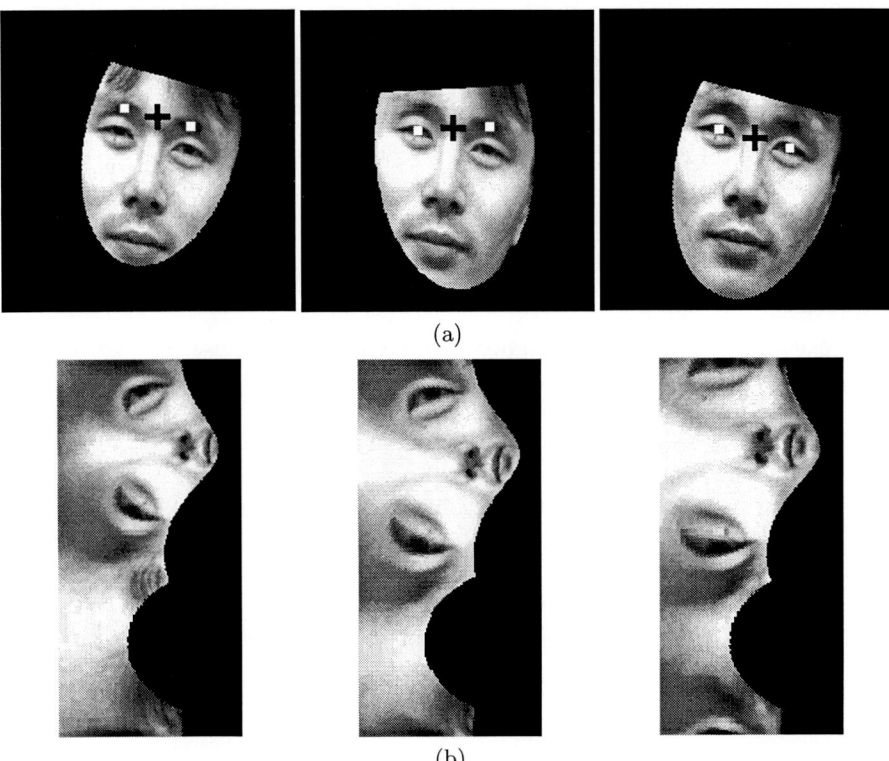

Fig. 5. (a) Examples of the candidate faces with three fixation points. Two small white boxes represent candidate eye pairs and black crosses represent fixation points. (b) Their corresponding log-polar images.

or by employing shift-invariant techniques, for instance, using the magnitude of Fourier transform or second-order local auto-correlation coefficient [12]. However, since scale information is available from the interocular distance between eyes and rotation angle from the slope of their connecting line, the amounts of shift Δu and Δv in each direction can be estimated by

$$\Delta u = \lfloor (\log_b d_i - \log_b d_t) + 0.5 \rfloor, \tag{3}$$
$$\Delta v = \lfloor (\theta_i - \theta_t) v_{max} + 0.5 \rfloor. \tag{4}$$

Here, d_i and d_t are respectively the interocular distances of the input candidate pairs and that of template face, θ_i and θ_t are their rotation angles, and v_{max} is the maximum value of v. The estimation of these shifting parameters allows us to avoid the further use of special shift-invariant techniques often introduced elsewhere [8]. It is also possible to normalize the input face to a template by interpolation using the interoccular distances and rotation angles in the Cartesian coordinates. The fine-tuning dithering mechanism needed to estimate its size and rotation angle more accurately is merely accomplished by simple horizontal shifts and vertical cyclic shifts in the log-polar plane while interpolation involving intensive computation is repeatedly needed for such a matching in the Cartesian coordinates. The dc values are removed from the templates as well as input log-polar images so as to further reduce illumination variations. The template correctly shifted is then overlapped and dithered over the mapped candidate faces. The candidate face showing minimum matching error is finally declared as accepted one. The elliptic model of a face designed to be fully specified by the two spatial positions of centers of pupils as shown in Fig. 1a is used to locate the remaining facial landmarks: a nose and a mouth. These location results are illustrated in Fig. 6.

We actually implemented and tested three templates whose sizes are 32×64, 64×128, and 128×256 (this corresponds to Fig. 2b). Their face location rates with varying fixation points are included in Table 1. The best performance is obtained at fixation point **a**, though the performance is only slightly degraded at fixation point **b** which corresponds to the center of a face ellipse. This result also confirms the previous findings [9] that the best candidates of fixation points lie within the central area around the eyes and the nose. Successful location is declared only when detected regions are really pupils and matching error remains within the predefined tolerance. To estimate accuracy of pupil detection, we marked the centers of pupils manually and compared them to the finally

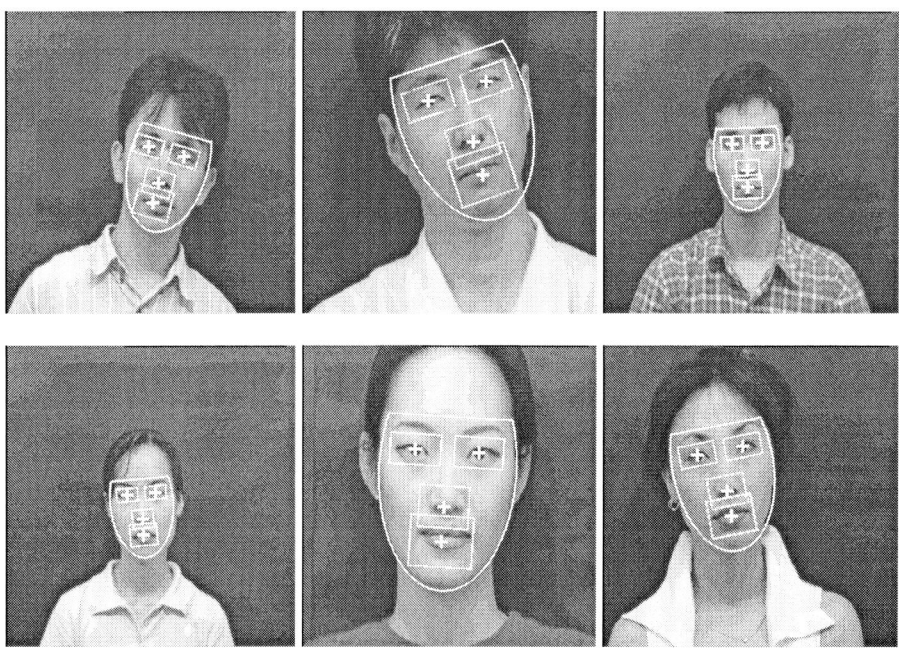

Fig. 6. Results of face and facial landmarks location with large variations of size and rotation.

Table 1. Table 1 Face location rates with varying fixation points.

Fixation Point	Template Size		
	32×64	64×128	128×256
a	98.1%	98.7%	98.7%
b	96.7%	98.4%	98.6%
c	91.0%	91.0%	91.0%

detected centers of labeled eye regions. The mean of location error was about 3 pixels with the variance of 0.25. We believe this result to be fairly good, considering that the original image size is 512×480. For a given image, the average run time of the proposed method is within 1 second on the PC with 333MHz Pentium CPU.

3 Conclusions

Scale and rotation invariant location of a face and its landmarks has been achieved based on log-polar mapping. The face model specified by two centers of pupil is employed to locate the positions of a nose and a mouth and also estimate their sizes and orientations. The developed method is quite attractive and efficient, since no special algorithm or architecture is needed to adapt multiple templates to scale and rotation variations of faces.

References

1. R. Chellappa, C. L. Wilson, and S. Sirohey: Human and machine recognition of faces: A survey. Proc. of IEEE, vol. 83, no. 5, 1995.
2. R. Brunelli and T. Poggio: Face recognition: features versus templates. IEEE Trans. Pattern Analysis and Machine Intelligence, vol. 15, no. 10, pp. 1042-1052, 1993.
3. B. Moghaddam and A. Pentland: Probabilistic visual learning for object representation. IEEE Trans. Pattern Analysis and Machine Intelligence, vol. 19, no. 7, pp. 696-710, 1993.
4. B. Takacs and H. Wechsler: Detection of faces and facial landmarks using iconic filter banks. Pattern Recognition, vol. 30, no. 10, pp. 1623-1636, 1997.
5. R. A. Messner and H. H. Szu: An image processing architecture for real time generation of scale and rotation invariant patterns. Computer Vision, Graphics and Image Processing, vol. 31, pp. 50-66, 1985.
6. B. Fischl, M. A. Cohen, and E. L. Schwartz: The local structure of space-variant images. Neural Networks, vol. 10, no. 5, pp. 815-831, 1997.
7. N. Otsu: A Threshold selection method from gray-level histogram. IEEE Trans. Systems, Man, and Cybernetics, vol. SMC-9, no. 1, pp. 62-66, Jan. 1979.
8. B. S. Srinivasa and B. N. Chatterji: An FFT-based Techniques for translation, rotation, and scale invariant image registration. IEEE Trans. Image Processing, vol. 5, no. 8, pp. 1266-1271, Aug. 1996.

9. M. Tistarelli: Recognition by using an active/space-variant sensor. Proc. IEEE Conf. Computer Vision and Pattern Recognition, pp. 833-837, June 1994.
10. D. Reisfeld, H. Wolfson, and Y. Yeshurun: Context-Free Attentional Operators: The Genarialized Symmetry Transform. Intl. Journal of Computer Vision, 14, pp. 119-130, 1995.
11. S-H Lin, S-Y Kung, and L-J Lin: Face Recognition/Detection by Probabilistic Decision-Based Neural Network. IEEE Trans. Neural Networks, vol. 8, no. 1, pp. 114-132, Jan. 1997.
12. K. Hotta, T. Kurita, and T. Mishima: Scale Invariant Face Detection Method using Higher-Order Local Autocorrelation Features extracted from Log-Polar Image. Proc. 2nd Intl. Conf. on Automatic Face and Gesture Recognition, pp. 70-75, April 14-16 1998.

Biology-Inspired Early Vision System for a Spike Processing Neurocomputer

Jörg Thiem, Carsten Wolff, and Georg Hartmann

Heinz Nixdorf Institute, Paderborn University, Germany
{thiem,wolff,hartmann}@get.uni-paderborn.de

Abstract. Examinations of the early vision system of mammals have shown GABOR-like behaviour of the simple cell responses. Furthermore, the advantages of GABOR-like image filtering are evident in computer vision. This causes strong demand to achieve GABOR-like responses in biology inspired spiking neural networks and so we propose a neural vision network based on spiking neurons with GABOR-like simple cell responses. The GABOR behaviour is theoretically derived and demonstrated with simulation results. Our network consists of a cascaded structure of photoreceptors, ganglion and horizontal neurons and simple cells. The receptors are arranged on a hexagonal grid. One main advantage of our approach compared to direct GABOR filtering is the availability of valuable intersignals. The network is designed as the preprocessing stage for pulse-coupled neural vision networks simulated on the SPIKE neurocomputer architecture. Hence, the simple cells are implemented as ECKHORN neurons.

1 Introduction

Commonly, there are two points of view for biology motivated computer vision. One research goal is the better understanding of the mechanisms in brains, especially in their vision part. The other challenge is to implement at least some of these capabilities into technical vision systems. In this contribution a model for the early stages of an image processing system is described which combines several features from biological vision systems also known to be useful in computer vision tasks.

The system is used as a preprocessing stage for a neurocomputer architecture and consists of a three step neural vision network composed of receptors, horizontal and ganglion cells and simple cells. Several examinations of the simple cell responses in brains have shown GABOR-like behaviour of these cells [1,11]. GABOR filtering is also proved to be advantageous in computer vision tasks. Especially the significant response to lines and edges and the robustness to illumination changes are widely used for image processing. So one feature of the presented vision network is the GABOR-like behaviour of the simple cells. As shown in [18], the photoreceptors in the mammal fovea are nearly hexagonally arranged. A hexagonal grid yields a superior symmetry, definite neighbourhood and less samples compared to a rectangular sampling while complying with SHANNON.

The information processing up to the ganglion cells is usually modelled with continuous signals. However, most of the brain's information processing is supposed to be pulse-coupled (these pulses are also known as spikes). Several examinations have shown that in addition to the pulse rate of these signals especially the phase information is used for vision tasks [3,4]. Neurons which represent a coherent feature - e. g. a continuous line in their receptive field - synchronize their pulses. This mechanism is supposed to be advantageous for many perception tasks, like object segmentation, and for this sake the presented network is adapted as a preprocessing stage for spiking neural networks. The ganglions convert their activity into pulse rates and the simple cells are designed as pulse-coupled ECKHORN neurons [3,4]. The combination of GABOR-like behaviour on a hexagonal grid with pulse-coupled information processing and the adaptation of this network to the special simulation hardware are presented in several steps. First, the hexagonal sampling of images is described. Based on this sampling a network structure is derived which consists of horizontal, ganglion and simple cells. For this network structure connection masks are estimated that cause GABOR behaviour for the simple cell responses. The GABOR impulse response is proved theoretically and documented with an example. The implemented network not only calculates signals of simple cells, but also gives access to valuable intersignals. Like in biology, these signals could be used for other visual tasks. The ganglion signal should be mentioned here as an example, because it is used in nature for gaze control or adaptation methods. As described in [7], it is also plausible, that ganglion signals could be used to realize similar ganglion receptive fields of bigger size for composition of simple cells with different resolutions and hence to easily build up a resolution pyramid. These mentioned intersignals are calculated automatically in the presented cascaded architecture and could improve a technical imaging system as well. In a further step the ganglions are combined with spike encoders similar to those of ECKHORN neurons [3,4]. The simple cell neurons are complete ECKHORN neurons without using the special linking capability. For these neurons a reasonable setup has to be found. Because spikes and also spike rates cannot be negative a separation between negative and positive ganglion responses has to be made which is similar to the on- and off-behaviour in brains. This leads to a special structure for the pulse-coupled network part. The presented preprocessing stage is designed for pulse-coupled vision systems and has to be calculated in real world scenes near to real time. This requirement cannot be fulfilled with standard computers because of the complex calculations for the individual neurons. Several neurocomputer architectures have been designed to offer the required simulation power. The presented network is designed for the SPIKE [5,6,17] neurocomputer architecture, especially for the ParSPIKE design [17]. ParSPIKE is based on a parallel DSP array and the presented network is adapted to this architecture. Preliminary results of this parallel approach are presented in the conclusion of this contribution.

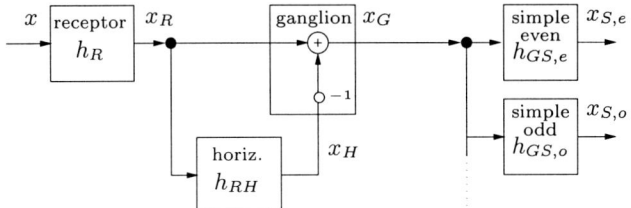

Fig. 1. Signal Circuit of the Filter Bank.

2 Topology and Design of the Network

The information processing in the retina is rather complex and not fully understood. Therefore, the main focus here will be the design and approximation of impulse responses of spatial digital filters – in neural aspects referred to as so called receptive fields (RFs). Neither behaviour in time domain, nor nonlinear effects should be considered. In Fig. 1 the cascaded architecture of the complete filter bank is shown. According to [15] the major subsystems concerning the retina are the receptor, the horizontal, and the ganglion filter. These systems should adopt the synaptical connections, respectively the spatial filtering characteristics of the retinal cells. There is a direct synaptical connection between the bipolar and the ganglion cells and only few influence by amacrin cells in the foveal region. For this reason, bipolar and amacrin cells are neglected in this specification. Before we go into details here, we first have to consider, that it is a digital system as far as the discrete spatial argument $\boldsymbol{k} \in \mathbb{Z}^2$ is concerned. The photoreceptors, lying on a discrete grid in the retina, are sampling the continous intensity distribution $x_c(t)$ which is formed by the lens of the eye. By several optical effects, i. e. diffraction and aberration, the eye behaves as a radial symmetric low-pass filter with 20 dB cut-off frequency below 50 cpd (cycles per degree) [9]. Therefore, the retinal signal is circularly bounded in the frequency domain. In this case, among all periodic sampling schemes $x(\boldsymbol{k}) = x_c(\boldsymbol{Tk})$ with a sampling matrix $\boldsymbol{T} \in \mathbb{R}^{2\times 2}$ the hexagonal sampling $\boldsymbol{T} = [\boldsymbol{t_1}, \boldsymbol{t_2}]$ with $\boldsymbol{t_{1,2}} = [1, \pm\sqrt{3}]^\mathrm{T}$ is the most efficient one, because there are fewest samples needed to describe the whole information of the continous signal [2]. And in fact, as shown in [18], the photoreceptors in the mammal fovea are nearly hexagonally arranged. The average cone spacing (csp) has been estimated to about 1 csp $= \|\boldsymbol{t_{1,2}}\| \approx 2.5\ldots 2.8$ μm. The sampling theorem of SHANNON [2] then would postulate a maximal frequency of $53\ldots 56$ cpd in the signal leading to the presumption that the sampling in the fovea of the human eye nearly complies with the sampling theorem. Indeed this conclusion seems to be correct and gets along with WILLIAMS [16].

Although the mathematical description on a hexagonal grid is quite more difficult than a conventional rectangular one, it offers many advantages, e. g. less samples while complying with SHANNON, less computation time for processing, definite neighbourhood, and superior symmetry. Hence, the presented filter design will be based upon a hexagonal sampling scheme with a spacing of 1 csp = $\|\boldsymbol{t}_{1,2}\| := 2.5\ \mu m$.

2.1 Photoreceptors

In accordance with [15] we use the features of the red-green cone system in the central fovea – where blue cones are nonexistent – to form the model's receptors. In this way we catch the highest spatial resolution in contrast to the rod-system or cones in peripheral regions. Because of the large spectral overlap and the wide spectral range of the red and green cones [14] we can regard red and green cones as one sampling system in this foveal region. The impulse response (or RF) of the receptors can be described with a differerence of two gaussians (DoG), but in most cases a gaussian distribution

$$h_R(\boldsymbol{k}) = g(\sigma_R, \boldsymbol{t} = \boldsymbol{Tk}) = \frac{1}{2\pi\sigma_R^2}\, e^{-\frac{\|\boldsymbol{Tk}\|^2}{2\sigma_R^2}} \qquad (1)$$

will be sufficient. It is well known that this field widens when average lighting conditions decrease to get a better signal-to-noise ratio. As described in [15], the standard deviation lies in the range $\sigma_R = 1.5\ldots 12$ csp, but for simplicity, adaptation will not be implemented yet and we use $\sigma_R := 1.5$ csp for the best spatial resolution.

2.2 Horizontal and Ganglion Cells

The family of ganglion cells can be subdivided into three classes [14]. For our purpose, we will use the X-ganglion cells (β-cells, parvocellular cells). They behave linear in time and space, provide the best spatial resolution and transmit information of color and shape.

As seen in Fig. 1, the ganglion cell gets input from one receptor $x_R(\boldsymbol{k})$ reduced by the signal from one horizontal cell $x_H(\boldsymbol{k})$. Obviously, the output signal of the ganglion cell can be calculated by the convolution sum $x_G(\boldsymbol{k}) = h_G(\boldsymbol{k}) * x(\boldsymbol{k})$, while

$$h_G(\boldsymbol{k}) = [1 - h_{RH}(\boldsymbol{k})] * h_R(\boldsymbol{k}) = h_{RG}(\boldsymbol{k}) * h_R(\boldsymbol{k}) \qquad (2)$$

is the ganglion RF. As physiological experiments indicate, this field shows a center-surround characteristic, which on the one hand can be approximated by a DoG-function, as used in existing models [15]. But in contrast to known models, here, the desired RF should be the more signal theoretical laplacian derivation of a gaussian (LoG) [12]

$$h_{G,\text{des}}(\boldsymbol{k}) = \Delta_t g(\sigma_G, t)\Big|_{t=\boldsymbol{Tk}} = -\frac{1}{\sigma_G^2}\left[2 - \frac{\|\boldsymbol{Tk}\|^2}{\sigma_G^2}\right]\cdot g(\sigma_G, t = \boldsymbol{Tk}). \qquad (3)$$

The central part measures a minimal diameter of 10 μm (4 csp) for best lighting conditions [8]. This hint constitutes the zero crossing of the LoG-function at 5 μm (2 csp) and a standard deviation of $\sigma_G = \frac{5\ \mu m}{\sqrt{2}} = \sqrt{2}$ csp.

Finally, we have to design the unknown system $h_{RG}(\boldsymbol{k})$ (or in equivalence $h_{RH}(\boldsymbol{k}) = 1 - h_{RG}(\boldsymbol{k})$) in a way, that the resulting RF of the ganglion cell is reached in an "optimum" sense. For this aim, we can write the double sum of the 2D-convolution as a single sum over $j = 1 \ldots N_G$ by indexing the vectors $\boldsymbol{n}$. In the same way, we describe $\boldsymbol{k}$ by a set of indices $i = 1 \ldots M$ to obtain

$$h_{G,\text{des}}(\boldsymbol{k}_i) \approx \sum_{j=1}^{N_G} h_{RG}(\boldsymbol{n}_j) \cdot h_R(\boldsymbol{k}_i - \boldsymbol{n}_j). \tag{4}$$

For $M > N_G$ we get an overdetermined linear system of equations, which can be solved in a least square sense.

2.3 Cortical Simple Cells

The output signals of ganglion cells are mostly transmitted to the primary visual cortex (Area 17 or V1). Here, scientists discovered the simple cells, whose RFs can be approximated with either even or odd GABOR elementary functions [10,13]. Therefore, these cells distinctively respond to either local oriented lines or edges of different spatial wideness in the image. In addition, there always exist two neighbouring cells having this opponent behaviour, which can be combined into one complex-valued filter with approximately quadrature characteristic. The signal circuit of the cortical part with one cell pair of even and odd behaviour is shown in Fig. 1. In the present model, several ganglion cells converge onto one simple cell, whose output signal is $x_S(\boldsymbol{k}) = h_S(\boldsymbol{k}) * x(\boldsymbol{k})$ with $h_S(\boldsymbol{k}) = h_{GS}(\boldsymbol{k}) * h_G(\boldsymbol{k})$. The mathematical expression for the desired RF of one simple cell is given by $h_{S,\text{des}}(\boldsymbol{k}) = \operatorname{Re} h_{S,\text{des}}^Q(\boldsymbol{k})$ for an even, and $h_{S,\text{des}}(\boldsymbol{k}) = \operatorname{Im} h_{S,\text{des}}^Q(\boldsymbol{k})$ for an odd RF with the complex impulse response of the quadrature filter

$$h_{S,\text{des}}^Q(\boldsymbol{k}) = K \cdot e^{-\frac{1}{2}(\boldsymbol{T}\boldsymbol{k})^T \boldsymbol{R}^T \boldsymbol{P} \boldsymbol{R}(\boldsymbol{T}\boldsymbol{k})} \cdot e^{j \boldsymbol{w}_0^T (\boldsymbol{T}\boldsymbol{k})}, \tag{5}$$

where $K = 1/(\sqrt{2\pi\sigma_1\sigma_2})^2$, $\boldsymbol{R}$ is a rotation matrix, and $\boldsymbol{P} = \operatorname{diag}\{1/\sigma_1^2, 1/\sigma_2^2\}$ is a parameter matrix. The RFs of cat's simple cells are investigated in great detail [10]. The measured modulation frequencies $\boldsymbol{w}_0$ are quite small compared to the other model components. However, firstly the highest spatial resolution will be selected in our design for technical image processing tasks. To determine the subsystem $h_{GS}(\boldsymbol{k})$ of one simple cell, we again solve an approximation problem

$$h_{S,\text{des}}(\boldsymbol{k}_i) \approx \sum_{j=1}^{N_S} h_{GS}(\boldsymbol{n}_j) \cdot h_G(\boldsymbol{k}_i - \boldsymbol{n}_j) \tag{6}$$

for $j = 1 \ldots N_S$, $i = 1 \ldots M$. We have to take into account, that $h_G(\boldsymbol{k})$ is not the desired ideal RF of the ganglion cell, but the result of the prior approximation. In Table 1 the mean squared errors for several conditions are shown ($R_i \subset \{\mathbb{Z}^2\}$ indicates the neighbourhood of the regarded simple cell). We conclude, that for our purpose $N_G = 7$ and $N_S = 37$ might be sufficient. It still have to be examined, in how far further processing steps depend on the achieved accuracy.

Table 1. Approximation of the Simple Cell's RF.

N_G	N_S	$n \in$	F_{MSE}
7	37	$\{R_0; R_1; R_2; R_3\}$	$4.4015 \cdot 10^{-7}$
	61	$\{R_0; R_1; R_2; R_3; R_4\}$	$2.8677 \cdot 10^{-8}$
19	37	$\{R_0; R_1; R_2; R_3\}$	$2.7315 \cdot 10^{-7}$
	61	$\{R_0; R_1; R_2; R_3; R_4\}$	$1.3635 \cdot 10^{-8}$

3 Implementation with Eckhorn Neurons

After designing the required filter bank, the adaptation to pulse-coupled information processing has to be considered. A spike or pulse in a pulse-coupled neural network is a single event and shows no sign or value. The presented network handles with negative or positive values in the range of -256 up to 255. This leads to the necessity to adapt the pulse-processing stages of the network. Pulse-processing in the context of this contribution means a timeslot simulation of the network and neurons which can emit only one spike for each timeslot. First, the ganglion cell responses have to be converted into pulse streams. To represent negative ganglion responses, two layers of pulse-emitting ganglion cells are established. One layer converts the positive values of the ganglion responses into a pulse-stream, the other layer converts the negative values by representing the absolute value with an appropriate pulse-stream. For the simple cells the same kind of representation for the positive and negative values is chosen. Furthermore, the responses of the simple cells are represented in even and odd parts. For six different orientations ($0°, 30°, \ldots, 150°$) this leads to 24 layers of simple cells and 2 layers of ganglion cells. For the simple cells which represent the positive part of the response, the positive ganglion cells are connected via the derived connection schemes to the excitatory input of the simple neurons and the negative ganglion cells are connected to the inhibitory input. For the negative simple cells the ganglions are connected in the opposite way.

ECKHORN neurons allow this type of network architecture by providing different dendrite trees for different dendritic potentials. In our implementation a model neuron with two dendritic trees (EP1 and IP) is chosen. One of these trees

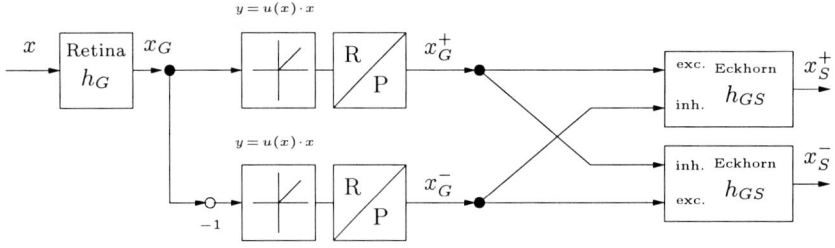

Fig. 2. Adaptation of the Network Structure to Pulse-Coupled Processing.

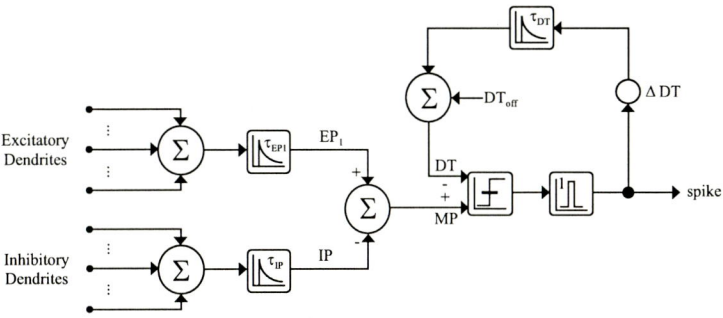

Fig. 3. Simple Cell Model Neuron (Derived from [3,4]).

has an excitatory effect (EP1) on the membrane potential (MP) and the other has an inhibitory effect (IP). The dendritic potentials are accumulated via leaky integrators. For spike emission the membrane potential is fed into a spike encoding circuitry similar to that of FRENCH and STEIN with a dynamic threshold (DT). This spike encoding circuitry is also used for the ganglion output.

Within this network architecture several parameters have to be adjusted. First an appropriate setup for the spike encoder of the ganglions has to be found which converts the ganglion activity into a pulse frequency in an approximately linear way. The parameters of the dynamic threshold (DT) are the time constant τ_{DT} of the leaky integrator, the threshold increment ΔDT and the offset of the threshold DT_{off}. A $\tau_{DT} = 32$ combined with $\Delta DT = 16$ leads to an almost linear conversion of the ganglion activity into a pulse frequency. By cutting off very low activities with $DT_{off} = 4$, the conversion is approximately linear.

The next step is the setup of the leaky integrators for the dendritic potentials of simple neurons. These dendritic potentials have to converge to the same value as the activity represented by the input pulse frequency. It has to be considered that in the case of input from different ganglions the spikes usually arrive at different times. This leads to a long integration time for the dendritic leaky

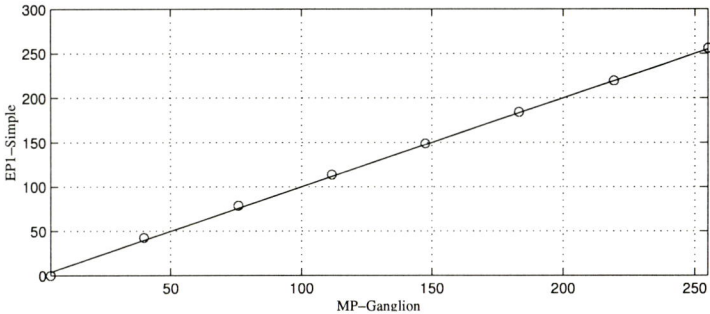

Fig. 4. Conversion of the Ganglion MP via Spikes to Simple Cell EP1 (circles: achieved, line: desired).

integrator. With a time constant $\tau_{EP1} = 1000$ this long integration time is achieved and the dendritic potential converges to the input value represented by the pulse frequency.

The simple cells use the same spike encoder circuitry as the ganglions. Internally, fixed point calculations are used for the neurons with an accuracy of s9.5 for the dendritic potentials and s3.5 for the connection weights. For simulation of dynamic input scenes an important fact has to be considered. These scenes are represented by an image stream delivered by a camera. Typically, the frame rate is much smaller than the maximum pulse frequency, which means that one image is input for several simulation timeslots. Frame updates may cause rapid changes in the pulse frequencies. This rhythmic changes have to be avoided if spike synchronization is used in the vision network. There are several possibilities to avoid rhythmic changes, like interpolation between images.

4 Adaptation to the ParSPIKE

The SPIKE architecture is a neurocomputer design for accelerated simulation of large pulse-coupled neural networks. One implementation is the SPIKE128K system [5,6] which allows the close to real time simulation of networks with 131072 neurons and up to 16 million connections. In this context, real time means an execution time of 1 ms for one simulation timeslot. The ParSPIKE design [17] extends the architecture by a parallel approach to networks with more than one million neurons. In both systems the neurons are represented by the values of their dendritic potentials and of the dynamic threshold. These values can be stored in a memory with a special neuron address for each neuron. The simulation algorithm working on this neuron memory has to update these values in each timeslot - if required. In vision networks only some neurons representing remarkable features in the scene are involved in the information processing. The other neurons are on their rest values. This leads directly to an event-driven simulation algorithm simulating only the active neurons. This algorithm can be implemented in hardware like SPIKE128K or in software on a parallel DSP array like ParSPIKE. For the parallel implementation the neurons are distributed onto several DSPs by storing their neuron memories in the dedicated memory of these processors. ParSPIKE is based on Analog Devices SHARC DSPs with large on-chip-memories. The system offers two type of DSP boards. One type of board is specialized to preprocessing and calculates the connections on-chip from connection masks. To communicate, the parallel DSPs only have to exchange spike lists. These regular connection (rc) boards are well suited for the presented preprocessing network. As described, the network consists of two ganglion layers and 24 simple cells layers. The ParSPIKE rc board offers 32 DSPs for simulation and their on-chip memory can be accessed via VME bus to feed in the input images. These input images are divided into four parts with an appropriate overlap and distributed onto eight DSPs. Each DSP calculates the hexagonal sampling and the ganglion filtering four one fourth of the image. Four DSPs generate the negative ganglion pulse streams and the other four DSPs create

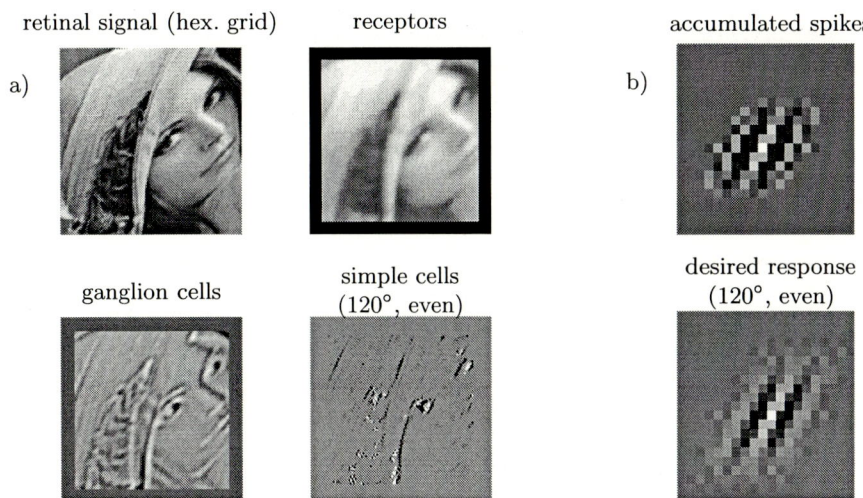

Fig. 5. a) Simulation Results: Due to the Hexagonal Sampling the Figures are Distorted. b) Simple Cell Impulse Response Compared to Desired Gabor Response.

the positive outputs. The 24 simple cell layers are distributed to one DSP each. Due to the hexagonal sampling, input images are processed by 14161 neurons in each layer (119x119). The whole network consists of 368186 neurons including the ganglions and can be processed on one ParSPIKE VME board.

5 Results and Conclusion

In this contribution a neural network has been presented which provides a GABOR-like image preprocessing for pulse-coupled neural vision networks. The network is adapted to a special neurocomputer architecture. The following results show this GABOR-like behaviour and give an estimation for the simulation performance with the proposed accelerator architecture. The example in Fig. 5 a) shows the processing of an input image via the different stages of the network. The ganglion cell responses are calculated from accumulated spikes from 10000 timeslots. The simple cells responses are shown for one orientation (120°) and for the even part of the GABOR response. Due to the several quantization steps in the presented network the GABOR responses differ from ideal GABOR responses. These quantization errors are demonstrated by the comparison of the desired GABOR impulse response with the results from the accumulated spikes of the simple cell impulse response in Fig. 5 b).

The simulation performance depends highly on the speed-up due to the parallel distribution of workload. In the example above the workload of the busiest processor is less than 4% higher than the average workload which means a very good load-balancing. The workload is calculated from simulations with a Sun workstation. Combined with the results of a DSP implementation of the simulation software on a Analog Devices evaluation kit a quite exact estimation for the simulation speed can be given. Simulation on a single processor Sun Ultra60 requires 1600 ms for one timeslot. For the ParSPIKE rc VME board an execution time of 36 ms for one timeslot is estimated.

References

1. Daugman, J.G.: Two-Dimensional Spectral Analysis of Cortical Receptive Field Profiles. Vision Research, Vol. 20 (1980) 847-856
2. Dudgeon, D.E., Mersereau, R.M.: Multidimensional Digital Signal Processing. Prentice Hall (1984)
3. Eckhorn, R., Reitbck, H. J., Arndt, M., Dicke, P.: Feature Linking via Stimulus - Evoked Oscillations: Experimental Results for Cat Visual Cortex and Functional Implications from a Network Model. Proc. IJCNN89, Vol. I (1989) 723-730
4. Eckhorn, R., Reitbck, H. J., Arndt, M., Dicke, P.: Feature Linking via Synchronization among Distributed Assemblies: Simulations of Results from Cat Visual Cortex. Neural Computations 2 (1990) 293-307
5. Frank, G., Hartmann, G., Jahnke, A., Schfer, M.: An Accelerator for Neural Networks with Pulse-Coded Model Neurons. IEEE Trans. on Neural Networks, Special Issue on Pulse Coupled Neural Networks, Vol. 10 (1999) 527-539
6. Hartmann, G., Frank, G., Schfer, M., Wolff, C.: SPIKE128K - An Accelerator for Dynamic Simulation of Large Pulse-Coded Networks. Proc. of the 6th MicroNeuro (1997) 130-139
7. Hartmann, G.: Recursive Features of Circular Receptive Fields. Biological Cybernetics, Vol. 43 (1982) 199-208
8. Hubel, D.: Auge und Gehirn. Spektrum d. Wissenschaft (1990)
9. Iglesias, I., Lopez-Gil, N., Arttal, P.: Reconstruction of the Point-Spread Function of the Human Eye from two Double-Pass Retinal Images by Phases-Retrieval Algorithms. Journal Optical Society of America 15(2) (1998) 326-339
10. Jones, J. P., Palmer, L. A.: An Evaluation of the Two Dimensional Gabor Filter Model of Simple Receptive Fields in Cat Striate Cortex. Journ. of Neurophys. 58(6) (1987) 1233-1258
11. Marcelja, S.: Mathematical Description of the Responses of Simple Cortical Cells. Journal Optical Society of America, Vol 70 (1980) 1297-1300
12. Marr, D., Hildreth, E.: Theory of Edge Detection. Proc. of the Royal Society of London, B 207 (1980)
13. Pollen, D. A., Ronner, S. F.: Visual Cortical Neurons as Localized Spatial Frequency Filters. IEEE Trans. on System, Man and Cybern., 13(5) (1983) 907-916
14. Schmidt, R. F., Thews, G.: Physiologie des Menschen. Springer (1997)
15. Shah, S., Levine, M. D.: Visual Information Processing in Primate Cone Pathways - Part I: A Model. IEEE Trans. on System, Man and Cybernetics, 26(2) (1996) 259-274
16. Williams, D. R.: Aliasing in Human Foveal Vision. Vision Research, 25(2) (1985) 195-205
17. Wolff, C., Hartmann, G., Rckert, U.: ParSPIKE - A Parallel DSP-Accelerator for Dynamic Simulation of Large Spiking Neural Networks. Proc. of the 7th MicroNeuro (1999) 324-331
18. Yellot, J. I.: Spectral Consequences of Photoreceptor Sampling in the Resus Retina. Science, Vol. 212 (1981) 382-385

A New Line Segment Grouping Method for Finding Globally Optimal Line Segments

Jeong-Hun Jang and Ki-Sang Hong

Image Information Processing Lab, Dept. of E.E., POSTECH, Korea
{jeonghun,hongks}@postech.ac.kr

Abstract. In this paper we propose a new method for extracting line segments from edge images. Our method basically follows a line segment grouping approach. This approach has many advantages over a Hough transform based approach in a practical situation. However, since its process is purely local, it does not provide a mechanism for finding more favorable line segments from a global point of view. Our method overcomes the local nature of the conventional line segment grouping approach, while retaining most of its advantages, by incorporating some useful concepts of the Hough transform based approach into the line segment grouping approach. We performed a series of tests to compare the performance of our method with those of other six existing methods. Throughout the tests our method ranked almost highest both in detection rate and computation time.

1 Introduction

A line segment is a primitive geometric object that is frequently used as an input to higher-level processes such as stereo matching, target tracking, or object recognition. Due to its importance, many researchers have devoted themselves to developing good line segment detection methods. The methods proposed up to date can be categorized into following three approaches:

Approach 1: Hough transform based approach [1,2,3]. The biggest advantage of this approach is to enable us to detect collinear edge pixels even though each of them is isolated. Therefore, this approach is useful in finding lines in noisy images where local information around each edge pixel is unreliable or unavailable. Another advantage of this approach is that it has a global nature since the score assigned to each detected line is computed by considering all the edge pixels lying on that line. However, there are several problems in this approach. It requires relatively a large amount of memory and long computation time, and raises the so-called *connectivity problem*, where illusionary lines that are composed of accidentally collinear edge pixels are also detected. To find lines with high accuracy, the approach needs fine quantization of parameter space, but the fine quantization makes it difficult to detect edge pixels that are not exactly collinear. The original Hough transform gives us only information on existence of lines, but not line segments. Therefore, we have to group pixels along each detected line, but this process is local, i.e., it does not guarantee the detection of globally optimal line segments.

Approach 2: Line segment grouping approach [4,5,6]. Elementary line segments (ELSs) are obtained by linking edge pixels and approximating them to piecewise straight line segments. These ELSs are used as an input to this approach. Adjacent line segments are grouped according to some grouping criteria and replaced by a new line segment. This process is repeated until no new line segment occurs. This approach overcomes many weaknesses of Approach 1. However, when most of the edge pixels are isolated or the ELSs are perturbed severely by noises so that information on them is almost useless, the approach does not work. Another disadvantage of this approach is that its process is purely local. Repetition of locally optimal grouping of line segments does not guarantee their globally optimal grouping. For example, consider an optimal grouping criterion in which an ELS needs to be grouped only into the longest line segment among detectable ones that can share the ELS. Such a criterion cannot be handled properly with this approach.

Basically, our method belongs to Approach 2. ELSs are used as an input to our method. However, the grouping mechanism of our method is different from that of the conventional approach. While retaining the advantages of Approach 2, in order to overcome its local nature, we adopted the concept of "line detection by voting" from Approach 1. By combining the concepts of both approaches, a more powerful method could be constructed.

Our method consists largely of three steps. In the first step, ELSs are grouped iteratively by measuring the orientation difference and proximity between an ELS and a line formed by ELSs grouped already. This step yields *base lines*, and for each base line, information on the ELSs contributing to the construction of the base line is also given. It should be noted that an ELS is allowed to belong to several base lines simultaneously. Therefore, a base line does not represent a real line, but represents a candidate line around which true line segments possibly exist. In the second step, ELSs around each base line are grouped into line segments along the base line with some grouping criteria. We designed the algorithm of this step so that closely positioned parallel line segments are effectively separated. In the third step, redundant line segments are removed by applying a restriction that an ELS should be grouped only into the longest line segment among ones sharing the ELS. The restriction can be relaxed so that an ELS is allowed to be shared by several, locally dominant line segments. Note that the criterion we use here for optimal grouping is the *length* of a line segment. Other criteria (e.g., the fitness of a line segment) can be applied easily by making a few modifications to the proposed algorithm. After the third step has been completed, if there exist ELSs remaining ungrouped, the above three steps are repeated, where the remaining ELSs are used as an input to the next iteration.

This paper is organized as follows. In Section 2, the proposed algorithms for the above three steps are described in detail. Section 3 shows experimental results on a performance comparison between our method and other conventional ones. Conclusions are given in Section 4.

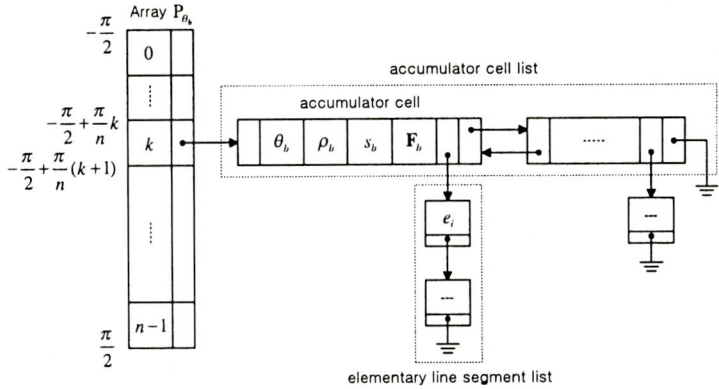

Fig. 1. Basic structure of an accumulator array.

2 The Proposed Method

An input to our method is a set of *elementary line segments*. If an edge image is given, the ELSs are obtained by first linking edge pixels using the *eight-directional following* algorithm, and then approximating them to piecewise straight segments using the *iterative endpoint fit-and-split algorithm* [7].

2.1 Finding Base Lines

Base lines are candidate lines near which true line segments may exist. They are obtained by grouping collinear ELSs. Figure 1 shows a data structure, called an *accumulator array*, which is used to manage base lines. The contents of the accumulator array are updated as the grouping goes on. The accumulator array is composed of *accumulator cells*, each of which contains four parameters (θ_b, ρ_b, s_b, and $\mathbf{F}_b$) concerning the corresponding base line, and an *elementary line segment list* whose elements are ELSs that contribute to the construction of the base line. The parameters θ_b and ρ_b represent two polar parameters of a base line that is given by

$$x \cos \theta_b + y \sin \theta_b = \rho_b, \quad -\frac{\pi}{2} \le \theta_b < \frac{\pi}{2} \quad (1)$$

and the parameter s_b represents a voting score, which is computed by summing all the lengths of the ELSs registered in the ELS list of an accumulator cell. The 3×3 matrix $\mathbf{F}_b$ is used to compute a weighted least-squares line fit to the ELSs in an ELS list. In Figure 1, the array P_{θ_b} is an array of n pointers, where the kth pointer indicates the first accumulator cell of the *accumulator cell list* whose accumulator cells have the values of θ_b in the range of $-\pi/2 + k\pi/n \le \theta_b < -\pi/2 + (k+1)\pi/n$. The array P_{θ_b} is used to find fast the accumulator cells whose values of θ_b are in a required range.

The detailed algorithm for finding base lines is given as follows. Initially, all the pointers of P_{θ_b} are set to null and the accumulator array has no accumulator cells. Let e_i denote the ith ELS ($i = 1, \ldots, N_e$) and let $\theta(e_i)$ and $\rho(e_i)$ represent two polar parameters of a line passing through two end points of e_i. For each e_i, accumulator cells satisfying the condition

$$|\theta_b - \theta(e_i)| \leq \Delta\theta_b + \frac{\Delta\theta_d}{l(e_i)} \qquad (2)$$

are searched for, where $\Delta\theta_b$ and $\Delta\theta_d$ are user-specified parameters and $l(e_i)$ denotes the length of e_i. If no such accumulator cells exist, a new accumulator cell is created and the parameters of the cell are set as follows:

$$\theta_b \leftarrow \theta(e_i), \quad \rho_b \leftarrow \rho(e_i), \quad s_b \leftarrow l(e_i), \quad \text{and} \quad \mathbf{F}_b \leftarrow \frac{l(e_i)}{2}(\mathbf{f}_1^T \mathbf{f}_1 + \mathbf{f}_2^T \mathbf{f}_2)$$

where

$$\mathbf{f}_1 = [x_1(e_i), y_1(e_i), -1], \quad \mathbf{f}_2 = [x_2(e_i), y_2(e_i), -1],$$

and $(x_1(e_i), y_1(e_i))$ and $(x_2(e_i), y_2(e_i))$ represent two end points of e_i. The created cell is registered in an appropriate accumulator cell list according to the value of θ_b. If accumulator cells satisfying Equation (2) exist, a proximity measure $p(e_i, b)$ between e_i and a base line b of each of the accumulator cells is computed, where $p(e_i, b)$ is defined by

$$p(e_i, b) = \sqrt{\frac{1}{l(e_i)} \int_0^1 (x(t) \cos \theta_b + y(t) \sin \theta_b - \rho_b)^2 dt} \qquad (3)$$

where

$$x(t) = x_1(e_i) + t(x_2(e_i) - x_1(e_i)) \quad \text{and} \quad y(t) = y_1(e_i) + t(y_2(e_i) - y_1(e_i)).$$

If $p(e_i, b)$ is less than a threshold T_p and e_i is not registered in the ELS list of the accumulator cell corresponding to the base line b, e_i is registered in the ELS list and the parameters of the accumulator cell are updated. First, $\mathbf{F}_b$ is updated by $\mathbf{F}_b \leftarrow \mathbf{F}_b + \frac{l(e_i)}{2}(\mathbf{f}_1^T \mathbf{f}_1 + \mathbf{f}_2^T \mathbf{f}_2)$ and then the normalized eigenvector corresponding to the smallest eigenvalue of $\mathbf{F}_b$ is computed. The elements of the eigenvector represent the parameters (a, b, and c) of a line ($ax + by = c$) that is obtained by weighted least-squares line fitting of the end points of the ELSs in the ELS list. New values of θ_b and ρ_b are calculated from the eigenvector. Finally, s_b is updated by $s_b \leftarrow s_b + l(e_i)$. Note that if the updated value of θ_b of an accumulator cell is outside the range of θ_b of the accumulator cell list the accumulator cell currently belongs to, the accumulator cell should be moved to a new appropriate accumulator cell list so that the value of θ_b of the accumulator cell is always kept inside the range of θ_b of a current accumulator cell list.

If the accumulator array has been updated for all e_i ($i = 1, \ldots, N_e$), the first iteration of our base line finding algorithm is now completed. The above procedure is repeated until no accumulator cell is updated. It should be noted that the creation of a new accumulator cell is allowed only during the first iteration of the algorithm. The repetition of the above procedure is necessary because for some

e_i and b, the orientation difference $|\theta_b - \theta(e_i)|$ and the proximity measure $p(e_i, b)$ may become less than given threshold values as b is continuously updated. The repetition has another important role. It allows an ELS to be shared by several base lines simultaneously. This property, with the two algorithms that will be explained in the next two subsections, makes it possible for our method to find more favorable line segments from a global point of view.

2.2 Grouping ELSs along Each Base Line

Once base lines have been obtained, we have to group ELSs into actual line segments along each base line. According to the algorithm of Subsection 2.1, each base line carries a list of ELSs that participate in the construction of that line. Therefore, when grouping ELSs along a base line, we take into account only ones registered in the ELS list of the line.

Let b_c represent a line segment that is obtained by clipping a base line b with an image boundary rectangle. If a point on b_c is represented by a single parameter s ($0 \leq s \leq l(b_c)$), the value of s of a point obtained by projecting a point (x, y) onto b_c is given by

$$s(x,y) = \frac{x_2(b_c) - x_1(b_c)}{l(b_c)}(x - x_1(b_c)) + \frac{y_2(b_c) - y_1(b_c)}{l(b_c)}(y - y_1(b_c)). \quad (4)$$

Let e_i^b denote the ith ELS ($i = 1, \ldots, N_b$) registered in the ELS list of the base line b and let $s_1(e_i^b)$ and $s_2(e_i^b)$ ($s_1 < s_2$) represent two values of s that are obtained by substituting two end points of e_i^b into Equation (4). Then our algorithm for grouping ELSs along the base line b is given as follows.

Procedure:

1. Create a 1D integer array I of length $N_I + 1$, where $N_I = \lfloor l(b_c) \rfloor$. Each element of I is initialized to 0.
2. Compute $p(e_i^b, b)$ for $i = 1, \ldots, N_b$.
3. Compute $s_1(e_i^b)$ and $s_2(e_i^b)$ for $i = 1, \ldots, N_b$.
4. Sort p's computed in Step 2 in a decreasing order of p. Let p_j denote the jth largest value of p ($j = 1, \ldots, N_b$) and let $e^b(p_j)$ represent an ELS corresponding to p_j.
5. Do $\{I[k] \leftarrow$ index number of $e^b(p_j)\}$ for $j = 1, \ldots, N_b$ and $k = \lfloor s_1(e^b(p_j)) + 0.5 \rfloor, \ldots, \lfloor s_2(e^b(p_j)) + 0.5 \rfloor$.
6. Create a 1D integer array S of length $N_b + 1$. Each element of S is initialized to 0.
7. Do $\{S[I[k]] \leftarrow S[I[k]] + 1\}$ for $k = 0, \ldots, N_I$.
8. Remove e_i^b from the ELS list if $S[i]/l(e_i^b) < T_m$, where T_m is a user-specified parameter ($0 \leq T_m \leq 1$). By doing so, we can separate closely positioned parallel line segments effectively.
9. Clear the array I and perform Step 4 once again.
10. Group $e_{I[i]}^b$ and $e_{I[j]}^b$ into the same line segment for $i, j = 0, \ldots, N_I$ ($I[i] \neq 0$ and $I[j] \neq 0$) if $|i - j| < T_g$, where T_g is a user-specified gap threshold ($T_g > 0$).

2.3 Removing Redundant Line Segments

According to the algorithm of Subsection 2.1, an ELS is allowed to be shared by several base lines. This property of the algorithm causes an ELS to be shared by several line segments after the grouping of Subsection 2.2. This property is very useful as will be shown in Section 3, but it may produce redundant line segments. For example, if most of the ELSs that belong to some line segment g are shared by more dominant line segments (i.e., longer line segments), the line segment g is redundant and should be removed.

Assume that a total of N_g line segments have been found with the grouping algorithm of Subsection 2.2 and let g_i denote the ith line segment ($i = 1, \ldots, N_g$). Here is our algorithm for redundant line segment removal.

Notations:

- $S(g)$: sum of the lengths of ELSs belonging to a line segment g.
- $L(e)$: label assigned to an ELS e.
- $G(e)$: a set of line segments sharing e as a common member ELS.
- $E(g)$: a set of ELSs belonging to a line segment g.

Procedure:

1. Do $\{L(e_i) \leftarrow$ index number of $g(e_i)\}$ for $i = 1, \ldots, N_e$, where $g(e_i) = \mathrm{argmax}_{g \in G(e_i)} S(g)$.
2. Create a 1D array M of length $N_g + 1$. Each element of M is initialized to 0.
3. Do $\{\mathrm{M}[L(e)] \leftarrow \mathrm{M}[L(e)] + l(e)\}$ for all $e \in E(g_i)$ and $i = 1, \ldots, N_g$.
4. Discard g_i if $\mathrm{M}[i]/S(g_i) \leq T_r$, where T_r is a user-specified parameter ($0 \leq T_r \leq 1$).

The procedures of Subsections 2.1, 2.2, and 2.3 are repeated if there exist ELSs remaining ungrouped. The remaining ELSs are used as an input to the next iteration. This iteration is necessary because ELSs that are not grouped into any line segment may appear during the processes of Subsections 2.2 and 2.3.

3 Experimental Results

In this section, experimental results on a comparison between our method and other line segment detection methods are shown. The conventional methods we chose for the purpose of comparison are as follows: Boldt et al.'s [4], Liang's [8], Nacken's [5], Princen et al.'s [10], Risse's [9], and Yuen et al.'s [11] methods, where Boldt et al.'s and Nacken's methods are based on the line segment grouping approach and the others are related with the Hough transform based approach. Note that originally, Liang's and Risse's methods are not designed for detecting line segments, but for lines. To make use of their methods for line segment detection, we appended a line segment extraction routine, where two adjacent edge pixels on each detected line are grouped into the same line segment if the distance between them is smaller than a given gap threshold.

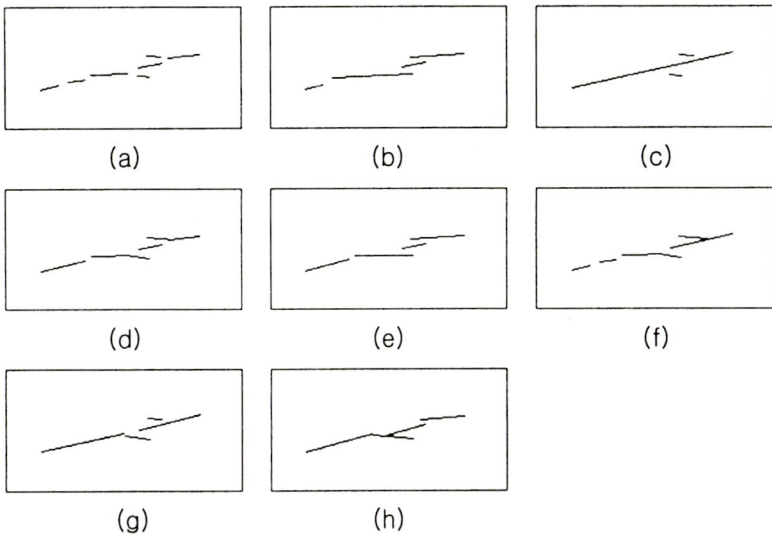

Fig. 2. Line segment detection results of conventional methods and ours for the simple test image: (a) input image, (b) Boldt et al.'s method, (c) our method ($T_r = 0.5$ and $\Delta\theta_b = 7.0°$), (d) Liang's method, (e) Nacken's method, (f) Princen et al.'s method, (g) Risse's method, and (h) Yuen et al.'s method.

To see the characteristics of the methods mentioned above, we made a simple test image as shown in Figure 2(a), where a sequence of ELSs, which looks like a single long line segment that is broken into several fragments, is given from upper right to lower left, and two short line segments are added. Figures 2(b)-(h) show the results yielded by the line segment detection methods. Note that all the methods failed to detect the long line segment except ours. The result of Figure 2(c) illustrates the global nature of our method well. In Boldt et al.'s and Nacken's methods, as can be seen in Figures 2(b) and (e), initial wrong grouping of ELSs (though it might be locally optimal) prohibited the methods from proceeding to group the ELSs further into the long segment. Such a behavior is a fundamental limitation of the conventional line segment grouping methods. In Figure 2(a), edge pixels constituting the long segment are not exactly collinear, which is a main reason for the failure of the Hough transform based methods. In such a case, locally collinear pixels may dominate the detection process of a Hough transform based approach. To solve the problem, coarse quantization of parameter space may be needed, but it causes another problem of the inaccurate detection of lines.

Figure 3 shows other detection results of our method for the same test image of Figure 2(a), obtained by adjusting some threshold values. In Figure 3(a), one can see that some ELSs are shared by more than one line segment. Allowing an ELS to be shared by more than one line segment is necessary when two or more real line segments are closely positioned so that some parts of them become

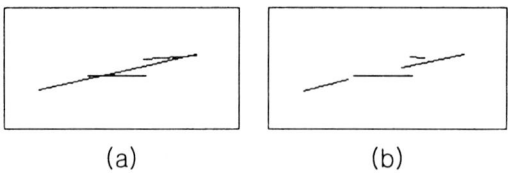

Fig. 3. Other detection results of our method for the same test image of Figure 2(a): (a) $T_r = 0.2$ and (b) $\Delta\theta_b = 3.0°$.

shared after a digitization process. Our method has a capability of allowing such kind of grouping and the degree of sharing is determined by the threshold T_r. The result of Figure 3(b) was obtained by decreasing the value of $\Delta\theta_b$, which has an effect of emphasizing collinearity of ELSs when grouping them.

We carried out four experiments to see the performance of each method. The size of test images used in the experiments is 300×300 and each image contains 18 true line segments that are generated randomly with a length of 40 to 200 pixels. In each experiment, a total of 50 images are generated and tested. In Experiment 1, each test image contains 18 line segments without any fragmentation of them and any noise added (see Figure 4(a)). In Experiment 2, each true line segment is broken into short ones, where the length of each short line segment is 6 to 20 pixels and the gap width is 2 to 6 pixels (see Figure 4(b)). In Experiment 3, a

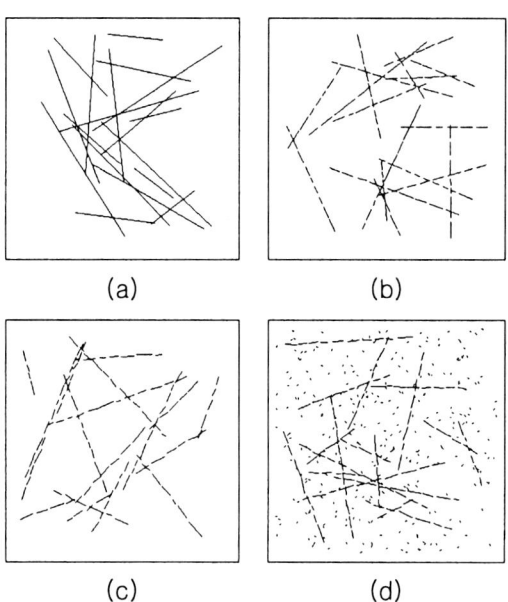

Fig. 4. Sample test images used in the experiments: (a) Experiment 1, (b) Experiment 2, (c) Experiment 3, and (d) Experiment 4.

Table 1. The result of Experiment 1.

match score	Boldt	Ours	Liang	Nacken	Princen	Risse	Yuen
0.0 – 0.9	1.9	1.9	7.1	1.4	3.0	2.4	2.4
0.9 – 1.0	16.5	16.4	13.8	16.3	16.3	16.5	16.0
0.98 – 1.0	14.9	15.2	11.7	14.0	14.0	14.6	13.0
# of line segments detected	18.4	18.3	20.9	17.7	19.3	19.0	18.4
execution time (msec)	153	178	784	1467	4299	6029	737

Table 2. The result of Experiment 2.

match score	Boldt	Ours	Liang	Nacken	Princen	Risse	Yuen
0.0 – 0.9	2.6	1.5	9.4	2.0	3.9	3.7	3.5
0.9 – 1.0	16.0	16.6	11.8	16.0	15.4	15.4	15.3
0.98 – 1.0	14.6	15.5	10.0	14.4	13.4	13.6	13.1
# of line segments detected	18.5	18.2	21.3	18.1	19.3	19.2	18.8
execution time (msec)	202	226	634	6164	3797	5757	701

Table 3. The result of Experiment 3.

match score	Boldt	Ours	Liang	Nacken	Princen	Risse	Yuen
0.0 – 0.9	3.8	2.4	22.7	5.1	8.8	5.5	17.4
0.9 – 1.0	15.1	16.2	5.3	14.4	12.7	14.2	8.5
0.98 – 1.0	13.7	14.8	3.8	12.7	9.1	12.0	5.1
# of line segments detected	18.9	18.5	28.0	19.5	21.5	19.8	26.0
execution time (msec)	200	231	624	6319	3716	5756	915

Table 4. The result of Experiment 4.

match score	Boldt	Ours	Liang	Nacken	Princen	Risse	Yuen
0.0 – 0.9	5.3	11.2	20.0	18.0	11.5	5.8	18.2
0.9 – 1.0	13.8	14.6	5.0	13.8	11.2	13.7	6.9
0.98 – 1.0	10.9	11.0	3.0	11.1	6.74	9.2	3.6
# of line segments detected	19.1	25.8	25.0	31.8	22.7	19.5	25.2
execution time (msec)	1400	833	700	50712	4482	6694	3141

small random perturbation is added to the orientation and the position of each short line segment (see Figure 4(c)). Finally, 300 noisy line segments, each of which is 3 pixels long and has an arbitrary orientation, are added to each test image in Experiment 4 (see Figure 4(d)). Note that the test images were not made incrementally as the experiments proceeded. A completely new set of test images was created for each experiment. We repeated the above experiments several times to obtain good parameter values for each method. It should be noted that the parameters of each method were set equal throughout all the experiments. The experiments were performed on a Pentium II(266MHz) PC and a programming tool used was Microsoft Visual C++ 6.0. The experimental results are shown in Tables 1-4, where given values are average values over 50 test images. In the tables, "match score" represents how much a detected line segment resembles the true line segment. Thus, "match score = 1.0" means a perfect match between a detected line segment and the corresponding true one. Since a test image always contains 18 true line segments, a good line segment detector would give values close to 18.0 in the third to the fifth row of the tables. The tables show that the performance of our method is almost highest among all the methods tested in terms of detection capability and execution time throughout the entire experiments.

4 Conclusions

In this paper we have proposed a new method for finding globally optimal line segments. Our method overcomes the local nature of a conventional line segment grouping approach while retaining most of its advantages. Experimental results showed that our method is fast and has a good detection capability compared to the existing line segment detection methods.

References

1. J. Illingworth and J. Kittler, "A Survey of the Hough Transform," *Computer Vision, Graphics, and Image Processing*, Vol. 44, pp. 87-116, 1988.
2. V. F. Leavers, "Survey - Which Hough Transform?" *Computer Vision, Graphics, and Image Processing*, Vol. 58, No. 2, pp. 250-264, 1993.
3. H. Kälviäinen, P. Hirvonen, L. Xu, and E. Oja, "Probabilistic and Non-probabilistic Hough Transforms: Overview and Comparisons," *Image and Vision Computing*, Vol. 13, No. 4, pp. 239-252, 1995.
4. M. Boldt, R. Weiss, and E. Riseman, "Token-Based Extraction of Straight Lines," *IEEE Transactions on System, Man, and Cybernetics*, Vol. 19, No. 6, pp. 1581-1594, 1989.
5. P. F. M. Nacken, "A Metric for Line Segments," *IEEE Transactions on Pattern Analysis and Machine Intelligence*, Vol. 15, No. 12, pp. 1312-1318, 1993.
6. E. Trucco and A. Verri, *Introductory Techniques for 3-D Computer Vision*, Prentice Hall, pp. 114-117, 1998.
7. R. Haralick and L. Shapiro, *Computer and Robot Vision*, Addison-Wesley, Vol. 1, pp. 563-565, 1992.
8. P. Liang, "A New Transform for Curve Detection," *Proc. International Conference on Computer Vision*, Osaka, Japan, pp. 748-751, 1990.
9. T. Risse, "Hough Transform for Line Recognition: Complexity of Evidence Accumulation and Cluster Detection," *Computer Vision, Graphics, and Image Processing*, Vol. 46, pp. 327-345, 1989.
10. J. Princen, J. Illingworth, and J. Kittler, "A Hierarchical Approach to Line Extraction Based on the Hough Transform," *Computer Vision, Graphics, and Image Processing*, Vol. 52, pp. 57-77, 1990.
11. S. Y. K. Yuen, T. S. L. Lam, and N. K. D. Leung, "Connective Hough Transform," *Image and Vision Computing*, Vol. 11, No. 5, pp. 295-301, 1993.

A Biologically-Motivated Approach to Image Representation and Its Application to Neuromorphology

Luciano da F. Costa[1], Andrea G. Campos[1], Leandro F. Estrozi[1], Luiz G. Rios-Filho[1]
and Alejandra Bosco[2]

[1] Cybernetic Vision Research Group, IFSC – University of São Paulo
Caixa Postal 369, São Carlos, SP, 13560-970, Brazil
{costa, campos, lfestroz, lgrios}@if.sc.usp.br
[2] Neuroscience Research Institute, University of California at Santa Barbara Santa Barbara, CA
93106, USA
bosco@lifesci.ucsb.edu

Abstract. A powerful framework for the representation, characterization and analysis of two-dimensional shapes, with special attention given to neurons, is presented. This framework is based on a recently reported approach to scale space skeletonization and respective reconstructions by using label propagation and the exact distance transform. This methodology allows a series of remarkable properties, including the obtention of high quality skeletons, scale space representation of the shapes under analysis without border shifting, selection of suitable spatial scales, and the logical hierarchical decomposition of the shapes in terms of basic components. The proposed approach is illustrated with respect to neuromorphometry, including a novel and fully automated approach to automated dendrogram extraction and the characterization of the main properties of the dendritic arborization which, if necessary, can be done in terms of the branching hierarchy. The reported results fully corroborate the simplicity and potential of the proposed concepts and framework for shape characterization and analysis.

1 Introduction

The problem of shape representation and analysis corresponds to a particularly interesting and important sub-area of computer vision, which has attracted growing interest. Although no consensus has been reached regarding the identification of the best alternatives for shape representation, the following three biologically motivated approaches have been identified as particularly promising and relevant: (i) the estimation of the curvature along the contours of the analyzed shapes [1, 2, 3] – motivated by the importance of curvature in biological visual perception (e.g. [4]); (ii) the derivation of skeletons [5, 6] – motivated by the importance of such representations [7] (and relationship with curvature [8]) for biological visual perception [9]; (iii) scale space representations allowing a proper spatial scale to be selected [10] – also of biological inspiration; and (iv) development of general shape representations capable of making explicit the hierarchical structure of the shape components, and that also allow a sensible decomposition of the shapes in terms of more basic elements.

The present article describes a powerful framework for representation and analysis of general shapes including the four above identified possibilities and based on the recently introduced family of scale space skeletons and respective reconstructions [5, 6] which is based in label propagation, avoiding the sophisticated straight line adjacency analysis implied by [11]. This approach allows a series of remarkable possibilities, including but not limited to: (a) high quality, connected-of-8 and unitary width skeletons are obtained; (b) the spatial scale can be properly selected in order to filter smaller detail out; (c) the position of the skeletons and the contours of the respective reconstructions do not shift while smaller detail are removed; and (d) a logical and powerful decomposition of the two-dimensional shapes in terms of segments can be easily obtained by "opening" the skeletons at the branch points. This approach was originally motivated by efforts aimed at the characterization of the spatial coverage by neural cells in terms of spatially quantized Minkowski sausages [12]. Indeed, it is argued here that neurons are amongst the most sophisticated and elaborated versions of general shapes and, as such, allows a particularly relevant perspective for addressing the problem of shape analysis and classification, as well as respective validations. As a matter of fact, the henceforth adopted skeletonization approach makes it clear that any two-dimensional shape can be represented in terms of dendrograms, i.e. a hierarchical branching structure of basic segments. The proposed representations and measures are illustrated with respect to a highly relevant problem in computational neuroscience, namely neuromorphometry. It should be observed that many neuronal cells in biological systems, such as Purkinje cells from Cerebellum and retinal ganglion cells, exhibit a predominantly planar organization.

The article starts by presenting the concepts of exact dilations and its use in order to obtain scale space skeletons and respective reconstructions, and follows by presenting how dendrograms can be easily extracted from the skeletons. In addition, it is shown how a logical decomposition of general shapes in terms of segments can be readily achieved by opening the skeletons at their branching points and reconstructing the portions of the shapes in terms of the distance transform. Next, the proposed concepts and techniques are illustrated with respect to the relevant problems of neural cell characterization and classification.

2 Exact Dilations, Scale Space Skeletonization and Reconstructions

Both the distance transform and the label propagation operations necessary for the extraction of scale space skeletons and respective reconstructions have been implemented in terms of exact dilations. As described in [5], exact dilations correspond to the process of producing all the possible dilations of a spatially quantized shape in an orthogonal lattice. A possible, but not exclusive, means for obtaining exact dilations consists in following the shape contour a number of times, respectively to every possible distance in an orthogonal lattice, in ascending order. For each of such subsequent distances, the relative positions around the shape contours are checked and updated in case they are still empty. In case the distance value is updated into that cell, the exact dilation will implement the exact Euclidean distance transform of the shape. On the other hand, in case the pixels along the contour of the original shape are labeled with subsequent integer values, and these values are updated into the respectively implied empty cells, the exact dilation algorithm will fully accurately implement the interesting process known as label propagation, which can be understood as the broadcasting, along the direction normal to the expanding contours, of the labels assigned to the contour of the original image (see Figure 1).

Fig. 1. A stage during label propagation. The original shape is shown in gray

Label propagation, often implemented in an approximated way, has been applied to the estimation of Voronoi diagrams [13]. In such approaches, each connected object in the image is assigned a unique label, which are then propagated. The positions where the propagating waves of labels collapse can be verified to correspond to the limiting frontiers of the respective generalized Voronoi diagram. As reported recently [5, 6], this process can be taken to its ultimate consequence and applied to each of the pixels along the contour of the objects in the image. Here, we report a slightly distinct approach which guarantees one-pixel-wide skeletons, though at the expense of eventual small displacements (no larger than one pixel) of some portions of the skeletons. Once the labels have been propagated, the maximum values of the differences between each the value at each pixel and its immediate 4-neighborhood is determined and used to construct an auxiliary image P. The warping problem implied by mapping the labels along closed contours (i.e. the starting point, labeled with 1, becomes adjacent to a pixel marked with the last label), can be easily circumvented by using the following criterion:

if label_difference > M/2
then label_difference = M-label_difference
else label_difference = label_difference

It can be shown that a family of scale space skeletons of the original shape can be obtained simply by thresholding P at distinct values T, which acts as a spatial scale parameter. The higher the value of T, the higher the scale and the more simplified the shape becomes. Give a specific value of T, the reconstruction of the shape for that specific skeleton is immediately achieved by making logical "ORs" between filled disks centered at each of the skeletons points having as radii the value of the distance transform at that point. Illustrations of such concepts can be found in Figure 2. The fact that the borders of larger detail do not shift while smaller detail portions of the shape are removed [6] is implied by the fact that the portions of the skeletons corresponding to the coarse scale detail do not move, but only change their lengths as the spatial scale is varied.

3 Automated Extraction of Dendrograms and Shape Decomposition

A dendrogram is a data structure expressing the hierarchical organization of the branching patterns of the respective shape components. Indeed, one of the facts highlighted in the present article is the possibility of representing not only neurons, but any two-dimensional shapes in terms of dendrograms, a possibility that is immediately suggested by the skeletons and respective reconstructions. Once the distance transform and the multiscale skeletons of the shape under analysis are obtained, and a suitable spatial scale (i.e. the threshold value T) is selected, the extraction of the respective dendrogram is immediate, as allowed by the good quality of the obtained skeletons, i.e. one-pixel-width and 8-connectivity. Indeed, the dendrogram can be obtained simply by following the binary representation of the skeletons at a specific spatial scale, while storing into a stack the coordinates of the branching points. The determination of such points is also straightforwardly achieved by adding the amount of non-zero pixels in the 8-neighborhood of each pixel along the skeleton; in case the obtained value is larger than 2, the central pixel is understood as a branching point. In the present article, a simple recursive program is used to follow the skeletons and to derive the dendrogram representation. To our best knowledge, this methodology represents the simplest and most effective means for automated dendrogram extraction. Indeed, it is the only fully automated solution to that problem (see [2] for the only alternative, semi-automated approach known to us). At the same time, the branching orders (i.e. the hierarchical levels in the dendrogram) and the angles at branching points are stored and later used to produce statistics describing the overall properties of the shape.

A particularly interesting property allowed by the skeletonization of general shapes is their representation and understanding in terms of a hierarchically organized composition of basic "segments". Basically, a "segment" is understood as any portion of the obtained skeletons that are comprised between to subsequent branching points or between a branching point and the origin of the skeleton or one of its terminations. Figure 2 illustrates the decomposition of a two-dimensional shape with respect to a specific segment. It can be verified that the original shape can be reconstructed by making logical "or" combinations of the respective segmented portions, considering the distance transform, as described in Section 2. Thus, such basic elements provide a natural and effective means of hierarchically decomposing and representing the original shape in terms of an intermediate set of geometrical elements.

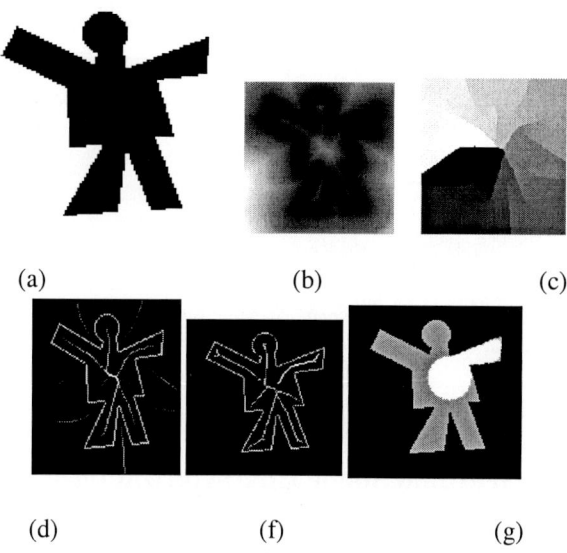

Fig. 2. Original shape (a) and respective distance transform (b), propagated labels (c), and difference image (d). The internal skeleton for $T = 10$ with original image (e), a selected segment (f) and its respective reconstruction (g).

The decomposition of a two-dimensional shape allows morphological properties - such as the length, average width, and curvature – to be measured separately along each segment. For instance, the estimation of the average curvature along a dendritic segment (we use the numerical approach to curvature estimation proposed in [1], representing the whole neuronal shape in terms of its external contour) can be easily performed by reconstructing each segment of the skeletons by combining disks having as radiuses the distance transform at the specific point - see Figure 2(g).

It should be observed that the dendrogram representation of shapes is not complete, in the sense that some information about the original shape is lost. If a complete representation of the shapes is needed, the skeletons instead of the dendrograms should be considered. However, dendrogram representations are particularly interesting for modeling, analyzing, comparing, and classifying shapes where the segments can appear bent, but preserving its length. It is clear that, provided the topology of the shape is maintained, the dendrogram representation will be invariant to the illustrated distortions, Figure 3, which can be useful in pattern recognition and analysis.

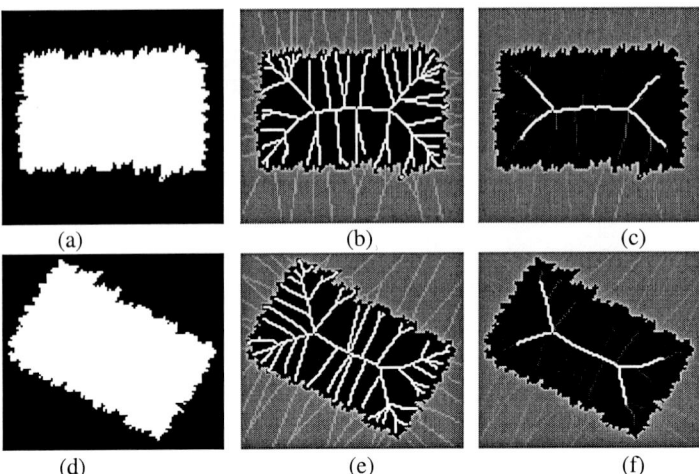

Fig. 3. Illustration of distortions invariance for skeleton. (d), (e) and (f) show, respectively, rotated version of original image(a), large scale skeleton (b) and fine scale skeleton (c).

4 Application: Neuronal Shape Characterization

The above presented framework to shape representation and analysis has been applied to the important problem of neuronal shape characterization. As above observed, neurons are naturally treated by the proposed framework as a particularly complex type of shapes characterized by intrincated branching patterns. Indeed, one of the few issues peculiar to this type of shapes is the possibility of characterizing the measures along the shape in terms of statistics conditioned to the branching order. Given a collection of neuronal cells, possibly from a same class or tissue, it is important to obtain meaningful measures characterizing their main properties, in order that the function and morphology of such cells can be properly related. Although much effort has been focused on such an endeavor (e.g. [14, 15, 16, 17, 18, 19, 20, 21, 22, 23]), there has been a lack of agreement about which measures and representations are most suited to neuronal cell characterization. As a matter of fact, it is important to note that such a characterization is often inherently related to the problem being investigated [3]. For instance, in case one wants to investigate the receptive fields of retinal ganglion cells by considering their respective morphology, it is important to select measures that express the spatial coverage and complexity of the cells.

One of the contributions of the present article is to argue that the scale space skeletonization and respective reconstructions of two-dimensional shapes, and especially the decomposition of such shapes in terms of basic elements (i.e. "segments"), represent a natural and powerful means for characterizing most neuronal shapes with respect to most typical problems.

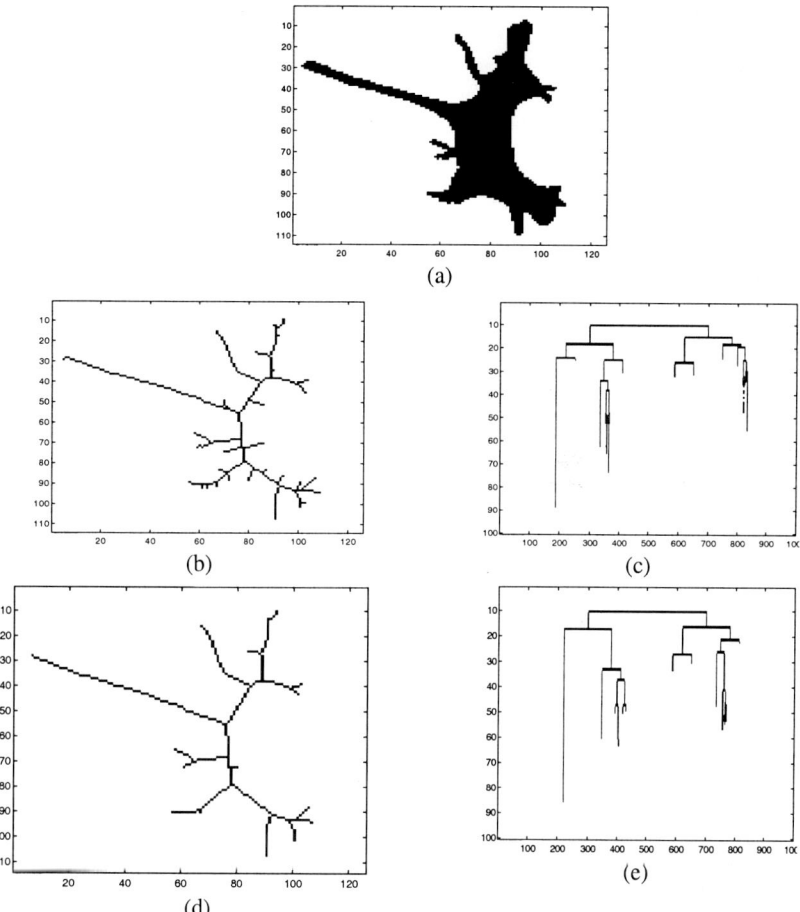

Fig. 4. One of the neuronal cells considered in the present article (a); its skeleton for $T = 3$ (b); respective dendrogram (c); skeleton for $T = 5$ (d); and respective dendrogram (e)

Firstly, each neuronal cell is segmented (i.e. its borders were extracted and organized as a list), and its respective distance transform and scale space skeletons obtained by using exact dilations and the procedures described in Section 2. The distance transform and skeletons are obtained, and the dendrograms are extracted (see Figure 4). While deriving the dendrograms, the values of a number of morphological features can be estimated and stored for further analysis, which can be done with respect to the branching order. Figure 5 illustrates histograms for several of possible features obtained from a specific class of neuronal cells. Such statistical characterizations provide exceedingly valuable information for neuronal shape characterization, analysis, comparison, and synthesis [2].

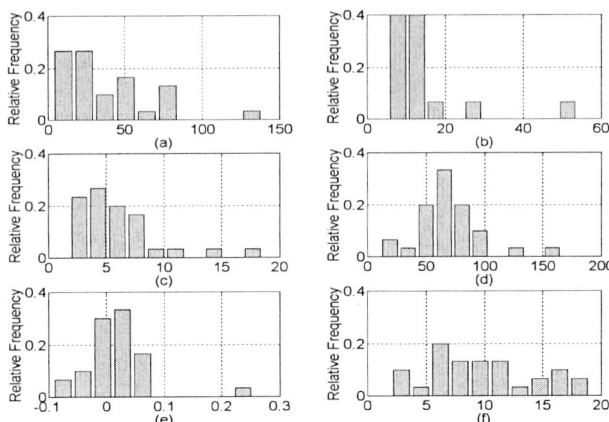

Fig. 5. Histograms characterizing the number of branches (a); average segment lengths (b), average width along segments (c), the angles at branching points (d); and average curvature along segments (e) with respect to all orders of dendritic segments in the considered neuronal cells.

5 Concluding Remarks

The present article has presented a novel biological inspired and comprehensive framework for the representation, analysis, comparison, and classification of general shapes. The key element in such a framework is the use of the recently introduced scale space representation of shapes, and respective reconstructions. It is proposed that such representations allow a natural, logical and general way to represent shapes which is invariant to several geometric transformations, including relative bending of the segments. A series of remarkable properties are allowed by such an approach, including the selection of a proper spatial scale (leaving out too small details, but preserving other features of interest); the preservation of the position of larger contours while smaller scale detail are removed; as well as the decomposition of two-dimensional shapes in terms of segments which, in addition to allowing morphological measures to be done respective to branching orders, presents great potential as subsidy for pattern recognition and shape understanding. The proposed framework and concepts have been illustrated with respect to the characterization and analysis of neuronal cells. Indeed, it is also proposed in the current article that the shape exhibited by neurons correspond to some of the richest and most elaborated versions of a morphological structure, nicely captured by the skeletons, which is inherent to any two-dimensional shapes. The vast potential of the proposed methodologies are illustrated with respect to the comprehensive characterization of morphological properties of the neuronal cells.

Acknowledgments

Luciano da Fontoura Costa is grateful to FAPESP (Procs 94/3536-6 and 94/4691-5) and CNPq (Proc 301422/92-3) for financial support. Andrea Gomes Campos and Luiz Gonzaga Rios-Filho are grateful to FAPESP (Procs 98/12425-4 and 98/13427-0 respectively). Leandro Farias Estrozi is grateful to Capes.

References

1. Cesar Jr., R.M. and L. da F. Costa. Towards Effective Planar Shape Representation with Multiscale Digital Curvature Analysis based on Signal Processing Techniques, Pattern Recognition, **29** (1996) 1559-1569
2. Costa, L. da F., R. M. Cesar Jr, R. C. Coelho, and J. S. Tanaka. 1998. Perspective on the Analysis and Synthesis of Morphologically Realistic Neural Networks, in Modeling in the Neurosciences (R. Poznanski Ed). Invited Paper, (1999) 505-528, Harwood Academic Publishers
3. Costa, L. da F. and T. J. Velte. Automatic characterization and classification of ganglion cells from the salamander retina, Journal of Comparative Neurology, **404(1)** (1999) 33-51
4. F. Attneave. Some Informational Aspects of Visual Perception, Psychological Review, **61** (1954) 183-193
5. Costa, L. da F. Multidimensional scale-space shape analysis. Santorini, In: Proceedings International Workshop on Synthetic-Natural Hybrid Coding and Three Dimensional Imaging, Santorini-Greece, (1999) 214-7
6. Costa, L. da F. & Estrozi, L. F. Multiresolution Shape Representation without Border Shifting, Electronics Letters, **35** (1999) 1829-1830
7. Wright, M.W. The Extended Euclidean Distance Transform - Dissertation submitted for the Degree of Doctor of Philosophy at the University of Cambridge, June 1995
8. Leyton, M. Symmetry-Curvature Duality, Computer Vision, Graphics and Image Processing, **38** (1987) 327-341
9. Blum, H. Biological Shape and Visual Science (Part I), J. Theor. Biol., **38** (1973) 205-287
10. Lindberg, T. Scale-space theory in computer vision. Kluwer Academic Publishers, 1994
11. Ogniewicz, R. L. Discrete Voronoi Skeletons, Hartung-Gorre Verlag, Germany, 1993
12. Tricot, C. 1995. Curves and Fractal Dimension, Springer-Verlag, Paris
13. Lantuéjoul, C. Skeletonization in quantitative metallography, In Issues of Digital Image Processing, R. M. Haralick and J.-C. Simon Eds., Sijthoff and Noordhoff, 1980
14. Fukuda, Y.; Hsiao, C. F.; Watanabe; M. and Ito, H.; Morphological Correlates of Physiologically Identified Y-, X-, and W-Cells in Cat Retina, Journal of Neurophysiology, **52(6)** (1984) 999-1013
15. Leventhal, A. G. and Schall, D., Structural Basis of Orientation Sensitivity of Cat Retinal Ganglion Cells, The Journal of Comparative Neurology, **220** (1983) 465-475
16. Linden, R. Dendritic Competition in the Developing Retina: Ganglion Cell Density Gradients and Laterally Displaced Dendrites, Visual Neuroscience, **10** (1993) 313-324
17. Morigiwa, K.; Tauchi, M. and Fukuda, Y., Fractal analysis of Ganglion Cell Dendritic Branching Patterns of the Rat and Cat Retinae, Neurocience Research, Suppl. 10, (1989) S131-S140
18. Saito, H. A., Morphology of Physiologically Identified X-, Y-, and W-Type Retinal Ganglion Cells of the Cat, The Journal of Comparative Neurology, **221** (1983) 279-288
19. Smith Jr., T. G.; Marks, W. B.; Lange, G. D.; Sheriff Jr., W. H. and Neale, E. A.; A Fractal Analysis of Cell Images, Journal of Neuroscience Methods, **27** (1989) 173-180

20. Van Ooyen, A. & Van Pelt, J., Complex Periodic Behavior in a Neural Network Model with Activity-Dependent Neurite Outgrowth, Journal of Theoretical Biology, **179**, (1996) 229-242
21. Van Oss, C. & van Ooyen, A., 1997. Effects of Inhibition on Neural Network Development Through Active-Dependent Neurite Outgrowth, Journal of Theoretical Biology, **185** (1997) 263-280
22. Vaney, D. I. Territorial Organization of Direction-Selective Ganglion Cells in Rabbit Retina, The Journal of Neuroscience, **14**(11) (1994) 6301-6316
23. Velte, T.J. and Miller, R. F., Dendritic Integration in Ganglion Cells of the Mudpuppy Retina, Visual Neuroscience, **12** (1995) 165-175

A Fast Circular Edge Detector for the Iris Region Segmentation

Yeunggyu Park, Hoonju Yun, Myongseop Song, and Jaihie Kim

I.V. Lab. Dept. of Electrical and Computer Engineering, Yonsei University,
134 Shinchon-dong, Seodaemun-gu, 120-749 Seoul, Korea
parkyk@seraph.yonsei.ac.kr
http://cherup.yonsei.ac.kr

Abstract. In this paper, we propose a fast circular edge detector for the iris region segmentation by detecting the inner and outer boundaries of the iris. In previous work, the circular edge detector which John G.Daugman proposed, searches the radius and the center of the iris to detect its outer boundary over an eye image. To do so, he used Gaussian filter to smooth texture patterns of the iris which cause its outer boundary to be detected incorrectly. Gaussian filtering requires much computation, especially when the filter size increases, so it takes much time to segment the iris region. In our algorithm, we could avoid procedure for Gaussian filtering by searching the radius and the center position of the iris from a position being independent of its texture patterns.
In experimental results, the proposed algorithm is compared with the previous ones, the circular edge detector with Gaussian filter and the Sobel edge detector for the eye images having different pupil and iris center positions.

1 Introduction

As the importance of the personal identification increases, the researches on the biometric technologies which use person's unique physical features are actively performed. The major biometric technologies are iris recognition, fingerprint verification, face recognition, and so on [1]. The iris recognition guarantees higher security than any other biometric, since the iris has highly detailed and unique textures which remain unchanged [2][3][4]. The iris recognition system verifies and identifies a person using these characteristics.

In general, the texture images have many small edge components representing local information. In the iris region segmentation, there may be obstacles because of them, so they should be removed to detect the correct inner and outer boundaries of the iris. So, John G.Daugman proposed the circular edge detector with Gaussian filter. However, it requires heavy computational complexity because of using Gaussian filter for smoothing texture patterns of the iris which cause its outer boundary to be detected incorrectly [2][5][6]. So we propose a fast circular edge detector in which Gaussian filtering is not necessary by searching the radius and the center position of the iris from a position being independent of the texture patterns of the iris.

2 Iris Recognition Procedure

The general iris recognition procedure is shown in Fig. 1. In the above procedure,

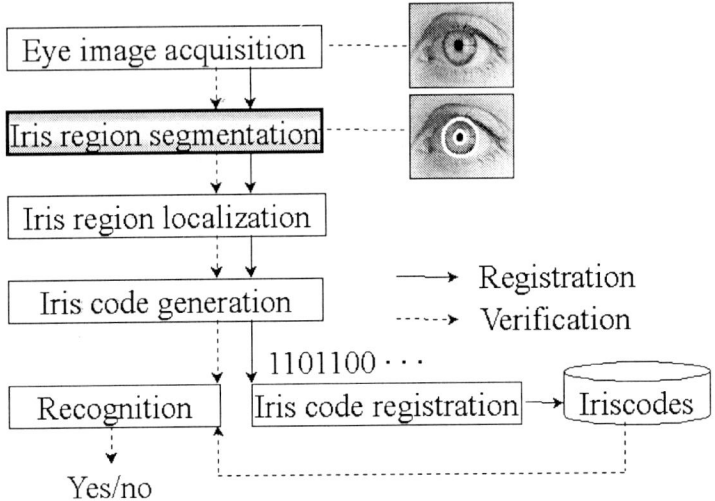

Fig. 1. Iris recognition procedure

it is important to segment the iris region, exactly and fast. After all, it affects the generation of the iris code and the recognition performance.

3 Iris Region Segmentation

The iris region segmentation is locating the iris region from an eye image like Fig. 2.

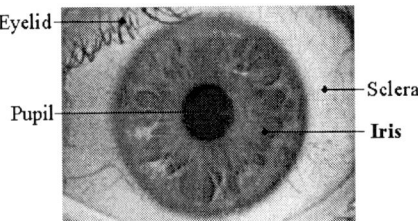

Fig. 2. Eye image

At first, we should detect the inner boundary between the iris and the pupil and then outer one between the iris and the sclera. In general, an iris has not only the texture patterns including its local characteristics but also many edge components inside it as shown in Fig.2. More features affecting the performance of the iris region detection exist close to the inner region than the outer one [2], especially, in case of brown color iris. Therefore, the typical edge detection operators such as Sobel or Prewitt operator may not be good to segment the iris region.

3.1 Circular Edge Detector with Gaussian Filtering

John G.Daugman proposed the circular edge detector with Gaussian filter to segment the iris region. The role of Gaussian filtering is to smooth the texture patterns inside it to avoid detecting the outer boundary of the iris, incorrectly such as shown in Fig. 3 [2]. However, Gaussian filtering requires much computa-

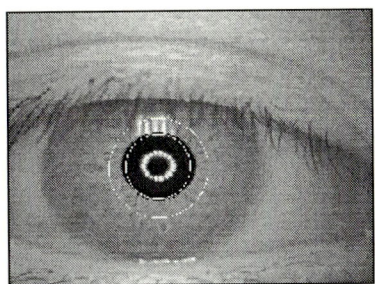

Fig. 3. Example of locating incorrect outer boundary of the iris

tion time for segmenting the iris region, especially when filter size increases. His algorithm searches the radius and the center position of the iris for detecting its outer boundary over an eye image by using Eq. (1). That requires much computation time, too. This method was implemented by integrodifferential operators that search over the image domain (x, y) for the maximum in the blurred partial derivative, with respect to increasing radius r of the normalized contour integral of I(x, y) along a circular arc ds of radius r and center coordinates (x_0, y_0):

$$max(r, x_0, y_0) \left| G_\sigma * \frac{\partial}{\partial r} \int_{r, x_0, y_0} \frac{I(x, y)}{2\pi r} ds \right| \quad (1)$$

where * denotes convolution and $G_\sigma(r)$ is smoothing function such as a Gaussian of scale σ [2][3][4][5][6][7].

3.2 Sobel Edge Operator

The Sobel operator is one of the most commonly used edge detectors [8][9][10][11]. In order to evaluate our algorithm, we applied the 8-directional Sobel operator to segment the iris region.

3.3 The Proposed Fast Circular Edge Detector

This method consists of five steps to segment the iris region as shown in Fig 4.

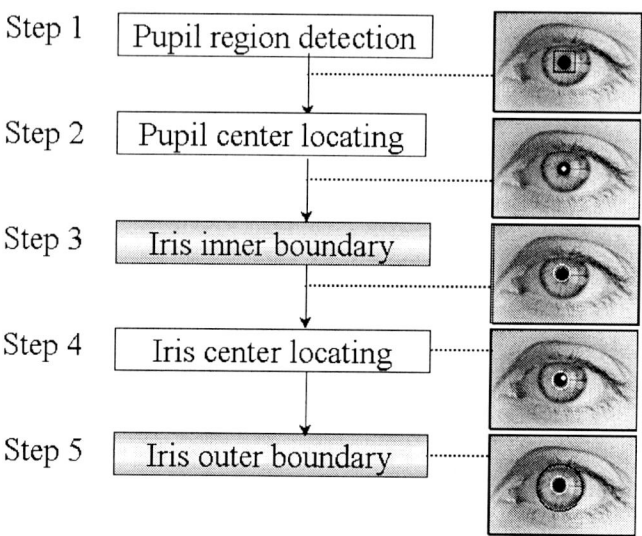

Fig. 4. Segmentation steps

At first, we detect a pupil region to locate its center position. Detecting the pupil region is achieved by thresholding an eye image.

In step 2, we locate the pupil center position based on the histogram and long axis of the pupil.

In step 3, we can easily find the iris inner boundary by using Eq. (2).

$$max(n\Delta r, x_0, y_0) \left| \frac{1}{\Delta r} \sum_k \sum_m I(x,y) \right| \qquad (2)$$

$$x = k\Delta r \, cos(m\Delta\theta + x_0)$$
$$y = k\Delta r \, sin(m\Delta\theta + x_0)$$

where Δr and $\Delta \theta$ are radius interval and angular interval, respectively.

In step 4 and step 5, we locate each iris center and its outer boundary. In these steps, Gaussian filtering may be necessary, since the iris region has the texture patterns similar to random noise. If we carefully analyze the pattern of the iris region, we could cognize it to be the smoother, the more outward. Using this trait, we need to start at radius r being independent of its texture patterns to detect the iris outer boundary. The start position r_s is computed from Eq. (3).

$$x = 2r_i \tag{3}$$

$$\begin{aligned} r_s &= 2r_p & If \quad x = 2r_p \\ &= 2r_i + d'(=r_p - r_i) & otherwise \end{aligned}$$

where r_i, r_p and x are the ideal pupil radius, the real pupil radius and the

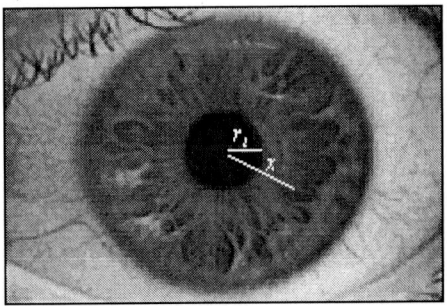

Fig. 5. Ideal pupil radius and reference point

reference point being independent of the texture patterns of the iris, respectively. Therefore, in our algorithm, Gaussian filtering requiring much computation time is not necessary.

4 Experimental Results

To evaluate the performance of algorithms for the iris region segmentation, we experiment in the following environment and for the eye images having the different pupil and iris center positions.
  * Input image size
     640 X 480 pixels gray level image.
  * Camera focus : fixed
  * Center of iris and pupil : not concentric
  * Camera : black and white CCD camera
  * CPU : Pentium II Xeon 450MHz

We could acquire the result images using the iris segmentation algorithms as shown in Fig. 6. As you can see in Fig. 6, the accuracy of the proposed algorithm is nearly equal to Daugman's one, but Sobel edge operator could not locate the correct iris region. It is why the edge of the outer boundary of the iris faints, and the iris region has many edge components.

In processing time, our algorithm is superior to any other method like Table 1. As you can see from the experimental results, because Daugman's algorithm

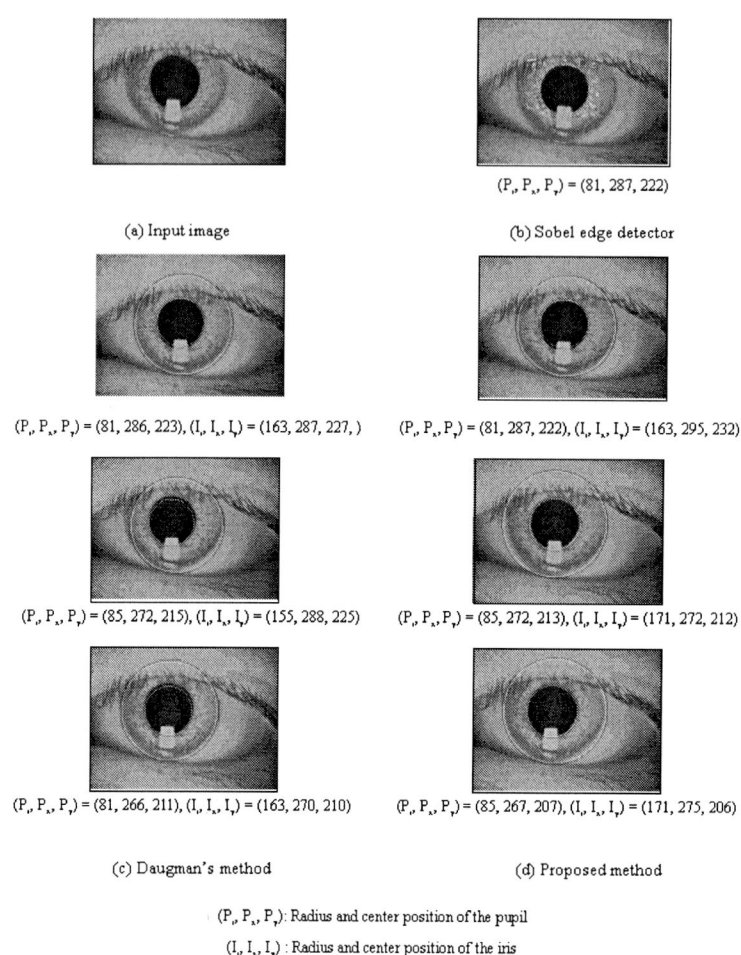

Fig. 6. Input and result images

uses Gaussian filter and searches the radius and the center position of the iris for detecting its outer boundary over an eye image, his algorithm takes much processing time than ours in detecting the iris region.

5 Conclusions

In this paper, we proposed a fast circular edge detector to segment the iris region from a person's eye image. In previous work, Daugman proposed circular edge detector with Gaussisn filter. His method has some factors increasing the computational complexity. We proposed a fast algorithm which Gaussian filter-

Table 1. Processing time(sec)

Center	Sobel edge operator			Daugman's method			Proposed method		
	Pupil	Iris	Total	Pupil	Iris	Total	Pupil	Iris	Total
1	0.13	x	x	0.02	5.13	5.15	0.02	2.97	2.99
2	0.11	x	x	0.02	4.94	4.96	0.02	1.89	1.91
3	0.11	x	x	0.02	4.22	4.24	0.02	1.34	1.36
Average	0.11	x	x	0.02	4.76	4.78	0.02	2.07	2.09

1: 1st pupil and iris center position in Fig. 5.
2: 2nd pupil and iris center position in Fig. 5.
3: 3rd pupil and iris center position in Fig. 5.
x: not available

ing was not necessary by using reference point being independent of the texture patterns of the iris. As results, our algorithm is faster than Daugman's one.

References

1. Despina Polemi.: Biometric Techniques: Review and Evaluation of Biometric Techniques for Identification and Authentication, Including an Appraisal of the Areas where They Are Most Application, Institute of Communication and Computer Systems National Technical University of Athens (1999) 5-7, 24-33
2. J. G. Daugman.: High Confidence Visual Recognition of persons by a Test of Statistical Independence, IEEE Transaction on Pattern Analysis and Machine Intelligence, Vol. 15, NO. 11 (1993) 1148-1160
3. J. G. Daugman.: Iris Recognition for Persons Identification (1997)
4. R. P. Wildes.: Iris Recognition-An Emerging Biometric Technology, Proceedings of the IEEE, Vol. 85, NO. 9 (1997) 1348-1363
5. N. Chacko, C. Mysen, and R. Singhal.: A Study in Iris Recognition (1999) 1-19
6. J. G. Daugman.: Recognizing Persons by Their Iris Patterns, Cambridge University (1997) 1-19
7. D. McMordie.: Texture Analysis of the Human Iris, McGill University (1997)
8. R. Jain, R. Kasturi, and B. G. Schunk.: Machine Vision, McGrow-Hill (1995) 145-153
9. E. Gose, R. Johnsonbaugh, and S. Jost.: Pattern Recognition and Image Analysis, Prentice Hall PRT, Inc (1996) 298-303
10. M. Nadler and E. P. Smith.: Pattern Recognition Engineering, Jonh Wiles & Sons, Inc (1993) 107-142
11. M. Sonka, V. Hlavac, and R. Boyle.: Image Processing, Analysis, and Machine Vision, Brooks/Cole Publishing Company (1999) 77-83

Face Recognition Using Foveal Vision

Silviu Minut[1], Sridhar Mahadevan[1], John M. Henderson[2], and Fred C. Dyer[3]

[1] Department of Computer Science and Engineering, Michigan State University
(minutsil, mahadeva)@cse.msu.edu
[2] Department of Psychology, Michigan State University
(john@eyelab.psy.msu.edu)
[3] Department of Zoology, Michigan State University, East Lansing, MI 48823
(fcdyer@pilot.msu.edu)

Abstract. Data from human subjects recorded by an eyetracker while they are learning new faces shows a high degree of similarity in saccadic eye movements over a face. Such experiments suggest face recognition can be modeled as a sequential process, with each fixation providing observations using both foveal and parafoveal information. We describe a sequential model of face recognition that is incremental and scalable to large face images. Two approaches to implementing an artificial fovea are described, which transform a constant resolution image into a variable resolution image with acute resolution in the fovea, and an exponential decrease in resolution towards the periphery. For each individual in a database of faces, a hidden-Markov model (HMM) classifier is learned, where the observation sequences necessary to learn the HMMs are generated by fixating on different regions of a face. Detailed experimental results are provided which show the two foveal HMM classifiers outperform a more traditional HMM classifier built by moving a horizontal window from top to bottom on a highly subsampled face image.

1 Introduction

A popular approach to face recognition in computer vision is to view it as a problem of clustering high-dimensional vectors representing constant resolution subsampled face images. Methods from linear algebra, such as principal components, are used to reduce the high dimensionality of the face image vector. Although the use of such methods for applications such as face recognition has resulted in impressive performance (e.g. [12]), these systems still require significant subsampling of the original image, and are usually not incremental.

This high dimensional pattern recognition approach works poorly when the vision problem requires simultaneously high resolution as well as a wide field of view. However, a wide range of vision problems solved by biological organisms, from foraging to detecting predators and prey, do require high resolution and wide field of view. Consequently, in virtually all vertebrates and invertebrates with well developed eyes, the visual field has a high acuity, high resolution center (the *fovea*). In humans, the fovea covers about 2° of the field of view. The resolution decreases gradually, but rapidly, from the fovea to the periphery [1,16].

The human vision system rapidly reorients the eyes via very fast eye movements [7,15,22] called *saccades*. These are ballistic motions, of up to 900°/s, during which visual information processing is severely limited [21]. Foveation reduces the dimensionality of the input data, but in turn generates an additional sequential decision problem: an efficient gaze control mechanism is needed to choose the next fixation point, in order to direct the gaze at the most visually salient object, and makes up for any potential loss of information incurred by the decrease in resolution in the periphery.

In computer vision, there has been much interest in "active vision" approaches including foveal vision [2,17,18,20], but there is as yet no widely accepted or successful computational model of saccadic eye movement that matches human performance. Our goal is to use foveated vision to build a biologically plausible face recognizer, as part of an effort towards the development of a general saccadic gaze control system for natural scenes. Our research is interdisciplinary, building simultaneously on computational modeling and experimental psychology. We conducted experiments with human subjects (see Fig. 1) who were presented with a number of face images and were requested to learn them for later recognition. An eye tracker recorded the fixation points, the fixation durations and the fixation order for each face. The resulting scan patterns were highly correlated across subjects, suggesting that there exists a common underlying policy that people use in gazing at faces while learning them or recalling them (e.g. eyes, nose and mouth regions represent a disproportionate number of the total number of fixations).

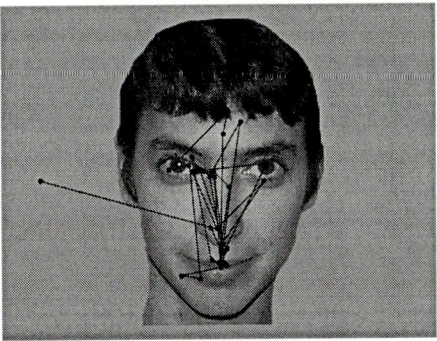

Fig. 1. Human scan pattern of eye movements recorded while a subject was given 10 seconds to learn a new face. Note majority of fixations tend to land on a relatively small region of the face.

Motivated by the human data, we want to develop a sequential model of face perception that uses foveal processing of the input images, and is incremental and scalable. We will use a stochastic sequential model of gaze control based on the framework of Hidden Markov Models (HMM), where the observations are generated stochastically by an underlying sequence of "hidden" states. Our long-term goal is to extend this approach using the general framework of Partially Observable Markov Decision Processes (POMDP) [8], which allows decisions at each fixation on where to move the fovea next.

2 Foveal Imaging

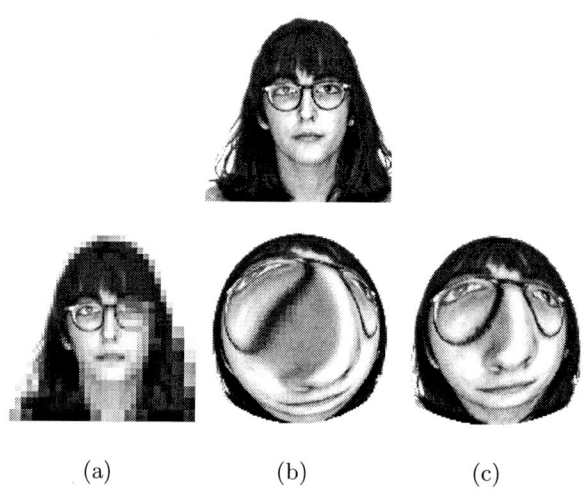

Fig. 2. Foveal processing of the original image (top) using (a) superpixels and (b) logpolar transformation with $r = log(R)$ (c) logpolar transformation with $r = log^2(R)$.

We describe two ways of simulating foveal vision in software, as shown in Fig. 2. In the first approach, following [3], we construct the simulated fovea as follows. Suppose we have a (large) constant resolution image C and we want to transform it into an image V with variable resolution, where the resolution decreases exponentially from the center to the periphery. The fovea is a central square in V of size p and is an identical copy of the same region in C. Then we surround the fovea with rings of superpixels. A superpixel is a square consisting of a few physical pixels in V. All pixels that make a superpixel have the same gray level, namely the average of the corresponding region in C. Within a ring, all superpixels have the same size. Each ring is obtained recursively from the previous one, by a dilation of a factor of 2. The construction is illustrated in Fig. 3.

The second method of building a simulated fove is based on the *log-polar* transform (see e.g. [20]). This is a nonlinear transformation of the complex plane to itself, which in polar coordinates is given by

$$(R, \Theta) \longrightarrow (r, \theta) \qquad (2.1)$$

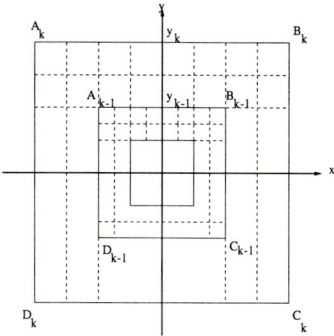

Fig. 3. This figure illustrates the construction of a rectangular fovea.

$$r = \log R \qquad (2.2)$$
$$\theta = \Theta \qquad (2.3)$$

Under this transformation each point is moved radially towards the origin, from polar distance R to $r = \log R$. Note that by differentiating (2.2) we get

$$dr = \frac{1}{R} dR \qquad (2.4)$$

Equation (2.4) is the key to understand how the reduction in resolution occurs. At distance $R = 20$ pixels from the origin, it takes a variation of $dR = 20$ pixels in the original image to produce a variation of $dr = 1$ pixel in the transformed image. However, near the origin, at distance, say, $R = 1$ pixel, it takes a variation of $dR = 1$ pixel to produce $dr = 1$ pixel change in the transformed image. Certainly, the rate of change of the resolution can be controlled by introducing some suitable coefficients, or by trying $r = f(R)$ for various functions f. Figure 2 shows an example of this transformation. The original image has size 512×512, as usual. The transformed images normally have size of the order of $f(R)$, but we rescale them to some reasonable size (e.g. 64×64 or to 32×32). In image (b) $r = \log R$ has been used, whereas in (c), $r = \log^2 R$.

2.1 Hidden Markov Models

In this paper, we view the observations at each fixation on a face as being generated by an underlying Hidden Markov Model (HMM). Although this approach allows for observations to be generated by "hidden" states, it is limited to fixed scan patterns. Our goal is to extend this approach to the POMDP model, which would allow for individual optimization in scan patterns across different faces.

An HMM is defined by a finite set of states, transition probabilities a_{ij} from state i to state j, and observation densities $b_i(o)$ for each i, which describe the probability of getting an observation vector o when the system is in state i. For an overview of HMMs see [14]. One assumes that the system starts in state i

with some (known) probability π_i. If the a_{ij}s, the b_is and the π_is are known it is straightforward (mathematically) to compute the probability of obtaining a certain observation sequence $O = O_1 O_2 \ldots O_n$ given that the system starts in state i. The *forward-backward* procedure is an efficient algorithm for estimating this probability.

One crucial problem for using HMMs is to infer the state it is in, based on the observation perceived. Specifically, given an observation sequence $O = O_1 O_2 \ldots O_n$ determine what was the most likely sequence of underlying states $i_1 i_2 \ldots i_n$ that produced that sequence of observations. The solution to this problem is *the Viterbi algorithm*, again under the assumption that the a's, the b's and π are known. In general, however, the HMM parameters are not known a priori, and we employ the well-known Baum-Welch expectation maximization algorithm to estimate these parameters. To this end, if we denote $\lambda = (a_{ij}, b_i, \pi)$ the set of all the parameters of an HMM, then for a given observation sequence $O = O_1 O_2 \ldots O_n$ we maximize the function $\lambda \to P(O|\lambda)$. This will provide the HMM that is most likely to have produced the observation sequence O, and then, if necessary, the Viterbi algorithm can be used for state estimation.

2.2 Description of the Recognizer

We now describe the construction of a face recognizer combining HMM techniques and foveated vision. Given a database of face images, with several images per person, an HMM is learned for each person in the database which can be viewed as an internal representation of that person. In this paper, we used a linear HMM model, in which the only nonzero transitions in a state are to the same state or to the next state. We varied the number of states to determine the best linear model. This idea has been used successfully in speech recognition for quite some time [14], and more recently also in face recognition [11,19].

The principal difference between our approach and previous HMM based face recognizers is in the generation of observation sequences. In [19] the images are divided into horizontal stripes roughly corresponding to semantic regions in a human face: forehead, eyes, nose, mouth, chin, etc. An observation then consists essentially of the graylevel values on each stripe. The observation sequence is produced by sliding a stripe from top to bottom. The system cannot tell deterministically whether an observation represents the eyes, or the mouth, or any other feature of a face, and for this reason the states are hidden. This technique was later refined in [11], where the horizontal stripes were subdivided into smaller blocks. This later system also used an algorithm that learned simultaneously a low level HMM that describes the transitions between the blocks within a stripe, and a higher level HMM that describes the transitions between the horizontal stripes.

We generate the observation sequences differently, motivated by the human experiments shown earlier in Fig. 1. The face images were divided into 10 regions corresponding to left eye, right eye, nose, mouth, chin, left cheek, right cheek, left ear, right ear and forehead. The 10 regions were the same across all the images and were defined by hand on the mean face, with the intent to capture the informative regions for most faces. Given a constant resolution image of a human face, we produce a sequence of foveated images S_i, each centered at one of the 10 regions, by using either foveation method described in the previous section. As observation vectors we take the Discrete Cosine Transform of the S_i's.

The HMMs learned are *independent* of each other and so if a new person is added to the database, we only need to learn that person. This is in sharp contrast with other face recognition techniques (notably Principal Component Analysis based methods such as [12]) where one has to compute certain invariants specific to the whole database (database mean, eigenvalues, eigenvectors, etc.). In the latter case, addition of a new image to the database typically requires re-learning the whole database.

3 Experiments

We now test the effectiveness of our approach to foveal face recognition. In all experiments we used gray level images of 58 women and 75 men from the Purdue AR database. For each person in the database we used 6 images. We used for testing both the resubstitution method, under which all 6 images were used for testing and for training, and the leave-one-out method, under which 5 images out of 6 were used for training and the remaining image for testing, resulting in 6 trials. Under leave-one-out, the results were averaged over all 6 trials.

3.1 Experiment 1: Foveal HMM Models

1. For each image C in the database, foveate 3 times in each of the 10 predefined regions to get 30 variable resolution images V_i.[1] Then collapse each V_i to get a small image S_i and compute the Discrete Cosine Transform on S_i to get the observation vector O_i. This way, each image C in the database produces an observation sequence $O = O_1 \ldots O_{30}$, with one observation per fixation.
2. *Training.* For each person in the database learn an HMM with s states by using the observation sequences selected for training.
3. *Testing.* Classify the testing observation sequences for each person by matching against all HMMs previously learned. Under leave-one-out, repeat steps 2 and 3 by selecting each image as a test sample.

Starting with a 512×512 constant resolution image and $d = 4$ layers of superpixels in each ring, we get an observation vector of only 198 components. The length of the observation sequence does not depend on the original image, and is in fact only equal to the number of fixations in the image.

Figure 4 shows the results of Experiment 1 using the resubstitution method. The two graphs show the recognition rate as a function of the number of states. Surprisingly, we got substantially different recognition rates on female images

---
[1] We use 30 fixations per image to roughly correspond to the number of fixations in the human experiments.

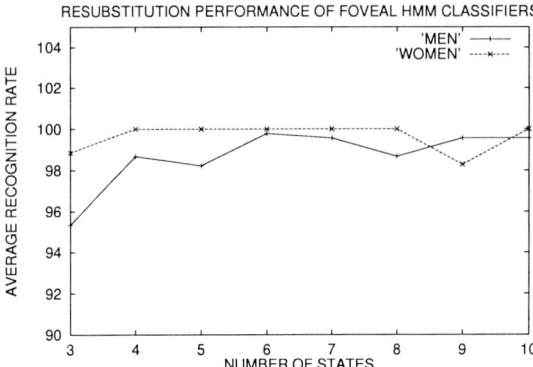

Fig. 4. Recognition results using superpixel-based foveally processed images. All images of each face were used in both training and testing.

than on male images, and we present the graphs individually. Notice that the best performance was achieved when the number of states was set to be equal to 6, which is significantly smaller than the number of regions used to collect observations (10).

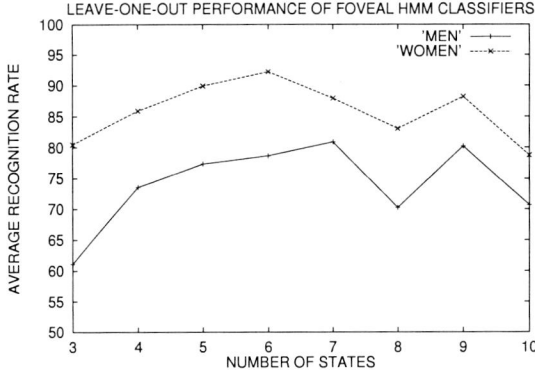

Fig. 5. Recognition results using superpixel-based foveally processed images. Recognition rates are averaged using the "leave-one-out" method on 6 images of each face.

Similar considerations apply for the leave-one-out method (Fig. 5). Here, the difference in recognition rates among the men and women is even more striking. For the sake of completeness, we include the results of this experiment when the log-polar foveation method was employed, instead of the superpixels method. The performance is roughly the same, as in the superpixel method, with a peak of 93.1% when 9 states were used with log-polar images, vs. a peak of about

92.2% when 6 state HMMs were built from superpixel images (see Fig. 6). The log-polar method has the advantage that it is easier to implement and faster.

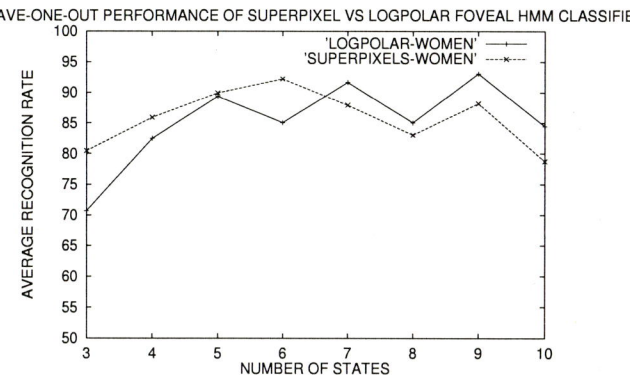

Fig. 6. Recognition rates achieved using the logpolar foveation method and the superpixel method on the female faces.

3.2 Experiment 2: Foveation vs. Stripes

We compare the performance of foveally generated HMMs to the performance of previously used HMM methods (see section 2.2). For each image C in the database, we produce the observation sequence by computing the Discrete Cosine Transform on a horizontal window sliding from top to bottom. An initial subsampling (by a factor of 4) is necessary, or else, the dimension of the resulting observation vectors would be too large. The remaining steps are identical to those in Experiment 1.

Figure 7 shows the results of the foveation method in comparison with the sliding window method. We repeated this experiment for different values of the width of the sliding stripe and of the overlap between the stripes and consistently, the recognition rates were substantially lower than the recognition rates for the foveation method.

The performance of our foveal HMM-based classifier is still not as good as pattern recognition based methods, such as eigen-faces (e.g. [12]). For comparison, we implemented a face recognizer which computes the eigen-faces of the database and retains the top ones. The performance of the eigen-face recognizer was extremely good with recognition rates between 98-100%. Of course, we would like to improve the performance, which is discussed below, but the emphasis of our work is to understand the role of foveal vision in face recognition, and more generally, in understanding natural scenes.

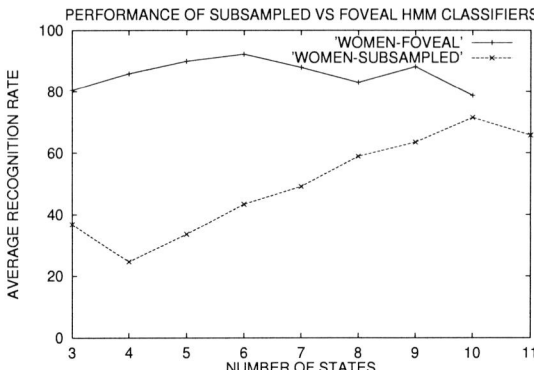

Fig. 7. Comparing recognition rates using subsampled images vs. foveally imaged faces.

4 Future Work

In this paper we describe the use of foveal vision to build a face recognition system. The observations needed to build a face model are generated by saccading over high importance regions. The approach scales to arbitrary-sized images, the learning is incremental, and it is motivated by eyetracking behavior in face recognition experiments.

The HMM approach is limited to a fixed scanning policy. The natural generalization would be to move to the POMDP framework, which allows the agent to select among several saccadic actions in each state to optimize the recognition policy. This would also better model data from human subjects, which show major differences in fixation location and duration between training (learning a face) and testing (recognizing a previously seen face).

Currently, the regions of interest must be predefined by the programmer, and are not relative to the person in the image. So, the system foveates in one region or another regardless of the position and the orientation of the face in the image, and expects to see certain features in that region (eyes, nose, mouth, etc.). A coordinate system relative to the object would perform better. We are also investigating building a *saliency map* [6] which can automatically select the next fixation point.

Acknowledgements

This research is supported in part by a Knowledge and Distributed Intelligence (KDI) grant from the National Science Foundation ECS-9873531. The data on human subjects was generated at Henderson's EyeLab in the Department of Psychology at Michigan State University. The computational experiments were carried out in Mahadevan's Autonomous Agents laboratory in the Department of Computer Science and Engineering.

References

1. Anstis, S. M. (1974). *A chart demonstrating variations in acuity with retinal position.* Vision Research, 14, pp. 589-592.
2. Ballard, D. H., (1991) *Animate Vision*, Artificial Intelligence journal, vol. 48, pp. 57-86.
3. Bandera, Cesar et. al. (1996) *Residual Q-Learning Applied to Visual Attention*, Proceedings of the International Conference on Machine Learning, Bari, Italy.
4. Desimone, R. (1991). *Face-selective cells in the temporal cortex of monkeys.* Journal of Cognitive Neuroscience, 3, pp. 1-8.
5. Friedman, A. (1979). *Framing pictures: The role of knowledge in automatized encoding and memory for gist.* Journal of Experimental Psychology: General, 108, pp. 316-355.
6. Henderson, J. M., & Hollingworth, A.. (1998). *Eye movements during scene viewing: an overview.* In G. W. Underwood (Ed.), Eye guidance while reading and while watching dynamic scenes, pp. 269-295. Amsterdam: Elsevier.
7. Henderson, J. M., & Hollingworth, A. (1999).*High-level scene perception.* Annual Review of Psychology, 50, pp. 243-271.
8. Kaelbling, L., Littman, M., & Cassandra, T. (1998). *Planning and Acting in Partially Observable Stochastic Domains,* Artificial Intelligence.
9. Matin, E. (1974). *Saccadic suppression: A review and an analysis.* Psychological Bulletin, 81, pp. 899-917.
10. Moghaddam, B. (1998). Beyond Eigenfaces: Probabilistic Face Matching for Face Recognition, International Conference on Automatic Face and Gesture Recognition.
11. Nefian, A., and Hayes, M. 1999. *Face recognition using an embedded HMM.* IEEE Conference on Audio and Video-based Biometric Person Authentication, Washington, D.C. 1999.
12. Pentland, A. et al., View-based and Modular Eigenfaces for Face Recognition, IEEE Conference on Computer Vision and Pattern Recognition, 1994.
13. Puterman, M. (1994). *Markov Decision Processes: Discrete Stochastic Dynamic Programming,* Wiley.
14. Rabiner, Lawrence R. (1989) *A Tutorial on Hidden Markov Models and Selected Applications in Speech Recognition*
15. Rayner, K. (1998). *Eye movements in reading, visual search and scene perception: 20 years of research.* Psychological Bulletin.
16. Riggs, L. A. (1965). *Eye movements. In C. H. Graham (Ed). Vision and Visual Perception,* pp. 321-349, New York: Wiley.
17. Rimey, R. D., & Brown, C. M. (1991). *Controlling eye movements with hidden Markov models.* International Journal of Computer Vision, November, pp. 47-65.
18. Rybak, I. A. et al. (1998) *A Model of Attention-Guided Visual Perception and Recognition.* Vision Research 38, pp. 2387-2400.
19. Samaria, F., & Young, S. (1994), *HMM based architecture for face identification.* Image and Computer Vision 12
20. Sela, G & Levine, D. M., (1997) *Real-Time Attention for Robotic Vision,* Real-Time Imaging 3, pp. 173-194, 1997.
21. Volkmann, F. C. (1986). *Human visual suppression.* Vision. Research, 26, pp. 1401-1416.
22. Yarbus, A. L. (1967). *Eye movements and vision.* New York: Plenum Press.

Fast Distance Computation with a Stereo Head-Eye System

Sang-Cheol Park and Seong-Whan Lee*

Center for Artificial Vision Research, Korea University,
Anam-dong, Seongbuk-ku, Seoul 136-701, Korea
{scpark, swlee}@image.korea.ac.kr

Abstract. In this paper, a fast method is presented for computing the 3D Euclidean distance with a stereo head-eye system using a disparity map, a vergence angle, and a relative disparity. Our approach is to use the dense disparity for an initial vergence angle and a fixation point for its distance from a center point of two cameras. Neither camera calibration nor prior knowledge is required. The principle of the human visual system is applied to a stereo head-eye system with reasonable assumptions. Experimental results show that the 3D Euclidean distance can be easily estimated from the relative disparity and the vergence angle. The comparison of the estimated distance of objects with their real distance is given to evaluate the performance of the proposed method.

1 Introduction

We propose a fast computational method of the 3D Euclidean distance from two images captured by a stereo head-eye system. In general, the relative distance has been known to be more feasible than the Euclidean distance in stereo vision. In robotic-vision applications[1], however, 3D Euclidean distance information can be used to find a target object and to avoid obstacles, to estimate the exact focus on the target object, and to detect a collision. Besides, the distance of a target object from the robot can be used in velocity computation for robot navigation and 3D distance information is useful to track an object occluded and moving along the optical axis. Therefore, 3D Euclidean distance can be more feasible than a relative distance in robot vision and navigation system.

Among many binocular vision systems, a stereo head-eye system is used as a model of the human visual system[5]. The difference in displacement of the two cameras, when fixating on a target object, allows us to find its image correspondences and to establish a disparity map from a stereo image pair. Distance information, which is lost when a 3D object is projected onto a 2D image plane, can be acquired from this differences[7]. In a stereo head-eye system, in particular, a disparity map, and a vergence angle are used for distance estimation. The vergence angle can be obtained from a simple trigonometric method. The disparity map can be obtained from the correspondence of an image pair.

---

* To whom all correspondence should be addressed. This research was supported by Creative Research Initiatives of the Ministry of Science and Technology, Korea.

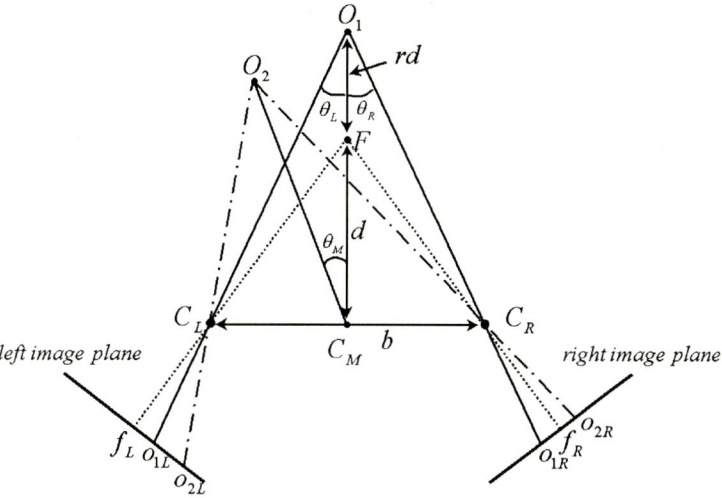

Fig. 1. Geometry of a stereo head-eye system. A 3D fixation point and two other 3D points are presented as well as two focal center points and a cyclopean point

In a stereo head-eye system configured as in Figure 1, the left and right focal points of the two cameras are depicted as C_L and C_R, respectively, and the center point of both cameras - the Cyclopean point of C_L and C_R - is C_M. A fixation point, F, is defined by the intersection of two optical rays passing through their focal center points and perpendicular to their image planes. The vergence angle of the left and right cameras fixating on the same 3D point, F, for instance, is represented as $\theta_L + \theta_R$. If a fixation point, F, is perpendicular on the baseline, the angle of θ_L equal to that of θ_R, respectively. The baseline of the two cameras is shown as b and the distance between C_M and the fixation point F, as d. In Figure 1, the 3D fixation point F is projected to f_L and f_R while other 3D points O_1 and O_2 are projected to o_{1L}, o_{1R} and o_{2L}, o_{2R}, respectively. Upper case letters are used to denote 3D world coordinate points and lower case letters to denote 2D image coordinate points.

The recovery of the relative distance has recently been considered in many related works as computer vision[1]. Nearly exact distance can be used in navigation, visual servoing, object segmentation, velocity computation, reconstruction, and so on. The 8-point algorithm is one of the best-known methods for 3D reconstruction[4]. However, the camera calibration for this method is a very expensive, and time-consuming process, and, moreover, is very sensitive to noise.

The proposed method requires neither camera calibration nor prior knowledge. Stereo matching and disparity estimation methods are classified into three groups: area-based, feature-based, and pixel-based method[6]. The present work used an area-based method.

We estimate the 3D Euclidean distance only from the geometry of a stereo head-eye system and a captured stereo image pair. This paper is organized as follows. In section 2, the computation of disparity between a stereo image pair and the vergence angle of the two cameras is described. In section 3, the estimation of the 3D distance using a relative disparity and a vergence angle are described. In section 4, experimental results and analysis are provided. Finally, in section 5, conclusions and future works are presented.

2 Disparity and Vergence Angle

We first estimate disparity variation for the estimation of a vergence angle and fixation point with an area-based method.

2.1 Construction of a Disparity Map

Stereo vision refers to the ability to infer information on the 3D structure and distance of a scene from two or more images taken from different viewpoints. 3D perception of the world is due to the interpretation that the brain gives for the computed differences in retinal position, named disparity, then the disparities of all the image points form the so-called disparity map[3].

Horizontal positional differences between corresponding features of the two image, i.e. horizontal disparities, are a powerful stimulus for the perception of distance even in the absence of other cues. Although horizontal disparity can by itself induce a percept of distance, it is not a direct measure of distance. The relationship between horizontal disparity and distance varies with the distance at which the disparity stimulus is viewed. Horizontal disparity has to be scaled for distance, which requires knowledge of distance provided by other sources of information[2].

Here, the simple SSD(Sum of Squared Differences) method among correlation-based and gradient-based methods is used. Correlation-based method is a common tool for the correspondence problem. Since gradient-based method is robust against the whole intensity variation[8], Figure 2 shows an example of a disparity map for which the disparities are estimated by the SSD method.

In general stereo systems, assumptions for computations are made as follows; two eyes are located at the same height and they fixate at the same 3D point, and so forth[4]. According to these assumptions, in stereo head-eye systems, distance information is directly extracted from a stereo image pair.

2.2 Vergence Control

The eye movement of human and other primates consist of saccade, pursuit, tremor and vergence movements. In particular, the vergence movements play a significant role in the stable fixation on the visual target of interest. In stereo vision, the control of vergence angles is defined as the motion of both cameras of a stereo head-eye system. The angle between two optical axes of the cameras is called the vergence angle. Any distance cues can be used to control this angle[5].

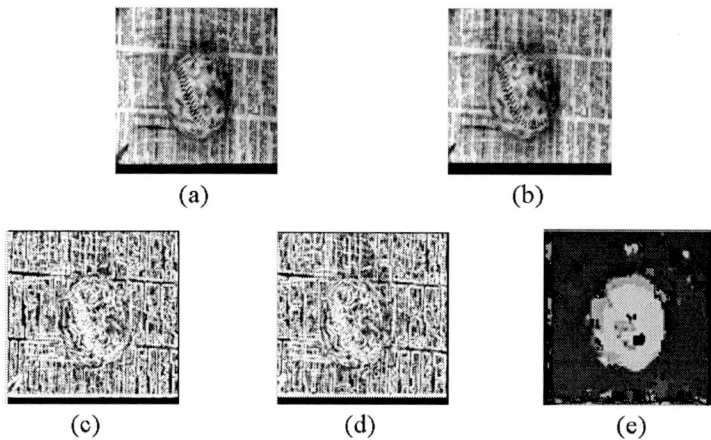

Fig. 2. Stereo image pair (a) Left image and (b) Right image (c) Left edge image (d) Right edge image (e) Disparity map

In Figure 3, R_l and R_r are the initial optical rays, which can be chosen to be either parallel to each other or some radian value of two rays. The cameras at the initial position are drawn by dotted line while the cameras after vergence movement are depicted by solid line. There are two constraints on the fixation point. First, the fixation point is projected onto the center of the retinal plane of each camera. Second, the vergence angle is $\theta_L+\theta_R$, when the cameras fixate at the fixation point. In Figure 3, R'_l and R'_r are the optical rays after fixation, when cameras move for focusing. Therefore, the vergence of the cameras is controlled to have the vergence angle where the object is located in the front-center of disparity map.

3 Relative Disparity and 3D Distance Computation

3.1 3D Distance Computation

In Figure 1, the distance(d) of the fixation point(F) from the Cyclopean point(C_M) is obtained by using the trigonometric method, as follows:

$$d = \frac{b}{2\tan(\theta_L)} . \qquad (1)$$

The relative distance between F and O_1 can be calculated by the disparity value of the fixation point, F, and an object point, O_1. Given the fixation point constraints that the cameras are located at the same height and that the angles of θ_L and θ_R are identical, the disparity value can be considered as the displacement of the corresponding points along the X-axis of two image planes. Therefore, the horizontal disparity is used. The computation of the relative disparity between

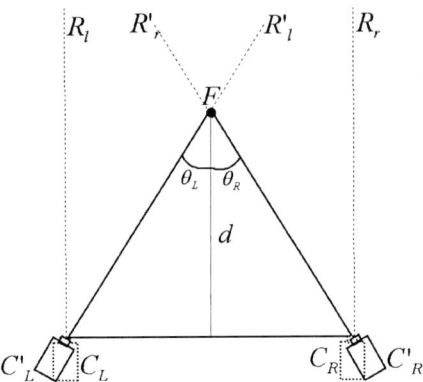

Fig. 3. The vergence angle and optical rays

F and O_1 about the X-axis is as follows:

$$redis = \frac{(o_{1Lx} - f_{Lx}) + (f_{Rx} - o_{1Rx})}{2} , \qquad (2)$$

where $redis$ is the value of relative disparity, (f_{Lx}, f_{Ly}) and (f_{Rx}, f_{Ry}) represent the projected coordinate of F in the left and right image planes, respectively, and (o_{1Lx}, o_{1Ly}) and (o_{1Rx}, o_{1Ry}) are those of O_1. The left-handed coordinate system for both cameras is used.

Above all, we need to calculate the transformation of the perspective projection from the 3D coordinate to the 2D image plane. Using the transformation of image coordinates to camera coordinates[7],

$$\begin{pmatrix} u \\ v \\ s \end{pmatrix} = \begin{pmatrix} -efk_u & 0 & u_0 & 0 \\ 0 & -efk_v & v_0 & 0 \\ 0 & 0 & 1 & 0 \end{pmatrix} \begin{pmatrix} x \\ y \\ z \\ 1 \end{pmatrix} , \qquad (3)$$

where ef is the effective focal length, k_u, k_v are the pixel scale factors to the retinal plane and u_0 and v_0 are the principal point. It is assumed that the focal lengths of two cameras are same in Equation 3.

The computation of the $(o_{1Lx} - f_{Lx})$ in Equation 2 using Equation 3 is as follows:

$$redis_L = O_{1Lx} - f_{Lx} = (-efk_u \frac{O_{1x}}{O_{1z}} + u_0 - (-efk_u \frac{F_x}{F_z} + u_0)) , \qquad (4)$$

where $redis_L$ is the disparity on the left image plane. In Equation 4, an assumption is that the principal point u_0 and v_0 is zero is used.

The 3D world coordinate system is newly chosen so that it has its origin at the Cyclopean point and its X-axis along the baseline with the direction toward the right camera being positive, that is, in the left-handed coordinate. The 3D world coordinate must be transformed to the left camera coordinate since the relative disparity is obtained from the left camera coordinate. Therefore, the 3D world coordinate can be transformed as follows:

$$C_L = \mathbf{R} C_M + \mathbf{T} , \qquad (5)$$

where the rotation matrix, $\mathbf{R}$, is obtained from the vergence angle and the translation vector, $\mathbf{T}$, is obtained from the baseline($b/2$) and C_L and C_L shown in Figure 1.

Equation 5 is rewritten as follows:

$$\begin{pmatrix} C_{LX} \\ C_{LY} \\ C_{LZ} \end{pmatrix} = \begin{pmatrix} cos\theta_L & 0 & sin\theta_L \\ 0 & 1 & 0 \\ -sin\theta_L & 0 & cos\theta_L \end{pmatrix} \begin{pmatrix} C_{MX} \\ C_{MY} \\ C_{MZ} \end{pmatrix} + \begin{pmatrix} -b/2 \\ 0 \\ 0 \end{pmatrix} , \qquad (6)$$

where $(C_{LX}, C_{LY}, C_{LZ})^T$ is the original point of the left camera coordinate and $(C_{MX}, C_{MY}, C_{MZ})^T$ is that of Cyclopean point coordinate.

The fixation point and the object point can be transformed as follows, when the 3D world coordinate is in the left camera coordinate, using a vergence angle and a baseline:

$$F' = \mathbf{R}' F + \mathbf{T}' , \qquad (7)$$

where F' obtained from the left camera coordinate, is used for 3D world coordinate value. According to the basic assumptions that the coordinate of F is $(0, 0, d)^T$ and that of an object point(O_1) is $(0, 0, rsd)^T$, let rsd be $d + rd$, when rsd is the distance from the fixation point to the object point(O_1) in Figure 1. The left camera coordinate is used for the coordinate of 3D world points of the following equation, using Equation 7,

$$\begin{pmatrix} x' \\ y' \\ z' \end{pmatrix} = \begin{pmatrix} cos(-\theta_L) & 0 & sin(-\theta_L) \\ 0 & 1 & 0 \\ -sin(-\theta_L) & 0 & cos(-\theta_L) \end{pmatrix} \begin{pmatrix} x \\ y \\ z \end{pmatrix} + \begin{pmatrix} b/2 \\ 0 \\ 0 \end{pmatrix} , \qquad (8)$$

where the coordinate value $[x', y', z']^T$ is transformed from $[x, y, z]^T$, according to $\mathbf{R}$ and $\mathbf{T}$. After the fixation point(F) and an object point(O_1) is transformed, Equation 4 is applied to following equations.

$$dor_L = (-efk_u \frac{O'_{1X}}{O'_{1Z}} + efk_u \frac{F'_X}{F'_Z}) , \qquad (9)$$

where $redis = (redis_L + redis_R)/2$ and the coordinate of a 3D point is $O1(O_{1X}, O_{1Y}, O_{1Z})$, and that of the fixation point is $F(F_X, F_Y, F_Z)$, $redis_R$ is calculated through the same way as Equation 4. From Equation 9, $redis_L = redis$ if we assume that $redis_L$ and $redis_R$ are the same. Using equation 9,

$$O'_{1Z} = \frac{efk_u O'_{1X}}{(efk_u \frac{F'_X}{F'_Z} - dor_L)} , \qquad (10)$$

where the coordinate value of O_{1x} and that of F_x is the same. Now, 3D Euclidean distance is easily estimated.

The relative distance between F and an object point O_2 can be computed in the same way. The same method of computation can be applied to the computation of the relative distance between F and any object on the horopter.

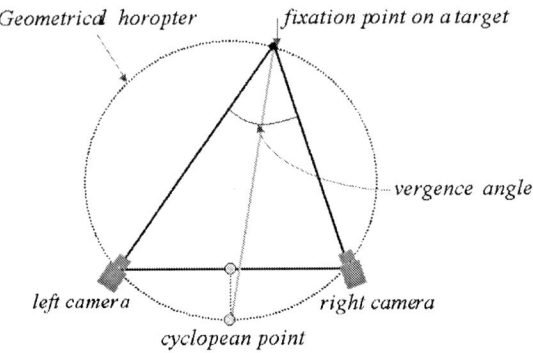

Fig. 4. Binocular geometry and geometrical horopter

The horopter is defined as the locus of a 3D object on which it has the same disparity value in Figure 4. Therefore, in Figure 1, when an object point is O_2, the computation of *redis* equals to $(redis_L + redis_R)/2$ using the horopter property, then the computational errors are so small that it can be ignored. In addition, the larger the absolute value of *redis*, the farther the fixation point is from the object point. If the sign of *redis* is negative, the object point is closer to the Cyclopean point than to the fixation point.

4 Experimental Results and Analysis

Experiments were conducted to verify the property of 3D Euclidean distance. Each objects were manually located at the real distances and the computation between the real 3D distances and estimated 3D distances of each objects were used for the verification of the property of interest of our proposed method.

Matching method used in experiments is a feature-based and gradient-based correlation method and feature points extracted from the *Sobel* edge operator and the thresholding operator for binary image are used.

In Experiment 1, the baseline is 25cm, the vergence angle is about $18°$ and efk_u is almost 162 in Equation 9. The fixation point of the scene is the center part of the first numbered object in Figure 5. The measurement unit of distance used was centi-meter(cm) and that of relative disparity was the pixel difference, as shown in Table 1. The values estimated by the proposed method were conducted to verify the property of the computation of 3D distance.

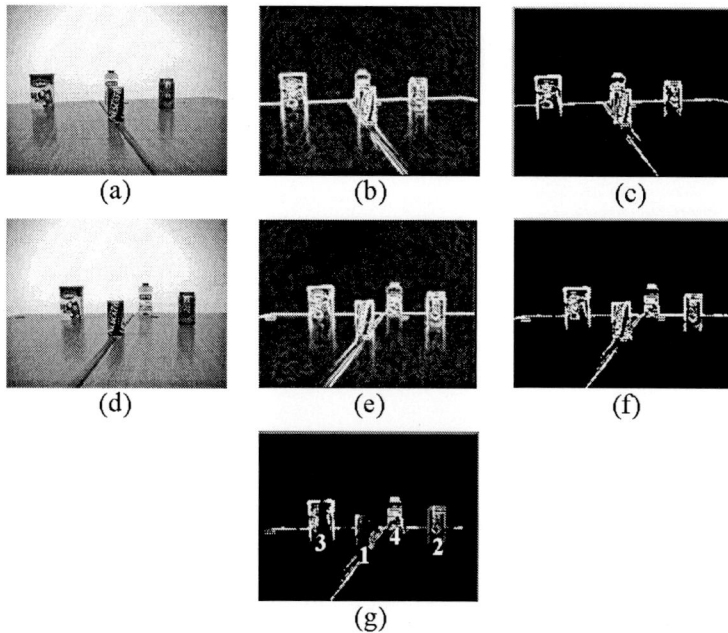

Fig. 5. Left and right images used in Experiment 1

Table 1. The comparison of real data and estimated data in Experiment 1

experiment1	real distance	estimated distance	relative disparity	computation error
object 1	80	79	0	1
object 2	100	98	-19	2
object 3	120	123	-25	3
object 4	140	136	-31	-4

In Experiment 2, object 1 and object 2 were located in front of the fixation point from the Cyclopean point and the same baseline as in Experiment 1 was used, the vergence angle was about $14°$ and efk_u was almost 308 in Equation 9. The relative disparity value of object 2 was -17 and that of object 3 was -18. In the case of object 3, a property of the horopter was used. The approximate values which were estimated by the proposed method were considerably similar to the real distances and can be feasible for many applications.

5 Conclusions and Future Works

In this paper, we proposed an efficient method for directly computing 3D Euclidean distances of points from a stereo image pair with respect to a fixation

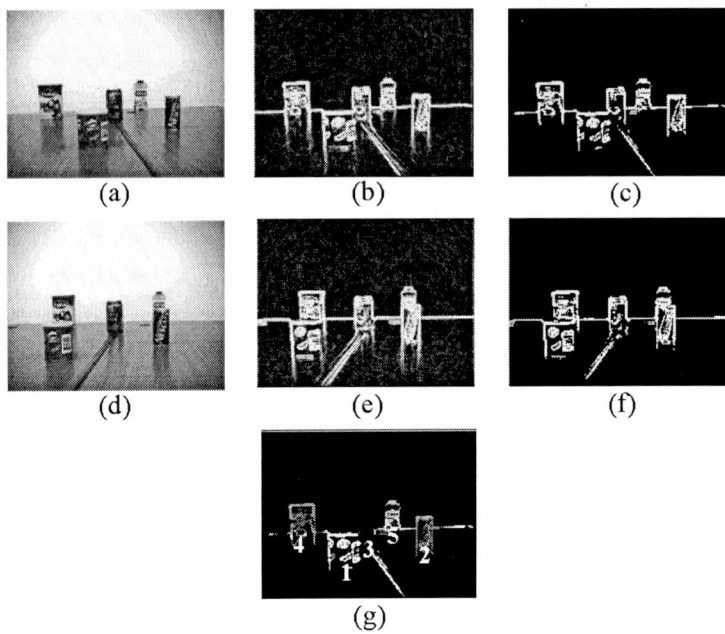

Fig. 6. Left and right images used in Experiment 2

Table 2. The comparison of real data and estimated data in Experiment 2

experiment2	real distance	estimated distance	relative disparity	computation error
object 1	60	78	39	18
object 2	80	87	12	7
object 3	100	102	0	2
object 4	120	118	-10	-2
object 5	140	147	-18	+7

point. The method is feasible for the navigation system and can also be used to detect and track a target object as well as to segment objects in the scene. Furthermore, no camera calibration is required.

However, an improved disparity estimation method robust to noise is required, since the SSD method employed here is sensitive to noise and intensity variation. The less it is influenced by errors, the more correctly the vergence can be controlled and fore ground segmentation is needed. For real-time implementation, a fast method for disparity estimation and vergence control is needed as well.

References

1. W.Y. Yau, H. Wang: Fast Relative Distance Computation for an Active Stereo Vision System. Real Time Imaging. **5** (1999) 189–202
2. C. J. Erkelens, R. van Ee: A Computational Model of Depth Perception based on Headcentric Disaprity. Vision Research. **38** (1998) 2999–3018
3. E. Trucco, A. Verri.: Introductory Techniques for 3D Computer Vision. Prentice Hall. Upper Saddle River. NJ (1998)
4. M. Chan, Z. Pizlo, D. Chelberg.: Binocular Shape Reconstruction:Psychological Plausibility of the 8-point Algorithm. Computer Vision and Image Understanding. **74** (1999) 121–137
5. G. S. M. Hansen.: Real-time Vergence Control using Local Phase Differences. Machine Graphics and Vision. **5** (1996)
6. G.-Q. Wei, W. Brauer, G. Hirzinger.: Intensity- and Gradient-based Stereo Matching using Hierarchical Gaussian Basis Functions. IEEE Trans. on Pattern Analysis and Machine Intelligence. **11** (1999) 1143–1160
7. O. Faugeras.: Three-Dimensional Computer Vision-A Geometric Viewpoint. MIT Press. Cambridge. MA. (1993)
8. A. Crouzil, L. Massip-Pailhes, S. Castan.: A New Correlation Criterion Based on Gradient Fields Similarity. Proc. Int. Conf. on Pattern Recognition. Vienna, Austria. (1996) 632–636

Bio-inspired Texture Segmentation Architectures

Javier Ruiz-del-Solar and Daniel Kottow

Dept. of Electrical Eng., Universidad de Chile
Casilla 412-3, 6513027 Santiago, CHILE
jruizd@cec.uchile.cl

Abstract. This article describes three bio-inspired Texture Segmentation Architectures that are based on the use of Joint Spatial/Frequency analysis methods. In all these architectures the bank of oriented filters is automatically generated using adaptive-subspace self-organizing maps. The automatic generation of the filters overcomes some drawbacks of similar architectures, such as the large size of the filter bank and the necessity of a priori knowledge to determine the filters' parameters. Taking as starting point the ASSOM (Adaptive-Subspace SOM) proposed by Kohonen, three growing self-organizing networks based on adaptive-subspace are proposed. The advantage of this new kind of adaptive-subspace networks with respect to ASSOM is that they overcome problems like the a priori information necessary to choose a suitable network size (the number of filters) and topology in advance.

1 Introduction

Image Processing plays today an important role in many fields. However, the processing of images of real-world scenes still presents some additional difficulties, compared with the processing of images in industrial environments, or in general, in controlled environments. Some examples of these problems are the wider range of variety of possible images and the impossibility to have control over external conditions. Natural images may contain incomplete data, such as partially hidden areas, and ambiguous data, such as distorted or blurred edges, produced by various effects like variable or irregular lighting conditions. On the other hand, the human visual system is adapted to process the visual information coming from the external world in an efficient, robust and relatively fast way. In particular, human beings are able to detect familiar and many times unfamiliar objects in variable operating environments. This fact suggest that a good approach to develop algorithms for the processing of real-world images consists of the use of some organizational principles present in our visual system or in general in biological visual systems. The processing of textures, and especially of natural textures, corresponds to an application field where this approach can be applied.

The automatic or computerized segmentation of textures is a long-standing field of research in which many different paradigms have been proposed. Among them, the *Joint Spatial/Frequency* paradigm is of great interest, because it is biologically based and because by using it, is possible to achieve high resolution in both the spatial (adequate for local analysis) and the frequency (adequate for global analysis) domains. Moreover, computational methods based on this paradigm are able to decompose textures into different orientations and frequencies (scales), which allows one to characterize them. This kind of characterization is very useful in the analysis of natural textures.

The segmentation of textures using Joint Spatial/Frequency analysis methods has been used by many authors (see partial reviews in [7, 8, 9]). These methods are based on the use of a bank of oriented filters, normally Gabor Filters, to extract a set of invariant features from the input textures. Then, these invariant features are used to classify the textures.

The main drawbacks of this kind of methods are the necessity of a large number of filters (normally more than sixteen), which slows down the segmentation process, and the a priori knowledge required to determine the filters' parameters (frequencies, orientations, bandwidths, etc.).

Based on the use of the ASSOM for the automatic generation of the oriented filters, recently was developed the TEXSOM-Architecture, an architecture for the segmentation of textures [9]. The TEXSOM architecture is described in section 2.

However, the use of the ASSOM model in the automatic generation of filters still presents some drawbacks because a priori information is necessary to choose a suitable network size (the number of filters) and topology in advance. Moreover, in some cases the lack of flexibility in the selection of the network topology (only rectangular or hexagonal grids are allowed) makes it very difficult to cover some areas of the frequency domain with the filters. On the other hand, growing self-organizing networks overcome this kind of problems. Taking these facts in consideration, the ASGCS, ASGFC and SASGFC networks, which correspond to growing networks based on adaptive-subspace concepts, were developed. These networks and two new texture segmentation architectures, TEXGFC and TEXSGFC, which are based on them, are presented in section 3. Finally, preliminary results and conclusions are given in section 4.

2 TEXSOM-Architecture

As it was pointed out, the TEXSOM-Architecture follows the Joint Spatial/Frequency paradigm, but it automatically generates the feature-invariant detectors (oriented filters) using the ASSOM.

The ASSOM corresponds to a further development of the SOM architecture, which allows one to generate invariant-feature detectors. In this network, each neuron is not described by a single parametric reference vector, but by basis vectors that span a linear subspace. The comparison of the orthogonal projections of every input vector into the different subspaces is used as the matching criterion by the network. If one wants these subspaces to correspond to invariant-feature detectors, one must define an episode (a group of vectors) in the training data, and then locate a representative winner for this episode. The training data is made of randomly displaced input patterns. The generation of the input vectors belonging to an episode is different, depending on translation, rotation, or scale invariant feature detectors needed to be obtained. The learning rules of the ASSOM and also more details about the generation of the input vectors can be found in [3-4].

2.1 Training Phase of the TEXSOM

The training is carried out using samples taken from all possible textures under consideration. If the architecture is used to perform defect identification on textured images, the samples must be taken from both defect and defect-free areas. In any case, the training is performed using two stages. In the first stage, or Filter Generation Stage, the feature-invariant detectors (Gabor-like filters) are generated using unsupervised learning. In the second stage, or Classifier Training Stage, these detectors generate invariant feature vectors, which are then used to train a supervised neural classifier. A detailed description can be found in [9].

Filter Generation. Before the training samples are sent to the ASSOM network they are pre-processed to obtain some degree of luminance invariance (the local mean value of the vectors is subtracted from them). The input parameters of the network are the size of the filters' mask and the number of neurons (or filters) in the network. Figure 1 shows a block diagram this stage.

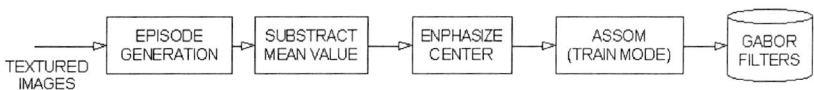

Fig. 1. Block diagram of the Filter Generation Stage.

Classifier Training. In this stage a LVQ network is trained with invariant feature vectors, which are obtained using the oriented filters generated in the Filter Generation Stage (see figure 2). The number of oriented filters gives the dimension of each feature vector. The vector components are obtained by taking the magnitude of the complex-valued response of the filters. This response is given by the convolution between an input pattern and the even and odd components of the filters.

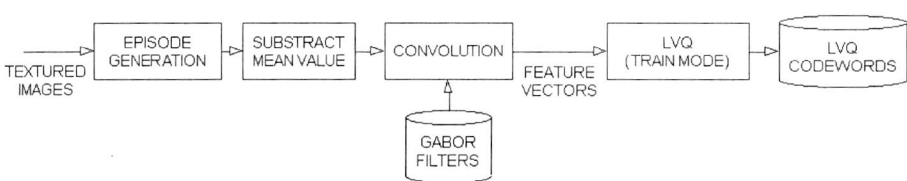

Fig. 2. Block diagram of the Classifier Training Stage.

2.2 Recall Phase of the TEXSOM

Figure 3 shows a simplified block diagram of the whole architecture working in recall-mode. The TEXSOM architecture includes a Non-linear Post-processing Stage that is not shown in this simplified diagram. The function of this stage is to improve the results of the pre-segmentation process. This post-processing stage performs Median Filtering over the pre-segmented images and then applies the Watershed Transformation over the median-filtered images (see description in [9]).

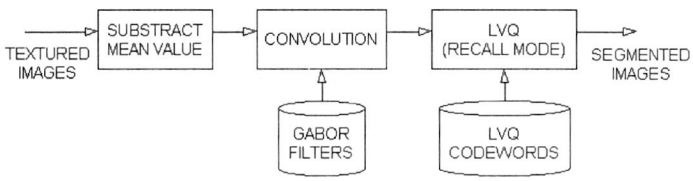

Fig. 3. Block diagram of the architecture in recall-mode.

3 TEXGFC and TEXSGFC Architectures

We will introduce two architectures that are strongly related to TEXSOM but improve it mainly on two aspects. First, a new Filter Generation Stage, implemented using ASGCS, ASGFC or SASGFC, allows an incremental growth of the number of filters during the training, and does not impose restrictions on the topology of the net. Second, the Classifier Training Stage is not unrelated to the generation of the filters anymore. Instead, a feedback path from the classification results to the filter generation process during training (see figure 4), influences the type of filters generated, in order to improve the overall performance of the architecture.

Fig. 4. Feedback path from classification results to the filter generation process during training.

3.1 Filter Generation: ASGCS, ASGFC and SASGFC

One of the main drawbacks of SOM in covering input spaces of rather complex morphology seems to be its rigid grid of fixed size and regular rectangular or hexagonal topology. The Growing Cell Structures (GCS) and shortly after the Growing Neural Gas (GNG), proposed in [1] and [2], precisely address this issue. Both networks start with a very small number of neurons and, while adapting the underlying prototype vectors, add neurons to the network. In order to determine where to insert a neuron, on-line counters for each neuron are used. These counters are updated every time a neuron results the winner neuron. Accordingly, zones of high input activity result in neurons having high counters where consequentially new neurons are inserted. The same idea is used to delete neurons from the network. We have borrowed this growing scheme for the development of ASGCS (Adaptive-Subspace GCS), ASGFC (Adaptive-Subspace Growing Filter Clusters) and SASGFC (Supervised ASGFC). An example of automatic filter generation using ASGCS and six natural images is shown in figures 5 and 6.

Fig. 5. Natural images used to train an ASGCS network.

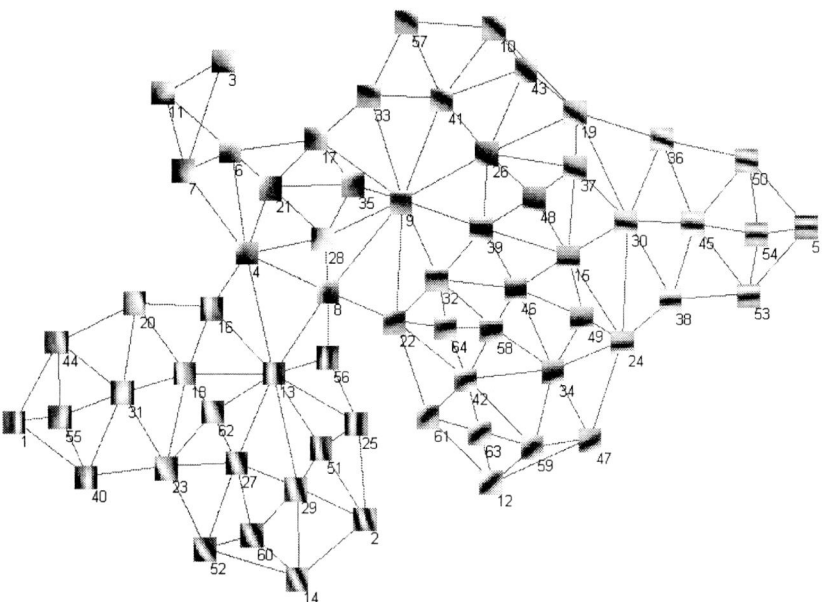

Fig. 6. The ASGCS filter network generated automatically using the images shown in figure 5.

While a large number of natural images or even artificial bi-dimensional sinusoids generate a nice array of Gabor Filters using either ASSOM or ASGCS (as an example see figure 6, or [6]), using a limited number of textures to grow a small number of filters seems to be a more difficult task. One way to understand this is to look at the textures in the frequency domain. Typically, each texture has pronounced clusters scattered over the Fourier domain. This suggests a strongly disconnected input space, where the number of filters and their freedom to move around without altering neighboring filters may be critical. ASGCS adds neurons to its network while adapting the underlying subspaces. Filters are inserted where the relationship between input activity and the number of filters is most critical [6]. The topology of the network formed by the neurons and their connections is a regular two-dimensional graph where each neuron belongs at least to one triangle (see figure 6 again).

ASGFC is very similar to ASGCS. The main difference is that filters are inserted where the relationship between matching error and the number of filters is most critical, which gives a more uniform distribution of the filters in the input space and generally a greater variety of filters. Also, the topology of ASGFC is looser; only pair-wise connections between neurons are used, resulting in network graphs with no closed paths. Moreover, connections might break down according to an aging scheme borrowed from GNG, which may create disconnected subgraphs and even isolated neurons. SASGFC is almost identical to ASGFC, but the filters carry a label relating the filter to one texture a priori and, accordingly, only adapt themselves to input episodes belonging to their texture. SASGFC starts with one filter for each texture and grows more filters just like ASGFC does, but the decision where to insert a filter is dramatically stressed since the new filter necessarily will belong to a certain texture. ASGCS, ASGFC and SASGFC are detailed described in [5].

3.2 The TEXGFC-Architecture

Figure 7 shows the complete training phase of TEXGFC. Once we generate the input samples and episodes, four things occur. 1. We adapt the existing filters of the ASGFC network (ASGFC in training mode). 2. Afterwards, we use the actual state of the ASGFC Filters to generate the feature vectors (convolution). 3. Then, we use the actual state of ASGFC to generate the winner filters information (ASGFC in recall mode). 4. The feature vectors are used to train the classifier (LVQ) just like in TEXSOM; and after a stabilization of the LVQ-training process we use the trained LVQ to classify the input samples. We now have a threefold information of the input samples. For each sample we know its class, the winner filter, and how it was classified by LVQ. We call this information the classification histogram because these results are summed up according to their attributes just like a histogram (see example in table 1).

ASGFC consults the histogram to know where and when to insert a new filter, based on which texture was badly classified and which filters are particularly responsible for it. We may stop this process once the classification results on the training samples are stabilized or are sufficiently satisfactory. Without any further changes the ASGFC network may be replaced by ASGCS, resulting in another variant of TEXGFC. In any case, the recall-phase of TEXGFC is the same as the one from TEXSOM (see figure 3).

3.3 The TEXSGFC-Architecture

TEXSGFC replaces ASGFC by its supervised version, the SASGFC (see block diagram in figure 8). This fact allows to skip the LVQ training during the overall filter-generation training phase, while retaining the possibility of reinforcing the filter generation by the histogram classification. Actually, the recall of SASGFC itself delivers the information of the winner filter and a preliminary classification result, given by the label of the winner neuron. When the filter-generation process is finished, a Classifier Training Phase using LVQ may be added to improve the final classification results. The recall-phase of this architecture is the same as the one from TEXSOM (see figure 3).

Table 1. An example of a classification histogram for 3 classes {C1, C2, C3} and four filters {F1, F2, F3, F4}. Each input sample is summed up to an histogram field according to the information of its class (Ci), the winner filter (Fi) and the result of the classification process {NS, S} (S: Success; NS: No Success).

	C_1 (NS)	C_2 (NS)	C_3 (NS)	C_1 (S)	C_2 (S)	C_3 (S)
F_1	12	0	3	3	0	4
F_2	2	1	7	3	1	4
F_3	0	15	0	0	0	0
F_4	0	3	9	1	4	6

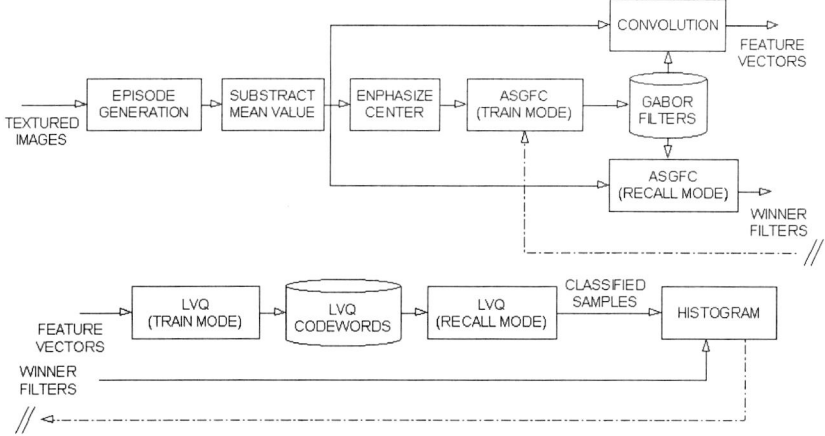

Fig. 7. TEXGFC. Block diagram of the training phase.

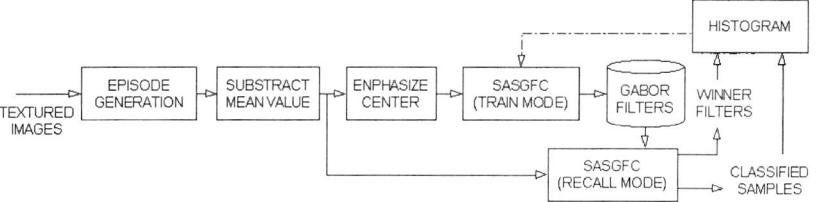

Fig. 8. TEXSGFC. Block diagram of the training phase.

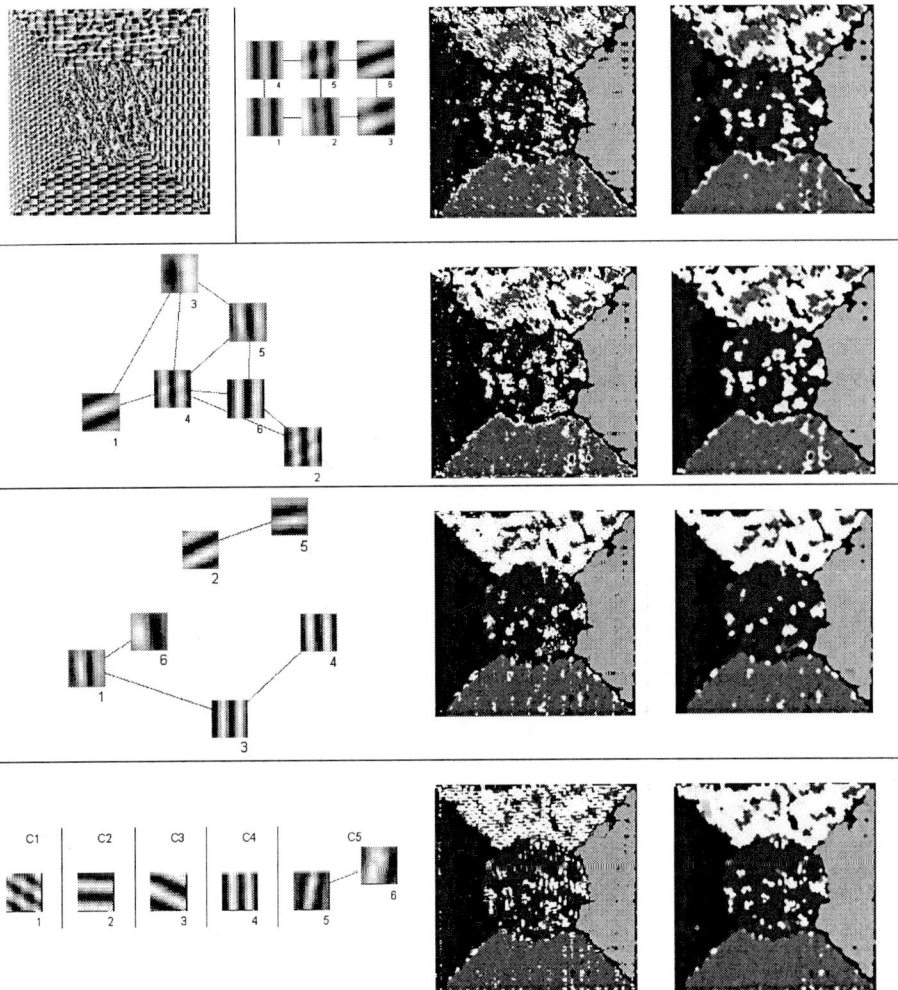

Fig. 9. Segmentation results of a textured image (left corner). From top to bottom: the results for TEXSOM, TEXGFC (using ASGCS), TEXGFC (using ASGFC) and TEXSGFC. Left: the generated filter networks; Middle: the segmented images; Right: Median-filtered images.

4 Preliminary Results and Conclusions

Some preliminary results of the segmentation capabilities of the architectures are presented here. The training of the architectures was performed using five textures, and a validation image was constructed using these textures (see left corner of figure 9). The results of the segmentation of this image applying TEXSOM, TEXGFC (using ASGCS), TEXGFC (using ASGFC), and TEXSGFC are shown in figure 9 (six filters are used in each case). In the left side the generated filter networks are displayed, while in the middle the segmented images are shown. By comparing these images can be seen that better results are obtained using TEXGFC (with ASGFC) and TEXSGFC. It should be pointed out, that these results are obtained without any post-processing. The use of a 3x3 median-filtering post-processing is shown in the right side of figure 9. As it can be seen, this very simple post-processing operation improves the result of the segmentation. The use of a more refined post-processing stage (as the one used in [9]) should improve largely the segmentation results.

Acknowledgements

This research was supported by FONDECYT (Chile) under Project Number 1990595.

References

1. Fritzke, B.: Growing Cell Structures - A self-organizing network for unsupervised and supervised learning. Neural Networks, Vol. 7, No. 9, 1994, pp. 1441-1460.
2. Fritzke, B.: A Growing Neural Gas Network Learns Topologies. Advances in Neural Information Processing Systems, 7, MIT Press, 1995.
3. Kohonen, T.: The Adaptive-Subspace SOM (ASSOM) and its use for the implementation of invariant feature detection. Proc. Int. Conf. on Artificial Neural Networks - ICANN '95, Oct. 9-13, Paris, 1995.
4. Kohonen, T.: Emergence of invariant-feature detectors in the adaptive-subspace self-organizing map. Biol. Cyber. 75 (4), 1996, pp. 281-291.
5. Kottow, D.: Dynamic topology for self-organizing networks based on adaptive subspaces, Master Degree Thesis, University of Chile, 1999 (Spanish).
6. Kottow, D., Ruiz-del-Solar, J.: A new neural network model for automatic generation of Gabor-like feature filters. Proc. Int. Joint Conf. On Neural Networks – IJCNN '99, Washington, USA, 1999.
7. Navarro, R., Tabernero, A., Cristóbal, G.: Image representation with Gabor wavelets and its applications. In: Hawkes, P.W. (ed.): Advances in Imaging and Electron Physics 97, Academic Press, San Diego, CA, 1996.
8. Reed, T., Du Buf, J.: A review of recent texture segmentation and feature techniques, CVGIP: Image Understanding 57 (3), 1993, pp. 359-372.
9. Ruiz-del-Solar, J.: TEXSOM: Texture Segmentation using Self-Organizing Maps, Neurocomputing (21) 1-3 1998, pp. 7-18.
10. Van Sluyters, R.C., Atkinson, J., Banks, M.S., Held, R.M., Hoffmann, K.-P., Shatz, C.J.: The Development of Vision and Visual Perception. In: Spillman L., Werner, J. (eds.): Visual Perception: The Neurophysiological Foundations, Academic Press (1990), 349-379.

3D Facial Feature Extraction and Global Motion Recovery Using Multi-modal Information

Sang-Hoon Kim[1] and Hyoung-Gon Kim[2]

Dept. of Control and Instrumentation, Hankyong National University
67 Seokjeong-dong,Ansung-City,Kyonggy-do,Korea
kimsh@hnu.hankyong.ac.kr
Imaging Media Research Center, KIST
39-1 Hawolgok-dong, Seongbuk-gu, Seoul, Korea
hgk@imrc.kist.re.kr

Abstract. Robust extraction of 3D facial features and global motion information from 2D image sequence for the MPEG-4 SNHC face model encoding is described. The facial regions are detected from image sequence using multi-modal fusion technique that combines range, color and motion information. 23 facial features among the MPEG-4 FDP(Face Definition Parameters) are extracted automatically inside the facial region using morphological processing. The extracted facial features are used to recover the 3D shape and global motion of the object using paraperspective factorization method. Stereo view and averaging technique are used to reduce the depth estimation error caused by the inherent paraperspective camera model. The recovered 3D motion information is transformed into global motion parameters of FAP(Face Animation Parameters) of the MPEG-4 to synchronize a generic face model with a real face.

1 Introduction

Recently, the focus on video coding technology has shifted to real-time object-based coding at rates of 8kb/s or lower. To meet such specification, MPEG-4 standardizes the coding of 2D/3D Audio/Visual hybrid data from natural and synthetic sources[1]. Specifically, to provide synthetic image capabilities, MPEG-4 SNHC AHG[13] has focused on the interactive/synthetic model hybrid encoding in the virtual space using real facial object perception technique. Although face detection is a prerequisite for the facial feature extraction, still too many assumptions are required. Generally, the informations of range, color and motion of human object has been used independently to detect faces based on the biologically motivated human perception technique. The range information alone is not enough to separate a general facial object from background due to its frequent motion. Also the motion information alone is not sufficient to distinguish a facial object from others. Recently, facial colors have been regarded as a remarkable clue to distinguish human objects from other regions[9]. But facial color has still many problems like similar noises in the complex background and color variation under the lighting conditions. Recent papers have suggested algorithms that combine range and color information[10][11], and that use multi-modal fusion range, color and motion information[15] to detect faces exactly. Two approaches are in use generally to recover object's motion and shape information, such as information flow-based method[2][3] and feature-based method[4][5][6]. The information flow-based method uses plane model[2] or oval face model[3], which accumulates errors as the number of image frame is increasing. The feature based method has a problem of diverging to an uncertain range according to the initial values[4][5], or works only under the assumption that the 3D structure of objects is already known[6]. The factorization-based method using paraperspective camera model shows the robustness in that field. The batch-type SVD computation method[7] is enhanced further to operate sequentially for the real-time execution[8]. However, it still has a problem in automatic feature extraction and has a large error in depth estimation owing to the inherent error of the paraperspective camera model.

The goal of this paper is 1) robust extraction of the 3D facial features from image sequence with complex background, and 2) recovery of the global motion information from the extracted facial features for the MPEG-4 SNHC face model encoding. Facial regions are detected from a image sequence using a multi-modal fusion technique that combines range, color and motion information. The 23 facial features suggested by the MPEG-4 FDP(Face Definition Parameters) are extracted automatically. The extracted facial features are used to recover the object's 3D shape and global motion sequentially based on the paraperspective factorization method. The proposed stereo view and averaging technique reduce a depth estimation error of the single paraperspective camera model. Finally, the recovered facial 3D motion and shape information is transformed into the global FAP motion parameters of the MPEG-4 to handle the synthetic face model synchronized with the real facial motion.

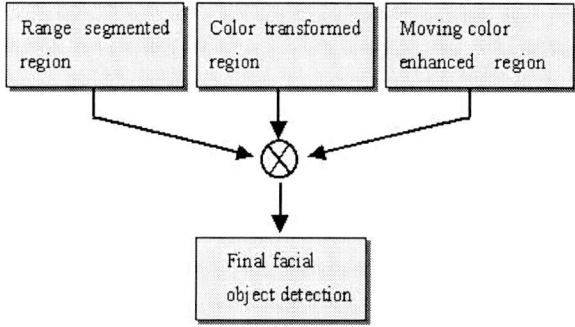

Fig. 1. Block diagram of the proposed face detection algorithm

2   The Multi-modal Fusion Technique

For the robust detection of the facial regions in real environment with various noises, a multi-modal fusion technique is proposed, where range, color and motion information is combined. The concept of multi-modal fusion technology is shown in figure 1. The information of the range segmented region, facial color transformed region and moving color region are extracted and these infomations are combined with AND operation in all pixels. The resulting image is transformed into a new gray level image whose intensity value is proportion to the probability to be a face. Then pixels of high probability are grouped and segmented as facial objects. It is described in [15] in detail.

3   Facial Features Extraction

3.1   Facial Features Position

The position of facial features is decided to be satisfied for the MPEG-4 SNHC FDP describing practical 3D structure of a face, as shown in figure 2. In our work, 23 important features among the 84 FDP are selected to be suitable as a stable input values for motion recovery. To extract facial features, normalized color information and color transform technique are used.

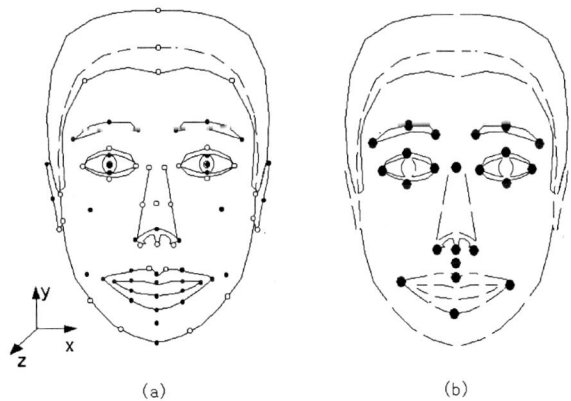

Fig. 2. Facial feature points (a)proposed by MPEG-4 SNHC and (b)proposed in this paper

3.2   Extraction of Eyes and Eyebrows

The most important facial features are eyes and eyebrows because of inherent symmetrical shape and color. To detect eyes and eyebrows within a face,

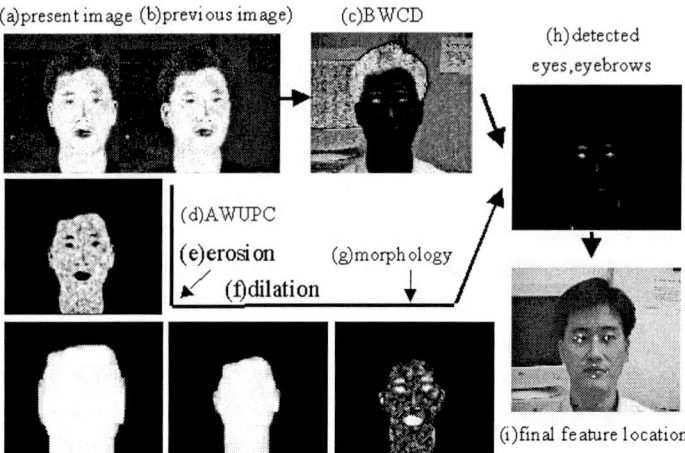

Fig. 3. A method of detecting eyes and eyebrows by top-hat transform

both the BWCD(Black and White Color Distribution) color transform and the AWUPC(Adaptive Weighted Unmatched Pixel Count) moving color enhancing technique[15] are combined together. Firstly, input image is transformed to enhance the region with BWCD as in figure 3 (c), which indicates eyes and eyebrows with normalized color space (r,g) = (85,85). Then, the motion detected image(figure 3(d)) is transformed by Top-hat operation[14] which enhances only the detail facial components. This Top- hat transformed image is combined with the prepared BWCD enhanced image to detect final eyes and eyebrows(figure 3(h)). All procedures including Top- hat transform and all procedures are shown in figure 3.

4 Shape and Motion Recovery Using a Paraperspective Stereo Camera Model

4.1 Paraperspective Camera Model

The paraperspective camera model is a linear approximation of the perspective projection by modeling both the scaling effect and the position effect, while retaining the linear properties with a scaling effect, near objects appear larger than the ones with distance, and with a position effect, objects in the periphery of the image are viewed from a different angle than those near the center of projection. As illustrated in figure 4, the paraperspective projection of an object onto an image involves two steps.

Step 1: An object point, $\vec{s_p}$, is projected along the direction of the line connecting the focal point of the camera and the object's center of mass, $\vec{c}$, onto a hypothetical image plane parallel to the real image plane and passing through the object's center of mass. In frame f, each object points $\vec{s_p}$ is projected along the direction of $\vec{c} - \vec{t_f}$ onto the hypothetical image plane where $\vec{t_f}$ is the camera origin with respect to the world origin. If the coordinate unit components of the camera origin are defined as i_f, j_f and k_f, the result of this projection, $\vec{s_{fp}}$, is given as follows,

$$\vec{s_{fp}} = \vec{s_p} - \frac{(\vec{s_p}\vec{k_f}) - (\vec{c} - \vec{k_f})}{(\vec{c} - \vec{t_f})\vec{k_f}} \vec{c} - \vec{t_f}. \quad (1)$$

Step 2: The point is then projected onto the real image plane using perspective projection. Since the hypothetical plane is parallel to the real image plane, this is equivalent to simply scaling the point coordinates by the ratio of the camera focal length and the distance between the two planes. Subtracting $\vec{t_f}$ from $\vec{s_{fp}}$, the position of the $\vec{s_{fp}}$ is converted with the camera coordinate system. Then, by scaling the result with the ratio of the camera's focal length l and the depth to the object's center of mass $\vec{z_f}$ results paraperspective projection of s_{fp} onto image plane. By placing the world origin at the object's center of mass, the equation can be simplified without loss of generality and the general paraperspective equation is given as follows,

$$u_{fp} = \frac{1}{z_f}\{\left[\vec{i_f} + \frac{\vec{i_f}\vec{t_f}}{z_f}\vec{k_f}\right]\vec{s_f} - \vec{t_f}\vec{i_f}\}, v_{fp} = \frac{1}{z_f}\{\left[\vec{j_f} + \frac{\vec{j_f}\vec{t_f}}{z_f}\vec{k_f}\right]\vec{s_f} - \vec{t_f}\vec{j_f}\}, \quad (2)$$

where u_{fp} and v_{fp} represents i and j component of the projected point.

The general paraperspective model equation can be rewritten simply as follows,

$$u_{fp} = \vec{m_f}\vec{s_p} + x_f, \quad v_{fp} = \vec{n_f}\vec{s_p} + y_f, \quad (3)$$

where

$$\vec{m_f} = \frac{\vec{i_f} - x_f\vec{k}}{z_f}, \vec{n_f} = \frac{\vec{j_f} - x_f\vec{k}}{z_f}, \quad (4)$$

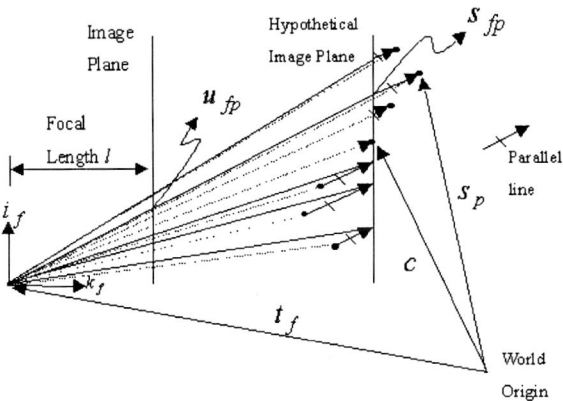

Fig. 4. Paraperspective camera model

$$z_f = -\vec{t_f}\vec{k_f}, x_f = -\frac{\vec{t_f}\vec{i_f}}{z_f}, y_f = -\frac{\vec{t_f}\vec{j_f}}{z_f}. \tag{5}$$

In equation (3), the term $\vec{m_f}$ and $\vec{n_f}$ represent the camera orientation information which means motion information, while $\vec{s_p}$ containing the coordinate of the feature points represents shape information. The x_f, y_f represent the translation information between world origin and camera origin. Notice that the motion information and the shape information can be separated in paraperspective camera model. The 3D feature points, $\vec{S_P}$, and motion information($\vec{m_f}, \vec{n_f}$) are calculated by solving equation (3) using Singular Value Decomposition(SVD). Details of the procedure are described in [7]. Although the motion and shape recovering algorithm using this paraperspective model shows good result[7][8], there is large difference between the measured object's depth information and calculated one using a paraperspective camera model which is due to the nonlinearity of practical perspective camera model. The depth error is about 30% of the measured value. The shape recovering average technique using stereo camera is suggested and evaluated.

4.2 Recovering Depth Error of the Paraperspective Camera Model

Figure 5 shows that there are differences between the images projected by objects(line objects a, b) through the perspective and the paraperspective camera. When certain points $b1, b2$ in the space is projected into the real image plane as $P1, P2$ using perspective camera model, the length of projected line($P1P2$) is presented as follows

$$b'' = P_2 - P_1, \tag{6}$$

where b'' means the real length of image reflected by the object b with the length of $(b2 - b1)$. When the same line b is projected into the real image plane using the paraperspective camera model, the length of projected object is expressed as follows,

$$b' = P_2' - P_1'. \tag{7}$$

The difference due to two independent camera model is given by $\Delta b = b' - b''$. The paraperspective factorization method that uses a linear camera model assumes the input values(u_{fp}, v_{fp}) are linearly proportional to the depth information, but those inputs are acquired through a general nonlinear perspective camera. This causes the error represented by Δa, Δb. Also the same error is reflected to the image plane by an object a(a1a2). If we assume (a = b), the size of projected images a' and b' using paraperspective camera model is always the same, while there exists a difference in the case of using the perspective camera model. If the distance between two cameras is long enough, it can be assumed that $\Delta a = \Delta b$. Therefore the error that should be corrected is defined as follows,

$$S_{fa} = S_{fb} = \frac{S_{fa} + S_{fb}}{2} = \frac{S_{fa''} + S_{fb''}}{2} = \frac{(S_{fa'} + \Delta a) + (S_{fb'} - \Delta b)}{2} = \frac{S_{fa'} + S_{fb'}}{2}, \tag{8}$$

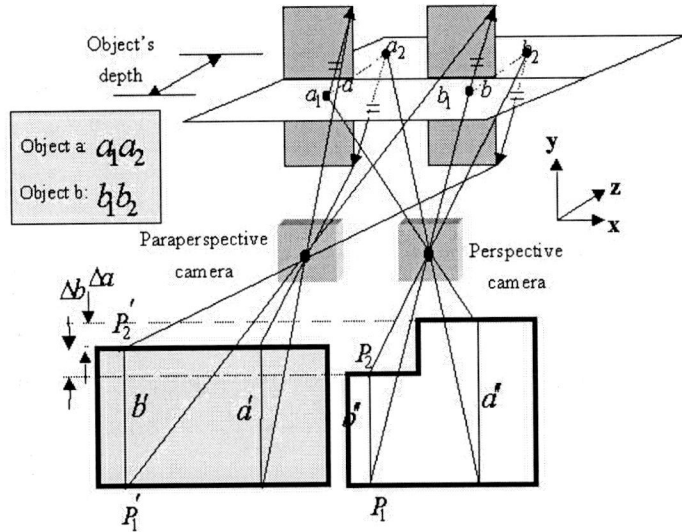

Fig. 5. Analysis of the depth error caused by the paraperspective camera model

where S_{fa}, S_{fb} are assumed to be the shape information of line objects a and b at frame f, respectively. Equation (8) means that the calculated average value of a, b approximates the measured value. Using this averaging technique, it is possible to recover the shape of 3D objects exactly. The accuracy of the motion information depends on that of the shape information.

5 Experimental Results

Figure 6 shows the results of multi-modal fusion technique to extract the facial regions. Test images include various skin color noises, complex background and different ranged objects. Figure 6(a)(b) show stereo image pairs at present time(t=0) for range information and figure 6(e) shows color image sequence at previous time(t=-1) for skin color information. Figure 6(f) shows inserted noises within the figure 6(a) by marking 'a','b','c'. Mark 'a' represents a object with motion, but without skin color. Mark 'b','c' represent object with skin color, but without motion. Figure 6(c) shows disparity map of figure 6(a)(b) stereo pairs and 6(d) is a range segmented image from figure 6(c). Figure 6(g) shows skin color enhanced image using GSCD and figure 6(h) represents the AWUPC result image enhancing only the region having both the motion and skin color, where skin color noises around human body and skin area having slight motion are removed by using a small motion variable during AWUPC operation. Final face detected regions on input image are shown in figure 6(i). Table 1 compares the performance of suggested multi-modal method with various uni-modal methods.

To prove the accuracy of the shape and motion recovery, the synthetic pyramid shaped object having 21 feature points along the four edges on the object

Table 1. performance comparison of multi-modal fusion face detection

information	No. of test images	No. of successful detection	success ratio
Range	100	46	46%
Skin color	100	82	82%
Moving color	100	88	88%
Multi-modal	100	96	96%

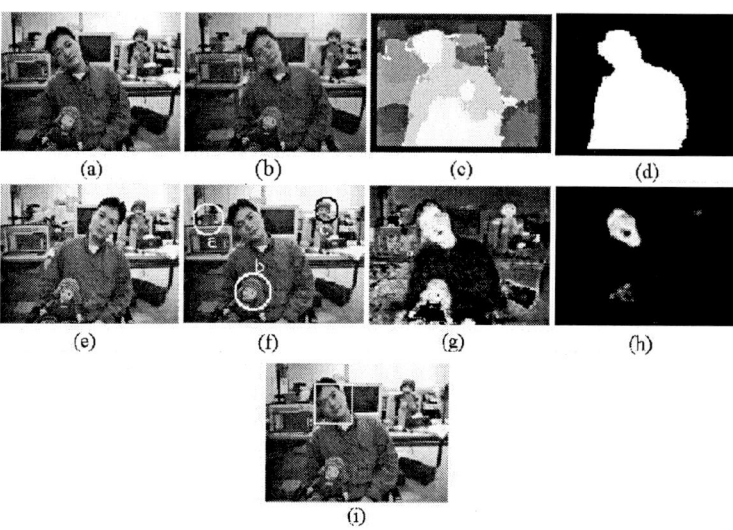

Fig. 6. face region detection result using range, skin color and motion information (a)(b) stereo image pairs(t=0) (c)MPC disparity map(d)range segmented image (e)color image sequence(t=-1) (f)inserted noises (g)skin color transformed image (h)AWUPC transform image (i)final face area

was used. The synthetic feature points were created by rotating the object, and the test consists of 60 image frames. Rotation of object was presented by three angles, α, β, γ. The coordinate of center point is defined as (0,0,0). The location of left camera is (0,-40,-400) and that of right one is (0,40,-400). The pyramid object rotates by 15 degree to the left and right around the x-axis and up and down around of the y-axis respectively. The distance between the neighboring features is 20 pixels on the x-axis, and 5 pixels on the y-axis. The result of motion recovery is represented by rotation angle, α, β, γ as shown in figure 7(b) and shape recovery on the x, y, z axis for all test frames is shown in figure 7(a).

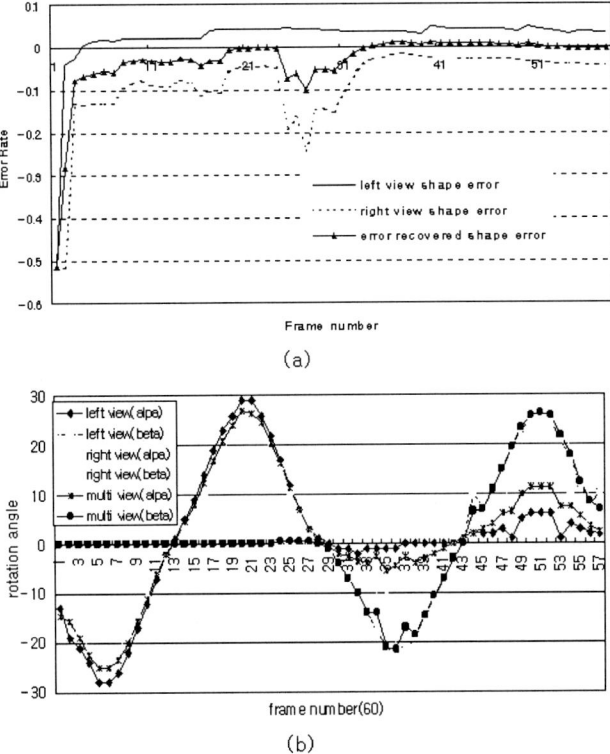

Fig. 7. Comparison of (a)shape recovery(for all frame)(b)motion recovery

6 Conclusions

New algorithm that recovers 3D facial features and motion information from 2D stereo image sequences for the MPEG-4 SNHC face model encoding has been proposed. The facial regions are detected using multi-modal fusion technique that combines range, color and motion information. The experiment shows that the success rate of the detection of facial region is over 96% for 100 image frames. The 23 facial features among the MPEG-4 FDP are extracted automatically using morphological processing and color transform algorithm. The facial features from 2D image sequences are used to recover the object's 3D shape and global motion sequentially based on the paraperspective factorization method. The stereo view and averaging technique are proposed to reduce a depth estimation error(about 30% of the measured value) caused by the inherent paraperspective camera model. Finally recovered facial 3D motion and shape information is transformed into the global motion parameters of FAP of the MPEG-4 to synchronize a generic face model with a real face.

References

1. MPEG-4 System Sub-group.: MPEG-4 System Methodology and Work Plan for Scene Description, ISO/IEC/JTC1/SC29/WG11/ N1786, Jul. 1997.
2. A. Pentland and B. Horowitz.: Recovery of Non-rigid Motion and Structure, IEEE Trans. Pattern Analysis and Machine Intelligence, vol. 13, no.7, pp. 730-742, 1991.
3. M.J. Black and Y. Yaccob.: Tracking and Recognizing Rigid and Non-rigid Facial Motion using Local Parametric Model of Image Motion, Proc. Intl Conf. Computer Vision, pp.374-381, 1995.
4. A. Azarbayejani and A. Pentland.: Recursive Estimation of Motion, Structure and Focal Length, IEEE Trans. Pattern Analysis and Machine Intelligence, vol. 7, no.6, pp. 562-575, Jun. 1995.
5. J. Weng , N. Ahuja and T. S. Huang.: Optimal Motion and Structure Estimation, IEEE Trans. Pattern Analysis and Machine Intelligence, vol. 15, no.9, Sept. 1993.
6. T.S.Huang and O.D.Faugeras.: Some Properties of the E-matrix in Two-view Motion Estimation, IEEE Trans. Pattern Analysis and Machine Intelligence, vol. 11, no.12, pp. 1310-1312, Dec. 1989.
7. C.J.Poelman and T. Kanade.: A Paraperspective Factorization Method for Shape and Motion Recovery, Technical Report CMU-CS-93-219, Carnegie Mellon University, 1993.
8. B.O.Jung.: A Sequential Algorithm for 3-D Shape and Motion Recovery from Image Sequences, Thesis for the degree of master, Korea University, Jun.1997.
9. Jibe Yang and Alex Waybill.: Tracking Human Faces in Real Time, Technical Report CMU-CS-95-210, Carnage Melon University, 1995.
10. S.H.Kim and H.G.Kim.: Object-oriented Face Detection using Range and Color Information, Proc. Intl Conf. Face and Gesture Recognition, Nara(Japan), pp. 76-81, April.1998.
11. S.H.Kim, H.G.Kim and K.H. Tchah.: Object-oriented Face Detection using Colour Transformation and Range Segmentation, IEE Electronics Letters, vol.34, no.10, 14th, pp.979-980, May. 1998.
12. S.H.Kim and H.G.Kim.: Facial Region Detection using Range Color Information, IEICE Trans. Inf. and Syst., vol.E81- D, no.9, pp.968-975, Sep.1998.
13. MPEG-4 SNHC Group.: Face and Body Definition and Animation Parameter, ISO/IEC JTC1/SC29/WG11 N2202, March 1998.
14. R.C. Gonzalez and R.E. Woods.: Digital Image Processing, Addison-Wesley, pp.225-238, 1992.
15. S.H.Kim and H.G.Kim.: Face Detection using Multi-Modal Information, Proc. Intl Conf. Face and Gesture Recognition, France, March.2000.

Evaluation of Adaptive NN-RBF Classifier Using Gaussian Mixture Density Estimates

Sung Wook Baik[1], SungMahn Ahn[2], and Peter W. Pachowicz[3]

[1] Datamat Systems Research Inc., McLean, VA, 22102, USA,
sbaik@dsri.com
[2] SsangYong Information & Communication Corp., Seoul, Korea,
sahn@sicc.co.kr
[3] Dept. of Electrical and Computer Engineering, George Mason University, Fairfax, VA 22030, USA,
ppach@gmu.edu

Abstract. This paper is focused on the development of an adaptive NN-RBF classifier for the recognition of objects. The classifier deals with the problem of continuously changing object characteristics used by the recognition processes. This characteristics change is due to the dynamics of scene-viewer relationship such as resolution and lighting. The approach applies a modified Radial-Basis Function paradigm for model-based object modeling and recognition. On-line adaptation of these models is developed and works in a closed loop with the object recognition system to perceive discrepancies between object models and varying object characteristics. The on-line adaptation employs four model evolution behaviors in adapting the classifier's structure and parameters. The paper also proposes that the models modified through the on-line adaptation be analyzed by an off-line model evaluation method (Gaussian mixture density estimates).

1 Introduction

Autonomous vision systems have ability to self-train so that it can adapt to new perceptual conditions under dynamic environments. One of approaches for such adaptation is to modify perceptual models of objects according to changes happened due to discrepancy between the internal models and object features found under new conditions. The systems are required to internally manipulate the perceptual models of objects in a coherent and stable manner. Humans are extremely good in adapting to a complex and changing perceptual environment. For as good an adaptation as humans, an on-line learning process is required to provide the adaptability feature to an object recognition system. A learning process improves the competence/performance of the system through a self-modification of object models. This modification of object models is executed through an interaction with the environment. Such on-line modification for adaptive object recognition need to be evaluated by delicate model representation functions (e.g. gaussian mixture density estimates) that can be applied off-line. The model representation using gaussian mixture density estimates is described in Section 6.

Most relevant research in the area of adaptive neural networks is focused on the dynamic pattern recognition problems. This includes, for example, the competitive self-organizing neural network [3] which is an unsupervised neural network system for pattern recognition and classification, a hybrid system of neural networks and statistical pattern classifiers using probabilistic neural network paradigm [13], and intelligent control, using neural networks, of complex dynamical systems [10].

2 On-line Model Evolution and Off-line Model Evaluation Paradigms

We introduce a computer vision system which can adapt to dynamic environments. The system involves P_{cv}, P_M and P_R for adaption to the environment by updating models over times from M_0 acquired off-line. P_{cv} is a vision process which uses object models to formulate decision (O). P_M is the model modification process to update the model to the perceived changes in the environment. P_R is the reinforcement process to define control and data used to modify models. This reinforcement is a process executed in an inner loop between P_M and P_R. The system is represented by the (1).

$$S_{cv} = < I_t, M_0, P_{cv}, P_M, P_R, O_t > \qquad (1)$$

For a computer vision system working in the time-varying environment, we introduce an optimal process $\hat{P}_{cv}$ (2).

$$\hat{P}_{cv} : I_t \times \hat{M}_t \longrightarrow \hat{O}_t \qquad (2)$$

$\hat{P}_{cv}$ means that we get the best decisions/output with the perfect models($\hat{M}_t$) representing current conditions as if it is processed by a human supervisor. The process is theoretically possible if $\hat{M}_t$ is available. There are, however, problems with the availability of the $\hat{M}_t$ due to the following reasons:

1. Models $\hat{M}_t$ for any condition t are usually not given a-priori because the set of conditions is infinite for dynamic systems expressed by I_t.
2. Models are always constructed with limited perfection. It means that much more resources (e.g. CPU time) are required for construction of more delicate models.

Therefore, $\hat{M}_t$ is used for off-line model evaluation by comparing with the realistic models constructed on-line. The method to build $\hat{M}_t$ is described in Section 6.

For on-line model evolution, we introduce a quasi optimal process $\tilde{P}_{cv}$ with $\tilde{M}_t$ a quasi optimal form of $\hat{M}_t$ (3). This process generates quasi optimal output $\tilde{O}_t$ when using $\tilde{M}_t$ under assumption that $\tilde{M}_t$ is gradually updated to be available at any t. The problem is that $\tilde{M}_t$ is not given at time t. Instead $\tilde{M}_{t-\Delta t}$ can be given at time t by being updated at just previous time in (4).

However, if the time interval is decreasing, i.e. $\Delta t \to 0$, P_{cv} (4) will converge to $\tilde{P}_{cv}$ (3) in (5). Considering further technicality of $\Delta t \to 0$, the time interval of system dynamics (Δt) must be much smaller than that of dynamics of environment (Δt_{env}) and it is greater than zero due to disparity between models and environments (6).

$$\tilde{P}_{cv} : I_t \times \tilde{M}_t \longrightarrow \tilde{O}_t \ where \ | \ \delta(\hat{O}_t, \tilde{O}_t) \ | < \epsilon \tag{3}$$

$$P_{cv} : I_t \times \tilde{M}_{t-\Delta t} \longrightarrow O_t \tag{4}$$

$$\lim_{\Delta t \to 0} P_{cv} = \tilde{P}_{cv} \tag{5}$$

Because of $\lim_{\Delta t \to 0} I_t \times \tilde{M}_{t-\Delta t} \longrightarrow \tilde{O}_t$ is true since $\lim_{\Delta t \to 0} \tilde{M}_{t-\Delta t} = \tilde{M}_t$

$$0 < \Delta t \ll \Delta t_{env} \tag{6}$$

3 RBF Classifier

A modified RBF classifier [9,4], with Gaussian distribution as a basis, was chosen for texture data modeling and classification. This is a well-known classifier widely used in pattern recognition and well-suited for engineering applications. Its well-defined mathematical model allows for further modifications and on-line manipulation with its parameters and structure. The RBF classifier models a complex multi-modal data distribution through its decomposition into multiple independent Gaussians. Sample classification provides a class membership along with a confidence measure of the membership. The classifier is implemented using the neural network approach [8].

Most of real texture images are represented by multi-modality of their characteristics rather than single modality. When it is perceived under variable perceptual condition, an object is registered with changing its texture images slightly. Therefore, in order to cover texture characteristics obtained from such slightly changing images of the object together, multi-modality of their characteristics is absolutely required. The RBF classifier deals with multi-modality of such texture characteristics. It is necessary to deal with multi-modality in a controllable and organized way for situations where subsequent different image frames are obtained under variable perceptual conditions. The structure of the NN-RBF classifier is shown in Figure 1. Each group of nodes corresponds to a different class. The combination of nodes is weighted. Each node is a Gaussian function with a trainable mean and spread. Classification decision yields class C_i of the highest $F_i(x)$ value for a sample vector x.

4 Reinforcement Formulation

A feedback reinforcement mechanism is designed to provide feedback information and control for the on-line adaptation of the NN-RBF classifier. This feedback

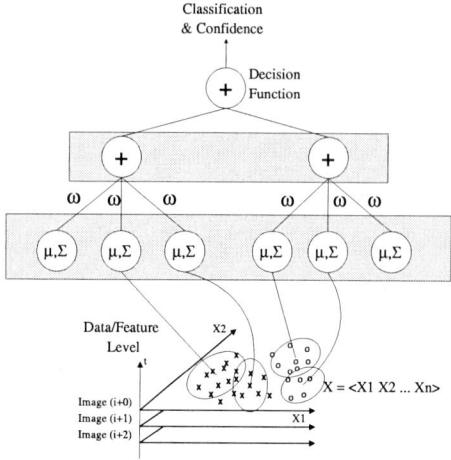

Fig. 1. A modified NN-RBF Classifier

exploits classification results of the NN-RBF classifier on the next segmented image of a sequence. Reinforcement is generated over the following three steps:

1. Sample Selection
2. Sample Categorization
3. Reinforcement Parameters Generation

Sample Selection is performed in an unsupervised manner. A given size window (15×15 pixels - meaningful size of a texture patch) randomly moves over the segmented image (see Figure 2). When all pixels in the window have the same class membership, the system understands that the window is located within a homogeneous area. Whenever such a window is found, a pixel position corresponding to the center of the window is picked up for feature data extraction. Redundant multiple overlapping windows are eliminated by rejecting samples of the highest deviation of classification confidence over the window [11]. Sample Categorization allocates selected data samples into groups of different similarity levels based on their confidences. Samples within each similarity group are generalized and described by reinforcement parameters expressing the direction and magnitude of a shift from the current RBF model.

5 Dynamic Modification and Verification of the Classifier

Reinforcement parameters are analyzed in relation to the structure and parameters of the NN-RBF classifier. First, the system selects strategies (called behaviors) for the classifier modification. Second, it binds reinforcement data to the selected behaviors. Finally, the behaviors are executed. There are four behaviors for the NN-RBF classifier modification that can be selected and executed

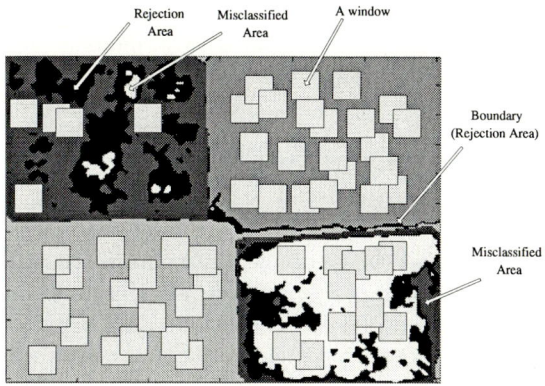

Fig. 2. Sample selection using a moving window located within homogeneous areas

independently: (1) Accommodation, (2) Translation, (3) Generation, and (4) Extinction. Each behavior is implemented separately using mathematical rules transposing reinforcement parameters onto actions of RBF modification. Accommodation and Translation behaviors modify the classifier parameters only. This modification is performed over selected nodes of the net. The basis for Accommodation is to combine reinforcement parameters with the existing node parameters. The result of Accommodation is adjusted function spread. The node center does not change/shift through the feature space. The goal for Translation is to shift the node center in the direction of reinforcement without modifying the spread of the function. Combining Accommodation and Translation, the system can fully modify an existing node of the NN-RBF classifier. Generation and Extinction behaviors modify the NN structure by expanding or pruning the number of nodes. The basic idea of Generation is to create a new node. A node is generated (allocated) when there is (1) a significant progressive shift in function location and/or (2) an increase in complexity of feature space, for example, caused by the increase in the multi-modality of data distribution. The goal of Extinction is to eliminate (de-allocate) useless nodes from a network. Extinction is activated by the utilization of network nodes in the image classification process. Components, which constantly do not contribute to the classifier, are disposed gradually. This allows for controlling the complexity of the neural net over time.

Classifier verification is (1) to confirm the progress of classifier modification and (2) to recover from eventual errors. Classifier verification is absolutely required because behavioral modification of the classifier is performed in an unsupervised manner. If errors occur and are not corrected, they would seriously confuse the system when working over the next images of a sequence. There are two possible causes of errors: 1) incorrect reinforcement generation, and 2) incorrect selection of modification behavior. Classifier verification compares the classification and image segmentation results on the same image. If the expected improvement is not reached, then the classifier structure and parameters are restored. Classifier modification is repeated with a different choice of behaviors and/or less progressive reinforcement.

6 Gaussian Mixture Density Estimates for Model Evaluation

Density estimation techniques can be divided into parametric and nonparametric approaches. A widely used parametric approach is the finite mixture model in combination with the Expectation-Maximization (EM) algorithm [6]. The EM algorithm is proved to converge to at least a local maximum in the likelihood surface.

In the finite mixture model, we assume that the true but unknown density is of the form in equation (1), where g is known, and that the nonnegative mixing coefficients, π_j, sum to unity.

$$f(x; \pi, \mu, \Sigma) = \sum_{j=1}^{g} \pi_j N(x; \mu_j, \Sigma_j) \qquad (7)$$

Traditional maximum likelihood estimation leads to a set of normal equations, which can only be solved using an iterative procedure. A convenient iterative method is the EM algorithm [6], where it is assumed that each observation, x_i, $i = 1, 2, \cdots, N$, is associated with an unobserved state, z_i, $i = 1, 2, \cdots, N$, and z_i is the indicator vector of length g, $z_i = (z_{i1}, z_{i2}, \cdots, z_{ig})'$, and z_{ij} is 1 if and only if x_i is generated by density j and 0 otherwise. The joint distribution of x_i and z_i under Gaussian mixture assumption is [18].

$$f(x_i, z_i | \Theta) = \prod_{j=1}^{g} [\pi_j N(x_i; \mu_j, \Sigma_j)]^{z_{ij}}. \qquad (8)$$

Therefore, the log likelihood for the complete-data is

$$l_c(\Theta) = \sum_{i=1}^{N} \sum_{j=1}^{g} z_{ij} \log[\pi_j N(x_i; \mu_j, \Sigma_j)]. \qquad (9)$$

Since z_{ij} is unknown, the complete-data log likelihood cannot be used directly. Thus we instead work with its expectation, that is, we apply the EM algorithm.

In the finite mixture model, the number of components is fixed. But in practice, it is not realistic to assume knowledge of the number of components ahead of time. Thus it is necessary to find the optimum number of components with given data. This is the problem of model selection and there are some approaches in the context of density estimation. Solka et al [16] used AIC (Akaike Information Criterion [2]), and Roeder and Wasserman [14] used BIC (Bayesian Information Criterion [15]) for the optimum number of components. But here we use the method of regularization [5,12]. A basic assumption we use here is that we begin with an overfitted model in terms of the number of components. That is, some of the components in the model are redundant and, thus, can be removed with the current likelihood level maintained. So the regularization term is chosen such that the solution resulting from the maximization of the new likelihood function favors less complex models over the one that we begin with. The goal is achieved by adjusting the mixing coefficients, π, and the resulting form of regularization function is as follows.

$$\sum_{i=1}^{N}\sum_{j=1}^{g} z_{ij} \log[\pi_j N(x_i; \mu_j, \Sigma_j)] + \lambda N \sum_{j=1}^{g}(\alpha - 1) \log \pi_j. \qquad (10)$$

A simple explanation about how the regularization term works is this. The value of the second term increases as the values of π decrease. Thus $\pi=0$ maximizes the regularization term. Since, however, we have to satisfy the constraint that the nonnegative mixing coefficients, π, sum to unity as well as to maximize the original likelihood, we expect some (not all) of π's drop to zero. It turns out that the method produces reasonable results with proper choices of λ and α. More details about the method including a choice of an overfitted model, the algorithm, and simulation results can be found in [1].

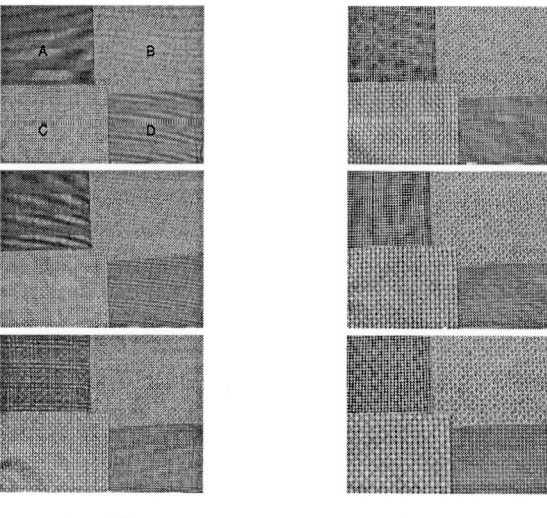

Image 1,5,9 Image 13,17,21

Fig. 3. Selected six images of a sequence of 22 images (In the first image, label A,B,C, and D represent different texture class areas).

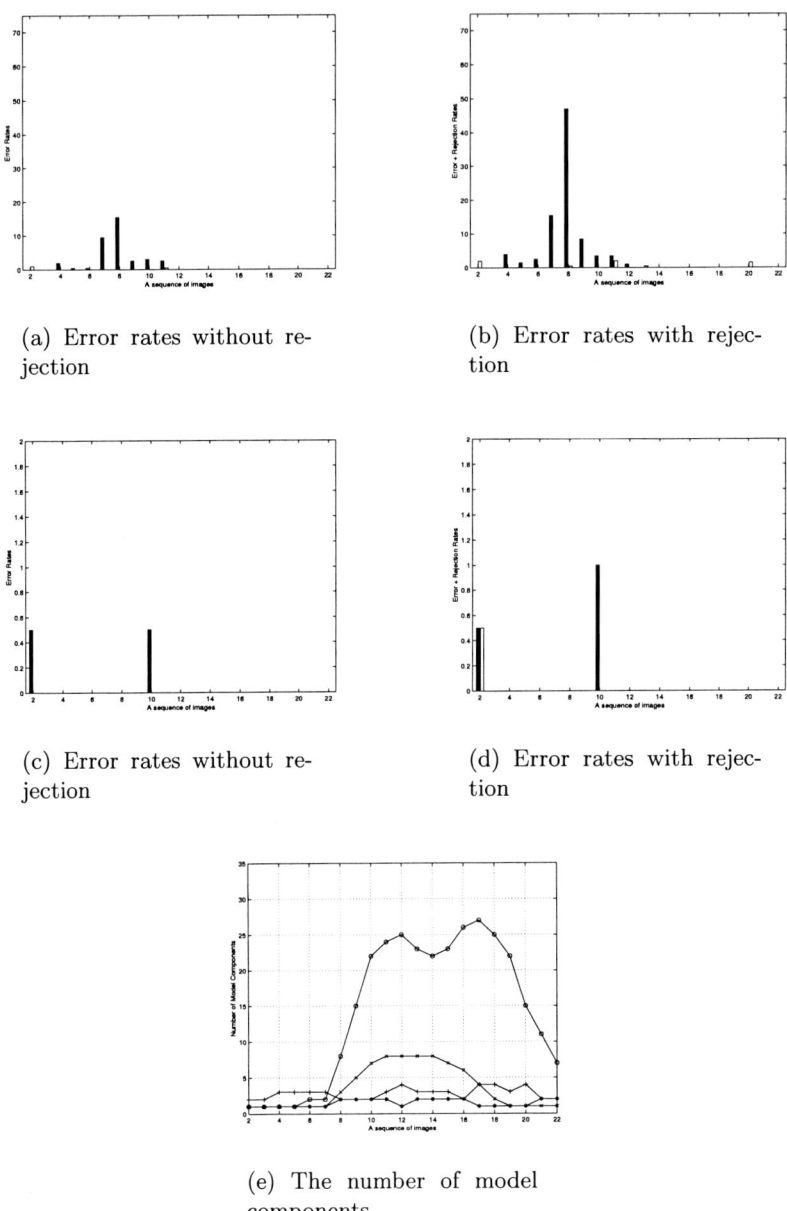

(a) Error rates without rejection

(b) Error rates with rejection

(c) Error rates without rejection

(d) Error rates with rejection

(e) The number of model components

Fig. 4. Experimental results for image sequence.(Black bar: Class A and White bar: Class D for (a) - (d); 'o': class A, '+': class B, '*': class C, 'x': class D for (e).

7 Experimental Results

In this section, the developed approach is experimentally validated through experiments with a sequence of indoor images. Figure 3 shows the selected images of a sequence for an indoor texture scene used for experimentation. Image sequences were acquired by a b&w camera. The distance was gradually decreased between the camera and the scene. The sequence has 22 images of the scene containing four fabrics (class A, B, C and D).

Each incoming image is processed to extract texture features through the following three steps:

1. Gabor spectral filtering [7].
2. Local 7×7 averaging of filter responses to estimate local energy response of the filter.
3. Local non-linear spatial filtering.

Non-linear filtering is used to eliminate a smoothing effect between distinctive homogenous areas. The filter computes standard deviation over five windows spread around a central pixel. The mean for the lowest deviation window is returned as the output. Values of each texture feature are subject to a normalization process [17] to eliminate negative imbalances in feature distribution.

Figure 4 shows experimental results with the indoor image sequence. There are two types of error rates registered: 1) error rate without rejection, and 2) error rate with rejection. Error rates with rejection provide a better analysis of experimental results. Classification errors are registered for each new incoming image I(i+1) before the NN-RBF classifier is modified (see diagrams a-b) and after it is modified over the I(i+1) image (see Figure 4(a) - 4(d)). Because the system goes through every image of a sequence, the modified classifier over the I(i+1) image is then applied to the next image. The results show a dramatic improvement in both error rates. Both error rates achieve almost zero level after the classifier is evolved over images of a sequence. Figure 4(e) shows the change in classifier complexity over the evolution process. The number of nodes for class A rapidly increases beginning from image 8 and reaches a maximum of 27 nodes when going through image 17. After that, it rapidly decreases to 9 nodes at the end of the image sequence. Other classes have relatively simpler structure than class A.

References

1. Ahn, S. M., Wegman, E. J.: A Penalty Function Method for Simplifying Adaptive Mixture Density. Computing Science and Statistics. **30** (1998) 134–143
2. Akaike, H.: A New Look at the Statistical Model Identification. IEEE Trans. Auto. Control. **19** (1974) 716–723
3. Athinarayanan, R.: A biologically based competition model for self-organizing neural networks. Proceedings of Systems, Man, and Cybernetics. **2** (1998) 1806–1811
4. Baik, S.: Model Modification Methodology for Adaptive Object Recognition in a Sequence of Images. Ph.D. Thesis, George Mason University. (1999)

5. Bishop, C. M.: Neural Networks for Pattern Recognition. Oxford. (1995)
6. Dempster, A. P., Laird, N. M., and Rubin, D. B.: Maximum Likelihood from Incomplete Data via the EM Algorithm. Journal of Royal Statistical Society(B). **39** (1977) 1-39
7. Jain, A.K., Bhattacharjee, S.: Text Segmentation Using Gabor Filters for Automatic Document Processing. Machine Vision and Applications. **5** (1992) 169-184
8. Haykin, S.. Neural Networks, Prentice Hall. (1999)
9. Musavi, M. T., et al.: A Probabilistic Model For Evaluation Of Neural Network Classifiers. Pattern Recognition. **25** (1992) 1241–1251
10. Narendra, K.S., Mukhopadhyay, S.: Intelligent control using neural networks. IEEE Control Systems Magazine. **12** (1992) 11–18
11. Pachowicz, P.W.: Invariant Object Recognition: A Model Evolution Approach. Proceedings of DARPA IUW. (1993) 715–724
12. Poggio, T. and Girosi. F.: Networks for approximation and Learning. Proceedings of the IEEE 78. **9** (1990) 1481–1497
13. Rigoll, G.: Mutual information neural networks for dynamic pattern recognition tasks. Proceedings of the IEEE International Symposium. **1** (1996) 80–85
14. Roeder, K., Wasserman, L.: Practical Bayesian Density Estimation Using Mixtures of Normals. Journal of American Statistical Association. **92** 894–902
15. Schwarz, G.: Estimating the Dimension of a Model. Annals of Statistics. **6** (1978) 461–464
16. Solka, J. L., Wegman, E. J., Priebe, C. E., Poston, W. L., Rogers, G. W.: A Method to Determine the Structure of an Unknown Mixture using Akaike Information Criterion and the Bootstrap. Statistics and Computing. (1998)
17. Theodoridis, S., Koutroumbas, K.: Pattern Recognition. (1999)
18. Titterington, D. M., Smith, A. F. M., Makov, U. E.: Statistical Analysis of Finite Mixture Distributions, Willey (1985)

Scene Segmentation by Chaotic Synchronization and Desynchronization

Liang Zhao

Laboratório de Integração e Testes, Instituto Nacional de Pesquisas Espaciais,
Av. dos Astronautas, 1758, CEP: 12227-010, São José dos Campos – SP, Brasil.
zhao@lit.inpe.br

Abstract. A chaotic oscillatory network for scene segmentation is presented. It is a two-dimensional array with locally coupled chaotic elements. It offers a mechanism to escape from the synchrony-desynchrony dilemma. As a result, this model has unbounded capacity of segmentation. Chaotic dynamics and chaotic synchronization in the model are analyzed. Desynchronization property is guaranteed by the definition of chaos. Computer simulations confirm the theoretical prediction

1 Introduction

Sensory segmentation is the ability to attend to some objects in a scene by separating them from each other and from their surroundings [12]. It is a problem of theoretical and practical importance. For example, it is used in situations where a person can separate a single, distinct face from a crowd or keep track of a single conversation from the overall noise at a party. Pattern recognition and scene analysis can be substantially simplified if a good segmentation is available.

Neuronal oscillation and synchronization observed in the cat visual cortex have been suggested as a mechanism to solve feature binding and scene segmentation problems [3, 4, 6]. Scene segmentation solution under this suggestion can be described by the following rule: the neurons which process different features of the same object oscillate with a fixed phase (synchronization), while neurons which code different objects oscillate with different phases or at random (desynchronization). This is the so-called Oscillatory Correlation [11]. The main difficulty encountered in these kinds of models is to deal with two totally contrary things at the same time: synchrony and desynchrony. The stronger the synchronization tendency among neurons, the more difficult it is to achieve desynchronization and vice versa. Then, segmentation solutions are tradeoffs between these two tendencies. We call this situation the Synchrony-Desynchrony Dilemma.

The segmentation capacity (number of objects that can be separated in a given scene) is directly related to this dilemma, i.e., the capacity will always be limited if the synchrony-desynchrony dilemma cannot be escaped from. This is because the enhancement of synchronization or desynchronization tendency will inevitably weaken another tendency. Usually desynchronization is weakened because a coherent object should not be broken up. Segmentation capacity is decided exactly by the model's desynchronization ability since the desynchronization mechanism serves to distinguish one group of synchronized neurons (an object) from another. Thus, the segmentation capacity decreases as the desynchronization tendency is weakened.

Many oscillatory correlation models for scene segmentation have been developed (see [2, 7, 8, 10, 11, 12, 13] and references there in). Most of them employ non-chaotic oscillator as each element. Because of the periodic nature, the chances of wrongly grouping two or more objects significantly increase as the number of objects to be segmented increases. Moreover, the synchronization and desynchronization analyses in this kind of models are usually based on a limit cycle solution of the system [2, 11]. Although each oscillator used in their models cannot be chaotic, the coupled system may be chaotic. Then, their analyses do not always valid since chaotic range and stable oscillating range in the parameter space have not been identified. On the other hand, few chaotic models have been proposed. However, long-range coupling is used in their networks. As pointed out by Wang [11], these long-range connections, globally coupling in particular, lead to indiscriminate segmentation since the network is dimensionless and loses critical geometrical information about the objects. To our knowledge, the maximal number of objects can be segmented by oscillatory correlation model is less than 12 [2, 11].

In this paper, an oscillator network with diffusively coupled chaotic elements is presented. We show numerically that a large number of elements in such a lattice can be synchronized. Each object in a given scene is represented by a synchronized chaotic trajectory, and all such chaotic trajectories can be easily distinguished by utilizing the sensitivity and aperiodicity properties of chaos. Therefore, the synchrony-desynchrony dilemma can be avoided. As a consequence, our model has unbounded capacity of object segmentation. The differential equations are integrated by using the fourth-order Runge-Kutta method. Lyapunov exponents are calculated by using the algorithm presented in [14].

This paper is organized as follows. Sec. 2 describes chaotic dynamics in a single periodically driven Wilson-Cowan oscillator. Sec. 3 gives the model definition and the segmentation strategy. Computer simulations are shown in Sec. 4. Sec. 5 concludes the paper.

2 Chaotic Dynamics in a Periodically Driven Wilson-Cowan Oscillator

A Wilson-Cowan oscillator is considered not a single neuron but a meanfield approximation of neuronal groups. It is modeled as a feedback loop between an excitatory unit and an inhibitory unit [1]:

$$\dot{x} = -ax + G(cx + ey + I - \theta_x)$$
$$\dot{y} = -by + G(dx + fy - \theta_y) \qquad (1)$$
$$G(v) = \frac{1}{1 + e^{-(v/T)}}$$

where a and b are decay parameters (positive numbers) of x and y, respectively; c and f are self-excitatory parameters; e is the strength of coupling from the inhibitory unit y to excitatory unit x, it is a negative value to assure that the variable y acts as inhibitory. The corresponding coupling strength from x to y is given by d. θ_x and θ_y are thresholds of unit x and y, respectively. $G(\bullet) \in [0, 1]$ is a sigmoid function and T defines its steepness. I is an external stimulus. If I is a constant, no chaos can appear since it is a two-dimensional continuous flow. In order to get a chaotic oscillator, the external stimulus is defined as a periodic function: $I(t) = A\cos(t)$, where A is the amplitude of the driven function.

In this paper, the amplitude of external stimulation A is considered a bifurcation parameter. Other parameters are held constant at: $a = 1.0$, $b = 0.01$, $c = 1.0$, $d = 0.6$, $e = -2.5$, $f = 0.0$, $\theta_x = 0.2$, $\theta_y = 0.15$, $T = 0.025$, in all the examples that appear in this text.

For characterizing the bifurcating and chaotic behavior, we consider the bifurcation diagram in Fig. 1-a, which shows the stroboscopic section of x at the fixed phase $A\cos(t) = 0$ against A.

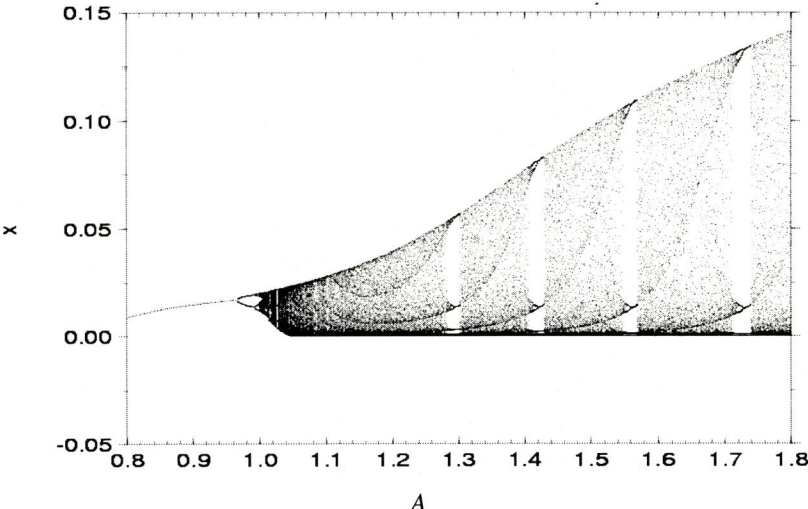

Fig. 1. Bifurcation diagram of periodically driven Wilson-Cowan oscillator by varying parameter A. The stepsize $\Delta A = 0.001$.

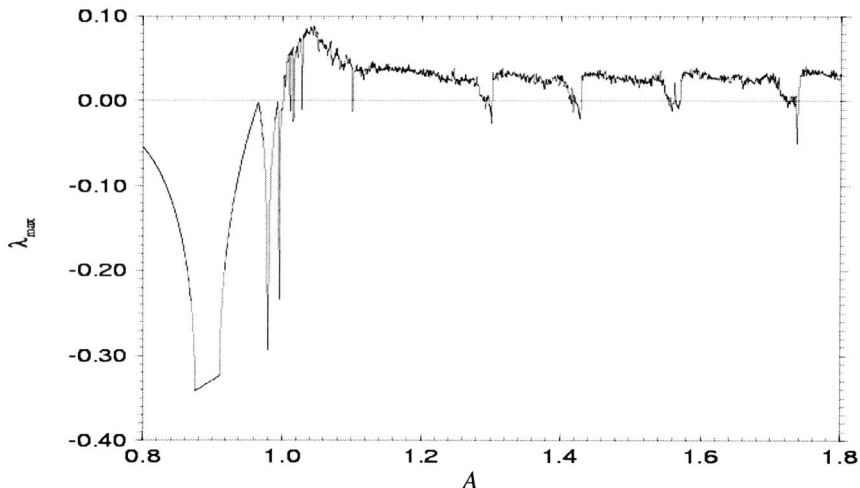

Fig. 2. Largest Lyapunov exponent against A. The stepsize $\Delta A = 0.001$.

From Fig. 1, we can see that periodic oscillation with period one occurs, when A is small. As A increases, a sequence of period-doubling bifurcation is observed. This bifurcation sequence accumulates at $A = 1.011$. For $A \geq 1.011$, chaotic behavior is observed. Fig. 2 shows the largest Lyapunov exponents corresponding to Fig. 1-a. Especially, when $A = 1.2$, the largest Lyapunov exponent $\lambda_{max} = 0.037 > 0$, which indicates that the oscillator is chaotic at this parameter value. Fig. 2 Shows the time series and phase trajectory of a chaotic attractor for $A = 1.2$.

3 Model Description

Our model is a two dimensional network. Each node is a chaotic Wilson-Cowan oscillator. It is governed by the following equations:

$$\dot{x}_{i,j} = -ax_{i,j} + G(cx_{i,j} + ey_{i,j} + I_{i,j} - \theta_x) + k\Delta x_{i,j} + \sigma_{i,j}$$
$$\dot{y}_{i,j} = -by_{i,j} + G(dx_{i,j} + fy_{i,j} - \theta_y) + k\Delta y_{i,j} \quad (3)$$
$$G(v) = \frac{1}{1 + e^{-(v/T)}}$$

where (i, j) is a lattice point with $1 \leq i \leq M$, $1 \leq j \leq N$, M and N are the lattice dimensions. $\sigma_{i,j}$ is a small perturbation term, which will be discussed below. k is the coupling strength. $\Delta x_{i,j}$ is the coupling term from other excitatory units. $\Delta y_{i,j}$ is the coupling term from other inhibitory units. These terms are given by

$$\Delta u_{i,j} = \Delta^+ u_{i,j} + \Delta^\times u_{i,j} \quad (4)$$

where Δ^+ and $\Delta^\times$ are two discrete two-dimensional Laplace operators given respectively by + and × shaped stencils on the lattice, that is, by

$$\Delta^+ u_{i,j} = u_{i+1,j} + u_{i-1,j} + u_{i,j+1} + u_{i,j-1} - 4u_{i,j}$$
$$\Delta^\times u_{i,j} = u_{i+1,j+1} + u_{i-1,j+1} + u_{i+1,j-1} + u_{i-1,j-1} - 4u_{i,j} \quad (5)$$

So, each lattice element is connected to its 8 nearest neighbors. The boundary condition is free-end, that is,

$$x_{i,j} = 0, \; y_{i,j} = 0, \text{ if either } i \leq 0, \text{ or } j \leq 0, \text{ or } i \geq M+1, \text{ or } j \geq N+1. \quad (6)$$

One can easily see that the interaction terms will vanish i.e., $\Delta x_{i,j} = 0$ and $\Delta y_{i,j} = 0$, when the oscillators are synchronized. Thus, the synchronous trajectory will remain once the synchronization state is achieved.

The segmentation strategy is described below. Considering a binary scene image containing p non-overlapped objects. The network is organized that each element corresponds to a pixel of the image. The parameters can be chosen so that the stimulated oscillators (receiving a proper input, corresponding to a figure pixel) is chaotic. The unstimulated oscillators (receiving zero or a very small input, corresponding to a background pixel) remain silent ($x_{i,j} = 0$). If each group of connected, stimulated oscillators synchronize in a chaotic trajectory, then each object is represented by a synchronized chaotic orbit, namely $X_1, X_2, ..., X_p$. Due to the sensitive dependency on initial condition, which is the main characteristic of chaos, if we give a different (or random) small perturbation to each trajectory of $X_1, X_2, ..., X_p$, i.e., $X_1+\delta_1, X_2+\delta_2, ..., X_p+\delta_2$, all these chaotic orbits will be exponentially distant from each other after a while. This is equivalent to giving a perturbation to each stimulated oscillator as shown in Eqn. (3). On the other hand, the synchronization state would not be destroyed by a small perturbation if it was an asymptotically stable state. In this way, all the objects in the scene image can be separated. From the above description, one can see that the segmentation mechanism is irrespective of the number of objects in a given scene. Thus, our model has unbounded capacity of segmentation. Computer simulations show that objects in a given scene can be separated (resulting synchronized chaotic trajectories are distinct) even without the perturbation mechanism.

Fujisaka & Yamada [5] and Heagy et al. [9] have proved analytically that a one-dimensional array of diffusively coupled chaotic elements can be synchronized and the synchronization state is asymptotically stable by providing large enough coupling strength. Moreover, under this coupling scheme, the synchronized orbit will be periodic if all the elements are periodic; it will be chaotic if at least one element is chaotic.

4 Computer Simulations

To show our segmentation method, each stimulated oscillator (corresponds to a figure pixel) receives an external input with amplitude $A = 1.2$. Unstimulated oscillators (corresponds to a background pixel) have $A = 0.0$. From the analysis of a single oscillator in Sec. 2, we know that each stimulated oscillator will be chaotic, while those without stimulation will remain silent.

In order to verify the superior segmentation capacity of this model, a computer simulation is performed with an input pattern containing 16 objects as shown in Fig. 3.

Some objects have identical size and shape. The input image is again presented on a 25×25 oscillator grid. The coupling strength is $k = 6.5$. Fig. 4 shows the temporal activities of the oscillator blocks. We can see the appearance of 16 synchronized chaotic orbits each of which represents an isolated object. The 16 orbits have mutually different temporal behavior.

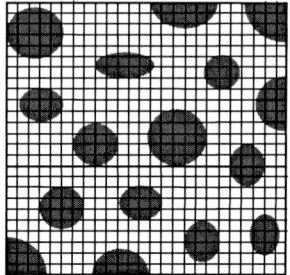

Fig. 3. An input pattern with 16 objects (black figures) mapped to a 25×25 oscillator network.

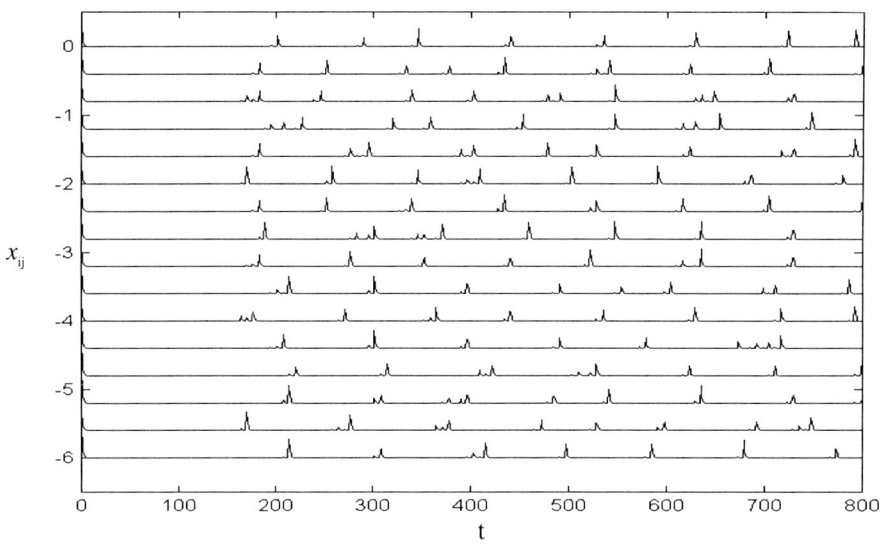

Fig. 4. Temporal activities of oscillator blocks (unperturbed solution). Each trace in the figure is a synchronized chaotic orbit, which contains elements corresponding to an object in the input pattern. Vertical scale of second to twentieth oscillator blocks are shifted downwards by 0.4.

Different groups of oscillators can reach the active phase at the same time. Therefore, usually we cannot correctly extract objects by only evaluating their activities at an instant. A formal method for extracting objects is to calculate cross-correlation among neurons in a certain time window. However, this is computationally demanding because a large number of neurons are involved. Here, we propose a simple method incorporating the characteristics of chaos.

The method works as follows: we know that different chaotic trajectories may cross a predefined Poincaré section simultaneously at some instant. However, because of the sensitive dependency property of chaos, they cannot keep together for a long time. Moreover, they separate rapidly once they have approached. Then, these chaotic trajectories can be distinguished by keeping observing their second, third or more crossing of the Poincaré section. If a set of oscillators cross the Poincaré section simultaneously on several successive times, they are considered to be a segmented object. For this purpose, a proper time interval is chosen within which each chaotic trajectory can cross the Poincaré section several times (3 or 4 times are sufficient). In this method, one object is extracted in each time interval.

Now, let the oscillators synchronize and desynchronize (round off the transient phase). During the first time window, a set of oscillators that first crosses the Poincaré section is denoted as E_1. E_1 may be composed of one or several objects. Anyway, an object included in E_1 will be extracted. At another instant, a set of oscillators E_2 crosses the Poincaré section. Note that individually E_1 or E_2 may be associated to different objects. Now, we check the intersection of E_1 and E_2, if $E_1 \cap E_2 = \phi$, i.e., E_1 and E_2 are two sets of oscillators containing completely different objects, that is, the object which will be extracted in this time interval is not contained in E_2, then E_2 is ignored and we continue observing the next crossing. Otherwise, if $E_1 \cap E_2 \neq \phi$, then, the content of E_1 is replaced by the intersection of E_1 and E_2, i.e., $E_1 \leftarrow E_1 \cap E_2$. As the system is running, a third set of oscillators will cross the Poincaré section, say E_3. Again, we check the intersection between the possibly modified E_1 and the new set E_3: if $E_1 \cap E_3 = \phi$, we only wait for the next crossing; if $E_1 \cap E_3 \neq \phi$, then, $E_1 \leftarrow E_1 \cap E_3$. This process continues until the end of the current time interval is reached. Now, $E_1 = E_1 \cap E_2 \cap ... \cap E_L$, where $E_1, E_2, ..., E_L$ are the crossing sets which have common elements. The markers associated with these selected oscillators are set to a special symbol and, these oscillators will not be processed further. Just after the first time interval, the second starts. The same process is repeated and the second object is extracted. This process continues until all activated oscillators are marked with the special symbol. Finally, the whole process terminates and all objects in the scene are extracted.

Fig. 5 shows the result of object extraction by utilizing the above introduced method. The input pattern is again shown by Fig. 6. The Poincaré section is defined as $x_{ij} = 0.05$. The time interval of length 300 is chosen. Elements represented by a same symbol means that they are extracted at the same time interval. Then, they are considered as a single object; elements represented by different symbols means they are extracted at different time intervals and, consequently, they are considered as elements in different objects. One can easily see that the 16 objects are correctly extracted by comparing with the input pattern.

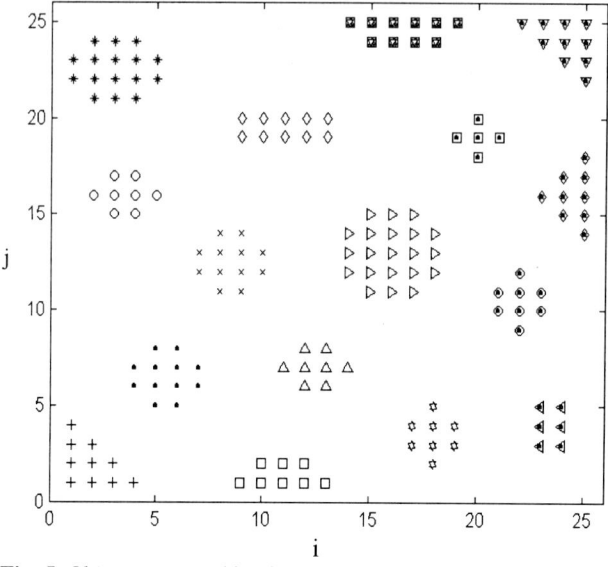

Fig. 5. Objects extracted by the simple method presented in the text.

5 Conclusions

Besides the complexity of chaotic systems, they can synchronize if proper conditions are held. On the other hand, sensitive dependence on initial condition implies that two nearby trajectories will diverge exponentially in time. Thus, chaos is a suitable solution to escape from the synchrony-desynchrony dilemma.

For the segmentation of a gray level image with overlapped objects, the amplitude of external stimulation of each oscillator can be arranged to take different values, i.e., oscillators represent different gray level pixels will be in different chaotic states. The coupling strengths among neurons are taken as an all-or-nothing filtering scheme, i.e. the coupling between pixels with small gray level difference will be maintained, while the coupling between pixels with great gray level difference will be cut. Other techniques are the same as used in the model for binary image segmentation. The results will be reported elsewhere.

Finally, we think that the simulation results of the model are consistent with our everyday experience. Let's consider a visual scene. We can see many stars in the sky on a summer night, but not only few of them. Why? It is only possible that all the stars we see have been firstly separated from one another, then recognized by our visual and central neural systems. Although we cannot pay attention to many things at a given instant, the underlying capacity of visual segmentation of human (animal) is unidentifiably large.

References

1. Baird, B.: Nonlinear dynamics of pattern formation and pattern recognition in the rabbit olfactory bulb, Physica 22D, (1986) 150-175.
2. Campbell, S. & Wang, D. L.: Synchronization and Desynchronization in a Network of Locally Coupled Wilson-Cowan Oscillators, IEEE Trans. Neural Networks, 7(3), (1996) 541-554.
3. Eckhorn, R., Bauer, R., Jordan, W., Brosch, M., Kruse, W., Munk, M. & Reitboeck, H. J.: Coherent oscillation: A mechanism of feature linking in the visual cortex?, Biol. Cybern. 60, (1988) 121-130.
4. Engel, A. K., König, P., Kreiter, A. K & Singer, W.: Interhemispheric Synchronization of Oscillatory Neuronal Responses in Cat Visual Cortex, Science, 252, (1991) 1177-1178.
5. Fujisaka, H. & Yamada, T.: Stability theory of synchronized motion in coupled-oscillator systems, Progress of Theoretical Physics, 60(1), (1983) 32-47.
6. Grey, C. M., König, P., Engel, A. K. & Singer, W.: Oscillatory responses in cat visual cortex exhibit inter-columnar synchronization which reflects global stimulus properties, Nature, 338, (1989) 334-337.
7. Grossberg, S. & Grunewald, A.: Cortical Synchronization and Perceptual Framing, Tecnical Report, CAS/CNS-TR-94-025, (1996)
8. Hansel, D. & Sompolinsky, H.: Synchronization and Computation in a Chaotic Neural Network, Phys. Rev. Lett., 68(5), (1992) 718-721.
9. Heagy, J. F., Carroll, T. L. & Pecora, L. M.: Synchronous chaos in coupled oscillator systems", Phys. Rev. E, 50(3), (1994) 1874-1885.
10. Kaneko, K.: Relevance of dynamic clustering to biological networks, Physica D, 75, (1994) 55-73.
11. Terman, D. & Wang, D.-L.: Global competition and local cooperation in a network of neural oscillators, Physica D, 81, (1995) 148-176.
12. von der Malsburg, Ch. & Buhmann, J.: Sensory segmentation with coupled neural oscillators, Biol. Cybern., 67, (1992) 233-242.
13. von der Malsburg, Ch. & Schneider, W.: A Neural Cocktail-Party Processor, Biol. Cybern., 54, (1986) 29-40.
14. Wolf, A., Swift, J. B., Swinney, H. L. & Vastano, J. A.: Determining Lyapunov Exponents From a Time Series, Physica 16D, (1985) 285-317.

Electronic Circuit Model of Color Sensitive Retinal Cell Network

Ryuichi Iwaki and Michinari Shimoda

Kumamoto National College of Technology, Kumamoto 861-1102, Japan
iwaki@ee.knct.ac.jp, shimoda@ee.knct.ac.jp

Abstract. An equivalent electronic circuit model is developed to analyze the response of color sensitive retinal cell network. Gap junction between adjacent cells is reduced by bi-directional conductance, and chemical synaptic junction between layers is reduced to uni-directional trans-admittance with time lag of first order. On the basis of the previous physiological studies, we estimate parameter values in the electronic circuit model. It is appreciated for the model to perform adequately the properties of spectral response and spatio-temporal response. Frequency response of the network are also calculated.

1 Introduction

Retina performs sensing and pre-processing of chromatic signals in biological vision. Especially in lower vertebrate, retinal cell networks have been studied physiologically and morphologically. In both layers of cone and horizontal cell, adjacent cells of the same type are coupled electrically through gap junction, while the cells are not coupled to the other types [1] [2]. Interconnections between layers have been observed in cone pedicle morphologically and the model for cone-horizontal cell organization has been proposed [3]. Properties of receptive field and spectral response have been recorded and classified [4] [5].

Computational approaches have been developed. The ill-posed problems of early vision have been solved in the framework of regularization theory [6]. Two layered architectures have been proposed and implemented for solving problems of image processing [7] [8]. These chips are intelligent sensors of monochromatic signal.

The purpose of this paper is to develop three-layered network model of color sensitive retinal network. In the electronic circuit model, gap junction is reduced by conductance, and chemical synaptic junction by trans-admittance with time lag of first order. Circuit parameter values should be estimated to match the response of cells. We simulate to appreciate the properties of spectral response and dynamic response, and calculate frequency response of the network.

2 Electronic Circuit Model

Network of outer retinal cells consists of three layers: photoreceptor, horizontal cell and bipolar cell layer. In the present model, three types of cones, namely

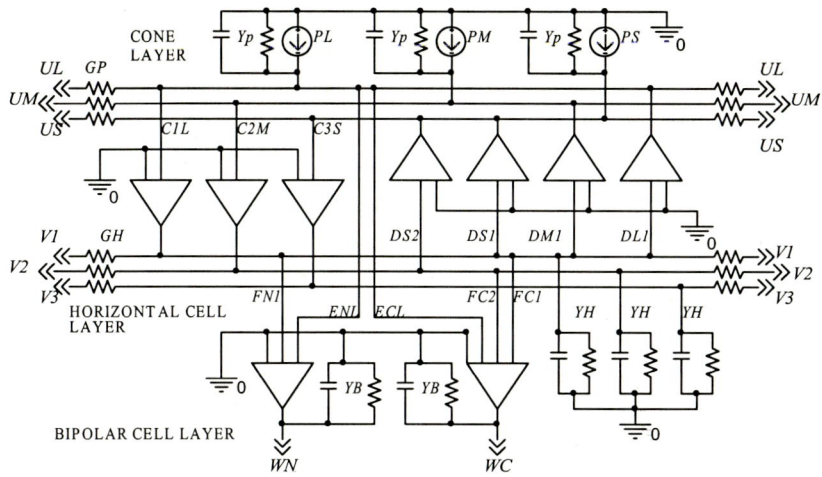

Fig. 1. Equivalent Electronic Circuit of Retina.

L-, M- and S-cone [9] are only taken into account in the photoreceptor layer. Horizontal cells are classified: mono-phasic L type, bi-phasic R-G type, bi-phasic Y-B type and triphasic type [10]. In bipolar cell layer, two types are taken into account : opponent color cell and cell without color cording [4] [5].

It has been studied that mono-phasic L horizontal cell receives directly from only L-cones. Similarly, bi-phasic R-G cell receives directly from M-cones, Y-B type and triphasic type directly from S-cones. There are feedback pathways from monophasic L type to L-cones, M-cones and S-cones, in addition from bi-phasic R-G type to S-cones [3]. It has been deduced that specific types of bipolar cells receive signals from specific types of cones and horizontal cells [5].

On the basis of these previous works, an equivalent electronic circuit of retina is developed as shown in Fig.1. Main feature of the model is the following;

1) Gap junction between same types of adjacent cells is reduced by bi-directional coupling conductance.

2) Chemical synaptic junction between cone layer and horizontal layer is reduced by uni-directional trans-admittance with time lag of first order. Feed forward and feedback pathways coincide with the previous work on the retinal network organization [3].

3) Pathways from cone layer and horizontal cell layer to bipolar cell layer are reduced also by the same type of trans-admittance stated above.

Applying Kirchhoff's law, we derive the following equations presented in the form of Laplace transform. Where notation is the following ; P's: input photo induced current of the cones, U's: voltage of the cones, V's: voltage of the horizontal cells, W's: voltage of the bipolar cells, G's: equivalent conductance of the gap junctions, Y's: membrane admittance of the cells, C's,D's,E's,F's: equivalent trans-admittance of chemical synaptic junctions, $[I]$: unit matrix.

Membrane conductance and capacitance have been measured in situ and in vitro, and estimated roughly 1[nS] and 50[pF]. Coupling conductance has been estimated roughly 50[nS] in cone layer, and 10^3[nS] in horizontal cell compartment. It has been appreciated that these estimations result the consistent properties of spatial response in the sense of half-decay [11]. Synaptic junction is described by trans-admittance with first order low pass filter, in which time constant has been estimated 16[msec] in the previous work [12].

Gain of each junction should be estimated to give consistent spectral response of the cells with the previous physiological studies. From the point of views, we estimate here typical values of parameters as shown in Table 1. To appreciate equivalency to biological retina, properties of spectral response to stationary light and dynamic response to flash light will be simulated with the model and compared with the previous physiological works in the following two sections.

$$G_P[\boldsymbol{A}]U_L + D_{L1}[\boldsymbol{I}]V_1 + [\boldsymbol{I}]P_L = 0 \tag{1}$$

$$G_P[\boldsymbol{A}]U_M + D_{M1}[\boldsymbol{I}]V_1 + [\boldsymbol{I}]P_M = 0 \tag{2}$$

$$G_P[\boldsymbol{A}]U_S + D_{S1}[\boldsymbol{I}]V_1 + D_{S2}[\boldsymbol{I}]V_2 + [\boldsymbol{I}]P_S = 0 \tag{3}$$

$$G_H[\boldsymbol{B}]V_1 + C_{1L}[\boldsymbol{I}]U_L = 0 \tag{4}$$

$$G_H[\boldsymbol{B}]V_2 + C_{2M}[\boldsymbol{I}]U_M = 0 \tag{5}$$

$$G_H[\boldsymbol{B}]V_3 + C_{3S}[\boldsymbol{I}]U_S = 0 \tag{6}$$

$$W_N = \pm(E_{NL}U_L + F_{N1}V_1)/Y_B \tag{7}$$

$$W_C = \pm(E_{CL}U_L + F_{C1}V_1 + F_{C2}V_2)/Y_B \tag{8}$$

where,

$$[\boldsymbol{A}] = \begin{bmatrix} a+1 & -1 & 0 & 0 & 0 \\ -1 & -a & -1 & 0 & 0 \\ & & \ddots & & \\ 0 & 0 & -1 & a & -1 \\ 0 & 0 & 0 & -1 & a+1 \end{bmatrix} \tag{9}$$

$$[\boldsymbol{B}] = \begin{bmatrix} b+1 & -1 & 0 & 0 & 0 \\ -1 & -b & -1 & 0 & 0 \\ & & \ddots & & \\ 0 & 0 & -1 & b & -1 \\ 0 & 0 & 0 & -1 & b+1 \end{bmatrix} \tag{10}$$

$$a = -(Y_P/G_P + 2), \quad b = -(Y_H/G_H + 2) \tag{11}$$

Table 1. Typical Value of Parameters [nS]

G_P	50	G_H	1×10^3
Y_P	$1 + 50 \times 10^{-3} s$	Y_H	$1 + 50 \times 10^{-3} s$
Y_B	$1 + 50 \times 10^{-3} s$	C_{1L}	$1/(1 + 16 \times 10^{-3} s)$
C_{2M}	$1/(1 + 16 \times 10^{-3} s)$	C_{3S}	$1/(1 + 16 \times 10^{-3} s)$
D_{C1}	$-1/(1 + 16 \times 10^{-3} s)$	D_{M1}	$-1/(1 + 16 \times 10^{-3} s)$
D_{S1}	$-0.3/(1 + 16 \times 10^{-3} s)$	D_{S2}	$-0.7/(1 + 16 \times 10^{-3} s)$
E_{NL}	$-1/(1 + 16 \times 10^{-3} s)$	E_{CL}	$-1/(1 + 16 \times 10^{-3} s)$
F_{N1}	$1/(1 + 16 \times 10^{-3} s)$	F_{C1}	$-0.7/(1 + 16 \times 10^{-3} s)$
F_{C2}	$-0.3/(1 + 16 \times 10^{-3} s)$		

3 Spectral Response of the Network

On the basis of the previous work [9], assume that normalized spectral response curves of photo currents in L-, M- and S-cone are represented as shown in Fig.2. Simulated response curves of horizontal cells are shown in Fig.3(a). Response curves of V_1 and V_2 coincide with these of mono-phasic L and bi-phasic R-G horizontal cell respectively, compared with the S-potentials which have measured in carp retina by Mitarai et al.[10]. Both response curves of Y-B and tri-phasic type are gained in V_3 for different values of trans-admittance ratio D_{1S}/D_{2S}. The increase of the ratio D_{1S}/D_{2S} shifts the neutral point to the right in abscissa. Critical ratio is $-0.45/-0.55$. It should be appreciated that only the three feed forward pathways and the four feed back pathways are essential to perform the well-known function of retina which converts tri-chromatic image to opponent color image, though there could be nine possible feed forward pathways and also nine feed back ones.

Spectral response curves of bipolar cells are shown in Fig.3 (b), in which response curves of bipolar cells W_C are drawn with triangles and voltages of bipolar cells W_N are drawn with circles.

The open symbol and the close one represent the response voltage value of the cells which are illuminated at the center and surround of the receptive field respectively. The response voltage curves of W_N and W_C coincide with these of the bipolar cell without color cording and the opponent color cell, which have been measured in goldfish retina by Kaneko [4] [5].

4 Dynamic Response of the Network

To simulate dynamic response of the network, the following empirical equations are applied for time course of photo induced currents which have been deduced by Baylor et al.[13], for step input illumination,

$$P_S = AI\{1 - \exp(-t/\tau)\}^n, \qquad (12)$$

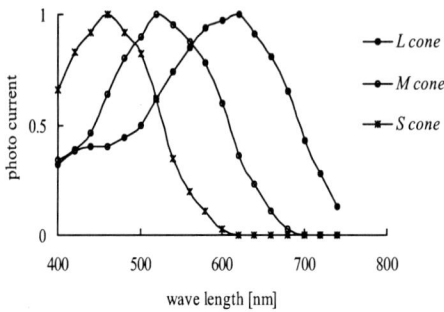

Fig. 2. Spectral Response Curves of Photo Induced Currents.

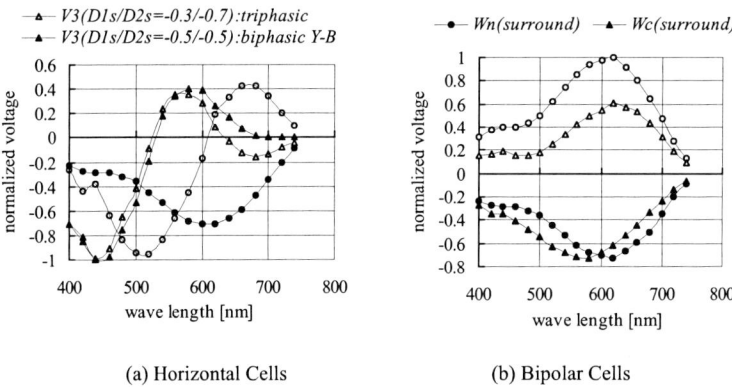

(a) Horizontal Cells   (b) Bipolar Cells

Fig. 3. Spectral Response Curves of Horizontal Cells and Bipolar cells.

for 5[msec] flash of light,

$$P_F = A'I \exp(-t/\tau)\{1 - \exp(-t/\tau)\}^{n-1} \tag{13}$$

where A, A', τ and n are empirical constants, I is strength of input light. The value of n has been deduced as 6 or 7, and value of τ should be deduced for each input light.

4.1 Dynamic Response to Diffuse Light

At first, we simulate the network property of dynamic response to the step input of diffused light, wave length of the light is 540[nm]. The values of τ and n in (12) are set to 10[msec] and 6 respectively. Time courses of the simulated photo current and voltage of cones are shown in Fig.4 (a) and (b) respectively.

The negative feed back pathways from horizontal cell layer to cone layer result the damped oscillation. It takes roughly 350[msec] to reach steady state after the step input has applied. Steady state voltage values of U_L and U_M are negative, whereas the value of U_S is positive, because of the negative feed back

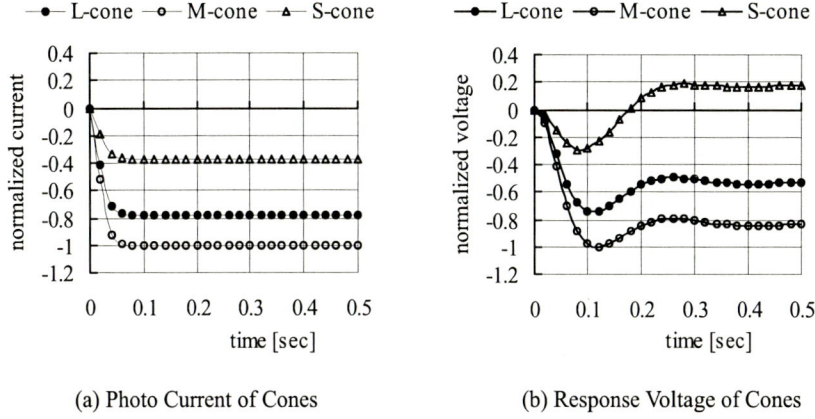

Fig. 4. Dynamic Response Curves to Step Input of Diffused Light, 540[nm].

from horizontal cell layer to cone layer. The steady state values coincide with spectral response at 540[nm] in Fig.3 (a) in the preceding section. Delay of cone voltage depends dominantly on the time constant of membrane admittance.

4.2 Dynamic Response to Slit of Light

Consider that a displaced slit of light illuminates the sequence of cones. Width of slit is assumed 20[μ m], and the period of flash is 5[msec]. Time course of photo currents is given by (13) above mentioned, in which the values of τ and n are set to 40[msec] and 6 respectively. Estimating the spacing of cones, horizontal cell compartments and bipolar cell compartments to be 10[μ m], we simulate the dynamic response of the cells located at the center of the slit, 40[μ m] and 200[μ m] apart from the center. All of the voltage amplitudes are normalized by the peak value of L-cone at the center of slit.

As cones are electrically coupled weakly to adjacent ones, the amplitudes of response voltage decay in a short distance. Moreover the small amplitudes of contradict voltage begin to appear at the peripheral cells after roughly 100[msec], because of negative feed back from horizontal cell layer. The amplitudes of voltage in horizontal calls are less than those of cones but decay little, therefor horizontal cells have much wider receptive fields, because of roughly 20-fold stronger coupling conductance.

Bipolar cells receive signals from cone and horizontal cell layer through negative and positive trans-admittance with delay of first order. At the center of the slit, the signal from L-cone gives dominant effect through the negative junction. At periphery of receptive fields, as the signal from horizontal cell is dominant and the sign of cone voltage is contradict , so the contradict sign of voltage appears at peripheral cells. Delay of response is dependant on the time constant of trans-admittance between layers. Results of simulations are shown in Fig.5.

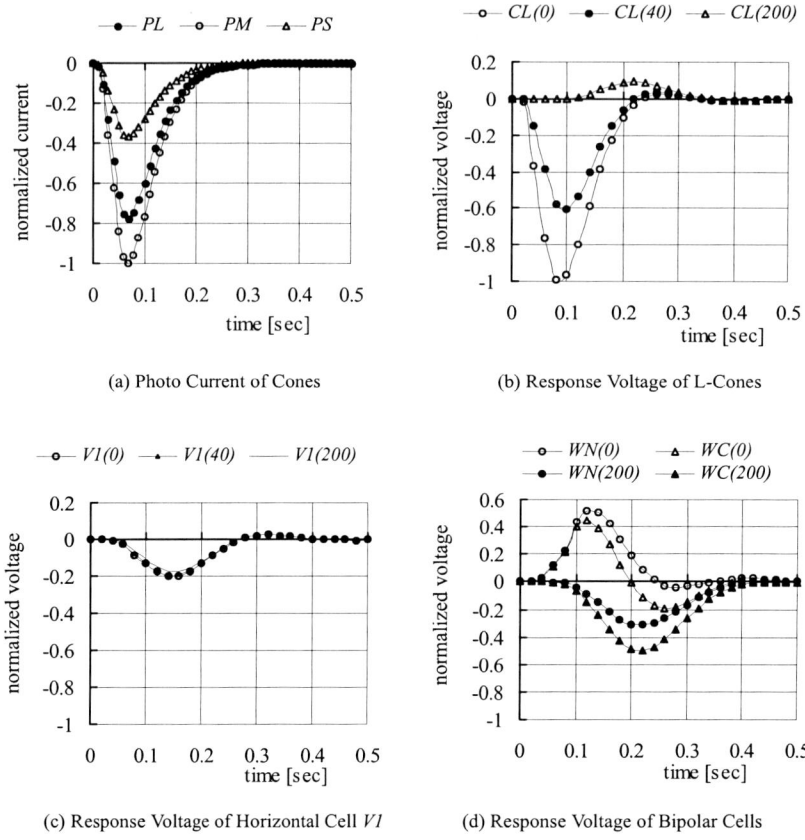

Fig. 5. Dynamic Response Curves to Slit of Flash Light, $20[\mu\,m]$, $5[\text{msec}]$.

5 Frequency Response of Gain

The frequency response of the network is simulated to appreciate the effect of parameters. The gains in horizontal cell layer and bipolar cell layer depend little on the coupling conductance between adjacent cells, though it has been revealed that the value of the conductance in horizontal layer G_H depends on the level of input light [14] [15]. The effect of coupling conductance is dominant on space-frequency response [16] rather than time-frequency response.

The frequency response curves of horizontal cells are shown in Fig.6 (a) for various values of membrane conductance: real part of Y_H which has typical value $1[\text{nS}]$. The network has low pass filter property and the knee points are roughly at $3\,[\text{Hz}]$ in all cases, being estimated the values of the other parameters as shown in Table 1. Overshoot of gain is prominent with decrease of membrane conductance, if it should vary with environmental conditions. In the high frequency region, gains have $-60[\text{dB/decade}]$ for various values of membrane conductance. In low frequency region, gain is lowered with increase of membrane conductance.

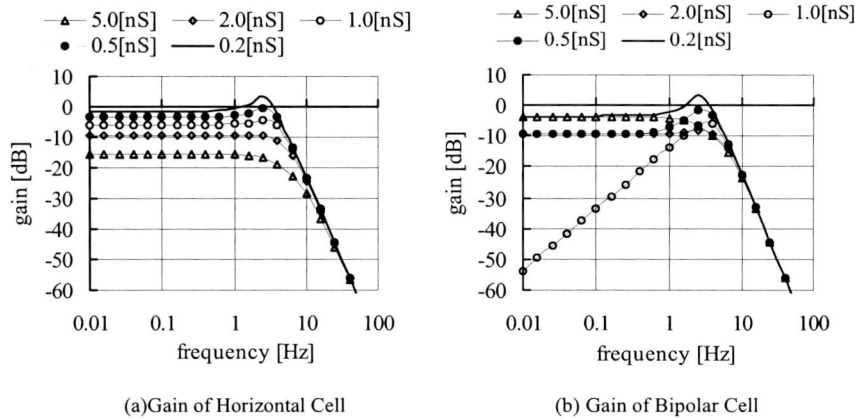

Fig. 6. Frequency Response Curves of Horizontal Cell and Bipolar Cell.

The frequency response curves of bipolar cells are shown in Fig.6 (b). It has the narrow band pass filter characteristic when the value of membrane conductance is 1[nS]. The peak of gain is given roughly at 3[Hz], and -20[dB/decade] in the low frequency region, -60[dB/decade] in the high frequency region. The property depends critically on membrane conductance of horizontal cells, if it should be varied with environmental conditions. Either increase or decrease of membrane conductance result the low pass filter characteristic, which have knee point frequency roughly at 3[Hz], -60[dB/decade] in high frequency region. Decrease of membrane conductance results also prominent overshoot, which is similar to the frequency response of horizontal cells.

6 Conclusion

We developed the three layered electronic circuit to perform equivalent response to the color sensitive retinal network of lower vertebrate. From this study, we conclude the followings.

1.Gap junction should be reduced by bi-directional conductance, chemical synaptic junction by uni-directional trans-admittance with time lag of first order. Only the three feed forward and four feedback pathways are essential to perform the consistent response with the retinal cell network which has been studied biologically. Values of the circuit parameters can be estimated adequately.

2.The model developed here performs the well-known function of retina; Conversion of tri-chromatic image in cone layer to opponent color image in horizontal layer, and contrary response of center/surround receptive field in both cone layer and especially bipolar layer.

3.Behavior of the network should be low path filter, and the frequency response of horizontal cell has the knee point at roughly 3[Hz] when parameters take the typical values shown in Table 1. Gain in low frequency region depends on membrane conductance, whereas it depends little on coupling conductance in horizontal layer. It has been founded the fact that dopamine is released from interplexiform cells to lower the conductance in horizontal layer at high level of input light[14][15]. We have appreciated in the preceding study [16] that the increase of conductance should enhance the low pass filtering characteristics in space-frequency response. It should be preferable that the space-frequency low pass filtering enhances to eliminate random noises at low level of inputs and high space-frequency signals should be passed to enhance the spatial sensitivity at high level of inputs.

4. The frequency response of bipolar cell is sensitive on membrane conductance of horizontal cell, if it should be varied. In the present simulations, the network has narrow band pass filtering characteristics having maximum gain roughly at frequency 3[Hz], when the parameters take the typical values. It belongs to future works to discuss the biological meaning of it.

The approach presented here combines the biological organization and computational analysis. It should contribute numerical analysis and synthetic neurobiology of early color vision.

Acknowledgments: The author would like to thank the anonymous reviewers for their helpful suggestions and comments. This work was supported by a Grant-in-Aid 11650365 from the Ministry of Education, Science, Sports and Culture, Japan.

References

1. P.B.Detwiler, and A.L.Hodgkin, "Electrical Coupling between Cones in Turtle Retina," Nature, vol.317, pp.314-319, 1985.
2. K.I.Naka, and W.A.H.Rushton, "The Generation and Spread of S-Potentials in Fish (Cyprinidae)," J. Physiol., vol. 192, pp.437-461, 1967.
3. W.K.Stell, and D.O.Lightfoot, "Color-Specific Interconnection of Cones and Horizontal Cells in the Retina of Goldfish," J. Comp. Neur., vol.159, pp.473-501
4. A.Kaneko, "Physiological and Morphological Identification of Horizontal, Bipolar and Amacrine Cells in the Goldfish Retina," J. Physiol., vol.207, pp.623-633, 1970.
5. A.Kaneko, "Receptive Field Organization of Bipolar and Amacrine Cells in the Goldfish Retina," J. Physiol., vol.235, pp.133-153, 1973.
6. T.Poggio, and V.Torre, and C.Koch, "Computational Vision and Regularization Theory," Nature, vol.317, pp.314-319, 1985.
7. C.Mead, and M.Machwald, "A Silicon Model of Early Visual Processing," Neural Networks, vol.1, pp.91-97, 1988.
8. H.Kobayashi, T.Matsumoto, T.Yagi, and T.Shimmi, " Image Processing Regularization Filters on Layered Architecture ," Neural Networks, vol.6, pp.327-350, 1993.
9. T.Tomita, A.Kaneko, M.Murakami, and E.L.Pautler, "Spectral Response Curves of Single Cones in the Carp," Vision Research, vol.7, pp.519-531, 1967.
10. G.Mitarai, T.Asano, and Y.Miyake, "Identification of Five Types of S-Potential and Their Corresponding Generating Sites in the Horizontal Cells of the Carp Retina," Jap. J. Ophthamol., vol.18, pp.161-176, 1974.

11. S.Ohshima, T.Yagi, and Y.Funahashi, "Computational Studies on the Interaction Between Red Cone and H1 Horizontal Cell", Vision Research, 35, No.1, pp.149-160, 1995
12. J.L.Schnapf and D.R.Copenhagen, "Differences in the kinetics of rod and cone synaptic transmission", Nature, 296, pp.862-864, 1982
13. D.A.Baylor, A.L.Hodgkin, and T.D.Lamb, "The Electrical Response of Turtle Cones to Flashes and Step of Light", J. of Physiol., 242, pp.685-727, 1974
14. T.Teranishi, K.Negishi, and S.Kato, "Dopamine Modulates S-Potential Amplitude and Dye-Coupling between External Horizontal Cells in Carp Retina", Nature, 301,pp.234—246, 1983
15. M.Kirsch and H.J.Wagner, "Release Pattern of Endogenous Dopamine in Teleost Retina during Light Adaptation and Pharmacological Stimulation", Vision Research, vol.29, no.2, pp.147-154, 1989
16. R.Iwaki and M.Shimoda, "Electronic Circuit Model of Retina in Color Vision", 36th International ISA Biomedical Sciences Instrumentation Symposium, vol.35, pp.373-378, 1999

The Role of Natural Image Statistics in Biological Motion Estimation

Ron O. Dror[1], David C. O'Carroll[2], and Simon B. Laughlin[3]

[1] MIT, 77 Massachusetts Ave., Cambridge, MA 02139, USA, `rdror@mit.edu`
[2] University of Washington, Box 351800, Seattle, WA 98195-1800, USA
[3] University of Cambridge, Downing Street, Cambridge CB2 3EJ, UK

Abstract. While a great deal of experimental evidence supports the Reichardt correlator as a mechanism for biological motion detection, the correlator does not signal true image velocity. This study examines the accuracy with which physiological Reichardt correlators can provide velocity estimates in an organism's natural visual environment. Both simulations and analysis show that the predictable statistics of natural images imply a consistent correspondence between mean correlator response and velocity, allowing the otherwise ambiguous Reichardt correlator to act as a practical velocity estimator. A computer vision system may likewise be able to take advantage of natural image statistics to achieve superior performance in real-world settings.

1 Introduction

The Reichardt correlator model for biological motion detection [15] has gained widespread acceptance in the invertebrate vision community. This model, which is mathematically equivalent to the spatiotemporal energy models popular for vertebrate motion detection [2], has also been applied to explain motion detection in humans, birds, and cats [21,22,7]. After forty years of physiological investigation, however, a fundamental issue raised by Reichardt and his colleagues remains unanswered. While both insects and humans appear capable of estimating image velocity [18,12], the output of a basic Reichardt correlator provides an inaccurate, ambiguous indication of image velocity. Correlator response to sinusoidal gratings depends on contrast (brightness) and spatial frequency (shape) as well as velocity; since the correlator is a nonlinear system, response to a broad-band image may vary erratically as a function of time. Some authors have concluded that velocity estimation requires either collections of differently tuned correlators [2], or an alternative motion detection system [18].

Before discarding the uniquely tuned Reichardt correlator as a velocity estimator, we consider the behavior of a physiologically realistic correlator in a natural environment. Previous experimental and modeling studies have typically focused on responses to laboratory stimuli such as sinusoidal or square gratings. We examine the responses of a Reichardt correlator to motion of natural broad-band images ranging from forests and animals to offices and city streets. In simulations, the correlator functions much better as a velocity estimator for

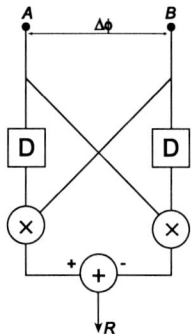

Fig. 1. A minimal Reichardt correlator.

motion of real-world imagery than for motion of traditional gratings. We develop a mathematical relationship between image power spectra and correlator response which shows that a system based on Reichardt correlators functions well in practice because natural images have predictable statistics and because the biological system is optimized to take advantage of these statistics. While this work applies generally to Reichardt correlators and mathematically equivalent models, we have chosen the fly as a model organism for computational simulations and experiments due to the abundance of behavioral, anatomical, and electrophysiological data available for its motion detection system.

The implication for machine vision is that the extensive recent body of work on statistics of natural images and image sequences (e.g., [8,3,17]) can be exploited in a computer vision system. Such a system may be superior to existing systems in practice despite inferior performance in simple test environments.

2 Correlator Response to Narrow-Band Image Motion

Figure 1 shows a simplified version of the correlator model. Receptors A and B are separated by an angular distance $\Delta\phi$. The signal from A is temporally delayed by the low-pass filter D before multiplication with the signal from B. This multiplication produces a positive output in response to rightward image motion. In order to achieve similar sensitivity to leftward motion and in order to cancel excitation by stationary stimuli, a parallel delay-and-multiply operation takes place with a delay on the opposite arm. The outputs of the two multiplications are subtracted to give a single time-dependent correlator output R.

Although the correlator is nonlinear, its response to sinusoidal stimuli is of interest. If the input is a sinusoidal grating containing only a single frequency component, the oscillations of the two subunits cancel and the correlator produces a constant output.[1] If the delay filter D is first-order low-pass with time

---

[1] A physical luminance grating must have positive mean luminance, so it will contain a DC component as well as an oscillatory component. In this case, the output will oscillate about the level given by (1).

constant τ, as in most modeling studies (e.g., [6]), a sinusoid of amplitude C and spatial frequency f_s traveling to the right at velocity v produces an output

$$R(t) = \frac{C^2}{2\pi\tau} \frac{f_t}{f_t^2 + 1/(2\pi\tau)^2} \sin(2\pi f_s \Delta\phi) , \qquad (1)$$

where $f_t = f_s v$ is the temporal frequency of the input signal [6]. The output level depends separably on spatial and temporal frequency. At a given spatial frequency, the magnitude of correlator output increases with temporal frequency up to an optimum $f_{t,opt} = \frac{1}{2\pi\tau}$, and then decreases monotonically as velocity continues to increase. Output also varies with the square of C, which specifies grating brightness or, in the presence of preprocessing stages, grating contrast.

3 Correlator Response to Broad-Band Images

Since the correlator is a nonlinear system, its response to a generic stimulus cannot be represented as a sum of responses to sinusoidal components of the input. In particular, the response to a broad-band image such as a natural scene may vary erratically with time.

3.1 Evaluation of Correlator Performance

In order to compare the performance of various velocity estimation systems, one must first establish a quantitative measure of accuracy. Rather than attempt to measure the performance of a motion detection system as a single number, we quantify two basic requirements for an accurate velocity estimation system:

1. Image motion at a specific velocity should always produce the same response.
2. The response to motion at a given velocity should be unambiguous; that is, it should differ from the response to motion at other velocities.

We restrict the range of potential input stimuli by focusing on responses to rigid, constant-velocity motion as observed by an eye undergoing rotational motion.

Given a large image moving at a particular constant velocity, consider an array of identically oriented correlators sampling the image at a dense grid of points in space and time. Define the mean response value $\overline{R}$ as the average of the ensemble outputs, and the relative error as the standard deviation of the ensemble divided by the mean response value. We call the graph of $\overline{R}$ as a function of velocity the velocity response curve. In order to satisfy requirement 1, different images should have similar velocity response curves and relative error should remain small. Requirement 2 implies that the velocity response curve should be monotonic in the relevant range of motion velocities.

Figure 2 shows velocity response curves for two simulated gratings of different spatial frequencies. The curves for the two gratings differ significantly, so that mean response level indicates velocity only if spatial frequency is known. In addition, the individual velocity response curves peak at low velocities, above which their output is ambiguous.

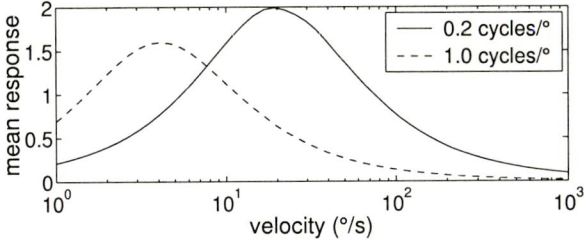

Fig. 2. Velocity response curves for the simple correlator model in response to sinusoidal gratings of two different spatial frequencies. Units on the vertical axis are arbitrary. The correlator in this simulation had a first-order low-pass delay filter with $\tau = 35$ ms, which matches the temporal frequency tuning observed experimentally in typical large flies such as *Calliphora*, *Eristalis*, and *Volucella* [9]. We set the inter-receptor angle to 1.08°, near the center of the physiologically realistic range for flies. These parameter choices do not critically influence our qualitative results.

3.2 Simulation with Natural Images

One can perform similar simulations with natural images. In view of the fact that the characteristics of "natural" images depend on the organism in question and its behavior, we worked with two sets of images. The first set consisted of panoramic images photographed from favored hovering positions of the hoverfly *Episyrphus balteatus* in the woods near Cambridge, U.K. The second set of photographs, acquired by David Tolhurst, includes a much wider variety of imagery, ranging from landscapes and leaves to people, buildings, and an office [19]. Figure 3 displays images from both sets. We normalized each image by scaling the luminance values to a mean of 1.0, both in order to discount differences in units between data sets and to model photoreceptors, which adapt to the mean luminance level and signal the contrast of changes about that level [11].

Figure 4A shows velocity response curves for the images of Fig. 3. The most notable difference between the curves is their relative magnitude. When the curves are themselves normalized by scaling so that their peak values are equal (Fig. 4B), they share not only their bell shape, but also nearly identical optimal velocities. We repeated these simulations on most of the images in both sets, and found that while velocity response curves for different images differ significantly in absolute magnitude, their shapes and optimal velocities vary little.

This empirical similarity implies that if the motion detection system could normalize or adapt its response to remove the difference in magnitude between these curves, then the spatially or temporally averaged mean correlator response would provide useful information on image velocity relatively independent of the visual scene. One cannot infer velocity from the mean response of a correlator to a sinusoidal grating, on the other hand, if one does not know the spatial frequency in advance.

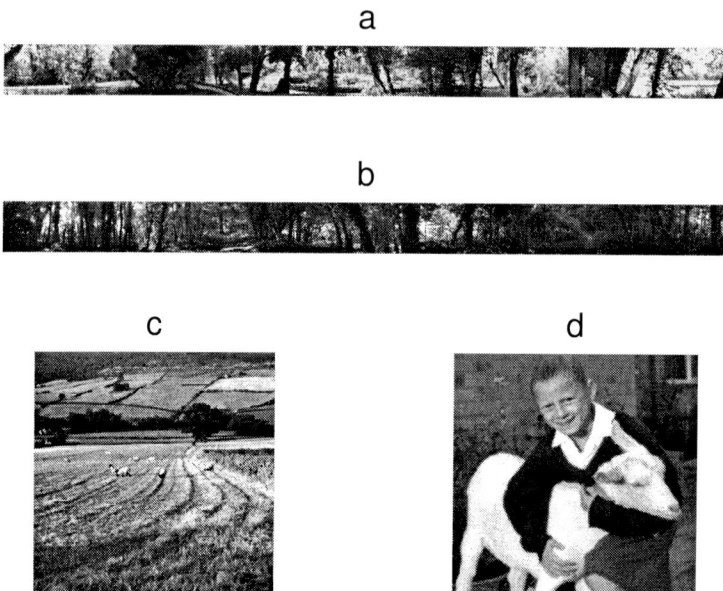

Fig. 3. Examples of the natural images used in simulations throughout this work. Images (a) and (b) are panoramic images from locations where *Episyrphus* chooses to hover. Images (c) and (d) are samples of the image set acquired by David Tolhurst. Details of image acquisition can be found in [5,19].

3.3 Mathematical Analysis of Mean Response to Broad-Band Images

This section develops a general mathematical relationship between the power spectrum of an image and mean correlator response, explaining the empirical similarities in shape and differences in magnitude between velocity response curves for different images. Natural images differ from sinusoidal gratings in that they possess energy at multiple non-zero spatial frequencies, so that they produce broad-band correlator input signals. As an image moves horizontally across a horizontally-oriented correlator, one row of the image moves across the two correlator inputs. One might think of this row as a sum of sinusoids representing its Fourier components. Because of the nonlinearity of the multiplication operation, the correlator output in response to the moving image will differ from the sum of the responses to the individual sinusoidal components. In particular, the response to a sum of two sinusoids of different frequencies f_1 and f_2 consists of the sum of the constant responses predicted by (1) to each sinusoid individually, plus oscillatory components of frequencies $f_1 + f_2$ and $|f_1 - f_2|$. Sufficient spatial or temporal averaging of the correlator output will eliminate these oscillatory components. The correlator therefore exhibits pseudolinearity

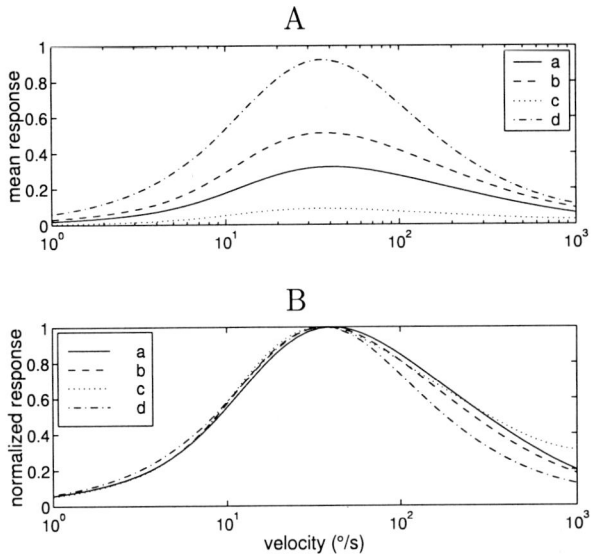

Fig. 4. Response of the simple correlator model to the natural images shown in Fig. 3. (A) Velocity response curves showing mean response to motion at different velocities, computed as in Fig. 2. (B) The same curves, normalized so that their maximum values are identical. Peak response velocities range from 35–40°/s.

or linearity in the mean [14], in that the mean response to a broad-band image is equal to the sum of the responses to each sinusoidal input component.

This pseudolinearity property implies that the mean response of a simple Reichardt correlator to a single row of an image depends only on the power spectrum of that row. Using (1) for correlator response to a sinusoid and the fact that $f_t = f_s v$, we can write the mean correlator output as

$$\overline{R} = \frac{1}{2\pi\tau} \int_0^\infty P(f_s) \frac{f_s v}{(f_s v)^2 + 1/(2\pi\tau)^2} \sin(2\pi f_s \Delta\phi) df_s , \qquad (2)$$

where $P(f_s)$ represents the power spectral density of one row of the image at spatial frequency f_s. Each velocity response curve shown in Fig. 4 is an average of the mean outputs of correlators exposed to different horizontal image rows with potentially different power spectra. This average is equivalent to the response of the correlator to a single row whose power spectrum $P(f_s)$ is the mean of the power spectra for all rows of the image.

If $P(f_s)$ were completely arbitrary, (2) would provide little information about the expected shape of the velocity response curve. A large body of research suggests, however, that power spectra of natural images are highly predictable. According to a number of studies involving a wide range of images, the two-dimensional power spectra are generally proportional to $f^{-(2+\eta)}$, where f is the modulus of the two-dimensional spatial frequency and η is a small constant (e.g., [8,19]). If an image has an isotropic two-dimensional power spectrum pro-

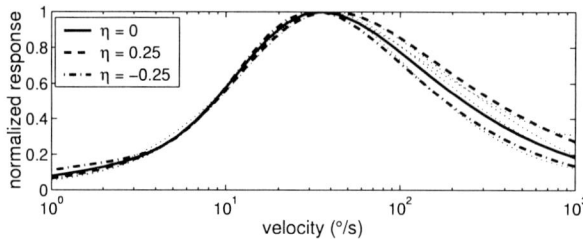

Fig. 5. Velocity response curves computed theoretically using (2), assuming row power spectra $P(f_s)$ of the form $f_s^{-(1+\eta)}$ for several values of η. Model parameters are as in Fig. 2. Simulated velocity response curves from Fig. 4B are shown in thin dotted lines for comparison. All curves have been normalized to a maximum value of 1.0. The predicted peak response velocities are 32, 35, and 40°/s for $\eta = -0.25$, 0, and 0.25, respectively.

portional to $f^{-(2+\eta)}$, the one-dimensional power spectrum of any straight-line section through the image is proportional to $f^{-(1+\eta)}$.

Overall contrast, which determines overall amplitude of the power spectrum, varies significantly between natural images and between orientations [20]. The best value of η also depends on image and orientation, particularly for images from different natural environments. Van der Schaaf and van Hateren [20] found, however, that a model which fixes $\eta = 0$ while allowing contrast to vary suffers little in its fit to the data compared to a model which allows variation in η.

The similarities in natural image power spectra lead to predictable peak response velocities and to similarities in the shapes of the velocity response curves for different images. Figure 5 shows velocity response curves predicted from hypothetical row power spectra $P(f_s) = f_s^{-1}$, $f_s^{-1.25}$, and $f_s^{-0.75}$, corresponding to $\eta = 0$, 0.25, and −0.25, respectively. The theoretical curves match each other and the simulated curves closely below the peak response value; in this velocity range, the velocity response is insensitive to the value of the exponent in the power spectrum.

Contrast differences between images explain the primary difference between the curves, their overall amplitude. Figure 6 shows horizontal power spectral densities for the images of Fig. 3, computed by averaging the power spectral densities of the rows comprising each image. On log-log axes, the spectra approximate straight lines with slopes close to −1, although the spectrum of image (d) has noticeable curvature. The relative magnitudes of the spectra correspond closely to the relative magnitudes of the velocity response curves of Fig. 4, as predicted by (2). Differences in the magnitude of the velocity response curves correspond to differences in overall contrast, except that image (d) has the largest response even though its contrast is only larger than that of (b) for frequencies near 0.1 cycles/°. This reflects the fact that some spatial frequencies contribute more than others to the correlator response.

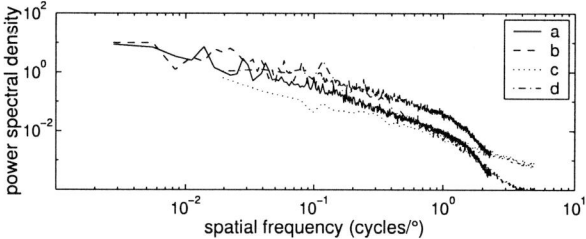

Fig. 6. Horizontal power spectral densities of the images in Fig. 3. Each spectrum is an average of power spectral densities of the rows comprising the image. Images (a) and (b) roll off in power at frequencies above 1.2 cycles/° due to averaging in the image acquisition process, but optical blur effects in the fly's eye reject almost all spatial frequency content above 1 cycle/°.

In order to use mean correlator response as a reliable indicator of velocity, the visual system needs to compensate for these contrast variations. One possibility is that contrast saturation early in the motion detection pathway eliminates significant differences in contrast. Alternatively, some form of contrast normalization akin to that observed in vertebrate vision systems [10] may work to remove contrast differences between images. Ideal localized contrast normalization would remove the dependence of correlator response on the spatial frequency of a sinusoidal grating [1], but this dependence has been documented experimentally, and our work suggests that a more global form of contrast normalization is likely. Our experimental results [13] confirm the relationships between the power spectrum and the velocity response curve predicted in this section and suggest that the response of wide-field neurons reflects image velocity consistently even as image contrast changes.

3.4 Limitations and Further Work

While the simple correlator model of Fig. 1 produces more meaningful estimates of velocity for natural images than for arbitrary sinusoids, it suffers from two major shortcomings. First, the standard deviation of the correlator output is huge relative to its mean, with relative error values ranging from 3.3 to 76 for the images and velocity ranges of Fig. 4. Second, mean correlator response for most natural images peaks at a velocity of 35 to 40°/s. While this velocity significantly exceeds the peak response velocity of 19.6°/s for a sinusoidal grating of optimal spatial frequency, it still leads to potential ambiguity since flies may turn and track targets at velocities up to hundreds of degrees per second. In further work [5,4], we found that a more physiologically realistic correlator model raises velocity response and lowers relative error dramatically through the inclusion of experimentally described mechanisms such as input prefiltering, output integration, compressive nonlinearities, and adaptive effects. Equation 2 generalizes naturally to predict the quantitative and qualitative effects of spatial and temporal prefiltering. This work examines only responses to constant velocity rigid motion; further work should consider natural image sequences.

4 Conclusions

While natural images appear more complicated than the gratings typically used in laboratory experiments and simulations, Reichardt correlators respond more predictably to motion of natural images than to gratings. The general structure and detailed characteristics of the physiological correlator suggest that it evolved to take advantage of natural image statistics for velocity estimation. While we worked with models based on data from insect vision, these conclusions also apply to models of vertebrate vision such as the Elaborated Reichardt Detector [21] and the spatiotemporal energy model [2], both of which are formally equivalent to the Reichardt correlators discussed here.

These results could be applied directly to hardware implementations of correlator-based motion-detection systems (e.g., [16]). A more important implication is that a machine vision system designed to perform a task involving real-world imagery would do well to take advantage of recent results in the field of natural image statistics. These results include statistics of image sequences [3] and extend well beyond power spectra [17]. A Bayesian approach to computer vision should take these statistics into account, as does the biological motion detection system.

Acknowledgments

We would like to thank David Tolhurst for sharing his set of images and Miranda Aiken for recording the video frames which formed the panoramic images. Rob Harris, Brian Burton, and Eric Hornstein contributed valuable comments. This work was funded by a Churchill Scholarship to ROD and by grants from the BBSRC and the Gatsby Foundation.

References

1. E. H. Adelson and J. R. Bergen. The extraction of spatio-temporal energy in human and machine vision. In *Proceedings from the Workshop on Motion: Representation and Analysis*, pages 151–55, Charleston, SC, 1986.
2. E. H. Adelson and J.R. Bergen. Spatiotemporal energy models for the perception of motion. *J. Opt. Soc. Am. A*, 2:284–99, 1985.
3. D. W. Dong and J. J. Atick. Statistics of natural time-varying images. *Network: Computation in Neural Systems*, 6:345–58, 1995.
4. R. O. Dror. Accuracy of velocity estimation by Reichardt correlators. Master's thesis, University of Cambridge, Cambridge, U.K., 1998.
5. R. O. Dror, D. C. O'Carroll, and S. B. Laughlin. Accuracy of velocity estimation by Reichardt correlators. Submitted.
6. M. Egelhaaf, A. Borst, and W. Reichardt. Computational structure of a biological motion-detection system as revealed by local detector analysis in the fly's nervous system. *J. Opt. Soc. Am. A*, 6:1070–87, 1989.

7. R. C. Emerson, M. C. Citron, W. J. Vaughn, and S. A. Klein. Nonlinear directionally selective subunits in complex cells of cat striate cortex. *J. Neurophysiology*, 58:33–65, 1987.
8. D. J. Field. Relations between the statistics of natural images and the response properties of cortical cells. *J. Opt. Soc. Am. A*, 4:2379–94, 1987.
9. R. A. Harris, D. C. O'Carroll, and S. B. Laughlin. Adaptation and the temporal delay filter of fly motion detectors. *Vision Research*, 39:2603–13, 1999.
10. D. J. Heeger. Normalization of cell responses in cat striate cortex. *Visual Neuroscience*, 9:181–97, 1992.
11. S. B. Laughlin. Matching coding, circuits, cells and molecules to signals: general principles of retinal design in the fly's eye. *Prog. Ret. Eye Res.*, 13:165–95, 1994.
12. Suzanne P. McKee, Gerald H. Silverman, and Ken Nakayama. Precise velocity discrimination despite random variations in temporal frequency and contrast. *Vision Research*, 26:609–19, 1986.
13. D. C. O'Carroll and R. O. Dror. Velocity tuning of hoverfly HS cells in response to broad-band images. In preparation.
14. T. Poggio and W. Reichardt. Visual control of orientation behaviour in the fly. Part II. Towards the underlying neural interactions. *Quarterly Reviews of Biophysics*, 9:377–438, 1976.
15. W. Reichardt. Autocorrelation, a principle for the evaluation of sensory information by the central nervous system. In A. Rosenblith, editor, *Sensory Communication*, pages 303–17. MIT Press and John Wiley and Sons, New York, 1961.
16. R. Sarpeshkar, W. Bair, and C. Koch. An analog VLSI chip for local velocity estimation based on Reichardt's motion algorithm. In S. Hanson, J. Cowan, and L. Giles, editors, *Advances in Neural Information Processing Systems*, volume 5, pages 781–88. Morgan Kauffman, San Mateo, 1993.
17. E.P. Simoncelli. Modeling the joint statistics of images in the wavelet domain. In *Proc SPIE, 44th Annual Meeting*, volume 3813, Denver, July 1999.
18. M. V. Srinivasan, S. W. Zhang, M. Lehrer, and T. S. Collett. Honeybee navigation *en route* to the goal: visual flight control and odometry. *J. Exp. Biol.*, 199:237–44, 1996.
19. D. J. Tolhurst, Y. Tadmor, and T. Chao. Amplitude spectra of natural images. *Ophthalmology and Physiological Optics*, 12:229–32, 1992.
20. A. van der Schaaf and J. H. van Hateren. Modelling the power spectra of natural images: statistics and information. *Vision Research*, 36:2759–70, 1996.
21. J. P. H. van Santen and G. Sperling. Elaborated Reichardt detectors. *J. Opt. Soc. Am. A*, 2:300–21, 1985.
22. F. Wolf-Oberhollenzer and K. Kirschfeld. Motion sensitivity in the nucleus of the basal optic root of the pigeon. *J. Neurophysiology*, 71:1559–73, 1994.

Enhanced Fisherfaces
for Robust Face Recognition

Juneho Yi, Heesung Yang, and Yuho Kim

School of Electrical and Computer Engineering
Sungkyunkwan University
300, Chunchun-dong, Jangan-gu
Suwon 440-746, Korea

Abstract. This research features a new method for automatic face recognition robust to variations in lighting, facial expression and eyewear. The new algorithm named SKKUfaces (Sungkyunkwan University faces) employs PCA (Principal Component Analysis) and FLD (Fisher's Linear Discriminant) in series similarly to Fisherfaces. The fundamental difference is that SKKUfaces effectively eliminates, in the reduced PCA subspace, portions of the subspace that are responsible for variations in lighting and facial expression and then applies FLD to the resulting subspace. This results in superb discriminating power for pattern classification and excellent recognition accuracy. We also propose an efficient method to compute the between-class scatter and within-class scatter matrices for the FLD analysis. We have evaluated the performance of SKKUfaces using YALE and SKKU facial databases. Experimental results show that the SKKUface method is computationally efficient and achieves much better recognition accuracy than the Fisherface method [1] especially for facial images with variations in lighting and eyewear.

1 Introduction

In face recognition, a considerable amount of research has been devoted to the problem of feature extraction for face classification that represents the input data in a low-dimensional feature space. Among representative approaches are Eigenface and Fisherface methods. Eigenface methods [7] [9] are based on PCA and use no class specific information. They are efficient in dimensionality reduction of input image data, but only provides us with feature vectors that represent main directions along which face images differ the most. On the other hand, Fisherface methods [1] [5] are based both PCA and FLD [10]. They first use PCA to reduce the dimension of the feature space and then applies the standard FLD in order to exploit class specific information for face classification. It is reported that the performance of Fisherface methods is far better in recognition accuracy than that of Eigenface methods.

The analysis of our method is similar to the Fisherface method suggested in [1]. The fundamental difference is that we apply FLD to a reduced subspace that is more appropriate for classification purpose than the reduced PCA subspace that Fisherface methods use. It has been suggested in the PCA based methods such as Eigenfaces that by discarding the three most significant principal components, the variation due to lighting is reduced [1]. However, this idea in concert with FLD has not been employed. We apply FLD to the reduced subspace that is computed by ignoring the first few eigenvectors from PCA corresponding to the top principal eigenvalues as illustrated in Figure 2. The effect is that, in this reduced subspace, portions of the vector space that are responsible for variations in lighting and facial expression are effectively eliminated. The reduced subspace is more appropriate for the FLD analysis than the reduced PCA subspace that Fisherfaces employ. That is, class separability is improved, and applying FLD to this reduced subspace can improve the discriminating power for pattern classification. Another important contribution of SKKUfaces is an efficient method to compute the between-class scatter and within-class scatter matrices.

We have evaluated our method using YALE and SKKU (Sungkyunkwan University) facial databases and have compared the performance of SKKUfaces with that of Fisherfaces. Experimental results show that our method achieves much better recognition accuracy than the Fisherface method especially for facial images with variations in lighting. In addition, a class separability measure computed for SKKUfaces and Fisherfaces shows that SKKUfaces has more discriminating power for pattern classification than Fisherfaces.

This paper is organized as follows. The following section briefly reviews Eigenface and Fisherface approaches. In section 3, we present our approach to feature extraction for robust face recognition and also describe a computationally very efficient method to compute within-class scatter and between-class scatter matrices. Section 4 presents experimental results using YALE and SKKU (Sungkyunkwan University) facial databases.

2 Related Works

2.1 Eigenface Method

Eigenface methods are based on PCA (or Karhunen-loeve transformation) that generates a set of orthonormal basis vectors. These orthonormal basis vectors are known as principal components that capture the main directions which face images differ the most. A face image is represented as a coordinates in the orthonormal basis. Kirby and Sirovish [7] first employed PCA for representing face images and PCA was used for face recognition by Turk and Pentland [2]. Eigenface methods are briefly described as follows.

Let a face image be a two-dimensional M by N array of intensity values. This image can be represented a vector $\boldsymbol{X}_i$ of dimension MN. Let $\mathbf{X} = [\boldsymbol{X}_1, \boldsymbol{X}_2, \cdots, \boldsymbol{X}_T]$ be the sample set of the face images. T is the total number of the face images. After subtracting the total mean denoted by $\boldsymbol{\Phi}$ from each face image, we get a new vector set $\boldsymbol{\Phi} = [\boldsymbol{X}_1 - \boldsymbol{\Phi}, \boldsymbol{X}_2 - \boldsymbol{\Phi}, \cdots, \boldsymbol{X}_T - \boldsymbol{\Phi}]$. Let $\boldsymbol{\Phi}_i$ denote $\boldsymbol{X}_i - \boldsymbol{\Phi}$. Then the covariance matrix is defined as:

$$S_T = \sum_{i=1}^{T} \boldsymbol{\Phi}_i \boldsymbol{\Phi}_i^T = \boldsymbol{\Phi}\boldsymbol{\Phi}^T. \tag{1}$$

The eigenvector and eigenvalue matrices, Ψ, Λ are computed as:

$$S_T \Psi = \Psi \Lambda. \qquad (2)$$

The size of the matrix, S_T is $MN \text{x} MN$ and determining the MN eigenvectors and eigenvalues is an intractable task for typical image sizes. A computationally feasible method that employs the eigenanalysis of $\boldsymbol{\Phi}^T \boldsymbol{\Phi}$ instead of $\boldsymbol{\Phi}\boldsymbol{\Phi}^T$ is used [2]. The size of $\boldsymbol{\Phi}^T \boldsymbol{\Phi}$ is $T \text{x} T$.

$$(\boldsymbol{\Phi}^T \boldsymbol{\Phi})V = V \Lambda' \qquad (3)$$

$\mathbf{V} = [\mathbf{V}_1, \mathbf{V}_2, \cdots, \mathbf{V}_T]$ and $\Lambda' = diag(\lambda_1, \lambda_2, \cdots, \lambda_T)$. Premultiplying $\boldsymbol{\Phi}$ on both sides, we have

$$\boldsymbol{\Phi}(\boldsymbol{\Phi}^T \boldsymbol{\Phi})V = (\boldsymbol{\Phi}\boldsymbol{\Phi}^T)(\boldsymbol{\Phi} V) = (\boldsymbol{\Phi} V)\Lambda' \qquad (4)$$

and $\boldsymbol{\Phi} V$ is the eigenvector matrix of $\boldsymbol{\Phi}\boldsymbol{\Phi}^T$. Assuming λ_i's are sorted as $\lambda_1 \geq \lambda_2 \geq \cdots \geq \lambda_T$, we obtain eigenvectors of $\boldsymbol{\Phi}\boldsymbol{\Phi}^T$ corresponding to the first largest m eigenvalues as follows. These eigenvectors constitute the projection matrix W_{pca}

$$W_{pca} = [\boldsymbol{\Phi} \mathbf{V}_1, \boldsymbol{\Phi} \mathbf{V}_2, \cdots, \boldsymbol{\Phi} \mathbf{V}_m]. \qquad (5)$$

$\boldsymbol{\Phi} \mathbf{V}_1, \boldsymbol{\Phi} \mathbf{V}_2, \cdots, \boldsymbol{\Phi} \mathbf{V}_m$ are refered to as eigenfaces. Refer to Figure 1 for an example of eigenfaces. A vector $\boldsymbol{X}_i$ that represents a face image is projected to a vector $\boldsymbol{Y}_i$ in a vector space of dimension, m using the following equation.

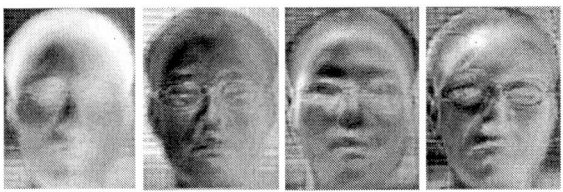

Fig. 1. The first four eigenfaces computed from SKKU facial images

$$\boldsymbol{Y}_i = W_{pca}^T(\boldsymbol{X}_i - \boldsymbol{\Phi}) \qquad (6)$$

A new face image $\boldsymbol{X}_i$ is recognized by comparison of $\boldsymbol{Y}_i$ with the projected vectors of the training face images that are computed off-line. Since PCA maximizes for all the scatter, it is more appropriate for signal representation rather than for recognition purpose.

2.2 Fisherface Method

The idea of the Fisherface method is that one can perform dimensionality reduction using W_{pca} and still preserve class separability. It applies FLD to the reduced PCA subspace to achieve more reliability for classification purpose. The Fisherface method is briefly described as follows. Let $\omega_1, \omega_2, \cdots, \omega_c$ and $N_1, N_2, \cdots, N_c$ denote the classes and the number of face images in each class, respectively. Let $M_1, M_2, \cdots, M_c$ and M be the means of the classes and the total mean in the reduced PCA subspace. Since $Y_{ij} = W_{pca}^T X_{ij}$, we can then have $M_i = \frac{1}{N_i} \sum_{j=1}^{N_i} Y_{ij} = W_{pca}^T (\frac{1}{N_i} \sum_{j=1}^{N_i} X_{ij})$. X_{ij} denotes the jth face image vector belonging to the ith class (i. e. subject). The between-class scatter and within-class scatter matrices S_b' and S_w' of Y_{ij}'s are expressed as follows.

$$S_b' = \sum_{i=1}^{C} N_i (M_i - M)(M_i - M)^T = W_{pca}^T S_b W_{pca} \qquad (7)$$

$$S_w' = \sum_{i=1}^{C} \frac{1}{N_i} \sum_{j=1}^{N_i} (Y_{ij} - M_i)(Y_{ij} - M_i)^T = W_{pca}^T S_w W_{pca} \qquad (8)$$

S_b and S_w denote the between-class scatter and within-class scatter matrices of X_{ij}'s, respectively. The projection matrix W that maximizes the ratio of the determinant, $\frac{|W^T S_b' W|}{|W^T S_w' W|}$ is chosen as the optimal projection, W_{fld}. The columns of W_{fld} are computed as the (C-1) leading eigenvectors of the matrix $(S_w')^{-1} S_b'$ [11] where C denotes the number of classes. For recognition, given an input face image X_k, it is projected to $\Omega_k = W_{fld}^T W_{pca}^T X_k$ and classified by comparison with the vectors Ω_{ij}'s that were computed off-line from a set of training face images.

3 SKKUfaces

3.1 SKKUface Method

The SKKUface method proposed in this research is illustrated in Figure 2. It is similar to Fisherface methods in that it applies PCA and FLD in series. Our algorithm is different from Fisherface methods in that face variations due to lighting, facial expression and eyewear are effectively removed by discarding the first few eigenvectors from the results of PCA, and then apply FLD to the reduced subspace to get the most class separability for face classification. The result is an efficient feature extraction that carries only features inherent in each face, excluding other artifacts such as changes in lighting and facial expression. Classification of faces using the resulting feature vectors leads to a considerably improved recognition accuracy than Fisherface methods.

As illustrated in Figure 2, we apply FLD to the reduced subspace that is computed by ignoring the first few eigenvectors corresponding to the top principal eigenvalues. For the experimental results, we have only discarded the first eigenvector. Another important contribution of SKKUfaces is the efficient computation of the between-class scatter and within-class scatter matrices S_b' and S_w' of Y_{ij}. The following section describes the method.

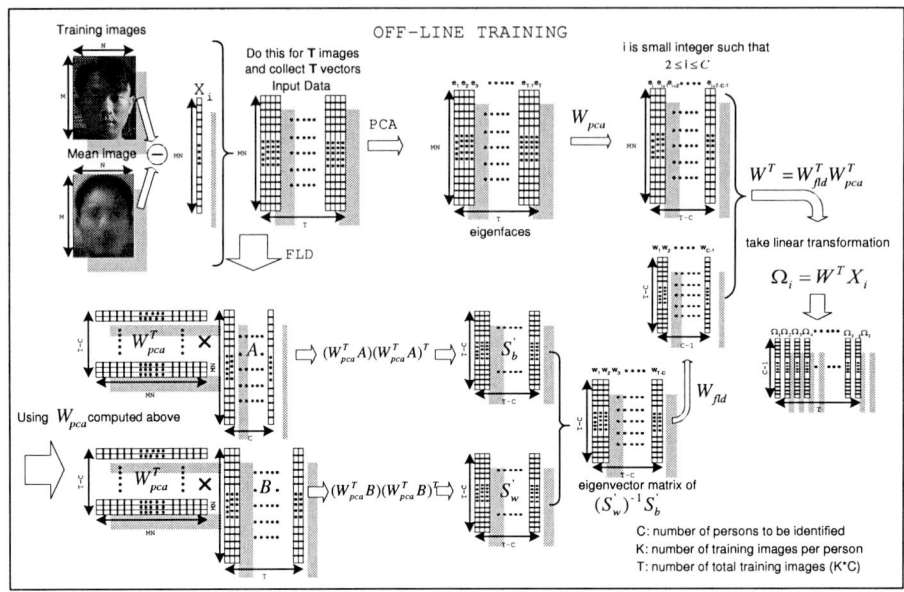
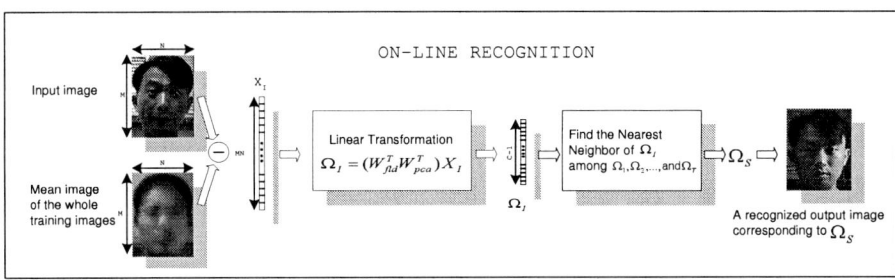

Fig. 2. The overview of the SKKUface method

3.2 Efficient Computation of Within-class Scatter and Between-class Scatter Matrices

After dimensionality reduction of the face vector space by the linear projection, W_{pca}, we need to compute the within-class scatter and between-class scatter matrices, S'_w and S'_b to apply the Fisher linear discriminant analysis to the reduced subspace. The resulting projection matrix, W_{fld} consists of columns of eigenvectors of $(S'_w)^{-1}S'_b$ corresponding to the largest (C-1) leading eigenvalues. In computing S'_w and S'_b represented by $W^T_{pca}S_wW_{pca}$ and $W^T_{pca}S_bW_{pca}$, respectively, we do not explicitly evaluate S_w and S_b. The size of the matrices, S_w and S_b, is $MN \times MN$ and it is an intractable task to compute them for typical image sizes. On the other hand, S_b can be expressed using equation (9) assuming the same size of each class.

$$S_b = \sum_{i=1}^{C} \frac{1}{C}(M_i - M)(M_i - M)^T$$
$$= \frac{1}{C}(M_1 - M)(M_1 - M)^T + \cdots + \frac{1}{C}(M_c - M)(M_c - M)^T$$
$$= \frac{1}{C}[M_1 - M, M_2 - M, \cdots M_C - M] \begin{bmatrix} (M_1 - M)^T \\ (M_2 - M)^T \\ \vdots \\ (M_C - M)^T \end{bmatrix} \quad (9)$$
$$= \frac{1}{C} A A^T$$

where $A = [M_1 - M, M_2 - M, \cdots M_c - M]$ and M_i, M denote the ith class mean and the total mean, respectively. $\frac{1}{C}$ is *prior* probability that represents the size of each class.

Since $MN \gg C$, we can save a huge amount of computation by using the matrix A of size $MNxC$ matrix rather than directly dealing with S_b of size $MNxMN$. Finally, S_b' is obtained using the following equation.

$$S_b' = W_{pca}^T S_b W_{pca} = W_{pca}^T A A^T W_{pca} = (W_{pca}^T A)(A^T W_{pca}) \quad (10)$$

Notice that S_b' is simply computed by multiplication of $W_{pca}^T A$ and its transpose. Similarly, S_w' can be written as follows.

$$S_w = \sum_{i=1}^{C} \sum_{j=1}^{N_i} (X_{ij} - M_i)(X_{ij} - M_i)^T$$
$$= [K_{11}, \cdots, K_{21}, \cdots, K_{CN_C}] \begin{bmatrix} K_{11}^T \\ \vdots \\ K_{21}^T \\ \vdots \\ K_{CN_C}^T \end{bmatrix} \quad (11)$$
$$= BB^T$$

$K_{ij} = X_{ij} - M_{ij}$ and $B = [K_{11}, \cdots, K_{21}, \cdots, K_{CN_C}]$. S_w' is computed as:

$$S_w' = W_{pca}^T S_w W_{pca} = W_{pca}^T B B^T W_{pca} = (W_{pca}^T B)(B^T W_{pca}) \quad (12)$$

The size of matrix B is $MNxT$ and $MN \gg T$. We could save a lot of computational effort using the matrix, B. Similarly to S_w', S_b' is simply computed by multiplication of $W_{pca}^T B$ and its transpose.

Suppose $M = N = 256, C = 10, K = 15$. The explicit computation of S_b and S_w involves matrices of size 65536 x 65536. Employing the proposed methods involves computation using a 65536 x 10 matrix for S_b' and a 65536 x 150 matrix for S_w'. This achieves about 6,500 times and 43 times less computation for S_b' and S_w', respectively.

4 Experimental Results

To assess the performance of SKKUfaces, the recognition rate of SKKUfaces is compared with that of Fisherfaces [1] using Yale facial database and SKKU facial database. The recognition rates were determined by the "leaving-one-out" method [11]. A face image is taken from the database for classification and all the images except this image are used for training the classifier. Classification was performed using a nearest neighbor classifier.

SKKU facial database contains ten different images of each of ten different subjects. The size of an image is 50 x 40. For a subject, five images out of ten images were taken first and the rest five images at a different time. All the images are frontal views of upright faces with changes in illumination, facial expression (open/closed eyes, smiling/nonsmiling/surprised), facial details (glasses/no glasses) and hair style. Refer to Figure 3 for the whole set of SKKU face images. In Yale facial database, each of sixteen different subjects have ten images which consist of three images under illumination changes, six with changes in facial expression and one with glasses worn. Figure 4 shows a set of images of a subject in Yale facial database.

Figures 5 and 6 show the relative performance of the algorithms when applied to SKKU facial database and Yale facial database, respectively. As can be seen in Figures 5 and 6, the performance of SKKUfaces is far better than that of Fisherfaces in the cases of variations in illumination and eyewear. This experimentally proves our claim that we apply FLD to a reduced subspace that is more appropriate for classification purpose than the reduced PCA subspace that Fisherface methods use. Application of FLD to this reduced subspace yields the better discriminating power for pattern classification and the recognition accuracy is far improved. The amount of computational saving we could benefit in computing S'_w and S'_b from the method proposed in section 3.2 is as follows.

Since $M = 50$, $N = 40$, $C = 10$, $K = 10$ in the case of SKKU facial database, directly evaluating with S_b and S_w should involve matrices of size 2000 x 2000. However, employing the proposed method only deals with a 2000 x 10 matrix for S'_b and a 2000 x 100 matrix for S'_w, respectively. The saving amounts to about 200 times and 20 times less computation for S'_b and S'_w, respectively.

5 Conclusion

We have proposed SKKUfaces for automatic face recognition robust to variations in lighting, facial expression and eyewear. In the reduced PCA subspace, SKKUfaces effectively removes portions of the vector space that are responsible for variations in lighting and facial expression, and applies FLD to this reduced subspace. The experimental results show that the discriminating power for pattern classification is considerably improved and excellent recognition accuracy is achieved. A study on the relationship between the number of eigenvectors to be discarded in the reduced PCA subspace and the degree of variations in lighting or facial expression will enable us to achieve the optimum performance of SKKUfaces.

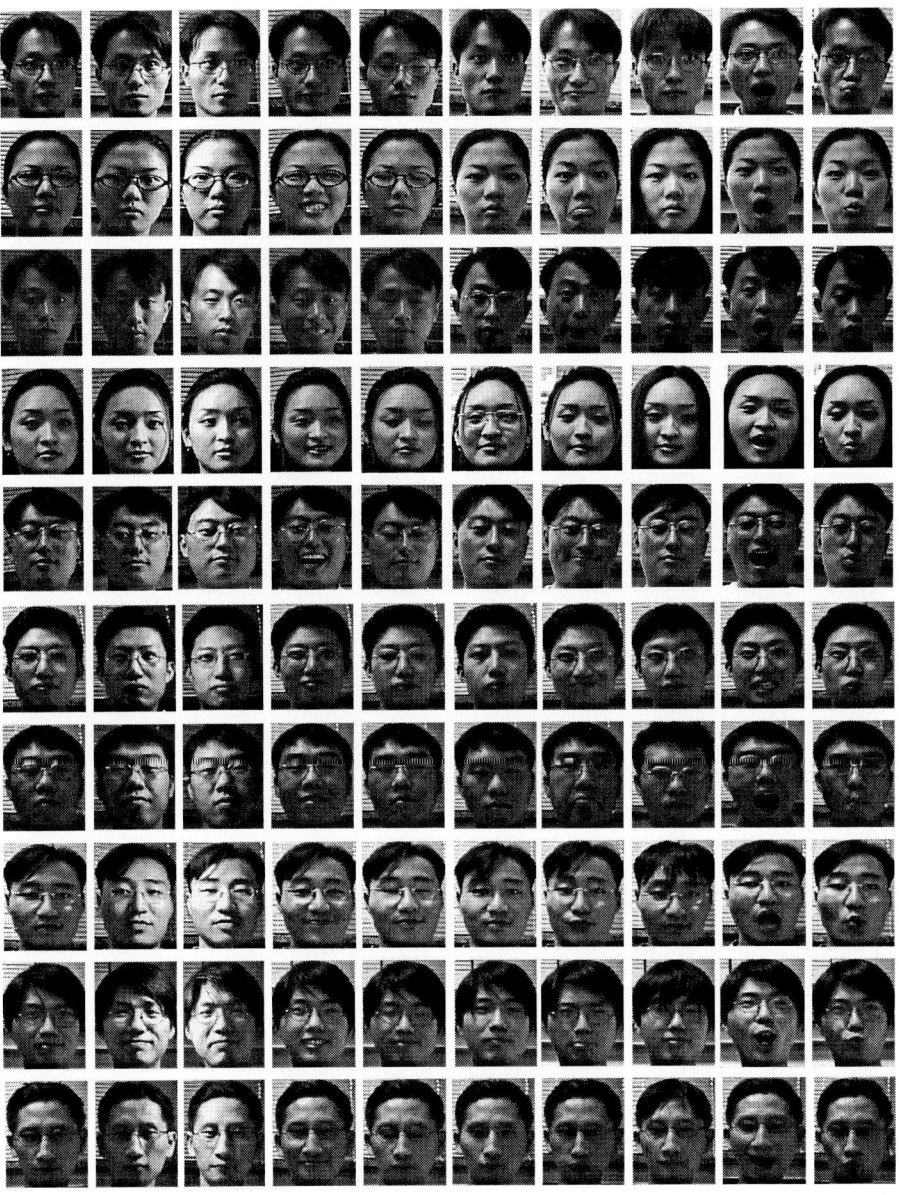

Fig. 3. The whole set of SKKU facial images [13]

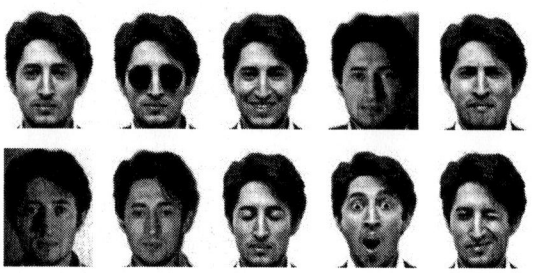

Fig. 4. Example images from Yale facial database [12]

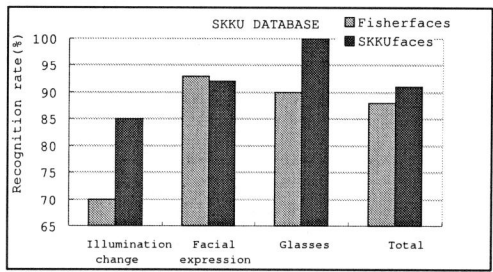

Fig. 5. The relative performance of the SKKUface and the Fisherface methods for SKKU facial images

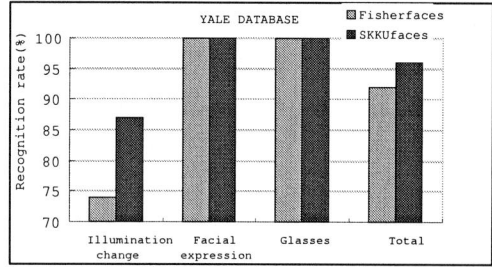

Fig. 6. The relative performance of the SKKUface and the Fisherface methods for Yale facial images

Acknowledgement

This work was supported by grant number 1999-2-515-001-5 from interdisciplinary research program of the KOSEF.

References

1. P. Belhumeur, J. Hespanha, and D. Kriegman, "Eigenfaces vs. Fisherfaces: Recognition Using Class Specific Linear Projection," *IEEE Trans. on PAMI*, vol. 19, no. 7, pp. 711-720, 1997.
2. M. Turk and A. Pentland, "Eigenfaces for Recognition," *Journal of Cognitive Neuroscience*, vol. 3, no. 1, pp. 71-86, 1991.
3. R. Brunelli and T. Poggio, "Face Recognition: Features vs. Templates," *IEEE Trans. on PAMI*, vol. 15, no. 15, pp. 1042-1052, 1993.
4. Shang-Hung Lin et al., "Face Recognition and Detection by Probabilistic Decision Based Neural Network," *IEEE Trans. on Neural Network*, vol. 8, no. 1, pp. 114-132, 1997.
5. Chengjun Liu and Harry Wechsler, "Enhanced Fisher Linear Discriminant Models for Face Recognition," Proceedings of the 14th International Conference on Pattern Recognition, vol. 2, pp. 1368-1372, 1998.
6. Rama Chellappa, Charles L. Wilson, and Saad Sirohey, "Human and Machine Recognition of Faces: A Survey," *Proceedings of IEEE*, vol. 83, no. 5, 1995.
7. M. Kirby and L. Sirovich, "Application of the Karhunen-Loeve Procedure for the Characterization of Human Faces," *IEEE Trans. on PAMI*, vol. 12, no. 1, pp. 103-108, 1990.
8. K. Etemad and R. Chellappa, "Discriminant Analysis for Recognition of Human faces image," *Journal of Optical Society of America*, vol. 14, no. 8, pp. 1724-1733, 1997.
9. A. Pentland, B. Moghaddam, T. Starner, and M. Turk, "View Based and Modular Eigenspaces for Face Recognition," Proceedings of the IEEE Conference on Computer Vision and Pattern Recognition, pp. 84-91, 1994.
10. R. A. Fisher, "The Use of Multiple Measures in Taxonomic Problems," *Ann. Eugenics*, vol. 7, pp. 179-188, 1936.
11. K. Fukunaga, Introduction to Statistical Pattern Recognition. Academic Press, second edition, 1991.
12. http://cvc.yale.edu/projects/yalefaces/yalefaces.html
13. http://vulcan.skku.ac.kr/research/skkufaces.html

A Humanoid Vision System for Versatile Interaction

Yasuo Kuniyoshi[1], Sebastien Rougeaux[2], Olivier Stasse[1], Gordon Cheng[1], and Akihiko Nagakubo[1]

[1] Humanoid Interaction Lab., Intelligent Systems Division, Electrotechnical Laboratory (ETL), 1-1-4 Umezono, Tsukuba, Ibaraki 305-8568 Japan.
{kuniyosh,stasse,gordon,nagakubo}@etl.go.jp,
WWW home page: http://www.etl.go.jp/~kuniyosh/
[2] RSISE, The Australian National University, Canberra ACT 0200, Australia

Abstract. This paper presents our approach towards a humanoid vision system which realizes real human interaction in a real environment. Requirements for visual functions are extracted from a past work on human action recognition system. Then, our recent development of biologically inspired vision systems for human interaction is presented as case studies on how to choose and exploit biological models, mix them with engineering solutions, and realize an integrated robotic system which works in real time in a real environment to support human interaction. A binocular active vision system with foveated wide angle lenses, a real time tracking using velocity and disparity cues, a real-time multi-feature attentional system, and a human motion mimicking experiment using a humanoid robot are presented.

1 Introduction

Observing human behavior and generating appropriate response behavior is becoming a more and more important issue in the trend of human-friendly robotics. It is obvious that vision plays a crucial role in such applications. It has to operate robustly in real time, in an unstructured environment. Moreover, the design of the vision system should always be done in the context of overall integration of the entire system.

Considering about required functions and constraints, we support that biologically inspired design is suitable for the above applications. On the other hand, if we want the system to operate in the real world and to establish real interactions, we have to meet tough constraints. Some of the essential constraints are common to humans and machines, which defines the principles of information processing. But there are other constraints which are not shared by the two kinds. Therefore, we should carefully examine which aspects of biological models should be adopted, which parts of the system should employ engineering models, and how they all fit together to realize the overall system.

This paper discusses the required functions, performance, and design issues for a vision system for human interaction, through case studies of some of our past and recent works in the area of vision based interactive robot systems.

2 Visual Functions for Human Interaction

2.1 Qualitative Action Recognition – A Case Study

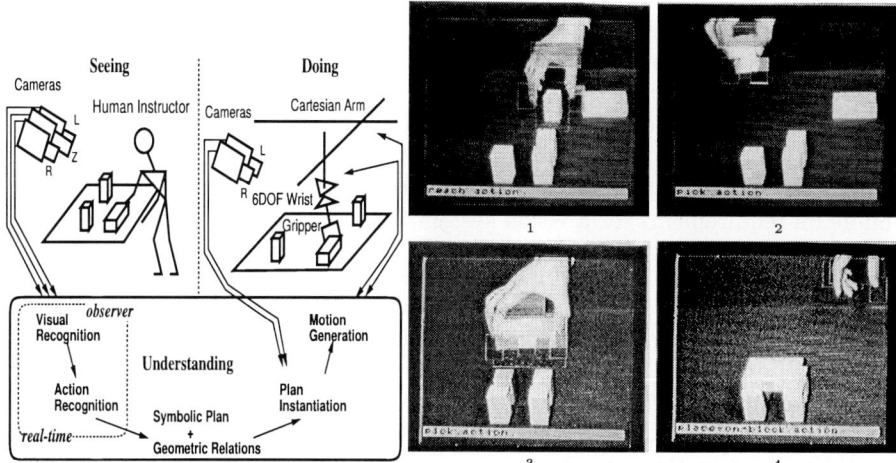

Fig. 1. Learning by Watching system ([1])(Left). A human performance of an arbitrary block stacking task is observed by a trinocular (stereo + zoom) camera system. Action units are identified in real time. The resulting symbolic action sequence is fed to a robot manipulator system which re-generates the learned actions to achieve the same task goal in a different workspace with a different initial block placements. Some snapshots of monitor display during task recognition (Right). Picking up the fourth pillar (top row) and placing the top plate (bottom row) in a "table" building task. The real time recognition results are displayed at the bottom; (1) reach, (2) pick, (3) pick, (4) place-on-block.

Kuniyoshi et al. [1] built an experimental system which recognizes pick and place sequences performed by a person in real time (Fig. 1). Their analysis [2] clarifies the following principles of action recognition.

- Action recognition is detecting causal relationship connecting the subject of action, its motion, the target object, and its motion.
- Temporal segmentation of actions is done when the causal relationship changes.
- The causal relationship is affected by the ongoing context of overall task.

The historical system above had many limitations: (1) it used only binary image processing and therefore assumed a black background, a white hand, and white rectangular blocks, (2) the cameras were fixed and the field of view was very narrow, covering only about 0.4m × 0.4m × 0.3m desktop space, (3) it assumed that there is always a single moving entity (the hand, sometimes with held objects).

In the following discussions, we first list up necessary vision functions for a more realistic system, then present our recent systems which realize the functions.

2.2 Realistic Visual Functions for Human Interaction

In human action recognition, the subject of action is primarily humans who are moving. The target of an action must be detected by attentional search with the help of basic knowledge about causality and the ongoing context. Motion, or more generally, changes, in these relevant objects must be detected timely. And all these information must be integrated in time to decide the current action as well as the ongoing context. To summarize, the following elements are the minimum requirements for a vision system for human action recognition.

- Complicated wide space: Humans generally move around in a wide cluttered environment and do things here and there. The system should have a very wide view angle as well as high magnification to monitor small localized motion and items. Space variant sensing combined with active gaze control is strongly supported.
- Human detection: The system should robustly detect and track humans and their limbs in the complicated background. Motion, skin color, and body structure are useful features.
- Attentional shift and memory: The system should be able to keep track of at least two targets. This requires spatial memory. The second (or further) target is often acquired by a guided visual search, such as searching from the primary target in the direction of its motion. This search process must be controlled by combining motion or other cues from the current target with previously learned knowledge about causal relationships between two or more events separated in space and time.
- Object detection: The minimum requirement for object recognition is to detect its existence as a "blob" robustly, assuming that it has an arbitrary color, shape and texture. Zero disparity filtering and kinetic depth are primarily useful. Flow-based time to impact may not be useful since the observer is not necessarily moving in the direction of the target. In addition, simple identification using 2D features are useful in combination with spatial memory. This is for re-acquiring the previously seen target or bind information from different times, e.g. "He did A to O then did B to O."
- Movement analysis: Either from optical flow or target tracking, movement patterns over a period of time should be classified.

The above considerations suggest the following system configuration as the minimum requirement:

- A binocular active vision system with space variant sensors.
- Optical flow and zero disparity filtering.
- Spatial attention mechanism and multi-feature integration.
- Simple spatial inference and its feedback to attentional shift.

In the following sections, we present our series of development of vision-based systems which capture parts of the above requirements. They emphasize different aspects of the above points and presented only briefly to constitute an overview of our approach. Interested readers are invited to refer to individual papers.

3 ESCHeR: A Binocular Head with Foveated Wide Angle Lens

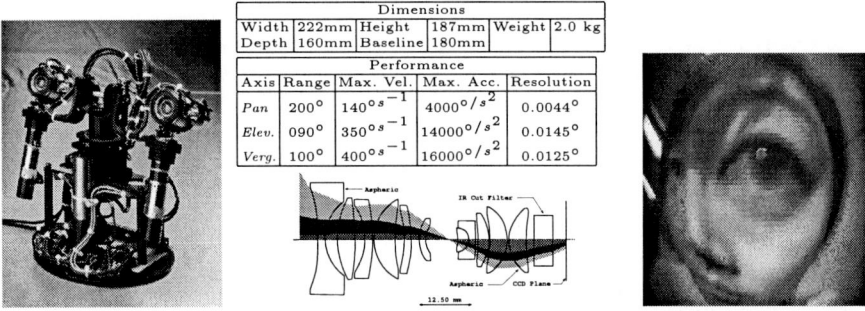

Fig. 2. *ESCHeR:* ETL Stereo Compact Head for Robot Vision with foveated wide angle lenses. The mechanism (left) and its spec. (middle-top). The optics design (middle-bottom) and an image (right) through the foveated wide angle lens.

ESCHeR [3] (E[tl] S[tereo] C[ompact] He[ad for] R[obot vision], Fig. 2) is a binocular active vision system developed at ETL as an essential component for our interactive robot system. Many aspects of the system are biologically inspired, e.g. the space variant sensing, active gaze control, optical flow and zero disparity filtering.

It has two CCD cameras which rotate independently ("vergence") in a common horizontal plane which can be tilted, and the whole platform can rotate around the vertical axis ("pan"). All joints are driven by DC servo motors equipped with rotary encoders. The mechanism partially mimics the eye mobility of human vision system, which is sufficient for tracking a moving object ("smooth pursuit") or to quickly change the focus of attention ("saccade").

Image processing is done at frame rate on a DataCube MaxVideo system with a Shamrock quad DSP system. Recently we are switching to a Intel Pentium III PC system. Exploiting the MMX and SSE acceleration, we achieved greater performance, e.g. 60Hz optical flow based target tracking. Servo control is done at 500Hz, interpolating the frame rate position/velocity data.

3.1 Space-Variant Sensors

The most significant part of ESCHeR is its "foveated wide angle" lenses[4]. Our lens simulates human visual system's compromise between the need for a wide field of view for peripheral detection and the need for high resolution for precise observation under limited number of pixels. Several implementations of space-variant visual sensors have been presented in the past. They include custom-designed CCD/CMOS sensors [5, 6], digital warping [7, 8, 9], or a combination of wide/tele cameras [10, 11], but suffer from problems such as continuity, efficiency or co-axial parallelism[1].

$$r(\theta) = \begin{cases} f_1 \tan \theta & , 0 \leq \theta \leq \theta_1, \\ \log_a (f_2 \theta) - p & , \theta_1 \leq \theta \leq \theta_2, \\ f_3 \theta + q & , \theta_2 \leq \theta \leq \theta_{max}. \end{cases} \quad (1)$$

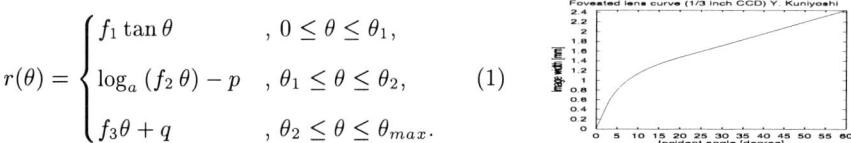

Fig. 3. Projection curve. The equation (left) and the incident-angle vs. image height graph (right). p and q are constants set to solve the continuity of the curve at the connection points.

Our system avoids these problems by an alternative approach pioneered by Suematsu[12]; designing a new optics. This method deals with the optical projection process which cannot be fully treated by the two dimensional methods such as CCD/CMOS design or the warping method. It is critical when we want to achieve a wide view angle. Even if we use the space variant CCD, we must use some wide angle lens, which generally deviates substantially from the standard (non-distorted) projection. Therefore, if we ignore the optical process, we cannot obtain a desired image. Integrating the optical process and the 2D image transformation is very important in achieving a foveated wide angle vision.

Our lens is carefully designed to provide a *projection curve*[2] which has useful intrinsic properties that help computer vision algorithms[4], unlike [12]. As seen in (1), the lens exhibits a standard projection in the fovea (0° – 2.5° half view angle) and a spherical one in the periphery (20° – 60°) which are smoothly linked using a log of spherical curve (2.5° – 20°). An example of image obtained with our lens is presented in Fig. 2.

4 Robust Real Time Tracking of a Human in Action

The most basic visual functions for versatile human-robot interaction are target detection and tracking, without even recognizing the content of what it sees. The target should be an open-ended assortment of moving objects (human heads, toys, etc.), with possibly deformable shapes (hands or body limbs). And the background would be highly cluttered scenes (usual laboratory environments).

---

[1] The latest Giotto sensor solved the continuity problem.
[2] A *projection curve* maps the incident angle Θ of a sight ray entering the lens system to an image height $r(\Theta)$ on the CCD surface.

A model based or template matching strategy would then be quite costly and even inappropriate. As an alternative, Rougeaux and Kuniyoshi [13, 14], showed how the integration of optical flow and binocular disparity, two simple cues commonly found in biological visual systems and well-known in image processing, can lead to very robust performance in the tracking of arbitrary targets in complex environment.

4.1 Optical Flow

Following an extensive review on the performance of optical flow algorithms [15], we decided to implement a gradient-based method with local smoothness constraints that is inspired from the work of [16] on stereo vision systems, because of its computational simplicity and the good results it yields in comparison with other techniques. We also added two modifications to the proposed scheme: an IIR recursive filter introduced by [17] for computing a reliable time gradient while respecting real-time constraints, and a Bayesian approach suggested by [18] for estimating flow reliabilities. Our first implementation extracts 36×36 flow vectors with confidence values from 324×324 pixels faster than frame rate[3].

4.2 Velocity Cues

The detection and pursuit of moving objects in the scene can be performed by making the distinction, in the motion field, between velocity vectors due to the camera egomotion and the others due to independent motion. In the case of ESCHeR, since the body is fixed to the ground, the egomotion flow can be computed from known camera rotations. However, because of the high distortions introduced by the foveated wide-angle lens, a simple background flow subtraction method [19] can not be directly implemented. We found that the egomotion flow in ESCHeR can be approximated in the velocity domain by an ellipsoid distribution (Fig. 4, [14]), whose center coordinates, orientation and axis length can be directly derived from the instantaneous rotation vector of the camera.

4.3 Disparity Cues

Disparity estimation plays another key role in binocular active vision systems [20, 14]. It not only ensures that we are fixating on the same object with both eyes, but provides also important cues for depth perception and focus control.

We adopt the phase disparity method which is considered to account for a Gabor filter based disparity detection [21], a biologically inspired model.It is also practically useful because it is robust, produces dense disparity maps. and can be implemented as a real-time process. It requires no explicit feature detection or matching process.

---

[3] Our latest PC version extracts 80×60 flow with confidence values from 640×480 pixels in less than 4 msec.

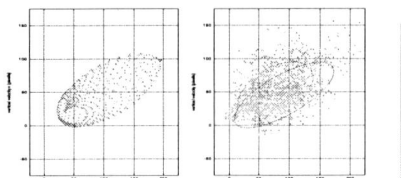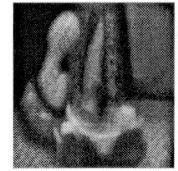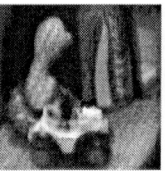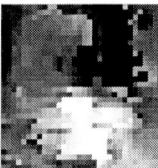

Fig. 4. Simulated and real egomotion flows in the velocity space (left two plots). Disparity estimation (right three photos): The original stereo pair and the phase-based disparity map.

From phase to disparity Let $I_l(\omega)$ and $I_r(\omega)$ be Fourier transforms of the left and right images, and $\phi_i(\omega) \equiv arg(I_i(\omega))$. The Fourier shift theorem states that if $\omega \neq 0$, the global shift Δx can then be recovered with

$$\Delta x = \frac{[\phi_l(\omega) - \phi_r(\omega)]_{2\pi}}{\omega} \qquad (2)$$

where $[\phi]_{2\pi}$ denotes the principal part of ϕ that lies between $-\pi$ and π.

Equation 2 supports that the disparity can be directly derived from the phase and frequency of the signals. This approach exhibits clear advantages over correspondence or correlation methods; First, because phase is amplitude invariant, varying illumination conditions between the left and right images do not perturb significantly the computation. Second, local phase is also robust against small image distortions [22], which makes it very suitable to handle the small optical deformations of our space-variant resolution in the foveal area (The disparity is only computed in the foveal area of the image).

However, the Fourier shift theorem cannot be directly applied for the analysis of stereo images, at least not in its present from: first, it requires a *global* shift between the two signals, whereas pixel displacements in a stereo pair are fundamentally *local*. Intuitive solutions, like applying the Fourier transform on small patches of the images, have been suggested [23], but the computational complexity becomes important and the precision decreases drastically with the size of the local window. A more interesting approach is to recover the local phase and frequency components in the signal using the output of complex-valued band-pass filters [21]. Fig. 4 shows a binocular disparity map obtained from the phase information after convolution with a complex-valued Gaussian difference filter [24]. In target tracking and fixation, we are interested in small range of disparity around zero, therefore the above method fits very well.

4.4 Integrating Velocity and Disparity Cues

To summarize, the overall processing diagram is shown in Fig. 5. The flow and disparity processing are done in parallel and integrated using confidence value. The entire processing including tracking control is done at frame rate.

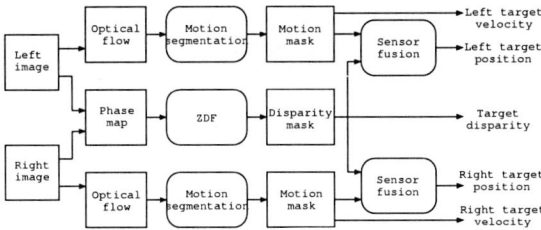

Fig. 5. Combining disparity and velocity information to obtain the target position and velocity in both left and right images, as well as its binocular disparity.

The target position is estimated as follows:

$$\mathbf{x}_t = \frac{\mu_v \mathbf{x}_v + \lambda \mu_d \mathbf{x}_d}{\mu_v + \lambda \mu_d}, \qquad (3)$$

where $\mathbf{x}_v$ is the position estimation from the velocity cue, $\mathbf{x}_d$ is the position from disparity, and μ_v and μ_d are respective confidence values. While the target is moving, velocity cue provides more information and when the target stops, disparity cue becomes more reliable. Fig. 6 shows such continuous and automatic integration.

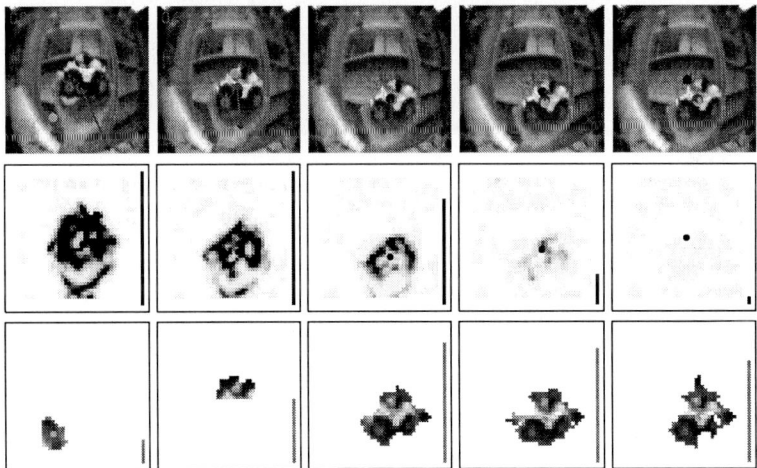

Fig. 6. Cue integration. The moving target stopped at third frame. The flow cue gradually disappears (middle row) and the disparity cue takes over (bottom row). Vertical bars in each frame indicates the confidence values.

The integrated system can find a walking person in a cluttered laboratory room, quickly saccade to the target, robustly track, and always attend to significantly moving target. Fig. 7 shows such an example. Although the tracking

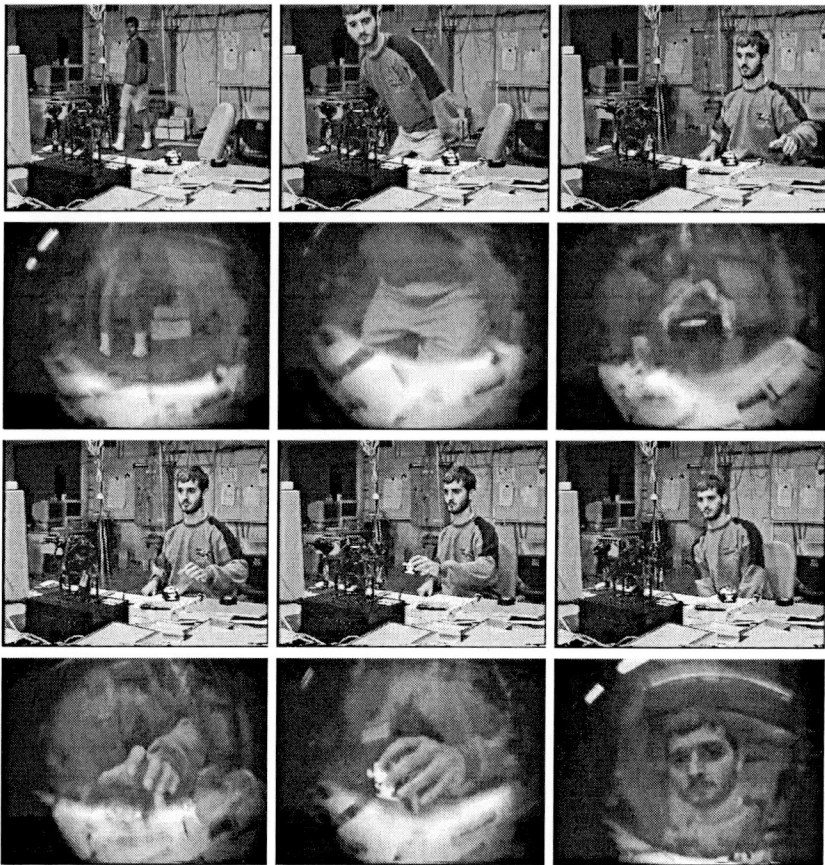

Fig. 7. Finding and tracking a human walking in a wide space as well as doing desktop actions. Bottom row is the view through ESCHeR's lens under tracking and saccading motion. ESCHeR is working completely autonomously through the sequence.

target is simply chosen based on flow intensity by the system, it picked up quite meaningful targets in experiments. Thanks to the space-variant sensor, the system could find and track target over a wide range of distances.

With ESCHeR's space variant sensor, the system usually converges onto a single target even when multiple flow patterns are present within the field of view. This is because the pattern closest to the fovea is magnified and dominate the center of mass estimation of the entire flow intensity values over the field of view. However, there is no top-down control on the choice of the target flow pattern.

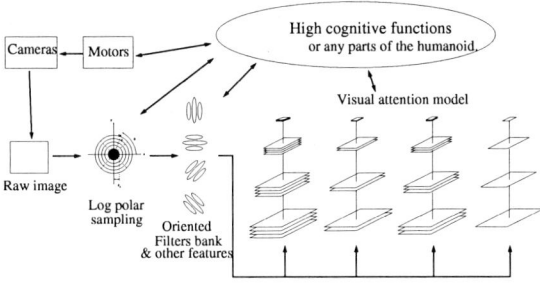

Fig. 8. Structure of our multi-feature visual attention system

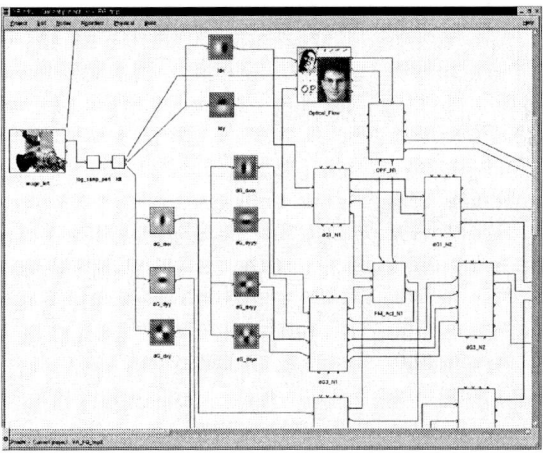

Fig. 9. XPredN: Friendly user interface to edit and modify a distributed real-time application. The above display shows the multi-feature attentional mechanism software.

5 A Real Time Multi-feature Attentional Mechanism

The target acquisition and tracking in the previous section cannot handle attentive processes, such as recognizing the previously seen object, choosing a particular target from several candidates with a top-down control, etc. According to the consideration in section2, a multiple feature based attentional mechanism would be appropriate for the above purpose.

Stasse and Kuniyoshi [25] developed a real time attentional mechanism. It computes a set of visual features in parallel, such as progressive orders of partial derivatives of Gaussian filters which constitute steerable filters, temporal filters, etc. The features are processed though FeatureGate model [26, 27], which detects salient points by combining bottom up and top down information. Moreover, optical flow extraction and phase-based disparity can be efficiently computed from steerable filter output. For details and experimental results, see [25].

In the past, such system as above required prohibitingly large computational resources and was not suitable for a real application. However, today, a cluster of PC's provides substantial computational power. Our system runs on a cluster with 16 PC's connected with ATM network. ATM was chosen from the viewpoint of strict real time requirement, i.e. no jitter in transmission delay. We make intensive use of SSE on Intel Pentium III processors to accelerate floating point computation. For example, it provides over 1 GFlops on a single 600MHz processor in floating point convolution.

On the above platform, we developed a highly efficient and very easy to program parallel real time software system, called PredN (Parallel Real time Event and Data driven Network) [28]. The application software is modeled as a network of nodes. And each node represents a computing thread with input and output data ports. A node is executed by a data-flow and/or periodic schemes as specified by the programmer. The network is automatically compiled and the codes are distributed over the PC cluster. The execution is very efficient, which is an order of magnitude faster than other real time distributed software platform, such as RT-CORBA.

The philosophy behind PredN is to facilitate flexible design of a complex real time intelligent systems by allowing the programmer to mix various processing models such as control theoretic filters, different types of neural networks, signal/image processing algorithms, etc. and connect them up with a data-flow network, which then runs in real time as an integrated system. This is important for a biologically inspired approach to real interactive systems, because often we have to mix engineering solutions (e.g. for fast image filtering) and biologically inspired models (e.g. neural networks for attentional models and adaptive modules) and integrate into one real time system.

6 A Humanoid Mimicking Human Motion

Gordon and Kuniyoshi [29] presented an early experiment on human-humanoid interaction as shown in Fig. 11. It notices a human either visually or by sound, orients towards the person, and mimics the arm motion at the direct perceptual mapping level [30]. All the multi-modal sensory-motor flows are continuously active, and the response motion (ballistic and constrained) is generated as a result of competition of activations. The system can engage in continuous interaction with humans, coming in and out of sight, without system initialization for more than 30 minutes.

6.1 ETL-Humanoid

In our current phase of development, the upper body of our humanoid robot has been completed 10. This upper body provides 24 degrees of freedom: 12 d-o-f for the arms, 3 d-o-f for the torso, 3 d-o-f for head/neck and 6 d-o-f for the eyes [31, 32]. Other parts of the body are still under construction. The processing for our system is currently performed over a cluster of six PCs [29].

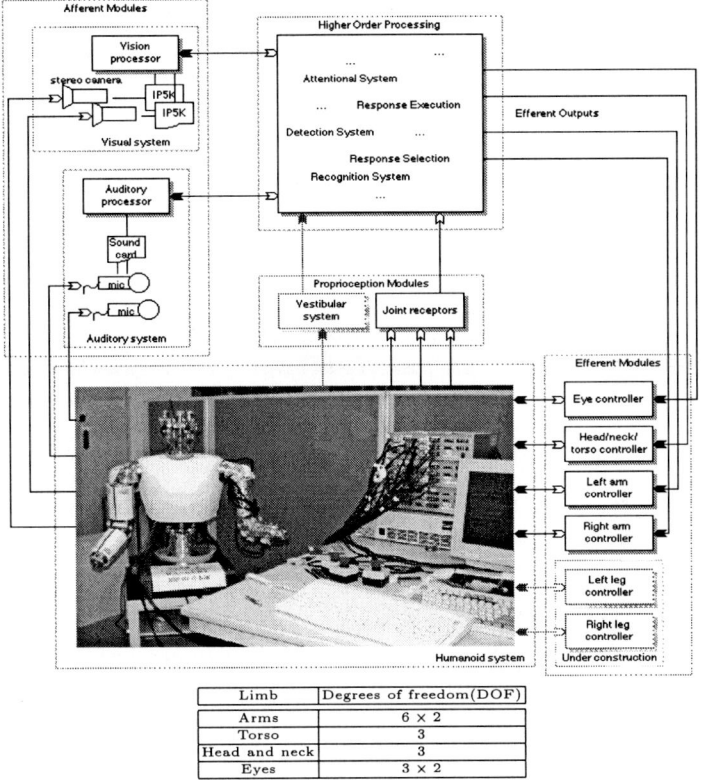

Fig. 10. ETL-Humanoid: Currently, the upper body has been completed. An overview of the processing system is shown. From left to right: Afferent Modules, process the external inputs; through to the Higher Order Processing, providing Response Selection, etc. These processes influence the internal sub-systems as well as the overall response of the system. The effect generates motion via the Efferent Modules.

6.2 Humanoid Mimicking

Figure 11 shows the experimental results of our system. The humanoid robot first orients toward, and track a person. When the upper body has been fully sighted, i.e. head and two arms have been detected, the humanoid robot mimics the upper body motion of the person. When the upper body can not be fully sighted, the robot continues to track the person. When the system loses sight of the person, it stays idle. Then an auditory response to a sound made by the person regains the attention of the robot. The system continues to track and mimic in a robust and continuous manner. Some experiments and demonstrations of this system have lasted continuously over 20–30 minutes. For details of the integration of this system see [29].

The basic information used in this experiment is the position of the head in the scene and the motion of the arms. The detection of an upper body of

Detect and track a person   Notice upper body   Start Mimicking

 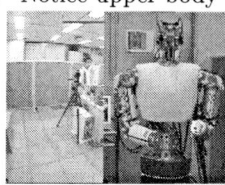

Stop Mimicking   Person, out of sight   Regain attention via sound

Fig. 11. Humanoid Interaction: Visual Attention and Tracking, Auditory Attention and Tracking, Mimicking of a person. This experiment shows the detection of a person, while estimating the upper body motion of the person. Detection of sound, allows our system to determine the spatial orientation of the sound source – spatial hearing.

a person within the environment is based on segmentation of skin color. The process of skin color detection is based on color distance and hue extraction. The overall process is as follows:

1. segmentation is made between the environment and the person.
2. select head, based on a set of attributes (e.g. colour, aspect ratio, depth etc.)
3. extract the arms from the scene based on a set of attributes (e.g. colour, aspect ratio, depth, etc.)

7 Conclusions

The series of system development presented above provide examples of how to choose and efficiently implement biologically inspired processing models for real human interaction systems. The viewpoint of global integration is always important. The hardware design of ESCHeR was done in the context of gaze movement, optics, and vision algorithms. The special lens affects the motion and stereo algorithms. The two major biologically inspired processes, flow and phase disparity, can be computed from a common steerable filter results, which also supports object identification and bottom up attention control. All these aspects are currently being integrated into a coherent humanoid system which interacts with humans in real environment.

Acknowledgments

The present work has been supported partly by the COE program and the Brain Science program funded by the Science and Technology Agency (STA) of Japan and ETL.

References

1. Y. Kuniyoshi, M. Inaba, and H. Inoue. Learning by watching: Extracting reusable task knowledge from visual observation of human performance. *IEEE Trans. Robotics and Automation*, 10(5), 1994.
2. Y. Kuniyoshi and H. Inoue. Qualitative recognition of ongoing human action sequences. In *Proc. IJCAI93*, pages 1600–1609, 1993.
3. Y. Kuniyoshi, N. Kita, S. Rougeaux, and T. Suehiro. Active stereo vision system with foveated wide angle lenses. In S.Z. Li, D.P. Mital, E.K. Teoh, and H. Wang, editors, *Recent Developments in Computer Vision*, Lecture Notes in Computer Science 1035. Springer-Verlag, 1995. ISBN 3-540-60793-5.
4. Y. Kuniyoshi, N. Kita, K. Sugimoto, S. Nakamura, and T. Suehiro. A foveated wide angle lens for active vision. In *Proc. IEEE Int. Conf. Robotics and Automation*, pages 2982–2988, 1995.
5. G. Sandini and V. Tagliaso. An anthropomorphic retina-like structure for scene analysis. *Computer Graphics and Image Processing*, 14(3):365–372, 1980.
6. G. Sandini et al. Giotto: Retina-like camera. Technical report, DIST - University of Genova, 1999.
7. R.S. Wallace, B.B. Bederson, and E.L. Schwartz. A miniaturized active vision system. *International Conference on Pattern Recognition*, pages 58–61, 1992.
8. W. Klarquist and A. Bovik. Fovea: a foveated vergent active stereo system for dynamic three-dimensional scene recovery. In *Proc., IEEE International Conference on Robotics and Automation, Leuven, Belgium*, 1998.
9. M. Peters and A. Sowmya. A real-time variable sampling technique for active vision: Diem. In *Proc., International Conference on Pattern Recognition, Brisbane, Australia*, 1998.
10. A. Wavering, J. Fiala, K. Roberts, and R. Lumia. Triclops: A high performance trinocular active vision system. In *Proc., IEEE International Conference on Robotics and Automation*, pages 3:410–417, 1993.
11. B. Scassellati. A binocular, foveated, active vision system. Technical Report MIT AI Memo 1628, Massachussetts Institute of Technology, January 1998.
12. Y. Suematsu and H. Yamada. A wide angle vision sensor with fovea - design of distortion lens and the simulated inage -. In *Proc., IECON93*, volume 1, pages 1770–1773, 1993.
13. S. Rougeaux and Y. Kuniyoshi. Robust real-time tracking on an active vision head. In *Proceedings of IEEE International Conference on Intelligent Robot and Systems (IROS)*, volume 2, pages 873–879, 1997.
14. S. Rougeaux and Y. Kuniyoshi. Velocity and disparity cues for robust real-time binocular tracking. In *International Conference on Computer Vision and Pattern Recognition , Puerto Rico*, pages 1–6, 1997.
15. J. L. Barron, D. J. Fleet, and S. S. Beauchemin. Performance of optical flow techniques. *International Journal of Computer Vision*, 12(1):43–77, February 1994.
16. B. Lucas and T. Kanade. An iterative image registration technique with an application to stereo vision. In *Proc. DARPA Image Understanding Workshop*, pages 121–130, 1981.
17. D. J. Fleet and K. Langley. Recursive filters for optical flow. *IEEE Transaction on Pattern Analysis and Machine Intelligence*, 17(1):61–67, January 1995.
18. E. P. Simoncelli, E. H. Adelson, and D. J. Heeger. Probability distribution of optical flow. In *Proc. IEEE Conf. on Computer Vision and Pattern Recognition*, pages 310–315, 1991.

19. K.J. Bradshaw, P.F. McLauchlan, I.D. Reid, and D.W. Murray. Saccade and pursuit on an active head/eye platform. *Image and Vision Computing journal*, 12(3):155–163, 1994.
20. D. Coombs and C. Brown. Real-time smooth pursuit tracking for a moving binocular robot. *Computer Vision and Pattern Recognition*, pages 23–28, 1992.
21. T. D. Sanger. Stereo disparity computation using gabor filters. *Biol. Cybern.*, 59:405–418, 1988.
22. D. J. Fleet. Stability of phase information. *Transaction on Pattern Analysis and Machine Intelligence*, 15(12):1253–1268, 1993.
23. J. Y. Weng. Image matching using the windowed fourier phase. *International Journal of Computer Vision*, 11(3):211–236, 1993.
24. C. J. Westelius. *Focus of Attention and Gaze Control for Robot Vision*. PhD thesis, Linkoping University, 1995.
25. O. Stasse, Y. Kuniyoshi, and G. Cheng. Development of a biologically inspired real-time visual attention system. In *Proceedings of IEEE International Workshop on Biologically Motivated Computer Vision (BMCV)*, 2000.
26. K. R. Cave. The featuregate model of visual selection. *(in review)*, 1998.
27. J. A. Driscoll, R. A. Peters II, and K. R. Cave. A visual attention network for a humanoid robot. In *IROS 98*, 1998.
28. O. Stasse and Y. Kuniyoshi. Predn: Achieving efficiency and code re-usability in a programming system for complex robotic applications. In *IEEE International Conference on Robotics and Automation*, 2000.
29. G. Cheng and Y. Kuniyoshi. Complex continuous meaningful humanoid interaction: A multi sensory-cue based approach. In *Proceedings of IEEE International Conference on Robotics and Automation*, April 2000. (to appear).
30. J. Piaget. *La Formation du Symbole chez l'Enfant*. 1945.
31. Y. Kuniyoshi and A. Nagakubo. Humanoid Interaction Approach: Exploring Meaningful Order in Complex Interactions. In *Proceedings of the International Conference on Complex Systems*, 1997.
32. Y. Kuniyoshi and A. Nagakubo. Humanoid As a Research Vehicle Into Flexible Complex Interaction. In *Proceedings of IEEE/RSJ International Conference on Intelligent Robots and Systems (IROS'97)*, 1997.

The Spectral Independent Components of Natural Scenes

Te-Won Lee[1,2], Thomas Wachtler[3], and Terrence J. Sejnowski[1,2]

[1] Institute for Neural Computation, University of California, San Diego, La Jolla, California 92093, USA
[2] Howard Hughes Medical Institute, Computational Neurobiology Laboratory, The Salk Institute, La Jolla, California 92037, USA
[3] Universität Freiburg, Biologie III, Neurobiologie und Biophysik, 79104 Freiburg, Germany

Abstract. We apply independent component analysis (ICA) for learning an efficient color image representation of natural scenes. In the spectra of single pixels, the algorithm was able to find basis functions that had a broadband spectrum similar to natural daylight, as well as basis functions that coincided with the human cone sensitivity response functions. When applied to small image patches, the algorithm found homogeneous basis functions, achromatic basis functions, and basis functions with overall chromatic variation along lines in color space. Our findings suggest that ICA may be used to reveal the structure of color information in natural images.

1 Learning Codes for Color Images

The efficient encoding of visual sensory information is an important task for image processing systems as well as for the understanding of coding principles in the visual cortex. Barlow [1] proposed that the goal of sensory information processing is to transform the input signals such that it reduces the redundancy between the inputs. Recently, several methods have been proposed to learn grayscale image codes that utilize a set of linear basis functions. Olshausen and Field [10] used a sparseness criterion and found codes that were similar to localized and oriented receptive fields. Similar results were obtained in [3,8] using the infomax ICA algorithm and a Bayesian approach respectively. In this paper we are interested in finding efficient color image codes. Analysis of color images have mostly focused on coding efficiency with respect to the postreceptoral signals [4,12]. Buchsbaum et al. [4] found opponent coding to be the most efficient way to encode human photoreceptor signals. In an analysis of spectra of natural scenes using PCA, Ruderman et al. [12] found principal components close to those of Buchsbaum. While cone opponency may give an optimal code for transmitting chromatic information through the bottleneck of the optic nerve, it may not necessarily reflect the chromatic statistics of natural scenes. For example, how the photoreceptor signals should be combined depends on their spectral properties. These however may not be determined solely by the spectral statistics in the environment, but by other functional (effects of infrared or UV sensitivity) or evolutionary (resolution) requirements. Therefore, opponent coding may not be the ultimate goal the visual system wants to achieve. And in fact, it is known that, while neurons in the Lateral Geniculate Nucleus (LGN) of trichromatic primates show responses along the coordinate axes ('cardinal directions') of cone-opponent color space [6], cortical cells do not adhere to these directions [7,13]. This suggest that a different coding scheme may be more appropriate to encode the chromatic structure of natural images. Here, we use ICA to analyze the spectral and spatial properties of natural images.

2 Independent Component Analysis

ICA is a technique for finding a linear non-orthogonal coordinate system in multivariate data. The directions of the axes of this coordinate system are determined by the data's second- and higher-order statistics. The goal of the ICA is to linearly transform the data such that the transformed variables are as statistically independent from each other as possible [2,5]. We assume that a data vector $\mathbf{x}$ can be modeled as a linear superposition of statistically independent source components $\mathbf{s}$ ($p(\mathbf{s}) = \prod_{i=1}^{M} \mathbf{p_i(s_i)}$) such that

$$\mathbf{x} = \mathbf{As}, \quad (1)$$

where $\mathbf{A}$ is a $N \times M$ scalar matrix. The columns of $\mathbf{A}$ are called the basis functions. The learning algorithm can be derived using the information maximization principle [2] or the maximum likelihood estimation formulation. The data likelihood can be expressed as:

$$p(\mathbf{x}|\mathbf{A}) = \frac{\mathbf{p(s)}}{|\det(\mathbf{A})|} \quad (2)$$

Maximizing the log-likelihood with respect to $\mathbf{A}$ and using the natural gradient gives

$$\Delta \mathbf{A} \propto -\mathbf{A}\left[\mathbf{I} - \varphi(\mathbf{s})\mathbf{x}^{\mathbf{T}}\right] \quad (3)$$

where $\varphi(\mathbf{s}) = -\frac{\partial \mathbf{p(s)}/\partial \mathbf{s}}{\mathbf{p(s)}}$. Our primary interest is to learn efficient codes, and we choose a Laplacian prior ($p(s) \propto \exp(-|s|)$) because it captures the sparse structure of coefficients ($\mathbf{s}$) for natural images. This leads to the simple learning rule which we used for our analysis

$$\Delta \mathbf{A} \propto -\mathbf{A}\left[\mathbf{I} - \mathrm{sign}(\mathbf{s})\mathbf{s}^{\mathbf{T}}\right]. \quad (4)$$

3 Independent Components of Single Hyperspectral Pixels

We analyzed a set of four hyperspectral images of natural scenes. The dataset was provided by Parraga et al. (http://www.crs4.it/gjb/ftpJOSA.html). A detailed

Fig. 1. Four hyperspectral color images of natural scenes.

description of the images is given in Parraga et al (1998). Briefly, the data set consists of 29 images. Each image has a size of 256x256 pixels, for which radiance values are given for 31 wavebands, sampled in 10 nm steps between 400 and 700 nm. Pixel size is 0.056x0.056 deg of visual angle. The images were recorded around Bristol, either outdoors, or inside the glass houses of Bristol Botanical Gardens. We chose four of these images which had been obtained outdoors under apparently different illumination conditions (figure 1). Training was done in 1000 sweeps, each using a set of spectra of 40000 pixels, which were chosen randomly from the four images. We used the logarithm of radiance values, as in the study by [12]. The data were not preprocessed otherwise. The resulting basis functions for the pixel spectra are shown in figure 2. The basis functions are plotted in order of decreasing L_2 norm. Figure 2 shows the corresponding relative contributions of the basis functions to the pixel spectra. The first basis function has a broadband spectrum, with a higher contribution in the short wavelength range. Its overall shape resembles typical daylight spectra [14]. Basis functions two to five show properties related to photoreceptor sensitivities: A comparison between the first five basis functions and human cone sensitivities is shown in figure3. Basis functions two and four have peaks that coincide with the peak of the M cone sensitivity. Note that basis functions are rescaled and sign-corrected to have positive peaks in this figure. Basis function three aligns with the short wavelength flank of the L cone sensitivity, and with the long wavelength flank of the M cone sensitivity. Finally, basis function five has a peak beyond the wavelength of the L cone sensitivity, where the difference between L and M cones is largest. These basis functions may represent object reflectances. Osorio et al. [11] showed that the human cone spectra are related to naturally occurring reflectance spectra. The remaining basis functions are mostly very narrow band, and their contributions are small.

4 Independent Components of Hyperspectral Natural Images

To analyze the spatial properties of the data set, we converted the spectra of each pixel to a vector of 3 cone excitation values (long-, medium-, and short-wavelength sensitive). This was done by multiplying the radiance value for each

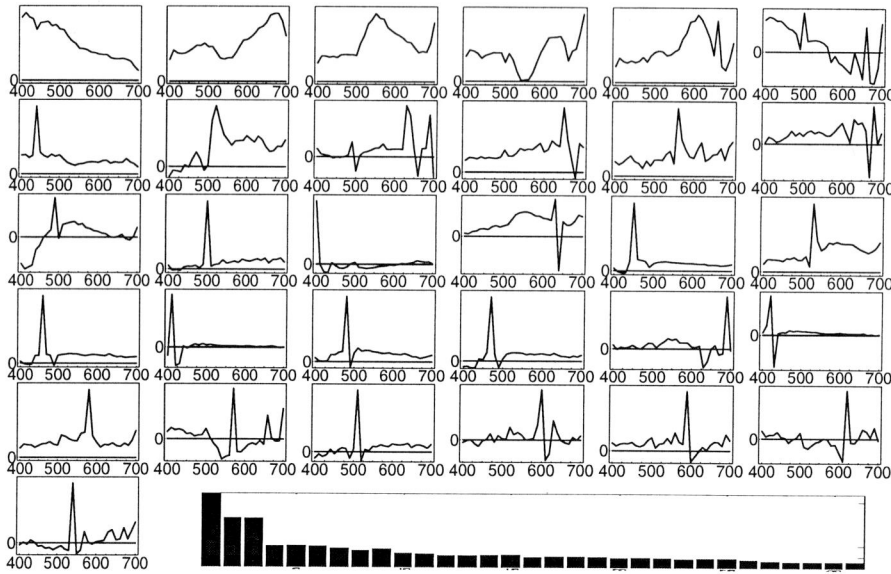

Fig. 2. The spectra of the learned basis functions ordered in decreasing L_2-norm their relative contributions (inset).

wavelength with the corresponding values for the human cone sensitivities as provided by Stockman et al (1993) (http://www-cvrl.ucsd.edu), and summing over the resulting values. From these data, 7x7 image patches were chosen randomly, yielding 7x7x3 = 147 dimensional vectors. Training was done in 500 sweeps, each using a set of spectra of 40000 image patches, which were chosen randomly from the four images. To visualize the resulting components, we plot the 7x7 pixels, with the color of each pixel indicating the combination of L, M, and S cone responses as follows. The values for each patch were normalized to values between 0 and 255, with 0 cone excitation corresponding to a value of 128. This method was used in [12]. Note that the resulting colors are not the colors that would be seen through the corresponding filters. Rather, the red, green, and blue components of each pixel represents the relative excitations of L, M, and S cones, respectively. In figure 4 A, we show the first 100 of the 147 components, ordered by decreasing L_2 norm. The first three, homogeneous, basis functions contribute on average 25% to the intensity of the images. Most of the remaining basis functions are achromatic, localized and oriented filters similar to those found in the analysis of grayscale natural images [3]. There are also many basis functions with color modulated between light blue and dark yellow. For both types of components, low spatial frequency components tend to have higher norm than components with higher spatial frequency. To illustrate the chromatic properties of the filters, we convert the L, M, S values for each pixel to its projection onto the isoluminant plane of cone-opponent color space. This space has been introduced by [9] and generalized to include achromatic colors by

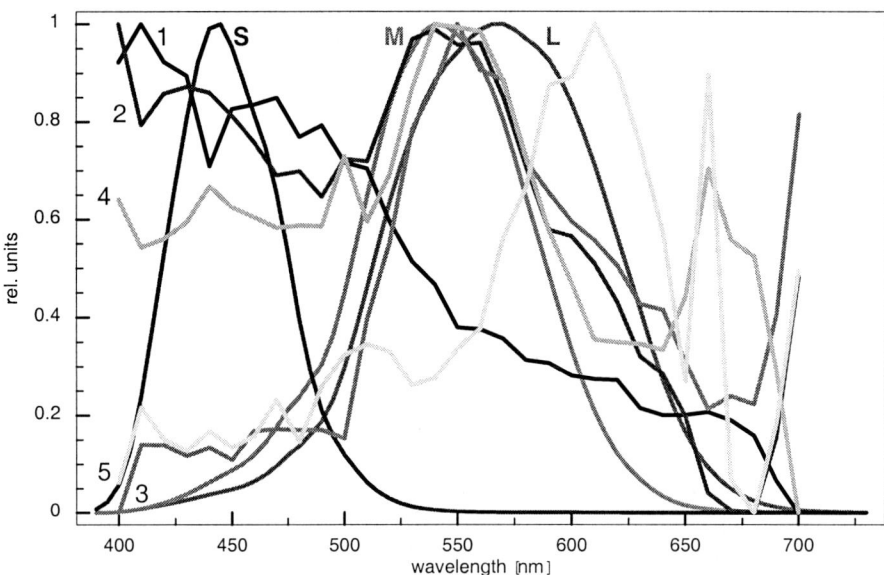

Fig. 3. Comparison between the first five basis functions and the human cone sensitivity response function.

[6]. In our plots, the x axis corresponds to the response of a L cone versus M cone opponent mechanism, the y axis corresponds to S cone modulation. Note that these axes do not coincide with colors we perceive as pure red, green, blue and yellow. For each pixel of the basis functions, we plot a point at its corresponding location in that color space. The color of the points are the same as used for the pixels in figure 4 (top). Thus, although only the projection onto the isoluminant plane is shown, the third dimension can be inferred by the brightness of the points. Interestingly, almost all components show chromatic variation along a line in color space. Only a few, weak, basis functions show color coordinates which do not form a line. The blue-yellow basis functions lie almost perfectly along the vertical S cone axis. The achromatic basis functions lie along lines that are slightly tilted away from this axis. This reflects the direction of variation of natural daylight spectra, whose coordinates in this color space lie along a line which is tilted counterclockwise with respect to the vertical axis. Notably, the yellow end of this line correlates with brighter colors (objects lit by sunlight), the blue end to darker colors (objects in shadow, lit by bluish skylight). The chromatic basis functions except the S cone modulated ones tend to lie along lines with orientations corresponding to (greenish) blue versus yellow/orange. There are no basis functions in the direction of L versus M cone responses (horizontal axis).

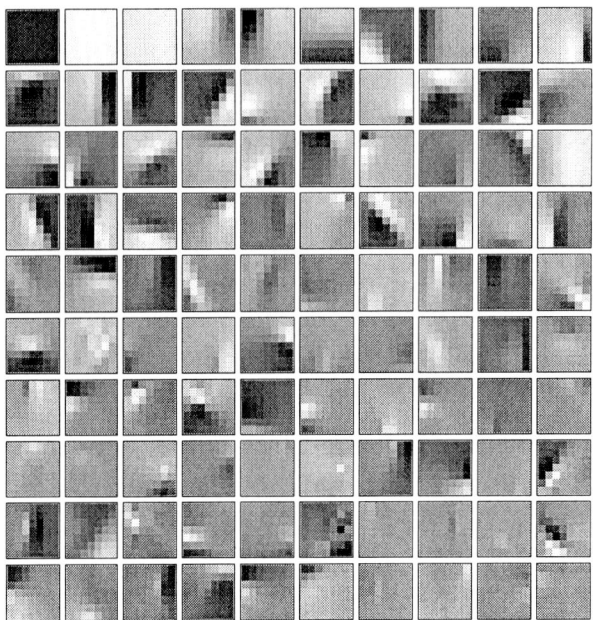

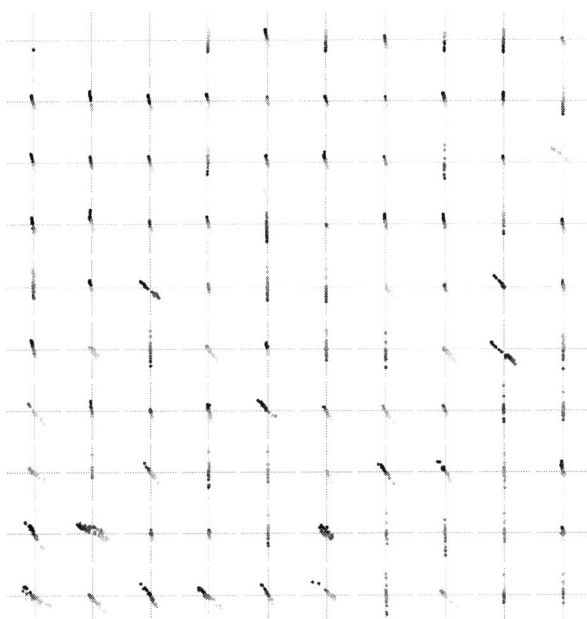

Fig. 4. (Top) 100 of 147 total learned basis functions (7 by 7 pixels and 3 colors) ordered in decreasing L_2-norm. (Bottom) Corresponding color-space diagrams for the 100 basis functions.

5 Discussion

We used ICA to analyze the spectral and spatial properties of natural images. In single hyperspectral pixels, we found that ICA was able to find basis functions with broadband spectra and basis functions related to human cone sensitivities. When applied to small image patches, ICA found homogeneous basis functions, achromatic and chromatic basis functions. Most basis functions showed pronounced opponency, i.e. their components form lines through the origin of color space. However, the directions of these lines do not always coincide with the coordinate axes. While it is known that chromatic properties of neurons in the LGN corresponds to variation restricted to these axes [6], cortical neurons show sensitivities for intermediate directions [7]. This suggests that the opponent coding along the 'cardinal directions' is used by the visual system to transmit visual information to the cortex, where the information is recoded, maybe to better reflect the statistical structure of the visual environment. Interestingly, ICA found only few basis functions with strong red-green opponency. The reason for this may lie in the fact that our images did not contain flowers or other strongly colored objects. Also, chromatic signals that are ecologically important [11] may not be typical or frequent in natural scenes. Using PCA, Ruderman et al. (1998) found components which reflect the opponent mechanisms to decorrelate chromatic signals, given the human cone sensitivities. The basis functions found by PCA are a result of the correlations introduced by the overlapping sensitivities of human cones. In contrast to PCA, ICA tries to discover the underlying statistical structure of the images. Our results are consistent with previously reported results on gray-scale images and we suggest that ICA may be used to reveal the structure of color information in natural images.

Acknowledgments

The authors would like to thank Mike Lewicki for fruitful discussions. We thank C. Parraga, G. Brelstaff, T. Troscianko, and I. Moorehead for providing the hyperspectral image data set.

References

1. H. Barlow. *Sensory Communication*, chapter Possible principles underlying the transformation of sensory messages, pages 217–234. MIT press, 1961.
2. A. J. Bell and T. J. Sejnowski. An Information-Maximization Approach to Blind Separation and Blind Deconvolution. *Neural Computation*, 7:1129–1159, 1995.
3. A. J. Bell and T. J. Sejnowski. The 'independent components' of natural scenes are edge filters. *Vision Research*, 37(23):3327–3338, 1997.
4. G. Buchsbaum and A. Gottschalk. Trichromacy, opponent colours coding and optimum colour information transmission in the retina. *Proceedings of the Royal Society London B*, 220:89–113, 1983.
5. J-F. Cardoso and B. Laheld. Equivariant adaptive source separation. *IEEE Trans. on S.P.*, 45(2):434–444, 1996.

6. A. M. Derrington, J. Krauskopf, and P. Lennie. Chromatic mechanisms in lateral geniculate nucleus of macaque. *Journal of Physiology*, 357:241–265, 1984.
7. P. Lennie, J. Krauskopf, and G. Sclar. Chromatic mechanisms in striate cortex of macaque. *Journal of Neuroscience*, 10:649–669, 1990.
8. M.S. Lewicki and B. Olshausen. A probablistic framwork for the adaptation and comparison of image codes. *J. Opt.Soc., A: Optics, Image Science and Vision*, in press, 1999.
9. D. I. A. MacLeod and R. M. Boynton. Chromaticity diagram showing cone excitation by stimuli of equal luminance. *Journal of the Optical Society of America*, 69:1183–1186, 1979.
10. B. Olshausen and D. Field. Emergence of simple-cell receptive field properties by learning a sparse code for natural images. *Nature*, 381:607–609, 1996.
11. D. Osorio and T. R. J. Bossomaier. Human cone-pigment spectral sensitivities and the reflectances of natural scenes. *Biological Cybernetics*, 67:217–222, 1992.
12. D. L. Ruderman, T. W. Cronin, and C.-C. Chiao. Statistics of cone responses to natural images: Implications for visual coding. *Journal of the Optical Society of America A*, 15:2036–2045, 1998.
13. T. Wachtler, T. J. Sejnowski, and T. D. Albright. Interactions between stimulus and background chromaticities in macaque primary visual cortex. *Investigative Ophthalmology & Visual Science*, 40:S641, 1999. ARVO abstract.
14. G. Wyszecki and W. S. Stiles. *Color Science: Concepts and Methods, Quantitative Data and Formulae.* Wiley, New York, 2nd ed. edition, 1982.

Topographic ICA as a Model of Natural Image Statistics

Aapo Hyvärinen, Patrik O. Hoyer, and Mika Inki

Neural Networks Research Centre
Helsinki University of Technology
P.O. Box 5400, FIN-02015 HUT, Finland
http://www.cis.hut.fi/projects/ica/

Abstract. Independent component analysis (ICA), which is equivalent to linear sparse coding, has been recently used as a model of natural image statistics and V1 receptive fields. Olshausen and Field applied the principle of maximizing the sparseness of the coefficients of a linear representation to extract features from natural images. This leads to the emergence of oriented linear filters that have simultaneous localization in space and in frequency, thus resembling Gabor functions and V1 simple cell receptive fields. In this paper, we extend this model to explain emergence of V1 topography. This is done by ordering the basis vectors so that vectors with strong higher-order correlations are near to each other. This is a new principle of topographic organization, and may be more relevant to natural image statistics than the more conventional topographic ordering based on Euclidean distances. For example, this topographic ordering leads to simultaneous emergence of complex cell properties: neighbourhoods act like complex cells.

1 Introduction

A fundamental approach in signal processing is to design a statistical generative model of the observed signals. Such an approach is also useful for modeling the properties of neurons in primary sensory areas. Modeling visual data by a simple linear generative model, Olshausen and Field [12] showed that the principle of maximizing the sparseness (or supergaussianity) of the underlying image components is enough to explain the emergence of Gabor-like filters that resemble the receptive fields of simple cells in mammalian primary visual cortex (V1). Maximizing sparseness is in this context equivalent to maximizing the independence of the image components [3,1,12].

We show in this paper that this same principle can be extended to explain the emergence of both topography and complex cell properties as well. This is possible by maximizing the sparseness of *local energies* instead of the coefficients of single basis vectors. This is motivated by a generative model in which a topographic organization of basis vectors is assumed, and the coefficients of near-by basis vectors have higher-order correlations in the form of dependent variances. (The coefficients are linearly uncorrelated, however.) This gives a topographic map where the distance of the components in the topographic representation is a function of the dependencies of the components. Moreover, neighbourhoods have invariance properties similar to those of complex cells. We derive a learning rule for the estimation of the model, and show its utility by experiments on natural image data.

2 ICA of Image Data

The basic models that we consider here express a static monochrome image $I(x,y)$ as a linear superposition of some features or basis functions $a_i(x,y)$:

$$I(x,y) = \sum_{i=1}^{m} a_i(x,y) s_i \qquad (1)$$

where the s_i are stochastic coefficients, different for each image $I(x,y)$. The crucial assumption here is that the s_i are nongaussian, and mutually independent. This type of decomposition is called independent component analysis (ICA) [3,1,9], or from an alternative viewpoint, sparse coding [12].

Estimation of the model in Eq. (1) consists of determining the values of s_i and $a_i(x,y)$ for all i and (x,y), given a sufficient number of observations of images, or in practice, image patches $I(x,y)$. We restrict ourselves here to the basic case where the $a_i(x,y)$ form an invertible linear system. Then we can invert the system as

$$s_i = <w_i, I> \qquad (2)$$

where the w_i denote the inverse filters, and $<w_i, I> = \sum_{x,y} w_i(x,y) I(x,y)$ denotes the dot-product. The $w_i(x,y)$ can then be identified as the receptive fields of the model simple cells, and the s_i are their activities when presented with a given image patch $I(x,y)$. Olshausen and Field [12] showed that when this model is estimated with input data consisting of patches of natural scenes, the obtained filters $w_i(x,y)$ have the three principal properties of simple cells in V1: they are localized, oriented, and bandpass. Van Hateren and van der Schaaf [15] compared quantitatively the obtained filters $w_i(x,y)$ with those measured by single-cell recordings of the macaque cortex, and found a good match for most of the parameters.

3 Topographic ICA

In classic ICA, the independent components s_i have no particular order, or other relationships. The lack of an inherent order of independent components is related to the assumption of complete statistical independence. When applying ICA to modeling of natural image statistics, however, one can observe clear violations of the independence assumption. It is possible to find, for example, couples of estimated independent components such that they are clearly dependent on each other.

In this section, we define topographic ICA using a generative model that is a hierarchical version of the ordinary ICA model. The idea is to relax the assumption of the independence of the components s_i in (1) so that components that are close to each other in the topography are not assumed to be independent in the model. For example, if the topography is defined by a lattice or grid, the dependency of the components is a function of the distance of the components on that grid. In contrast, components that are not close to each other in the topography *are* independent, at least approximately; thus most pairs of components are independent.

3.1 What Kind of Dependencies Should Be Modelled?

The basic problem is then to choose what kind of dependencies are allowed between near-by components. The most basic dependence relation is linear correlation. However, allowing linear correlation between the components does not seem very useful. In fact, in many ICA estimation methods, the components are constrained to be uncorrelated [3,9,6], so the requirement of uncorrelatedness seems natural in any extension of ICA as well.

A more interesting kind of dependency is given by a certain kind of higher-order correlation, namely correlation of energies. This means that

$$\text{cov }(s_i^2, s_j^2) = E\{s_i^2 s_j^2\} - E\{s_i^2\}E\{s_j^2\} \neq 0 \qquad (3)$$

if s_i and s_j are close in the topography. Here, we assume that this covariance is positive. Intuitively, such a correlation means that the components tend to be active, i.e. non-zero, at the same time, but the actual values of s_i and s_j are not easily predictable from each other. For example, if the variables are defined as products of two zero-mean independent components z_i, z_j and a common "variance" variable σ:

$$s_i = z_i \sigma$$
$$s_j = z_j \sigma \qquad (4)$$

then s_i and s_j are uncorrelated, but their energies are not. In fact the covariance of their energies equals $E\{z_i^2 \sigma^2 z_j^2 \sigma^2\} - E\{z_i^2 \sigma^2\}E\{z_j^2 \sigma^2\} = E\{\sigma^4\} - E\{\sigma^2\}^2$, which is positive because it equals the variance of σ^2 (we assumed here for simplicity that z_i and z_j are of unit variance). This kind of a dependence has been observed, for example, in linear image features [13,8]; it is illustrated in Fig. 1.

3.2 The Generative Model

Now we define a generative model that implies correlation of energies for components that are close in the topographic grid. In the model, the observed image patches are generated as a linear transformation of the components s_i, just as in the basic ICA model in (1). The point is to define the joint density of **s** so that it expresses the topography.

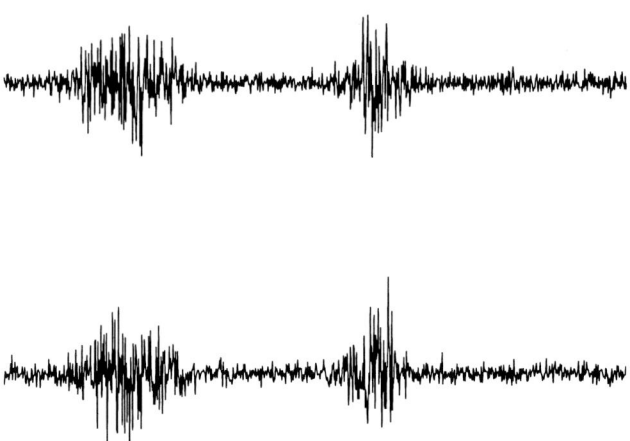

Fig. 1. Illustration of higher-order dependencies. The two signals in the figure are uncorrelated but they are not independent. In particular, their energies are correlated. The signals were generated as in (4), but for purposes of illustration, the variable σ was replaced by a time-correlated signal.

We define the joint density of **s** as follows. The variances σ_i^2 of the s_i are not constant, instead they are assumed to be random variables, generated according to a model to be specified. After generating the variances, the variables s_i are generated independently from each other, using some conditional distributions to be specified. In other words, the s_i are *independent given their variances*. Dependence among the s_i is implied by the dependence of their variances. According to the principle of topography, the variances corresponding to near-by components should be (positively) correlated, and the variances of components that are not close should be independent, at least approximatively.

To specify the model for the variances σ_i^2, we need to first define the topography. This can be accomplished by a neighborhood function $h(i,j)$, which expresses the proximity between the i-th and j-th components. The neighborhood function can be defined in the same ways as with the self-organizing map [10]. The neighborhood function $h(i,j)$ is thus a matrix of hyperparameters. In this paper, we consider it to be known and fixed.

Using the topographic relation $h(i,j)$, many different models for the variances σ_i^2 could be used. We prefer here to define them by an ICA model followed by a nonlinearity:

$$\sigma_i = \phi(\sum_{k=1}^{n} h(i,k) u_k) \qquad (5)$$

where u_i are the "higher-order" independent components used to generate the variances, and ϕ is some scalar nonlinearity. The distributions of the u_i and the actual form of ϕ are additional hyperparameters of the model; some suggestions will be given below. It seems natural to constrain the u_k to be non-negative. The function ϕ can then be constrained to be a monotonic transformation in the set of non-negative real numbers. This assures that the σ_i's are non-negative. The resulting topographic ICA model is summarized in Fig. 2.

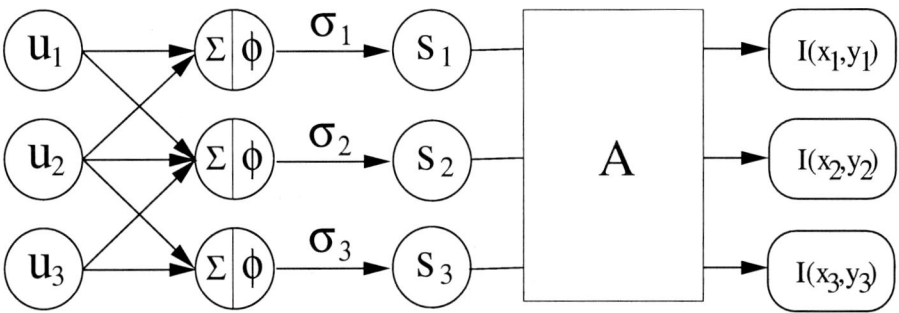

Fig. 2. An illustration of the topographic ICA model. First, the "variance-generating" variables u_i are generated randomly. They are then mixed linearly inside their topographic neighborhoods. The mixtures are then transformed using a nonlinearity ϕ, thus giving the local variances σ_i. Components s_i are then generated with variances σ_i. Finally, the components s_i are mixed linearly to give the observed image patches I.

3.3 Basic Properties of the Topographic ICA Model

The model as defined above has the following properties. (Here, we consider for simplicity only the case of sparse, i.e. supergaussian, data.) The first basic property is that all the components s_i are uncorrelated, as can be easily proven by symmetry arguments [7]. Moreover, their variances can be defined to be equal to unity, as in classic ICA. Second, components s_i and s_j that are near to each other, i.e. such that $h(i, j)$ is significantly non-zero, tend to be active (non-zero) at the same time. In other words, their energies s_i^2 and s_j^2 are positively correlated. This is exactly the dependence structure that we wanted to model in the first place. Third, latent variables that are far from each other are practically independent. Higher-order correlation decreases as a function of distance. For details, see [7].

4 Estimation of the Model

4.1 Approximating Likelihood

To estimate the model, we can use maximum likelihood estimation. The model is, however, a missing variables model in which the likelihood cannot be obtained in closed form. To simplify estimation, we can derive a tractable approximation of the likelihood.

To obtain the approximation, we make the following assumptions. First, we fix the conditional density p_i^s of the components s_i to be gaussian. Note that this does not imply that the unconditional density of s_i were gaussian; in fact, the unconditional density is supergaussian [7]. Second, we define the nonlinearity ϕ as $\phi(y) = y^{-1/2}$. Third, we assume for simplicity that the p_i^s are the same for all i. The main motivation for these assumptions is algebraic simplicity that makes a simple approximation possible. These choices are compatible with natural image statistics, if the variable u_i are taken from distributions with very heavy tails.

Thus we obtain [7] the following approximation of the log-likelihood, for T observed image patches $I_t, t = 1, ..., T$:

$$\log \tilde{L}(w_i, i=1,...,n) = \sum_{t=1}^{T} \sum_{j=1}^{n} G(\sum_{i=1}^{n} h(i,j) < w_i, I_t >^2) + T \log |\det \mathbf{W}|. \quad (6)$$

where the scalar function G is obtained from the pdf's of p_u, the variance-generating variables, by:

$$G(y) = \log \int \frac{1}{\sqrt{2\pi}} \exp(-\frac{1}{2}uy) p_u(u) \sqrt{h(i,i)u} \, du, \quad (7)$$

and the matrix $\mathbf{W}$ contains the vectors $w_i(x,y)$ as its columns.

Note that the approximation of likelihood is a function of local energies. Every term $\sum_{i=1}^{n} h(i,j) < w_i, I >^2$ could be considered as the energy of a neighborhood, possibly the output of a higher-order neuron as in visual complex cell models [8].

The function G has a similar role as the log-density of the independent components in classic ICA. If the data is sparse (like natural image data), the function $G(y)$ needs to be chosen to be *convex* for non-negative y [7]. For example, one could use the function:

$$G_1(y) = -\alpha_1 \sqrt{y} + \beta_1, \quad (8)$$

This function is related to the exponential distibution [8]. The scaling constant α_1 and the normalization constant β_1 are determined so as to give a probability density that is compatible with the constraint of unit variance of the s_i, but they are irrelevant in the following.

4.2 Learning Rule

The approximation of the likelihood given above enables us to derive a simple gradient learning rule. First, we assume here that the data is preprocessed by whitening and that the w_i, are constaned to form an orthonormal system [3,9,6]. This implies that the estimates of the components are uncorrelated. Such a simplification is widely used in ICA, and it is especially useful here since it allows us to concentrate on higher-order correlations.

Thus we can simply derive a gradient algorithm in which the i-th (weight) vector w_i is updated as

$$\Delta w_i(x,y) \propto E\{I(x,y) < w_i, I > r_i)\} \quad (9)$$

where

$$r_i = \sum_{k=1}^{n} h(i,k) g(\sum_{j=1}^{n} h(k,j) < w_j, I >^2). \quad (10)$$

The function g is the derivative of G. After every step of (9), the vectors w_i are normalized to unit variance and orthogonalized, as in [9,6], for example.

In a neural interpretation, the learning rule in (9) can be considered as "modulated" Hebbian learning, since the learning term is modulated by the term r_i. This term could be considered as top-down feedback as in [8], since it is a function of the local energies which could be the outputs of higher-order neurons (complex cells).

4.3 Connection to Independent Subspace Analysis

Another closely related modification of the classic ICA model was introduced in [8]. As in topographic ICA, the components s_i were not assumed to be all mutually independent. Instead, it was assumed that the s_i can be divided into couples, triplets or in general n-tuples, such that the s_i inside a given n-tuple could be dependent on each other, but dependencies between different n-tuples were not allowed. A related relaxation of the independence assumption was proposed in [2,11]. Inspired by Kohonen's principle of feature subspaces [10], the probability densities for the n-tuples of s_i were assumed in [8] to be *spherically symmetric*, i.e. depend only on the norm.

In fact, topographic ICA can be considered a generalization of the model of independent subspace analysis. The likelihood in independent subspace analysis can be expressed as a special case of the approximative likelihood (6), see [7]. This connection shows that topographic ICA is closely connected to the principle of invariant-feature subspaces. In topographic ICA, the invariant-feature subspaces, which are actually no longer independent, are completely overlapping. Every component has its own neighborhood, which could be considered to define a subspace. Each of the terms $\sum_{i=1}^{n} h(i,j) < w_i, I >^2$ could be considered as a (weighted) projection on a feature subspace, i.e. as the value of an invariant feature. Neurophysiological work that shows the same kind of connection between topography and complex cells is given in [4].

5 Experiments

We applied topographic ICA on natural image data. The data was obtained by taking 16×16 pixel image patches at random locations from monochrome photographs depicting wild-life scenes (animals, meadows, forests, etc.). For details on the experiments, see [7]. The mean gray-scale value of each image patch (i.e. the DC component) was subtracted. The data was then low-pass filtered by reducing the dimension of the data vector by principal component analysis, retaining the 160 principal components with the largest variances, after which the data was whitened by normalizing the variances of the principal components. These preprocessing steps are essentially similar to those used in [8,12,15]. In the results shown above, the inverse of these preprocessing steps was performed.

The neighborhood function was defined so that every neighborhood consisted of a 3×3 square of 9 units on a 2-D torus lattice [10]. The choice of a 2-D grid was here motivated by convenience of visualization only; further research is needed to see what the "intrinsic dimensionality" of natural image data could be. The function G was chosen as in (8). The approximation of likelihood in Eq. (6) for 50,000 observations was maximized under the constraint of orthonormality of the filters in the whitened space, using the gradient learning rule in (9).

The obtained basis vectors a_i are shown in Fig. 3. The basis vectors are similar to those obtained by ordinary ICA of image data [12,1]. In addition, they have a clear topographic organization.

The connection to independent subspace analysis [8], which is basically a complex cell model, can also be found in these results. Two neighboring basis vectors in Fig. 3 tend to be of the same orientation and frequency. Their locations are near to each other as well. In contrast, their phases are very different. This means that a neighborhood of such basis vectors, i.e. simple cells, functions as a complex cell: The local energies that are summed in the approximation of the likelihood in (6) can be considered as the outputs of complex cells. Likewise, the feedback r_i in the learning rule could be considered as coming from complex cells. For details, see [8,7].

6 Discussion

6.1 Comparison with Some Other Topographic Mappings

Our method is different from ordinary topographic mappings [10] in several ways. First, according to its definition, topographic ICA attempts to find a decomposition into components that are independent. This is because only near-by components are not independent, at least approximately, in the model. In contrast, most topographic mappings choose the representation vectors by principles similar to vector quantization and clustering [5,14].

Most interestingly, the very principle defining topography is different in topographic ICA and most topographic maps. Usually, the similarity of vectors in the data space is defined by Euclidean geometry: either the Euclidean distance, or the dot-product, as in the "dot-product Self-Organizing Map" [10]. In topographic ICA, the similarity of two vectors in the data space is defined by their higher-order correlations, which cannot be expressed as Euclidean relations. In fact, our principle makes it possible to define a topography even among a set of orthogonal vectors, whose Euclidean distances are all equal.

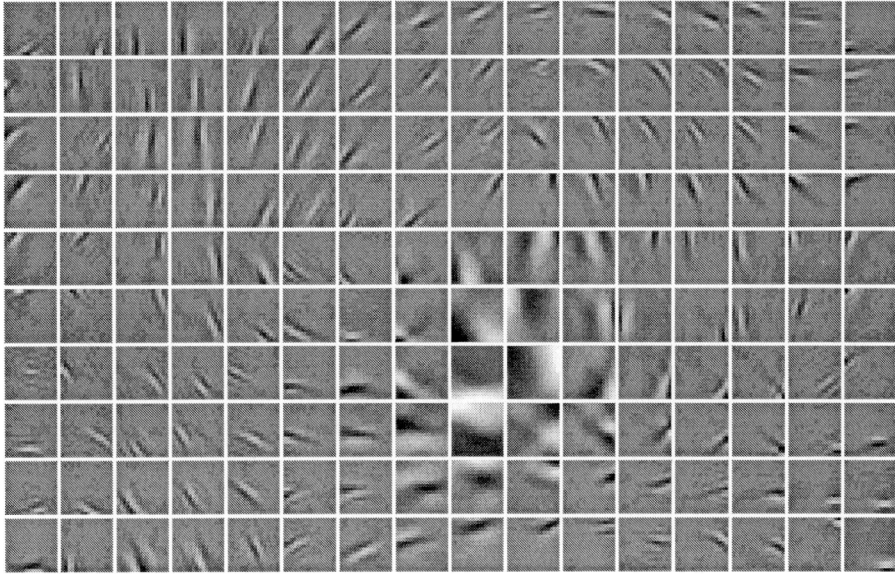

Fig. 3. Topographic ICA of natural image data. The model gives Gabor-like basis vectors for image windows. Basis vectors that are similar in location, orientation and/or frequency are close to each other. The phases of nearby basis vectors are very different, giving the neighborhoods properties similar to those of complex cells.

6.2 Conclusion

To avoid some problems associated with modelling natural image statistics with independent component analysis, we introduced a new model. Topographic ICA is a generative model that combines topographic mapping with ICA. As in all topographic mappings, the distance in the representation space (on the topographic "grid") is related to the distance of the represented components. In topographic ICA, the distance between represented components is defined by the mutual information implied by the higher-order correlations, which gives the natural distance measure in the context of ICA.

Applied on natural image data, topographic ICA gives a linear decomposition into Gabor-like linear features. In contrast to ordinary ICA, the higher-order dependencies that linear ICA could not remove define a topographic order such that near-by cells tend to be active at the same time. This implies also that the neighborhoods have properties similar to those of complex cells. The model thus shows simultaneous emergence of complex cell properties and topographic organization. These two properties emerge from the very same principle of defining topography by simultaneous activation of neighbours.

References

1. A.J. Bell and T.J. Sejnowski. The 'independent components' of natural scenes are edge filters. *Vision Research*, 37:3327–3338, 1997.
2. J.-F. Cardoso. Multidimensional independent component analysis. In *Proc. IEEE Int. Conf. on Acoustics, Speech and Signal Processing (ICASSP'98)*, Seattle, WA, 1998.
3. P. Comon. Independent component analysis – a new concept? *Signal Processing*, 36:287–314, 1994.
4. G. C. DeAngelis, G. M. Ghose, I. Ohzawa, and R. D. Freeman. Functional microorganization of primary visual cortex: Receptive field analysis of nearby neurons. *Journal of Neuroscience*, 19(10):4046–4064, 1999.
5. G. J. Goodhill and T. J. Sejnowski. A unifying objective function for topographic mappings. *Neural Computation*, 9(6):1291–1303, 1997.
6. A. Hyvärinen. Fast and robust fixed-point algorithms for independent component analysis. *IEEE Trans. on Neural Networks*, 10(3):626–634, 1999.
7. A. Hyvärinen and P. O. Hoyer. Topographic independent component analysis. 1999. Submitted, available at http://www.cis.hut.fi/~aapo/.
8. A. Hyvärinen and P. O. Hoyer. Emergence of phase and shift invariant features by decomposition of natural images into independent feature subspaces. *Neural Computation*, 2000. (in press).
9. A. Hyvärinen and E. Oja. A fast fixed-point algorithm for independent component analysis. *Neural Computation*, 9(7):1483–1492, 1997.
10. T. Kohonen. *Self-Organizing Maps*. Springer-Verlag, Berlin, Heidelberg, New York, 1995.
11. J. K. Lin. Factorizing multivariate function classes. In *Advances in Neural Information Processing Systems*, volume 10, pages 563–569. The MIT Press, 1998.
12. B. A. Olshausen and D. J. Field. Natural image statistics and efficient coding. *Network*, 7(2):333–340, May 1996.
13. E. P. Simoncelli and O. Schwartz. Modeling surround suppression in V1 neurons with a statistically-derived normalization model. In *Advances in Neural Information Processing Systems 11*, pages 153–159. MIT Press, 1999.
14. N. V. Swindale. The development of topography in the visual cortex: a review of models. *Network*, 7(2):161–247, 1996.
15. J. H. van Hateren and A. van der Schaaf. Independent component filters of natural images compared with simple cells in primary visual cortex. *Proc. Royal Society ser. B*, 265:359–366, 1998.

Independent Component Analysis of Face Images

Pong C. Yuen[1] and J. H. Lai[1,2]

[1] Department of Computer Science
Hong Kong Baptist University, Hong Kong
{pcyuen, jhlai}@comp.hkbu.edu.hk

[2] Department of Mathematics
Zhongshan University, China
stsljh@zsu.edu.cn

Abstract. This paper addresses the problem of face recognition using independent component analysis. As the independent components (IC) are not orthogonal, to represent a face image using the determined ICs, the ICs have to be orthogonalized, where two methods, namely Gram-Schmit Method and Householder Transformation, are proposed. In addition, to find a better set of ICs for face recognition, an efficient IC selection algorithm is developed. Face images with different facial expressions, pose variations and small occlusions are selected to test the ICA face representation and the results are encouraging.

1 Introduction

Bartlett and Sejnowski [1] first proposed to use independent component analysis (ICA) for face representation. They found that the recognition accuracy using ICA basis vectors is higher that from the principal component analysis (PCA) basis vectors using 200 face images. They also found that ICA representation of faces had greater invariance to change in pose.

From the theoretical point of view, ICA offers two additional advantages over PCA. First, ICA is a generalization of the PCA approach. ICA [2] decorrelates higher-order statistics from the training images, while PCA decorrelates up to 2nd order statistics only. Barlett and Sejnowski [1] demonstrated that much of the important information for image recognition is contained in high-order statistics. As such, representational bases that decorrelate high-order statistics provide better accuracy than those that decorrelate low-order statistics.

Secondly, ICA basis vectors are more spatially local than the PCA basis vectors. This property is particular useful for recognition. As face is a non-rigid object, local representation of faces will reduce the sensitivity of the face variations such as facial expressions, small occlusion and pose variations. That means, some independent components (IC) are invariant under such variations.

Follows the direction of Barlett and Sejnowski, we further develop an IC selection algorithm. Moreover, in [1], all the testing images are within training set. In order to represent images outside training set, the independent components have to be orthogonalized.

2 Review on ICA

The objective of ICA is to represent a set of multidimensional measurement vectors in a basis where the components are statistically independent or as independent as possible. In the simplest form of ICA [2], there are m scalar random variables x_1, x_2, ..., x_m to be observed, which are assumed to be linear combinations of n unknown independent components s_1, s_2, ..., s_n. The independent components are mutually statistically independent and zero-mean. We will denote the observed variables x_i as a observed vector $\mathbf{X}=(x_1, x_2, ..., x_m)^T$ and the component variables s_i as a vector $\mathbf{S}=(s_1, s_2, ..., s_n)$, respectively. Then the relation between $\mathbf{S}$ and $\mathbf{X}$ can be modeled as

$$\mathbf{X}=\mathbf{AS} \qquad (1)$$

where, $\mathbf{A}$ is an unknown m×n matrix of full rank, called the mixing matrix. The columns of $\mathbf{A}$ represent features, and s_i signals the amplitude of the i-th feature in the observed data x. If the independent components s_i have unit variance, it will make independent components unique, up to their signs.

The current implementation algorithms for ICA can be divided into two approaches. One [2] rely on minimizing or maximizing some relevant criteria functions. The problem in this approach is that it requires very complex matrix or tensor operations. Another approach [3] contains adaptive algorithms often based on stochastic gradient methods. The limitation of this approach is slow convergence. Hyvärinen et al. [4] introduce an algorithm using a very simple, yet highly efficient, fixed-point iteration scheme for finding the local extrema of the kurtosis of a linear combination of the observed variables. This fixed-point algorithm is adopted in this paper.

3 Selection of Independent Components for Face Representation

According to the ICA theory, the matrix $\mathbf{S}$ contains all the independent components, which are calculated from a set of training face images, $\mathbf{X}$. The matrix $\mathbf{AS}$ can reconstruct the original signals $\mathbf{X}$. So $\mathbf{A}$ only represents a crude weighted components of the original images. If $\mathbf{X}$ contains individuals with different variations, we can select some independent components (IC) using $\mathbf{A}$. The ICs should be able to reflect the similarity for the same individual while they posses the discriminative power for face images with different persons. To achieve this goal, we have the following two criteria in selecting ICs from $\mathbf{A}$.

The distance within the same class (same person under different variations) should be minimized

The distance between different classes (different persons) should be maximized.

If the matrix **X** is constructed by n individual persons and each person has m variations, the mixing matrix **A** shows the "weighting" of each independent component in **S**. Let a_{ij} represents the entry at the ith column and the jth row. The mean of within-class distance is then given by,

$$W = \frac{1}{nm(m-1)} \sum_{i=1}^{n} \sum_{u=1}^{m} \sum_{v=1}^{m} (a_{i,ixm+u} - a_{i,ixm+v}) \qquad (2)$$

The mean of between-class distance:

$$\mathbf{B} = \frac{1}{n(n-1)} \sum_{s=1}^{n} \sum_{t=1}^{n} (\overline{a}_s - \overline{a}_t) \qquad (3)$$

where,

$$\overline{a}_i = \frac{1}{m} \sum_{j=1}^{m} a_{ij}$$

In this paper, we employ the ratio of within-class distance and between-class distance to select stable mixing features of **A**. The ratio γ is defined as

$$\gamma = \frac{W}{B} \qquad (4)$$

From the definition γ, the smaller the ratio γ is, the better effect the classifier will be.

Using equation (3), each column feature in A is ranked accounting to the γ value. We will only select the top r ICs on the list for recognition, where r < n.

4 Face Recognition Using ICA

This section is divided into two parts. In the first part, we present how the ICA can be employed to recognize images inside training set. The second part discusses the recognition of images outside training set.

4.1 Recognition of Face Image Inside Training Set

The performance for recognition of face image inside training set is used as a reference index. Given n individuals and each individual has m variations, the matrix **X** can be constructed. Each individual is represented by a row in **X**. Using the fixed-point algorithm [3], the mixing matrix **A** can be calculated. Each row in **A** represents the ICA features of the corresponding row in **X**. For the sake of recognition, a test image is recognized by assigning it the label of the nearest of the other (nxm − 1) images in Euclidean distance.

4.2 Recognition of Face Image Outside Training Set

In practice, we are going to recognize face images outside training set. What we need to do is to represent face image using the determined independent components **S** in the training stage as described in section 4.1. However, the independent components (IC) in **S** are not orthogonal. To represent face images outside training set, the ICs have to be orthogonalized. To do this, two methods are proposed in this paper, namely Gram-Schmit method and Householder Transformation. A brief description on these two methods is as follows.

Gram-Schmit Method

Given a matrix **S** which contains a set of independent components $s_1, s_2, s_3,...,s_n$. Gram-Schmit method [7] is to find an orthonormal set of bases, $\beta_1, \beta_2, \beta_3,..., \beta_n$ from **S** and described as follows.

$$\beta_i = \begin{cases} s_1 / \|s_1\| & i = 1 \\ \alpha_i / \|\alpha_i\| & i > 1 \end{cases}$$

where

$$\alpha_i = s_i - \sum_{k=1}^{i-1} <s_i, \beta_k> \beta_k$$

To represent a face image x_i using the orthonormalized ICs, the face image is projected into the space that is spaned by orthonormal basis $\{\beta_j : j=1,2,...,n\}$. We will get the representation vector $a_i = (a_{i1}, a_{i2},..., a_{in})$, where

$$a_{ij} = x \cdot \beta_j \qquad j=1,2,...,n$$

The feature vector a_i represents the image x_i and is used for distance measurement with other reference images in the domain of interest.

Householder Transformation

Householder transformation is another method to determine a set of orthonormal bases from the independent matrix **S**. For each face image $x_i = (x_{i1}, x_{i2},..., x_{ir})$, the equation $x_i = a_i S$ is an over-determined equation, where $a_i = (a_{i1}, a_{i2},..., a_{ir})$ is a vector. Although the equation has no exact solution, it has a single least square solution. The least square solution of the equation $x_i = a_i S$, $a_i \in R^n$ is obtained as follows,

$$\|a_i S' - x_i\|_2 = \min\{\|b_i S' - x_i\|_2 : b_i \in R^{n_1}\}$$

The least square solution can be acquired using Householder Transformation [7]. Again, the feature vector a_i represents the image x_i and will be used for recognition.

5 Experimental Results

The preliminary experimental results presented in this section are divided into three parts. The first part shows how to select the independent components for recognition. The second part demonstrates that the ICA representation is less sensitive to the face pose variations. The final part demonstrates that the ICA representation is less sensitive to the illuminations.

5.1 Implementation Details

Two standard databases from Olivetti research laboratory and Yale University are selected to evaluate the recognition accuracy of the proposed method. Yale database is selected because this database contains all frontal view face images with different facial expression and illuminations. Yale database consists of 15 persons and each person has 11 different views that represent various expressions, illumination conditions and small occlusion as shown in Figure 1. Olivetti database contains face images with different orientations, which are used to demonstrate the performance of the proposed method on pose variations. There are 40 persons Olivetti database and each person consists of 10 different facial views that represent various expressions, small occlusion (by glasses), different scales and orientations as shown in Figure 2.

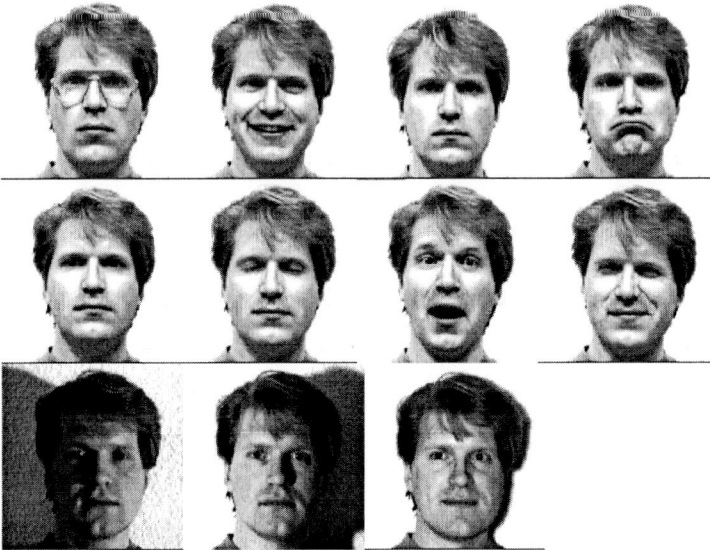

Fig. 1. Various view images of one person in the Yale database

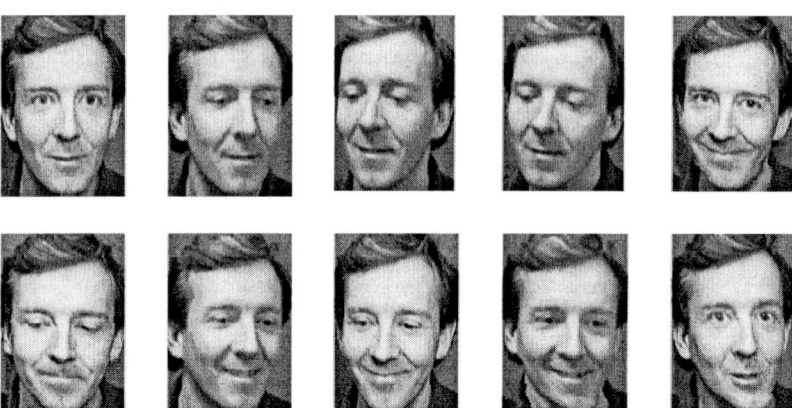

Fig. 2. Various view images of one person in Olivetti database

In some experiments, we also want to evaluate the effect of image resolution against the accuracy. To reduce the resolution, wavelet decomposition is employed. Generally speaking, all wavelet transforms with smooth, compactly support, orthogonality (or biorthogonality) can be used. Through out this paper, the well-known Daubechies wavelet D4 is adopted.

5.2 Selection of Independent Components

We have developed a criterion based on equation (3) to determine a subset of independent components (IC) which have high interclass distance and low intraclass distance. However, Using this criterion, the optimal case is to eliminate all the independent components, as the ratio will be equal to zero. Instead, we want to find a lowest ratio value while the performance is good. We select the Olivetti database to do an experiment. In this experiment, we calculate the recognition accuracy for each ratio from 0.5 to 1.5. The results are plotted in Figure 3. It is found that the accuracy increases when the ratio of the selected ICs increases. The accuracy reaches the maximum when the ratio is around 0.8 and then decreases when the ratio further increases. From this experiment, we can find out the optimal value of the ratio (0.8).

5.3 Sensitivity to Pose Variations

The objective of this section is to demonstrate that ICA representation is less sensitive to pose variations. Continue the experiment in section 5.2, we select 5 images per persons to determine the independent components. Then we select those ICs with the ratio values smaller than or equal to 0.8 for recognition. Using the Gram-Schmit Method and Householder Transformation, these ICs are orthonormalized. The remaining five face images are used for testing. These images are projected onto the orthonormalized IC bases. An Euclidean distance measurement is performed for classification. The results are shown in Table 1.

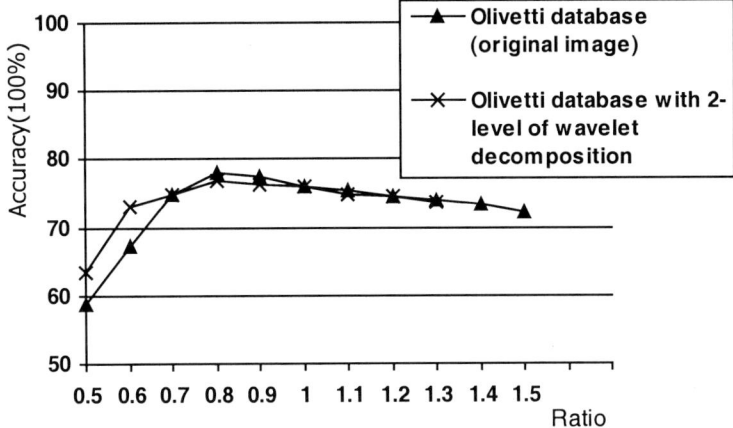

Fig. 3. Selection of independent components

Table 1. Olivetti Database

	Household transformation (selected Ics)	Gram-Schmit Method (selected ICs)	PCA
Accuracy	75.28%	76.39%	70.83%

Table 1 shows that using either household transform or Gram-Schmit method, the performance on the ICA with selected ICs are better than that of PCA. The recognition accuracy of the proposed method is around 6% higher than that of PCA method. This result shows that ICA is less sensitive to the pose variations. The first five independent components and the principal components are shown in Figures 4 and 5 respectively. It can be seen that the ICs bases are more spatially local than that of PCs.

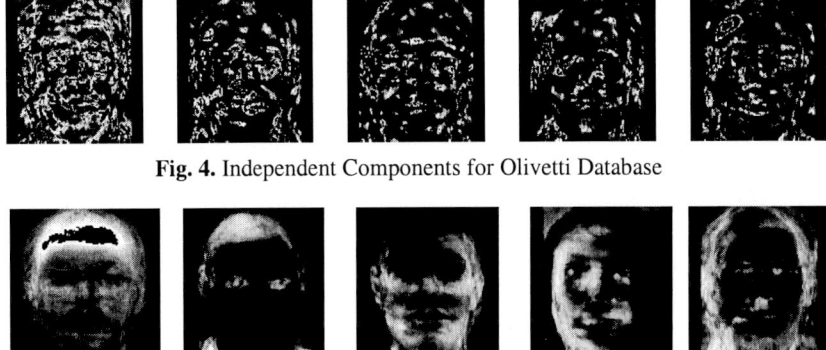

Fig. 4. Independent Components for Olivetti Database

Fig. 5. Eigenfaces for Olivetti Database

5.4 Sensitivity to the Illuminations

To demonstrate the ICA method is less sensitive to the illumination, the Yale database is selected as this database contains two poor and non-linear illumination images as shown in Figure 1. Using the same steps as described in Section 5.3, the results are recorded and shown in Table 2. It can be seen that the recognition accuracy of the proposed method is around 84% while it is only 74% for PCA method. These results show that the proposed method is insensitive to the illumination. Again, the first five independent components and the principal components are shown in Figures 6 and 7 respectively.

Table 2. Yale Database

	Household transformation (selected ICs)	Gram-Schmit Method (selected ICs)	PCA
Accuracy	84.00%	83.33%	74%

Fig. 6. Independent Components for Yale Database

  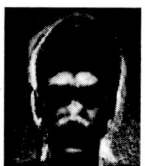  

Fig. 7. Eigenfaces for Yale Database

6 Conclusions

An independent component analysis of face images has been discussed and reported in this paper. We have proposed (1) a selection criterion and (2) two methods to orthogonalize the independent components such that the images outside training set can be represented by the IC bases. Experimental results show that the proposed method is

  Less sensitive to the pose variations
  Less sensitive to the illumination changes

Acknowledgments

This project was supported by the Faculty Research Grant, Hong Kong Baptist University. The second author is partly supported by Natural Science Foundation of Guangdong.

References

1. M. S. Bartlett and T. J. Sejnowski, "Viewpoint invariant face recognition using independent component analysis and attractor networks", *Neural Information Processing Systems – Natural and Synthetic*, Vol. 9, pp. 817-823, 1997
2. P. Comon, "Independent component analysis - a new concept?" *Signal Processing*, Vol. 36, pp. 287-314, 1994.
3. N. Defosse and P. Loubaton, "Adaptive blind separation of independent sources: adeflation approach", *Signal Processing*, Vol. 45, pp. 59-83, 1995.
4. A. Hyvärinen. "Fast and Robust Fixed-Point Algorithms for Independent Component Analysis", *IEEE Transactions on Neural Networks* 10(3):626-634, 1999.
5. A. Bell and T. J. Sejnowski, "Edges are the independent components of natural scenes", *in NIPS 96 (Denver Colorado)*, 1996.
6. P. S. Penev and J. J. Atick, "Local feature analysis: a general statistical theory for object representation", *Network: Computation in Neural Systems*, Vol. 7, pp. 477-500, 1996.
7. D. S. Watkins, *Fundamentals of matrix computations*, John Wiley & Sons, 1991.
8. M. Turk and A. Pentland, "Eigenfaces for recognition", *Journal of Cognitive Neuroscience*, Vol. 3, No.1, pp. 71-86, 1991.

Orientation Contrast Detection in Space-Variant Images*

Gregory Baratoff, Ralph Schönfelder, Ingo Ahrns, and Heiko Neumann

Dept. of Neural Information Processing, University of Ulm, 89069 Ulm, Germany

Abstract. In order to appropriately act in a dynamic environment, any biological or artificial agent needs to be able to locate object boundaries and use them to segregate the objects from each other and from the background. Since contrasts in features such as luminance, color, texture, motion and stereo may signal object boundaries, locations of high feature contrast should summon an agent's attention. In this paper, we present an orientation contrast detection scheme, and show how it can be adapted to work on a cortical data format modeled after the retino-cortical remapping of the visual field in primates. Working on this cortical image is attractive because it yields a high resolution, wide field of view, and a significant data reduction, allowing real-time execution of image processing operations on standard PC hardware. We show how the disadvantages of the cortical image format, namely curvilinear coordinates and the hemispheric divide, can be dealt with by angle correction and filling-in of hemispheric borders.

1 Introduction

One of the primary tasks of a visual observer is to analyze images of the environment with the aim of extracting visible object boundaries and features. Object features - both direct ones like luminance and color, as well as indirect ones like texture, motion, and stereo disparity - are useful in *identifying* objects, whereas object boundaries provide the information necessary for *segmenting* the scene into individual objects and for determining the spatial relations between them. Where the projections of two objects abut in the image, *feature contrasts* are generally present in at least some of the feature dimensions due to differences in surface properties, motions and depths of the objects. Even though feature contrasts can occur within object boundaries, e.g. due to irregular texturing, every location of high feature contrast potentially indicates an object boundary. These locations should therefore attract the attention of the observer. Closer inspection can then allow the observer to determine whether a particular feature contrast arises from an object boundary or not.

---

* This research was performed in the collaborative research center devoted to the "*Integration of symbolic and sub-symbolic information processing in adaptive sensory-motor systems*" (SFB-527) at the University of Ulm, funded by the German Science Foundation (DFG).

A visual observer needs a wide field of view to monitor its surroundings, and a high resolution to discriminate objects. In primates, these two competing requirements are satisfied by a space-variant remapping of the visual field onto the cortex, which features a highly-resolved central region and a coarsely resolved periphery. Additionally, the mapping realizes a significant data reduction, which helps reduce the amount of processing power required for the analysis of the visual information. A formal model of this mapping has been proposed by Schwartz[10,11] in the form of the (displaced) complex logarithmic mapping (DCLM).

In this article, we combine a newly developed mechanism for detection of orientation contrast with a space-variant image representation. In section 2 we review psychophysical evidence concerning the role of feature contrast in attentional capture, and present an algorithm for orientation contrast detection compatible with this data. In section 3 we give a detailed description of the retino-cortical mapping in primates. This nonlinear mapping introduces curvilinear coordinates in the cortical image and a splitting of the visual field along the vertical midline into two cortical hemispheres. These two properties potentially introduce artifacts when standard image processing operations are applied to the cortical image. In section 4 we discuss how some basic image processing operations need to be adapted to avoid these artifacts. Based on this, we present an implementation of our orientation contrast detection scheme in space-variant images. In section 5, we summarize our contributions and identify research problems that we plan to work on.

2 A Model of Orientation Contrast Detection

2.1 Role of Feature Contrast in Preattentive Vision

Support for the view of feature *contrast* as opposed to feature *coherence* as the determinant of attentional capture - at least as far as bottom-up information is concerned - comes from psychophysical experiments[6,7]. These experiments showed that human performance in the detection of salient targets and segmentation of texture fields is determined by feature (orientation, motion, or color) contrast, and not by the feature coherence. For orientation and motion, this result was also obtained in visual search and figure-ground discrimination tasks, whereas for color performance was also found to be determined by feature coherence. A general result valid across all feature dimensions is that psychophysically measured salience increases with feature contrast[7].

2.2 Computational Model

Our orientation contrast detection scheme is realized as a nonlinear multi-stage operation over a local neighborhood, and consists of the following steps :

1. *Orientation-selective filtering* : The original image is processed using a fixed set of N orientation-selective filters, yielding orientation feature maps $o_i, i =$

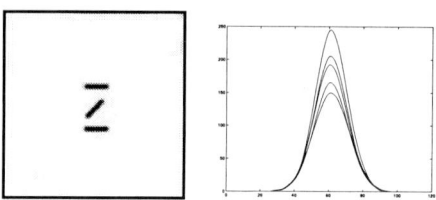

Fig. 1. Orientation contrasts in line segment stimuli. Left : stimulus consisting of central line segment (target) at an orientation of 45°, with two flanking horizontal line segments. Right : orientation contrast responses for target orientations of $0°, 22.5°, 45°, 67.5°, 90°$ (bottom to top). Each response curve represents a horizontal slice of the orientation contrast image at the height of the center of the target.

1,..., N. In our implementation we first computed horizontal and vertical image derivatives I_x and I_y using a Sobel filters (central difference operator), then interpolated the N=8 orientation responses as $o_i = |\cos(\frac{i}{N}\pi)I_x + \sin(\frac{i}{N}\pi)I_y|$. Absolute values are taken to render the result independent of contrast polarity.

2. *Accumulation of spatial support for each orientation* : The orientation response images are individually lowpass-filtered, yielding N accumulated orientation feature maps :

$$\bar{o}_i = G_\sigma * o_i \quad (1)$$

where G_σ is an isotropic Gaussian filter with standard deviation σ, and $*$ denotes convolution.

3. *Computation of orientation-specific orientation contrasts* : For each orientation i, the following pointwise operation is performed, yielding N contrast response feature maps c_i that are selective to orientation :

$$c_i = \bar{o}_i \cdot (\sum_{j \neq i} w_j \bar{o}_j) \quad (2)$$

The w_i form a Gaussian-shaped filter mask with peak at the orthogonal orientation.

4. *Pooling of orientation contrasts across orientations* : The orientation-specific contrast feature maps c_i are merged by maximum selection to yield a non-specific orientation contrast feature map :

$$c = \max_i c_i \quad (3)$$

This processing scheme yields results compatible with the psychophysical data discussed above, as illustrated in Fig. 1. In a stimulus made up of three vertically stacked line segments, the orientation of the middle line segment (target) is systematically varied. The target appears most salient when its orientation is orthogonal to the other two line segments and least salient when its orientation matches it. In between these two extremes, the perceived salience varies monotonically.

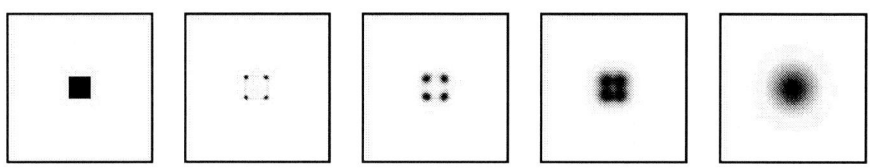

Fig. 2. Orientation contrast responses to a square test image. From left to right : original image, orientation contrast responses for $\sigma = 1, 2, 4, 8$.

An important parameter in the proposed scheme is the size of the neighborhood (determined by σ, the standard deviation of the isotropic Gaussian filter mask) over which the individual orientation responses are accumulated in step 2 of the algorithm. As shown in Fig. 2, choosing a small neighborhood results in orientation contrast maxima at the corners of the square. With increasing σ the regions of high orientation contrast grow, and finally merge to a single blob representing the entire square.

3 Space-Variant Retino-Cortical Mapping

3.1 Architecture of the Primate Visual System

An important characteristic of the primate visual system is the space-variant remapping of the visual field onto the cortex. At the center of the visual field there is a small area of highest resolution, called the fovea, spanning about 5° of visual angle. From there the resolution falls off towards the periphery. This space-variant sampling is realized by the photoreceptor and ganglion cell arrangement in the retina [13]. A fast eye movement subsystem allows rapid deployment of the high-resolution fovea to interesting regions of the environment.

On its way from the retina to the cortex the information from the right visual field is routed to the left brain hemisphere, and vice versa. Here, a potential problem arises for cells with receptive fields covering the vertical midline region of the visual field, since they need access to information from both the left and the right visual field. The brain addresses this problem by a double representation of the vertical stripe surrounding the central meridian. The width of this doubly-covered midline region corresponds to about 1° of visual angle, which expands to 3° in the fovea [2].

3.2 A Model of the Retino-Cortical Projection in Primates

A model of the retino-cortical projection in primates based on anatomical data was first proposed by Schwartz[10] in the form of the complex logarithmic mapping (CLM). In this model, the image plane (retina) is identified with the complex plane :

$$z = x + i\,y = re^{i\phi} \tag{4}$$

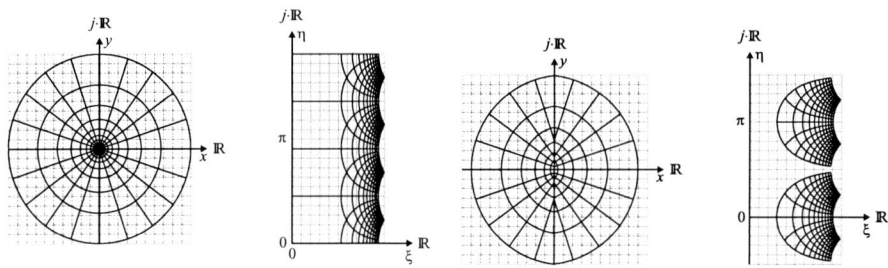

Fig. 3. Coordinate lines of the CLM (left pair) and the DCLM (right pair). Left image of each pair : retinal image, right image : cortical image.

where $r = \sqrt{x^2 + y^2}$ and $\tan\phi = y/x$. The CLM maps the retinal position represented by the complex variable z onto the cortical position represented by the complex variable $\zeta = \xi + i\,\eta$:

$$\zeta = \log z = \log r + i\,\phi \tag{5}$$

It transforms the cartesian representation into a log-polar one consisting of an angular component and a logarithmically compressed radial component. These properties are illustrated in Fig. 3 (left), which shows the coordinate lines of the CLM. This form of the mapping has seen extensive application in the computational vision community[9,14,5,12,3], in particular for optical flow based obstacle avoidance and rotation and scale invariant object recognition. One disadvantage of the complex logarithmic mapping is, however, the singularity of the logarithm at the origin of the coordinate system, which is located at the image center. In a practical implementation, this problem is circumvented by cutting out a disk around the image center.

Schwartz[11]'s second model, the displaced complex logarithmic map (DCLM) - which he found to be a better description of the retino-cortical projection in different primate species - introduces a displacement of the origin that eliminates the singularity of the plain logarithm :

$$\zeta = \begin{cases} \log(z+a)\,, Re\{z\} > 0 \\ \log(z-a)\,, Re\{z\} < 0 \end{cases} \tag{6}$$

where a is a small positive constant, whose introduction causes the singularity at the origin to disappear. Note that the DCLM exhibits the hemispheric divide present in the primate cortical representation, whereas the plain CLM does not. These properties are illustrated in Fig. 3 (right), which shows the coordinate lines of the DCLM. In the DCLM polar separability is lost, especially in the central region. Indeed, for small $|z| \ll a$ in the right hemisphere ($Re\{z\} > 0$), (5) yields $\zeta \propto z$, i.e. the mapping is approximately linear. For large $|z| \gg a$ on the other hand, the plain logarithmic mapping becomes dominant, i.e. $\zeta \propto \log(z)$. When the natural logarithm is used, both the CLM and the DCLM are conformal mappings, i.e. angles are locally preserved. Furthermore, both can be parametrized to be length- and area- preserving.

4 Space-Variant Image Processing

4.1 Implementation of the DCLM

We realize the retino-cortical image transformation with the help of a look-up-table. For each pixel in the cortical image we store its preimage in the retinal image as well as the size of its receptive field. In the simplest case the intensity of the cortical pixel is computed by averaging all retinal pixels that map onto it. Alternatives involving weighted averaging with overlapping receptive fields are discussed in [1]. The inverse transformation is also sometimes needed, for example when reconstructing the retinal image corresponding to a given cortical image, or for the operation of border filling-in introduced in section 4.2. For this purpose we also store the inverse transformation in a look-up-table.

The cortical image representation is discretized as follows:

$$\zeta = \xi + i \cdot \eta \mapsto \frac{\xi - \xi_0}{\Delta_\xi} + i \cdot \frac{\eta}{\Delta_\eta} \tag{7}$$

where ξ_0 is the smallest value of ξ, and Δ_ξ and Δ_η are sizes of the cortical pixels in ξ- and η-direction. Given the number of cortical pixels in the angular dimension, N_η, the pixel size in the angular dimension is:

$$\Delta_\eta = \frac{2\pi}{N_\eta} \tag{8}$$

The radial pixel size is obtained by dividing the range of logarithmically compressed radii by the number of cortical pixels in the radial dimension N_ξ:

$$\Delta_\xi = \frac{ln(r) - ln(a)}{N_\xi} = \frac{ln(r/a)}{N_\xi} \tag{9}$$

where r is the maximal radius measured from the displaced origin. The discretized version of the mapping remains conformal if the cortical pixel sizes are chosen to be equal: $\Delta_\xi = \Delta_\eta$. Using this equality, and letting N_η be the main parameter, the number of pixels in the ξ-direction is determined as:

$$N_\xi = \frac{N_\eta}{2\pi} ln(r/a) \tag{10}$$

The remaining degree of freedom, the constant a, is set by requiring the mapping to be 1:1, i.e. area- and length-preserving, in the image center. For this purpose, the Jacobian of the discretized mapping at the image center should be the identity mapping[8]. This is achieved when $a = 1/\Delta_\eta$.

For the examples presented in this paper, we have chosen $N_\eta = 180$, corresponding to an angular resolution of 2° per pixel. For an input image of size $w^2 = 256 \times 256$ this yields $N_\xi = 56$ pixels for the radial dimension. We thus obtain a data reduction of $256^2/(180 \cdot 56) \approx 6.5$. For image resolutions typically obtained from standard CCD-cameras (512×512 pixels) the reduction factor increases to about 20.

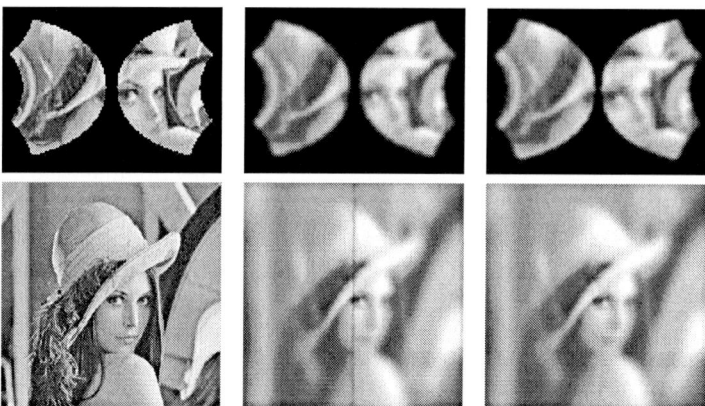

Fig. 4. Effect of filling-in of interhemispheric border on smoothing in the cortical image. Lower left : original image, upper left : cortical image obtained by DCLM. Smoothing involved four iterations of filtering with the 3×3 binomial filter applied to the cortical image. Middle column shows result *without* filling-in of midline regions prior to smoothing (top : cortical image, bottom : its retinal reconstruction). Right column shows result *with* prior filling-in.

4.2 Filling-in of the Interhemispheric Border

Several steps in the orientation contrast detection scheme involve linear filtering, i.e. convolution of the image with a discrete mask of fixed size. If this operation is applied blindly to the cortical image, artifacts occur at the hemispheric borders due to the fact that some of the pixels under the convolution mask are located in the other hemisphere. Here, we solve this problem by "growing" the hemispheric border regions by the number of pixels required by the convolution mask. The intensity values of the new border pixels are filled in from the other hemisphere. For each pixel p in the border region, its preimage $p' = \tau^{-1}(p)$ under the DCLM in the retinal image is first computed. Then, the image of p' is computed, yielding $p'' = \tau(p')$, a point in the other hemisphere. Since this point usually lands between pixels, we calculate its intensity value by bilinear interpolation from its four neighboring pixels. Finally, this value is copied into the border pixel p. Hemispheric border filling-in is performed before each linear filtering operation. It can be performed very efficiently by keeping in a look-up table the position p'' for each border pixel p. Furthermore, the width of the border region requiring filling-in can be reduced for filtering operations that can be decomposed into a cascade of operations with smaller filter masks. This can for example be done for Gaussian or binomial filters. Fig. 4 illustrates the importance of the filling-in operation for the example of smoothing using cascade of binomial filters. Blind application of the convolution operation leads to artifacts along the hemispheric divide. When the borders are filled in before the convolution no artifacts are introduced.

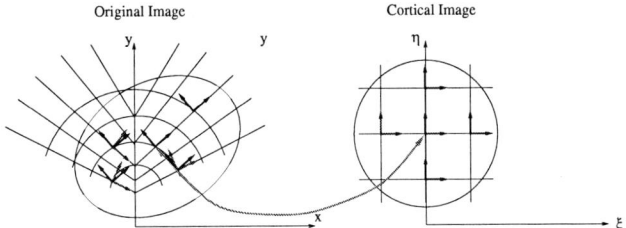

Fig. 5. Variation of local coordinate system orientation. Within a hemisphere the orientation changes slowly along $\eta = const$ lines. Drastic orientation changes occur across hemispheric divide.

4.3 Angle Correction

In section 4.1 we parameterized the DCLM to be conformal. This means that angles formed by image structures are preserved by the DCLM. However, angles with respect to the absolute horizontal or vertical axis are not preserved. In fact, the orientation of the local coordinate system varies in the cortical image, as shown in Fig. 5. In order to express angles in the same global coordinate system as that of the retinal image, the angles obtained in the cortical image must be corrected by subtracting η, the angle that the local ξ-axis forms with the global x-axis [4]. Fig. 6 illustrates the angle-correction operation for edge (gradient) computations.

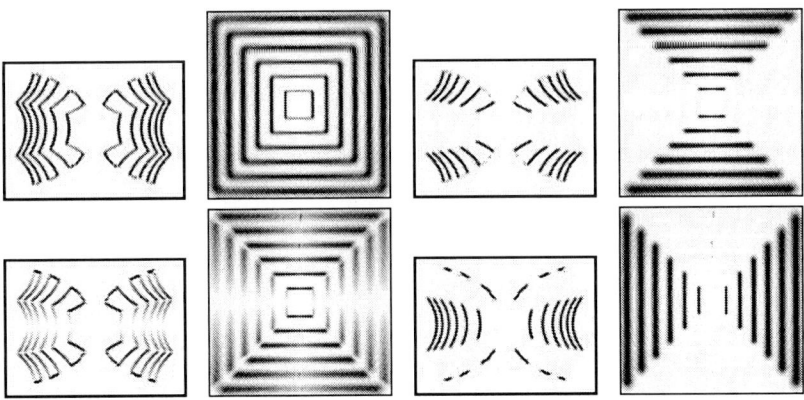

Fig. 6. Gradient computation in the cortical image. The original image and its cortical image are the same as in Fig. 4, left column. Top row shows x-gradients, middle row y-gradients. First column shows gradients in cortical image *without* angle correction, second column shows retinal reconstructions. Third column shows gradients in cortical image *with* angle correction, fourth column shows retinal reconstructions.

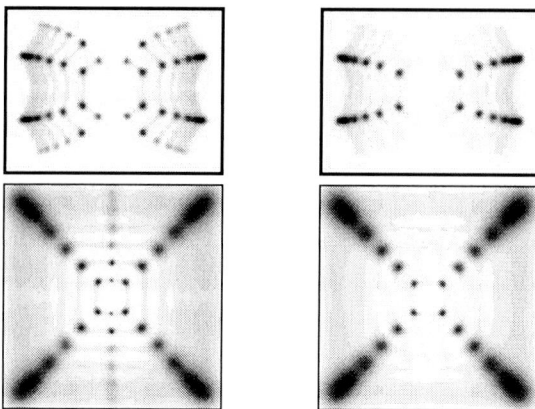

Fig. 7. Orientation contrast detection in the cortical image. The original image is the same as in Fig. 4. Left column : orientation contrast computed *without* angle correction. Right column : orientation contrast computed *with* angle correction. Top row : cortical image, bottom row : retinal reconstruction.

4.4 Space-Variant Orientation Contrast Detection

By incorporating border filling-in and angle correction the orientation contrast detection scheme gives correct results when applied to the cortical image, as shown in Fig. 7. One also notices that the characteristic scale of salient features increases towards the periphery. Whereas the corners of the nested squares are detected individually in the center of the image, they merge into one blob in the periphery. (Compare this to Fig. 2.) Because of the space-variant nature of the cortical image, features at different scales are detected using the same accumulation scale : fine details are detected in the fovea, and coarse features in the periphery.

The execution times on a Pentium II 300 Mhz were as follows : 0.05 sec for the DCLM, and 0.25 sec for the orientation contrast detection. In the latter, the majority of the time was spent on step 2 (accumulation of spatial support), which involved 10 iterations per orientation of the 3x3 binomial filter. For the original image of size 256×256, the equivalent computation would have taken 1.6 seconds. This corresponds to an efficiency gain factor of $1.6/0.3 \approx 5$. It approximately doubles when the DCLM resolution is halved ($N_\eta = 90$), or when an input image twice the size is considered.

5 Conclusions

The work presented in this article is a first step in the development of an attentive visual observer that analyzes its environment and reacts to external events in real-time. In order to achieve real-time execution, we employed a highly data-reducing space-variant mapping modelled after the retino-cortical projection in primates, which features a central region of high resolution and a wide field of view. In trying to apply an orientation contrast detection scheme to this space-variant representation, one notices that a blind application of standard image processing techniques introduces artifacts into the computations. These artifacts arise because of the curvilinear coordinates and the discontinuity at the hemispheric border introduced by the mapping. In order to eliminate these artifacts, we developed two techniques : an angle correction operation which aligns local coordinate systems with a global one, and a hemispheric border filling-in operation, which allows linear filtering across the hemispheric discontinuity. It should be noted that these techniques are not restricted to the particular situation studied in this article, but are general and are applicable to other image processing operations and to other forms of space-variant mappings. In future work we plan to address the following issues : (1) implementation of feature contrast maps for motion, stereo and color, (2) prioritized selection of salient image locations by integration of contrast information from different feature maps, and (3) real-time system integration on an active stereo vision head.

References

1. M. Bolduc and M.D. Levine. A review of biologically motivated space-variant data reduction models for robotic vision. *CVIU*, 69(2):170–184, 1999.
2. A.H. Bunt and D.S. Minckler. Foveal sparing. *AMA Archives of Ophtalmology*, 95:1445–1447, 1977.
3. K. Daniilidis. Computation of 3d-motion parameters using the log-polar transform. In V. Hlavac and R. Sara, editors, *Proc CAIP'95*. Springer, 1995.
4. B. Fischl, M.A. Cohen, and E.L. Schwartz. The local structure of space-variant images. *Neural Networks*, 10(5):815–831, 1997.
5. R. Jain, S. L. Bartlett, and N. O'Brian. Motion stereo using ego-motion complex logarithmic mapping. *IEEE Trans. PAMI*, 9:356–369, 1987.
6. H.C. Nothdurft. Texture segmentation and pop-out from orientation contrast. *Vis. Res.*, 31(6):1073–1078, 1991.
7. H.C. Nothdurft. The role of features in preattentive vision: Comparison of orientation, motion and color cues. *Vis. Res.*, 33(14):1937–1958, 1993.
8. A.S. Rojer and E.L. Schwartz. Design considerations for a space-variant visual sensor with complex-logarithmic geometry. In *Proc. ICPR'90*, pages 278–284, 1990.
9. G. Sandini and V. Tagliasco. An anthropomorphic retina-like structure for scene analysis. *CVGIP*, 14:365–372, 1980.
10. E. Schwartz. Spatial mapping in the primate sensory projection: Analytic structure and relevance to perception. *Biol. Cyb.*, 25:181–194, 1977.
11. E. Schwartz. Computational anatomy and functional architecture of striate cortex : A spatial mappping approach to perceptual coding. *Vis. Res.*, 20:645–669, 1980.
12. M. Tistarelli and G. Sandini. On the advantages of polar and log-polar mapping for direct estimation of time-to-impact from optical flow. *IEEE Trans. PAMI*, 15:401–410, 1992.
13. H. Waessle, U. Gruenert, J. Roehrenbeck, and B.B. Boycott. Retinal ganglion cell density and cortical magnification factor in the primate. *Vis. Res.*, 30(11):1897–1911, 1990.
14. C.F.R. Weiman and G. Chaikin. Logarithmic spiral grids for image processing and display. *CGIP*, 11:197–226, 1979.

Multiple Object Tracking in Multiresolution Image Sequences

Seonghoon Kang and Seong-Whan Lee*

Center for Artificial Vision Research, Korea University,
Anam-dong, Seongbuk-ku, Seoul 136-701, Korea
{shkang, swlee}@image.korea.ac.kr

Abstract. In this paper, we present an algorithm for the tracking of multiple objects in space-variant vision. Typically, an object-tracking algorithm consists of several processes such as detection, prediction, matching, and updating. In particular, the matching process plays an important role in multiple objects tracking. In traditional vision, the matching process is simple when the target objects are rigid. In space-variant vision, however, it is very complicated although the target is rigid, because there may be deformation of an object region in the space-variant coordinate system when the target moves to another position. Therefore, we propose a deformation formula in order to solve the matching problem in space-variant vision. By solving this problem, we can efficiently implement multiple objects tracking in space-variant vision.

1 Introduction

In developing an active vision system, there are three main requirements imposed on the system: high resolution for obtaining details about the regions of interest, a wide field of view for easy detection of a looming object or an interesting point, and the fast response time of the system[1]. However, a system that uses a traditional image representation which has uniformly distributed resolution cannot satisfy such requirements. There have been many research works on the active vision algorithms based on the biological vision system that satisfies all of the requirements, among which is space-variant vision using the multi-resolution property of the biological vision system.

Space-variant vision has recently been applied to many active vision applications such as object tracking, vergence control, etc. In a typical active tracking system, there is only one target to track because the camera head continuously fixates on one object at a time. In order to transfer its fixation to another object, it is important to keep track of the positions of the objects moving in the background. This gives rise to the necessity for multiple objects tracking. In multiple objects tracking, new problems are introduced, such as multiple objects detection, and matching between the current target model and the detected object in the space-variant coordinate system.

---

* To whom all correspondence should be addressed. This research was supported by Creative Research Initiatives of the Ministry of Science and Technology, Korea.

In space-variant vision, multiple objects detection is difficult to achieve, because a motion vector in the space-variant coordinate system is represented differently from that of the Cartesian one in size and direction. Consequently, it is difficult to segment a moving region directly in the space-variant coordinate system. The matching problem is also very difficult, because there may be deformation of an object region in the space-variant coordinate system when it moves to another position, although the target object is rigid.

In this paper, we propose an efficient algorithm that overcomes the difficulties mentioned above.

2 Related Works

A representative of the space-variant models is Schwartz's $log(z)$ model[2]. This model and its variations have been applied to various areas, such as active vision, visual perception, etc. Panerai et al.[3] developed a technique for vergence and tracking in log-polar images. Lim et al.[4] proposed a tracking algorithm for space-variant active vision. Recently, Jurie[5] proposed a new log-polar mapping procedure for face detection and tracking.

There have also been many works on multiple objects tracking. Recently, Bremond and Thonnat[6] presented a method to track multiple non-rigid objects in a video sequence, and Haritaoglu et al.[7] developed *Hydra*, a real-time system for detecting and tracking multiple people.

3 Space-Variant Vision

3.1 Space-Variant Representation Model

In this section, we describe in detail our space-variant model used in this paper. We use the property of a ganglion cell whose receptive field size is small in the fovea and becomes larger toward its periphery. The fovea and the periphery are considered separately for the model, as shown in Figure 1, since the distribution of a receptive field is uniform in the fovea and gets sparser toward the periphery.

Following equations (1) and (2) are the radius of the nth eccentricity and the number of receptive fields at each ring in the periphery, respectively.

$$R_n = \frac{2(1-o)w_0 + (2+w_0 k)R_{n-1}}{2-(1-2o)w_0 k} \qquad (1)$$

$$K = round\left(\frac{2\pi R_0}{(1-o)(w_0 + w_0 k R_0)}\right) \qquad (2)$$

Equations (3) and (4), given below, are the radius of the nth eccentricity and the number of receptive fields at the nth ring in the fovea, respectively.

$$R_n = R_{n-1} - \frac{R_0}{round(R_0/(1-o)/w_o)} \qquad (3)$$

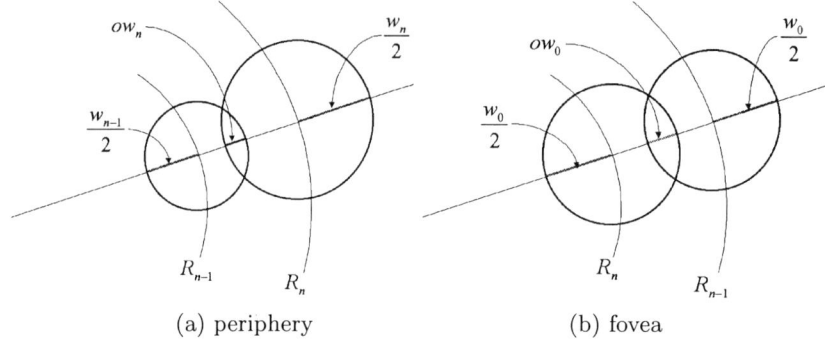

Fig. 1. Receptive fields in fovea and periphery

$$K_n = round\left(\frac{2\pi R_n}{(1-o)w_0}\right) \quad (4)$$

In the equations (1) ∼ (4), w_n, w_0 and R_0 are the size of the receptive field at the nth ring, the size of the receptive field at the fovea, and the size of the fovea, respectively. k is constant and o is an overlapping factor. When an overlapping factor is one, the receptive fields are completely overlapped, and when it is zero, there is no overlap whatsoever. The function $round(\cdot)$ performs a 'rounding off' operation.

Figure 2 shows a mapping template generated by the above equations.

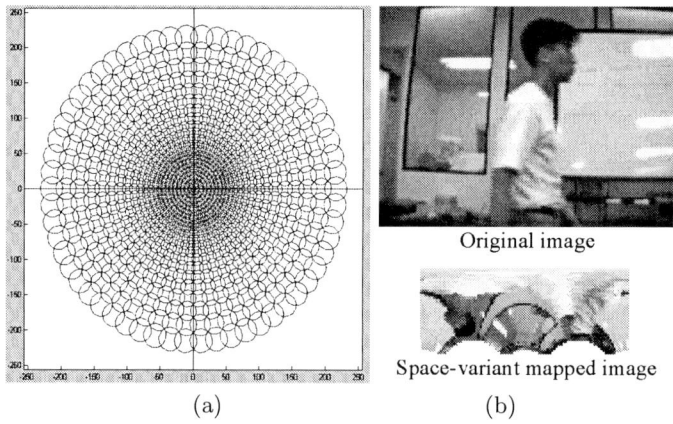

Fig. 2. (a) A space-variant mapping template, and (b) an example of mapping

3.2 Deformation Formula in Space-Variant Vision

There may be a deformed movement in the space-variant coordinate system, which corresponds to a simple translation in the Cartesian coordinate system, as shown in Figure 3. We should be aware of the deformation formula in order to analyze the movement of a region in the space-variant coordinate system.

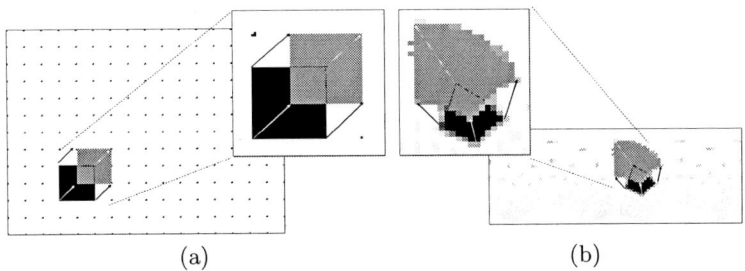

Fig. 3. (a) A movement in the Cartesian coordinate system, and (b) the deformed movement in the space-variant coordinate system, which corresponds to (a)

In Figure 4, we know the deformed position of a certain point in the space-variant coordinate system that corresponds to the translated position in the Cartesian coordinate system. The radius R and the angle θ of the given point (ξ, η) in the space-variant coordinate system can be obtained using the mapping functions (1) and (2) given below:

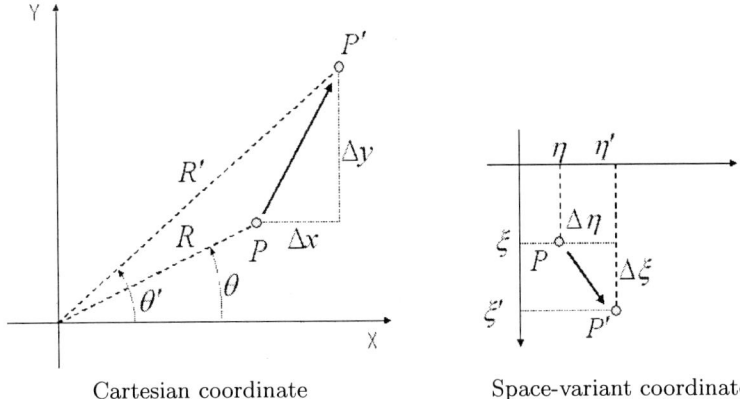

Cartesian coordinate   Space-variant coordinate

Fig. 4. Deformed movement in the space-variant coordinate corresponding to movement in the Cartesian coordinate

$$R = SM_R(\xi) = \sum_{n=1}^{\xi} ab^{n-1} + R_0 b^{\xi}$$
$$= \frac{(a + bR_0 - R_0)b^{\xi} - a}{b - 1}, \quad (5)$$

$$\theta = SM_\theta(\eta) = \frac{2\pi}{K}\eta, \quad (6)$$

where $a = \frac{2(1-o)w_0}{2-(1-2o)w_0k}$ and $b = \frac{2+w_0k}{2-(1-2o)w_0k}$. Then, the position (R', θ') of the point in the Cartesian coordinate system, after the movement of the point by Δx and Δy, can be found easily, as follows:

$$R' = \sqrt{(x + \Delta x)^2 + (y + \Delta y)^2}, \quad (7)$$
$$\theta' = \arctan\left(\frac{y + \Delta y}{x + \Delta x}\right), \quad (8)$$

where $x = R\cos\theta$, $y = R\sin\theta$. From the equations given above, the deformed position (ξ', η') in the space-variant coordinate system can be derived as shown below:

$$\xi' = SM_R^{-1}(R')$$
$$= \log_b\left(\frac{(b-1)R' + a}{(a + bR_0 - R_0)}\right), \quad (9)$$
$$\eta' = SM_\theta^{-1}(\theta') = \frac{K}{2\pi}\theta'. \quad (10)$$

Finally, the deviation $(\Delta\xi, \Delta\eta)$ in the space-variant coordinate system can be obtained as follows:

$$\Delta\xi = SM_R^{-1}(R') - \xi, \quad (11)$$
$$\Delta\eta = SM_\theta^{-1}(\theta') - \eta. \quad (12)$$

As we have shown above, the deformed deviation $(\Delta\xi, \Delta\eta)$ of a point (ξ, η), which comes from a translation $(\Delta x, \Delta y)$ in the Cartesian coordinate system, can be found when the point (ξ, η) and the translation $(\Delta x, \Delta y)$ are given.

4 Moving Object Detection

4.1 Motion Estimation in Space-Variant Vision

Horn and Schunk's optical flow method[8] is used for motion estimation. In addition, the optical flow vectors in the space-variant coordinate system are transformed to the Cartesian coordinate system in order to segment a moving region and to calculate a mean flow vector easily, as shown on Figure 5.

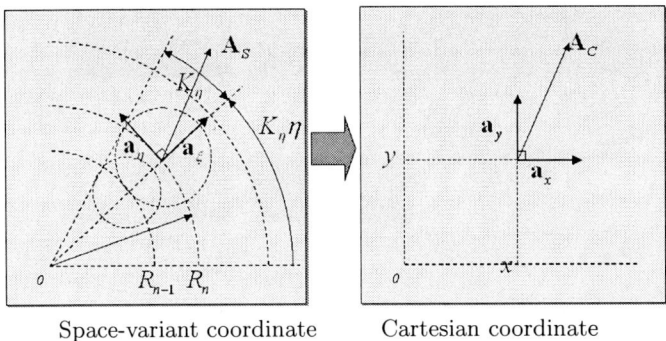

Fig. 5. Vector transformation from the space-variant coordinate system to the Cartesian coordinate system

A vector $\mathbf{A}_S = A_\xi \mathbf{a}_\xi + A_\eta \mathbf{a}_\eta$ in the space-variant coordinate system can be represented as a polar coordinate vector $\mathbf{A}_S = A_\xi(R_n - R_{n-1})\mathbf{a}_\rho + A_\eta K_\eta \mathbf{a}_\phi$. Then, it is easily transformed to a Cartesian coordinate vector $\mathbf{A}_C = A_x \mathbf{a}_x + A_y \mathbf{a}_y$ where

$$\begin{aligned}
A_x &= \mathbf{A}_S \cdot \mathbf{a}_x \\
&= A_\xi(R_n - R_{n-1})\mathbf{a}_\rho \cdot \mathbf{a}_x + A_\eta K_\eta \mathbf{a}_\phi \cdot \mathbf{a}_x \\
&= A_\xi(R_n - R_{n-1})\cos(K_\eta \eta) - A_\eta K_\eta \sin(K_\eta \eta), \quad (13) \\
A_y &= \mathbf{A}_S \cdot \mathbf{a}_y \\
&= A_\xi(R_n - R_{n-1})\mathbf{a}_\rho \cdot \mathbf{a}_y + A_\eta K_\eta \mathbf{a}_\phi \cdot \mathbf{a}_y \\
&= A_r(R_n - R_{n-1})\sin(K_\eta \eta) + A_\eta K_\eta \cos(K_\eta \eta). \quad (14)
\end{aligned}$$

4.2 Moving Object Region Segmentation

For moving region segmentation, a region-based segmentation and labeling are employed. As shown in Figure 6, we construct a binary optical flow map, then morphological post-filtering is applied. Finally, a labeled region map is obtained by using a connected component analysis.

5 Multiple Objects Tracking

5.1 Dynamic Target Model

For multiple objects tracking, we construct a dynamic target model which consists of the texture of an object region. This target model is updated continuously in time during the tracking process. It is defined by:

$$\Psi^t(\xi, \eta) = wI(\xi, \eta) + (1-w)D_{\Psi^{t-1}, \Delta P}(\xi, \eta), \quad (15)$$

where $\Psi^t(\xi, \eta)$ is the intensity value at (ξ, η) in a texture template, $I(\xi, \eta)$ is that of the input image, $D_{\Psi^{t-1}, \Delta P}(\xi, \eta)$ is that of a deformed template at (ξ, η) and $w(0 < w < 1)$ is a weight value.

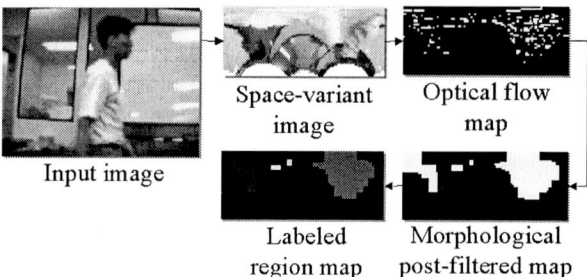

Fig. 6. Region based moving region segmentation

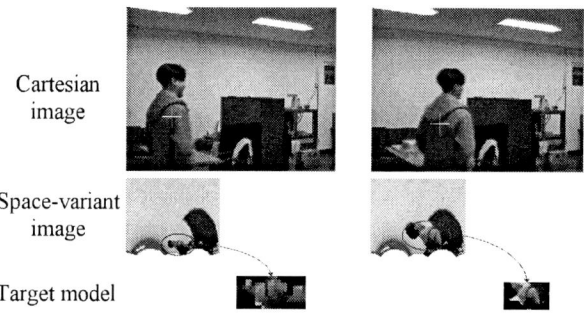

Fig. 7. Examples of extracted target model

5.2 Deformable Matching of Moving Targets

For matching between the target model and a detected object, we use a correlation technique, the quadratic difference, defined by:

$$C_{ij} = \frac{\sum_{\xi,\eta \in \Psi} \left(I(\xi+i, \eta+j) - D_{\Psi,\Delta P}(\xi,\eta) \right)^2}{N^2}, \quad (16)$$

where N is the number of pixels in the target template Ψ and $|i,j| < W$. W is the size of a search window. When C_{ij} is smaller than a certain threshold value, the detected object is matched with the current target. The target model is then updated to the detected object.

5.3 Selection of a Target to Track

Now, we can track multiple objects by using the detection, prediction, matching, and updating loop. A target is selected for tracking with a camera head. For the target selection, we introduce a selection score defined by:

$$S_{target} = w_1 s + w_2 v + w_3 d, \quad (17)$$

where $w_1 + w_2 + w_3 = 1$, s, v and d are the number of pixels in the target region, the velocity of the target, and the inverse of the distance from the center of the target, respectively. A selection score represents a weighted consideration for the target size, velocity, and the distance.

6 Experimental Results and Analysis

Our experimental system consists of a Pentium III 500MHz PC with a Matrox Meteor II image grabber, and a 4-DOF (pan, tilt, and vergence of the left and right cameras) head-eye system, Helpmate BiSight. Figure 8 depicts a brief configuration of our experimental system.

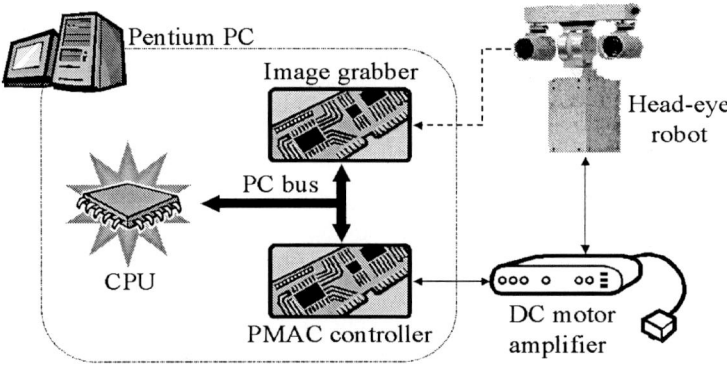

Fig. 8. Configuration of our experimental system

The performance of multiple objects tracking in space-variant active vision is shown. In our experiment, people moving around in an indoor environment are chosen as target objects to track. During the experiment, they are walking across the field of view in various situations.

Figure 9 shows the result images of each sub-process of the tracking process. The images in the third row are binary optical flow maps, each of which is constructed by transformation of the directions of the optical flow vectors into a grey-level image with an appropriate threshold. The images in the forth row are morphological post-filtered maps, each of which is constructed by a morphological closing operation. Each of these maps is used as an extracting mask for target model construction. Finally, the images in the last row are segmented target models. The target model is used for matching between the tracked objects in the previous frame and the detected objects in the current frame. Throughout these processes, multiple objects tracking is accomplished.

Figure 10 shows a sequence of images taken by the head-eye camera. As shown in the figure, the system did not lose the targets. However, it detected only one target for two objects upon occlusion, since occlusion is not considered in our system. Future research will take this into account for a better tracking performance.

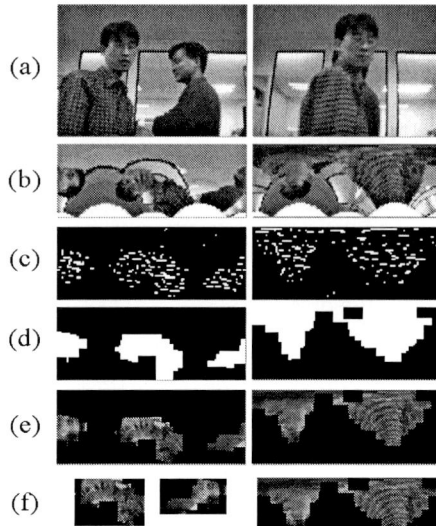

Fig. 9. Result images of sub processes of our system:(a) is a Cartesian image, (b) is a space-variant image, (c) is an optical flow map, (d) is a morphological post-filtered map, (e) is an extracted target region and (f) is an extracted target model.

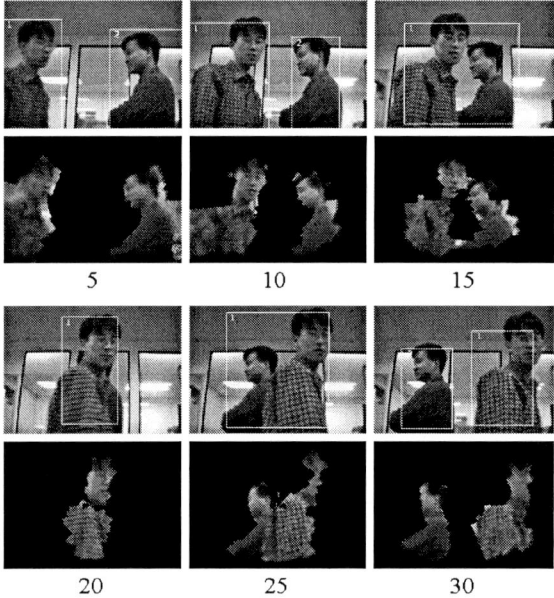

Fig. 10. Tracking sequence: Images on odd row are tracking sequence and images on even row are reconstructed images from detected target region.

7 Conclusions and Further Researches

In this paper, we have shown multiple targets tracking in space-variant active vision in a low cost PC environment. Motion detection is very efficient because of using optical flows in the space-variant coordinate system. Considering deformation in the space-variant coordinate system, caused by the movement of the target region, matching in space-variant vision becomes very efficient and simple. Nevertheless, the proposed algorithm does not consider occlusion problems-solutions to which should be included for better tracking performance.

References

1. Bolduc, M., Levin, M. D.: A Real-time foveated Sensor with Overlapping Receptive Fields. Real-Time Imaging **3:3** (1997) 195–212
2. Schwartz, E. L.: Spatial Mapping in the Primate Sensory Projection: Analytic Structure and Relevance to Perception. Biological Cybernetics **25** (1977) 181–194
3. Panerai, F., Capurro, C., Sandini, G.: Space Variant Vision for an Active Camera Mount. Technical Report TR1/95, LIRA-Lab-DIST University of Genova (1995)
4. Lim, F. L., West, G. A. W., Venkatesh, S.: Tracking in a Space Variant Active Vision System. Proc. of 13th International Conference on Pattern Recognition, Vienna, Austria (1996) 745-749
5. Jurie, F.: A new log-polar mapping for space variant imaging: Application to face detection and tracking. Pattern Recognition **32** (1999) 865–875
6. Bremond, F., Thonnat, M.: Tracking Multiple Nonrigid Objects in Video Sequences. IEEE Trans. on Circuits and Systems for Video Technology **8:5** (1998) 585–591
7. Haritaoglu, I., Harwood, D., Davis, L. S.: Hydra: Multiple People Detection and Tracking Using Silhouettes. Proc. of 2nd IEEE Workshop on Visual Surveillance, Fort Collins, Colorado (1999) 6–13
8. Tekalp, A. M.: Digital Video Processing. Prentice Hall (1995)

A Geometric Model for Cortical Magnification

Luc Florack

Utrecht University, Department of Mathematics, PO Box 80010,
NL-3508 TA Utrecht, The Netherlands.
Luc.Florack@math.uu.nl

Abstract. A Riemannian manifold endowed with a conformal metric is proposed as a geometric model for the cortical magnification that characterises foveal systems. The eccentricity scaling of receptive fields, the relative size of the foveola, as well as the fraction of receptive fields involved in foveal vision can all be deduced from it.

1 Introduction

Visual processing in mammalians is characterised by a foveal mechanism, in which resolution decreases roughly in proportion to eccentricity, *i.e.* radial distance from an *area centralis* (*e.g.* humans, primates) or vertical distance from an elongated *visual streak* (*e.g.* rabbit, cheetah). As a consequence, it is estimated that in the case of humans about half of the striate cortex is devoted to a foveal region covering only 1% of the visual field [15].

In this article we aim for a theoretical underpinning of retinocortical mechanisms. In particular we aim to quantify the *cortical magnification* which characterises foveal systems. The stance adopted is the conjecture pioneered by Koenderink that the visual system can be understood as a "geometry engine" [11]. More specifically, we model the retinocortical mechanism in terms of a metric transform between two manifolds, $\mathcal{M}$ and $\mathcal{N}$, representing the visual space as it is embodied in the *retina* (photoreceptor output), respectively *lateral geniculate nucleus* (LGN) and *striate cortex* (ganglion, simple and complex cell output).

We consider the rotationally symmetric case only, for which we propose a conformal metric transform which depends only on eccentricity. A simple geometric constraint then uniquely establishes the form of the metric up to a pair of integration constants, which can be matched to empirics quite reasonably.

2 Preliminaries

We introduce two base manifolds, $\mathcal{M}$ and $\mathcal{N}$, representing the neural substrate of visual space as embodied by the photoreceptor tissue in the retina, respectively the ganglion, simple and complex cells in the LGN and the striate cortex (V1).

We furthermore postulate the existence of a smooth, reversible *retinocortical mapping* $\varrho : \mathcal{M} \to \mathcal{N}$ which preserves neighbourhood relations.

We introduce some additional geometric structure by considering the *total spaces* $(\mathcal{E}, \pi_{\mathcal{E}}, \mathcal{M})$, respectively $(\mathcal{F}, \pi_{\mathcal{F}}, \mathcal{N})$, comprising so-called *fibres* over $\mathcal{M}$, respectively $\mathcal{N}$, and corresponding projection maps $\pi_{\mathcal{E}} : \mathcal{E} \to \mathcal{M}$, respectively $\pi_{\mathcal{F}} : \mathcal{F} \to \mathcal{N}$. The totality of fibres constitutes a so-called *fibre bundle*, cf. Spivak [16]. For brevity we will denote the total spaces by $\mathcal{E}$ and $\mathcal{F}$; no confusion is likely to arise. The physiological significance of all geometric concepts involved is (loosely) explained in the appendix. Heuristically, each fibre is a linearisation of all neural activity associated with a *fixed* point in visual space; the associated projection map allows us to retrieve this point given an arbitrary visual neuron in the fibre bundle.

The reason for considering a fibre bundle construct is that if we monitor the output of different cells, even if their receptive fields have the same base point, we generally find different results. Likewise, the output of any given photoreceptor in the retina contributes to many receptive fields simultaneously, and therefore it is not sufficient to know only the "raw" retinal signal produced by the photoreceptor units, even though this contains, strictly speaking, all optical information. See Fig. 1. We will not elaborate on the internal structure of the fibres, but instead concentrate on the Riemannian nature of the base manifolds.

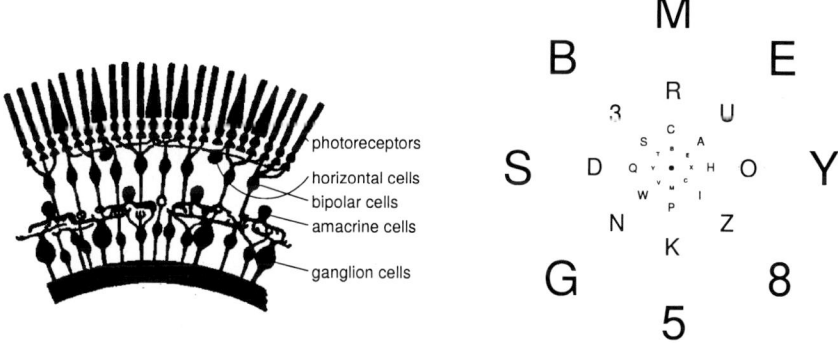

Fig. 1. Left: Sketch of retinal layers. Right: Spatial acuity depends on eccentricity; all characters appear more or less equally sharp when focusing on the central dot.

3 The Scale-Space Paradigm

Our point of departure for modeling retinocortical processing is the *causality principle* introduced by Koenderink [10], which entails that extrema of the retinal irradiance distribution f should not be enhanced when resolved at finite resolution. The simplest *linear* partial differential equation (p.d.e.) that realises this is the isotropic diffusion equation (boundary conditions are implicit):

$$\begin{cases} u_s = \Delta u \\ \lim_{s \to 0} u = f \; ; \end{cases} \tag{1}$$

Δu is the Laplacean of u, and $\sigma = \sqrt{2s} > 0$ the observation scale. The function u is known as the *scale-space representation* of f [5, 9, 10, 12, 13, 17, 19, 20].

However, one must be cautious when relating physical and psychophysical quantities, since the parametrisation of the former is essentially arbitrary. The parametrisation of the latter is whatever evolution has accomplished, which depends on the species of interest and its natural environment. Therefore, although Eq. (1) serves as the point of departure, we will account for all transformation degrees of freedom that preserve scale causality, notably reparametrisations of scale, space, and neural response. Applying the respective reparametrisations, with $u = \gamma(v)$ subject to $\gamma' > 0$, say, yields the following, generalised p.d.e.:

$$v_t = \frac{1}{\sqrt{h}} \sum_{\alpha,\beta=1}^{2} \nabla_\alpha \left(\sqrt{h}\, h^{\alpha\beta} \nabla_\beta v \right) + \mu(v) \sum_{\alpha,\beta=1}^{2} h^{\alpha\beta} \nabla_\alpha v \nabla_\beta v . \tag{2}$$

Here $h_{\alpha\beta}$ are the components of the metric tensor relative to a coordinate basis, $h^{\alpha\beta}$ is the inverse of the matrix $h_{\alpha\beta}$, and $h = \det h_{\alpha\beta}$. The scale parameter has been renamed t to distinguish it from s, of which it is some monotonic function. The linear term on the right hand side of Eq. (2) is the so-called "Laplace-Beltrami" operator, the nonlinear one accounts for nonlinear phototransduction. Indeed, the coefficient function $\mu = (\ln \gamma')'$ can be related to the Weber-Fechner law [6, 7]. In general, the flexibility of choosing γ so as to account for threshold, saturation and other nonlinear transfer phenomena, justifies linearity, which is the *raison d'être* of differential geometry. Here we concentrate on the construction of a suitable metric that can be related to foveal vision.

4 Foveation

Consider a conformal metric transform, or "metric rescaling", induced by the previously introduced retinocortical mapping $\varrho : (\mathcal{M}, \mathbf{h}) \longrightarrow (\mathcal{N}, \mathbf{g})$, and define $\mathbf{g}(q) = \varrho^{\text{inv}*} \mathbf{h}(\varrho^{\text{inv}}(q))$, such that (using sloppy shorthand notation):

$$\mathbf{g} = e^{2\zeta}\, \mathbf{h}, \tag{3}$$

in which $\zeta(r)$ is a radially symmetric, smooth scalar function defined everywhere on the retinal manifold except at the fiducial origin, $r = 0$, *i.e.* the foveal point; diagrammatically:

$$\begin{array}{ccc} \mathbf{h} \in T^*\mathcal{M}_p \otimes T^*\mathcal{M}_p & \xleftarrow{\varrho^*} & \mathbf{g} \in T^*\mathcal{N}_q \otimes T^*\mathcal{N}_q \\ \downarrow & & \downarrow \\ p \in \mathcal{M} & \xrightarrow{\varrho} & q \in \mathcal{N} \end{array} \tag{4}$$

If one departs from an unbiased Euclidean metric such a transform induces a spatial bias which depends only on radial distance from the foveal point[1].

A special property that holds only in $n=2$ dimensions is that if we take

$$v_s = \frac{1}{\sqrt{g}} \sum_{\alpha,\beta=1}^{2} \nabla_\alpha \left(\sqrt{g}\, g^{\alpha\beta} \nabla_\beta v\right) + \mu(v) \sum_{\alpha,\beta=1}^{2} g^{\alpha\beta} \nabla_\alpha v \nabla_\beta v, \tag{5}$$

as our basic equation, then using Eq. (3) this can be rewritten as

$$v_s = e^{-2\zeta} \left\{ \frac{1}{\sqrt{h}} \sum_{\alpha,\beta=1}^{2} \nabla_\alpha \left(\sqrt{h}\, h^{\alpha\beta} \nabla_\beta v\right) + \mu(v) \sum_{\alpha,\beta=1}^{2} h^{\alpha\beta} \nabla_\alpha v \nabla_\beta v \right\}. \tag{6}$$

In other words, we can interpret the metric rescaling as an eccentricity dependent size rescaling, for if

$$t = e^{-2\zeta} s + t_0, \tag{7}$$

for some constant t_0, then Eqs. (6–7) will take us back to Eq. (2). Furthermore, the assumption that $(\mathcal{M}, \mathbf{h})$ is flat[2] (or equivalently, has a vanishing Riemann curvature tensor) implies that $(\mathcal{N}, \mathbf{g})$ is flat as well, provided ζ is a harmonic function satisfying the Laplace equation. This follows from the fact that for $n=2$ the Riemann curvature tensor of $(\mathcal{M}, \mathbf{h})$ is completely determined in terms of the Riemann curvature scalar; if we denote the Riemann curvature scalar for $(\mathcal{M}, \mathbf{h})$ and $(\mathcal{N}, \mathbf{g})$ by $\mathbf{R}_\mathcal{M}$, respectively $\mathbf{R}_\mathcal{N}$, then the metric rescaling of Eq. (3) yields

$$e^{2\zeta} \mathbf{R}_\mathcal{N} = \mathbf{R}_\mathcal{M} + 2(n-1)\Delta\zeta + (n-1)(n-2)\|\nabla\zeta\|^2. \tag{8}$$

The reader is referred to the literature for details [2,8]. Thus if $\Delta\zeta = 0$ Eq. (3) relates flat metrics defined at various eccentricities $0 < r < R$. The solution is

$$\zeta_q(r) = q \ln \frac{r_0}{r}, \tag{9}$$

in which $r_0 > 0$ is some constant radius satisfying $0 < r_0 < R$, R is the radius of our *"geometric retina"*, and $q \in \mathbb{R}$ is a dimensionless constant. With a modest amount of foresight we will refer to r_0 as the radius of the *geometric foveola*. Physical considerations lead us to exclude $q \leq 0$, as this is inconsistent with actual cortical magnification, but an unambiguous choice remains undecided here on theoretical grounds. However, in order to connect to the well-known log-polar paradigm consistent with psychophysical and neurophysiological evidence (*v.i.*) we shall henceforth assume that $q=1$, writing $\zeta_1 = \zeta$ for simplicity. In this case, a Euclidean metric $\mathbf{h} = r^2 d\theta \otimes d\theta + dr \otimes dr$ implies

$$\mathbf{g} = \left(\frac{r_0}{r}\right)^2 \left(r^2 d\theta \otimes d\theta + dr \otimes dr\right). \tag{10}$$

---

[1] In practice one considers *equivalent eccentricity* to effectively enforce isotropy [15].
[2] This is basically an *ad hoc* assumption, but it is in some sense the simplest *a priori* option, and will turn out consistent with empirical findings.

Indeed, switching to log-polar coordinates[3], $(\overline{\theta}, \overline{r}) \in [0, 2\pi r_0) \times \mathbb{R}$, defined by

$$(\overline{\theta}, \overline{r}) = \chi(\theta, r) : \begin{cases} \overline{\theta} = r_0 \theta & \theta \in [0, 2\pi) \\ \overline{r} = r_0 \ln \dfrac{r}{r_0} & r > 0, \end{cases} \quad (11)$$

we obtain $\mathbf{g} = d\overline{\theta} \otimes d\overline{\theta} + d\overline{r} \otimes d\overline{r}$. This shows that relative to a log-polar coordinate chart the conformal metric becomes homogeneous. We can interpret the two-form

$$\chi^* \epsilon_{\mathcal{N}} = \chi^* \left(d\overline{\theta} \wedge d\overline{r} \right) = \left(\frac{r_0}{r} \right)^2 r d\theta \wedge dr = \left(\frac{r_0}{r} \right)^2 \epsilon_{\mathcal{M}}, \quad (12)$$

with unit two-form $\epsilon_{\mathcal{M}} \in \Lambda^2(T^*\mathcal{M}_p)$, as the eccentricity-weighted elementary area element at retinal position (r, θ). This implies that, resolution limitations apart, typical receptive field size should scale in direct proportion to eccentricity. There is ample evidence for this scaling phenomenon from psychophysics [1,18] as well as neurophysiology; *e.g.* typical dendritic field diameter of retinal ganglion cells appears to be directly proportional to eccentricity [4,15].

However, the log-polar transform exhibits a spurious singularity at $r=0$, and thus fails to be a viable model for the foveal centre. Since the problem pertains to a point of measure zero, it is better to consider Eq. (9) in conjunction with the physically meaningful scales of Eq. (7):

$$\sigma_\lambda(r) = \sigma_0 \sqrt{e^{2\lambda} \left(\frac{r}{r_0} \right)^2 + 1}. \quad (13)$$

Here we have reparametrised $s = t_0 e^{2\lambda}$ with $\lambda \in \mathbb{R}_0^+$; $\sigma_0 = \sqrt{2t_0}$ (corresponding to $\lambda = r = 0$) denotes the finest scale represented, something in the order of photoreceptor spacing, say. Eq. (13) reveals resolution limitations, unlike conformal metric or log-polar transform, Eqs. (10–11). It predicts receptive field sizes approximately proportional to eccentricity in the periphery, and approximately constant in the foveola, $r < r_0$: Fig. 2. Receptive fields satisfying this scaling behaviour for fixed λ constitute natural spatial assemblies, since they reflect the *a priori* acuity bias of the visual system. At each retinal location, moreover, we encounter scale assemblies of nested receptive fields parametrised by λ; these encode the *deep structure* of the retinal irradiance function at a fixed point [10].

5 Cortical Magnification

If we consider the smallest receptive fields at eccentricity r, and cover the retina with discs of radius $\sigma(r) = \sigma_0(r)$, Eq. (13), then we may count the number of discs in the immediate neighbourhood of the fovea, $0 \leq r < \rho$, and compare it to that in the periphery $\rho \leq r < R$. The fractional covering of $\sigma(r)$-discs inside a ring

---

[3] Strictly speaking neither polar nor log-polar coordinates are genuine coordinates in the technical sense of being locally diffeomorphic to Cartesian coordinates, due to the singularity at the foveal centre [16].

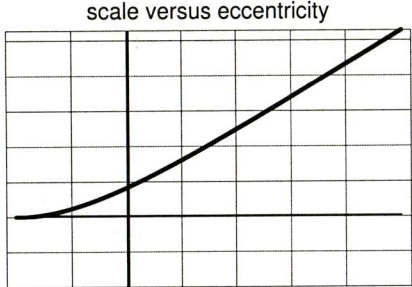

Fig. 2. Scale (vertical axis) *versus* eccentricity (horizontal axis). The graph is approximately linear in the periphery ($r \gg r_0$), and approximately constant in the geometric foveola ($r \ll r_0$). The vertical line indicates the radius of the geometric foveola, $r = r_0$. The horizontal asymptote, $\sigma = \sigma_0$, reveals the resolution limitation of the visual system.

of width dr at eccentricity r clearly equals $2\pi r dr$ divided by $\pi\sigma^2(r)$. Therefore, the respective numbers in central and peripheral regions are given by

$$\mathcal{N}_c = \left(\frac{r_0}{\sigma_0}\right)^2 \int_0^\rho \frac{2r\,dr}{r^2 + r_0^2} = \left(\frac{r_0}{\sigma_0}\right)^2 \ln\frac{\rho^2 + r_0^2}{r_0^2}, \quad (14)$$

$$\mathcal{N}_p = \left(\frac{r_0}{\sigma_0}\right)^2 \int_\rho^R \frac{2r\,dr}{r^2 + r_0^2} = \left(\frac{r_0}{\sigma_0}\right)^2 \ln\frac{R^2 + r_0^2}{\rho^2 + r_0^2}. \quad (15)$$

Let us now define ρ via the equality $\mathcal{N}_c = \mathcal{N}_p$, i.e.

$$\left(\frac{\rho}{r_0}\right)^2 + 1 = \sqrt{\left(\frac{R}{r_0}\right)^2 + 1}, \quad (16)$$

which is independent of σ_0 (therefore of any fixed relative overlap), but does depend on the size of the geometric foveola, or rather on the dimensionless ratio

$$\epsilon \stackrel{\text{def}}{=} \frac{r_0}{R}. \quad (17)$$

We shall refer to ρ as the *geometric equipartitioning radius*, as it separates retinal regions containing equally many receptive fields. Eq. (16) expresses an intimate relation between this radius and that of the geometric foveola:

$$\frac{\rho}{R} = \epsilon^{\frac{1}{2}}\left((1+\epsilon^2)^{\frac{1}{2}} - \epsilon\right)^{\frac{1}{2}} = \epsilon^{\frac{1}{2}} + \mathcal{O}(\epsilon^{\frac{3}{2}}). \quad (18)$$

Note that equipartitioning does *not* refer to areas, for if $\mathcal{A}_c = \pi\rho^2$ is the area occupied by half of all receptive fields, and $\mathcal{A}_r = \pi R^2$ the entire retinal area, then

$$\frac{\mathcal{A}_c}{\mathcal{A}_r} = \left(\frac{\rho}{R}\right)^2 \approx \epsilon, \quad (19)$$

radius of	in mm
whole retina[a]	21
central retina	6.0
equipartitioning[b]	2.0
parafovea	1.25
macula lutea[c]	0.75
foveola	0.20

[a] half distance from ora to ora across horizontal meridian
[b] radius of central retina serving 50 % of all ganglion cells
[c] half distance from foveal rim to foveal rim

Table 1. Some figures on human retina; 288 μm corresponds to 1° of visual angle. (Sources: Polyak [14] and Rodieck [15].)

where the latter approximation holds if $\varepsilon \ll 1$, recall Eq. (18).

Table 1 presents some figures on the human retina. It is estimated that about half of the striate cortex is devoted to a foveal region covering roughly 7° of visual angle, *i.e.* 1 % of the visual field [15]. Thus Eq. (19) yields $\varepsilon \approx 1\%$, from which we deduce that $r_0/R \approx 1\%$, cf. Eq. (17). For a retinal radius $R \approx 21$ mm this means that $r_0 \approx 210$ μm, which is indeed roughly the correct radius of the human foveola[4] [14]. Moreover, from Eq. (18) we find that $\rho/R \approx 10\%$, or $\rho \approx 2$ mm, the equipartitioning radius in the human retina by construction.

In addition, we can estimate the fraction of receptive fields involved in foveal vision (*i.e.* which have their base point in the foveola $r < r_0$) to the total number of receptive fields. If $\mathcal{N}_f$ and $\mathcal{N}_r$ denote the respective numbers (note that these follow from Eq. (14) by formal substitution $\rho \to r_0, R$ respectively), then

$$\frac{\mathcal{N}_f}{\mathcal{N}_r} = \frac{\ln 2}{\ln(1+\kappa^2)} = \log_2^{-1}(1+\kappa^2), \quad (20)$$

in which we have introduced

$$\kappa \stackrel{\text{def}}{=} \varepsilon^{-1} \gg 1, \quad (21)$$

the boundary curvature of the geometric foveola. For $\varepsilon \approx 1\%$ this ratio is approximately 7.5 %. To get a feeling for all these scales, cf. Fig. 3.

Finally, let us turn to a quantification of cortical magnification. It is usually defined as the size of visual cortex in mm that is devoted to 1° of visual angle [3]. In order to get rid of the somewhat arbitrary length scale, let us redefine it here as the size increment of visual cortex $d\ell(\alpha)/L$ per eccentricity increment $d\alpha = dr/R$, in which $L = \ell(1)$ denotes total size, and $\alpha \in [0,1]$ relative eccentricity. Reasoning similar to that which led to Eqs. (14–15) then yields, using Eqs. (17–21),

$$\ell'(\alpha) = \frac{1}{\ln(1+\kappa^2)} \frac{2\alpha\kappa^2}{1+\alpha^2\kappa^2} L. \quad (22)$$

---
[4] Note that it is not a circular argument to deduce the size of the foveola from cortical magnification; these have been introduced as independent concepts.

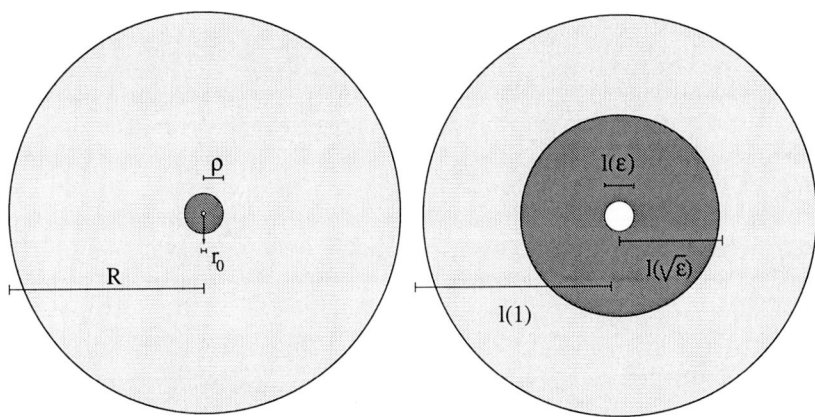

Fig. 3. Left: Decomposition of the geometric retina into three concentric regions: R is the radius of the entire retina, ρ is the radius of the central region containing 50% of all receptive fields, and r_0 is the radius of the geometric foveola. Right: Corresponding cortical regions after cortical magnification.

By integration it follows that a foveal neighbourhood of relative radius $\alpha = r/R$ claims a cortical area of size

$$\ell(\alpha) = \frac{\ln(1+\alpha^2\kappa^2)}{\ln(1+\kappa^2)} L , \qquad (23)$$

see Fig. 4.

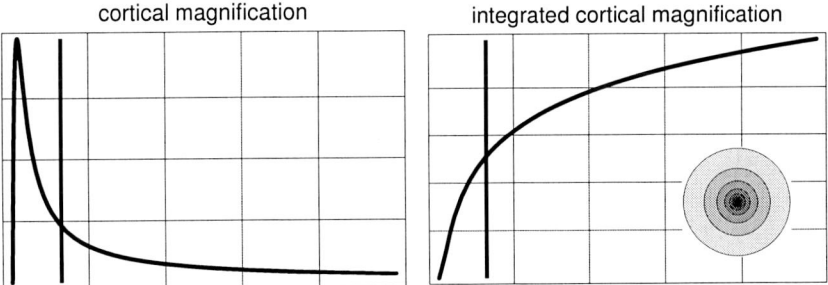

Fig. 4. Cortical magnification $\ell'(\alpha)/L$ and its integral $\ell(\alpha)/L$ as a function of eccentricity $\alpha \in [0,1]$. The peak in the left graph is found at the boundary of the geometric foveola, $\alpha = \epsilon$, i.e. $r = r_0$. The vertical line indicates the equipartitioning radius, $\alpha = \sqrt{\epsilon}$, i.e. $r = \rho$, defined as the (approximate) solution to the equation $\ell(\alpha) = L/2$. The inset shows some α-equidistant contours $\ell(\alpha) =$ constant.

6 Conclusion and Discussion

Retinocortical mechanisms of foveal vision can be understood in terms of a conformal metric transform in combination with scale-space axiomatics. It provides a realistic model for the typical eccentricity scaling of receptive field size. The familiar log-polar construct arises as an approximation valid outside the foveola. A remarkable conclusion is that knowledge of cortical magnification determines the relative size of the foveola, as well as the relative number of receptive fields involved. These theoretical predictions are in reasonable agreement with known facts on human retinocortical mechanisms.

Of course the human visual front-end is a lot more complicated than the way it has been modelled here. But in view of the descriptive power of differential geometry the interesting conjecture that it can be understood as a "geometry engine" does appear feasible.

A Geometric Terminology Loosely Explained

Here we list some of the geometric concepts introduced in the theory, with a brief explanation of their relation to visual modules in retina, LGN and striate cortex according to the proposed model:

$\mathcal{M}$: The spatial loci of photoreceptor cells within the retinal tissue, represented as a smooth manifold. This deemphasizes the discrete structure in favour of a continuum, which is closer to our percept of a non-discrete optical world.

$\mathcal{N}$: Similar to $\mathcal{M}$, but representing the spatial substrate of ganglion, simple and complex cells in LGN and striate cortex, i.e. the collection of base points receptive fields "are looking at".

$\mathcal{E}$: The "fibres over $\mathcal{M}$" consisting of all photoreceptor units stratified according to spatial position in the retina (for given point resolution).

$\mathcal{F}$: The "fibres over $\mathcal{N}$", the totality of ganglion, simple and complex cells in LGN and striate cortex stratified according to "cortical" position, i.e. the base point in $\mathcal{N}$ to which they are attached.

ϱ: Retinocortical mapping which couples a unique point in $\mathcal{N}$ retinotopically to any given one in $\mathcal{M}$, vice versa, roughly identifiable with the optic tract.

$\pi_{\mathcal{E}}(\phi) = \mathrm{p} \in \mathcal{M}$: The base point in the retina that forms the "centre of mass" for the photoreceptor $\phi \in \mathcal{E}$. This may be a many-to-one mapping depending on point resolution. Reversely, $\pi_{\mathcal{E}}^{\mathrm{inv}}(\mathrm{p}) \subset \mathcal{E}$ is the subset of all photoreceptors located at p (for given point resolution).

$\pi_{\mathcal{F}}(\psi) = \mathrm{q} \in \mathcal{N}$: The base point in the LGN or striate cortex that forms the "centre of mass" for the receptive field $\psi \in \mathcal{F}$ (obviously a many-to-one mapping). This can be related to a unique corresponding point in the retina by virtue of the retinocortical mapping ϱ, viz. $\mathrm{p} = \pi(\psi) \in \mathcal{M}$ with backprojection map $\pi = \varrho^{\mathrm{inv}} \circ \pi_{\mathcal{F}}(\psi)$. Reversely, $\pi_{\mathcal{F}}^{\mathrm{inv}}(\mathrm{q}) \subset \mathcal{F}$ (or $\pi^{\mathrm{inv}}(\mathrm{p}) \subset \mathcal{F}$) is the subset of all receptive fields with "centre of mass" at $\mathrm{q} \in \mathcal{N}$ (respectively $\mathrm{p} \in \mathcal{M}$).

References

1. P. Bijl. *Aspects of Visual Contrast Detection*. PhD thesis, University of Utrecht, Department of Physics, Utrecht, The Netherlands, May 8 1991.
2. Y. Choquet-Bruhat, C. DeWitt-Morette, and M. Dillard-Bleick. *Analysis, Manifolds, and Physics. Part I: Basics*. Elsevier Science Publishers B.V. (North-Holland), Amsterdam, 1991.
3. P. M. Daniel and D. Whitteridge. The representation of the visual field on the cerebral cortex in monkeys. *Journal of Physiology*, 159:203–221, 1961.
4. B. Fischer and H. U. May. Invarianzen in der Katzenretina: Gesetzmäßige Beziehungen zwischen Empfindlichkeit, Größe und Lage rezeptiver Felder von Ganglienzellen. *Experimental Brain Research*, 11:448–464, 1970.
5. L. M. J. Florack. *Image Structure*, volume 10 of *Computational Imaging and Vision Series*. Kluwer Academic Publishers, Dordrecht, The Netherlands, 1997.
6. L. M. J. Florack. Non-linear scale-spaces isomorphic to the linear case. In B. K. Ersbøll and P. Johansen, editors, *Proceedings of the 11th Scandinavian Conference on Image Analysis (Kangerlussuaq, Greenland, June 7–11 1999)*, volume 1, pages 229–234, Lyngby, Denmark, 1999.
7. L. M. J. Florack, R. Maas, and W. J. Niessen. Pseudo-linear scale-space theory. *International Journal of Computer Vision*, 31(2/3):247–259, April 1999.
8. L. M. J. Florack, A. H. Salden, B. M. ter Haar Romeny, J. J. Koenderink, and M. A. Viergever. Nonlinear scale-space. In B. M. ter Haar Romeny, editor, *Geometry-Driven Diffusion in Computer Vision*, volume 1 of *Computational Imaging and Vision Series*, pages 339–370. Kluwer Academic Publishers, Dordrecht, 1994.
9. B. M. ter Haar Romeny, L. M. J. Florack, J. J. Koenderink, and M. A. Viergever, editors. *Scale-Space Theory in Computer Vision: Proceedings of the First International Conference, Scale-Space'97, Utrecht, The Netherlands*, volume 1252 of *Lecture Notes in Computer Science*. Springer-Verlag, Berlin, July 1997.
10. J. J. Koenderink. The structure of images. *Biological Cybernetics*, 50:363–370, 1984.
11. J. J. Koenderink. The brain a geometry engine. *Psychological Research*, 52:122–127, 1990.
12. J. J. Koenderink and A. J. van Doorn. Receptive field families. *Biological Cybernetics*, 63:291–298, 1990.
13. T. Lindeberg. *Scale-Space Theory in Computer Vision*. The Kluwer International Series in Engineering and Computer Science. Kluwer Academic Publishers, Dordrecht, The Netherlands, 1994.
14. S. L. Polyak. *The Retina*. University of Chicago Press, Chicago, 1941.
15. R. W. Rodieck. *The First Steps in Seeing*. Sinauer Associates, Inc., Sunderland, Massachusetts, 1998.
16. M. Spivak. *Differential Geometry*, volume 1–5. Publish or Perish, Berkeley, 1975.
17. J. Sporring, M. Nielsen, L. M. J. Florack, and P. Johansen, editors. *Gaussian Scale-Space Theory*, volume 8 of *Computational Imaging and Vision Series*. Kluwer Academic Publishers, Dordrecht, The Netherlands, 1997.
18. F. W. Weymouth. Visual sensory units and the minimal angle of resolution. *American Journal of Ophthalmology*, 46:102–113, 1958.
19. A. P. Witkin. Scale-space filtering. In *Proceedings of the International Joint Conference on Artificial Intelligence*, pages 1019–1022, Karlsruhe, Germany, 1983.
20. R. A. Young. The Gaussian derivative model for machine vision: Visual cortex simulation. *Journal of the Optical Society of America*, July 1986.

Tangent Fields from Population Coding

Niklas Lüdtke, Richard C. Wilson, and Edwin R. Hancock

Department of Computer Science, University of York, York YO10 5DD, UK

Abstract. This paper addresses the problem of local orientation selection or tangent field estimation using population coding. We use Gabor filters to model the response of orientation sensitive units in a cortical hypercolumn. Adopting the biological concept of *population vector decoding* [4], we extract a continuous orientation estimate from the discrete set of responses in the Gabor filter bank which is achieved by performing vectorial combination of the broadly orientation-tuned filter outputs. This yields a population vector the direction of which gives a precise and robust estimate of the local contour orientation. We investigate the accuracy and noise robustness of orientation measurement and contour detection and show how the certainty of the estimated orientation is related to the shape of the response profile of the filter bank. Comparison with some alternative methods of orientation estimation reveals that the tangent fields resulting from our population coding technique provide a more perceptually meaningful representation of contour direction and shading flow.

1 Introduction

Although much of the early work on low-level feature detection in computer vision has concentrated on the characterisation of the whereabouts of intensity features [12], [2], the problem is a multi-faceted one which also involves the analysis of orientation and scale. In fact, one of the criticisms that can be levelled at many of these approaches is that they are goal directed [2] and fail to take into account the richness of description that appears to operate in biological vision systems. Only recently have there been any serious attempts to emulate this descriptive richness in building image representations.

To illustrate this point, we focus on the topic of orientation estimation. Here there is a consiserable body of work which suggests that tangent field flows play an important role in the interpretation of 2D imagery and the subsequent perception of shape [20], [21], [14], [1]. Despite some consolidated efforts, however, the extraction of reliable and consistent tangent fields from 2D intensity images has proved to be computationally elusive. One of the most focussed and influential efforts to extract and use tangent fields for computational vision has been undertaken by Zucker and his co-workers. Here the contributions have been to show how to characterise tangent directions [20], to improve their consistency using relaxation operations and fine-spline coverings [21] and how to use the resulting information for the analysis of shading flows [1].

Although it is clear that cortical neurons are capable of responding to a diversity of feature contrast patterns, orientations and scales, computational models of how to combine these responses have also proved quite elusive. Recently, however, there has been a suggestion that neuronal ensembles of the sort encountered in cortical hypercolumns can be conveniently encoded using a *population vector* [4]. In fact, population coding has become an essential paradigm in cognitive neuroscience over the past decade and is increasingly studied within the neural network community [18], [17], [15], [13], [10]. In the vision domain, Vogels [18] examined a model of population vector coding of visual stimulus orientation by striate cortical cells. Based on an ensemble of broadly orientation-tuned units, the model explains the high accuracy of orientation discrimination in the mammalian visual system. Gilbert and Wiesel [5] used a very similar approach to explain the context dependence of orientation measurements and related this to physiological data and the psychophysical phenomenon of "tilt illusion".

Our aim in this paper is to make some first steps in the development of a computational model for feature localisation and tangent field estimation in a setting provided by population coding. Following the work of Heitger, von der Heydt and associates, we take as our starting point an approach in which the orientation selective complex cells found in a cortical hypercolumn are modelled by the reponse moduli of a bank of complex Gabor filters [6]. We show how simple vectorial addition of the individual filter responses produces a population vector which accurately encodes the orientation response of the hypercolumnar ensemble. Although each individual filter has very broad orientation tuning, the vectorial combination of the set of responses can yield a much better estimate of local stimulus orientation. In particular, by investigating the resulting orientation measurements under noise, we demonstrate that the population coding method is accurate and robust.

Moreover, from the distribution of responses in the filter-bank, a certainty measure can be calculated, which characterizes the reliability of the orientation measurement *independent of contour contrast*. This allows us to distinguish points of high local anisotropy (contour points) from noisy regions, corner points and junctions.

2 Orientation Tuning of Gabor Filters

Gabor filters are well-known models of orientation selective cells in striate cortex [3] that have found numerous applications in the computer vision literature including edge detection [16], texture analysis and object recognition [8]. In this paper we use the moduli of the quadrature pair responses ("local energy"), since their response properties resemble those of complex cells where responses do not depend on contrast polarity (edges vs. lines) and are robust against small stimulus translations within the receptive field [6].

Like simple and complex cells in striate cortex, Gabor filters have rather broad orientation tuning. The essential control parameter for the shape of the filter kernels is the width of the Gaussian envelope which we will refer to as

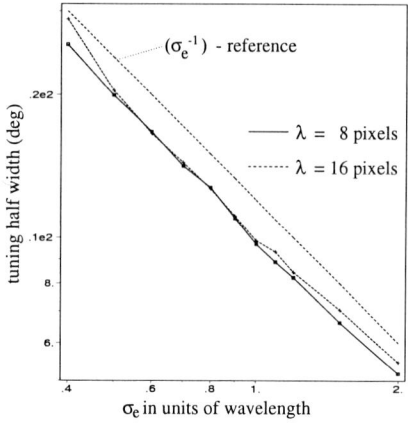
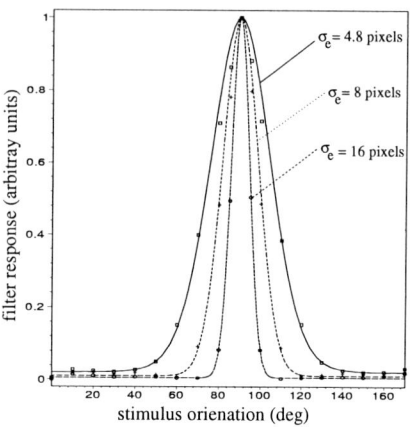

Fig. 1. The log-log-plot of the tuning half-width as a function of the kernel width illustrates the uncertainty relation between tuning width and spatial width: $w \propto (\sigma_e)^{-1}$.

Fig. 2. Normalized empirical tuning curve for the moduli of three Gabor-filters ($\lambda = 8$ pixels, vertical orientation). The half-widths are $w = 17.2°$, $9.7°$ and $5.2°$.

σ_e. A value of three times σ_e can be considered the *radius of the "receptive field"* since the Gaussian envelope virtually vanishes at greater distances from the center. To determine the tuning curve and to examine the influence of the envelope width on the tuning width, we have used synthetic images of straight lines with orientations ranging from 0° to 170° as test stimuli. Figure 2 shows the tuning curves for three filters which had a preferred orientation $\theta_{pref} = 90°$, a wavelength $\lambda = 8$ pixels and envelope widths $\sigma_e = 0.6$ (4.8 pixels), 1.0 (8 pixels) and 2.0 (16 pixels), respectively. The estimated half-widths of the tuning curves are $w = 17.2°$, $9.7°$ and $5.2°$. The first value is comparable to typical orientation tuning half-widths of striate cortical cells [18]. Interestingly, as will be demonstrated later, this turns out to be the most suitable tuning width for orientation measurement.

The data in Figure 2 has been fitted with a model tuning function. Here we use the von Mises function [11], since this is appropriate to angular distributions:

$$f(\theta; \theta_{pref}) = e^{\kappa \cos[2(\theta - \theta_{pref})]} - f_0. \tag{1}$$

θ_{pref} is the preferred orientation of the filter, κ is the width parameter which plays a similar role to the width σ of a Gaussian and f_0 is a "mean activity" which models the effect of discretization noise. In Figure 1 the tuning width is plotted as a function of the kernel width σ_e. The two quantities are inversely proportional to one-another due to the general uncertainty relation between orientational bandwidth (w) and spatial width (σ_e). In fact, Gabor filters have been shown to minimize the quantity $w \sigma_e$ [3].

The main conclusion to be drawn from this data is that Gabor filter banks provide rather coarse estimates of feature orientation unless the full range of orientations is sampled with a large number of filters, which is obviously highly inefficient. In the next section we demonstrate that when *population coding* is used to represent the convolution responses of the filter bank, the outputs of only a small number of filters need to be combined in order to achieve a considerable improvement of the precision of orientation estimation.

3 The Population Vector of a Gabor Filter Bank

The concept of a population vector has been introduced by Georgopoulos and collegues in order to describe the representation of limb movements by direction sensitive neurons in motor cortex [4]. According to their definition, the population vector for a set of n Gabor-filters is computed as follows [4]: Consider a wavelength λ. Let $G(x, y; \theta_i, \lambda)$ be the response modulus ("energy") of a quadrature pair of Gabor filters of orientation θ_i. Let $e_i = (\cos \theta_i, \sin \theta_i)^T$ be the unit vector in the direction θ_i. Then the population vector $\boldsymbol{p}$ is:

$$\boldsymbol{p}(x,y) = \sum_{i=1}^{n} G(x, y; \theta_i, \lambda) \, \boldsymbol{e}_i \,, \qquad (2)$$

meaning that each filter is represented by a two component vector. The vector orientation and magnitude are given by the preferred orientation θ_i and the response magnitude (modulus) $G(x, y; \theta_i)$ of the filter at location (x, y). The population vector is the sum of the n filter vectors.

However, equation (2) cannot be directly applied since we are dealing with filters sensitive only to orientation but not direction, i.e., there is a $180°$-ambiguity. To take this into account the population vector is computed using the scheme in Figure 3. This is the process of *encoding* the stimulus orientation. The *decoding* is achieved by determining the orientation of the population vector θ_{pop}, which is given by:

$$\theta_{pop}(x,y) = \arctan\left[p_y(x,y)/p_x(x,y)\right] \qquad (3)$$

The magnitude of the population vector $\|\boldsymbol{p}(x,y)\|$ is related to the response "energy" of the filter bank at position (x, y). If evaluated at contour locations, i.e. local maxima of $\|\boldsymbol{p}\|$, θ_{pop} gives an estimate of the local tangent angle. Theoretically, the coding error can be made as small as desired by applying more and more filters. However, computational cost and discretization errors in digital images put limitations on the optimal number of filters.

4 Performance of Orientation Estimation

In this section we investigate the accuracy of the orientation estimate in digital images delivered by filter-banks of different sizes. We examine the dependence

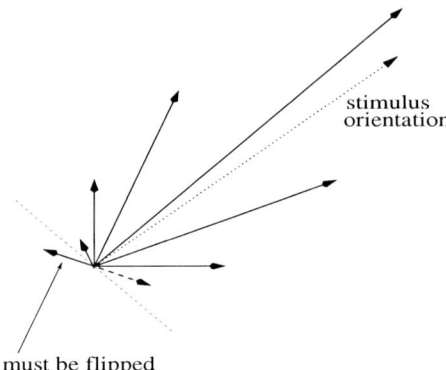

Fig. 3. Orientations are restricted to the range $0° - 180°$. Therefore, the vector components are computed with respect to a symmetry axis, in this case the orientation of maximum response. Components outside the $\pm 90°$ range around the axis have to be "flipped" back into that range to enforce a symmetrical arrangement. A component perpendicular to the symmetry axis (i.e. on the dashed line) would effectively cancel itself out and can thus be ignored.

of the error on tuning width and noise level. The test images and the filter wavelength are the same as previously described in Section 2. The filter-banks consist of 8, 16 and 32 Gabor filters. Figure 4 shows the root mean square (rms) error of the population angle $\langle \delta\theta_{pop} \rangle_{rms}$ as a function of the tuning half-width of the applied Gabor filters. The error increases when the tuning width is too small to guarantee sufficient filter overlap to cover the whole range of 180 degrees (given the number of filters). In our experiments this limit has never been reached for 32 filters. Here the error seems to decrease further for decreasing tuning width. However, the receptive field size then becomes large and the spatial resolution is no longer optimal. For large tuning widths the envelope parameter is so small that the whole receptive field consists of only a few pixels and discretization errors become noticeable. As a result, eight filters should be sufficient for practical purpose, since the computational cost is low and the precision only slightly smaller than with 16 filters.

Compared to the tuning width of a single Gabor filter, the population vector estimate of stimulus orientation is very accurate. The resulting rms-deviation of the angle of the population vector from the ground truth value of the stimulus orientation was only $\langle \delta\theta_{pop} \rangle_{rms} \approx 1°$. This should be compared with the half-width of the tuning curve for the most suitable filter ($w \approx 17°$). The error of the population coded orientation estimate consists of two components: the coding error due to the limited number of filters and the discretization error due to the lattice structure of digital images. Moreover, the measured rms-error is consistent with simulations by Vogels [18].

We also experimented with the sensitivity to additive Gaussian noise. Figure 5 shows the rms-error as a function of the noise variance for different numbers of

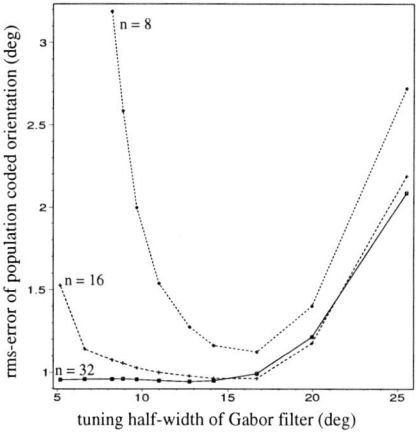 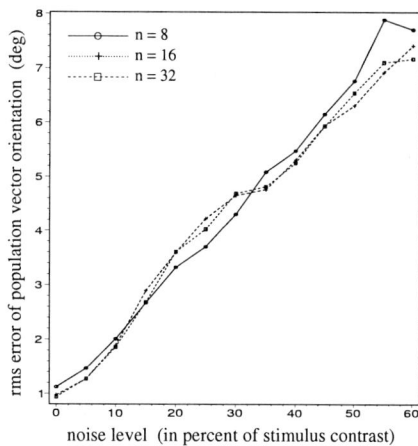

Fig. 4. Dependence of the rms-error of orientation of the population vector on the tuning half-witdth for 8, 16 and 32 filters. The wavelength is $\lambda = 8$ pixels. A minimum of rms-error occurs at a tuning width similar to that of cells in striate cortex.

Fig. 5. The rms-error as a function of the noise variance for additive Gaussian noise using 8, 16 and 32 filters. The method seems to be quite robust. The filter banks show no significant difference in noise sensitivity.

filters (8,16 and 32). The dependence is roughly linear for all three filter banks with no significant difference in noise sensitivity.

5 Analysis of Filter Bank Response Profiles

The response profile of the filter bank, i.e., the angular distribution of filter outputs at a given point in the input image, contains valuable information of the local contour structure. Zemel and collegues proposed to represent certainty of local information in terms of the sum of responses [19]. In our notation this yields: $C(x,y) = \sum_i^n G_i(x,y) / \left(\sum_i^n G_i\right)_{max}$, where the denominator is some global maximum of the summed reponses. However, this measure only depends on response energy (contour contrast) and cannot discriminate between low contrast contours and intense noise. Also, points of multimodal anisotropy, such as corners (points of high curvature) and junctions, can produce high responses in the filter bank, though local tangent orientation is ill-defined.

We argue that the "sharpness" of the response profile is more suitable to characterize the reliability of the local orientation estimate, as it is contrast independent. Thus, we have contour contrast and certainty as two separate pieces of information. In fact, there is evidence that perceived contrast and the appearance of contours is not so closely linked as is commonly assumed [7].

Figure 6 shows the response profile of the filter bank at a number of different points in a natural image. Despite the fact that the response profiles are nor-

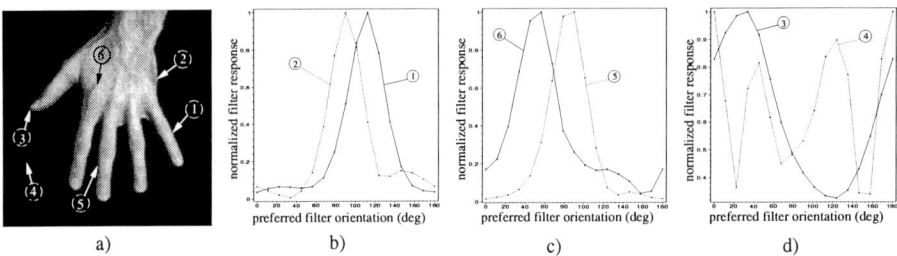

Fig. 6. Normalized response profiles of the filter bank (16 filters, $\lambda = 8$ pixels, $\sigma_e = 0.6 \times \lambda$) at different points in a natural image (a). Note that the distribution of filter responses reflects the quality of the edge. The width of the response profile allows the distinction of noisy regions from weak contours independent of their contrast level. The half-widths at well pronounced edges are very similar to the tuning width of the filter obtained from synthetic line images: compare $w = 17.2°$ with $w_1 = 19.8°$, $w_2 = 21.5°$, $w_5 = 21.3°$, $w_6 = 24.2°$ and $w_3 = 52.9°$.

malized, the quality of the edge (degree of anisotropy) and, thus, the expected reliability of orientation measurement, is well-reflected in the width of the profile. Accordingly, certainty should be measured in terms of the angular variance of the response energy distribution. At a contour the reponse energy of the filter bank can be assumed to be clustered around the contour angle. Therefore, we use the average of the cosines of orientation differences weighted by the responses:

$$C(x,y) = \left(\frac{\sum_i^n G_i(x,y) \, \cos\left[\frac{\pi}{2} - |\frac{\pi}{2} - \Delta\theta_i|\right]}{\sum_i^n G_i(x,y)} \right)^2, \quad \Delta\theta_i = |\theta_i - \theta_{pop}| \quad (4)$$

Here θ_i is the filter orientation and θ_{pop} the population coded contour angle. The average is squared to stretch the range of values. The certainty measure effectively has an upper bound less than unity since the maximum degree of response clustering is limited by the tuning width. Additionally, every certainty value below 0.5 means total unreliability since $C = 0.5$ corresponds to 45° and any response clustering further away than 45° from the measured orientation simply means that there is a multimodal distribution. A possible way of normalizing the certainty measure would be to divide it by the largest certainty value detected.

6 Tangent Fields

In this section we combine our population coding techniques for the measurement of orientation, response energy and certainty to obtain tangent fields. The results are promising, not only as far as contour detection is concerned but also for the purpose of representing tangent flow fields. Figure 7 shows a detail from the tangent flow field corresponding to an infra-red aerial image obtained with

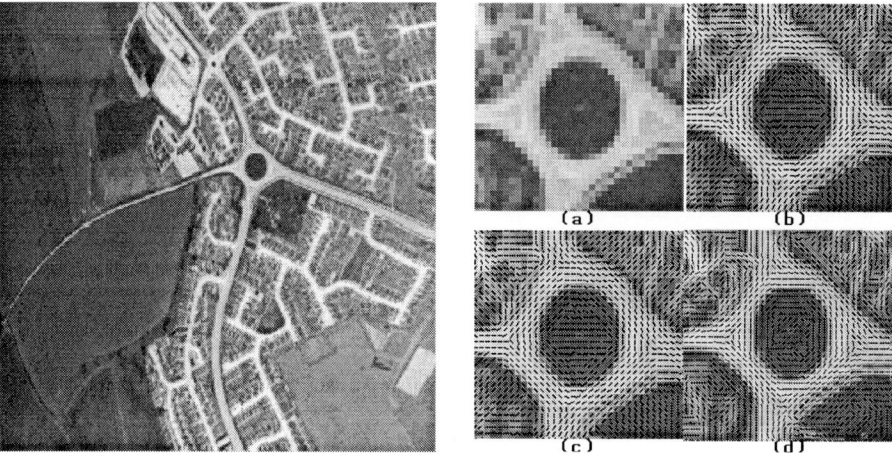

Fig. 7. Left: an Infra-red aerial image. Right: magnified detail (roundabout). (a) original, (b) tangent field following [21], (c) tangent field obtained by selecting the strongest response from 8 S-Gabor filters, (d) tangent field from population coding with 8 Gabor filters ($\lambda = 3$pixels). Some fine details in the flow field are better preserved than with the other methods. Images (a)-(c) after [9].

different methods of orientation measurement. We compare Zucker's method (which uses second Gaussian derivatives, [21]), selection of tangent orientation from S-Gabor filters ([9]) and population vector coding. Note how the population vector approach is able to recover some fine details in the flow that have been lost by the other algorithms due to smoothing.

6.1 Contour Representation

In order to represent contours by means of local line segments from the tangent field it is necessary to select those line segments on the crestlines of the response energy landscape given by the magnitude of the population vector. In the terminology of Parent and Zucker this problem is referred to as the search for *lateral maxima* [14]. Our algorithm first performs the local maximum search on the product of certainty and response energy, $\|\boldsymbol{p}(x,y)\| C(x,y)$. Thus, we exclude points of high curvature or junctions where orientation measurement is not well-defined as well as noisy regions where virtually no orientational structure is present. In a subsequent step, spurious parallels are eliminated through competition among neighbouring parallel line segments. The remaining points undergo thresholding. As a result, only points of high contrast and high certainty "survive".

Once these "key points" of reliable contour information and the corresponding tangent angles are determined, they provide a symbolic representation in terms of local line segments (figure 8). Moreover, they could serve as an ini-

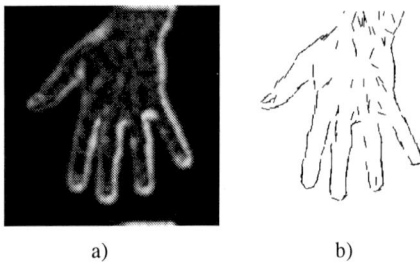

a)　　　　　　　　　b)

Fig. 8. Contour representation from population coded tangent fields. a) the directional energy map, b) the contour tangent field of a natural image (human hand, see fig. 6a). The tangents in Image b) represent the local orientation at "key points", i.e., local maxima of the product of directional energy and certainty.

tialization of knodes in a graph representation and be further updated by more global constraints but this is beyond the scope of this paper.

7   Conclusion and Discussion

We have shown how population coding can be used to accurarely estimate the orientation of grey-scale image features from Gabor filter responses. This raises the question of the biological plausibility of the population vector approach. The fact that a population vector interpretation allows read-out of the information encoded by a neural ensemble through the experimentor [4] does not mean that such decoding is actually performed in the brain [15], [13], [10]. It is more likely that distributed coding is maintained to secure robustness against noise and loss of neurons. We do *not* claim that our algorithm models the cortical processing of orientation information. However, all the operations necessary for computing the population vector could easily be realized by cortical neural networks. Also, it turns out that the optimal performance of orientation estimation by our system is reached when the tuning width of the filters resembles that of striate cortical cells.

Apart from the ongoing neurobiological debate, the popuation vector approach presented here has shown to be an efficient tool for computational vision. In the future we plan to extend research towards probabilistic interpretations of population coding [19] in order to recover full probability distributions not only of orientation, but also of more complex features such as curve segments or shading flow fields.

References

1. Breton, P., Iverson, L.A., Langer, M.S., Zucker, S.W. (1992) *Shading Flows and Scenel Bundles: A New Approach to Shape from Shading*, Proeceeding of the European Conf. on Comp. Vision (ECCV'92), lecture notes in computer science 588, Springer Verlag, pp. 135-150

2. Canny, J. (1986) *A Compuational Approach to Edge Detection* IEEE Transact. on Patt. Rec. and Machine Intell., (PAMI) 8(6), pp. 679-700
3. Daugman, J. (1985) *Uncertainty Relation for Resolution in Space, Spatial Freuqency and Orientation Optimized by Two-dimensional Visual Cortical Filters*, Journal of the Optical Society of America, 2, pp. 1160-1169
4. Georgopoulos A.P. , Schwarz, A.B. and Kettner, R.E. (1986) "Neural Population Coding of Movement Direction", *Science* 233, pp. 1416-1419
5. Gilbert, C. and Wiesel, T.N. (1990) *The Influence of Contextual Stimuli on the Orientation Selectivity of Cells in Primary Visual Cortex of the Cat*, Vis. Res., 30(11), pp. 1689-1701
6. Heitger, F. , Rosenthaler, L. , von der Heydt, R. , Peterhans, E. , Kübler, O. (1992) *Simulation of Neural Contour Mechanisms: From Simple to End-stopped Cells*, Vis. Res. 32(5), pp. 963-981
7. Hess, R.F. , Dakin, S.C. and Field, D.J. (1998) *The Role of "Contrast Enhancement" in the Detection and Appearance of Visual Contrast*, Vis. Res. 38(6), pp. 783-787
8. Lades, M., Vorbrüggen, J., Buhmann, J., Lange, J., von der Malsburg,C. , Würtz, R.P., and Konen, W. (1993) *Distortion Invariant Object Recognition in the Dynamic Link Architecture*, IEEE Trans. on Computers, 42(3), pp. 300-311
9. Leite, J.A.F. and Hancock, E. (1997) *Iterative Curve Organization*, Pattern Recognition Letters 18, pp. 143-155
10. Lehky, S.R. and Sejnowski, T.J.(1998) *Seeing White: Qualia in the Context of Population Decoding*, Neural Computation 11, pp. 1261-1280
11. Mardia, K.V. (1972), *Statistics of Directional Data*, Academic Press
12. Marr, D. (1982) Vision: A Computational Investigation into the Human Representtion and Processing of Visual Information. Freeman, San Francisco
13. Oram, M.W., Földiàk, P., Perrett, D.I. and Sengpiel, F. (1998) "The 'Ideal Homunculus': Decoding Neural Population Signals", *Trends in Neuroscience* 21 (6), pp. 259-265
14. Parent, P. and Zucker, S.W. (1989) *Trace Inference, Curvature Consistency, and Curve Detection*, IEEE Trans. on Patt. Anal. and Machine Intell., 11, pp. 823-839
15. Pouget, A., Zhang, K. (1997) "Statistically Efficient Estimation Using Cortical Lateral Connections", *Advances in Neural Information Processing* 9, pp. 97-103
16. Shustorovich, A. (1994) *Scale Specific and Robust Edge/Line Encoding with Linear Combinations of Gabor Wavelets*, Pattern Recognition, vol.27(5), pp. 713-725
17. Snippe, H.P. (1996) *Parameter Extraction form Population Codes: A Ciritcal Assessment*, Neural Computation 8, pp. 511-529
18. Vogels, R. (1990) "Population Coding of Stimulus Orientation by Striate Cortical Cells", *Biological Cybernetics* 64, pp. 25-31
19. Zemel, R. , Dayan, P. and Pouget, A. (1998) *Probabilistic Interpretation of Population Codes*, Neural Computation, 10(2), pp. 403-430
20. Zucker, S.W. (1985) *Early Orientation Selection: Tangent Fileds and the Dimensionality of their Support*, Comp. Vis., Graph., and Image Process., 32, pp. 74-103
21. Zucker, S.W., David, C., Dobbins, A., Iverson, L. (1988) *The Organization of Curve Detection: Coarse Tangent Fields and Fine Spline Coverings*, Proc. of the 2nd Int. Conf. on Comp. Vision (ICCV'88), pp. 577-586

Efficient Search Technique for Hand Gesture Tracking in Three Dimensions

Takanao Inaguma, Koutarou Oomura,
Hitoshi Saji, and Hiromasa Nakatani

Computer Science, Shizuoka University
Hamamatsu 432-8011, Japan
s4009@cs.inf.shizuoka.ac.jp

Abstract. We describe a real-time stereo camera system for tracking a human hand from a sequence of images and measuring the three dimensional trajectories of the hand movement. We incorporate three kinds of constraints into our tracking technique to put restrictions on the search area of targets and their relative positions in each image. The restrictions on the search area are imposed by continuity of the hand locations, use of skin color segmentation, and epipolar constraint between two views. Thus, we can reduce the computational complexity and obtain accurate three-dimensional trajectories. This paper presents a stereo tracking technique and experimental results to show the performance of the proposed method.

1 Introduction

Computer analysis of gestures has been widely studied to provide a natural interface in human-computer interaction [1], [2]. Especially, hand gesture recognition has many promising applications that provide new means of manipulation in artificial and virtual environments. Using three-dimensional hand gesture models, we can track the motions precisely [3]. Such methods, however, can be time-consuming, and automated model fitting is not easy. Color, background image, and difference between frames are also used as keys for tracking the targets [4,5,6,7,8,9,10,11]. Though their computational complexity is relatively low, tracking precision is not so high as the methods that use three-dimensional models.

Recovery of hand trajectory in three dimensions plays an important role in gesture recognition. However, methods using only a single camera hardly track the three-dimensional movements. We present a real-time stereo camera system for tracking a human hand from a sequence of images and measuring the three dimensional trajectories of the hand movement. We incorporate three kinds of constraints into our tracking technique to put restrictions on the search area of targets and their relative positions in each image, so that we can reduce the computational complexity and obtain accurate three-dimensional trajectories in real-time.

In the following, we introduce the new stereo tracking technique and describe our feature matching scheme. We also present experimental results to show the performance of the proposed method.

2 Experimental Arrangement

The essential elements of the experimental arrangement are two video cameras (SONY DCR-VX1000). The distance of the baseline between the two cameras is set to 0.25m. A subject is asked to move his/her hand and keep the palm in the direction of the cameras. The subject is positioned at a distance of 2.4m from the baseline. In this paper, we assume that the background is motionless and in uniform color so that we can segment the human region easily from the input image. The assumptions we made may seem unrealistic, but the technique we present in this paper has such a practical application like recognition of instructions by human gesture that will replace the conventional keyboards or mouses. The system tracks the motion of the palm in stereo and calculates the three-dimensional locations from a sequence of images; each image is captured at the video frame rate of 30Hz on the resolution of 320 × 240 pixel.

Many techniques have been investigated concerning human detection [12],[13]. We set aside the human detection for those studies, and we concentrate on the tracking of the hand gesture.

3 Tracking Procedure

The hand tracking technique consists of the following steps:

1. camera calibration and parameter estimation
2. initialization of template matching
3. computing the three-dimensional trajectory
4. restriction on the search area and tracking the hand motion
5. return to 3

Each step is described in the following subsections.

3.1 Camera Calibration and Parameter Estimates

For measuring the three dimensions of an object, we need the internal and external camera calibration before commencing three dimensional measurements. Therefore, from image correspondences between two views of an object with known size, we estimate the camera parameters such as the orientation and position of each camera.

Let (x, y, z) be an object point, (x', y') be the corresponding image point. Then, the transformation between the object and image point is represented by the camera parameters s_i [14].

$$x'(s_1 x + s_2 y + s_3 z + 1) + s_4 x + s_5 y + s_6 z + s_7 = 0 \qquad (1)$$

$$y'(s_1x + s_2y + s_3z + 1) + s_8x + s_9y + s_{10}z + s_{11} = 0 \qquad (2)$$

Thus, we can obtain the camera parameters by solving the above equations for each camera. Though six points are enough to obtain those 11 parameters, we may as well use more than 6 points to make the measurement more accurate. We use 7 points in this work.

3.2 Initialization of Template Matching

3.2.1 Initial Template in the Left Image

The subject is asked to keep their hands from their face so that the hand region and the face region in the image can be clearly separated. The initial shape of the template is determined as follows:

At first, we transform the input image into YCrCb color space to extract skin color regions [15]. The transformation from RGB to YCrCb is described as:

$$\begin{pmatrix} Y \\ Cr \\ Cb \end{pmatrix} = \begin{pmatrix} 0.2990 & 0.5870 & 0.1140 \\ 0.5000 & -0.4187 & -0.0813 \\ -0.1687 & -0.3313 & 0.5000 \end{pmatrix} \begin{pmatrix} R \\ G \\ B \end{pmatrix} + \begin{pmatrix} 0 \\ 128 \\ 128 \end{pmatrix} \qquad (3)$$

The values of Cr and Cb in skin color region are in the range of $133 \leq Cr \leq 173, 77 \leq Cb \leq 127$ [15]. We show the skin color segmentation in Fig.1, where skin color regions are depicted by white pixels and the remainder in black. Using this binary image, we search for the biggest square inside which all the pixels are white, and set the square as the initial template for the hand.

input image

skin color segmentation

Fig. 1. Skin color segmentation

3.2.2 Initial Template in the Right Image

After we locate the initial template in the left image, we set the initial template in the right image by considering the correspondence between the right and left images. If we set the templates independently in the two views, we might estimate inaccurate three-dimensional positions.

The initial position of the right template is determined under the epipolar constraint. The epipolar line is calculated by substituting the center coordinates of the left template into Equations (1) and (2). In Fig. 2, white lines are epipolar lines that pass the center of the palm. We locate the right template by template matching along the epipolar line. The details are described later in subsection 3.4.

left image   right image

Fig. 2. Epipolar line

3.3 Computing the Three-Dimensional Trajectory

From the tracking results and the camera parameters, we calculate the three-dimensional coordinates of the hand. Let t_i be the right camera parameters, s_i the left camera parameters, (x_R, y_R) the tracking results in the right image, and (x_L, y_L) in the left image. Then from Equations (1) and (2), we obtain the following equations:

$$x_L(s_1 x + s_2 y + s_3 z + 1) + s_4 x + s_5 y + s_6 z + s_7 = 0 \qquad (4)$$
$$y_L(s_1 x + s_2 y + s_3 z + 1) + s_8 x + s_9 y + s_{10} z + s_{11} = 0 \qquad (5)$$
$$x_R(t_1 x + t_2 y + t_3 z + 1) + t_4 x + t_5 y + t_6 z + t_7 = 0 \qquad (6)$$
$$y_R(t_1 x + t_2 y + t_3 z + 1) + t_8 x + t_9 y + t_{10} z + t_{11} = 0 \qquad (7)$$

By solving these equations, we can obtain the three-dimensional coordinates (x, y, z) of the hand location. We will use them for the restrictions on the search area.

3.4 Restrictions on the Search Area and Tracking the Hand Motion

We track the hand movement by template matching:

$$D = \sum_x \sum_y \sum_i \sum_j |f(x+i, y+j) - g(i,j)| \qquad (8)$$

where D stands for dissimilarity between the input image f and the template g. Since the input images are color images, we calculate D_r, D_g, D_b for each color component, red, green, blue, respectively and add them. We determine the location of the hand by finding such a location (x, y) that gives the minimum of $D_r + D_g + D_b$.

If we would search the whole image for the hand, the procedure should be inefficient and the possibility of mismatching could rise. In this paper, we incorporate three kinds of constraints into our tracking technique to put restrictions on the search area of targets and their relative positions in each image.

These constraints are as follows:

1. continuity of the hand locations,
2. use of the skin color segmentation, and
3. epipolar constraint between two views.

3.4.1 Continuity of the Hand Location

We limit the range of (x, y) in Equation (8) by the constraint of the continuity of the hand locations. Because the locations of the hand do not change all of a sudden between the consecutive frames that are captured at every 1/30 sec, we can restrict the search area of the next location of the hand to the neighbors of the current location. However, if we determined the search area by considering the neighbors only in the image-plane, we cannot search for the exact locations when the conditions like the hand size change in the image. Thus, we consider the neighbors in the three-dimensional trajectory. We set the search area to $15 \times 15 \times 15 cm^3$ in the three-dimensional space.

3.4.2 Use of the Skin Color Segmentation Image

By considering the results of the skin color segmentation, we decide whether or not we use Equation(8). Black pixels in the skin image are most likely to be the background or the clothes if we assume that the color of the background is not the same as the skin. Therefore, we will neglect matching inside those black areas; otherwise the matched area could gradually deviate from the hand area to the background.

In this work, we limit the search area to such regions that have more than 90% white pixels, so that we can deal with an oblique hand or a far going hand in that case the corresponding templates have some black pixels.

3.4.3 Epipolar Constraint between Two Views

Here, if we determined such locations independently in the right and left image, those two locations could not correspond in the space. Instead, we limit the search area along the epipolar line.

We calculate the epipolar line from the tracking results of the previous frame. Though each pixel has a different epipolar line, we use the epipolar line of the center of the hand for the matching. The epipolar line calculated from the result of the previous tracking is used over the whole search area.

Tracking the hand location is performed as follows:

1. Calculate the dissimilarity on the epipolar line in the left image.
2. Calculate the dissimilarity on the corresponding epipolar line in the right image.
3. Add two dissimilarities.

We iterate this procedure along the epipolar line until we find the minimum of the dissimilarity.

4 Experimental Results

We tracked two kinds of hand gestures, wave and circle (Fig. 3). These gestures show the same trajectory in the image-plane (Fig. 4), but we can distinguish them by comparing the three dimensional trajectories (Fig. 5) or the gesture eigenspaces which are obtained by KL expansion (Fig. 6).

As far as the efficiency of the proposed method, we compare the number of matching with and without three constraints. The number of matching is 10^3 per frame with three constraints, while the number is 1.4×10^5 without their constraints. By the three constraints, we can limit the search areas to less than 1/100 when it is compared to the exhaustive matching.

circle　　　　　　　　　　　　　　wave

Fig. 3. tracking result

5 Conclusions

We have developed a stereo camera system for tracking and measuring the three dimensional trajectories of the hand movement. We incorporate three kinds of constraints into our tracking technique to put restrictions on the search area

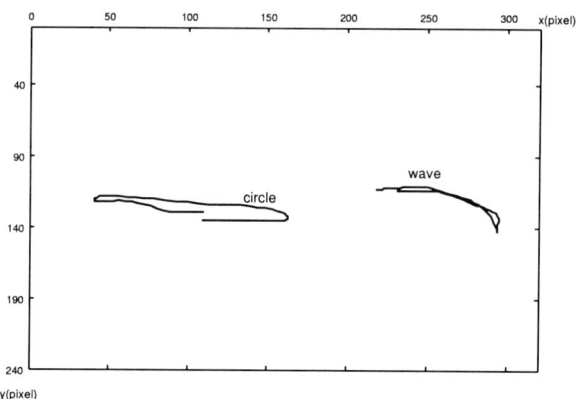

Fig. 4. Trajectry in the image-plane

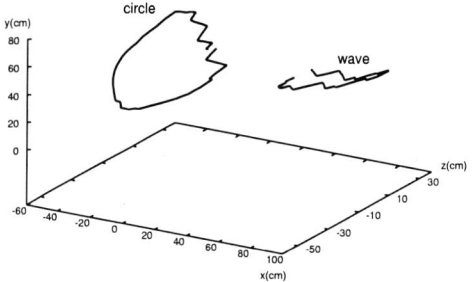

Fig. 5. Trajectry in the 3D position

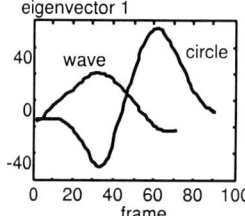 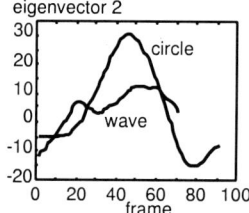 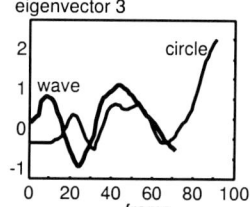

Fig. 6. KL expansion

of targets and their relative positions in each image, so that we can reduce the computational complexity and obtain accurate three-dimensional trajectories.

Our present system assumes a single subject in front of a motionless background. We need to explore tracking techniques that will work even when those assumptions are not satisfied. That is left to our future work.

References

1. Vladimir I. Pavlovic, Rajeev Sharma and Thomas S. Huang, "Visual Interpretation of Hand Gestures for Human-Computer Interaction: A Review," *IEEE Trans. Pattern Analysis and Machine Intelligence*, vol. 19, no. 7, pp. 677-695, 1997.
2. D. M. Gavrila, "The Visual Analysis of Human Movement: A Survey," *Computer Vision and Image Understanding*, vol. 73, no. 1, pp. 82-98, 1999.
3. James J. Kuch, Thomas S. Huang, "Virtual Gun: A Vision Based Human Computer Interface Using the Human Hand," *IAPR Workshop on Machine Vision Application*, pp. 196-199, 1994.
4. Ross Cutler, Matthew Turk, "View-based Imterpretation of Real-time Optical Flow for Gesture Recognition," *Proc. 3rd International Conference on Automatic Face and Gesture Recognition*, pp. 416-421, 1998.
5. Takio Kurita, Satoru Hayamizu, "Gesture Recognition using HLAC Features of PARCOR Images and HMM based recognizer," *Proc. 3rd International Conference on Automatic Face and Gesture Recognition*, pp. 422-427, 1998.
6. Takahiro Watanabe and Masahiko Yachida, "Real Time Gesture Recognition Using Eigenspace from Multi Input Image Sequences," *Proc. 3rd IEEE International Conference on Automatic Face and Gesture Recognition*, pp. 428-433, 1998.
7. Kazuyuki Imagawa, Shan Lu and Seiji Igi, "Color-Based Hands Tracking System for Sign Language Recognition," *Proc. 3rd International Conference on Automatic Face and Gesture Recognition*, pp. 462-467, 1998.
8. Nobutaka Shimada, Kousuke Kimura, Yoshiaki Shirai and Yoshinori Kuno, "3-D Hand Posture Estimation by Indexing Monocular Silhouette Images," *Proc. 6th Korea-Japan Joint Workshop on Computer Vision*, pp. 150-155, 2000.
9. Kiyofumi Abe,Hideo Saito and Shinji Ozawa, "3-D Drawing System via Hand Motion Recognition from Two Cameras," *Proc. 6th Korea-Japan Joint Workshop on Computer Vision*, pp. 138-143, 2000.
10. Yusuf Azoz, Lalitha Devi, Rajeev Sharma, "Tracking Hand Dynamics in Unconstrained Environments," *Proc. 3rd International Conference on Automatic Face and Gesture Recognition*, pp. 274-279, 1998.
11. Hyeon-Kyu Lee, Jin H. Kim, "An HMM-Based Threshold Model Approach for Gesture Recognition," *IEEE Trans. Pattern Analysis and Machine Intelligence*, vol. 21, no. 10, pp. 961-973, 1999.
12. Masanori Yamada,Kazuyuki Ebihara,Jun Ohya, "A New Robust Real-time Method for Extracting Human Silhouettes from Color Image," *Proc. 3rd International Conference on Automatic Face and Gesture Recognition*, pp. 528-533, 1998.
13. Christopher Richard Wren, Ali Azarbayejani, Trevor Darrell, and Alex Paul Pentland, "Pfinder:Real-Time Tracking of the Human Body," *IEEE Trans. Pattern Analysis and Machine Intelligence*, vol. 19, no. 7, pp. 780-785, 1997.
14. B. K. P. Horn, *Robot Vision*, MIT Press, 1986.
15. Douglas Chai, King N. Ngan, "Locating Facial Region of a Head-and-Shoulders Color Image," *Proc. 3rd International Conference on Automatic Face and Gesture Recognition*, pp. 124-129, 1998.

Robust, Real-Time Motion Estimation from Long Image Sequences Using Kalman Filtering*

J. A. Yang[1] and X. M. Yang[2]

[1] Institute of Artificial Intelligence, Hefei University of Technology
Hefei, Anhui Province 230009, P. R. China
jayang@mail.hf.ah.cn
[2] Department of Computer Sciences, Purdue University,
West Lafayette, In 47907-1398, U.S.A.
yang17@cs.purdue.edu

Abstract. This paper presents how to estimate the left and right monocular motion and structure parameters of two stereo image sequences including direction of translation, relative depth, observer rotation and rotational acceleration, and how to compute absolute depth, absolute translation and absolute translational acceleration parameters at each frame. For improving the accuracy of the computed parameters and robustness of the algorithm, A Kalman filter is used to integrate the parameters over time to provide a "best" estimation of absolute translation at each time.

1 Introduction

The monocular motion parameters can be estimated by solving simple linear systems of equations[1][2]. These parameters may be computed in a camera-centered coordinate system using adjacent 3-tuples of flow fields from a long monocular flow sequence and are then integrated over time using a Kalman filter[3]. The work extends that algorithm to use binocular flow sequences to compute, in addition to the monocular motion parameters for both left and right sequences, binocular parameters for absolute translational speed, U, acceleration in translational speed, δU and absolute depth, X_3 at each image pixel. The direction of translation is available from the monocular sequences and together with U and δU provides absolute observer translation.

Our algorithm does not require that the corrspondence problem be solved between stereo images or stereo flows, i.e. no features or image velocities have to be matched between left and right frames, and it does not require an *a priori* surface structure model[4][5]. The algorithm does require that the observer be rotating and that the spatial baseline be significant, otherwise only monocular parameters can be recovered[6][7].

---

* This work was supported by The National Natural Science Foundation of China under contracts No. 69585002 and No. 69785003.

In this paper, we assume the observer rotational motion is no more than "second order", in other words, observer motion is either *constant* or has at most *constant acceleration*. Allowable observer is a camera rigidly attached to the moving vehicle, which travels along a smooth trajectory in a stationary environment. As the camera moves it acquires images at some reasonable sampling rate. Given a sequence of such images we analyze them to recover the camera's motion and depth information about various surfaces in the environment. As the camera moves relative to some 3D environmental point, the 3D relative velocity that occurs is mapped (under perspective projection) onto the camera's image plane as 2D image motion[8]. Optical flow or image velocity is an infinitesimal approximation to this image motion[4][9]. As in the monocular case, we use a nested Kalman Filter to integrate the computed binocular motion parameters over time, thus providing a "best" estimate of the parameter values at each time.

2 The Projected Motion Geometry

2.1 The Projection Geometry from 3D to 2D in Monocular Images

The standard image velocity equation relates an image velocity measured at image location $\boldsymbol{Y} = (y_1, y_2, 1) = \boldsymbol{P}/X_3$, i.e. the perspective projection of a 3D point $\boldsymbol{P} = (X_1, X_2, X_3)$, to the 3D observer translation $\boldsymbol{U}$ and 3D observer rotation $\boldsymbol{\omega}$. Figure 1 shows this coordinates system setup. Assuming a focal length of 1 we can write the image velocity $\boldsymbol{v} = (v_1, v_2)$ as

$$\boldsymbol{v}(\boldsymbol{Y}, t) = \boldsymbol{v}_T(\boldsymbol{Y}, t) + \boldsymbol{v}_R(\boldsymbol{Y}, t) \tag{1}$$

where $\boldsymbol{v}_T$ and $\boldsymbol{v}_R$ are the translational and rotational components of image velocity:

$$\boldsymbol{v}_T(\boldsymbol{Y}, t) = A_1(\boldsymbol{Y})\boldsymbol{u}(\boldsymbol{Y}, t)\|\boldsymbol{Y}\|_2 \quad \text{and} \quad \boldsymbol{v}_R(\boldsymbol{Y}, t) = A_2(\boldsymbol{Y})\boldsymbol{\omega}(t) \tag{2}$$

where

$$A_1 = \begin{pmatrix} -1 & 0 & y_1 \\ 0 & -1 & y_2 \end{pmatrix} \quad \text{and} \quad A_2 = \begin{pmatrix} y_1 y_2 & -(1+y_1^2) & y_2 \\ (1+y_2^2) & -y_1 y_2 & -y_1 \end{pmatrix} \tag{3}$$

2.2 The Projection Geometry from 3D to 2D in Binocular Images

In our binocular setup a second camera is rigidly attached to the first camera with a known baseline s as shown in Figure 1. We subscript variables in the left image with L and variables in the right image with R. Our solution depends on using left and right monocular solutions computed from separate left and right long image sequences. That is, both left and right cameras have constant translational direction, $\hat{U}$, the same rotation and rotation acceleration, $\boldsymbol{\omega}$ and $\boldsymbol{\delta}$, but different relative depth, μ_L and μ_R, different translational accelerations, U_L and U_R, and rotation acceleration, $\boldsymbol{\omega}$ and $\delta\boldsymbol{\omega}$. Of course, $\boldsymbol{U}_L$ and $\boldsymbol{U}_R$ can be computed from U_L and U_R and $\hat{u}$. We use the relationships

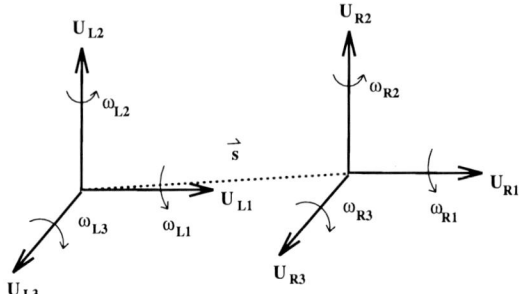

Fig. 1. The binocular setup. Subscripts L and R are used to indicate variables in the *left* and *right* image sequences. s is the spatial baseline and is assumed known.

between $\boldsymbol{U}_L$ and $\delta \boldsymbol{U}_L$ and $\delta \boldsymbol{U}_R$ and δU_R that use the values of $\boldsymbol{\omega}$ and $\delta \boldsymbol{\omega}$ to avoid the correspondencee problem. Given 3D translational information, $\boldsymbol{U}_L$ and $\boldsymbol{U}_R$, we can compute 3D depth, X_{3L} and X_{3R} or equivalently the 3D coordinates, $\boldsymbol{P}_L = X_{3L}\boldsymbol{Y}$ and $\boldsymbol{P}_R = X_{3R}\boldsymbol{Y}$. We give the main relationships below.

The physical stereo allows us to write:

$$\boldsymbol{U}_R = \boldsymbol{U}_L + \boldsymbol{\omega} \times \boldsymbol{s} \qquad (4)$$

Thus $\boldsymbol{u}_L$ and $\boldsymbol{u}_R$ can be written as

$$\boldsymbol{u}_L = \frac{\boldsymbol{U}_L}{X_{3L}\|\boldsymbol{Y}\|_2} \qquad (5)$$

$$\boldsymbol{u}_R = \frac{\boldsymbol{U}_R}{X_{3R}\|\boldsymbol{Y}\|_2} = \frac{\boldsymbol{U}_L + \boldsymbol{\omega} \times \boldsymbol{s}}{X_{3R}\|\boldsymbol{Y}\|_2}. \qquad (6)$$

Using $\boldsymbol{u}_L$ and $\boldsymbol{u}_R$ in the left and right image velocity equations, we can subtract the image velocity at $\boldsymbol{Y}$ to obtain

$$\boldsymbol{v}_R(\boldsymbol{Y}) - \boldsymbol{v}_L(\boldsymbol{Y}) = \begin{pmatrix} -1 & 0 & y_1 \\ 0 & -1 & y_2 \end{pmatrix} \boldsymbol{u}_L \|\boldsymbol{Y}\|_2\, \mathrm{a}(\boldsymbol{Y}) + \begin{pmatrix} -1 & 0 & y_1 \\ 0 & -1 & y_2 \end{pmatrix} (\boldsymbol{\omega} \times \boldsymbol{s})\, \mathrm{b}(\boldsymbol{Y}). \qquad (7)$$

These are 2 linear equations in 2 unknowns, $\mathrm{a}(\boldsymbol{Y})$ and $\mathrm{b}(\boldsymbol{Y})$, where $\mathbf{b} = \frac{1}{X_{3R}}$ and $\mathbf{a} = X_{3L}\mathbf{b} - 1$. Note that if there is no observer rotation, i.e. $\boldsymbol{\omega} = \boldsymbol{0}$, $\mathbf{b}$ cannot be recovered (only the parameter $\mathbf{a}$ can be computed), and as a result absolute depth cannot be computed. $\mathbf{a}$ yields the ratio of depth in the left and right images at each image point. Given the monocular parameters $\hat{\boldsymbol{u}}$, $\boldsymbol{\omega}$ and μ_L and μ_R (computed separately from the left and right image sequences) we can compute $\mathbf{a}$ and $\mathbf{b}$ for each image point (s is known). We compute absolute translations $\boldsymbol{U}_L$ and $\boldsymbol{U}_R$ as

$$\boldsymbol{U}_L = X_{3L}\|\boldsymbol{Y}\|_2 \boldsymbol{u}_L, \qquad \boldsymbol{U}_R = X_{3R}\|\boldsymbol{Y}\|_2 \boldsymbol{u}_R. \qquad (8)$$

Each image point potentially yields a different U value. We can then compute constant acceleration in translational speed in the left and right images using averaged U_L and U_R values as

$$\delta U_L = ||\boldsymbol{U}_L(t+\delta t)||_2 - ||\boldsymbol{U}_L(t)||_2, \quad \delta U_R = \delta U_L + ||\delta\boldsymbol{\omega}\times\boldsymbol{s}||_2. \quad (9)$$

Then $\delta\boldsymbol{U}_L = \delta U_L \hat{u}$ and $\delta\boldsymbol{U}_R = \delta U_R \hat{u}$.

Since we are dealing with a monocular observer we cannot recover the observer's absolute translation, $\boldsymbol{U}$, or the actual 3D coordinates, $\boldsymbol{P}$, of environmental points but rather the ratio of the two. We define the *relative* depth as the ratio of 3D translation and 3D depth as

$$\mu(\boldsymbol{Y},t) = \frac{||\boldsymbol{U}(t)||_2}{||\boldsymbol{P}(t)||_2} = \frac{||\boldsymbol{U}||_2}{X_3||\boldsymbol{Y}||_2} \quad (10)$$

where $\hat{u} = \hat{U} = (u_1, u_2, u_3)$ is the normalized direction of translation and $\boldsymbol{u}(\boldsymbol{Y},t) = \mu(\boldsymbol{Y},t)\hat{u}$ is the depth-scaled observer translation.

3 The Least Squares Solution

Modelling uniform rotational acceleration as:

$$\boldsymbol{\omega}(t+\delta t) = \boldsymbol{\omega}(t) + \delta\boldsymbol{\omega}\delta t \quad (11)$$

We can write

$$\frac{r_1}{r_2} = \frac{v_1(\boldsymbol{Y},t_1) + v_1(\boldsymbol{Y},t_{-1}) - 2v_1(\boldsymbol{Y},t_0)}{v_2(\boldsymbol{Y},t_1) + v_2(\boldsymbol{Y},t_{-1}) - 2v_2(\boldsymbol{Y},t_0)} \quad (12)$$

which is a linear equation in terms of velocity differences and the components of $\hat{u}$. Note we have assumed that $\delta t = |t_{i+1} - t_i| = |t_i - t_{i-1}|$, where t_i is the central frame, i.e. the flow is sampled at every δt. The ratios of the two components of $\boldsymbol{r}$ give us 1 linear equation in (u_1, u_2, u_3):

$$\frac{r_1}{r_2} = \frac{-u_1 + y_1 u_3}{-u_2 + Y_2 u_3}. \quad (13)$$

Since only 2 components of $\hat{u}$ are independent, then if u_3 is not zero, we can write

$$r_2\frac{u_1}{u_3} - r_1\frac{u_2}{u_3} = r_2 y_1 - r_1 y_2. \quad (14)$$

This is 1 linear equation in 2 unknowns, $(\frac{u_1}{u_3}, \frac{u_2}{u_3})$. Given m image velocities we obtain m equations of the form in (14) which we can write as

$$WM_2(\frac{u_1}{u_3}, \frac{u_2}{u_3}) = WB_2, \quad (15)$$

where M_2 is a $m\times 2$ matrix, B_2 is a $m\times 1$ matrix, and W is an $m\times m$ diagonal weight matrix that is based on confidence measures of the computed optical flow for the real data. To continue, we solve this system of equations in the least squares sense as

$$\left(\frac{u_1}{u_3}, \frac{u_2}{u_3}\right) = (M_2^T W^2 M_2)^{-1} M_2^T W^2 B_2, \tag{16}$$

which involves solving a simple 2×2 system of linear equations to obtain the direction of translation $(u_1/u_3, u_2/u_3, 1)$, which when normalized yields $\hat{u}$. In the event u_3 is zero (or near-zero) we can in principle solve for $(1, u_2/u_1, u_3/u_1)$ or $(u_1/u_2, 1, u_3/u_2)$ by appropriate manipulation of equation (14). Again, systems of equations of the form (16) can be set up for these cases. We compute $\hat{u}$ using all three methods and then choose the $\hat{u}$ from the system of equations having the smallest condition number, κ^1. If there is no observer translation, all three systems of equations for $(\frac{u_1}{u_3}, \frac{u_2}{u_3})$, $(\frac{u_1}{u_2}, \frac{u_3}{u_2})$, and $(\frac{u_2}{u_1}, \frac{u_3}{u_1})$, will be singular. We can detect this situation by choosing the result with the smallest condition number. If this κ is large enough (indicating singularity as $\hat{u} \approx (0,0,0)$, then only $\boldsymbol{\omega}$ can be recovered by solving a linear system of equations comprised of equations of the form

$$\boldsymbol{v}(\boldsymbol{Y}, t) = A_2(\boldsymbol{Y})\boldsymbol{\omega}(t). \tag{17}$$

Results in [5] indicate this system of equations is quite robust in the face of noisy data. Given $\hat{u}$ computed using this strategy we can then compute $\boldsymbol{\omega}$ and then the $\mu's$. Note that in order to compute $\hat{u}$ we have assumed we can measure image velocities at the same image locations at all three times. Since poor velocity measurements should have small confidence measures, an image location with one or more poor velocity measurements at any time should have little or no effect on to the computation of $\hat{u}$. Given $\hat{u}$, we can compute $\boldsymbol{\omega}$. Now we first compute the normalized direction of $\boldsymbol{v}_T$ as

$$\hat{d} = \frac{(A_1(\boldsymbol{Y})\hat{u})}{||A_1(\boldsymbol{Y})\hat{u}||_2}. \tag{18}$$

Given $\hat{d} = (d_1, d_2)$ we can compute $\hat{d}^\perp$ as $(d_2, -d_1)$. Hence we obtain one linear equation in the 3 components of $\boldsymbol{\omega}(t)$

$$\hat{d}^\perp \cdot \boldsymbol{v} = \hat{d}^\perp \cdot (A_2(\boldsymbol{Y})\boldsymbol{\omega}(t)) \tag{19}$$

Given m image velocities we obtain m equations of the form in (19) which we can write as

$$W M_3 \boldsymbol{\omega} = W B_3 \tag{20}$$

where M_3 is a $m \times 3$ matrix, B_3 is a $m \times 1$ matrix and W is the same $m \times m$ diagonal matrix whose diagonal elements are based on the confidence measures of the corresponding velocity measurements. We can solve (20) as

$$\boldsymbol{\omega} = (M_3^T W^2 M_3)^{-1} M_3^T W^2 B_3 \tag{21}$$

which involves solving a simple 3×3 linear system of equations. If $\hat{d}$ and $A_2(\boldsymbol{Y})\boldsymbol{\omega}$ are parallel, then $\boldsymbol{\omega}$ cannot be recovered. Now rotational acceleration can be

found at frame i as
$$\delta\boldsymbol{\omega}(t_i) = \frac{\boldsymbol{\omega}(t_{i+1}) - \boldsymbol{\omega}(t_{i-1})}{2}. \tag{22}$$
Finally, given $\hat{u}$ and $\boldsymbol{\omega}(t)$, each image velocity, $\boldsymbol{v}(\boldsymbol{Y},t)$ yields two equations for $\mu(\boldsymbol{Y}_i, t)$
$$\boldsymbol{r} = \boldsymbol{v} - A_2(\boldsymbol{Y}_i)\boldsymbol{\omega}(t) = A_1(\boldsymbol{Y})\hat{u}\mu(\boldsymbol{Y}_i,t)\|\boldsymbol{Y}_i\|_2 = s\mu(\boldsymbol{Y}_i,t), \tag{23}$$
which we solve for each image velocity measurement as:
$$\mu(\boldsymbol{Y}_i) = \frac{r_1}{s_1} = \frac{r_2}{s_2}. \tag{24}$$
If $\boldsymbol{v}$ is purely horizontal or vertical at some image location then one of s_1 or s_2 will be zero, but the other can still allow μ to be recovered. If both s_1 and s_2 are non-zero we average the two computed μ values:
$$\mu(\boldsymbol{Y}) = \frac{1}{2}\left(\frac{r_1 s_2 + r_2 s_1}{s_1 s_2}\right). \tag{25}$$
If both s_1 and s_2 are zero we have a singularity and μ cannot be recovered. Typically, these singularities arise at the focus of expansions(FOEs) where the velocity is zero. Note that while we have assumed the direction of translation in the observer's coordinate system is constant, there can be acceleration in the translational speed. Changing translational speed cannot be separated from changing depth values and the combined effect of both are reflected in changing μ values.

4 Kalman Filtering

The 3D depth and translation can, in principle, be determined for every $\boldsymbol{Y}$ location. Since speeds U_L and U_R should be the same for every location we compute the average and variance of these values at every image location, but noise can seriously corrupt individual speeds. Some were obviously wrong, i.e, $\|\boldsymbol{U}_L\|_2 > 1000.0$ or $\|\boldsymbol{U}_R\|_2 > 1000.0$, and so our algorithm classified them as outliers and removed from further consideration. We use the least squares variance $\sigma_{U_L}^2$ as the variance in our Kalman filter calculations. We subscript variables with M for measured quantities (computed from individual stereo flows), with C for quantities computed using a Kalman update equation and with P for predicted quantities. The steps of the Kalman filter are:

1. Initialize predicted parameters: $U_{LP} = 0$, $\sigma_{U_{LP}}^2 = \infty$, $i = 1$.
2. Compute X_3 values and U_{LM} from (7) and (8), average U_L and compute its standard deviation, we have
$$K_{U_L} = \frac{\sigma_{U_{LP}}^2}{\sigma_{U_{LP}}^2 + \sigma_{U_{LM}}^2}, \tag{26}$$
$$U_{LC} = U_{LP} + K_{U_L}(U_{LM} - U_{LP}), \tag{27}$$
$$\sigma_{U_{LC}}^2 = K_{U_L}\sigma_{U_{LM}}^2. \tag{28}$$

3. Update predicted quantities $U_{LP} = U_{LC}$, $\sigma^2_{U_{LP}} = K_{U_L}\sigma^2_{U_{LM}}$, $i = i + 1$.

These steps are performed for each pair of speeds recovered from each stereo flow pair. we also run the Kalman filter with the actual error squared as the variance for the purpose of comparison.

We use a series of nested Kalman filters to integrate these measurements over time. Below we outline the steps in our filter computation for $n+2$ images, numbered 0 to $n+1$, and show how we continually update the solutions for frames 1 to n. We subscript items by M if they are measured quantities, by C if they are computed quantities and by P if they are predicted quantities.

5 Experimental Results

5.1 The Experimental Results of One Stereo Sequence

A Kalman filter was applied to the U_L and V_R estimates at each time to recursively refine them and decrease their uncertainty. We present experimental results for one stereo sequence that consists of 22 stereo flows. The left sequences were generated for $\boldsymbol{U}_L = (0, 0, 40)$ with an angular rotation of 0.174 radians. The initial observer position was $(0, 0, 0)$ and the flow was sampled at every $\frac{1}{10}$ time unit. The right sequences were generated using a spatial baseline of $\boldsymbol{s} = (10, 0, 10)f$ units, yielding $\boldsymbol{U}_R = (-1.74, 0, 41.74)$. We could not use $\boldsymbol{s} = (10, 0, 0)$ because then both $\boldsymbol{U}_L$ and $\boldsymbol{U}_R$ were in the same direction and equation (2.7) is singular. As well, X_{3L} and X_{3R} have to be different at the same image location or equation (2.7) cannot be solved there. The images consisted of a number of ray-traced planes viewed in stereo at various depths. The 22 stereo flows allow 20 left and right monocular motion and structure calculations to be performed. Figure 2 show the measured error in speeds U_L and U_R for 1% random error in the left and roght image velocity fields (we can use up to 3% random error before speed error ranges from 2% to 15%. The performance for the actual variance and least squares variance was nearly identical.

5.2 Real Image Experimental Results

We have applied our algorithm to two long image sequences. The first sequence consists of 36 images with the camera translating directly towards the scene. The second sequence also consists of 36 images with general camera motion comprised of simple translation and simple rotation about the camera's vertical axis. Neither sequence has rotational acceleration ($\delta\omega = (0, 0, 0)$). Optical flow was measured using the differential method involves prefiltering the images with a spatio-temporal Gaussian filter with isotropic standard deviation $\sigma = 1.5$. Differentiation was performed using a 4-point central difference and integration of the spatio-temporal derivatives (normal image velocity) into full image velocities was performed in 5×5 neighbourhoods using standard least squares. The prefiltering and differentiation requirements meant that only 22 flow fields can be computed for the 36 input images. Confidence measures based on the smallest

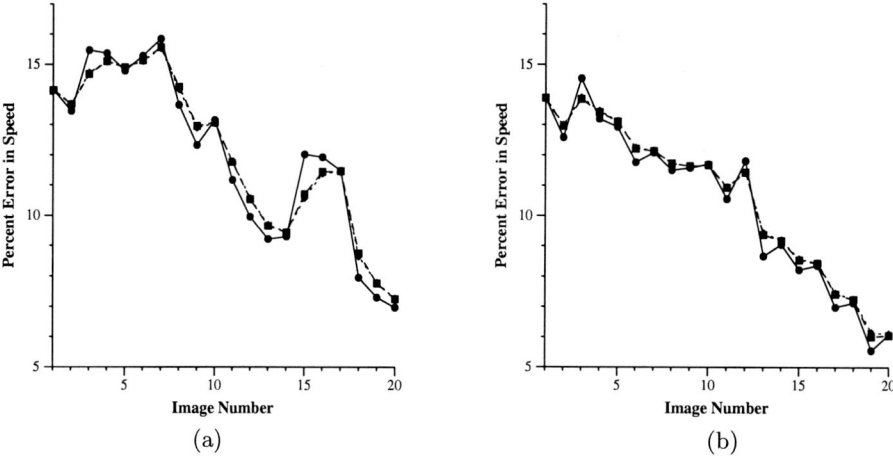

Fig. 2. (a) Measured (solid lines) and actual and least squares Kalman filtered (dashed and dotted lines) error for speed U_L; (b) Measured (solid lines) and actual and least squares Kalman filtered (dashed and solid lines) error for speed U_R.

eigenvalues of the least squares matrix, λ_1, were used to weight "good" image velocities, those velocities for which $\lambda_1 \leq 1.0$ were rejected outright. The confidence measures of surviving image velocities were used as the weights in the W matrix; since our algorithm requires tuples of image velocities at the same image location at three consecutive times, each weight, w_i (the diagonal elements of W), is computed as:

$$w_i = \min(\lambda_1(t_{k-1}), \lambda_1(t_k), \lambda_1(t_{k+1})), \tag{29}$$

where t_{k-1}, t_k and t_{k+1} are three consecutive times. Rejected velocities have an assigned confidence measure of 0.0 and so have no effect on the motion and structure calculation.

Figure 3 and Figure 4 show the 19th image and its flow field for both sequences. Eigenvalue (λ_1) images for each flow field are also shown. Here *white* indicates high eigenvalues ("good" velocities") while *black* indicates low eigenvalues ("poor" velocities).

At one meter with the camera focused, a ruler viewed horizontally was 32.375 inches long in a 512 pixel wide image row while the same ruler viewed vertically was 24.25 inches long in a 482 pixel high image column. This results in a 44.8° horizontal field of view and a 34.24° vertical field of view. Since our algorithm assumes a focal length of 1, we convert velocity positions from pixels to f units and image velocities from pixels/frame to f units/frame by applying the appropriate scaling to the pixel locations and their image velocities.

The first real image experiment involved pure translation. The correct translational direction was $\hat{u} = (0,0,1)$, the correct rotation was $\omega = (0,0,0)$ and the correct rotational acceleration was $\delta\omega = (0,0,0)$. Figure 5 shows the experimental results.

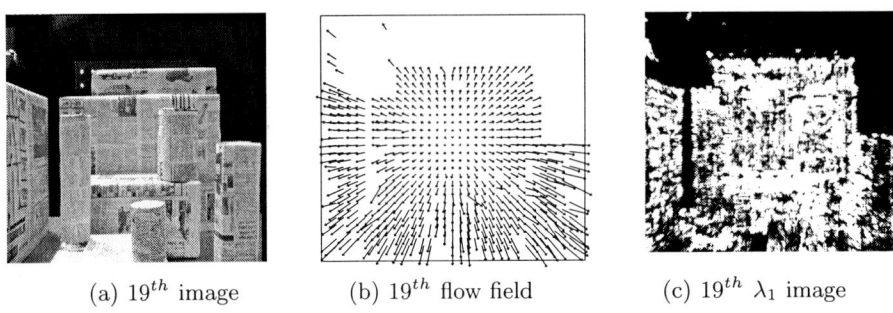

(a) 19th image (b) 19th flow field (c) 19th λ_1 image

Fig. 3. (a) The 19th image of the translating newspaper sequence; (b) its flow field thresholded using $\lambda_1 \leq 1.0$; (c) an image of the confidence values λ_1 for the flow field.

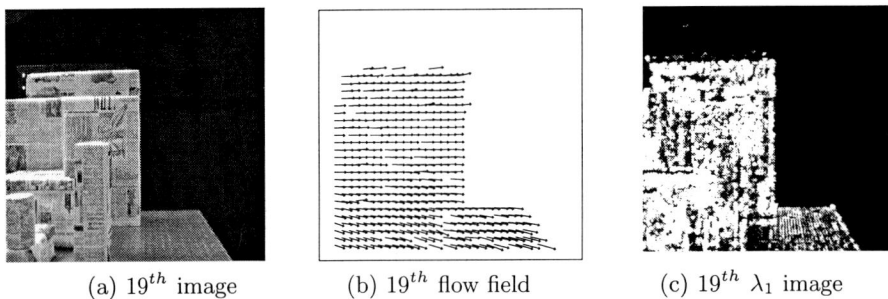

(a) 19th image (b) 19th flow field (c) 19th λ_1 image

Fig. 4. (a) The 19th image of the general newspaper sequence, (b) its flow field thresholded using $\lambda_1 \leq 1.0$ and (c) an image of the confidence values (λ_1) for the flow field.

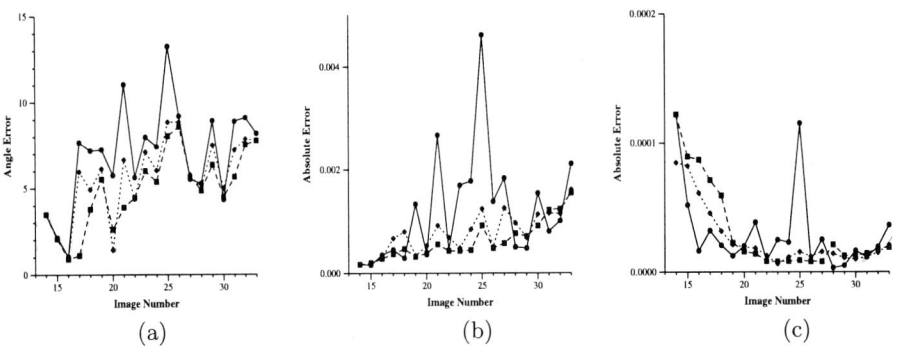

Fig. 5. Measured and Kalman filtered error for (a) $\hat{u}$, (b) $\boldsymbol{\omega}$ and (c) $\delta\boldsymbol{\omega}$ for 5% random error for the line-of-sight translation newspaper sequence.

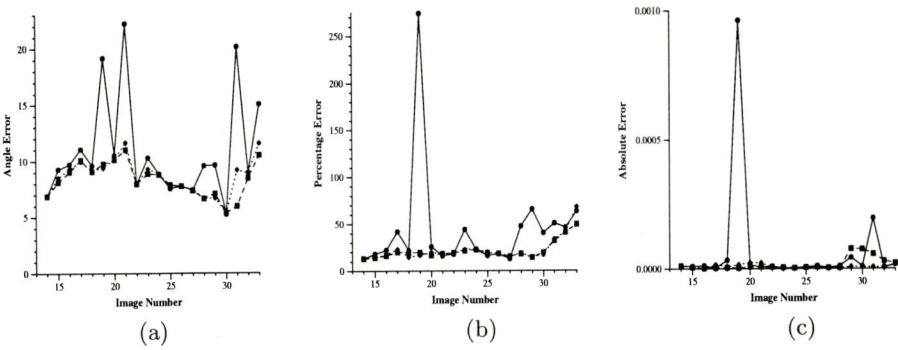

Fig. 6. Measured and Kalman filtered error for (a) $\hat{u}$, (b) ω and (c) $\delta\omega$ for 5% random error for the general motion (line-of-sight translation and vertical rotation) newspaper image sequence.

(a) The 19th frame of the Rubic image sequence (b) Unthresholded optical flow

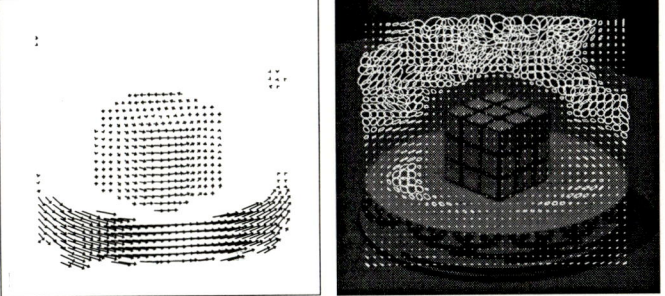

(c) Thresholded optical flow T_e=0.00(d) The uncertainty estimation of the recovery

Fig. 7. Optical flow of Rubic sequence and the uncertainty estimation of its recovery.

The second real image experiment involved line-of-sight translation and angular rotation. The correct translational direction was $\hat{u} = (0, 0, 1)$, rotation was $\omega = (0, 0.00145, 0)$ ($\frac{1}{12}^{\circ}$ per frame) and rotational acceleration was $\delta\omega = (0, 0, 0)$. Experimental results are given in Figure 6. Figure 7 shows the optical flow of

Rubic sequence and the uncertainty estimation of its recovery. As with the synthetic image data, the depth error depends on the accuracy of individual recovered image velocities and only slightly on the accuracy of the recovered motion parameters.

6 Discussion

The recovery of binocular parameters from stereo flow fields is very sensitive to noise in the flow vectors. This is because the differences in similar stereo image velocities can have large error even if the individual image velocity error is quite small. We are currently investigating the direct application of Kalman filtering on both relative and absolute depth calculations and the optical flow fields to produce more accurate depth values. A Kalman filtering was then used to integrate these calculations over time, resulting in real-time frame-by-frame best estimate of each unknown parameters and significant improvement in the recovered solution. Other pending work includes using a local depth model to make depth calculations more robust and testing our algorithms using image sequences acquired from a camera mounted on a robot arm.

References

1. Li L., Duncan, J. H.: 3D translational motion and structure from binocular image flows. IEEE PAMI, **15** (1993) 657-667
2. Hu, X., Ahuja, N.: Motion and structure estimation using long sequence motion models. Image and Vision Computing, **11** (1993) 549-569
3. Matthies, L., Szeliski, R., Kanade, T.: Kalman filter-based algorithms for estimating depth from image sequences. IJCV, **3** (1989) 209-238
4. Cui, N. et al.: Recursive-batch estimation of motion and structure from monocular image sequences. CVGIP: Image Understanding, **59** (1994) 154-170
5. Yang, J. A.: Computing general 3D motion of objects without correspondence from binocular image flow, Journal of Computer, **18** (1995) 849-857
6. Yang, J. A.: A neural paradigm of time-varying motion segmentation. Journal of Computer Science and Technology, **9** (1999) 238-251
7. Zhang, Z., Faugeras O. D.: Three-dimensional motion computation and object segmentation in a long sequence of stereo frames. IJCV, **7** (1992) 211-242
8. De Micheli, E. et al.: The accuracy of the computation of optical flow and the recovery of motion parameters. IEEE PAMI, **15** (1993) 434-447
9. Heeger, D. J., Jepson, A. D.: Subspace methods for recovering rigid motion 1: algorithm and implementation. IJCV, **7** (1992) 95-117

T-CombNET - A Neural Network Dedicated to Hand Gesture Recognition

Marcus V. Lamar[1], Md. Shoaib Bhuiyan[2], and Akira Iwata[1]

[1] Dept. of Electrical and Computer Eng., Nagoya Institute of Technology,
Showa, Nagoya, Japan
{lamar,iwata}@mars.elcom.nitech.ac.jp
http://mars.elcom.nitech.ac.jp

[2] Dept. of Information Science, Suzuka University of Medical Science and Technology,
Suzuka, Mie, Japan
s.bhuiyan@computer.org

Abstract. T-CombNET neural network structure has obtained very good results in hand gesture recognition. However one of the most important setting is to define an input space that can optimize the global performance of this structure. In this paper the Interclass Distance Measurement criterion is analyzed and applied to select the space division in T-CombNET structure. The obtained results show that the use of the IDM criterion can improve the classification capability of the network when compared with practical approaches. Simulations using Japanese finger spelling has been done. The recognition rate has improved from 91.2% to 96.5% for dynamic hand motion recognition.

1 Introduction

Gesture recognition is a promising sub-field of the discipline of computer vision. Researchers are developing techniques that allow computers to understand human gestures. Due to the parallelism existing with speech recognition, the use of techniques such as HMM and DTW [9] has given very good results in sign language recognition applications. The use of Neural Networks (NN) is not a new idea [4], but new structures have been proposed to handle more efficiently with these problems. The T-CombNET Neural Network structure is one promising example [10,11]. The main point of T-CombNET is the splitting of processing efforts in two stages, performed by Stem Network and Branches Networks, which analyze different input vectors extracted from an input feature space. A good sub-spaces selection is fundamental to the success of the T-CombNET.

In this paper we present the use of Interclass Distance Measurement (IDM) criterion in the space selection problem. The IDM is a very popular technique applied to *feature selection* problems. We modified the original IDM, creating the Joint IDM (JIDM) criterion which is dedicated to split an input space in

---

[1] Marcus V. Lamar is an Assistant Professor at the Federal University of Paraná/DELT/CIEL - Brazil

two complementary sub-spaces. Then, the vectors in these sub-spaces can be analyzed by any multi-stage classifier. in our case we apply it to the T-CombNET Neural Network model.

2 The T-CombNET Model

The T-CombNET model is inspired by Iwata's CombNET-II neural network [8] and was developed mainly to classify high dimensional time series into a very large number of categories, as required in speech, financial data series and gesture recognition problems [11]. The structure of T-CombNET is shown in Fig.1.

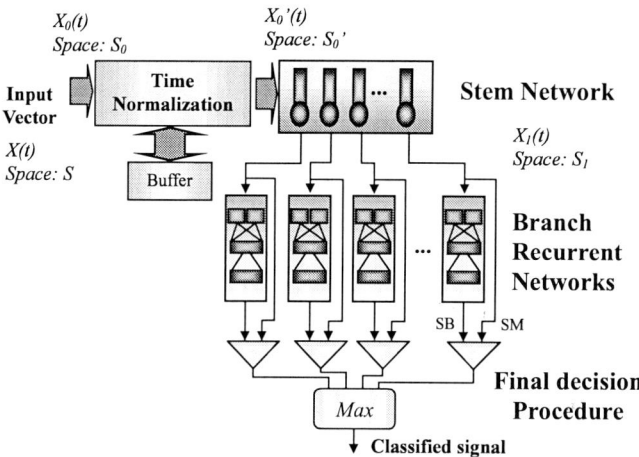

Fig. 1. The T-CombNET Structure

The T-CombNET model is composed of a Time Normalization (TN) preprocessing block, followed by Stem, and Branch network layers and by a Final Decision Procedure. The Stem network is composed of a Learning Vector Quantization (LVQ) based NN and the Branch Networks consisting of Elman Partially Recurrent Neural Networks (RNN) [6]. The Time Normalization procedure aims to fit the N_0 dimensional time series present in subspace S_0 into a fixed dimensional input vector X'_0 required by the LVQ NN.

The operation of the network is as follows: Initially the input vector X is projected into subspace S_0, generating the time series X_0. This time series is processed by TN procedure obtaining a time normalized vector X'_0 which is analyzed by LVQ NN in the Stem layer. The input vector X receives a score SM_i from each neuron i in Stem layer, calculated by Eq.(1).

$$SM_i = \frac{1}{1 + |\boldsymbol{X'_0} - \boldsymbol{N_i}|} \qquad (1)$$

where $|X'_0 - N_i|$ is the Euclidean Distance from the input vector X'_0 and the ith neuron template.

In a next step the input vector X is projected into subspace S_1 and the correspondent time series X_1 is analyzed by the Elman RNN in the Branch layers. The highest neuron output of each branch network i is chosen as being the scores SB_i. Finally, the Final Decision Procedure chooses the pair (SM_i, SB_i) that maximize the fitness function Z_i, defined by Eq.(2).

$$Z_i = (SB_i)^\lambda \times (SM_i)^{1-\lambda} \qquad 0 \leq \lambda \leq 1 \qquad (2)$$

where λ must be estimated from a training set to maximize the recognition rate.

The T-CombNET model uses the Divide-and-Conquer principle to simplify a problem. Using the TN procedure a rough description of time series in subspace S_0 is first analyzed by the Stem layer, and then the reminder information present in vector X is analyzed finely by the Branch layers. This division reduces the complexity of the input spaces in Stem and Branches, allowing non-correlated information in input space S be processed separately, and thus improving the analyzing capability and reducing the complexity in a training stage, by reduction of number of local minima in modeling error hyper-surface. It allows a problem with a large number of classes to be solved easily, reducing the training time and/or permitting a better solution, close to global minimum to be found, and thus increases the recognition rate.

2.1 Sub-spaces Selection

The selection of the subspaces S_0 and S_1 is a very important point in the T-CombNET structure. In practical applications, the selection of most effective features from a measurement set can be a very difficult task, and many methods have been proposed in the literature [7] to solve this *feature selection* problem.

One possible method for *feature selection*, is based on a class separability criterion that is evaluated for all of possible combinations of the input features. The Eq.(3) presents the Interclass Distance Measure (IDM) criterion based on the Euclidean Distance [2].

$$J_\delta = \frac{1}{2} \sum_{i=1}^{m} P(\omega_i) \sum_{j=1}^{m} P(\omega_j) \frac{1}{N_i N_j} \sum_{k=1}^{N_i} \sum_{l=1}^{N_j} \delta(\boldsymbol{\xi}_{ik}, \boldsymbol{\xi}_{jl}) \qquad (3)$$

where m is the number of classes; $P(\omega_i)$ is the probability of the ith class; N_i is the number of pattern vectors belonging to the class ω_i, and $\delta(\boldsymbol{\xi}_{ik}, \boldsymbol{\xi}_{jl})$ is the Euclidean Distance from the kth candidate pattern of the class i to the lth candidate pattern of the class j.

The maximization of J_δ in the Eq.(3) with respect to the candidate pattern ξ, defines the feature set to be used. The IDM separability criterion estimates the distances between classes by computing the distances of a statistically representative set of input patterns, in order to determine the best feature set which may improve the efficiency of a general classifier.

In the T-CombNET context, we apply the IDM based *feature selection* method to split the input space in two subspaces to be used by the stem and branch networks. In the T-combnet structure the class probabilities $P(\omega_i)$ are unknown *a priori* for the Stem and Branches networks, due to the dependence of the S_0 sub-space definition and the stem training algorithm used. Thus, we need to design the Stem NN first to know those probabilities.

To model these features required by the T-CombNET structure, we propose the following joint interclass distance measurement (JIDM) J_T

$$J_T = J_{\delta S} + \frac{1}{M}\sum_{p=1}^{M} J_{\delta Bp} \qquad (4)$$

where M is the total number of branches NN, $J_{\delta S}$ is the IDM for the stem network and $J_{\delta Bp}$ for the pth branch network.

As the originally proposed IDM defined by Eq.(3) includes not only the interclass distance measurements but also the intraclass measurements, we propose a modification in this formulation, in order to explicit the contributions of the interclass and intraclass measurements, generating more effective functions to be maximized.

We define the interclass distance measurement $Ji_{\delta S}$ and intraclass distance measurement $Jo_{\delta S}$ for the stem network as

$$Ji_{\delta S} = \sum_{i=1}^{m-1} P(\varphi_i) \sum_{j=i+1}^{m} P(\varphi_j) \cdot \frac{1}{N_i N_j} \sum_{k=1}^{N_i}\sum_{l=1}^{N_j} \delta(\boldsymbol{x}'_{0_{ik}}, \boldsymbol{x}'_{0_{jl}}) \qquad (5)$$

$$Jo_{\delta S} = \sum_{i=1}^{m} P(\varphi_i)^2 \frac{1}{N_i^2 - N_i} \sum_{k=1}^{N_i-1}\sum_{l=k+1}^{N_i} \delta(\boldsymbol{x}'_{0_{ik}}, \boldsymbol{x}'_{0_{il}}) \qquad (6)$$

where $P(\varphi_i)$ is the *a priori* probability obtained for the ith pseudo class designed by the stem network training algorithm; $\boldsymbol{x}'_{0_{ik}}$ is the kth input vector belonging to the ith pseudo class, in the S'_0 space; m is the total number of pseudo classes of the stem layer; and N_i is the total number of input vectors belonging to the ith pseudo class.

The interclass distance measurement $Ji_{\delta Bp}$ and the intraclass distance measurement $Jo_{\delta Bp}$ for the branch network p, can be written as

$$Ji_{\delta Bp} = \sum_{i=1}^{m_p-1} P(\omega_{ip}|\varphi_p) \sum_{j=i+1}^{m_p} P(\omega_{jp}|\varphi_p) \cdot \frac{1}{N_i N_j} \sum_{k=1}^{N_i}\sum_{l=1}^{N_j} \delta(\boldsymbol{x}'_{1_{ik}}, \boldsymbol{x}'_{1_{jl}}) \qquad (7)$$

$$Jo_{\delta Bp} = \sum_{i=1}^{m_p} P(\omega_{ip}|\varphi_p)^2 \cdot \frac{1}{N_i^2 - N_i} \sum_{k=1}^{N_i-1}\sum_{l=k+1}^{N_i} \delta(\boldsymbol{x}'_{1_{ik}}, \boldsymbol{x}'_{1_{kl}}) \qquad (8)$$

where $P(\omega_{ip}|\varphi_p)$ is the conditional probability obtained for the ith class of the pth branch network given the stem neuron p; $\boldsymbol{x}'_{1_i}$ are input vectors allocated by the stem NN for the class i in the Time Normalized S_1' space; and m_p is the number of classes allocated to the stem neuron p.

Considering only the interclass distances, substituting the Eqs.(5) and (7) in Eq.(4), we obtain the function to be maximized with respect to the candidate vectors x_0' and x_1', to reach the optimum space division S_0 and S_1.

3 Simulation Results

This section presents the results of the experiments performed using Silicon Graphics Indy workstation with a color camera, using frames of 160×120 pixels size and 24 bits/pixel for color.

The Japanese finger spelling consists of 42 static hand postures and 34 dynamic hand gestures. The static hand postures recognition task was subject of an earlier work [10]. Current work is dedicated to the analysis and the classification of the dynamic gestures. The six basic hand movements present in Japanese finger spelling are presented in Fig.2.

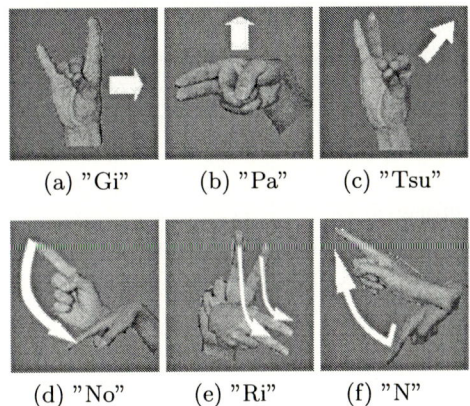

Fig. 2. Japanese Kana hand alphabet basic movements

In hand gesture set there are 20 symbols containing the horizontal, left to right hand motion shown by the Fig.2(a), where only the hand shape is changed; 5 symbols with motion type (b); 5 for type (c); 2 symbols for (d); and 1 symbol for each motion (e) and (f), totalizing the 34 categories of symbolic dynamic hand gestures.

A database is generated using the gestures performed by 5 non native sign language speakers. Each person was invited to perform each gesture 5 times, thus generating 170 video streams per person. The Training Set was generated from the samples of 4 persons, generating 680 video streams. The 5th person's gesture samples were reserved to use as an independent Test Set.

3.1 Space Division Analysis

In the joint IDM based criterion, the set of features x_0 and x_1 that maximize the class separability given by Eq.(4) is chosen to define the spaces S_0, S_1. This process requires a huge amount of computation time to obtain an optimum solution, because of the Eq.(4) which must be evaluated $n = \sum_{i=1}^{d-1} C_{d_i}^d$ times. In our application, the original input space dimension $d = 22$, resulting in $n = 4.194.303$ iterations, demanding a time estimated in months of continuous computation using a Pentium II 450MHz processor based computer. In the T-CombNET context, the previous design of the stem network for each iteration to evaluate the $P(\omega_{ip}|\varphi_p)$ probability, makes the optimization task infeasible. Thus, we need to use practical approaches or/and fast search methods in order to obtain a solution.

Practical Approach. A working solution can be found using the designer's feelings and, if required by trial and error method. If a physical interpretation of the vectors in the spaces S_0, S_0' and S_1 can be taken, the network model can be easily analyzed and better choices can be made. In this work, we tested the following practical approach: From the 22-D input vector presented in [10], the two components corresponding to the $P_v(t)$, hand movement direction on the screen, are chosen to define the S_0 subspace, generating two sub-spaces– a 2-D S_0 and its complementary 20-D S_1, to be used by the T-CombNET model. It seems to be efficient due to the natural non correlation existing between hand posture, described by S_1 space, and hand trajectory, described by S_0'. Using these definitions we assign the stem layer to analyze a normalized trajectory and the branch networks to analyze fine hand postures variation for a previously selected hand trajectory.

To demonstrate that the use of complementary subspaces S_0 and S_1 can improve the quality of the T-CombNET structure, we trained 3 networks. Table 1 shows the obtained results for the tested approaches.

Table 1. Recognition rates obtained for the practical approach

Structure	Sub-space S_0	Sub-space S_1	Recognition Rate
A	S	S	89.4 %
B	$\{P_v^x, P_v^y\}$	S	90.0 %
C	$\{P_v^x, P_v^y\}$	$\bar{S}_0$	91.2 %

Where S is the original 22-D dimensional input space, $\{P_v^x, P_v^y\}$ is the 2-D space defined by the hand location components, and $\bar{S}_0$ is the S_0 complementary 20-D space. Analyzing Table 1 we conclude that setting the stem layer to use all input space, that is setting $S_0 = S$ in Structure A, makes the steam layer design complex, because of the very high input dimension generated by preprocessing using the TN procedure. On the other hand, once that the information contained in the S_0 space is already processed by steam NN, its inclusion in the S_1 subspace, as occurs in Structure B, only increases the complexity of the branch training. Thus, it reduces the efficiency of the entire structure. Therefore, the best performance of the T-CombNET structure is reached using the input spaces for Structure C, that is complementary input spaces. A user dependent experiment using this approach is presented in [10].

Optimization Method. Sub-optimum solutions for the optimization of the joint IDM criterion can be found by using Fast Search (FS) methods [7]. In order to quickly obtain an approach for the input space division, we applied the Sequential Forward Selection Method presented in [2]. Applying the FS method, the time necessary to obtain a sub-optimum solution was reduced to a few hours.

Figure 3 shows the joint IDM values obtained during the maximization process and its reached maximum values, for each evaluated dimension of the input space S_0. We tested two stem NN training algorithms, a modified version of LBG algorithm [1] and the original self-growing algorithm proposed to the CombNET-II model training [8].

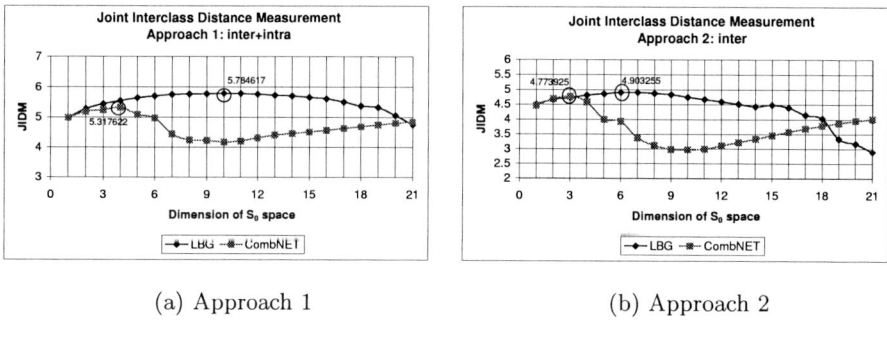

(a) Approach 1   (b) Approach 2

Fig. 3. Joint Interclass Distance Measurement

The optimization process is applied to two different approaches to define the joint IDM criterion. In the Fig.3(a), the original proposed definition of IDM, given by Eq.(3), is applied to stem and branches as indicated by Eq.(4), this means that the interclass and intraclass distances measurements are added to obtain the final IDM value. The Approach 2 uses only the interclass distances, defined by Eqs.(5) and (7), to calculate the IDM value. From the graphics we conclude that the original proposed IDM criterion generates a stem network with higher input vector dimension, once the maximum values are obtained for dimensions 4 and 10, for CombNET-II and LBG stem training algorithms respectively. Then incorporating the interclass measurement to the IDM to select the input space, a higher input dimension is needed to optimize the class separability.

From Fig.3 we can verify the strong dependence between the obtained joint IDM values and the training method used in the stem NN, it is due to the ability of IDM in capture the essence of the designed stem network. The objective of the LBG algorithm is to optimize the space division in order to obtain a minimum global distortion between neuron templates and training set vectors. On the other hand, the CombNET-II training algorithm aims to divide the space in order to minimize the classification error. As the IDM is a criterion based on the distance measurement, that is, based on distortion, then it is natural that the LBG algorithm results have a higher IDM value.

Table 2 presents a comparison of the IDM values resulting of the application of the practical approach, selecting $S_0 = \{P_v^x, P_v^y\}$, and the correspondent results obtained by the optimization method using the CombNET-II training algorithm.

Table 2. Joint IDM values for Practical Approach and Optimization Method

Sub-space S_0	Approach 1	Approach 2
Practical	4.514	3.661
Optimum	5.318	4.774

The values presented in Table 2 suggest that the practical approach does not define the best space division for the T-CombNET model.

Table 3 shows the obtained final recognition rates for the networks trained by approaches presented by Fig.3, for the maximum and minimum points of the CombNET-II trained IDM and maximum point for LBG trained IDM curves.

Table 3. Recognition Rates of T-CombNET

Optimization	Approach 1	Approach 2
CombNET-II Max	91.76%	96.47%
CombNET-II Min	87.06%	89.40%
LBG Max	73.53%	91.18%

Comparing the obtained recognition rates for the maximum points and minimum points of the CombNET-II training algorithm, from the Table 3 we conclude that the JIDM criterion is highly correlated with the obtainable recognition rate for the entire T-CombNET structure. Even if the LBG algorithm generates higher IDM values, the final recognition rates obtained is not so good as expected, demonstrating the low efficiency of distortion based algorithms when applied to classification problems. The application of joint IDM allows to find the optimum space division criterion to be applied to the T-CombNET structure. However, the results presented were obtained by using a sub-optimum space division, once that a Fast Search method was applied to maximize the JIDM, thus better results are expected if the global maximum can be found.

Evaluating the performance of the proposed structure, comparisons with classical network structures were performed. The T-CombNET structure designed using Approach 2 is compared with Kohonen's LVQ1 [5], Elman and Jordan [3] RNNs trained with the standard Backpropagation algorithm, single 3-layer feedforward backpropagation trained Multi-Layer Perceptron (MLP) and original CombNET-II structures. The LVQ1, MLP and CombNET-II are not recurrent neural networks, so they needed a pre-processing stage by the previously described Time Normalization algorithm in order to do the "spatialization" of the temporal information. The input space in this case is composed of all 22 dimensions of the input vector, and the Time Normalization parameter was set to 8 samples in this experiment. Elman and Jordan RNN are applied directly to the 22-D input vectors over the time. The best structures and recognition rates obtained from the experiments are presented in Table 4.

Table 4. Recognition Results

Network Type	Network Structure	Recognition Rate
Jordan	$22 \times 272 \times 34 + 34$	75.88%
Elman	$22 \times 272 + 272 \times 34$	86.47%
LVQ1	$176 \times 136 \times 34$	94.71%
MLP	$176 \times 136 \times 34$	94.71%
CombNET-II	stem: 176×3 branches: $176 \times 50 \times 20$ $176 \times 50 \times 11$ $176 \times 50 \times 9$	95.29%
T-CombNET	stem: 24×5 branches: $19 \times 100 + 100 \times 25$ $19 \times 88 + 88 \times 22$ $19 \times 60 + 60 \times 15$ $19 \times 32 + 32 \times 8$ $19 \times 32 + 32 \times 8$	96.47%

We have shown that the Recurrent Neural Networks like, Elman and Jordan RNN, are very good approaches for time series processing reaching high recognition rates in user dependent system [10], but such RNNs do not generalize as expected for the user independent problem treated here. The use of time "spatialization" techniques in classic NN approaches reaches a superior performance. However, the combined use of time "spatialization" and RNN in the proposed T-CombNET model overcomes the classical approaches, achieving a 96.47% of correct recognition rate.

4 Conclusions

The T-CombNET model is a multi-stage Neural Network that consists of a Time Normalization preprocessing procedure, followed by a LVQ NN based stem network and many Elman RNN in the branch networks. The input spaces of Stem and Branches NN are complementary sub-spaces selected from a input feature space. This work proposed the use of a *feature selection* method based on Interclass Distance Measurement to select the input spaces in T-CombNET structure. The optimal selection is a very computational time demanding problem, working approaches using a sub-optimum fast search method is discussed. The system was applied to an user independent, 34 classes, Japanese finger spelling recognition problem. The obtained results show that the use of Joint IDM criterion improves the recognition rate from 91.2%, reached by using the practical approach, to 96.5%; overcoming another classic neural network structures.

References

1. Linde, Y., Buzo, A., and Gray, R. M.: An Algorithm for Vector Quantizers Design. IEEE Trans. on Communications, Vol. COM-28 (1980) 84-95
2. Young, T. Y., and Fu, K. S.:Handbook of Pattern Recognition and Image Processing, Academic Press (1986)
3. Jordan, M. I.: Serial order: A parallel distributed processing approach. Technical Report Nr. 8604, Institute for Cognitive Science, University of California, San Diego (1986)
4. Tamura, S., and Kawasaki, S.: Recognition of Sign Language Motion Images. Pattern Recognition, Vol.21 (1988) 343-353
5. Kohonen, T.: Improved Versions of Learning Vector Quantization. International Joint Conference on Neural Networks, San Diego (1990) 545-550
6. Elman, J. L.: Finding Structure in Time. Cognitive Science, 14, (1990) 179-221
7. Fukunaga, K.: Introduction to Statistical Pattern Recognition, Academic Press (1990)
8. Iwata, A., Suwa, Y., Ino, Y., and Suzumura, N.: Hand-Written Alpha-Numeric Recognition by a Self-Growing Neural Network : CombNET-II. Proceedings of the International Joint Conference on Neural Networks, Baltimore (1992)
9. Starner, T., and Pentland, A.: Visual Recognition of American Sign Language Using Hidden Markov Models. International Workshop on Automatic Face and Gesture Recognition, Zurich, Switzerland (1995)
10. Lamar, M. V., Bhuiyan, M. S., and Iwata, A.: Hand Gesture Recognition using Morphological PCA and an improved CombNET-II. Proceedings of the 1999 International Conference on System, Man, and Cybernetics, Vol.IV, Tokyo, Japan (1999) 57-62
11. Lamar, M. V., Bhuiyan, M. S., and Iwata, A.: Temporal Series Recognition using a new Neural Network structure T-CombNET. Proceedings of the 6th International Conference on Neural Information Processing, Vol.III, Perth, Australia (1999) 1112-1117

Active and Adaptive Vision: Neural Network Models

Kunihiko Fukushima

The University of Electro-Communications,
Chofu, Tokyo 182-8585, Japan
`fukushima@ice.uec.ac.jp`

Abstract. To capture and process visual information flexibly and efficiently from changing external world, the function of active and adaptive information processing is indispensable. Visual information processing in the brain can be interpreted as a process of eliminating irrelevant information from a flood of signals received by the retina. Selective attention is one of the essential mechanisms for this kind of active processing. Self-organization of the neural network is another important function for flexible information processing. This paper introduces some neural network models for these mechanisms from the works of the author: such as "recognition of partially occluded patterns", "recognition and segmentation of face with selective attention", "binding form and motion with selective attention" and "self-organization of shift-invariant receptive fields".

1 Introduction

When looking at an object, we do not passively accept the entire information available within our visual field, but actively gather only necessary information. We move our eyes and change the focus of our attention to the places that attract our interest. We capture information from there and process it selectively. Visual information processing in the brain can be interpreted as a process of eliminating irrelevant information from a flood of signals received by the retina. Selective attention is one of the essential mechanisms for this kind of active processing of visual information. The interaction between the feedforward and feedback signals in neural networks plays an important role. Self-organization of the neural network is another important function for flexible information processing. The brain modifies its own structure to adapt to the environment.

This paper discusses the importance of active and adaptive processing in vision, showing our recent research on modeling neural networks.

2 Recognition of Occluded Patterns

We often can read or recognize a letter or word contaminated by stains of ink, which partly occlude the letter. If the stains are completely erased and the occluded areas of the letter are changed to white, however, we usually have difficulty in recognizing the letter, which now have some missing parts. For example, the patterns in Fig. 1(a), in which the occluding objects are invisible, are almost illegible, but the patterns in (b), in which the occluding objects are visible, are much easier to read. Even after having watched patterns in (b) and having known what the occluding objects are, we still have difficulty in recognizing patterns in (a) if the occluding objects are invisible.

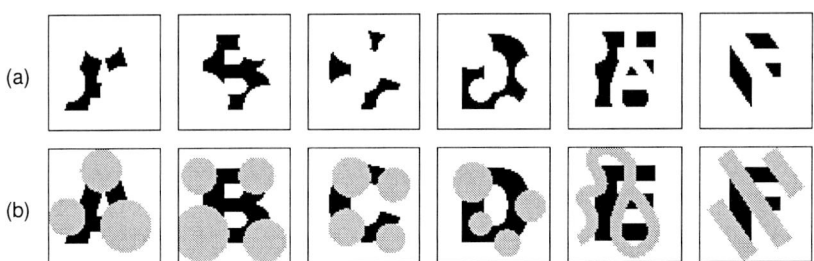

Fig. 1. Patterns partially occluded by (a) invisible and (b) visible masking objects. These patterns are used to test the proposed neural network model.

This section proposes a hypothesis explaining why a pattern is easier to be recognized when the occluding objects are visible. A neural network model is constructed based on the proposed hypothesis and is simulated on a computer.

A pattern usually contains a variety of visual features, such as edges, corners, and so on. The visual system extracts, at an early stage, these local features from the input pattern and then recognizes it. When a pattern is partially occluded, a number of new features, which did not exist in the original pattern, are generated near the contour of the occluding objects. When the occluding objects are invisible, the visual system will have difficulty in distinguishing which features are relevant to the original pattern and which are newly generate by the occlusion. These irrelevant features will hinder the visual system from recognizing the occluded pattern correctly.

When the occluding objects are visible, however, the visual system can easily discriminate relevant from irrelevant features. The features extracted near the contours of the occluding objects are apt to be irrelevant to the occluded pattern. If the responses of feature extractors in charge of the area covered by the occluding objects are suppressed, the signals related to the disturbing irrelevant features will be blocked and will not reach higher stages of the visual system. Since the visual system usually has some tolerance to partial absence of local features of the pattern, it can recognize the occluded pattern correctly if the disturbing signals from the irrelevant features disappear.

We have constructed a neural network model based on this hypothesis. The model is an extended version of the *neocognitron* [1]. The neocognitron is a neural network model of the visual system, and has a hierarchical multilayered architecture. It acquires an ability to robustly recognize visual patterns through learning. It consists of layers of S-cells, which resemble simple cells in the primary visual cortex, and layers of C-cells, which resemble complex cells. In the whole network, layers of S-cells and C-cells are arranged alternately in a hierarchical manner. S-cells are feature-extracting cells, and input connections to S-cells are variable and are modified through learning. C-cells, whose input connections are fixed and unmodifiable, exhibit an approximate invariance to the position of the stimuli presented within their receptive fields. The C-cells in the highest stage work as recognition cells indicating the result of pattern recognition.

Figure 2 shows the architecture of the proposed neural network model. It consists of four stages of S- and C-cell layers. It has a layer of contrast-extracting cells (U_G), which correspond to retinal ganglion cells or LGN cells, between the input layer U_0 (photoreceptor layer) and the S-cell layer of the first stage (U_{S1}). Layer U_G consists of two cell-planes: a cell-plane consisting of on-center cells, and a cell-plane consisting of off-center cells.

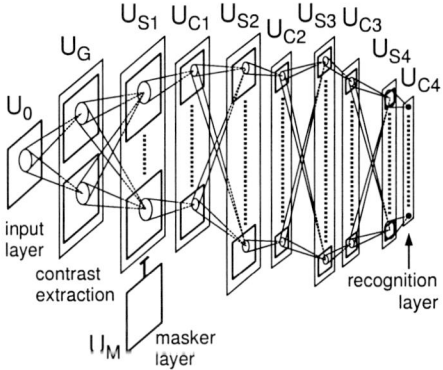

Fig. 2. The architecture of the neural network model that can recognize partially occluded patterns.

The output of layer U_{Sl} (S-cell layer of the lth stage) is fed to layer U_{Cl}, and a blurred version of the response of layer U_{Sl} appears in layer U_{Cl}. The density of the cells in each cell-plane is reduced between layers U_{Sl} and U_{Cl}.

The S-cells of the first stage (U_{S1}), which have been trained with supervised learning [1], extract edge components of various orientations from input image. The S-cells of the intermediate stages (U_{S2} and U_{S3}) are self-organized by unsupervised competitive learning in a similar way as for the conventional neocognitron. Layer U_{S4} at the highest stage is trained to recognize all learning patterns correctly through supervised competitive learning [2].

The model contains a layer (U_M) that detects and responds only to occluding objects. The layer, which has only one cell-plane, is called the *masker layer* in this paper. The shape of the occluding objects is detected and appears in layer U_M, in the same shape and at the same location as in the input layer U_0. There are slightly diverging, and topographically ordered, inhibitory connections from layer U_M to all cell-planes of layer U_{S1}. The response to features irrelevant to the occluded pattern are thus suppressed by the inhibitory signals from layer U_M. (The mechanism of segmenting occluding objects from the input image is not the main issue here. In the computer simulation below, occluding objects are segmented based on the difference in brightness between occluding objects and occluded patterns.)

The model is simulated on a computer. The network has initially been trained to recognize alphabetical characters. Figure 3 shows the learning patterns used to train the network.

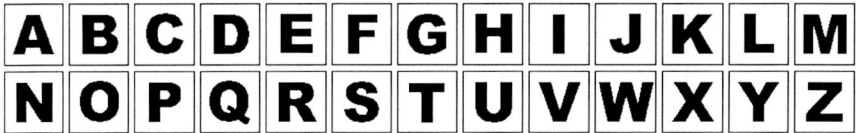

Fig. 3. The learning patterns used to train the network.

After finishing the learning, partially occluded alphabetical characters are presented to the input layer of the network. Figure 1 are some examples of the input images used for the test. Tests are made for two types of input images: the occluding objects are (a) invisible, and (b) visible.

When the occluding objects are visible as shown in (b), they are detected and appear in the masker layer U_M. When occluding objects are invisible as shown in (a), however, no response appears in the masker layer U_M, and no inhibitory signals come from the layer.

Table 1 shows how the network recognizes these test patterns, under two different conditions: namely, when the masker layer is working, and not working.

Table 1. Result of recognition by the network, when the masker layer is working (proposed model), and not working (neocognitron). This table shows how the images in Figure 1 are recognized.

model	occluding objects	recognized as
either model	none	A B C D E F
either model	invisible (Fig. 1(a))	I E Z E L G
with masker layer (proposed model)	visible (Fig. 1(b))	A B C D E F
without masker layer (neocognitron)	visible (Fig. 1(b))	Q B G Q E X

As can be seen from the table, the network with a masker layer recognizes all patterns correctly, if the occluding objects are visible. If the occluding objects are invisible, however, all patterns are recognized erroneously because the masker layer cannot detects occluding objects and cannot send inhibitory signals to layer U_{S1}. The patterns with invisible occluding objects are hardly recognized correctly even by human beings. Our network thus responds to occluded patterns like human beings.

If the masker layer is removed from the network, the network becomes almost like the conventional neocognitron, and fails to recognize most of the occluded patterns even when the occluding objects are visible.

3 Face Recognition

We have proposed an artificial neural network that recognizes and segments a face and its components (e.g., eyes and mouth) from a complex background [3][4]. The network can find out a face from an image containing numerous other objects. It then focuses attention on the eyes and mouth, and segment each of these facial components away from other objects.

We extend Fukushima's *selective attention model* [5] to have two channels of different resolutions. Figure 4 shows the architecture of the proposed network together with a typical example of the response of some layers in the network. Both high- and low-resolution channels have forward and backward paths. The forward paths manage the function of pattern recognition, and the backward paths manage the function of selective attention, segmentation and associative recall.

The layers in the forward paths are denoted by U, and those in the backward paths by W. The layers in the low- and high-resolution channels are represented with suffixes L and H, respectively.

The low-resolution channel mainly detects the approximate shape of a face, while the high-resolution channel detects the detailed shapes of the facial components. The high-resolution channel can analyze input patterns in detail, but usually lacks the ability to gather global information because of the small receptive fields of its cells. Furthermore, direct processing of high-resolution information is susceptible to deformation of the image. Although the processing of low-resolution information is tolerant of such variations, detailed information is lost. Therefore, the proposed network analyses objects using signals from both channels.

The left and right eyes, for example, have a similar shape. It is consequently difficult to discriminate from shape alone. In the proposed network, a cell (of layer U_{SH4}) extracting the left eye analyses the shape of the eye in detail using the signals from the high-resolution channel, but it also obtains from the low-resolution channel rough positional information to discriminate the left eye from the right eye. The three cell-planes of U_{SH4} in Fig. 4 thus respond to the left eye, right eye and mouth, respectively from the top.

A cell (of layer U_{S5}) recognizing a face also receives signals from both channels and can reject a non-facial oval object if the object does not have two eyes and mouth, which are extracted by the cells of the high-resolution channel.

The output of the recognition layer U_{C5} at the highest stage is fed back through backward paths of both channels. The backward signals through W_{CH4},

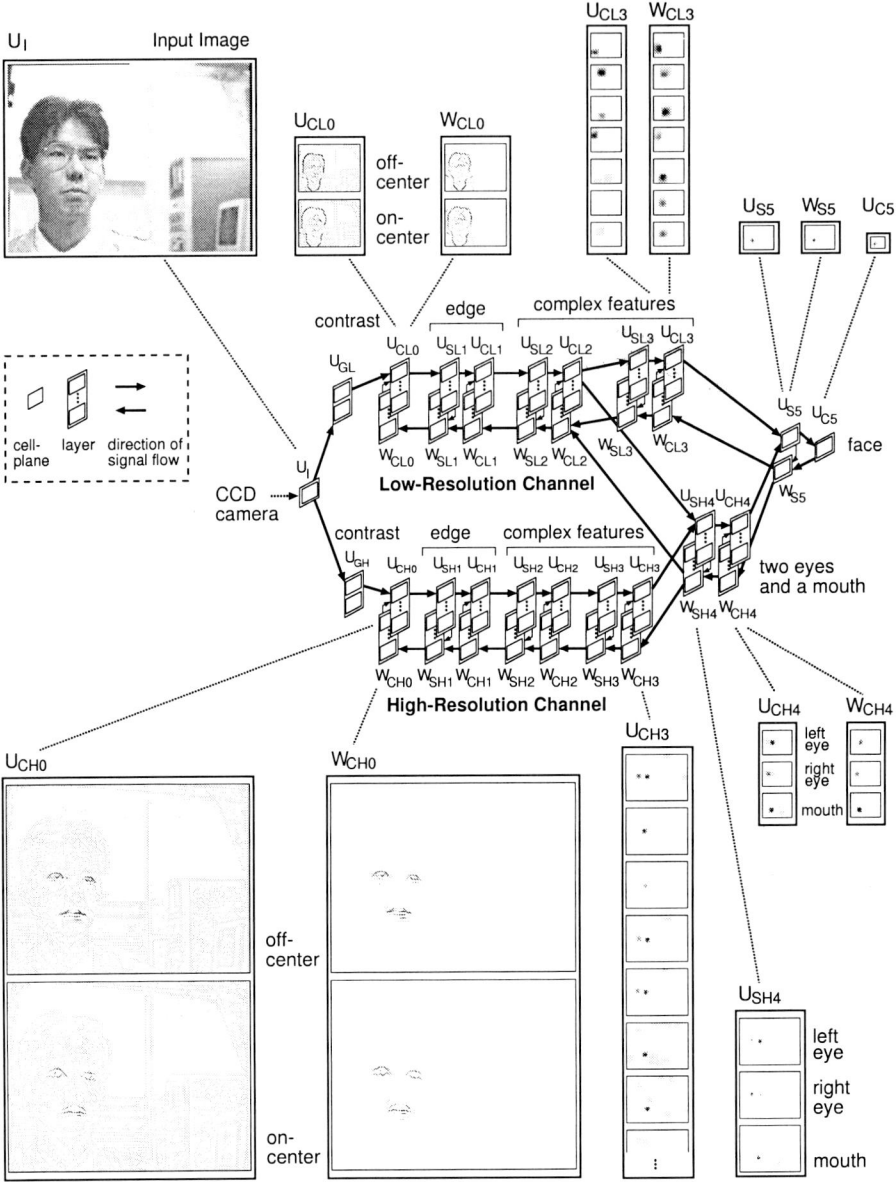

Fig. 4. The architecture of the face-recognition network, and a typical example of the response of some layers in the network.

which detects facial components, generate the image (or contours) of those facial components in W_{CH0} at the lowest stage of the high resolution channel. In a standard situation, three facial components, that is, two eyes and a mouth, are simultaneously segmented in W_{CH0}. If necessary, however, we can let the network segment only a single facial component, say, the left eye, by controlling the response of layer W_{CH4}.

4 Binding Form and Motion

In the visual systems of mammals, visual scenes are analyzed in parallel by separate channels. If a single object is presented to the retina, the form channel, or the occipito-temporal pathway, recognizes what it is. The motion channel, or the occipito-parietal pathway, detects the type of the motion, for example, movement direction of the object. In each channel, the input image is analyzed in a hierarchical manner. Cells in a higher stage generally have larger receptive fields and are more insensitive to the location of the object.

Then, the so-called "binding problem" has to be solved, when two or more objects are simultaneously presented to the retina. Different attributes extracted from the same object have to be bound together. Suppose, for example, a triangle moving downward and a square moving leftward are presented simultaneously. The form channel will recognizes both the triangle and square, and the motion channel will detect both the downward and leftward movements. How does the brain know that the triangle is not moving leftward but downward?

Anatomical observations have shown that the forward connections in the form and motion channels are segregated from each other, but the backward connections are more diffused. Suggested by this anatomical evidence, we have proposed a neural network model [6], which can solve the binding problem by selective attention.

Figure 5 shows the network architecture. The network has two separate channels for processing form and motion information. Both channels have hierarchical architectures, and the cells in the higher stages have larger receptive fields with more complex response properties. Both channels not only have forward but also backward signal flows. By the interactions of the forward and the backward signals, each channel exhibits the function of selective attention.

If we neglect the interaction of the two channels, the form channel has basically the same architecture and function as the Fukushima's selective-attention model [5]. The highest stage of this channel corresponds to the inferotemporal areas, and consists of recognition cells, which respond selectively to the categories of the attended object. The lowest stage of the backward path is the segmentation layer. The image of the attended object is segregated from that of other objects, and emerges in this layer.

A lower stage of the motion channel consists of motion-detecting cells with small receptive fields. A motion-detecting cell responds selectively to the movement in a particular direction. The highest stage of the motion channel, which corresponds to MST, has direction-indicating cells with large receptive fields: One direction-indicating cell exists for each different direction of movement. They only respond to the motion of the attended object, even if there are many other objects moving in other directions. Each direction-indicating cell receives

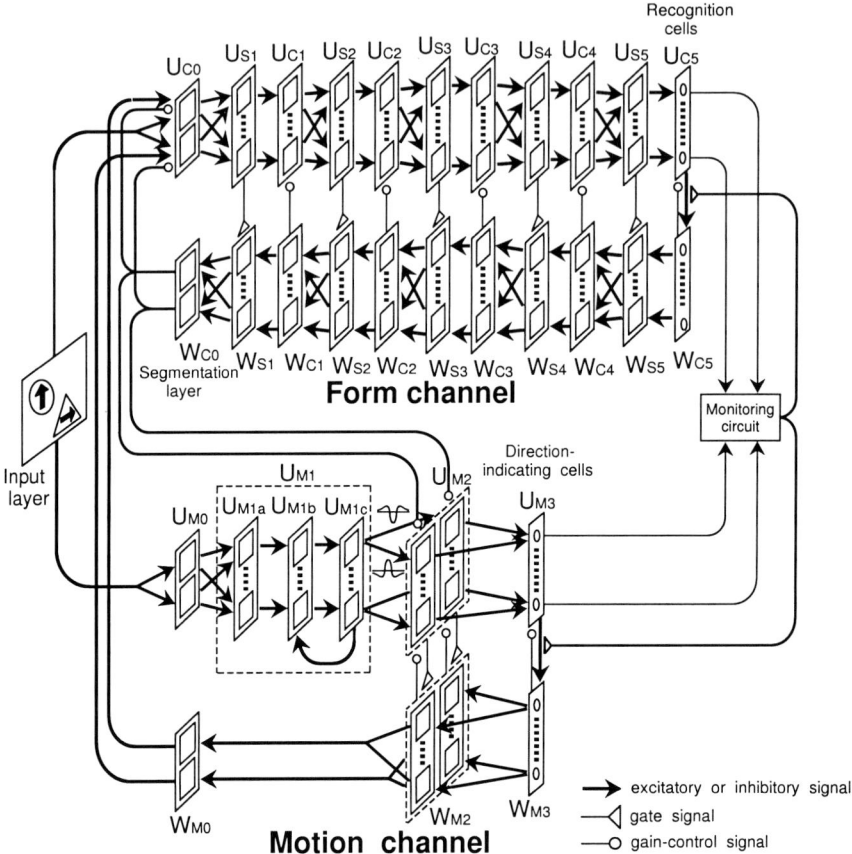

Fig. 5. The architecture of the model that has separate channels for processing form and motion information [6].

converging excitatory signals from lower-layer cells of the same preferred direction, but signals originating from non-attended objects are suppressed by the attention mechanism.

Backward signals from higher stages of each channel are fed back not only to the lower stages of the same channel but also to the lower stages of the other channel and controls the forward signal flows of both channels. Because of this interaction, both channels focus attention on the same object even if a number of objects are presented simultaneously to the retina.

Figure 6 shows an example of the response of the model. Three moving patterns made of random dots are presented to the input layer, as shown in the top of the figure. In the figure, the responses of the segmentation layer, the recognition cells, and the motion-indicating cells are shown in a time sequence.

In the form channel, the recognition cell representing a diamond responded first in the highest layer, and the contour of the diamond appeared in the seg-

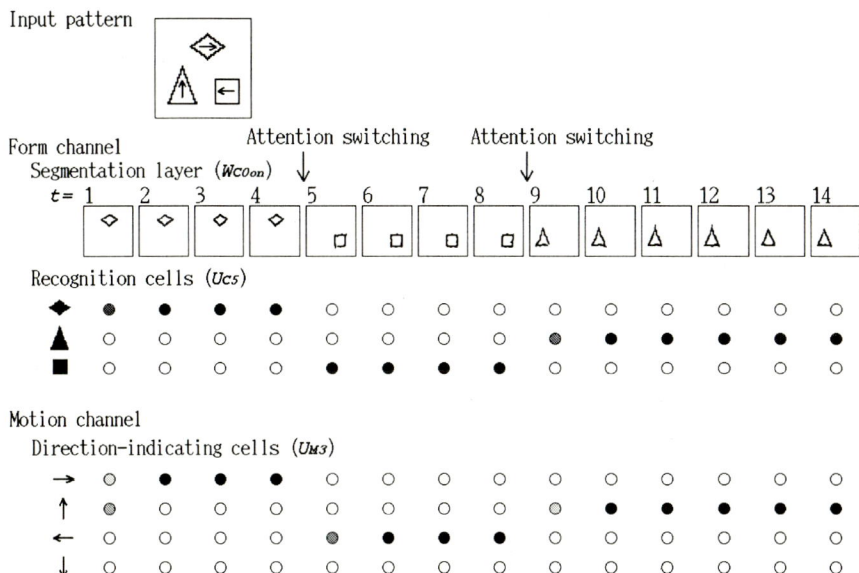

Fig. 6. Binding form and motion. The response of the model to random dot patterns moving in different directions [6].

mentation layer ($t \leq 4$). In the motion channel, the direction-indicating cell for rightward movement responded.

When the response of the network reached a steady state at $t=4$, attention was switched automatically. After switching attention ($t \geq 5$), the form channel recognized and segmented the square, and the motion channel detected its leftward movement. After the second switch of attention ($t \geq 9$), the triangle was recognized and its upward movement was detected. The model thus recognizes all objects in the visual scene in turn by a process of attention switching.

5 Self-Organization of Shift-Invariant Receptive Fields

This section discusses a learning rule by which cells with shift-invariant receptive fields are self-organized [7]. With this learning rule, cells similar to simple and complex cells in the primary visual cortex are generated in a network. Training patterns that are presented during the learning are long straight lines of various orientations that sweep across the input layer.

To demonstrate the learning rule, we simulate a three-layered network that consists of an input layer, a layer of S-cells (or simple cells), and a layer of C-cells (or complex cells). The network, which is illustrated in Fig. 7, resembles the first stage of the *neocognitron*. In contrast to the neocognitron, however, input connections to C-cells, as well as S-cells, can be created through learning. The network neither has the architecture of cell-planes nor requires the condition of shared connections to realize shift-invariant receptive fields of C-cells.

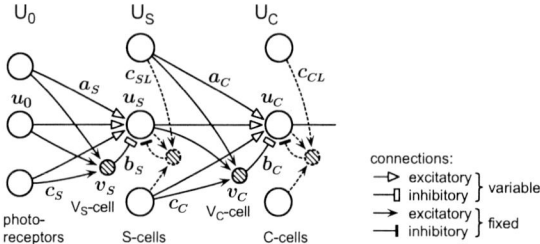

Fig. 7. Self-organization of shift-invariant receptive fields. Connections between cells in the network [7].

The learning rule of the S-cells is similar to the unsupervised learning for the conventional neocognitron. Each cell (S-cell) competes with other cells in its vicinity, and the competition depends on the instantaneous activities of the cells. Only winners of the competition have their input connections increased. The increment of each connection is proportional to the presynaptic activity. In other words, LTP (long term potentiation) is induced in the input synapses of the winner S-cells.

If a straight line is presented during the learning phase, winners in the competition are generally distributed along the line. When the line is shifted to another location, another set of winners is chosen. Thus, after finishing a sweep of the line, S-cells whose receptive fields have the same preferred orientation as the line are generated and become distributed over the layer U_S. S-cells of other preferred orientations are generated by sweeps of lines of other orientations.

C-cells also compete with each other for the self-organization. In contrast to S-cells, however, the competition is based on the *traces* of their activities, not on their instantaneous activities. A trace is a kind of temporal average (or moving average) [8]. Once the winners are determined by the competition, the excitatory connections to the winners increase (LTP). The increment of each connection is proportional to the instantaneous presynaptic activity. At the same time, the losers have their excitatory input connections decreased (LTD). The decrement of each connection is proportional to both the instantaneous presynaptic activity and the instantaneous postsynaptic activity.

We will now discuss how the self-organization of the C-cell layer (U_C) progresses under this learning rule. If a line stimulus sweeps across the input layer, S-cells whose preferred orientation matches the orientation of the line become active. The timings of becoming active, however, differ among S-cells. For the creation of shift-invariant receptive fields of C-cells, it is desired that a single C-cell obtain strong excitatory connections from all of these S-cells in its connectable area.

Since winners are determined based on traces of outputs of C-cells, the same C-cell has a large probability to continue to be a winner throughout the period when the line is sweeping across its receptive field. Namely, if a C-cell once becomes a winner, the same C-cell will be apt to keep winning for a while after that because the trace of an output lasts for some time after the extinction of the output. Hence a larger number of S-cells will become connected to the C-cell. The larger the number of connected S-cells becomes, the larger the chance of winning becomes. Thus, the C-cell will finally acquire connections from all relevant S-cells.

The increase of inhibitory connections from V_C-cell to C-cell is also important for preventing a single C-cell from coming to respond to lines of all orientations. The inhibitory connection to each C-cell is increased in such a way that the total amount of the excitatory inputs never exceeds the inhibitory input to the C-cell. In other words, the activity of a cell is always regulated so as not to exceed a certain value.

Figure 8 shows responses of the network that has finished self-organization. The responses of photoreceptors of U_0, S-cells of U_S and C-cells of U_C are displayed in the figure by the size of the small dots (filled squares).

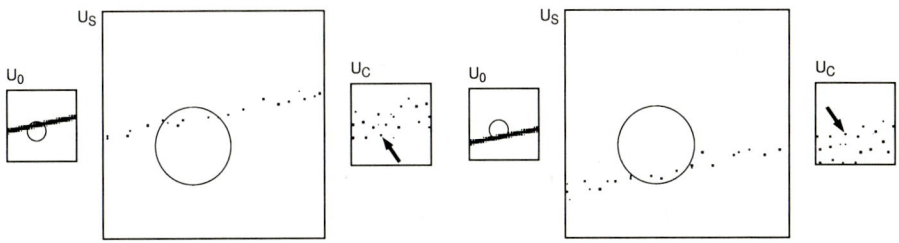

(a) The stimulus line is optimally oriented to the C-cell marked with an arrow.

(b) The same line as in (a) is presented at a different location.

Fig. 8. Responses of the network that has finished self-organization [7].

If a line is presented to a location on the input layer as shown in Fig. 8(a), a number of S-cells respond to the line, and consequently C-cells that receive excitatory connections from these S-cells become active. Let us watch, for example, an active C-cell marked with an arrow in layer U_C. The circles in U_S and U_0 show the connectable area and the effective receptive field of the C-cell, respectively. When comparing the sizes of these areas, note that layers U_S and U_C, that have a larger number of cells, are displayed in this figure on a larger scale than layer U_0.

If the line shifts to a new location as shown in Fig. 8(b), other S-cells become active because S-cells are sensitive to stimulus location. In layer U_C, however, several C-cells that were active in Fig. 8(a) are still active for this shifted stimulus. For example, the C-cell marked with an arrow continues responding. This shows that the C-cell exhibits a shift-invariance within the receptive field. When the line is rotated to another orientation, this C-cell is of course silent even if the line is presented in its receptive field.

Figure 9 shows responses of typical S- and C-cells. The orientation tuning curves of the cells are shown in (a). A stimulus line is rotated at various orientations, and is swept across the input layer. The abscissa of a curve is the line orientation, and the ordinate is the total response of the cell during a sweep of the line in each orientation. Figure (b) shows the response of the same cells to the shift of an optimally oriented line. It can be seen that the C-cell responds to the line within a wide range of locations and shows shift-invariance within its receptive field.

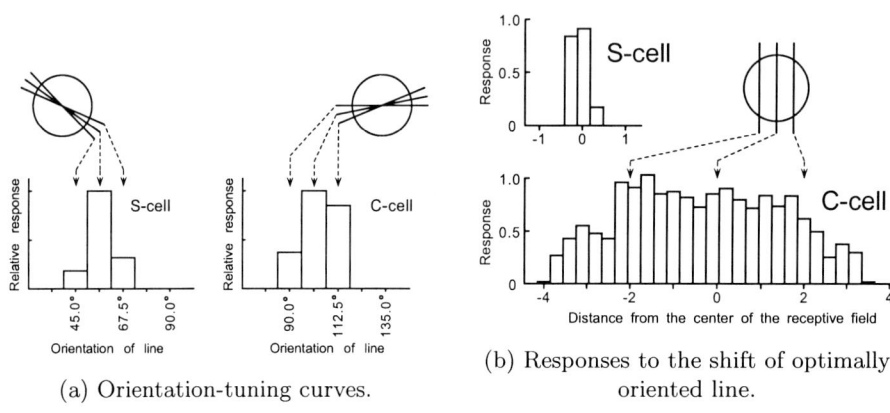

(a) Orientation-tuning curves.

(b) Responses to the shift of optimally oriented line.

Fig. 9. Typical responses of S- and C-cells [7].

References

1. K. Fukushima: "Neocognitron: a hierarchical neural network capable of visual pattern recognition", *Neural Networks*, **1**[2], pp. 119–130 (1988).
2. K. Fukushima, K. Nagahara, H. Shouno: "Training neocognitron to recognize handwritten digits in the real world", *pAs'97 (The Second Aizu International Symposium on Parallel Algorithms/Architectures Synthesis*, Aizu-Wakamatsu, Japan), pp. 292–298 (March, 1997).
3. H. Hashimoto, K. Fukushima: "Recognition and segmentation of components of a face with selective attention" (in Japanese), *Trans. IEICE D-II*, **J80-D-II**[8], pp. 2194–2202 (Aug. 1997).
4. K. Fukushima: "Active vision: neural network models", *Brain-Like Computing and Intelligent Information Systems*, eds.: N. Kasabov, S. Amari, pp. 3–24, Singapore: Springer-Verlag (1998).
5. K. Fukushima: "Neural network model for selective attention in visual pattern recognition and associative recall", *Applied Optics*, **26**[23], pp. 4985–4992 (Dec. 1987).
6. M. Kikuchi, K. Fukushima: "Neural network model of the visual system: Binding form and motion", *Neural Networks*, **9**[8], pp. 1417–1427 (Nov. 1996).
7. K. Fukushima: "Self-organization of shift-invariant receptive fields", *Neural Networks*, **12**[6], pp. 791–801 (July 1999).
8. P. Földiák: "Learning invariance from transformation sequences", *Neural Computation*, **3**, 194–200 (1991).

Temporal Structure in the Input to Vision Can Promote Spatial Grouping

Randolph Blake and Sang-Hun Lee

Vanderbilt Vision Research Center,
Vanderbilt University, Nashville TN 37240, USA
randolph.blake@vanderbilt.edu

1 Introduction

> *Humpty Dumpty sat on a wall*
> *Humpty Dumpty had a great fall*
> *All the King's horses and all the King's men*
> *Couldn't put Humpty together again.*

This familiar "Mother Goose" nursery rhyme captures the essence of what has become a central problem in the field of visual science: how can an aggregate of individuals working together reassemble a complex object that has been broken into countless parts? In the case of vision, the "horses and men" comprise the many millions of brain neurons devoted to the analysis of visual information, neurons distributed over multiple areas of the brain (Van Essen et al. 1992; Logothetis, 1998). And, in the case of vision, the source to be reassembled corresponds to the panorama of objects and events we experience upon looking around our visual world. In contemporary parlance, this "reassembly" process has been dubbed the *binding problem*, and it has become a major focus of interest in neuroscience (Gray, 1999) and cognitive science (Treisman 1999).

We have no way of knowing exactly why the King's men and horses failed to solve their particular version of the binding problem, but perhaps they possessed no "blueprint" or "picture" of Humpty Dumpty to guide their efforts. (Just think how difficult it would be to assemble a jig-saw puzzle without access to the picture on the box.) Of course, the brain doesn't have blueprints or pictures to rely on either, but it can -- and apparently does -- exploit certain regularities to constrain the range of possible solutions to the binding problem (Marr, 1982). These regularities arise from physical properties of light and matter, as well as from relational properties among objects. Among those relational properties are ones that arise from the temporal structure created by the dynamical character of visual events; in this chapter we present psychophysical evidence that the brain can exploit that temporal structure to make educated guesses about what visual features go with one another. To begin, we need to back up a step and review briefly how the brain initially registers information about the objects and events we see and then outline several alternative theories of feature binding.

2 Early Vision and Possible Binding Mechanisms

It is generally agreed that early vision entails local feature analyses of the retinal image carried out in parallel over the entire visual field. By virtue of the receptive-field properties of the neurons performing this analysis, visual information is registered at multiple spatial scales, ranging from coarse to fine, for different contour orientations (DeValois and DeValois, 1988). Of course, we are visually unaware of this multiscale dissection of the retinal image: from a perceptual standpoint, coarse and fine spatial details bind seemlessly into harmonious, coherent visual representations of objects. In addition to spatial scale, different qualitative aspects of the visual scene -- color, form, motion -- engage populations of neurons distributed among numerous, distinct visual areas (Zeki, 1993). But, again, the perceptual concomitants of those distributed representations are united perceptually; it's visually impossible to disembody the "red" from the "roundness" of an apple. Evidently, then, the process of binding -- wherein the distributed neural computations underlying vision are conjointly represented -- transpires automatically and efficiently.

Several possible mechanisms for feature binding have been proposed over the years. These include:
- coincidence detectors - these are neurons that behave like logical AND-gates, responding only when specific features are all present simultaneously in the appropriate spatial arrangement (Barlow, 1972). This idea is implicit in Hubel and Wiesel's serial model of cortical receptive fields (Hubel and Wiesel, 1962), in which a simple cell with an oriented receptive field is activated only when each of its constituent thalamic inputs is simultaneously active. Coincidence detection has also been proposed as the neural basis for registration of global, coherent motion (Adelson and Movshon, 1982) and for the integration of information within the "what" and "where" visual pathways (e.g., Rao et al, 1997). A principal objection to coincidence detection as a mechanism of binding is the combinatorial problem: there are simply not enough neurons to register the countless feature combinations that define the recognizable objects and events in our world (e.g., Singer and Gray, 1995; but see Ghose and Maunsell, 1999, who question this objection).
- attention - distributed representations of object features are conjoined by the act of attending to a region of visual space (Treisman and Gelade, 1980; Ashby et al, 1996). This cognitively grounded account, while minimizing the combinatorial problem, remains ill-defined with respect to the actual neural concomitants of attentional binding.
- temporal synchronization - grouping of object features is promoted by synchronization of neural activity among distributed neurons responsive to those various features (von der Malsburg, 1995; Milner, 1974; Singer, 1999). On this account, the coupling of activity among neurons can occur within aggregates of neurons within the same visual area, among neurons within different visual areas and, for that matter, among neurons located in the separate hemispheres of the brain.

This third hypothesis has motivated recent psychophysical work on the role of temporal structure in spatial grouping, including work in our laboratory using novel displays and tasks. In the following sections this chapter examines the extent to which visual grouping is jointly determined by spatial and temporal structure. From the outset we acknowledge that there is considerable skepticism within the field of biological vision about the functional significance of synchronized discharges among neurons. This skepticism exists for several reasons. Some have argued that synchronized activity may be an artifact that simply arises from the spiking behavior of retinal and/or thalamic cells unrelated to stimulus-driven phase locking (Ghose and Freeman 1997) or from the statistical properties of rapidly spiking cells (Shadlen and Movshon 1999). Others question whether the noisy spike trains from individual neurons possess the temporal fidelity for temporal patterning to be informationally relevant (Shadlen and Newsome 1998). Still others argue that neural synchrony, even if it were to exist, provides a means for *signalling* feature clusters but not a means for *computing* which features belong to an object, therefore leaving the binding problem unsolved (Shadlen and Movshon, 1999). Finally, some have found stimulus induced synchrony is just as likely between neurons responding to figure and background elements as between neurons responding to figure elements only (Lamme and Spekreijse 1998). Balanced overviews of these criticisms are provided in recent reviews by Gawne (1999) and by Usrey and Reid (1999).

While not ignoring these controversies, our approach has been to study the effect of externally imposed synchrony on visual perception. We reason that if temporal synchronization were to provide a means for binding and segmentation, psychophysical performance on perceptual grouping tasks should be enhanced when target features vary synchronously along some dimension over time, whereas performance should be impaired when features vary out-of-phase over time. These predictions are based on the assumption that temporal modulations of an external stimulus produce modulations in neural activity, within limits of course.[1] We give a more detailed exposition of this rationale elsewhere (Alais et al 1998). The following sections review evidence that bears on the question: "To what extent do features changing together over time tend to group together over space?" Next, we turn to our very recent studies using stochastic temporal structure, the results from which provide the most compelling demonstrations to date for the role of temporal structure in spatial grouping. The chapter closes with speculative comments about whether the binding problem really exists from the brain's perspective.

3 Temporal Fluctuation and Spatial Grouping

<u>Periodic Temporal Modulation</u> The simplest, most widely used means for varying temporal structure is to repetitively flicker subsets of visual features, either in-phase or out-of-phase. If human vision exploits temporal phase for spatial organization, features flickering in synchrony should group together and appear segregated from

---

[1] 1. Most versions of the temporal binding hypothesis posit the existence of <u>intrinsically</u> mediated neural synchrony engendered even by static stimulus features. On this model, synchrony is the product of neural circuitry, not just stimulation conditions. Hence, one must be cautious in drawing conclusions about explicit mechanisms of "temporal binding" from studies using externally induced modulations in neural activity.

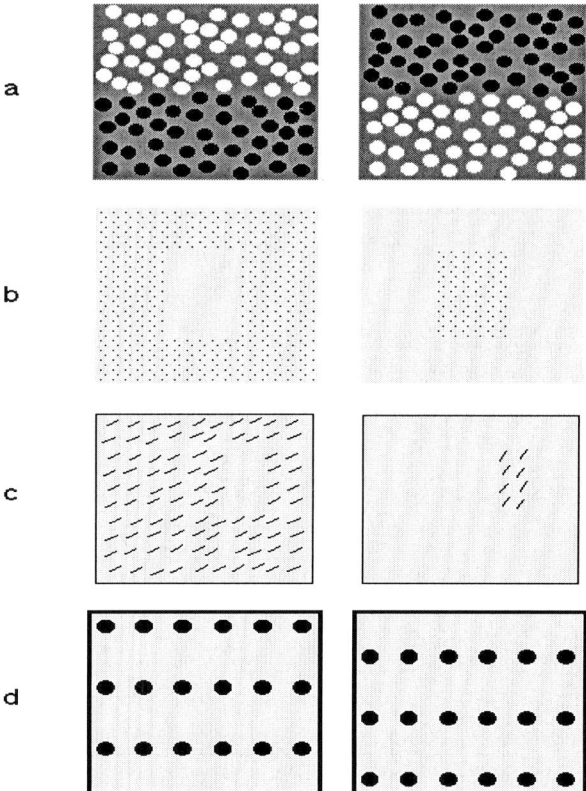

Fig. 1. Examples of visual displays used to assess the contribution of repetitive flicker on spatial grouping (redrawn from originals). A. Roger-Ramachandran and Ramachandran (1998). Black and white spots formed a texture border in the first and second frames of a two-frame "movie" -- the contrast of all spots reversed each frame. B. Fahle (1993). Rectangular arrays of small dots forming the "figure" and the "background" were rapidly interchanged. C. Kiper et al (1996). Oriented contours form successively presented "background" and "figure" frames that were rapidly interchanged. D. Usher & Donnelly (1998). A square lattice of elements (shown here as filled circles) was presented with alternating rows presented successively or with all elements presented simultaneously (not shown)

those flickering in different temporal phases. The following paragraphs briefly summarize studies that have utilized this form of temporal modulation.

Rogers-Ramachandran and Ramachandran (1991, 1998) created an animation consisting of two frames (Figure 1a). Black and white dots were spatially distributed against a grey background to create a texture border in the first frame. Then, in the second frame, the luminance of all spots was reversed (black turned to white and vice versa). Repetitively alternating the two frames created counterphase flicker of the two groups of dots. When the display flickered at 15 hz, that is, the phase difference between the two groups of spots was 15 msec, observers could still perceive the texture boundary but could not discern whether any given pair of spots was flickering in-phase or out-of-phase. Rogers-Ramachandran and Ramachandran termed this boundary a "phantom contour" because clear texture segregation was perceived even though texture elements themselves were indistinguishable.

Using a similar strategy, Fahle (1993) manipulated the stimulus onset asynchrony between two groups of flickering dots and measured the clarity of figure-ground segregation. He tested arrays of regularly and randomly spaced dots (Figure 1b). The dots within a small "target" region (denoted by a rectangle in Figure 1b) were flickered in synchrony while dots outside the target region flickered at the same rate but with temporal phase the flicker delayed relative to that of the target dots. Observers could judge the shape of the target region with temporal phase shifts as brief as 7 msec under optimal conditions. Fahle concluded that the visual system can segregate a visual scene into separate regions based on "purely temporal cues" because the dots in figure and those in ground were undifferentiated within any single frame, with temporal phase providing the only cue for shape. Kojima (1998) subsequently confirmed this general finding using band-passed random dot texture stimuli.

The results from these two studies imply that very brief temporal delays between "figure" and "background" elements provide an effective cue for spatial grouping and texture segregation. However, other studies using rather similar procedures have produced conflicting results. Kiper,

Gegenfurtner and Movshon (1991, 1996) asked whether onset asynchrony of texture elements influences performance on tasks involving texture segmentation and grouping. Observers discriminated the orientation (vertical vs horizontal) of a rectangular region containing line segments different in orientation from those in a surrounding region (Figure 1c). Kiper et al varied the angular difference in orientation between target and background texture elements, a manipulation known to affect the conspicuity of the target shape. They also varied the onset time between target and background elements, reasoning that if temporal phase is utilized by human vision for texture segmentation, performance should be superior when "target" texture elements are presented out of phase with the "background" elements. To the contrary, however, temporal asynchrony had no effect on the ease of texture segmentation; performance depended entirely on the spatial cue of orientation disparity.

Using a novel bistable display (Figure 2), Fahle and Koch (1995) also failed to find evidence that temporal cues promote spatial organization. When the two identical Kanizsa triangles formed by illusory contours partially overlapped, observers typically experienced perceptual rivalry: one triangle appeared nearer than the other, with the depth order reversing spontaneously every several seconds. When one triangle was made less conspicuous by misaligning slightly the inducing elements, the other, unperturbed triangle dominated perceptually. However, introducing temporal offsets between the different inducing elements of a triangle had no significant effect on its perceptual dominance.

How can we reconcile these conflicting results? In those studies where temporal phase mattered (Roger-Ramachandran and Ramachandran, 1991, 1998; Fahle, 1993; Kojima, 1998), there were no spatial cues for segmentation -- all elements in the displays were identical in form, orientation, disparity and other static properties. In contrast, obvious spatial cues were present in displays revealing little or no effect of

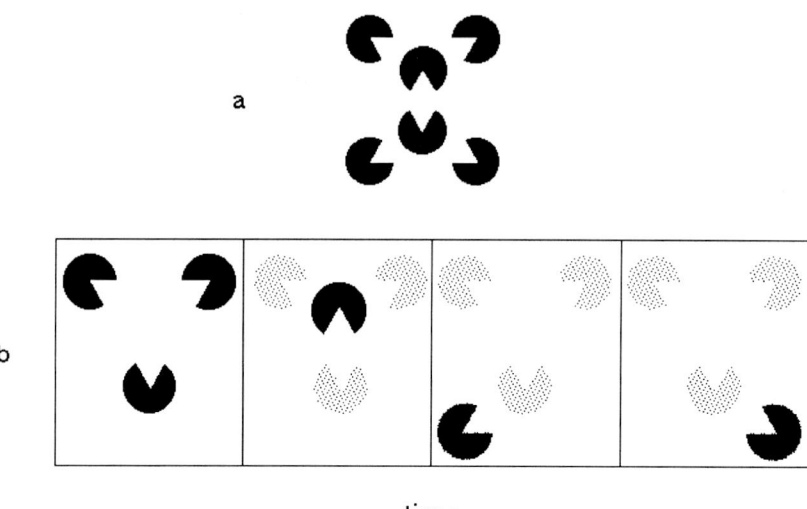

Fig. 2. Display used by Fahle & Koch (1995). When viewed without flicker, the two superimposed illusory Kanizsa triangles fluctuate in perceptual dominance, with one appearing in front of the other for several seconds at a time. The temporal configuration of the components of the illusory triangles was manipulated by briefly presenting all three pacmen for one triangle simultaneously followed by brief, sequential presentation of the three pacmen forming the other triangle. (In this schematic, the lightly stippled pacmen were not actually presented during the sequence and are shown here as reference for the positions of the single pacmen)

temporal phase (Kiper et al, 1991, 1996; Fahle and Koch, 1995). Perhaps, then, the salience of temporal structure on spatial grouping is modulated by the presence and strength of spatial cues. Leonards, Singer and Fahle (1996) explicitly tested this idea using texture arrays like those used by Kiper et al (1991; 1996). Figure and background could be defined by a difference in temporal phase alone, by a difference in orientation alone, or by both temporal phase and orientation differences (Figure 1c). In line with Roger-Ramachandran and Ramachandran (1991, 1998), Fahle (1993) and Kojima (1998), the effect of temporal cues on texture segmentation was significant when figures were defined only by temporal phase difference or when temporal cues and spatial cues defined the same figures. When figures were well defined by spatial cues, temporal phase had no influence on figure/ground segregation. Based on their results, Leonards et al proposed a flexible texture segmentation mechanism in which spatial and temporal cues interact.

Comparable results have been reported by Usher and Donnelly (1998), who used a square lattice display (Figure 1d) in which elements appear to group into either rows or columns. When the elements in alternating rows (or columns) of the lattice were flickered asynchronously, the display was perceived as rows (or columns) correspondingly. In the second experiment, Usher and Donnelly presented arrays of randomly oriented lines segments and asked observers to detect collinear target elements. Performance was better when target and background line segments flickered asynchronously than when they flickered in synchrony. Under this condition, it should be noted, temporal and spatial cues were congruent. However, the efficacy of temporal phase waned when the same target elements were randomly oriented and no longer collinear. Now the temporal lag had to be extended to about 36 msec before target elements were segregated from background elements. These results, together with those of earlier works, indicate that the potency of temporal information for spatial segmentation and grouping depends on the salience of available spatial cues.

Random Contrast Modulation The studies summarized above all used periodic temporal modulation in which luminance values fluctuated predictably between levels. In an effort to create more unpredictable temporal modulation, our laboratory developed displays in which the contrast levels of spatial frequency components comprising complex visual images (e.g., a face) are modulated over time. With this form of temporal modulation, the amplitude of the contrast envelope increases and decreases by random amounts over time without changing the space-average luminance of the display. In our initial work (Blake and Yang, 1997), we found that observers were better able to detect synchronized patterns of temporal contrast modulation within hybrid visual images composed of two components when those components were drawn from the same original picture. We then went on to show that "spatial structure" coincides with the phase-relations among component spatial frequencies (Lee and Blake, 1999a). These two studies set the stage for our more recent experiments examining the role of synchronous contrast modulations in perception of bistable figures.

In one study (Alais et al, 1998), we created a display consisting of four spatially distributed apertures each of which contained a sinusoidal grating that, when viewed alone, unambiguously appeared to drift in the direction orthogonal to its orientation (Figure 3). When all four gratings were viewed together, however, they intermittently grouped together to form a unique "diamond" figure whose global direction of motion corresponded to the vector sum of the component motions. Thus upon viewing this quartet of gratings, observers typically experience perceptual fluctuations over time between global motion and local motion. We found that the incidence of global motion increased when contrast modulations among the gratings were correlated and decreased when the component contrast modulations were uncorrelated. Similar results were obtained using a motion plaid in which two gratings drifting in different directions are spatially superimposed. Under optimal conditions, this display is also bistable, appearing either as two transparent gratings drifting in different directions or as one plaid moving in a direction determined by a vector sum of the velocities of component gratings (Adelson and Movshon, 1982).

Again, the incidence of coherent motion was enhanced by correlated contrast modulations and suppressed by uncorrelated contrast modulations. A second study assessed the role of temporal patterning of contrast modulation in grouping visual features using another bistable phenomenon, binocular rivalry. When dissimilar patterns are imaged on corresponding areas of the retinae in the two eyes, the patterns compete for perceptual dominance (Breese, 1899; Blake, 1989). Since binocular rivalry is strongly local in nature, however, predominance become piecemeal with small zones of suppression when rival targets are large or when small targets are distributed over space (Blake, O'Shea and Mueller, 1992). Alais and Blake (1999)

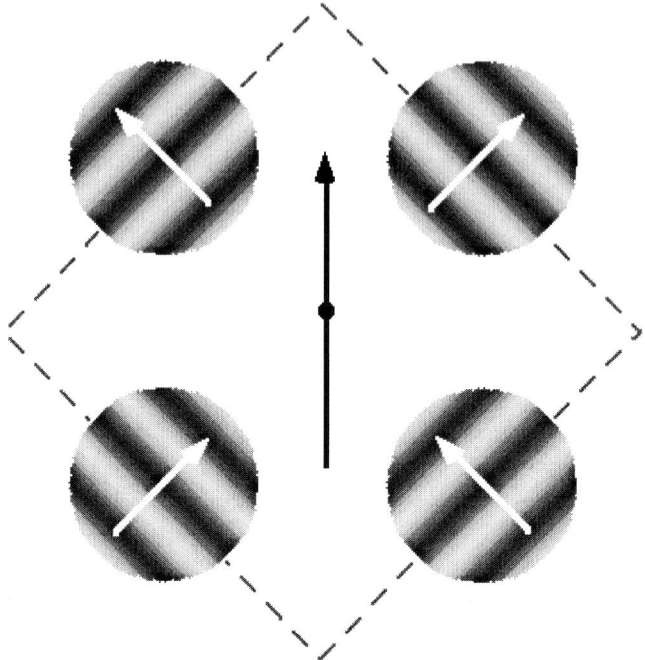

Fig. 3. Display used by Alais et al (1998). Gratings drifted smoothly in the direction indicated by the white arrows. When all four gratings are viewed simultaneously, the four occasionally group to form a partially occluded 'diamond' whose direction of motion corresponds to the vector sum of the component motions ("upward" in this illustration)

investigated the potency of correlated contrast modulation to promote conjoint dominance of two, spatially separated rival targets pitted against random-dot patches presented to corresponding portions of the other eye. The orientations of the two gratings were either collinear, parallel or orthogonal, and they underwent contrast modulations that were either correlated or uncorrelated. Correlated contrast modulation promoted joint grating predominance relative to the uncorrelated conditions, an effect strongest for collinear gratings. Joint predominance depended strongly on the angular separation between gratings and the temporal phase-lag in contrast modulations. Alais and Blake (1999) speculated that these findings may reflect neural interactions subserved by lateral connections between cortical hypercolumns.

Generalizing from the studies summarized in these last two sections, it appears that temporal flicker cannot overrule explicit spatial structure - flicker, in other words, does not behave like "super glue" to bind spatial structures that do not ordinarily form coherent objects. Nor, for that matter, does out-of-phase flicker destroy spatial structure defined by luminance borders. Based on the above results, one would conclude that temporal flicker promotes grouping primarily when spatial structure is ambiguous or weak. The generality and implications of these earlier studies must be qualified in two ways:

- All the studies cited above utilized local stimulus features whose luminance or contrast was modulated periodically, with only the rates and phases of flicker varying among different components of the display. The use of periodic flicker is grounded in linear systems analysis, which has played an important role in shaping techniques used in visual science. Still, periodic flicker constitutes a highly predictable, deterministic signal which could be construed as a somewhat artificial event. In the natural environment, there exists considerable irregularity in the temporal structure of the optical input to vision: objects can move about the visual environment unpredictably, and as observers we move our eyes and heads to sample that environment. In contrast, repetitively flashed or steadily moving stimuli, so often used in the laboratory, are highly predictable events which, from an information theoretical perspective, convey little information. In fact, periodically varying stimulation may significantly underestimate the temporal resolving power of neurons in primate visual cortex (eg Buracas et al 1998).
- In those studies where evidence for binding from temporal synchrony was found, successive, individual frames comprising a flicker sequence contained visible luminance discontinuities that clearly differentiated figure from background; spatial structure was already specified in given, brief "snapshots" of those dynamic displays. Thus these studies do not definitively prove that human vision can group spatial features based purely on temporal synchrony.

Stochastic Temporal Structure The two considerations presented in the previous paragraph motivated us to develop stochastic animations in which individual frames contained no static cue about spatial structure, leaving only temporal information to specify spatial grouping. How does one remove all static cues from individual frames? A hint for tackling this challenge came from an expanded notion of 'temporal structure' conveyed by time-varying signals. There are three alternative ways to define temporal structure (see Appendix, section A): (i) a time series of absolute quantity, (ii) a collection of distinctive times when a particular event occurs (point process), and (iii) a collection of times for events and magnitudes associated with those events (marked point process). Among of these representations, the point process contains information only about 'time' and not about magnitude. Thus we reasoned that if it's possible to create an animation display in which groups of local elements differ only in terms of their respective point processes, but do not differ in other stimulus properties when considered on a frame by frame basis, that display would be devoid of static spatial cues.

To create these conditions, we created animation displays in which each individual frame consisted of an array of many small sinusoidal gratings each windowed by a stationary, circular gaussian envelope -- such stimuli are termed 'Gabor patches'. All Gabor patches throughout the array had the same contrast, and their orientations were randomly determined. As the animation was played, grating contours within each small, stationary Gabor patch moved in one of two directions orthogonal to their orientation. Each grating reversed direction of motion irregularly over time according to random (Poisson) process. Temporal structure of each Gabor element was described by a point process consisting of points in time at which motion reversed direction (Figure 4). When all Gabor elements within a virtual 'figure' region reversed their direction of motion simultaneously while Gabor elements outside of

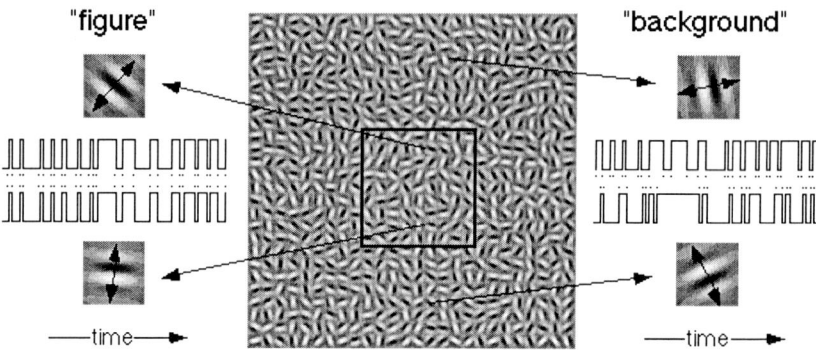

Fig. 4. Display used by Lee and Blake (1999). One frame from an animation sequence consisting of an array of small Gabor patches within which contours move in one of two directions orthogonal to their orientation; from frame-to-frame motion direction changes irregularly. Shown schematically on either side of the square array of Gabor patches are enlarged pictures of several Gabor patches with double-headed arrows indicating the two possible directions of motion. The time series indicate direction of motion, and the small dots associated with each time series denote that time series' point process - points in time at which direction changed. Gabor patches within a virtual region of the array (dotted outline) have point processes that are correlated while Gabor patches outside this virtual area are uncorrelated.(Reprinted by permission from Science, volume 5417, p. 1166. Copyright 1999 © American Association for the Advancement of Science)

this area changed direction independently of one another, the 'figure' region defined by temporal synchrony stood out conspicuously from the background. Here, the 'figure' region and the 'ground' region differed only in point process and they were not defined by static cues such as luminance, contrast and orientation in individual frames. Furthermore, there was no information about the figure in any two successive frames of the animation, for all contours moved from frame to frame. Therefore, only the difference in point process (which specifies 'when' direction of motion changes) distinguished figure from background.

Because all local elements throughout the display changed directions of motion 'irregularly' over time, we were able systematically to manipulate two variables that reveal just how sensitive human vision is to the temporal structure contained in these dynamic displays. One variable is the 'predictability' (or 'randomness') of temporal structure conveyed by individual elements -- this is easily manipulated by changing the probability distribution associated with the two directions of motion. Borrowing ideas from information theory, this 'predictability' was quantified by computing the entropy of temporal patterns of elements (see Appendix, section B). Since time-varying signals with high entropy convey more dynamic, fine temporal structure, the systematic manipulation of the entropy of all the elements made it possible to examine how accurately human vision can register fine temporal structure contained in contours irregularly changing direction of motion. The second variable was the temporal relationship among elements in the 'figure' region -- this was manipulated by varying the extent to which all possible pairs of those elements are correlated (see Appendix, section C). Since the time points at which the elements change direction of motion could be represented by point processes as described earlier, the index of temporal relationship among those elements could be quantified by computing the correlations among their point processes. By varying this index systematically, the efficiency of human vision in utilizing temporal structure was examined.

Lee and Blake (1999b) found that these two factors (entropy and correlation) both systematically affected the perceptual quality of spatial structure; the clarity of perceived figure/ground segregation increased with increases in the entropy of the entire display and with the correlation among elements within the 'figure' region. Our results clearly show that human vision can register fine temporal structure with high fidelity and can efficiently construct spatial structure solely based on the temporal relationship among local elements distributed over space.

From the outset of this work, we carefully tried to identify and evaluate possible stimulus artifacts that could underlie perception of shape in these displays. An artifact would be any cue that does not depend on correlations among point processes for shape definition. In the displays utilizing an array of Gabor patches, for example, reversals in direction of motion mean that contours return to the same positions they occupied one frame earlier. If contrast were summed over multiple frames, it is possible that these 2-frame change sequences, when they occurred, could produce enhanced contrast of that Gabor. When all Gabors in the figure region obey the same point process, these "pulse" increases in apparent contrast could define the figure against a background of irregular contrast pulses occurring randomly throughout the background where changes were unsynchronized. To counteract this cue, we randomly varied the contrast values of Gabor patches from frame-to-frame throughout the array, thereby eliminating contrast as a potential cue. Spatial form from temporal synchrony remains clearly visible. When the elements defining the figure are all synchronized but the background elements are not, the background contains a more varied pattern of temporal change, which means that the temporal frequency amplitude spectra of the background - considered across all background elements - may differ in detail from that of the figure, although both would be quite broad. However, this difference does not exist when background elements all obey the same point process (albeit one different from the figure) - still, shape discrimination performance remains quite good for this condition, ruling out differences in temporal frequency amplitude spectra as an alternative cue.

Another possible artifact arises from the possibility of temporal integration over some number of consecutive animation frames in which no change in direction occurs. If the grating contours were to step through one complete grating cycle without change in direction and if the integration time constant were to match the time taken for that event, the net luminance of pixels defining that grating would sum to levels near the mean and effectively produce a patch of approximately uniform brightness. For the synchronized condition, all Gabors in the figure would assume this brightness level simultaneously while those in the background would not because they are changing direction at different times on average - with time-integrated signals, the figure region could occasionally stand out from the background simply based on luminance. In fact, this is not at all what one sees when viewing these displays - one instead sees smooth motion continuously throughout the array of Gabor elements. Still, it could be argued that this putative integrated luminance cue is being registered by a separate, temporally sluggish mechanism. To formalize this hypothesis, we have simulated it in a MatLab program. In brief, the model integrates individual pixel luminance values over n successive animation frames, using a weighted temporal envelope. This procedure creates a modified sequence of animation frames whose individual pixels have been subjected to temporal integration, with a time constant explicitly designed to pick up n-frame cycles of no change. We then compute the standard deviation of luminance values within the figure region and within the background and use those values to compute a signal/noise index. To the extent that luminance mediates detection of spatial structure in these dynamic displays, psychophysical performance should covary with this "strength" index. We have created sequences minimizing the incidence of "no change" sequences by manipulating entropy and by selectively removing "no change" sequences in an animation - both manipulations affect the "strength" index. When we compare performance on trials where this cue is potentially present to trials where it is not, we find no difference between these two classes of trials.

Our work to date has focused on changes in direction of translational motion and changes in direction of rotational motion. These were selected, in part, because registration of motion signals requires good temporal resolution. There is no reason to believe, however, that structure from temporal synchrony is peculiar to motion. Indeed, to the extent that temporal structure is fundamentally involved in feature binding, any stimulus dimension for which vision possesses reasonable temporal resolution should in principle be able to support figure/ground segmentation from synchrony. There must, of course, be limits to the abstractness of change supporting grouping. Consider, for example, a letter array composed of vowels and consonants. It would be very unlikely that irregular, synchronized changes from vowels to consonants, and vice versa, in the array would support perceptual grouping (unless those changes were accompanied by prominent feature changes, such as letter size or font type). Grouping should be restricted to image properties signifying surfaces and their boundaries.

4 Concluding Remarks

Our results, in concert with earlier work, lend credence to the notion that temporal and spatial coherence are jointly involved in visual grouping. Biological vision, in other words, interprets objects and events in terms of the relations -- spatial and temporal -- among features defining those objects and events. Our results also imply that visual neurons modulate their responses in a time-locked fashion in response to external time-varying stimuli. Theoretically, this stimulus-locked variation in neural response could be realized either of two ways depending on how information is coded in the spike train. One class of models emphasizes temporal coding wherein fine temporal structure of dynamic visual input is encoded by exact locations in time of individual neural spikes in single neurons; in effect, the "code" is contained in the point processes associated with trains of neural activity.

Alternatively, a second class of models assumes that stimulus features are encoded by ensembles of neurons with similar receptive field properties. The average firing rate within a neural ensemble can fluctuate in a time-locked fashion to time-varying stimuli with temporal precision sufficient to account for the psychophysical data presented here (Shadlen and Newsome, 1994). Although the temporal coding scheme allows neurons to more efficiently transmit information about temporal variations in external stimuli than the rate coding scheme, psychophysical evidence alone does not definitively distinguish between the two models.

Regardless how this coding controversy is resolved, our results using stochastic animations convincingly show that temporal structure in the optical input to vision provides a robust source of information for spatial grouping. Indeed, one could argue that vision's chief job is extracting spatio-temporal structure in the interest of object and event perception. After all, the optical input to vision is rich in temporal structure, by virtue of the movement of objects and the movement of the observer through the environment. Consequently, our eyes and brains have evolved in a dynamic visual world, so why shouldn't vision be designed by evolution to exploit this rich source of information?

In closing, we are led to speculate whether the rich temporal structure characteristic of normal vision may, in fact, imprint its signature from the outset of neural processing. If this were truly the case, then concern about the binding problem would fade, for there would be no need for a mechanism to reassemble the bits and pieces comprising visual objects. Perhaps temporal structure insures that neural representations of object "components" remain conjoined from the very outset of visual processing. Construed in this way, the brain's job is rather different from that facing the King's horses and men who tried to put Humpty Dumpty back together. Instead of piecing together the parts of a visual puzzle, the brain may resonate to spatio-temporal structure contained in the optical input to vision.

References

Adelson EH and Movshon JA (1982) Phenomenal coherence of moving visual patterns. Nature, 300, 523-525.

Alais D and Blake R (1998) Interactions between global motion and local binocular rivalry Vision Research, 38, 637-644.

Alais D, Blake R and Lee S (1998) Visual features that vary over time group over space. Nature Neuroscience., 1, 160-164.

Ashby FG, Prinzmetal W, Ivry R, Maddox,WT (1996) A formal theory of feature binding in object perception. Psychological Review, 103, 165-192.

Barlow, HB (1972) Single units and sensation: a neuron doctrine for perceptual psychology? Perception, 1, 371-394.

Blake R (1989) A neural theory of binocular rivalry. Psychological Review 96, 145-167.

Blake R and Yang Y (1997) Spatial and temporal coherence in perceptual binding. Proceedings of the National Academy of Science, 94, 7115-7119.

Blake R, O'Shea RP and Mueller TJ (1992) Spatial zones of binocular rivalry in central and peripheral vision. Visual Neuroscience, 8, 469-478.

Breese, BB (1899) On inhibition. Psychological Monograph, 3, 1-65.

Brillinger, DR (1994) Time series, point processes, and hybrids. Canadian Journal of Statistics, 22, 177-206

Brook D and Wynne RJ (1988) Signal Processing: principles and applications. London: Edward Arnold.

De Coulon, F (1986). Signal Theory and Processing. Dedham MA: Artech House, Inc.

DeValois RL and DeValois KK (1988) Spatial vision. New York: Oxford University Press.

Fahle M (1993) Figure-ground discrimination from temporal information. Proceedings of the Royal Society of London, B, 254, 199-203.

Fahle M and Koch C (1995) Spatial displacement, but not temporal asynchrony, destroys figural binding. Vision Research, 35, 491-494.

Gawne T J (1999) Temporal coding as a means of information transfer in the primate visual system. Critical Reviews in Neurobiology 13, 83-101.

Ghose GM and Freeman RE (1997) Intracortical connections are not required for oscillatory activity in the visual cortex. Visual Neuroscience 14, 963-979.

Ghose GM and Maunsell J (1999) Specialized representations in visual cortex: a role for binding? Neuron 24 79-85.

Gray CM (1999) The temporal correlation hypothesis of visual feature integration: still alive and well. Neuron 24 31-47.

Hubel DH and Wiesel TN (1962) Receptive fields, binocular interaction and functional architecture in the cat's visual cortex. Journal of Physiology, London, 160, 106-154.

Kiper DC, Gegenfurtner KR and Movshon JA (1996) Cortical oscillatory responses do not affect visual segmentation. Vision Research, 36, 539-544.

Kiper DC and Gegenfurtner KR (1991) The effect of 40 Hz flicker on the perception of global stimulus properties. Society of Neuroscience Abstracts, 17, 1209.

Kojima H (1998) Figure/ground segregation from temporal delay is best at high spatial frequencies. Vision Research, 38, 3729-3734.

Lamme VAF and Spekreijse H (1998) Neuronal synchrony does not represent texture segregation. Nature 396, 362-366.

Lee SH & Blake R (1999a) Detection of temporal structure depends on spatial structure. Vision Research, 39, 3033-3048.

Lee SH & Blake R(1999b) Visual form created solely from spatial structure. Science, 284, 1165-1168.

Leonards U, Singer W, & Fahle M (1996) The influence of temporal phase differences on texture segmentation. Vision Research, 36, 2689-2697.

Logothetis,NK (1998) Single units and conscious vision. Philosophical Transactions of the Royal Society, London B, 353, 1801-1818.

Mainen ZF and Sejnowski TJ (1995) Reliability of spike timing in neocortical neurons. Science, 268, 1503-1506.

Mansuripur M (1987) Introduction to Information Theory. Englewood Cliffs NJ: Basic Books.

Marr D (1982) Vision: A computational Investigation into the human representation and processing of visual information. San Francisco: WH Freeman.

Milner, P.M. (1974) A model for visual shape recognition. *Psychological Review*, 81, 521-535.

Rogers-Ramachandran DC and Ramachandran VS (1998) Psychophysical evidence for boundary and surface systems in human vision. Vision Research, 38, 71-77.
Rogers-Ramachandran DC and Ramachandran VS (1998) Phantom contours: Selective stimulation of the magnocellular pathways in man. Investigative Ophthalmology and Visual Science, Suppl., 26.
Rao SC, Rainer G, Miller EK (1997) Integration of what and where in the primate prefrontal cortex. Science, 276, 821-824
Shadlen M and Movshon JA (1999) Synchrony unbound: a critical evaluation of the temporal binding hypothesis. Neuron, 24 67-77.
Shadlen MN and Newsome WT (1994) Noise, neural codes and cortical organization. Current Opinion in Neurobiology, 4, 569-579.
Shadlen MN and Newsome WT (1998) The variable discharge of cortical neurons: implications for connectivity, computation, and information coding. Journal of Neuroscience, 18, 3870-3896.
Shannon CE and Weaver W (1949) The mathematical theory of communication. Urbana: Univ. of Illinois Press.
Singer W and Gray CM (1995) Annual Review of Neuroscience, 18, 555-586.
Singer W.(1999) Striving for coherence. Nature, 397, 391-392.
Treisman A (1999) Solutions to the binding problem: progress through controversy and convergence. Neuron 24 105-110.
Treisman A and Gelade G (1980) A feature-integration theory of attention. Cognitive Psychology, 12, 97-136.
Usher M and Donnelly N (1998) Visual synchrony affects binding and segmentation in perception. Nature, 394, 179-182.
Usrey WM and Reid RC (1999) Synchronous activity in the visual system. Annual Review of Physiology, 61, 435-456.
Van Essen DC, Anderson CH and Felleman DJ (1992) Information processing in the primate visual system, an integrated systems perspective. Science 255, 419-423 .
von der Malsburg,C (1995) Binding in models of perception and brain function. Current Opinions in Neurobiology, 5, 520-526.
Zeki S (1993) A vision of the brain. Cambridge MA: Blackwell Scientific.

Appendix: Time-Varying Signals and Information Theory

Throughout this chapter we employ the term "temporal structure" in reference to spatial grouping. In this appendix, we define "temporal structure" using information theory and signal processing theory as conceptual frameworks.

A. Time-Varying Signals

According to signal processing theory, time-varying signals are defined by variations of some quantity over time which may or may not be detectable by a given system (Brook and Wynne, 1988). For the visual system, time-varying signals would be the temporal fluctuations of visual properties of external stimuli. Thus, the temporal structure of visual stimuli can be carried by any property which is potentially detectable by human vision, including luminance, color, stereoscopic disparity, and motion. When a vertically oriented sinusoidal grating shifts its spatial phase either right-ward or left-ward over time, for instance, time-varying signals can be defined by the temporal fluctuations of direction of motion.

<u>Deterministic vs Random Signals</u> Time-varying signals can be distinguished depending on whether they are deterministic or random. Deterministic signals are perfectly predictable by an appropriate mathematical model, while random signals are unpredictable and can generally be described only through statistical observation (de Coulon, 1986). Figure 5a shows an example of a deterministic signal and Figure 5b shows a random signals. In Figure 5a, a deterministic time-varying signal, directions of motion of a grating can be exactly predicted on the basis of time points where the grating shifts its spatial phase because the two opposite directions of motion alternate in a deterministic fashion. For the time-varying signals in Figure 5b, one cannot predict in which direction the grating will be moving at time $t + 1$ by knowing in which direction it is moving at time t, because direction of motion changes randomly over time. Instead, one only can make a statistical prediction (e.g., "the grating is more likely to move left-ward than right-ward") by estimating the probabilities of the two motion directions. Unlike most engineered communications systems that use well-defined frequencies to transmit information, many biological systems must deal with irregularly fluctuating signals.

<u>Representation of the temporal structure of time-varying signals</u> The temporal structure contained in time-varying visual signals can be described in any of several ways, depending upon what properties one wishes to emphasize.
(1) Time series of absolute quantity (Figure 5c). The time series plot of absolute quantity is the most direct and simple description of temporal signals (Brillinger, 1994). Here, simply the absolute quantity of visual stimuli is plotted against time;

$$Y(t), \; -4 < t < +4$$

where t is time and Y represents the quantity of stimulus. When a dot changes in luminance over time, for example, its temporal structure can be represented by plotting the absolute level of luminance over time (Figure 5c).
(2) Marked point process (Figure 5d). An irregular time series representing both the times at which events occur as well as the quantities associated with those times:

$$\{(\tau_j, M_j)\}, \; j = "\;1,\;"\;2,\;"\;3,...$$

where M_j represents the magnitude of change associated with the jth event. In Figure 5d, the locations of vertical lines represent time points when a stimulus changes in luminance and the directions and lengths of vertical lines represent the magnitude and direction (increase or decrease in luminance) of changes, respectively.
(3) Point process (Figure 5e). If we are interested only in 'when' a given stimulus quantity changes or 'when' events occur, we may specify a point process which is the collection of distinctive times when a stimulus changes its quantity;

$$\{\tau_j\}, \; j = 0,\;"\;1,\;"\;2,\;"\;3,...$$

where the t_j is a time point for the jth event. For instance, asterisks in Figure 5e denote the times when a stimulus element changes its luminance, without regard to the value of luminance itself. In a point process, the characteristics of temporal structure in time-varying signals can be described by analyzing the distribution of events in a given time period and also the distribution of time intervals between events.

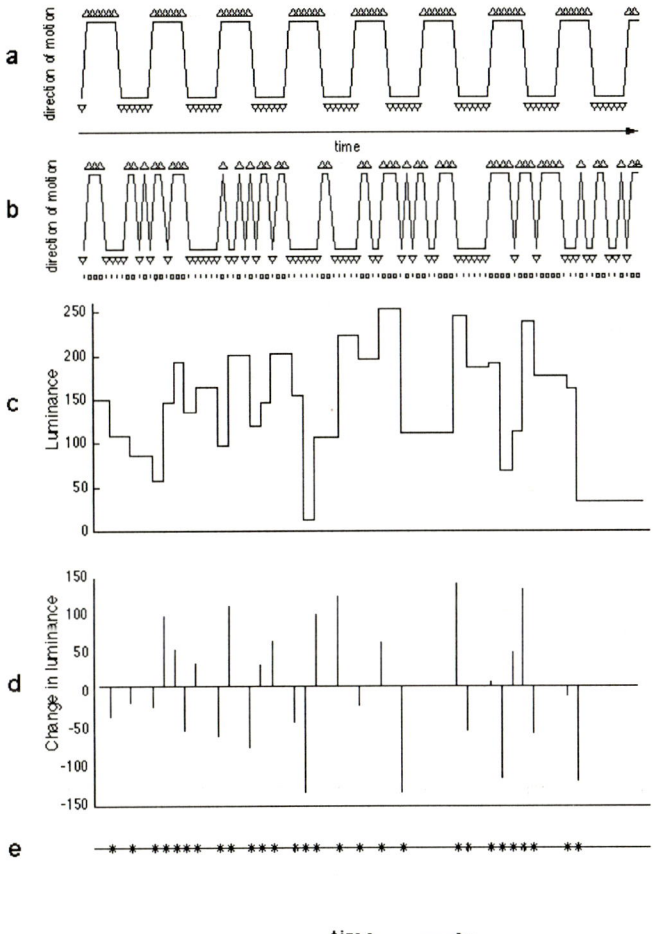

Fig. 5. Time series indicating direction of visual motion over time. (a) Deterministic time-varying signal. Direction of motion is entirely predictable and specifiable mathematically. (b) Random time-varying signal. Direction of motion reverses randomly over time according to a Poisson process. (c) - (e) show three types of representations of temporal structure. (c) Time series of absolute quantity. The absolute luminance level of a given visual stimulus is plotted against time. (d) Marked point process. The locations of vertical lines represent points in time when the stimulus changes in luminance, and the directions and lengths of vertical lines represent the magnitude and direction of change in luminance, respectively. (e) Point process. Asterisks denote times when the luminance value changes.

B. Time-Varying Signals as Information Flow

How well does human vision respond to the temporal structure generated by unpredictable time-varying visual signals? Information theory (Shannon and Weaver, 1949; Mansuripur, 1987) sheds light on how we can conceptualize this problem. According to information theory, 'time-varying visual signals' can be treated as information flow over time as long as they are an ensemble of mutually exclusive and statistically independent messages. For example, the time series of motion direction in Figure 5b can be understood as an ensemble of 'left-ward motion' messages and 'right-ward motion' messages. These messages are mutually exclusive because the grating cannot move in two opposite directions simultaneously. They are also statistically independent since the probability of one message does not depend on the probability of the other. If we assign a symbol '0' to the left-ward motion signal and '1' to the right-ward motion signal, the time plot in the top part of Figure 5b can be translated into an information flow composed of '0' and '1' as illustrated in the bottom portion of Figure 5b.

Furthermore, information theory allows us to have a measure of uncertainty or randomness about the temporal structure of visual stimuli. We can quantify the amount of information as 'entropy' if the probability distribution of messages comprising time-varying signals is known. Suppose that a time series of signals, S, is an ensemble of n different messages, $\{m_1, m_2, \ldots, m_i, \ldots, m_n\}$, and the probabilities for those messages are $\{p_1, p_2, \ldots, p_i, \ldots, p_n\}$. Then the information (expressed in binary units called "bits") attributed to each message is defined by

$$H_i = -\log_2(p_i)$$

and the averaged amount of information,

$$H(S) = -\sum_{i=1}^{N} p_i * \log_2(p_i)$$

is the probability-weighted mean of information contained in all the messages. This equation implies that the averaged amount of information is maximized when all of the possible messages are equally probable. In the earlier example of time-varying stimuli, the direction of motion is the most unpredictable when the two directions of motion are equally probable.

While H(S) represents the uncertainty of time-varying signals, the temporal complexity of signals can be evaluated by the rate of flow of information
$$H(S)/T$$
where T is the average duration of messages from the ensemble, weighted for frequency of occurrence, that is,

$$T = \sum_{i=1}^{N} p_i * T_i$$

The high value of H(S)/T indicates that the relatively large amount of information (high value of H(S)) flow in a short duration (low value of T). Thus, human vision's ability to process the temporal structure of signals can be evaluated by finding the maximum information flow rate which can be processed by human vision. If the visual system successfully processes time-varying signals with a specific information flow rate, it should (i) reliably generate identical outputs in response to identical input signals and (ii) accurately discriminate among different signals. By measuring these abilities, reliability and accuracy, while varying the rate of information flow, the capacity of human vision for time-varying signals can be determined (e.g., Mainen and Sejnowski, 1995).

C. Characterization of Relationship among Spatially Distributed Temporal Structures

In the visual environment, the sources of time-varying signals are often distributed over space. It seems reasonable to assume that time-varying visual signals arising from the same object are likely to have related temporal structures. Since this is the case, is the visual system able to detect important relations among temporal structures of visual signals? If so, what kinds of relationship among temporal structures is human vision sensitive to?

Once the temporal structures of time-varying signals are defined quantitatively by one of the ways mentioned above (time series of absolute quantity, point process and marked point process), the relationship among stochastic time series can be quantitatively expressed by computing correlation coefficients in the time domain. Suppose we have N observations on two series of temporal signals, x and y,

$$\{x_1, x_2, x_3, ..., x_N\}$$
$$\{y_1, y_2, y_3, ..., y_N\}$$

at unit time interval over the same time period. Then, the relationship between those series can be characterized by the cross-correlation function.

$$\rho_{xy}(k) = \gamma_{xy}(k)/\sqrt{[\gamma_{xx}(0)\gamma_{yy}(0)]}, \quad k = 1, 2, 3, ..., (N-1)$$

$$\gamma_{xy}(k) = Cov(x_t, y_{t+k})$$

$$\gamma_{xx}(0) = Cov(x_0, x_{0+k})$$

$$\gamma_{yy}(0) = Cov(y_0, y_{0+k})$$

where k is a temporal phase lag between two signals.

Author Index

Ahn, S.	463	Florack, L.	297, 574
Ahrns, I.	554	Fukushima, K.	623
Aloimonos, Y.	118		
Ameur, S.	258	Grupen, R. A.	52
Baek, K.	238	Hadid, A.	258
Baik, S. W.	463	Han, K.-H.	353
Baker, P.	118	Hancock, E. R.	584
Bang, S. Y.	316	Hartmann, G.	387
Baratoff, G.	554	Henderson, J. M.	424
Bhuiyan, M. S.	613	Hocini, H.	258
Blake, R.	635	Hong, J.-Y.	353
Blanz, V.	308	Hong, K.-S.	397
Bosco, A.	407	Hoyer, P. O.	535
Bülthoff, H. H.	10	Hwang, B.-W.	308
Byun, H.	286	Hyvärinen, A.	535
Campos, A. G.	407	Inaguma, T.	594
Chang, D.-H.	343	Inki, M.	535
Cheng, G.	150, 512	Ishihara, Y.	108
Chien, S.-I.	379	Iwaki, R.	482
Cho, K.-J.	353	Iwata, A.	613
Cho, Y.	286		
Cho, Y.-J.	160	Jang, D.-S.	248
Choi, H.-I.	248	Jang, J.-H.	397
Choi, I.	379	Jang, S.-W.	248
Choi, S.	42	Jeong, H.	227
Christensen, H.	209		
Christou, C. G.	10	Kang, S.	564
Costa, L. da F.	32, 407	Kim, C.-Y.	62
		Kim, H.-G.	453
da Costa, F. M. G.	32	Kim, H.-J.	179
Djeddi, M.	258	Kim, J.	417
Draper, B. A.	238	Kim, N.-G.	168
Dresp, B.	336	Kim, S.	268
Dror, R. O.	492	Kim, S.-H.	453
Dyer, F. C.	424	Kim, Y.	502
		Kopp, C.	336
Eklundh, J.-O.	209	Kottow, D.	444
Estrozi, L. F.	407	Kuniyoshi, Y.	150, 512
Fan, K.-C.	359	Kweon, I.-S.	62
Fermuller, C.	118		
Fischer, S.	336	Lai, J. H.	545

Lamar, M. V.	613	Roh, H.-K.	369
Laughlin, S. B.	492	Rougeaux, S.	512
Lee, H.-S.	286	Ruiz-del-Solar, J.	444
Lee, O.	42		
Lee, S.	268	Saji, H.	594
Lee, S.-H.	635	Sali, E.	73
Lee, S.-I.	129	Schönfelder, R.	554
Lee, S.-K.	268	Sehad, A.	258
Lee, S.-W.	179, 308, 369, 434, 564	Sejnowski, T. J.	527
		Seo, Y.-S.	62
Lee, S.-Y.	129	Sherman, E.	88
Lee, T.-W.	527	Shim, M.	326
Lim, J.	160	Shimoda, M.	482
Lin, C.	359	Song, M.	417
Lowe, D. G.	20	Spitzer, H.	88
Lüdtke, N.	584	Stasse, O.	150, 512
Mahadevan, S.	424	Tankus, A.	139
Minut, S.	424	ter Haar Romeny, B. M.	297
Morita, S.	108	Thiem, J.	387
Müller, V.	98		
		Ullman, S.	73
Nagakubo, A.	512		
Nakatani, H.	594	Vetter, T.	308
Nattel, E.	217	von der Malsburg, C.	276
Neumann, H.	554		
		Wachtler, T.	527
O'Carroll, D. C.	492	Wieghardt, J.	276
Oh, S.-R.	160	Wilson, R. C.	584
Oh, Y.	227	Wolff, C.	387
Oomura, K.	594		
		Yang, H.	502
Pachowicz, P. W	463	Yang, J. A.	602
Park, C. S.	316	Yang, X. M.	602
Park, J.-S.	343	Yeshurun, Y.	139, 217
Park, S.-C.	434	Yi, J.	502
Park, S.-G.	343	Yoo, J.	286
Park, S. H.	268	Yoo, M.-H.	179
Park, Y.	417	Yoon, K.-J.	62
Piater, J. H.	52	You, B.-J.	160
Pless, R.	118	Yuen, P. C.	545
Poggio, T.	1	Yun, H.	417
Pyo, H.-B.	268		
		Zhao, L.	473
Riesenhuber, M.	1	Zucker, S. W.	189
Rios-Filho, L. G.	407		

Lecture Notes in Computer Science

For information about Vols. 1–1727
please contact your bookseller or Springer-Verlag

Vol. 1728: J. Akoka, M. Bouzeghoub, I. Comyn-Wattiau, E. Métais (Eds.), Conceptual Modeling – ER '99. Proceedings, 1999. XIV, 540 pages. 1999.

Vol. 1729: M. Mambo, Y. Zheng (Eds.), Information Security. Proceedings, 1999. IX, 277 pages. 1999.

Vol. 1730: M. Gelfond, N. Leone, G. Pfeifer (Eds.), Logic Programming and Nonmonotonic Reasoning. Proceedings, 1999. XI, 391 pages. 1999. (Subseries LNAI).

Vol. 1731: J. Kratochvíl (Ed.), Graph Drawing. Proceedings, 1999. XIII, 422 pages. 1999.

Vol. 1732: S. Matsuoka, R.R. Oldehoeft, M. Tholburn (Eds.), Computing in Object-Oriented Parallel Environments. Proceedings, 1999. VIII, 205 pages. 1999.

Vol. 1733: H. Nakashima, C. Zhang (Eds.), Approaches to Intelligent Agents. Proceedings, 1999. XII, 241 pages. 1999. (Subseries LNAI).

Vol. 1734: H. Hellwagner, A. Reinefeld (Eds.), SCI: Scalable Coherent Interface. XXI, 490 pages. 1999.

Vol. 1564: M. Vazirgiannis, Interactive Multimedia Documents. XIII, 161 pages. 1999.

Vol. 1591: D.J. Duke, I. Herman, M.S. Marshall, PREMO: A Framework for Multimedia Middleware. XII, 254 pages. 1999.

Vol. 1624: J. A. Padget (Ed.), Collaboration between Human and Artificial Societies. XIV, 301 pages. 1999. (Subseries LNAI).

Vol. 1635: X. Tu, Artificial Animals for Computer Animation. XIV, 172 pages. 1999.

Vol. 1646: B. Westfechtel, Models and Tools for Managing Development Processes. XIV, 418 pages. 1999.

Vol. 1735: J.W. Amtrup, Incremental Speech Translation. XV, 200 pages. 1999. (Subseries LNAI).

Vol. 1736: L. Rizzo, S. Fdida (Eds.): Networked Group Communication. Proceedings, 1999. XIII, 339 pages. 1999.

Vol. 1737: P. Agouris, A. Stefanidis (Eds.), Integrated Spatial Databases. Proceedings, 1999. X, 317 pages. 1999.

Vol. 1738: C. Pandu Rangan, V. Raman, R. Ramanujam (Eds.), Foundations of Software Technology and Theoretical Computer Science. Proceedings, 1999. XII, 452 pages. 1999.

Vol. 1739: A. Braffort, R. Gherbi, S. Gibet, J. Richardson, D. Teil (Eds.), Gesture-Based Communication in Human-Computer Interaction. Proceedings, 1999. XI, 333 pages. 1999. (Subseries LNAI).

Vol. 1740: R. Baumgart (Ed.): Secure Networking – CQRE [Secure] '99. Proceedings, 1999. IX, 261 pages. 1999.

Vol. 1741: A. Aggarwal, C. Pandu Rangan (Eds.), Algorithms and Computation. Proceedings, 1999. XIII, 448 pages. 1999.

Vol. 1742: P.S. Thiagarajan, R. Yap (Eds.), Advances in Computing Science – ASIAN'99. Proceedings, 1999. XI, 397 pages. 1999.

Vol. 1743: A. Moreira, S. Demeyer (Eds.), Object-Oriented Technology. Proceedings, 1999. XVII, 389 pages. 1999.

Vol. 1744: S. Staab, Extracting Degree Information from Texts. X; 187 pages. 1999. (Subseries LNAI).

Vol. 1745: P. Banerjee, V.K. Prasanna, B.P. Sinha (Eds.), High Performance Computing – HiPC'99. Proceedings, 1999. XXII, 412 pages. 1999.

Vol. 1746: M. Walker (Ed.), Cryptography and Coding. Proceedings, 1999. IX, 313 pages. 1999.

Vol. 1747: N. Foo (Ed.), Advanced Topics in Artificial Intelligence. Proceedings, 1999. XV, 500 pages. 1999. (Subseries LNAI).

Vol. 1748: H.V. Leong, W.-C. Lee, B. Li, L. Yin (Eds.), Mobile Data Access. Proceedings, 1999. X, 245 pages. 1999.

Vol. 1749: L. C.-K. Hui, D.L. Lee (Eds.), Internet Applications. Proceedings, 1999. XX, 518 pages. 1999.

Vol. 1750: D.E. Knuth, MMIXware. VIII, 550 pages. 1999.

Vol. 1751: H. Imai, Y. Zheng (Eds.), Public Key Cryptography. Proceedings, 2000. XI, 485 pages. 2000.

Vol. 1752: S. Krakowiak, S. Shrivastava (Eds.), Advances in Distributed Systems. VIII, 509 pages. 2000.

Vol. 1753: E. Pontelli, V. Santos Costa (Eds.), Practical Aspects of Declarative Languages. Proceedings, 2000. X, 327 pages. 2000.

Vol. 1754: J. Väänänen (Ed.), Generalized Quantifiers and Computation. Proceedings, 1997. VII, 139 pages. 1999.

Vol. 1755: D. Bjørner, M. Broy, A.V. Zamulin (Eds.), Perspectives of System Informatics. Proceedings, 1999. XII, 540 pages. 2000.

Vol. 1757: N.R. Jennings, Y. Lespérance (Eds.), Intelligent Agents VI. Proceedings, 1999. XII, 380 pages. 2000. (Subseries LNAI).

Vol. 1758: H. Heys, C. Adams (Eds.), Selected Areas in Cryptography. Proceedings, 1999. VIII, 243 pages. 2000.

Vol. 1759: M.J. Zaki, C.-T. Ho (Eds.), Large-Scale Parallel Data Mining. VIII, 261 pages. 2000. (Subseries LNAI).

Vol. 1760: J.-J. Ch. Meyer, P.-Y. Schobbens (Eds.), Formal Models of Agents. Poceedings. VIII, 253 pages. 1999. (Subseries LNAI).

Vol. 1761: R. Caferra, G. Salzer (Eds.), Automated Deduction in Classical and Non-Classical Logics. Proceedings. VIII, 299 pages. 2000. (Subseries LNAI).

Vol. 1762: K.-D. Schewe, B. Thalheim (Eds.), Foundations of Information and Knowledge Systems. Proceedings, 2000. X, 305 pages. 2000.

Vol. 1763: J. Akiyama, M. Kano, M. Urabe (Eds.), Discrete and Computational Geometry. Proceedings, 1998. VIII, 333 pages. 2000.

Vol. 1764: H. Ehrig, G. Engels, H.-J. Kreowski, G. Rozenberg (Eds.), Theory and Application of Graph Transformations. Proceedings, 1998. IX, 490 pages. 2000.

Vol. 1765: T. Ishida, K. Isbister (Eds.), Digital Cities. IX, 444 pages. 2000.

Vol. 1767: G. Bongiovanni, G. Gambosi, R. Petreschi (Eds.), Algorithms and Complexity. Proceedings, 2000. VIII, 317 pages. 2000.

Vol. 1768: A. Pfitzmann (Ed.), Information Hiding. Proceedings, 1999. IX, 492 pages. 2000.

Vol. 1769: G. Haring, C. Lindemann, M. Reiser (Eds.), Performance Evaluation: Origins and Directions. X, 529 pages. 2000.

Vol. 1770: H. Reichel, S. Tison (Eds.), STACS 2000. Proceedings, 2000. XIV, 662 pages. 2000.

Vol. 1771: P. Lambrix, Part-Whole Reasoning in an Object-Centered Framework. XII, 195 pages. 2000. (Subseries LNAI).

Vol. 1772: M. Beetz, Concurrent Ractive Plans. XVI, 213 pages. 2000. (Subseries LNAI).

Vol. 1773: G. Saake, K. Schwarz, C. Türker (Eds.), Transactions and Database Dynamics. Proceedings, 1999. VIII, 247 pages. 2000.

Vol. 1774: J. Delgado, G.D. Stamoulis, A. Mullery, D. Prevedourou, K. Start (Eds.), Telecommunications and IT Convergence Towards Service E-volution. Proceedings, 2000. XIII, 350 pages. 2000.

Vol. 1775: M. Thielscher, Challenges for Action Theories. XIII, 138 pages. 2000. (Subseries LNAI).

Vol. 1776: G.H. Gonnet, D. Panario, A. Viola (Eds.), LATIN 2000: Theoretical Informatics. Proceedings, 2000. XIV, 484 pages. 2000.

Vol. 1777: C. Zaniolo, P.C. Lockemann, M.H. Scholl, T. Grust (Eds.), Advances in Database Technology – EDBT 2000. Proceedings, 2000. XII, 540 pages. 2000.

Vol. 1778: S. Wermter, R. Sun (Eds.), Hybrid Neural Systems. IX, 403 pages. 2000. (Subseries LNAI).

Vol. 1780: R. Conradi (Ed.), Software Process Technology. Proceedings, 2000. IX, 249 pages. 2000.

Vol. 1781: D.A. Watt (Ed.), Compiler Construction. Proceedings, 2000. X, 295 pages. 2000.

Vol. 1782: G. Smolka (Ed.), Programming Languages and Systems. Proceedings, 2000. XIII, 429 pages. 2000.

Vol. 1783: T. Maibaum (Ed.), Fundamental Approaches to Software Engineering. Proceedings, 2000. XIII, 375 pages. 2000.

Vol. 1784: J. Tiuryn (Ed.), Foundations of Software Science and Computation Structures. Proceedings, 2000. X, 391 pages. 2000.

Vol. 1785: S. Graf, M. Schwartzbach (Eds.), Tools and Algorithms for the Construction and Analysis of Systems. Proceedings, 2000. XIV, 552 pages. 2000.

Vol. 1786: B.H. Haverkort, H.C. Bohnenkamp, C.U. Smith (Eds.), Computer Performance Evaluation. Proceedings, 2000. XIV, 383 pages. 2000.

Vol. 1787: J. Song (Ed.), Information Security and Cryptology – ICISC'99. Proceedings, 1999. XI, 279 pages. 2000.

Vol. 1790: N. Lynch, B.H. Krogh (Eds.), Hybrid Systems: Computation and Control. Proceedings, 2000. XII, 465 pages. 2000.

Vol. 1792: E. Lamma, P. Mello (Eds.), AI*IA 99: Advances in Artificial Intelligence. Proceedings, 1999. XI, 392 pages. 2000. (Subseries LNAI).

Vol. 1793: O. Cairo, L.E. Sucar, F.J. Cantu (Eds.), MICAI 2000: Advances in Artificial Intelligence. Proceedings, 2000. XIV, 750 pages. 2000. (Subseries LNAI).

Vol. 1795: J. Sventek, G. Coulson (Eds.), Middleware 2000. Proceedings, 2000. XI, 436 pages. 2000.

Vol. 1794: H. Kirchner, C. Ringeissen (Eds.), Frontiers of Combining Systems. Proceedings, 2000. X, 291 pages. 2000. (Subseries LNAI).

Vol. 1796: B. Christianson, B. Crispo, J.A. Malcolm, M. Roe (Eds.), Security Protocols. Proceedings, 1999. XII, 229 pages. 2000.

Vol. 1800: J. Rolim et al. (Eds.), Parallel and Distributed Processing. Proceedings, 2000. XXIII, 1311 pages. 2000.

Vol. 1801: J. Miller, A. Thompson, P. Thomson, T.C. Fogarty (Eds.), Evolvable Systems: From Biology to Hardware. Proceedings, 2000. X, 286 pages. 2000.

Vol. 1802: R. Poli, W. Banzhaf, W.B. Langdon, J. Miller, P. Nordin, T.C. Fogarty (Eds.), Genetic Programming. Proceedings, 2000. X, 361 pages. 2000.

Vol. 1803: S. Cagnoni et al. (Eds.), Real-World Applications and Evolutionary Computing. Proceedings, 2000. XII, 396 pages. 2000.

Vol. 1805: T. Terano, H. Liu, A.L.P. Chen (Eds.), Knowledge Discovery and Data Mining. Proceedings, 2000. XIV, 460 pages. 2000. (Subseries LNAI).

Vol. 1806: W. van der Aalst, J. Desel, A. Oberweis (Eds.), Business Process Management. VIII, 391 pages. 2000.

Vol. 1807: B. Preneel (Ed.), Advances in Cryptology – EUROCRYPT 2000. Proceedings, 2000. XVIII, 608 pages. 2000.

Vol. 1811: S.W. Lee, H.. Bülthoff, T. Poggio (Eds.), Biologically Motivated Computer Vision. Proceedings, 2000. XIV, 656 pages. 2000.

Vol. 1815: G. Pujolle, H. Perros, S. Fdida, U. Körner, I. Stavrakakis (Eds.), Networking 2000 – Broadband Communications, High Performance Networking, and Performance of Communication Networks. Proceedings, 2000. XX, 981 pages. 2000.

Vol. 1816: T. Rus (Ed.), Algebraic Methodology and Software Technology. Proceedings, 2000. XI, 545 pages. 2000.

Vol. 1818: C.G. Omidyar (Ed.), Mobile and Wireless Communications Networks. Proceedings, 2000. VIII, 187 pages. 2000.

Vol. 1823: B. Hertzberger, H. Afsarmanesh, R. Williams, M. Bubak (Eds.), High Performance Computing and Networking. Proceedings, 2000. XVIII, 719 pages. 2000.